2ⁿᵈ Edition

Beginning Algebra: Early Graphing

John Tobey

North Shore Community College
Danvers, Massachusetts

Jeffrey Slater

North Shore Community College
Danvers, Massachusetts

Jamie Blair

Orange Coast College
Costa Mesa, California

Prentice Hall
is an imprint of

Upper Saddle River, NJ 07458

Library of Congress Cataloging-in-Publication Data

Tobey, John

Beginning algebra : early graphing / John Tobey, Jeffrey Slater, Jamie Blair. — 2nd ed.

p. cm.

Includes index.

ISBN-13: 978-0-321-57821-1 (annotated instructor's ed.)

ISBN-10: 0-321-57821-X (annotated instructor's ed.)

ISBN-13: 978-0-321-57796-2 (student ed.)

ISBN-10: 0-321-57796-5 (student ed.)

1. Algebra—Graphic methods—Textbooks.   I. Slater, Jeffrey.   II. Blair, Jamie.   III. Title.

QA219.T63 2010

512.9—dc22   2008046293

Editorial Director, Mathematics: *Christine Hoag*
Editor in Chief: *Paul Murphy*
Executive Project Manager: *Kari Heen*
Senior Project Editor: *Lauren Morse*
Assistant Editors: *Georgina Brown and Christine Whitlock*
Production Management: *Elm Street Publishing Services*
Senior Managing Editor: *Linda Mihatov Behrens*
Operations Specialist: *Ilene Kahn*
Senior Operations Supervisor: *Diane Peirano*
Marketing Manager: *Marlana Voerster*
Marketing Assistant: *Nathaniel Koven*
Art Director: *Heather Scott*
Interior/Cover Designer: *Tamara Newnam*
AV Project Manager: *Thomas Benfatti*
Executive Manager, Course Production: *Peter Silvia*
Media Producer: *Audra J. Walsh*
Associate Producer: *Emilia Yeh*
Manager, Content Development: *Rebecca Williams*
QA Manager: *Marty Wright*
Senior Content Developer: *Mary Durnwald*
Photo Research Development Manager: *Elaine Soares*
Image Permission Coordinator: *Kathy Gavilanes*
Photo Researcher: *Michael Downey*
Manager, Cover Visual Research and Permissions: *Karen Sanatar*
Cover Image: *Ryan McVay/Photodisc/Getty Images, Inc.*
Compositor: *Macmillan Publishing Solutions*
Art Studios: *Scientific Illustrators and Laserwords*

Prentice Hall
is an imprint of

PEARSON

© 2010, 2006 by Pearson Education, Inc.
Pearson Prentice Hall
Pearson Education, Inc.
Upper Saddle River, New Jersey 07458

Printed in the United States of America

10 9 8 7 6 5 4 3 2 1

ISBN 10: 0-321-57796-5
ISBN 13: 978-0-321-57796-2

Pearson Education Ltd., London
Pearson Education Singapore, Pte. Ltd.
Pearson Education Canada, Inc.
Pearson Education, Japan
Pearson Education Australia PTY, Limited

Pearson Education North Asia, Ltd., Hong Kong
Pearson Educación de Mexico, S.A. de C.V.
Pearson Education Malaysia, Pte. Ltd.
Pearson Education Upper Saddle River,
  New Jersey

Beginning Algebra: Early Graphing

STUDY SMARTER

CHAPTER
Test Prep
VIDEO CD

Step-by-step solutions on video for all chapter test exercises from the text

Beginning Algebra: Early Graphing

SECOND EDITION   John Tobey   Jeffrey Slater   Jamie Blair

CHAPTER Test Prep VIDEO CD   PEARSON

Main Menu   Exit

Chapter 3 - Graphing and Functions

Select an exercise from the Chapter Test in the book to view the step-by-step solution.

1	2	3	4	5	6	7
8	9	**10**	11	12	13	14
15	16	17				

$(2, 7)$ $(2, -2)$

$m = \dfrac{-2 - 7}{2 - 2} = \dfrac{-9}{0}$

$= \text{no slope}$

Well notice that our x values repeat, x is two here, x is two here; so we have the line x equals to two.

00:18 / 00:50   CC ESP

English and Spanish Subtitles Available

INCLUDED WITH EVERY NEW COPY OF THIS TEXTBOOK!

With this edition of the Tobey/Slater/Blair Developmental Mathematics series, we are committed to helping you get the most out of your learning experience by showing you how important math can be in your daily lives. For example, the new *Use Math to Save Money* feature presents practical examples of how math can help you save money, cut costs, and spend less. One of the best ways you can save money is to pass your course the first time you take it. This text and its features and MyMathLab can help you do that.

The best place to start is the *How Am I Doing? Guide to Math Success.* This clear path for you to follow is based upon how our successful students have utilized the textbook in the past. Here is how it works:

EXAMPLES and PRACTICE PROBLEMS: When you study an Example, you should immediately do the Practice Problem that follows to make sure you understand each step in solving a particular problem. The worked-out solution to every Practice Problem can be found in the back of the text, starting at page SP-1, so you can check your work and receive immediate guidance in case you need to review.

EXERCISE SETS—Practice, Practice, Practice: You learn math by *doing* math. The best way to learn math is to *practice, practice, practice*. Be sure that you complete every exercise your instructor assigns as homework. In addition, check your answers to the odd-numbered exercises in the back of the text to see whether you have correctly solved each problem.

QUICK QUIZ: After every exercise set, be sure to do the problems in the Quick Quiz. This will tell you immediately if you have understood the key points of the homework exercises.

CONCEPT CHECK: At the end of the Quick Quiz is a Concept Check. This will test your understanding of the key concept of the section. It will ask you to explain in your own words how a procedure works. It will help you clarify your knowledge of the methods of the section.

HOW AM I DOING? MID-CHAPTER REVIEW: This feature allows you to check if you understand the important concepts covered to that point in a particular chapter. Many students find that halfway through a chapter is a crucial point for review because so many different types of problems have been covered. This review covers each of the types of problems from the first half of the chapter. Do these problems and check your answers at the back of the text. If you need to review any of these problems, simply refer back to the section and objective indicated next to the answer.

HOW AM I DOING? CHAPTER TEST: This test (found at the end of every chapter) provides you with an excellent opportunity to both practice and review for any test you will take in class. Take this test to see how much of the chapter you have mastered. By checking your answers, you can once again refer back to the section and the objective of any exercise you want to review further. This allows you to see at once what has been learned and what still needs more study as you prepare for your test or exam.

HOW AM I DOING? CHAPTER TEST PREP VIDEO CD: If you need to review any of the exercises from the *How Am I Doing? Chapter Test,* this video CD found at the front of the text provides a clear explanation of how to do each step of every problem on the test. Simply insert the CD into a computer and watch a math instructor solve each of the Chapter Test exercises in detail. By reviewing these problems, you can study through any points of difficulty and better prepare yourself for your upcoming test or exam.

These steps provide a clear path you can follow in order to successfully complete your math course. More importantly, the *How Am I Doing? Guide to Math Success* is a tool to help you achieve an understanding of mathematics. We encourage you to take advantage of this new feature.

John Tobey and Jeffrey Slater
North Shore Community College

Jamie Blair
Orange Coast College

This book is dedicated to the memory of Lexie Tobey and John Tobey, Sr.
They have left a legacy of love, a memory of four decades of faithful teaching,
and a sense of helping others that will influence generations to come.
For their grandchildren they have left an inspiring model of a loving family,
true character, and service to God and community.

Contents

CHAPTER 0

Prealgebra Review   1

CHAPTER 1

Real Numbers and Variables   57

CHAPTER 4

Systems of Equations

CHAPTER 5

Exponents and Polynomials

CHAPTER 8

CHAPTER 9

TO THE INSTRUCTOR

One of the hallmark characteristics of *Beginning Algebra: Early Graphing* that makes the text easy to learn and teach from is the building-block organization. Each section is written to stand on its own, and every homework set is completely self-testing. Exercises are paired and graded and are of varying levels and types to ensure that all skills and concepts are covered. As a result, the text offers students an effective and proven learning program suitable for a variety of course formats—including lecture-based classes; discussion-oriented classes; distance learning classes; modular, self-paced courses; mathematics laboratories; and computer-supported centers. The book has been written to be especially helpful in online courses. The authors usually teach at least one course online each semester.

   Beginning Algebra: Early Graphing is part of a series that includes the following:

   Tobey/Slater/Blair, *Basic College Mathematics,* Sixth Edition

   Tobey/Slater/Blair, *Essentials of Basic College Mathematics,* Second Edition

   Blair/Tobey/Slater, *Prealgebra,* Fourth Edition

   Tobey/Slater, *Beginning Algebra,* Seventh Edition

   Tobey/Slater/Blair, *Beginning Algebra: Early Graphing,* Second Edition

   Tobey/Slater, *Intermediate Algebra,* Sixth Edition

   Tobey/Slater/Blair, *Beginning and Intermediate Algebra,* Third Edition

We have visited and listened to teachers across the country and have incorporated a number of suggestions into this edition to help you with the particular learning delivery system at your school. The following pages describe the key continuing features and changes in the second edition.

NEW! FEATURES IN THE SECOND EDITION

Quick Quiz

At the end of each problem section, there is a **Quick Quiz** for the student. The quiz contains three problems that cover all of the essential content of that entire section of the book. If a student can do those three problems, the student has mastered the mathematics skills of that section. If a student cannot do those three problems, the student is made aware that further study is needed to obtain mastery. At the end of the Quick Quiz is a **Concept Check** question. This question stresses a mastery of the concepts of each section of the book. The question asks the student to explain how and why a solution method actually works. It forces the student to analyze problems and reflect on the mathematical concepts that have been learned. The student is asked to explain in his or her own words the mathematical procedures that have been practiced in a given section.

Classroom Quiz

Adjacent to each Quick Quiz for the student, the Annotated Instructor's Edition of the book contains a **Classroom Quiz.** This quiz allows the instructor to give a short quiz in class that covers the essential content of each section of the book. Immediately, an instructor can find out if the students have mastered the material in this section or not. No more will an instructor have to rush to pick out the right balance of problems that test a student's knowledge of a section. The Classroom Quiz is instantly available to assist the instructor in assessing student knowledge.

Putting Your Skills to Work: Use Math to Save Money

Each chapter of the book presents a simple, down-to-earth, practical example of how to save money. Students are given a straightforward, realistic way to cut costs and spend less. They are shown logical ways to get out of debt. They are given motivating examples of how other students saved money. Students are very motivated to read these articles and soon begin to see how the course will actually help them in everyday life. Many of these activities were contributed by professors, and our thanks go out to Mary Pearce, Suellen Robinson, Mike Yarbrough, Betty Ludlum, Connie Buller, Armando Perez, and Maria Luisa Mendez for their insightful contributions.

Teaching Examples

Having in-class practice problems readily available is extremely helpful to both new and experienced instructors. These instructor examples, called **Teaching Examples,** are included in the margins of the Annotated Instructor's Edition for practice in class.

KEY FEATURES IN THE SECOND EDITION

The **How Am I Doing? Guide to Math Success** shows students how they can effectively use this textbook to succeed in their mathematics course. This clear path for them to follow is based upon how students have successfully utilized this textbook in the past.

The text has been designed so the **Examples and Practice Problems** are clearly connected in a cohesive unit. This encourages students to try the Practice Problem associated with each Example to ensure they understand each step in solving a particular problem. The worked-out solution to every Practice Problem can be found in the back of the text so students can check their work. **New!** to this edition, the answers are now included next to each Practice Problem in the Annotated Instructor's Edition.

Each **exercise set** progresses from easy to medium to challenging problems, with appropriate quantities of each, and has paired even and odd problems. All concepts are fully represented, with every Example from the section covered by a group of exercises. Exercise sets include **Mixed Practice** problems, which require students to identify the type of problem and the best method they should use to solve it, as well as **Verbal and Writing Skills** exercises, which allow students to explain new concepts fully in their own words. Throughout the text the application exercises have been updated. These **Applications** relate to everyday life, global issues beyond the borders of the United States, and other academic disciplines. Roughly 25 percent of the applications have been contributed by actual students based on scenarios they have encountered in their home or work lives.

Many students find that halfway through a chapter is a crucial point for review because so many different types of problems have been covered. The **How Am I Doing? Mid-Chapter Review** covers each of the types of problems from the first half of the chapter and allows instructors to check if students understand the important concepts covered to that point. Specific section and objective references are provided with each answer to indicate where a student should look for further review.

Developing Problem-Solving Abilities

As authors, we are committed to producing a textbook that emphasizes mathematical reasoning and problem-solving techniques as recommended by AMATYC, NCTM, AMS, NADE, MAA, and other bodies. To this end, the problem sets are built on a wealth of real-life and real-data applications. Unique problems have been developed and incorporated into the exercise sets that help train students in data interpretation, mental mathematics, estimation, geometry and graphing, number sense, critical thinking, and decision making.

The successful **Mathematics Blueprint for Problem Solving** strengthens problem-solving skills by providing a consistent and interactive outline to help students organize their approach to problem solving. Once students fill in the blueprint, they can refer back to their plan as they do what is needed to solve the problem. Because of its flexibility, this feature can be used with single-step problems, multi-step problems, applications, and nonroutine problems that require problem-solving strategies.

The **Developing Your Study Skills** boxes are integrated throughout the text to provide students with techniques for improving their study skills and succeeding in math courses.

Integration and Emphasis on Geometry Due to the emphasis on geometry on many statewide exams, geometry problems are integrated throughout the text. Examples and exercises that incorporate a principle of geometry are marked with a triangle icon for easy identification.

When students encounter mathematics in real-world publications, they often encounter data represented in a **graph, chart, or table** and are asked to make a reasonable conclusion based on the data presented. This emphasis on graphical interpretation is a continuing trend with today's expanding technology. In this text, students are asked to make simple interpretations, to solve medium-level problems, and to investigate challenging applied problems based on the data shown in a chart, graph, or table.

Mastering Mathematical Concepts

Text features that develop the mastery of concepts include the following:

Concise **Learning Objectives** listed at the beginning of each section allow students to preview the goals of that section.

To Think About questions extend the concept being taught, providing the opportunity for all students to stretch their minds, to look for patterns, and to make conclusions based on their previous experience. These critical-thinking questions may follow Examples in the text and appear in the exercise sets.

Almost every exercise set concludes with a section of **Cumulative Review** problems. These problems review topics previously covered, and are designed to assist students in retaining the material.

Calculator boxes are placed in the margin of the text to alert students to a scientific calculator application. In the exercise section a scientific calculator icon is used to indicate problems that are designed for solving with a calculator. There is also instruction on how to use a scientific calculator in an appendix.

Reviewing Mathematical Concepts

At the end of each chapter, we have included problems and tests to provide your students with several different formats to help them review and reinforce the ideas that they have learned. This assists them not only with that specific chapter, but reviews previously covered topics as well.

The concepts and mathematical procedures covered are reviewed at the end of each chapter in a unique **Chapter Organizer.** It lists concepts and methods, and provides a completely worked-out example for each type of problem.

Chapter Review Problems are grouped by section as a quick refresher at the end of the chapter. They can also be used by the student as a quiz of the chapter material.

Found at the end of the chapter, the **How Am I Doing? Chapter Test** is a representative review of the material from that particular chapter that simulates an actual testing format. This provides the students with a gauge of their preparedness for the actual examination.

At the end of each chapter is a **Cumulative Test.** One-half of the content of each cumulative test is based on the math skills learned in previous chapters. By completing these tests for each chapter, the students build confidence that they have mastered not only the contents of the chapter but those of previous chapters as well.

RESOURCES FOR THE STUDENT

Student Solutions Manual
(ISBNs: 0-321-57816-3, 978-0-321-57816-7)

- Solutions to all odd-numbered section exercises
- Solutions to every exercise (even and odd) in the Quick Quiz, mid-chapter reviews, chapter reviews, chapter tests, and cumulative reviews

Worksheets for Classroom or Lab Practice
(ISBNs: 0-321-57824-4, 978-0-321-57824-2)

- Extra practice exercises for every section of the text with ample space for students to show their work

Chapter Test Prep Video CD
Provides step-by-step video solutions to each problem in each How Am I Doing? Chapter Test in the textbook. Automatically included with every new copy of the text, inside the front cover.

Lecture Series on DVD
(ISBNs: 0-321-57831-7, 978-0-321-57831-0)

- Organized by section, contain problem-solving techniques and examples from the textbook
- Step-by-step solutions to selected exercises from each textbook section

MathXL® Tutorials on CD
(ISBNs: 0-321-57826-0, 978-0-321-57826-6)
This interactive tutorial CD-ROM provides:

- Algorithmically generated practice exercises correlated at the objective level
- Practice exercises accompanied by an example and a guided solution
- Tutorial video clips within the exercise to help students visualize concepts
- Easy-to-use tracking of student activity and scores and printed summaries of students' progress

RESOURCES FOR THE INSTRUCTOR

Annotated Instructor's Edition
(ISBNs: 0-321-57821-X, 978-0-321-57821-1)

- Complete student text with answers to all practice problems, section exercises, mid-chapter reviews, chapter reviews, chapter tests, cumulative tests, and practice final exam.
- Teaching Tips placed in the margin at key points where students historically need extra help
- New! Teaching Examples provide in-class practice problems and are placed in the margins accompanying each example.

Instructor's Solutions Manual
(ISBNs: 0-321-57818-X, 978-0-321-57818-1)

- Detailed step-by-step solutions to the even-numbered section exercises
- Solutions to every exercise (odd and even) in the Classroom Quiz, mid-chapter reviews, chapter reviews, chapter tests, cumulative tests, and practice final

Instructor's Resource Manual with Tests and Mini-Lectures
(ISBNs: 0-321-57819-8, 978-0-321-57819-8)

- For each section there is one Mini-Lecture with key learning objectives, classroom examples, and teaching notes.

- Two short group activities per chapter are provided in a convenient ready-to-use handout format.
- Three forms of additional practice exercises that help instructors support students of different ability and skill levels.
- Answers are included for all items.
- Alternate test forms with answers:
  - Two Chapter Pretests per chapter (1 free response, 1 multiple choice)
  - Six Chapter Tests per chapter (3 free response, 3 multiple choice)
  - Two Cumulative Tests per even-numbered chapter (1 free response, 1 multiple choice)
  - Two Final Exams (1 free response, 1 multiple choice)

TestGen®

- Enables instructors to build, edit, print, and administer tests.
- Features a computerized bank of questions developed to cover all text objectives.
- Creates multiple but equivalent versions of the same question or test with the click of a button.
- Instructors can modify questions or add new questions.
- Tests can be printed or administered online.

The software and testbank are available for download from Pearson Education's online catalog.

Pearson Adjunct Support Center

The Pearson Adjunct Support Center is staffed by qualified mathematics instructors with more than 50 years of combined experience at both the community college and university level. Assistance is provided for faculty in the following areas:

- Suggested syllabus consultation
- Tips on using materials packed with your book
- Book-specific content assistance
- Teaching suggestions including advice on classroom strategies

MEDIA RESOURCES

MathXL® www.mathxl.com

MathXL is a powerful online homework, tutorial, and assessment system that accompanies Pearson Education textbooks in mathematics and statistics. With MathXL, instructors can create, edit, and assign online homework and tests using algorithmically generated exercises correlated at the objective level to the textbook. They can also create and assign their own online exercises and import TestGen tests for added flexibility. All student work is tracked in MathXL's online gradebook. Students can take chapter tests in MathXL and receive personalized study plans based on their test results. The study plan diagnoses weaknesses and links students directly to tutorial exercises for the objectives they need to study and retest. Students can also access supplemental animations and video clips directly from selected exercises. MathXL is available to qualified adopters. For more information, visit our Web site at www.mathxl.com, or contact your sales representative.

MyMathLab® www.mymathlab.com

MyMathLab is a series of text-specific, easily customizable online courses for Pearson Education textbooks in mathematics and statistics. Powered by CourseCompass™ (our online teaching and learning environment) and MathXL® (our online homework, tutorial, and assessment system), MyMathLab gives instructors the tools they need to deliver all or a portion of their course online, whether students are in a lab or working at home or elsewhere. MyMathLab provides a rich and flexible set of course materials, featuring free-response exercises that are algorithmically generated for unlimited practice and mastery. Students can also use online tools, such as video lectures, animations, and a multimedia textbook, to independently improve their understanding and performance. Instructors can use MyMathLab's homework and test managers to select and assign online exercises correlated directly to the textbook, and they can create and assign their own online exercises and import TestGen tests for added flexibility. MyMathLab's online gradebook—designed specifically for mathematics and statistics—automatically tracks students' homework and test results and gives the instructor control over how to calculate final grades. Instructors can also add offline (paper-and-pencil) grades to the gradebook. MyMathLab also includes access to the Pearson Tutor Center, which provides students with tutoring via toll-free phone, fax, e-mail, and interactive Web sessions. MyMathLab is available to qualified adopters. For more information, visit our Web site at www.mymathlab.com, or contact your sales representative.

ACKNOWLEDGMENTS

This book is the product of many years of work and many contributions from faculty and students across the country. We would like to thank the many reviewers and participants in focus groups and special meetings with the authors in preparation of previous editions. Our deep appreciation to each of the following:

Carla Ainsworth, *Salt Lake Community College*

George J. Apostolopoulos, *DeVry Institute of Technology*

Mary Lou Baker, *Columbia State Community College*

Jana Barnard, *Angelo State University*

Katherine Barringer, *Central Virginia Community College*

Linda Beattie, *Western New Mexico University*

Mark Billiris, *St. Petersburg Junior College*

Jamie Blair, *Orange Coast College*

Larry Blevins, *Tyler Junior College*

Lydia Botsford, *Heald Business College, Milpitas*

Bonnie Brooks, *Sussex County Community College*

David Brown, *Heald Business College, Vacaville*

John Buckley, *Delaware Technical and Community College*

Robert Christie, *Miami-Dade Community College*

Gene Coco, *Heald Business College, Roseville*

Nelson Collins, *Joliet Junior College*

Mike Contino, *California State University at Heyward*

Kevin Cooper, *National American University*

Gregory Daubenmire, *Las Positas College*

Judy Dechene, *Fitchburg State University*

Floyd L. Downs, *Arizona State University*

Robert Dubuc, Jr., *New England Institute of Technology*

Barbara Edwards, *Portland State University*

Mahmoud El-Hashash, *Bridgewater State College*

Janice F. Gahan-Rech, *University of Nebraska at Omaha*

Naomi Gibbs, *Pitt Community College*

Mark Glines, *Salt Lake Community College*

Cynthia Gubitose, *Southern Connecticut State University*

Renu Gupta, *Louisiana State University, Alexandria*

Philip Hammersley, *Ivy Tech Community College*

Mary Beth Headlee, *Manatee Community College*

Autumn Hoover, *Angelo State University*

Laura Kaufmann, *Orange Coast College*

Carolyn Krause, *Delaware Technical and Community College*

Euniz Lochte, *Florida Metropolitan University, Orlando North*

Elizabeth Lucas, *North Shore Community College*

Jim Lynn, *Florida Metropolitan University, Orlando North*

Doug Mace, *Baker College*

Valerie Maley, *Cape Fear Community College*

Carl Mancuso, *William Paterson College*

James Matovina, *Community College of Southern Nevada*

Todd Mattson, *DeVry University, DuPage Campus*

Jim McKee, *Florida Metropolitan University, Tampa*

Janet McLaughlin, *Montclair State College*

Beverly Meyers, *Jefferson College*

Nancy Meyers, *University of Southern Indiana*

Wayne L. Miller, *Lee College*

Gloria Mills, *Tarrant County Junior College*

Norman Mittman, *Northeastern Illinois University*

Kathy Morgan, *Coastal Carolina Community College*

Christina Morian, *Lincoln University*

Sharon L. Morrison, *St. Petersburg Junior College*

Joe Nagengast, *Florida Career College*

Charlcie Neal, *Clayton College and State University*

Jim Osborn, *Baker College*

Linda Padilla, *Joliet Junior College*

Cathy Panik, *Manatee Community College*

Elizabeth A. Polen, *County College of Morris*

Larry Pontaski, *Pueblo Community College*

Joel Rappaport, *Miami-Dade Community College*

Juan Rivera, *Florida Metropolitan University, Orlando South*

Graciela Rodriguez, *Laredo Community College*

Ronald Ruemmler, *Middlesex County College*

Dennis Runde, *Manatee Community College*

Sally Search, *Tallahassee Community College*

Jeff Simmons, *Ivy Tech Community College*

Carolyn Gigi Smith, *Armstrong Atlantic State University*

Jed Soifer, *Atlantic Cape Community College*

Lee Ann Spahr, *Durham Technical Community College*

Richard Sturgeon, *University of Southern Maine*

Ara B. Sullenberger, *Tarrant County Community College*

Gwen Terwilliger, *University of Toledo*

Sharon Testone, *Onondaga Community College*

Margie Thrall, *Manatee Community College*

Michael Trappuzanno, *Arizona State University*

Georgina Vastola, *Middlesex Community College*

Jonathan Weissman, *Essex County College*

Alice Williamson, *Sussex County Community College*

Mike Wilson, *Florida Metropolitan University, Tampa*

Jerry Wisnieski, *Des Moines Community College*

Mary Jane Wolfe, *University of Rio Grande*

Michelle Younker, *Terra Community College*

Peter Zimmer, *West Chester University*

In addition, we want to thank the following individuals for providing splendid insight and suggestions for this new edition:

Suzanne Battista, *St. Petersburg Junior College, Clearwater*

Annette Benbow, *Tarrant County College, Northwest*

Karen Bingham, *Clarion College*

Nadine Branco, *Western Nevada Community College*

Connie Buller, *Metropolitan Community College*

Judy Carter, *North Shore Community College*

John Close, *Salt Lake Community College*

Cecil Ellard, *Ivy Tech Community College, Bloomington*

Colin Godfrey, *University of Massachusetts, Boston*

Shanna Goff, *Grand Rapids Community College*

Edna Greenwood, *Tarrant County College, Northwest*

Veronica Hupper, *Hesser College, Manchester*

Peter Kaslik, *Pierce College*

Joyce Keenan, *Horry-Georgetown Technical College*

Rajeesh Lal, *Pierce College*

Nam Lee, *Griffin Technical Institute*

Tanya Lee, *Career Technical College*

Betty Ludlum, *Austin Community College*

Phyllis Lurvey, *Hesser College, Manchester*

Mary Marlin, *West Virginia Northern Community College*

Carolyn T. McIntyre, *Horry-Georgetown Technical College*

Maria Luisa Mendez, *Laredo Community College*

Steven J. Meyer, *Erie Community College*

Marcia Mollé, *Metropolitan Community College*

Jay L. Novello, *Horry-Georgetown Technical College*

Sandra Lee Orr, *West Virginia State University*

Mary Pearce, *Wake Technical Community College*

Armando Perez, *Laredo Community College*

Regina Pierce, *Davenport University*

Anne Praderas, *Austin Community College*

Jose Rico, *Laredo Community College*

Suellen Robinson, *North Shore Community College*

Kathy Ruggieri, *Lansdale School of Business*

Randy Smith, *Des Moines Area Community College*

Dina Spain, *Horry-Georgetown Technical College*

Lori Welder, *PACE Institute*

Michael Yarbrough, *Cosumnes River College*

We have been greatly helped by a supportive group of colleagues who not only teach at North Shore Community College and Orange Coast College, but have also provided a number of ideas as well as extensive help on all of our mathematics books. Our special best wishes to our colleague Bob Campbell, who recently retired. He has given us a friendly smile and encouraging ideas for 35 years! Also, a special word of thanks to Wally Hersey, Judy Carter, Rick Ponticelli, Lora Connelly, Sharyn Sharaf, Donna Stefano, Nancy Tufo, Elizabeth Lucas, Anne O'Shea, Marsha Pease, Walter Stone, Evangeline Cornwall, Rumiya Masagutova, Charles Peterson, and Neha Jain.

Jenny Crawford provided major contributions to this revision. She provided new problems, new ideas, and great mental energy. She greatly assisted us during the production process. She made helpful decisions. Her excellent help was much appreciated. She has become an essential part of our team as we work to provide the best possible textbook.

A special word of thanks goes to Cindy Trimble and Associates for their excellent work in accuracy checking manuscript and page proofs.

Each textbook is a combination of ideas, writing, and revisions from the authors and wise editorial direction and assistance from the editors. We especially want to thank our editor at Pearson Education—Paul Murphy. He has a true vision of how authors and editors can work together as partners, and it has been a rewarding experience to create and revise textbooks together with him. We especially want to thank our Project Manager Lauren Morse for patiently answering questions and solving many daily problems. We also want to thank our entire team at Pearson Education—Marlana Voerster, Nathaniel Koven, Christine Whitlock, Georgina Brown, Linda Behrens, Tom Benfatti, Heather Scott, Ilene Kahn, Audra Walsh, and MiMi Yeh—as well as Allison Campbell and Karin Kipp at Elm Street Publishing Services for their assistance during the production process.

Nancy Tobey served as our administrative assistant. Daily she was involved with mailing, photocopying, collating, and taping. A special thanks goes to Nancy. We could not have finished the book without you.

Book writing is impossible for us without the loyal support of our families. Our deepest thanks and love to Nancy, Johnny, Melissa, Marcia, Shelley, Rusty, Abby, Jerry, Joe, and Wendy. Your understanding, your love and help, and your patience have been a source of great encouragement. Finally, we thank God for the strength and energy to write and the opportunity to help others through this textbook.

We have spent more than 37 years teaching mathematics. Each teaching day, we find that our greatest joy is helping students learn. We take a personal interest in ensuring that each student has a good learning experience in taking this course. If you have some personal comments, suggestions, or ideas for future editions of this textbook, please write to us at:

Prof. John Tobey, Prof. Jeffrey Slater, and Prof. Jamie Blair
Pearson Education
Office of the College Mathematics Editor
75 Arlington Street, Suite 300
Boston, MA 02116

or e-mail us at

jtobey@northshore.edu

We wish you success in this course and in your future life!

John Tobey
Jeffrey Slater
Jamie Blair

Follow the directions for each question. Simplify each answer.

Chapter 0

1. Add. $3\frac{1}{4} + 2\frac{3}{5}$

2. Multiply. $\left(1\frac{1}{6}\right)\left(2\frac{2}{3}\right)$

3. Divide. $\frac{15}{4} \div \frac{3}{8}$

4. Multiply. $(1.63)(3.05)$

5. Divide. $120 \div 0.0006$

6. Find 7% of 64,000.

Chapter 1

7. Add. $-3 + (-4) + (+12)$

8. Subtract. $-20 - (-23)$

9. Combine. $5x - 6xy - 12x - 8xy$

10. Evaluate $2x^2 - 3x - 4$ when $x = -3$.

11. Remove the grouping symbols. $2 - 3\{5 + 2[x - 4(3 - x)]\}$

12. Evaluate. $-3(2 - 6)^2 + (-12) \div (-4)$

Chapter 2

In questions 13 and 14, solve each equation for x.

13. $7(3x - 1) = 5 + 4(x - 3)$

14. $\frac{2}{3}x - \frac{3}{4} = \frac{1}{6}x + \frac{21}{4}$

15. Solve for x and graph the result. $42 - 18x < 48x - 24$

16. The length of a rectangle is 7 meters longer than twice the width. The perimeter is 46 meters. Find the dimensions.

17. Hector has four test scores of 80, 90, 83, and 92. What does he need to score on the fifth test to have an average of 86 on the five tests?

18. The drama club put on a play for Thursday, Friday, and Saturday nights. The total attendance for the three nights was 6210. Thursday night had 300 fewer people than Friday night. Saturday night had 510 more people than Friday night. How many people came each night?

1. _____

2. _____

3. _____

4. _____

5. _____

6. _____

7. _____

8. _____

9. _____

10. _____

11. _____

12. _____

13. _____

14. _____

15. _____

16. _____

17. _____

18. _____

19. _____

20. _____

21. _____

22. _____

23. _____

24. _____

25. _____

26. _____

27. _____

28. _____

Chapter 3

19. Graph. $y = 2x - 4$

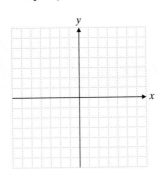

20. Graph. $3x + 4y = -12$

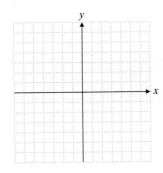

21. What is the slope of a line passing through $(6, -2)$ and $(-3, 4)$?

22. If $f(x) = 2x^2 - 3x + 1$, find $f(3)$.

23. Graph the region. $y \geq -\dfrac{1}{3}x + 2$

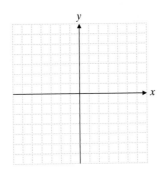

24. Find the equation of a line with a slope of $\frac{3}{5}$ that passes through the point $(-1, 3)$.

Chapter 4

Solve each system by the appropriate method.

25. Substitution method
$$x + y = 17$$
$$2x - y = -5$$

26. Addition method
$$-5x + 4y = 8$$
$$2x + 3y = 6$$

27. Any method
$$2(x - 2) = 3y$$
$$6x = -3(4 + y)$$

28. Any method
$$x + \frac{1}{3}y = \frac{10}{3}$$
$$\frac{3}{2}x + y = 8$$

29. Is $(2, -3)$ a solution for the following system?

$3x + 5y = -9$

$2x - 3y = 13$

30. A man bought three pairs of gloves and four scarves for $53. A woman bought two pairs of the same-priced gloves and three of the same-priced scarves for $38. How much did each item cost?

Chapter 5

31. Multiply. $(-2xy^2)(-4x^3y^4)$

32. Divide. $\dfrac{36x^5y^6}{-18x^3y^{10}}$

33. Raise to the indicated power. $(-2x^3y^4)^5$

34. Evaluate. $(-3)^{-4}$

35. Multiply. $(3x^2 + 2x - 5)(4x - 1)$

36. Divide. $(x^3 + 6x^2 - x - 30) \div (x - 2)$

Chapter 6

Factor completely.

37. $5x^2 - 5$

38. $x^2 - 12x + 32$

39. $8x^2 - 2x - 3$

40. $3ax - 8b - 6a + 4bx$

Solve for x.

41. $16x^2 - 24x + 9 = 0$

42. $\dfrac{x^2 + 8x}{5} = -3$

Chapter 7

43. Simplify. $\dfrac{x^2 + 3x - 18}{2x - 6}$

44. Multiply. $\dfrac{6x^2 - 14x - 12}{6x + 4} \cdot \dfrac{x + 3}{2x^2 - 2x - 12}$

45. Divide and simplify.

$\dfrac{x^2}{x^2 - 4} \div \dfrac{x^2 - 3x}{x^2 - 5x + 6}$

46. Add.

$\dfrac{3}{x^2 - 7x + 12} + \dfrac{4}{x^2 - 9x + 20}$

47. Solve for x.

$2 - \dfrac{5}{2x} = \dfrac{2x}{x + 1}$

48. Simplify.

$\dfrac{3 + \dfrac{1}{x}}{\dfrac{9}{x} + \dfrac{3}{x^2}}$

29. _____

30. _____

31. _____

32. _____

33. _____

34. _____

35. _____

36. _____

37. _____

38. _____

39. _____

40. _____

41. _____

42. _____

43. _____

44. _____

45. _____

46. _____

47. _____

48. _____

49. _____

50. _____

51. _____

52. _____

53. _____

54. _____

55. _____

56. _____

57. _____

58. _____

59. _____

60. _____

Chapter 8

49. Evaluate. $\sqrt{121}$

50. Simplify. $\sqrt{125x^3y^5}$

51. Multiply and simplify.

$$\left(\sqrt{2} + \sqrt{6}\right)\left(2\sqrt{2} - 3\sqrt{6}\right)$$

52. Rationalize the denominator.

$$\frac{\sqrt{5} - \sqrt{3}}{\sqrt{6}}$$

53. In the right triangle with sides a, b, and c, find side c if side $a = 4$ and side $b = 6$.

54. y varies directly with x. When $y = 56$, then $x = 8$. Find y when $x = 11$.

Chapter 9

In questions 55–58, solve for x.

55. $14x^2 + 21x = 0$

56. $2x^2 + 1 = 19$

57. $2x^2 - 4x - 5 = 0$

58. $x^2 - x + 8 = 5 + 6x$

59. Graph the equation. $y = x^2 + 8x + 15$

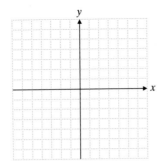

60. A rectangle is 4 inches longer in length than in width. The area of the rectangle is 96 square inches. Find the length and the width of the rectangle.

CHAPTER

0

What image comes to mind when you think of the Internal Revenue Service? A once-a-year challenge to get all your financial records together? Whether you file income taxes online or on paper, it is a difficult task for most people. However, if you master the mathematics in this chapter, you will be able to accurately complete your income tax returns. It is just one more way that mathematical knowledge will help you.

Prealgebra Review

Student Learning Objectives

After studying this section, you will be able to:

 1 Understand basic mathematical definitions.

2 Simplify fractions to lowest terms using prime numbers.

3 Convert between improper fractions and mixed numbers.

4 Change a fraction to an equivalent fraction with a given denominator.

Chapter 0 is designed to give you a mental "warm-up." In this chapter you'll be able to step back a bit and tone up your math skills. This brief review of prealgebra will increase your math flexibility and give you a good running start into algebra.

1 Understanding Basic Mathematical Definitions

Whole numbers are the set of numbers 0, 1, 2, 3, 4, 5, 6, 7, ... They are used to describe whole objects, or entire quantities.

Fractions are a set of numbers that are used to describe parts of whole quantities. In the object shown in the figure there are four equal parts. The *three* of the *four* parts that are shaded are represented by the fraction $\frac{3}{4}$. In the fraction $\frac{3}{4}$ the number 3 is called the **numerator** and the number 4, the **denominator.**

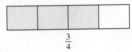

$$\frac{3}{4}$$

$3 \leftarrow$ *Numerator* is on the top

$4 \leftarrow$ *Denominator* is on the bottom

The *denominator* of a fraction shows the number of equal parts in the whole and the *numerator* shows the number of these parts being talked about or being used.

Numerals are symbols we use to name numbers. There are many different numerals that can be used to describe the same number. We know that $\frac{1}{2} = \frac{2}{4}$. The fractions $\frac{1}{2}$ and $\frac{2}{4}$ both describe the same number.

Usually, we find it more useful to use fractions that are simplified. A fraction is considered to be in **simplest form** or **reduced form** when the numerator (top) and the denominator (bottom) can both be divided exactly by no number other than 1, and the denominator is greater than 1.

$\frac{1}{2}$ is in simplest form.

$\frac{2}{4}$ is *not* in simplest form, since the numerator and the denominator can both be divided by 2.

If you get the answer $\frac{2}{4}$ to a problem, you should state it in simplest form, $\frac{1}{2}$. The process of changing $\frac{2}{4}$ to $\frac{1}{2}$ is called **simplifying** or **reducing** the fraction.

2 Simplifying Fractions to Lowest Terms Using Prime Numbers

Natural numbers or **counting numbers** are the set of whole numbers excluding 0. Thus the natural numbers are the numbers 1, 2, 3, 4, 5, 6, ...

When two or more numbers are multiplied, each number that is multiplied is called a **factor.** For example, when we write $3 \times 7 \times 5$, each of the numbers 3, 7, and 5 is called a factor.

Prime numbers are natural numbers greater than 1 whose only natural number factors are 1 and themselves. The number 5 is prime. The only natural number factors of 5 are 5 and 1.

$$5 = 5 \times 1$$

The number 6 is not prime. The natural number factors of 6 are 3 and 2 or 6 and 1.

$$6 = 3 \times 2 \qquad 6 = 6 \times 1$$

The first 15 prime numbers are

$$2, 3, 5, 7, 11, 13, 17, 19, 23, 29, 31, 37, 41, 43, 47.$$

Any natural number greater than 1 either is prime or can be written as the product of prime numbers. For example, we can take each of the numbers 12, 30, 14,

19, and 29 and either indicate that they are prime or, if they are not prime, write them as the product of prime numbers. We write as follows:

$$12 = 2 \times 2 \times 3 \qquad 30 = 2 \times 3 \times 5 \qquad 14 = 2 \times 7$$

19 is a prime number.   29 is a prime number.

To reduce a fraction, we use prime numbers to factor the numerator and the denominator. Write each part of the fraction (numerator and denominator) as a product of prime numbers. Note any *factors* that appear in both the *numerator* (top) and *denominator* (bottom) of the fraction. If we divide numerator and denominator by these values we will obtain an equivalent fraction in *simplest form*. When the new fraction is simplified, it is said to be in **lowest terms.** Throughout this text, to *simplify* a fraction will always mean to write the fraction in lowest terms.

EXAMPLE 1 Simplify each fraction.

(a) $\dfrac{14}{21}$ (b) $\dfrac{15}{35}$ (c) $\dfrac{20}{70}$

Solution

(a) $\dfrac{14}{21} = \dfrac{\cancel{7} \times 2}{\cancel{7} \times 3} = \dfrac{2}{3}$ We factor 14 and factor 21. Then we divide numerator and denominator by 7.

(b) $\dfrac{15}{35} = \dfrac{\cancel{5} \times 3}{\cancel{5} \times 7} = \dfrac{3}{7}$ We factor 15 and factor 35. Then we divide numerator and denominator by 5.

(c) $\dfrac{20}{70} = \dfrac{2 \times \cancel{2} \times \cancel{5}}{7 \times \cancel{2} \times \cancel{5}} = \dfrac{2}{7}$ We factor 20 and factor 70. Then we divide numerator and denominator by both 2 and 5.

Practice Problem 1 Simplify.

(a) $\dfrac{10}{16}$ (b) $\dfrac{24}{36}$ (c) $\dfrac{36}{42}$

Sometimes when we simplify a fraction, all the prime factors in the top (numerator) are divided out. When this happens, we must remember that a 1 is left in the numerator.

EXAMPLE 2 Simplify each fraction.

(a) $\dfrac{7}{21}$ (b) $\dfrac{15}{105}$

Solution

(a) $\dfrac{7}{21} = \dfrac{\cancel{7} \times 1}{\cancel{7} \times 3} = \dfrac{1}{3}$ (b) $\dfrac{15}{105} = \dfrac{\cancel{5} \times \cancel{3} \times 1}{7 \times \cancel{5} \times \cancel{3}} = \dfrac{1}{7}$

Practice Problem 2 Simplify.

(a) $\dfrac{4}{12}$ (b) $\dfrac{25}{125}$ (c) $\dfrac{73}{146}$

If all the prime numbers in the bottom (denominator) are divided out, we do not need to leave a 1 in the denominator, since we do not need to express the answer as a fraction. The answer is then a whole number and is not usually expressed as a fraction.

EXAMPLE 3   Simplify each fraction.

(a) $\dfrac{35}{7}$ **(b)** $\dfrac{70}{10}$

Solution

(a) $\dfrac{35}{7} = \dfrac{5 \times \cancel{7}}{\cancel{7} \times 1} = 5$ **(b)** $\dfrac{70}{10} = \dfrac{7 \times \cancel{5} \times \cancel{2}}{\cancel{5} \times \cancel{2} \times 1} = 7$

Practice Problem 3   Simplify.

(a) $\dfrac{18}{6}$ **(b)** $\dfrac{146}{73}$ **(c)** $\dfrac{28}{7}$

Sometimes the fraction we use represents how many of a certain thing are successful. For example, if a baseball player was at bat 30 times and achieved 12 hits, we could say that he had a hit $\frac{12}{30}$ of the time. If we reduce the fraction, we could say he had a hit $\frac{2}{5}$ of the time.

EXAMPLE 4   Cindy got 48 out of 56 questions correct on a test. Write this as a fraction in simplest form.

Solution   Express as a fraction in simplest form the number of correct responses out of the total number of questions on the test.

$$48 \text{ out of } 56 \rightarrow \dfrac{48}{56} = \dfrac{6 \times \cancel{8}}{7 \times \cancel{8}} = \dfrac{6}{7}$$

Cindy answered the questions correctly $\frac{6}{7}$ of the time.

Practice Problem 4   The major league pennant winner in 1917 won 56 games out of 154 games played. Express as a fraction in simplest form the number of games won in relation to the number of games played.

NOTE TO STUDENT: Fully worked-out solutions to all of the Practice Problems can be found at the back of the text starting at page SP-1

The number *one* can be expressed as $1, \frac{1}{1}, \frac{2}{2}, \frac{6}{6}, \frac{8}{8}$, and so on, since

$$1 = \dfrac{1}{1} = \dfrac{2}{2} = \dfrac{6}{6} = \dfrac{8}{8}.$$

We say that these numerals are *equivalent ways* of writing the number *one* because they all express the same quantity even though they appear to be different.

SIDELIGHT: The Multiplicative Identity
When we simplify fractions, we are actually using the fact that we can multiply any number by 1 without changing the value of that number. (Mathematicians call the number 1 the **multiplicative identity** because it leaves any number it multiplies with the same identical value as before.)
Let's look again at one of the previous examples.

$$\dfrac{14}{21} = \dfrac{7 \times 2}{7 \times 3} = \dfrac{7}{7} \times \dfrac{2}{3} = 1 \times \dfrac{2}{3} = \dfrac{2}{3}$$

So we see that

$$\dfrac{14}{21} = \dfrac{2}{3}$$

When we simplify fractions, we are using this property of multiplying by 1.

③ Converting Between Improper Fractions and Mixed Numbers

If the numerator is less than the denominator, the fraction is a **proper fraction.** A proper fraction is used to describe a quantity smaller than a whole.

Fractions can also be used to describe quantities larger than a whole. The following figure shows two bars that are equal in size. Each bar is divided into 5 equal pieces. The first bar is shaded completely. The second bar has 2 of the 5 pieces shaded.

The shaded-in region can be represented by $\frac{7}{5}$ since 7 of the pieces (each of which is $\frac{1}{5}$ of a whole box) are shaded. The fraction $\frac{7}{5}$ is called an improper fraction. An **improper fraction** is one in which the numerator is larger than or equal to the denominator.

The shaded-in region can also be represented by 1 whole added to $\frac{2}{5}$ of a whole, or $1 + \frac{2}{5}$. This is written as $1\frac{2}{5}$. The fraction $1\frac{2}{5}$ is called a mixed number. A **mixed number** consists of a whole number added to a proper fraction (the numerator is smaller than the denominator). The addition is understood but not written. When we write $1\frac{2}{5}$, it represents $1 + \frac{2}{5}$. The numbers $1\frac{7}{8}$, $2\frac{3}{4}$, $8\frac{1}{3}$, and $126\frac{1}{10}$ are all mixed numbers. From the preceding figure it seems clear that $\frac{7}{5} = 1\frac{2}{5}$. This suggests that we can change from one form to the other without changing the value of the fraction.

From a picture it is easy to see how to *change improper fractions to mixed numbers*. For example, suppose we start with the fraction $\frac{11}{3}$ and represent it by the following figure (where 11 of the pieces, each of which is $\frac{1}{3}$ of a box, are shaded). We see that $\frac{11}{3} = 3\frac{2}{3}$, since 3 whole boxes and $\frac{2}{3}$ of a box are shaded.

Changing Improper Fractions to Mixed Numbers   You can follow the same procedure without a picture. For example, to change $\frac{11}{3}$ to a mixed number, we can do the following:

$$\frac{11}{3} = \frac{3}{3} + \frac{3}{3} + \frac{3}{3} + \frac{2}{3}$$ By the rule for adding fractions (which is discussed in detail in Section 0.2)

$$= 1 + 1 + 1 + \frac{2}{3}$$ Write 1 in place of $\frac{3}{3}$, since $\frac{3}{3} = 1$.

$$= 3 + \frac{2}{3}$$ Write 3 in place of $1 + 1 + 1$.

$$= 3\frac{2}{3}$$ Use the notation for mixed numbers.

Now that you know how to change improper fractions to mixed numbers and why the procedure works, here is a shorter method.

> **TO CHANGE AN IMPROPER FRACTION TO A MIXED NUMBER**
> 1. Divide the denominator into the numerator.
> 2. The quotient is the whole-number part of the mixed number.
> 3. The remainder from the division will be the numerator of the fraction. The denominator of the fraction remains unchanged.

We can write the fraction as a division statement and divide. The arrows show how to write the mixed number.

$$\frac{7}{5} \qquad 5)\overline{7} \qquad \text{Whole-number part} \qquad \text{Numerator of fraction}$$
$$\frac{5}{2} \qquad \text{Remainder} \qquad 1\frac{2}{5}$$

Thus, $\frac{7}{5} = 1\frac{2}{5}$.

$$\frac{11}{3} \qquad 3\overline{)11} \qquad \text{Whole-number part} \qquad\qquad \text{Numerator of fraction}$$

$$\underline{9}$$
$$2 \qquad \text{Remainder}$$

$$\longrightarrow 3\frac{2}{3}\longleftarrow$$

Thus, $\dfrac{11}{3} = 3\dfrac{2}{3}$.

Sometimes the remainder is 0. In this case, the improper fraction changes to a whole number.

EXAMPLE 5 Change to a mixed number or to a whole number.

(a) $\dfrac{7}{4}$ **(b)** $\dfrac{15}{3}$

Solution

(a) $\dfrac{7}{4} = 7 \div 4 \qquad 4\overline{)7}$
$$\underline{4}$$
$$3 \quad \text{Remainder}$$

Thus $\dfrac{7}{4} = 1\dfrac{3}{4}$

(b) $\dfrac{15}{3} = 15 \div 3 \qquad 3\overline{)15}$
$$\underline{15}$$
$$0 \quad \text{Remainder}$$

Thus $\dfrac{15}{3} = 5$

Practice Problem 5 Change to a mixed number or to a whole number.

(a) $\dfrac{12}{7}$ **(b)** $\dfrac{20}{5}$

NOTE TO STUDENT: Fully worked-out solutions to all of the Practice Problems can be found at the back of the text starting at page SP-1

Changing Mixed Numbers to Improper Fractions It is not difficult to see how to change mixed numbers to improper fractions. Suppose that you wanted to write $2\frac{2}{3}$ as an improper fraction.

$$2\frac{2}{3} = 2 + \frac{2}{3} \qquad \text{The meaning of mixed number notation}$$

$$= 1 + 1 + \frac{2}{3} \qquad \text{Since } 1 + 1 = 2$$

$$= \frac{3}{3} + \frac{3}{3} + \frac{2}{3} \qquad \text{Since } 1 = \frac{3}{3}$$

When we draw a picture of $\frac{3}{3} + \frac{3}{3} + \frac{2}{3}$, we have this figure:

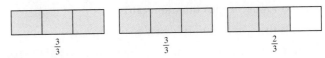

$$\frac{3}{3} \qquad\qquad \frac{3}{3} \qquad\qquad \frac{2}{3}$$

If we count the shaded parts, we see that

$$\frac{3}{3} + \frac{3}{3} + \frac{2}{3} = \frac{8}{3}. \qquad \text{Thus} \quad 2\frac{2}{3} = \frac{8}{3}.$$

Now that you have seen how this change can be done, here is a shorter method.

TO CHANGE A MIXED NUMBER TO AN IMPROPER FRACTION

1. Multiply the whole number by the denominator.
2. Add this to the numerator. The result is the new numerator. The denominator does not change.

EXAMPLE 6   Change to an improper fraction.

(a) $3\frac{1}{7}$ **(b)** $5\frac{4}{5}$

Solution

(a) $3\frac{1}{7} = \frac{(3 \times 7) + 1}{7} = \frac{21 + 1}{7} = \frac{22}{7}$

(b) $5\frac{4}{5} = \frac{(5 \times 5) + 4}{5} = \frac{25 + 4}{5} = \frac{29}{5}$

Practice Problem 6   Change to an improper fraction.

(a) $3\frac{2}{5}$ **(b)** $1\frac{3}{7}$ **(c)** $2\frac{6}{11}$ **(d)** $4\frac{2}{3}$

---

④ Changing a Fraction to an Equivalent Fraction with a Given Denominator

A fraction can be changed to an equivalent fraction with a different denominator by multiplying both numerator and denominator by the same number.

$$\frac{5}{6} = \frac{5 \times 2}{6 \times 2} = \frac{10}{12} \qquad \frac{3}{7} = \frac{3 \times 3}{7 \times 3} = \frac{9}{21}$$

So $\frac{5}{6}$ is equivalent to $\frac{10}{12}$. $\frac{3}{7}$ is equivalent to $\frac{9}{21}$.

We often multiply in this way to obtain an equivalent fraction with a *particular denominator*.

EXAMPLE 7   Find the missing number.

(a) $\frac{3}{5} = \frac{?}{25}$ **(b)** $\frac{4}{7} = \frac{?}{14}$ **(c)** $\frac{2}{9} = \frac{?}{36}$

Solution

(a) $\frac{3}{5} = \frac{?}{25}$ Observe that we need to multiply the denominator by 5 to obtain 25. So we multiply the numerator 3 by 5 also.

$\frac{3 \times 5}{5 \times 5} = \frac{15}{25}$ The desired numerator is 15.

(b) $\frac{4}{7} = \frac{?}{14}$ Observe that $7 \times 2 = 14$. We need to multiply the numerator by 2 to get the new numerator.

$\frac{4 \times 2}{7 \times 2} = \frac{8}{14}$ The desired numerator is 8.

(c) $\frac{2}{9} = \frac{?}{36}$ Observe that $9 \times 4 = 36$. We need to multiply the numerator by 4 to get the new numerator.

$\frac{2 \times 4}{9 \times 4} = \frac{8}{36}$ The desired numerator is 8.

Practice Problem 7   Find the missing number.

(a) $\frac{3}{8} = \frac{?}{24}$ **(b)** $\frac{5}{6} = \frac{?}{30}$ **(c)** $\frac{2}{7} = \frac{?}{56}$

Verbal and Writing Skills

1. In the fraction $\frac{12}{13}$, what number is the numerator?

2. In the fraction $\frac{13}{17}$, what number is the denominator?

3. What is a factor? Give an example.

4. Give some examples of the number 1 written as a fraction.

5. Draw a diagram to illustrate $2\frac{2}{3}$.

6. Draw a diagram to illustrate $3\frac{3}{4}$.

Simplify each fraction.

7. $\frac{18}{24}$

8. $\frac{20}{35}$

9. $\frac{12}{36}$

10. $\frac{12}{48}$

11. $\frac{60}{12}$

12. $\frac{45}{15}$

13. $\frac{24}{36}$

14. $\frac{32}{64}$

15. $\frac{30}{85}$

16. $\frac{33}{55}$

17. $\frac{42}{54}$

18. $\frac{63}{81}$

Change to a mixed number.

19. $\frac{17}{6}$

20. $\frac{19}{5}$

21. $\frac{47}{5}$

22. $\frac{54}{7}$

23. $\frac{38}{7}$

24. $\frac{41}{6}$

25. $\frac{41}{2}$

26. $\frac{25}{3}$

27. $\frac{32}{5}$

28. $\frac{79}{7}$

29. $\frac{111}{9}$

30. $\frac{124}{8}$

Change to an improper fraction.

31. $3\frac{1}{5}$

32. $4\frac{2}{5}$

33. $6\frac{3}{5}$

34. $5\frac{1}{12}$

35. $1\frac{2}{9}$

36. $1\frac{5}{6}$

37. $8\frac{3}{7}$

38. $6\frac{2}{3}$

39. $24\frac{1}{4}$

40. $10\frac{1}{9}$

41. $15\frac{2}{3}$

42. $13\frac{3}{5}$

Find the missing numerator.

43. $\dfrac{3}{8} = \dfrac{?}{64}$

44. $\dfrac{5}{9} = \dfrac{?}{54}$

45. $\dfrac{3}{5} = \dfrac{?}{35}$

46. $\dfrac{5}{9} = \dfrac{?}{45}$

47. $\dfrac{4}{13} = \dfrac{?}{39}$

48. $\dfrac{13}{17} = \dfrac{?}{51}$

49. $\dfrac{3}{7} = \dfrac{?}{49}$

50. $\dfrac{10}{15} = \dfrac{?}{60}$

51. $\dfrac{3}{4} = \dfrac{?}{20}$

52. $\dfrac{7}{8} = \dfrac{?}{40}$

53. $\dfrac{35}{40} = \dfrac{?}{80}$

54. $\dfrac{45}{50} = \dfrac{?}{100}$

Applications *Solve.*

55. *Basketball* During the regular 2006–07 WNBA basketball season, Asjha Jones of the Connecticut Sun scored 378 points in 24 games. Express as a mixed number in simplified form how many points she averaged per game.

56. *Kentucky Derby* In 2006, 440 horses were nominated to compete in the Kentucky Derby. Only 20 horses were actually chosen to compete in the Derby. What simplified fraction shows what portion of the nominated horses actually competed?

57. *Income Tax* Last year, my parents had a combined income of $64,000. They paid $13,200 in federal income taxes. What simplified fraction shows how much my parents spent on their federal taxes?

58. *Transfer Students* The University of California system accepted 14,664 out of 18,330 California Community College students who applied as transfers in the fall of 2003. What simplified fraction shows what portion of applications were accepted?

Trail Mix The following chart gives the recipe for a trail mix.

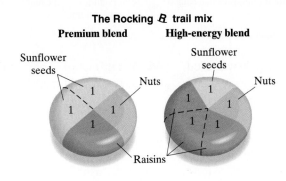

59. What part of the premium blend is nuts?

60. What part of the high-energy blend is raisins?

61. What part of the premium blend is not sunflower seeds?

62. What part of the high-energy blend does not contain nuts?

Theme Parks *The following chart provides some statistics about the number of visitors to the two most popular theme parks in the United States.*

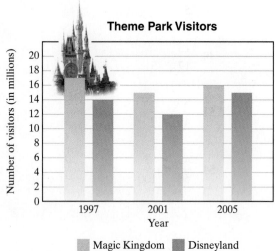

Theme Park Visitors

(*Source:* www.solarius.com, www.themeparkinsider.com)

63. What fractional part of the visitors to the two parks in 2001 went to Disneyland?

64. What fractional part of the visitors to the two parks in 2005 went to the Magic Kingdom?

Fish Catch *The following chart provides some statistics about the fish catch in the United States.*

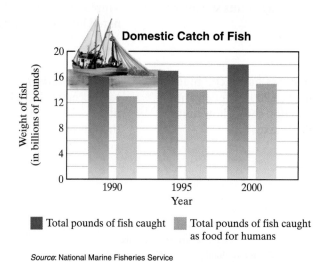

Domestic Catch of Fish

Source: National Marine Fisheries Service

65. What fractional part of the fish catch in 1990 was used as food for humans?

66. What fractional part of the fish catch in 2000 was used as food for humans?

Quick Quiz 0.1

1. Simplify. $\dfrac{84}{92}$

2. Write as an improper fraction. $6\dfrac{9}{11}$

3. Write as a mixed number. $\dfrac{103}{21}$

4. Concept Check Explain in your own words how to change a mixed number to an improper fraction.

0.2 Adding and Subtracting Fractions

 1 **Adding or Subtracting Fractions with a Common Denominator**

If fractions have the same denominator, the numerators may be added or subtracted. The denominator remains the same.

TO ADD OR SUBTRACT TWO FRACTIONS WITH A COMMON DENOMINATOR

1. Add or subtract the numerators.
2. Keep the same (common) denominator.
3. Simplify the answer whenever possible.

Student Learning Objectives

After studying this section, you will be able to:

1 Add or subtract fractions with a common denominator.

2 Use prime factors to find the least common denominator of two or more fractions.

3 Add or subtract fractions with different denominators.

4 Add or subtract mixed numbers.

EXAMPLE 1 Add the fractions. Simplify your answer whenever possible.

(a) $\dfrac{5}{7} + \dfrac{1}{7}$ **(b)** $\dfrac{2}{3} + \dfrac{1}{3}$ **(c)** $\dfrac{1}{8} + \dfrac{3}{8} + \dfrac{2}{8}$ **(d)** $\dfrac{3}{5} + \dfrac{4}{5}$

Solution

(a) $\dfrac{5}{7} + \dfrac{1}{7} = \dfrac{5+1}{7} = \dfrac{6}{7}$ **(b)** $\dfrac{2}{3} + \dfrac{1}{3} = \dfrac{2+1}{3} = \dfrac{3}{3} = 1$

(c) $\dfrac{1}{8} + \dfrac{3}{8} + \dfrac{2}{8} = \dfrac{1+3+2}{8} = \dfrac{6}{8} = \dfrac{3}{4}$ **(d)** $\dfrac{3}{5} + \dfrac{4}{5} = \dfrac{3+4}{5} = \dfrac{7}{5}$ or $1\dfrac{2}{5}$

Practice Problem 1 Add.

(a) $\dfrac{3}{6} + \dfrac{2}{6}$ **(b)** $\dfrac{3}{11} + \dfrac{8}{11}$

(c) $\dfrac{1}{8} + \dfrac{2}{8} + \dfrac{1}{8}$ **(d)** $\dfrac{5}{9} + \dfrac{8}{9}$

EXAMPLE 2 Subtract the fractions. Simplify your answer whenever possible.

(a) $\dfrac{9}{11} - \dfrac{2}{11}$ **(b)** $\dfrac{5}{6} - \dfrac{1}{6}$

Solution

(a) $\dfrac{9}{11} - \dfrac{2}{11} = \dfrac{9-2}{11} = \dfrac{7}{11}$ **(b)** $\dfrac{5}{6} - \dfrac{1}{6} = \dfrac{5-1}{6} = \dfrac{4}{6} = \dfrac{2}{3}$

Practice Problem 2 Subtract.

(a) $\dfrac{11}{13} - \dfrac{6}{13}$ **(b)** $\dfrac{8}{9} - \dfrac{2}{9}$

Although adding and subtracting fractions with the same denominator is fairly simple, most problems involve fractions that do not have a common denominator. Fractions and mixed numbers such as halves, fourths, and eighths are often used. To add or subtract such fractions, we begin by finding a common denominator.

Using Prime Factors to Find the Least Common Denominator of Two or More Fractions

Before you can add or subtract fractions, they must have the same denominator. To save work, we select the smallest possible common denominator. This is called the **least common denominator** or LCD (also known as the *lowest common denominator*).

The LCD of two or more fractions is the smallest whole number that is exactly divisible by each denominator of the fractions.

EXAMPLE 3 Find the LCD. $\frac{2}{3}$ and $\frac{1}{4}$

Solution   The numbers are small enough to find the LCD by inspection. The LCD is 12, since 12 is exactly divisible by 4 and by 3. There is no smaller number that is exactly divisible by 4 and 3.

NOTE TO STUDENT: Fully worked-out solutions to all of the Practice Problems can be found at the back of the text starting at page SP-1

Practice Problem 3   Find the LCD. $\frac{1}{8}$ and $\frac{5}{12}$

In some cases, the LCD cannot easily be determined by inspection. If we write each denominator as the product of prime factors, we will be able to find the LCD. We will use $(\cdot)$ to indicate multiplication. For example, $30 = 2 \cdot 3 \cdot 5$. This means $30 = 2 \times 3 \times 5$.

PROCEDURE TO FIND THE LCD USING PRIME FACTORS

1. Write each denominator as the product of prime factors.

2. The LCD is a product containing each different factor.

3. If a factor occurs more than once in any one denominator, the LCD will contain that factor repeated the greatest number of times that it occurs in any one denominator.

EXAMPLE 4   Find the LCD of $\frac{5}{6}$ and $\frac{1}{15}$ by this new procedure.

Solution

$$6 = 2 \cdot 3$$
$$15 = \begin{vmatrix} & 3 \cdot 5 \end{vmatrix}$$

Write each denominator as the product of prime factors.

$$\text{LCD} = 2 \cdot 3 \cdot 5$$
$$\text{LCD} = 2 \cdot 3 \cdot 5 = 30$$

The LCD is a product containing each different prime factor. The different factors are 2, 3, and 5, and each factor appears at most once in any one denominator.

Practice Problem 4   Use prime factors to find the LCD of $\frac{8}{35}$ and $\frac{6}{15}$.

Great care should be used to determine the LCD in the case of repeated factors.

EXAMPLE 5 Find the LCD of $\frac{4}{27}$ and $\frac{5}{18}$.

Solution

$$27 = 3 \cdot 3 \cdot 3$$

Write each denominator as the product of prime factors. We observe that the factor 3 occurs three times in the factorization of 27.

$$18 = \quad 3 \cdot 3 \cdot 2$$

$$LCD = 3 \cdot 3 \cdot 3 \cdot 2$$
$$LCD = 3 \cdot 3 \cdot 3 \cdot 2 = 54$$

The LCD is a product containing each different factor. The factor 3 *occurred most* in the factorization of 27, where it occurred *three* times. Thus the LCD will be the product of *three* 3's and *one* 2.

Practice Problem 5 Find the LCD of $\frac{5}{12}$ and $\frac{7}{30}$.

EXAMPLE 6 Find the LCD of $\frac{5}{12}$, $\frac{1}{15}$, and $\frac{7}{30}$.

Solution

$$12 = 2 \cdot 2 \cdot 3$$

Write each denominator as the product of prime factors. Notice that the only repeated factor is 2, which occurs twice in the factorization of 12.

$$15 = \quad 3 \cdot 5$$
$$30 = \quad 2 \cdot 3 \cdot 5$$

$$LCD = 2 \cdot 2 \cdot 3 \cdot 5$$
$$LCD = 2 \cdot 2 \cdot 3 \cdot 5 = 60$$

The LCD is the product of each different factor, with the factor 2 appearing twice since it occurred twice in one denominator.

Practice Problem 6 Find the LCD of $\frac{2}{27}$, $\frac{1}{18}$, and $\frac{5}{12}$.

3 Adding or Subtracting Fractions with Different Denominators

Before you can add or subtract them, fractions must have the same denominator. Using the LCD will make your work easier. First you must find the LCD. Then change each fraction to a fraction that has the LCD as the denominator. Sometimes one of the fractions will already have the LCD as the denominator. Once all the fractions have the same denominator, you can add or subtract. Be sure to simplify the fraction in your answer if this is possible.

TO ADD OR SUBTRACT FRACTIONS THAT DO NOT HAVE A COMMON DENOMINATOR

1. Find the LCD of the fractions.

2. Change each fraction to an equivalent fraction with the LCD for a denominator.

3. Add or subtract the fractions.

4. Simplify the answer whenever possible.

Let us return to the two fractions of Example 3. We have previously found that the LCD is 12.

EXAMPLE 7   Bob picked $\frac{2}{3}$ of a bushel of apples on Monday and $\frac{1}{4}$ of a bushel of apples on Tuesday. How much did he have in total?

Solution   To solve this problem we need to add $\frac{2}{3}$ and $\frac{1}{4}$, but before we can do so, we must change $\frac{2}{3}$ and $\frac{1}{4}$ to fractions with the same denominator. We change each fraction to an equivalent fraction with a common denominator of 12, the LCD.

$$\frac{2}{3} = \frac{?}{12} \qquad \frac{2 \times 4}{3 \times 4} = \frac{8}{12} \quad \text{so} \quad \frac{2}{3} = \frac{8}{12}$$

$$\frac{1}{4} = \frac{?}{12} \qquad \frac{1 \times 3}{4 \times 3} = \frac{3}{12} \quad \text{so} \quad \frac{1}{4} = \frac{3}{12}$$

Then we rewrite the problem with common denominators and add.

$$\frac{2}{3} + \frac{1}{4} = \frac{8}{12} + \frac{3}{12} = \frac{8 + 3}{12} = \frac{11}{12}$$

In total, Bob picked $\frac{11}{12}$ of a bushel of apples.

NOTE TO STUDENT: Fully worked-out solutions to all of the Practice Problems can be found at the back of the text starting at page SP-1

Practice Problem 7   Carol planted corn in $\frac{5}{12}$ of the farm fields at the Old Robinson Farm. Connie planted soybeans in $\frac{1}{8}$ of the farm fields. What fractional part of the farm fields of the Old Robinson Farm was planted in corn or soybeans?

Sometimes one of the denominators is the LCD. In such cases the fraction that has the LCD for the denominator will not need to be changed. If every other denominator divides into the largest denominator, the largest denominator is the LCD.

EXAMPLE 8   Find the LCD and then add. $\dfrac{3}{5} + \dfrac{7}{20} + \dfrac{1}{2}$

Solution   We can see by inspection that both 5 and 2 divide exactly into 20. Thus 20 is the LCD. Now add.

$$\frac{3}{5} + \frac{7}{20} + \frac{1}{2}$$

We change $\frac{3}{5}$ and $\frac{1}{2}$ to equivalent fractions with a common denominator of 20, the LCD.

$$\frac{3}{5} = \frac{?}{20} \qquad \frac{3 \times 4}{5 \times 4} = \frac{12}{20} \quad \text{so} \quad \frac{3}{5} = \frac{12}{20}$$

$$\frac{1}{2} = \frac{?}{20} \qquad \frac{1 \times 10}{2 \times 10} = \frac{10}{20} \quad \text{so} \quad \frac{1}{2} = \frac{10}{20}$$

Then we rewrite the problem with common denominators and add.

$$\frac{3}{5} + \frac{7}{20} + \frac{1}{2} = \frac{12}{20} + \frac{7}{20} + \frac{10}{20} = \frac{12 + 7 + 10}{20} = \frac{29}{20} \quad \text{or} \quad 1\frac{9}{20}$$

Practice Problem 8   Find the LCD and add.

$$\frac{3}{5} + \frac{4}{25} + \frac{1}{50}$$

Now we turn to examples where the selection of the LCD is not so obvious. In Examples 9 through 11 we will use the prime factorization method to find the LCD.

EXAMPLE 9   Add. $\dfrac{7}{18} + \dfrac{5}{12}$

Solution   First we find the LCD.

$$18 = 3 \cdot 3 \cdot 2$$
$$12 = \;\mid\; 3 \cdot 2 \cdot 2$$
$$\text{LCD} = 3 \cdot 3 \cdot 2 \cdot 2 = 9 \cdot 4 = 36$$

Now we change $\frac{7}{18}$ and $\frac{5}{12}$ to equivalent fractions that have the LCD.

$$\frac{7}{18} = \frac{?}{36} \qquad \frac{7 \times 2}{18 \times 2} = \frac{14}{36}$$
$$\frac{5}{12} = \frac{?}{36} \qquad \frac{5 \times 3}{12 \times 3} = \frac{15}{36}$$

Now we add the fractions.

$$\frac{7}{18} + \frac{5}{12} = \frac{14}{36} + \frac{15}{36} = \frac{29}{36} \qquad \text{This fraction cannot be simplified.}$$

Practice Problem 9   Add.

$$\frac{1}{49} + \frac{3}{14}$$

EXAMPLE 10   Subtract. $\dfrac{25}{48} - \dfrac{5}{36}$

Solution   First we find the LCD.

$$48 = 2 \cdot 2 \cdot 2 \cdot 2 \cdot 3$$
$$36 = \;\mid\mid\; 2 \cdot 2 \cdot 3 \cdot 3$$
$$\text{LCD} = 2 \cdot 2 \cdot 2 \cdot 2 \cdot 3 \cdot 3 = 16 \cdot 9 = 144$$

Now we change $\frac{25}{48}$ and $\frac{5}{36}$ to equivalent fractions that have the LCD.

$$\frac{25}{48} = \frac{?}{144} \qquad \frac{25 \times 3}{48 \times 3} = \frac{75}{144}$$
$$\frac{5}{36} = \frac{?}{144} \qquad \frac{5 \times 4}{36 \times 4} = \frac{20}{144}$$

Now we subtract the fractions.

$$\frac{25}{48} - \frac{5}{36} = \frac{75}{144} - \frac{20}{144} = \frac{55}{144} \qquad \text{This fraction cannot be simplified.}$$

Practice Problem 10   Subtract.

$$\frac{1}{12} - \frac{1}{30}$$

NOTE TO STUDENT: Fully worked-out solutions to all of the Practice Problems can be found at the back of the text starting at page SP-1

EXAMPLE 11 Combine. $\dfrac{1}{5} + \dfrac{1}{6} - \dfrac{3}{10}$

Solution   First we find the LCD.

$$5 = 5$$
$$6 = \quad 2 \cdot 3$$
$$10 = 5 \cdot 2 \ \Big|$$
$$\downarrow \ \downarrow \ \downarrow$$
$$\text{LCD} = 5 \cdot 2 \cdot 3 = 10 \cdot 3 = 30$$

Now we change $\frac{1}{5}$, $\frac{1}{6}$, and $\frac{3}{10}$ to equivalent fractions that have the LCD for a denominator.

$$\dfrac{1}{5} = \dfrac{?}{30} \qquad \dfrac{1 \times 6}{5 \times 6} = \dfrac{6}{30}$$

$$\dfrac{1}{6} = \dfrac{?}{30} \qquad \dfrac{1 \times 5}{6 \times 5} = \dfrac{5}{30}$$

$$\dfrac{3}{10} = \dfrac{?}{30} \qquad \dfrac{3 \times 3}{10 \times 3} = \dfrac{9}{30}$$

Now we combine the three fractions.

$$\dfrac{1}{5} + \dfrac{1}{6} - \dfrac{3}{10} = \dfrac{6}{30} + \dfrac{5}{30} - \dfrac{9}{30} = \dfrac{2}{30} = \dfrac{1}{15}$$

Note the important step of simplifying the fraction to obtain the final answer.

Practice Problem 11 Combine.

$$\dfrac{2}{3} + \dfrac{3}{4} - \dfrac{3}{8}$$

Adding or Subtracting Mixed Numbers

If the problem you are adding or subtracting has mixed numbers, change them to improper fractions first and then combine (add or subtract). As a convention in this book, if the original problem contains mixed numbers, express the result as a mixed number rather than as an improper fraction.

EXAMPLE 12   Combine. Simplify your answer whenever possible.

(a) $5\dfrac{1}{2} + 2\dfrac{1}{3}$ **(b)** $2\dfrac{1}{5} - 1\dfrac{3}{4}$ **(c)** $1\dfrac{5}{12} + \dfrac{7}{30}$

Solution

(a) First we change the mixed numbers to improper fractions.

$$5\dfrac{1}{2} = \dfrac{5 \times 2 + 1}{2} = \dfrac{11}{2} \qquad 2\dfrac{1}{3} = \dfrac{2 \times 3 + 1}{3} = \dfrac{7}{3}$$

Next we change each fraction to an equivalent form with the common denominator of 6.

$$\dfrac{11}{2} = \dfrac{?}{6} \qquad \dfrac{11 \times 3}{2 \times 3} = \dfrac{33}{6}$$

$$\dfrac{7}{3} = \dfrac{?}{6} \qquad \dfrac{7 \times 2}{3 \times 2} = \dfrac{14}{6}$$

Finally, we add the two fractions and change our answer to a mixed number.

$$\frac{33}{6} + \frac{14}{6} = \frac{47}{6} = 7\frac{5}{6}$$

Thus $5\frac{1}{2} + 2\frac{1}{3} = 7\frac{5}{6}$.

(b) First we change the mixed numbers to improper fractions.

$$2\frac{1}{5} = \frac{2 \times 5 + 1}{5} = \frac{11}{5} \qquad 1\frac{3}{4} = \frac{1 \times 4 + 3}{4} = \frac{7}{4}$$

Next we change each fraction to an equivalent form with the common denominator of 20.

$$\frac{11}{5} = \frac{?}{20} \qquad \frac{11 \times 4}{5 \times 4} = \frac{44}{20}$$

$$\frac{7}{4} = \frac{?}{20} \qquad \frac{7 \times 5}{4 \times 5} = \frac{35}{20}$$

Now we subtract the two fractions.

$$\frac{44}{20} - \frac{35}{20} = \frac{9}{20}$$

Thus $2\frac{1}{5} - 1\frac{3}{4} = \frac{9}{20}$.

Note: It is not necessary to use these exact steps to add and subtract mixed numbers. If you know another method and can use it to obtain the correct answers, it is all right to continue to use that method throughout this chapter.

(c) Now we add $1\frac{5}{12} + \frac{7}{30}$.

The LCD of 12 and 30 is 60. Why? Change the mixed number to an improper fraction. Then change each fraction to an equivalent form with a common denominator.

$$1\frac{5}{12} = \frac{17 \times 5}{12 \times 5} = \frac{85}{60} \qquad \frac{7}{30} = \frac{7 \times 2}{30 \times 2} = \frac{14}{60}$$

Then add the fractions, simplify, and write the answer as a mixed number.

$$\frac{85}{60} + \frac{14}{60} = \frac{99}{60} = \frac{33}{20} = 1\frac{13}{20}$$

Thus $1\frac{5}{12} + \frac{7}{30} = 1\frac{13}{20}$.

Practice Problem 12 Combine.

(a) $1\frac{2}{3} + 2\frac{4}{5}$ **(b)** $5\frac{1}{4} - 2\frac{2}{3}$

▲ **EXAMPLE 13** Manuel is enclosing a triangle-shaped exercise yard for his new dog. He wants to determine how many feet of fencing he will need. The sides of the yard measure $20\frac{3}{4}$ feet, $15\frac{1}{2}$ feet, and $18\frac{1}{8}$ feet. What is the perimeter of (total distance around) the triangle?

Solution

Understand the problem. Begin by drawing a picture.

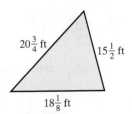

We want to add up the lengths of all three sides of the triangle. This distance around the triangle is called the **perimeter.**

$$20\frac{3}{4} + 15\frac{1}{2} + 18\frac{1}{8} = \frac{83}{4} + \frac{31}{2} + \frac{145}{8}$$

$$= \frac{166}{8} + \frac{124}{8} + \frac{145}{8} = \frac{435}{8} = 54\frac{3}{8}$$

He will need $54\frac{3}{8}$ feet of fencing.

NOTE TO STUDENT: Fully worked-out solutions to all of the Practice Problems can be found at the back of the text starting at page SP-1

▲ **Practice Problem 13** Find the perimeter of a rectangle with sides of $4\frac{1}{5}$ cm and $6\frac{1}{2}$ cm. Begin by drawing a picture. Label the picture by including the measure of *each* side.

Developing Your Study Skills

Class Attendance

A student of mathematics needs to get started in the right direction by choosing to attend class every day, beginning with the first day of class. Statistics show that class attendance and good grades go together. Classroom activities are designed to enhance learning and you must be in class to benefit from them. Vital information and explanations that can help you in understanding concepts are given each day. Do not be deceived into thinking that you can just find out from a friend what went on in class. There is no good substitute for firsthand experience. Give yourself a push in the right direction by developing the habit of going to class every day.

Class Participation

People learn mathematics through active participation, not through observation from the sidelines. If you want to do well in this course, be involved in classroom activities. Sit near the front where you can see and hear well and where your focus is on the instruction process and not on the students around you. Ask questions, be ready to contribute toward solutions, and take part in all classroom activities. Your contributions are valuable to the class and to yourself. Class participation requires an investment of yourself in the learning process, which you will find pays huge dividends.

Verbal and Writing Skills

1. Explain why the denominator 8 is the least common denominator of $\frac{3}{4}$ and $\frac{5}{8}$.

2. What must you do before you add or subtract fractions that do not have a common denominator?

Find the LCD (least common denominator) of each set of fractions. Do not combine the fractions; only find the LCD.

3. $\frac{4}{5}$ and $\frac{11}{16}$

4. $\frac{21}{25}$ and $\frac{8}{15}$

5. $\frac{7}{10}$ and $\frac{1}{4}$

6. $\frac{3}{16}$ and $\frac{1}{24}$

7. $\frac{5}{18}$ and $\frac{7}{54}$

8. $\frac{7}{24}$ and $\frac{5}{48}$

9. $\frac{1}{15}$ and $\frac{4}{21}$

10. $\frac{11}{12}$ and $\frac{7}{20}$

11. $\frac{17}{40}$ and $\frac{13}{60}$

12. $\frac{7}{30}$ and $\frac{8}{45}$

13. $\frac{2}{5}, \frac{3}{8},$ and $\frac{5}{12}$

14. $\frac{1}{7}, \frac{3}{14},$ and $\frac{9}{35}$

15. $\frac{5}{6}, \frac{9}{14},$ and $\frac{17}{26}$

16. $\frac{3}{8}, \frac{5}{12},$ and $\frac{11}{42}$

17. $\frac{1}{2}, \frac{1}{18},$ and $\frac{13}{30}$

18. $\frac{5}{8}, \frac{3}{14},$ and $\frac{11}{16}$

Combine. Be sure to simplify your answer whenever possible.

19. $\frac{3}{8} + \frac{2}{8}$

20. $\frac{2}{9} + \frac{5}{9}$

21. $\frac{5}{14} - \frac{1}{14}$

22. $\frac{9}{20} - \frac{3}{20}$

23. $\frac{3}{8} + \frac{5}{6}$

24. $\frac{7}{10} + \frac{8}{15}$

25. $\frac{5}{7} - \frac{2}{9}$

26. $\frac{7}{8} - \frac{2}{3}$

27. $\frac{1}{3} + \frac{2}{5}$

28. $\frac{2}{7} + \frac{1}{3}$

29. $\frac{5}{9} + \frac{5}{12}$

30. $\frac{11}{15} + \frac{1}{10}$

31. $\frac{11}{15} - \frac{31}{45}$

32. $\frac{21}{12} - \frac{23}{24}$

33. $\frac{16}{24} - \frac{1}{6}$

34. $\frac{11}{28} - \frac{1}{7}$

35. $\frac{3}{8} + \frac{4}{7}$

36. $\frac{7}{4} + \frac{5}{9}$

37. $\frac{2}{3} + \frac{7}{12} + \frac{1}{4}$

38. $\frac{4}{7} + \frac{7}{9} + \frac{1}{3}$

39. $\frac{5}{30} + \frac{3}{40} + \frac{1}{8}$

40. $\frac{1}{12} + \frac{3}{14} + \frac{4}{21}$

41. $\frac{1}{3} + \frac{1}{12} - \frac{1}{6}$

42. $\frac{1}{5} + \frac{2}{3} - \frac{11}{15}$

43. $\dfrac{5}{36} + \dfrac{7}{9} - \dfrac{5}{12}$

44. $\dfrac{5}{24} + \dfrac{3}{8} - \dfrac{1}{3}$

45. $4\dfrac{1}{3} + 3\dfrac{2}{5}$

46. $3\dfrac{1}{8} + 2\dfrac{1}{6}$

47. $1\dfrac{5}{24} + \dfrac{5}{18}$

48. $6\dfrac{2}{3} + \dfrac{3}{4}$

49. $7\dfrac{1}{6} - 2\dfrac{1}{4}$

50. $7\dfrac{2}{5} - 3\dfrac{3}{4}$

51. $8\dfrac{5}{7} - 2\dfrac{1}{4}$

52. $7\dfrac{8}{15} - 2\dfrac{3}{5}$

53. $2\dfrac{1}{8} + 3\dfrac{2}{3}$

54. $3\dfrac{1}{7} + 4\dfrac{1}{3}$

55. $11\dfrac{1}{7} - 6\dfrac{5}{7}$

56. $17\dfrac{1}{5} - 10\dfrac{3}{5}$

57. $3\dfrac{5}{12} + 5\dfrac{7}{12}$

58. $2\dfrac{13}{16} + 2\dfrac{3}{16}$

Mixed Practice

59. $\dfrac{7}{8} + \dfrac{1}{12}$

60. $\dfrac{19}{30} + \dfrac{3}{10}$

61. $3\dfrac{3}{16} + 4\dfrac{3}{8}$

62. $5\dfrac{2}{3} + 7\dfrac{2}{5}$

63. $\dfrac{16}{21} - \dfrac{2}{7}$

64. $\dfrac{15}{24} - \dfrac{3}{8}$

65. $5\dfrac{1}{5} - 2\dfrac{1}{2}$

66. $6\dfrac{1}{3} - 4\dfrac{1}{4}$

67. $25\dfrac{2}{3} - 6\dfrac{1}{7}$

68. $45\dfrac{3}{8} - 26\dfrac{1}{10}$

69. $1\dfrac{1}{6} + \dfrac{3}{8}$

70. $1\dfrac{2}{3} + \dfrac{5}{18}$

71. $8\dfrac{1}{4} + 3\dfrac{5}{6}$

72. $7\dfrac{3}{4} + 6\dfrac{2}{5}$

73. $32 - 1\dfrac{2}{9}$

74. $24 - 3\dfrac{4}{11}$

Applications

75. *Inline Skating* Nancy and Sarah meet three mornings a week to skate. They skated $8\dfrac{1}{4}$ miles on Monday, $10\dfrac{2}{3}$ miles on Wednesday, and $5\dfrac{3}{4}$ miles on Friday. What was their total distance for those three days?

76. *Marathon Training* Paco and Eskinder are training for the Boston Marathon. Their coach gave them the following schedule: run $10\dfrac{1}{2}$ miles on Thursday, a short run of $5\dfrac{1}{4}$ miles on Friday, a rest day on Saturday, and a long run of $18\dfrac{2}{3}$ miles on Sunday. How many miles did they run over these four days?

77. *MTV Video* Sheryl has $8\frac{1}{2}$ hours this weekend to work on her new video. She estimates that it will take $2\frac{2}{3}$ hours to lip sync the new song and $1\frac{3}{4}$ hours to learn the new dance steps. How much time will she have left over for her MTV interview?

78. *Aquariums* Carl bought a 20-gallon aquarium. He put $17\frac{3}{4}$ gallons of water into the aquarium, but it looked too low, so he added $1\frac{1}{4}$ more gallons of water. He then put in the artificial plants and the gravel but now the water was too high, so he siphoned off $2\frac{2}{3}$ gallons of water. How many gallons of water are now in the aquarium?

To Think About

Carpentry *Carpenters use fractions in their work. The picture below is a diagram of a spice cabinet. The symbol " means inches. Use the picture to answer exercises 79 and 80.*

79. Before you can determine where the cabinet will fit, you need to calculate the height, *A*, and the width, *B*. Don't forget to include the $\frac{1}{2}$-inch thickness of the wood where needed.

80. Look at the close-up of the drawer. The width is $4\frac{9}{16}''$. In the diagram, the width of the opening for the drawer is $4\frac{5}{8}''$. What is the difference?

Why do you think the drawer is smaller than the opening?

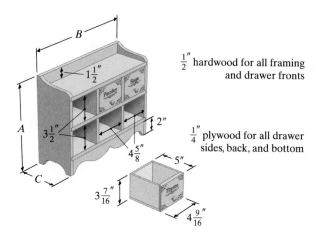

81. *Putting Green Care* The Falmouth Country Club maintains the putting greens with a grass height of $\frac{7}{8}$ inch. The grass on the fairways is maintained at $2\frac{1}{2}$ inches. How much must the mower blade be lowered by a person mowing the fairways if that person will be using the same mowing machine on the putting greens?

82. *Fairway Care* The director of facilities maintenance at the club discovered that due to slippage in the adjustment lever, the lawn mower actually cuts the grass $\frac{1}{16}$ of an inch too long or too short on some days. What is the maximum height that the fairway grass could be after being mowed with this machine? What is the minimum height that the putting greens could be after being mowed with this machine?

Cumulative Review

83. [0.1.2] Simplify. $\dfrac{36}{44}$

84. [0.1.3] Change to an improper fraction. $26\dfrac{3}{5}$

Quick Quiz 0.2 Perform the operations indicated. Simplify your answers whenever possible.

1. $\dfrac{3}{4} + \dfrac{1}{2} + \dfrac{5}{12}$

2. $2\dfrac{3}{5} + 4\dfrac{14}{15}$

3. $6\dfrac{1}{9} - 3\dfrac{5}{6}$

4. Concept Check Explain how you would find the LCD of the fractions $\frac{4}{21}$ and $\frac{5}{18}$.

1 Multiplying Fractions, Whole Numbers, and Mixed Numbers

Multiplying Fractions During a recent snowstorm, the runway at Beverly Airport was plowed. However, the plow cleared only $\frac{3}{5}$ of the width and $\frac{2}{7}$ of the length. What fraction of the total runway area was cleared? To answer this question, we need to multiply $\frac{3}{5} \times \frac{2}{7}$.

The answer is that $\frac{6}{35}$ of the total runway area was cleared.

The multiplication rule for fractions states that to multiply two fractions, we multiply the two numerators and multiply the two denominators.

TO MULTIPLY ANY TWO FRACTIONS

1. Multiply the numerators.
2. Multiply the denominators.

EXAMPLE 1 Multiply.

(a) $\dfrac{3}{5} \times \dfrac{2}{7}$ **(b)** $\dfrac{1}{3} \times \dfrac{5}{4}$ **(c)** $\dfrac{7}{3} \times \dfrac{1}{5}$ **(d)** $\dfrac{6}{5} \times \dfrac{2}{3}$

Solution

(a) $\dfrac{3}{5} \times \dfrac{2}{7} = \dfrac{3 \cdot 2}{5 \cdot 7} = \dfrac{6}{35}$ **(b)** $\dfrac{1}{3} \times \dfrac{5}{4} = \dfrac{1 \cdot 5}{3 \cdot 4} = \dfrac{5}{12}$

(c) $\dfrac{7}{3} \times \dfrac{1}{5} = \dfrac{7 \cdot 1}{3 \cdot 5} = \dfrac{7}{15}$ **(d)** $\dfrac{6}{5} \times \dfrac{2}{3} = \dfrac{6 \cdot 2}{5 \cdot 3} = \dfrac{12}{15} = \dfrac{4}{5}$

Note that we must simplify this fraction.

Practice Problem 1 Multiply.

(a) $\dfrac{2}{7} \times \dfrac{5}{11}$ **(b)** $\dfrac{1}{5} \times \dfrac{7}{10}$ **(c)** $\dfrac{9}{5} \times \dfrac{1}{4}$ **(d)** $\dfrac{8}{9} \times \dfrac{3}{10}$

NOTE TO STUDENT: *Fully worked-out solutions to all of the Practice Problems can be found at the back of the text starting at page SP-1*

It is possible to avoid having to simplify a fraction as the last step. In many cases we can divide by a value that appears as a factor in both a numerator and a denominator. Often it is helpful to write the numbers as products of prime factors in order to do this.

EXAMPLE 2 Multiply.

(a) $\dfrac{3}{5} \times \dfrac{5}{7}$ **(b)** $\dfrac{4}{11} \times \dfrac{5}{2}$ **(c)** $\dfrac{15}{8} \times \dfrac{10}{27}$

Solution

(a) $\dfrac{3}{5} \times \dfrac{5}{7} = \dfrac{3 \cdot 5}{5 \cdot 7} = \dfrac{3 \cdot \overset{1}{\cancel{5}}}{7 \cdot \underset{1}{\cancel{5}}} = \dfrac{3}{7}$ Note that here we divided numerator and denominator by 5.

If we factor each number, we can see the common factors.

(b) $\dfrac{4}{11} \times \dfrac{5}{2} = \dfrac{2 \cdot \overset{1}{\cancel{2}}}{11} \times \dfrac{5}{\underset{1}{\cancel{2}}} = \dfrac{10}{11}$ **(c)** $\dfrac{15}{8} \times \dfrac{10}{27} = \dfrac{\overset{1}{\cancel{3}} \cdot 5}{2 \cdot 2 \cdot \underset{1}{\cancel{2}}} \times \dfrac{5 \cdot \overset{1}{\cancel{2}}}{\underset{1}{\cancel{3}} \cdot 3 \cdot 3} = \dfrac{25}{36}$

After dividing out common factors, the resulting multiplication problem involves smaller numbers and the answers are in simplified form.

Practice Problem 2 Multiply.

(a) $\dfrac{3}{5} \times \dfrac{4}{3}$

(b) $\dfrac{9}{10} \times \dfrac{5}{12}$

SIDELIGHT: Dividing Out Common Factors

Why does this method of dividing out a value that appears as a factor in both numerator and denominator work? Let's reexamine one of the examples we solved previously.

$$\frac{3}{5} \times \frac{5}{7} = \frac{3 \cdot 5}{5 \cdot 7} = \frac{3 \cdot \overset{1}{\cancel{5}}}{7 \cdot \underset{1}{\cancel{5}}} = \frac{3}{7}$$

Consider the following steps and reasons.

$\dfrac{3}{5} \times \dfrac{5}{7} = \dfrac{3 \cdot 5}{5 \cdot 7}$ Definition of multiplication of fractions.

$\phantom{\dfrac{3}{5} \times \dfrac{5}{7}} = \dfrac{5 \cdot 3}{5 \cdot 7}$ Change the order of the factors in the numerator, since $3 \cdot 5 = 5 \cdot 3$. This is called the commutative property of multiplication.

$\phantom{\dfrac{3}{5} \times \dfrac{5}{7}} = \dfrac{5}{5} \cdot \dfrac{3}{7}$ Definition of multiplication of fractions.

$\phantom{\dfrac{3}{5} \times \dfrac{5}{7}} = 1 \cdot \dfrac{3}{7}$ Write 1 in place of $\frac{5}{5}$, since 1 is another name for $\frac{5}{5}$.

$\phantom{\dfrac{3}{5} \times \dfrac{5}{7}} = \dfrac{3}{7}$ $1 \cdot \frac{3}{7} = \frac{3}{7}$, since any number can be multiplied by 1 without changing the value of the number.

Think about this concept. It is an important one that we will use again when we discuss rational expressions.

Multiplying a Fraction by a Whole Number

Whole numbers can be named using fractional notation. $3, \frac{9}{3}, \frac{6}{2}$, and $\frac{3}{1}$ are ways of expressing the number *three*. Therefore,

$$3 = \frac{9}{3} = \frac{6}{2} = \frac{3}{1}.$$

When we multiply a fraction by a whole number, we merely express the whole number as a fraction whose denominator is 1 and follow the multiplication rule for fractions.

EXAMPLE 3 Multiply.

(a) $7 \times \dfrac{3}{5}$

(b) $\dfrac{3}{16} \times 4$

Solution

(a) $7 \times \dfrac{3}{5} = \dfrac{7}{1} \times \dfrac{3}{5} = \dfrac{21}{5}$ or $4\dfrac{1}{5}$

(b) $\dfrac{3}{16} \times 4 = \dfrac{3}{16} \times \dfrac{4}{1} = \dfrac{3}{4 \cdot 4} \times \dfrac{\cancel{4}}{1} = \dfrac{3}{4}$

Notice that in **(b)** we did not use *prime* factors to factor 16. We recognized that $16 = 4 \cdot 4$. This is a more convenient factorization of 16 for this problem. Choose the factorization that works best for each problem. If you cannot decide what is best, factor into primes.

Practice Problem 3 Multiply.

(a) $4 \times \dfrac{2}{7}$

(b) $12 \times \dfrac{3}{4}$

Multiplying Mixed Numbers

When multiplying mixed numbers, we first change them to improper fractions and then follow the multiplication rule for fractions.

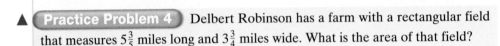

▲ **EXAMPLE 4**   How do we find the area of a rectangular field $3\frac{1}{3}$ miles long by $2\frac{1}{2}$ miles wide?

Solution   To find the area, we multiply length times width.

$$3\frac{1}{3} \times 2\frac{1}{2} = \frac{10}{3} \times \frac{5}{2} = \frac{\cancel{2} \cdot 5}{3} \times \frac{5}{\cancel{2}} = \frac{25}{3} = 8\frac{1}{3}$$

The area is $8\frac{1}{3}$ square miles.

▲ **Practice Problem 4**   Delbert Robinson has a farm with a rectangular field that measures $5\frac{3}{5}$ miles long and $3\frac{3}{4}$ miles wide. What is the area of that field?

EXAMPLE 5   Multiply. $2\frac{2}{3} \times \frac{1}{4} \times 6$

Solution

$$2\frac{2}{3} \times \frac{1}{4} \times 6 = \frac{8}{3} \times \frac{1}{4} \times \frac{6}{1} = \frac{\cancel{4} \cdot 2}{\cancel{3}} \times \frac{1}{\cancel{4}} \times \frac{2 \cdot \cancel{3}}{1} = \frac{4}{1} = 4$$

Practice Problem 5   Multiply.

$$3\frac{1}{2} \times \frac{1}{14} \times 4$$

② Dividing Fractions, Whole Numbers, and Mixed Numbers

Dividing Fractions   To divide two fractions, we invert the second fraction (that is, the divisor) and then multiply the two fractions.

> **TO DIVIDE TWO FRACTIONS**
> 1. Invert the second fraction (that is, the divisor).
> 2. Now multiply the two fractions.

EXAMPLE 6   Divide.

(a) $\frac{1}{3} \div \frac{1}{2}$ **(b)** $\frac{2}{5} \div \frac{3}{10}$ **(c)** $\frac{2}{3} \div \frac{7}{5}$

Solution

(a) $\frac{1}{3} \div \frac{1}{2} = \frac{1}{3} \times \frac{2}{1} = \frac{2}{3}$ Note that we always invert the *second* fraction.

(b) $\frac{2}{5} \div \frac{3}{10} = \frac{2}{5} \times \frac{10}{3} = \frac{2}{\cancel{5}} \times \frac{\cancel{5} \cdot 2}{3} = \frac{4}{3}$ or $1\frac{1}{3}$ **(c)** $\frac{2}{3} \div \frac{7}{5} = \frac{2}{3} \times \frac{5}{7} = \frac{10}{21}$

Practice Problem 6   Divide.

(a) $\frac{2}{5} \div \frac{1}{3}$ **(b)** $\frac{12}{13} \div \frac{4}{3}$

Dividing a Fraction and a Whole Number   The process of inverting the second fraction and then multiplying the two fractions should be done very carefully when one of the original values is a whole number. Remember, a whole number such as 2 is equivalent to $\frac{2}{1}$.

EXAMPLE 7   Divide.

(a) $\dfrac{1}{3} \div 2$

(b) $5 \div \dfrac{1}{3}$

Solution

(a) $\dfrac{1}{3} \div 2 = \dfrac{1}{3} \div \dfrac{2}{1} = \dfrac{1}{3} \times \dfrac{1}{2} = \dfrac{1}{6}$

(b) $5 \div \dfrac{1}{3} = \dfrac{5}{1} \div \dfrac{1}{3} = \dfrac{5}{1} \times \dfrac{3}{1} = \dfrac{15}{1} = 15$

Practice Problem 7   Divide.

(a) $\dfrac{3}{7} \div 6$

(b) $8 \div \dfrac{2}{3}$

NOTE TO STUDENT: Fully worked-out solutions to all of the Practice Problems can be found at the back of the text starting at page SP-1

SIDELIGHT: Number Sense

Look at the answers to the problems in Example 7. In part **(a)**, you will notice that $\frac{1}{6}$ is less than the original number $\frac{1}{3}$. Does this seem reasonable? Let's see. If $\frac{1}{3}$ is divided by 2, it means that $\frac{1}{3}$ will be divided into two equal parts. We would expect that each part would be less than $\frac{1}{3}$. $\frac{1}{6}$ is a reasonable answer to this division problem.

In part **(b)**, 15 is greater than the original number 5. Does this seem reasonable? Think of what $5 \div \frac{1}{3}$ means. It means that 5 will be divided into thirds. Let's think of an easier problem. What happens when we divide 1 into thirds? We get *three* thirds. We would expect, therefore, that when we divide 5 into thirds, we would get 5×3 or 15 thirds. 15 is a reasonable answer to this division problem.

Complex Fractions   Sometimes division is written in the form of a **complex fraction** with one fraction in the numerator and one fraction in the denominator. It is best to write this in standard division notation first; then complete the problem using the rule for division.

EXAMPLE 8   Divide.

(a) $\dfrac{\frac{3}{7}}{\frac{3}{5}}$

(b) $\dfrac{\frac{2}{9}}{\frac{5}{7}}$

Solution

(a) $\dfrac{\frac{3}{7}}{\frac{3}{5}} = \dfrac{3}{7} \div \dfrac{3}{5} = \dfrac{\cancel{3}}{7} \times \dfrac{5}{\cancel{3}} = \dfrac{5}{7}$

(b) $\dfrac{\frac{2}{9}}{\frac{5}{7}} = \dfrac{2}{9} \div \dfrac{5}{7} = \dfrac{2}{9} \times \dfrac{7}{5} = \dfrac{14}{45}$

Practice Problem 8   Divide.

(a) $\dfrac{\frac{3}{11}}{\frac{5}{7}}$

(b) $\dfrac{\frac{12}{5}}{\frac{8}{15}}$

SIDELIGHT: Invert and Multiply

Why does the method of "invert and multiply" work? The division rule really depends on the property that any number can be multiplied by 1 without changing the value of the number. Let's look carefully at an example of division of fractions:

$$\frac{2}{5} \div \frac{3}{7} = \frac{\frac{2}{5}}{\frac{3}{7}}$$

We can write the problem using a complex fraction.

$$= \frac{\frac{2}{5}}{\frac{3}{7}} \times 1$$

We can multiply by 1, since any number can be multiplied by 1 without changing the value of the number.

$$= \frac{\frac{2}{5}}{\frac{3}{7}} \times \frac{\frac{7}{3}}{\frac{7}{3}}$$

We write 1 in the form $\frac{\frac{7}{3}}{\frac{7}{3}}$, since any nonzero number divided by itself equals 1. We choose this value as a multiplier because it will help simplify the denominator.

$$= \frac{\frac{2}{5} \times \frac{7}{3}}{\frac{3}{7} \times \frac{7}{3}}$$

Definition of multiplication of fractions.

$$= \frac{\frac{2}{5} \times \frac{7}{3}}{1}$$

The product in the denominator equals 1.

$$= \frac{2}{5} \times \frac{7}{3}$$

Thus we have shown that $\frac{2}{5} \div \frac{3}{7}$ is equivalent to $\frac{2}{5} \times \frac{7}{3}$ and have also shown some justification for the "invert and multiply rule."

Dividing Mixed Numbers This method for division of fractions can be used with mixed numbers. However, we first must change the mixed numbers to improper fractions and then use the rule for dividing fractions.

EXAMPLE 9 Divide.

(a) $2\frac{1}{3} \div 3\frac{2}{3}$

(b) $\dfrac{2}{3\frac{1}{2}}$

Solution

(a) $2\frac{1}{3} \div 3\frac{2}{3} = \frac{7}{3} \div \frac{11}{3} = \frac{7}{\cancel{3}} \times \frac{\cancel{3}}{11} = \frac{7}{11}$

(b) $\dfrac{2}{3\frac{1}{2}} = 2 \div 3\frac{1}{2} = \frac{2}{1} \div \frac{7}{2} = \frac{2}{1} \times \frac{2}{7} = \frac{4}{7}$

NOTE TO STUDENT: Fully worked-out solutions to all of the Practice Problems can be found at the back of the text starting at page SP-1

Practice Problem 9 Divide.

(a) $1\frac{2}{5} \div 2\frac{1}{3}$

(b) $4\frac{2}{3} \div 7$

(c) $\dfrac{1\frac{1}{5}}{1\frac{2}{7}}$

EXAMPLE 10   A chemist has 96 fluid ounces of a solution. She pours the solution into test tubes. Each test tube holds $\frac{3}{4}$ fluid ounce. How many test tubes can she fill?

Solution   We need to divide the total number of ounces, 96, by the number of ounces in each test tube, $\frac{3}{4}$.

$$96 \div \frac{3}{4} = \frac{96}{1} \div \frac{3}{4} = \frac{96}{1} \times \frac{4}{3} = \frac{\cancel{3} \cdot 32}{1} \times \frac{4}{\cancel{3}} = \frac{128}{1} = 128$$

She will be able to fill 128 test tubes.

Pause for a moment to think about the answer. Does 128 test tubes filled with solution seem like a reasonable answer? Did you perform the correct operation?

Practice Problem 10   A chemist has 64 fluid ounces of a solution. He wishes to fill several jars, each holding $5\frac{1}{3}$ fluid ounces. How many jars can he fill?

Sometimes when solving word problems involving fractions or mixed numbers, it is helpful to solve the problem using simpler numbers first. Once you understand what operation is involved, you can go back and solve using the original numbers in the word problem.

EXAMPLE 11   A car traveled 301 miles on $10\frac{3}{4}$ gallons of gas. How many miles per gallon did it get?

Solution   Use simpler numbers: 300 miles on 10 gallons of gas. We want to find out how many miles the car traveled on 1 gallon of gas. You may want to draw a picture.

10 gallons

300 miles

Divide. $300 \div 10 = 30$.

Now use the original numbers given in the problem.

$$301 \div 10\frac{3}{4} = \frac{301}{1} \div \frac{43}{4} = \frac{301}{1} \times \frac{4}{43} = \frac{1204}{43} = 28$$

The car got 28 miles per gallon.

Practice Problem 11   A car traveled 126 miles on $5\frac{1}{4}$ gallons of gas. How many miles per gallon did it get?

Developing Your Study Skills

Why Is Homework Necessary?

Mathematics is a set of skills that you learn by doing, not by watching. Your instructor may make solving a mathematics problem look very easy, but for you to learn the necessary skills, you must practice them over and over again, just as your instructor once had to do. There is no other way. Learning mathematics is like learning to play a musical instrument, to type, or to play a sport. No matter how much you watch someone else do it, how many books you may read on "how to" do it, or how easy it may seem to be, the key to success is to practice on a regular basis. Homework provides this practice.

PRACTICE   WATCH   DOWNLOAD   READ   REVIEW

Verbal and Writing Skills

1. Explain in your own words how to multiply two mixed numbers.

2. Explain in your own words how to divide two proper fractions.

Multiply. Simplify your answer whenever possible.

3. $\dfrac{36}{7} \times \dfrac{5}{9}$

4. $\dfrac{3}{8} \times \dfrac{24}{5}$

5. $\dfrac{17}{18} \times \dfrac{3}{5}$

6. $\dfrac{21}{22} \times \dfrac{11}{6}$

7. $\dfrac{4}{5} \times \dfrac{3}{10}$

8. $\dfrac{5}{6} \times \dfrac{7}{11}$

9. $\dfrac{24}{25} \times \dfrac{5}{2}$

10. $\dfrac{15}{24} \times \dfrac{8}{9}$

11. $\dfrac{7}{12} \times \dfrac{8}{28}$

12. $\dfrac{6}{21} \times \dfrac{9}{18}$

13. $\dfrac{6}{35} \times 5$

14. $\dfrac{2}{21} \times 15$

15. $9 \times \dfrac{2}{5}$

16. $\dfrac{8}{11} \times 3$

Divide. Simplify your answer whenever possible.

17. $\dfrac{8}{5} \div \dfrac{8}{3}$

18. $\dfrac{13}{9} \div \dfrac{13}{7}$

19. $\dfrac{3}{7} \div 3$

20. $\dfrac{7}{8} \div 4$

21. $10 \div \dfrac{5}{7}$

22. $18 \div \dfrac{2}{9}$

23. $\dfrac{6}{14} \div \dfrac{3}{8}$

24. $\dfrac{8}{12} \div \dfrac{5}{6}$

25. $\dfrac{7}{24} \div \dfrac{9}{8}$

26. $\dfrac{9}{28} \div \dfrac{4}{7}$

27. $\dfrac{\frac{7}{8}}{\frac{3}{4}}$

28. $\dfrac{\frac{5}{6}}{\frac{10}{13}}$

29. $\dfrac{\frac{5}{6}}{\frac{7}{9}}$

30. $\dfrac{\frac{3}{4}}{\frac{11}{12}}$

31. $1\dfrac{3}{7} \div 6\dfrac{1}{4}$

32. $4\dfrac{1}{2} \div 3\dfrac{3}{8}$

33. $3\dfrac{1}{3} \div 2\dfrac{1}{2}$

34. $5\dfrac{1}{2} \div 3\dfrac{3}{4}$

35. $6\dfrac{1}{2} \div \dfrac{3}{4}$

36. $\dfrac{1}{4} \div 1\dfrac{7}{8}$

37. $\dfrac{15}{2\frac{2}{5}}$

38. $\dfrac{18}{4\frac{1}{2}}$

39. $\dfrac{\frac{2}{3}}{1\frac{1}{4}}$

40. $\dfrac{\frac{5}{6}}{2\frac{1}{2}}$

Mixed Practice *Perform the proper calculations. Simplify your answer whenever possible.*

41. $\dfrac{4}{7} \times \dfrac{21}{2}$

42. $\dfrac{12}{18} \times \dfrac{9}{2}$

43. $\dfrac{5}{14} \div \dfrac{2}{7}$

44. $\dfrac{5}{6} \div \dfrac{11}{18}$

45. $10\dfrac{3}{7} \times 5\dfrac{1}{4}$

46. $10\dfrac{2}{9} \div 2\dfrac{1}{3}$

47. $30 \div \dfrac{3}{4}$

48. $8 \div 1\dfrac{1}{3}$

49. $6 \times 4\dfrac{2}{3}$

50. $5\dfrac{1}{2} \times 10$

51. $2\dfrac{1}{2} \times \dfrac{1}{10} \times \dfrac{3}{4}$

52. $3\dfrac{1}{3} \times \dfrac{1}{5} \times \dfrac{2}{3}$

53. (a) $\dfrac{1}{15} \times \dfrac{25}{21}$
  (b) $\dfrac{1}{15} \div \dfrac{25}{21}$

54. (a) $\dfrac{1}{6} \times \dfrac{24}{15}$
  (b) $\dfrac{1}{6} \div \dfrac{24}{15}$

55. (a) $\dfrac{2}{3} \div \dfrac{12}{21}$
  (b) $\dfrac{2}{3} \times \dfrac{12}{21}$

56. (a) $\dfrac{3}{7} \div \dfrac{21}{25}$
  (b) $\dfrac{3}{7} \times \dfrac{21}{25}$

Applications

57. *Shirt Manufacturing* A denim shirt at The Gap requires $2\dfrac{3}{4}$ yards of material. How many shirts can be made from $71\dfrac{1}{2}$ yards of material?

58. *Pullover Manufacturing* A fleece pullover requires $1\dfrac{5}{8}$ yards of material. How many fleece pullovers can be made from $29\dfrac{1}{4}$ yards of material?

59. *Farm Field* Jesse purchased a large rectangular field so that he could farm the land. The field measures $11\dfrac{1}{3}$ miles long and 12 miles wide. What is the area of his field?

60. *Flower Garden* Sara must find the area of her flower garden so that she can determine how much fertilizer to purchase. What is the area of her rectangular garden, which measures 15 feet long and $10\dfrac{1}{5}$ feet wide?

Cumulative Review

Write as a mixed number.

61. **[0.1.3]** (a) $\dfrac{22}{7}$ (b) $\dfrac{18}{5}$

Write as an improper fraction.

62. **[0.1.3]** (a) $12\dfrac{1}{3}$ (b) $9\dfrac{7}{8}$

Quick Quiz 0.3 Perform the operations indicated. Simplify answers whenever possible.

1. $\dfrac{7}{15} \times \dfrac{25}{14}$

2. $3\dfrac{1}{4} \times 4\dfrac{1}{2}$

3. $3\dfrac{3}{10} \div 2\dfrac{1}{2}$

4. **Concept Check** Explain the steps you would take to perform the calculation $3\dfrac{1}{4} \div 2\dfrac{1}{2}$.

1. _____

2. _____

3. _____

4. _____

5. _____

6. _____

7. _____

8. _____

9. _____

10. _____

11. _____

12. _____

13. _____

14. _____

15. _____

16. _____

17. _____

18. _____

19. _____

How are you doing with your homework assignments in Sections 0.1 to 0.3? Do you feel you have mastered the material so far? Do you understand the concepts you have covered? Before you go further in the textbook, take some time to do each of the following problems.

0.1

In exercises 1 and 2, simplify each fraction.

1. $\dfrac{15}{55}$

2. $\dfrac{46}{115}$

3. Write $\dfrac{15}{4}$ as a mixed number.

4. Change $4\dfrac{5}{7}$ to an improper fraction.

Find the missing number.

5. $\dfrac{4}{5} = \dfrac{?}{30}$

6. $\dfrac{5}{9} = \dfrac{?}{81}$

0.2

7. Find the LCD, but do not add. $\dfrac{3}{8}, \dfrac{5}{6},$ and $\dfrac{7}{15}$

Perform the calculation indicated. Write the answer in simplest form.

8. $\dfrac{3}{7} + \dfrac{2}{7}$

9. $\dfrac{5}{14} + \dfrac{2}{21}$

10. $2\dfrac{3}{4} + 5\dfrac{2}{3}$

11. $\dfrac{17}{18} - \dfrac{5}{9}$

12. $\dfrac{6}{7} - \dfrac{2}{3}$

13. $3\dfrac{1}{5} - 1\dfrac{3}{8}$

0.3

Perform the calculations indicated. Write the answer in simplest form.

14. $\dfrac{25}{7} \times \dfrac{14}{45}$

15. $12 \times 3\dfrac{1}{2}$

16. $4 \div \dfrac{8}{7}$

17. $2\dfrac{1}{3} \div 3\dfrac{1}{4}$

18. $\dfrac{\frac{6}{25}}{\frac{9}{10}}$

19. Ranak owns a rectangular plot of land that measures $3\dfrac{3}{4}$ miles long and $5\dfrac{1}{3}$ miles wide. What is the area of Ranak's land?

Now turn to page SA-1 for the answer to each of these problems. Each answer also includes a reference to the objective in which the problem is first taught. If you missed any of these problems, you should stop and review the Examples and Practice Problems in the referenced objective. A little review now will help you master the material in the upcoming sections of the text.

0.4 USING DECIMALS

① Understanding the Meaning of Decimals

We can express a part of a whole as a fraction or as a decimal. A **decimal** is another way of writing a fraction whose denominator is 10, 100, 1000, and so on.

$$\frac{3}{10} = 0.3 \qquad \frac{5}{100} = 0.05 \qquad \frac{172}{1000} = 0.172 \qquad \frac{58}{10,000} = 0.0058$$

The period in decimal notation is known as the **decimal point.** The number of digits in a number to the right of the decimal point is known as the number of **decimal places** of the number. The place value of decimals is shown in the following chart.

Hundred-thousands	Ten-thousands	Thousands	Hundreds	Tens	Ones	← Decimal point	Tenths	Hundredths	Thousandths	Ten-thousandths	Hundred-thousandths
100,000	10,000	1000	100	10	1	•	$\frac{1}{10}$	$\frac{1}{100}$	$\frac{1}{1000}$	$\frac{1}{10,000}$	$\frac{1}{100,000}$

EXAMPLE 1 Write each of the following decimals as a fraction or mixed number. State the number of decimal places. Write out in words the way the number would be spoken.

(a) 0.6   **(b)** 0.29   **(c)** 0.527   **(d)** 1.38   **(e)** 0.00007

Solution

Decimal Form	Fraction Form	Number of Decimal Places	The Words Used to Describe the Number
(a) 0.6	$\frac{6}{10}$	one	six tenths
(b) 0.29	$\frac{29}{100}$	two	twenty-nine hundredths
(c) 0.527	$\frac{527}{1000}$	three	five hundred twenty-seven thousandths
(d) 1.38	$1\frac{38}{100}$	two	one and thirty-eight hundredths
(e) 0.00007	$\frac{7}{100,000}$	five	seven hundred-thousandths

Practice Problem 1 Write each decimal as a fraction or mixed number and in words.

(a) 0.9   **(b)** 0.09   **(c)** 0.731   **(d)** 1.371   **(e)** 0.0005

Student Learning Objectives

After studying this section, you will be able to:

① **Understand the meaning of decimals.**

② **Change a fraction to a decimal.**

③ **Change a decimal to a fraction.**

④ **Add and subtract decimals.**

⑤ **Multiply decimals.**

⑥ **Divide decimals.**

⑦ **Multiply and divide a decimal by a multiple of 10.**

You have seen that a given fraction can be written in several different but equivalent ways. There are also several different equivalent ways of writing the decimal form of fractions. The decimal 0.18 can be written in the following equivalent ways:

$$\text{Fractional form:} \quad \frac{18}{100} = \frac{180}{1000} = \frac{1800}{10,000} = \frac{18,000}{100,000}$$

$$\text{Decimal form:} \quad 0.18 = 0.180 = 0.1800 = 0.18000.$$

Thus we see that *any number of terminal zeros may be added to the right-hand side of a decimal* without changing its value.

$$0.13 = 0.1300 \qquad 0.162 = 0.162000$$

Similarly, *any number of terminal zeros may be removed from the right-hand side of a decimal* without changing its value.

② Changing a Fraction to a Decimal

A fraction can be changed to a decimal by dividing the denominator into the numerator.

EXAMPLE 2   Write each of the following fractions as a decimal.

(a) $\frac{3}{4}$ **(b)** $\frac{21}{20}$ **(c)** $\frac{1}{8}$ **(d)** $\frac{3}{200}$

Solution

(a) $\frac{3}{4} = 0.75$ since
$$\begin{array}{r} 0.75 \\ 4\overline{)3.00} \\ 2\,8 \\ \hline 20 \\ 20 \\ \hline 0 \end{array}$$

(b) $\frac{21}{20} = 1.05$ since
$$\begin{array}{r} 1.05 \\ 20\overline{)21.00} \\ 20 \\ \hline 1\,00 \\ 1\,00 \\ \hline 0 \end{array}$$

(c) $\frac{1}{8} = 0.125$ since
$$\begin{array}{r} 0.125 \\ 8\overline{)1.000} \\ 8 \\ \hline 20 \\ 16 \\ \hline 40 \\ 40 \\ \hline 0 \end{array}$$

(d) $\frac{3}{200} = 0.015$ since
$$\begin{array}{r} 0.015 \\ 200\overline{)3.000} \\ 2\,00 \\ \hline 1\,000 \\ 1\,000 \\ \hline 0 \end{array}$$

Practice Problem 2   Write as decimals.

(a) $\frac{3}{8}$ **(b)** $\frac{7}{200}$ **(c)** $\frac{33}{20}$

Calculator

Fraction to Decimal

You can use a calculator to change $\frac{3}{5}$ to a decimal.
Enter:

3 ÷ 5 =

The display should read

0.6

Try the following.

(a) $\frac{17}{25}$ **(b)** $\frac{2}{9}$

(c) $\frac{13}{10}$ **(d)** $\frac{15}{19}$

Ans:

(a) 0.68 **(b)** $0.\overline{2}$ **(c)** 1.3
(d) 0.7894737 (Rounded to seven decimal places.)

Sometimes division yields an infinite repeating decimal. We use three dots to indicate that the pattern continues forever. For example,

$$\frac{1}{3} = 0.3333\ldots \qquad \begin{array}{r} 0.333 \\ 3\overline{)1.000} \\ \underline{9} \\ 10 \\ \underline{9} \\ 10 \\ \underline{9} \\ 1 \end{array}$$

An alternative notation is to place a bar over the repeating digit(s):

$$0.3333\ldots = 0.\overline{3} \qquad 0.575757\ldots = 0.\overline{57}$$

EXAMPLE 3 Write each fraction as a decimal.

(a) $\dfrac{2}{11}$ **(b)** $\dfrac{5}{6}$

Solution

(a) $\dfrac{2}{11} = 0.181818\ldots$ or $0.\overline{18}$ **(b)** $\dfrac{5}{6} = 0.8333\ldots$ or $0.8\overline{3}$

$$\begin{array}{r} 0.1818 \\ 11\overline{)2.0000} \\ \underline{11} \\ 90 \\ \underline{88} \\ 20 \\ \underline{11} \\ 90 \\ \underline{88} \\ 2 \end{array} \qquad \begin{array}{r} 0.8333 \\ 6\overline{)5.0000} \\ \underline{4\,8} \\ 20 \\ \underline{18} \\ 20 \\ \underline{18} \\ 20 \\ \underline{18} \\ 2 \end{array}$$

Note that the 8 does not repeat. Only the digit 3 is repeating.

Practice Problem 3 Write each fraction as a decimal.

(a) $\dfrac{1}{6}$ **(b)** $\dfrac{5}{11}$

Sometimes division must be carried out to many places in order to observe the repeating pattern. This is true in the following example:

$$\frac{2}{7} = 0.285714285714285714\ldots \qquad \text{This can also be written as } \frac{2}{7} = 0.\overline{285714}.$$

It can be shown that the denominator determines the maximum number of decimal places that might repeat. So $\frac{2}{7}$ must repeat in the seventh decimal place or sooner.

3 Changing a Decimal to a Fraction

To convert from a decimal to a fraction, merely write the decimal as a fraction with a denominator of 10, 100, 1000, 10,000, and so on, and simplify the result when possible.

EXAMPLE 4 Write each decimal as a fraction.

(a) 0.2 **(b)** 0.35 **(c)** 0.516 **(d)** 0.74 **(e)** 0.138 **(f)** 0.008

Solution

(a) $0.2 = \dfrac{2}{10} = \dfrac{1}{5}$

(b) $0.35 = \dfrac{35}{100} = \dfrac{7}{20}$

(c) $0.516 = \dfrac{516}{1000} = \dfrac{129}{250}$

(d) $0.74 = \dfrac{74}{100} = \dfrac{37}{50}$

(e) $0.138 = \dfrac{138}{1000} = \dfrac{69}{500}$

(f) $0.008 = \dfrac{8}{1000} = \dfrac{1}{125}$

Practice Problem 4 Write each decimal as a fraction and simplify whenever possible.

(a) 0.8 **(b)** 0.88 **(c)** 0.45 **(d)** 0.148 **(e)** 0.612 **(f)** 0.016

All repeating decimals can also be converted to fractional form. In practice, however, repeating decimals are usually rounded to a few places. It will not be necessary, therefore, to learn how to convert $0.\overline{033}$ to $\frac{11}{333}$ for this course.

 Adding and Subtracting Decimals

Last week Bob spent $19.83 on lunches purchased at the cafeteria at work. During this same period, Sally spent $24.76 on lunches. How much did the two of them spend on lunches last week?

Adding and subtracting decimals is similar to adding and subtracting whole numbers, except that it is necessary to line up decimal points. To perform the operation $19.83 + 24.76$, we line up the numbers in column form and add the digits:

$$
\begin{array}{r}
19.83 \\
+\ 24.76 \\
\hline
44.59
\end{array}
$$

Thus Bob and Sally spent $44.59 on lunches last week.

> **ADDITION AND SUBTRACTION OF DECIMALS**
>
> **1.** Write in column form and line up the decimal points.
>
> **2.** Add or subtract the digits.

EXAMPLE 5 Perform the following operations.

(a) $3.6 + 2.3$ **(b)** $127.32 - 38.48$ **(c)** $3.1 + 42.36 + 9.034$ **(d)** $5.0006 - 3.1248$

Solution

(a)
$$
\begin{array}{r}
3.6 \\
+\ 2.3 \\
\hline
5.9
\end{array}
$$

(b)
$$
\begin{array}{r}
127.32 \\
-\ 38.48 \\
\hline
88.84
\end{array}
$$

(c)
$$
\begin{array}{r}
3.1 \\
42.36 \\
+\ 9.034 \\
\hline
54.494
\end{array}
$$

(d)
$$
\begin{array}{r}
5.0006 \\
-\ 3.1248 \\
\hline
1.8758
\end{array}
$$

NOTE TO STUDENT: Fully worked-out solutions to all of the Practice Problems can be found at the back of the text starting at page SP-1

Practice Problem 5 Add or subtract.

(a) $3.12 + 5.08 + 1.42$ **(b)** $152.003 - 136.118$

(c) $1.1 + 3.16 + 5.123$ **(d)** $1.0052 - 0.1234$

SIDELIGHT: Adding Zeros to the Right-Hand Side of the Decimal
When we added fractions, we had to have common denominators. Since decimals are really fractions, why can we add them without having common denominators? Actually, we have to have common denominators to add any fractions, whether they are in decimal form or fraction form. However, sometimes the notation does not show this. Let's examine Example 5(c).

Original Problem   We are adding the three numbers:

$$3.1$$
$$42.36$$
$$\underline{+\ 9.034}$$
$$54.494$$

$$3\tfrac{1}{10} + 42\tfrac{36}{100} + 9\tfrac{34}{1000}$$
$$3\tfrac{100}{1000} + 42\tfrac{360}{1000} + 9\tfrac{34}{1000}$$
$$3.100 + 42.360 + 9.034 \quad \text{This is the new problem.}$$

Original Problem   **New Problem**

3.1	3.100
42.36	42.360
+ 9.034	+ 9.034
54.494	54.494

We notice that the results are the same. The only difference is the notation. We are using the property that any number of zeros may be added to the right-hand side of a decimal without changing its value.

This shows the convenience of adding and subtracting fractions in decimal form. Little work is needed to change the decimals so that they have a common denominator. All that is required is to add zeros to the right-hand side of the decimal (and we usually do not even write out that step except when subtracting).

As long as we line up the decimal points, we can add or subtract any decimal fractions.

In the following example we will find it useful to add zeros to the right-hand side of the decimal.

EXAMPLE 6   Perform the following operations.

(a) $1.0003 + 0.02 + 3.4$

(b) $12 - 0.057$

Solution   We will add zeros so that each number shows the same number of decimal places.

(a)
$$1.0003$$
$$0.0200$$
$$\underline{+\ 3.4000}$$
$$4.4203$$

(b)
$$12.000$$
$$\underline{-\ 0.057}$$
$$11.943$$

Practice Problem 6   Perform the following operations.

(a) $0.061 + 5.0008 + 1.3$

(b) $18 - 0.126$

---

⑤ Multiplying Decimals

MULTIPLICATION OF DECIMALS

To multiply decimals, you first multiply as with whole numbers. To determine the position of the decimal point, you count the total number of decimal places in the two numbers being multiplied. This will determine the number of decimal places that should appear in the answer.

EXAMPLE 7 Multiply. 0.8×0.4

Solution

$$
\begin{array}{r}
0.8 \quad \text{(one decimal place)} \\
\underline{\times 0.4} \quad \text{(one decimal place)} \\
0.32 \quad \text{(two decimal places)}
\end{array}
$$

Practice Problem 7 Multiply. 0.5×0.3

Note that you will often have to add zeros to the left of the digits obtained in the product so that you obtain the necessary number of decimal places.

EXAMPLE 8 Multiply. 0.123×0.5

Solution

$$
\begin{array}{r}
0.123 \quad \text{(three decimal places)} \\
\underline{\times \quad 0.5} \quad \text{(one decimal place)} \\
0.0615 \quad \text{(four decimal places)}
\end{array}
$$

Practice Problem 8 Multiply. 0.12×0.4

Here are some examples that involve more decimal places.

EXAMPLE 9 Multiply.

(a) 2.56×0.003 **(b)** 0.0036×0.008

Solution

$$
\textbf{(a)} \quad
\begin{array}{r}
2.56 \quad \text{(two decimal places)} \\
\underline{\times \ 0.003} \quad \text{(three decimal places)} \\
0.00768 \quad \text{(five decimal places)}
\end{array}
$$

$$
\textbf{(b)} \quad
\begin{array}{r}
0.0036 \quad \text{(four decimal places)} \\
\underline{\times \ 0.008} \quad \text{(three decimal places)} \\
0.0000288 \quad \text{(seven decimal places)}
\end{array}
$$

Practice Problem 9 Multiply.

(a) 1.23×0.005 **(b)** 0.003×0.00002

NOTE TO STUDENT: Fully worked-out solutions to all of the Practice Problems can be found at the back of the text starting at page SP-1

SIDELIGHT: Counting the Number of Decimal Places

Why do we count the number of decimal places? The rule really comes from the properties of fractions. If we write the problem in Example 8 in fraction form, we have

$$
0.123 \times 0.5 = \frac{123}{1000} \times \frac{5}{10} = \frac{615}{10,000} = 0.0615.
$$

 Dividing Decimals

When discussing division of decimals, we frequently refer to the three primary parts of a division problem. Be sure you know the meaning of each term.

The **divisor** is the number you divide into another.

The **dividend** is the number to be divided.

The **quotient** is the result of dividing one number by another.

In the problem $6 \div 2 = 3$ we represent each of these terms as follows:

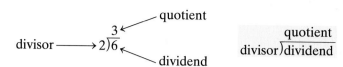

When dividing two decimals, count *the number of decimal places* in the divisor. Then *move the decimal point to the right* that *same number of places* in both *the divisor* and *the dividend*. Mark that position with a caret $(_\wedge)$. Finally, perform the division. Be sure to line up the decimal point in the quotient with the position indicated by the caret.

EXAMPLE 10   Four friends went out for lunch. The total bill, including tax, was $32.68. How much did each person pay if they shared the cost equally?

Solution   To answer this question, we must calculate $32.68 \div 4$.

$$
\begin{array}{r}
8.17 \\
4\overline{)32.68} \\
\underline{32} \\
06 \\
\underline{4} \\
28 \\
\underline{28} \\
0
\end{array}
$$

Since there are no decimal places in the divisor, we do not need to move the decimal point. We must be careful, however, to place the decimal point in the quotient directly above the decimal point in the dividend.

Thus $32.68 \div 4 = 8.17$, and each friend paid $8.17.

Practice Problem 10   Sally Keyser purchased 6 boxes of paper for an inkjet printer. The cost was $31.56. There was no tax since she purchased the paper for a charitable organization. How much did she pay for each box of paper?

Note that sometimes we will need to place extra zeros in the dividend in order to move the decimal point the required number of places.

EXAMPLE 11   Divide. $16.2 \div 0.027$

Solution

$$0.027\overline{)16.200}$$

three decimal places

There are **three** decimal places in the divisor, so we move the decimal point **three places** to **the right** in the **divisor** and **dividend** and mark the new position by a caret. Note that we must add two zeros to 16.2 in order to do this.

$$
\begin{array}{r}
600. \\
0.027\overline{)16.200} \\
\underline{16\ 2} \\
000
\end{array}
$$

Now perform the division as with whole numbers. The decimal point in the answer is directly above the caret.

Thus $16.2 \div 0.027 = 600$.

Practice Problem 11   Divide. $1800 \div 0.06$

Special care must be taken to line up the digits in the quotient. Note that sometimes we will need to place zeros in the quotient after the decimal point.

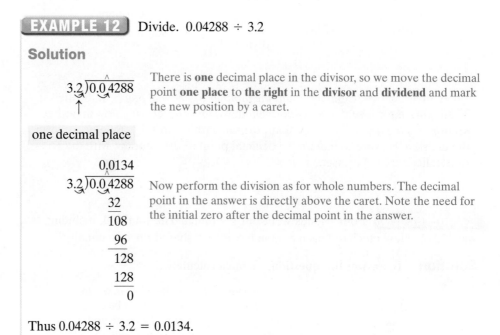

EXAMPLE 12 Divide. 0.04288 ÷ 3.2

Solution

$$3.2\overline{)0.0\,4288}$$

↑
one decimal place

There is **one** decimal place in the divisor, so we move the decimal point **one place** to **the right** in the **divisor** and **dividend** and mark the new position by a caret.

$$
\begin{array}{r}
0.0134 \\
3.2\overline{)0.0\,4288} \\
32 \\
\overline{108} \\
96 \\
\overline{128} \\
128 \\
\overline{0}
\end{array}
$$

Now perform the division as for whole numbers. The decimal point in the answer is directly above the caret. Note the need for the initial zero after the decimal point in the answer.

Thus 0.04288 ÷ 3.2 = 0.0134.

Practice Problem 12 Divide. 0.01764 ÷ 4.9

NOTE TO STUDENT: Fully worked-out solutions to all of the Practice Problems can be found at the back of the text starting at page SP-1

SIDELIGHT: Dividing Decimals by Another Method

Why does this method of dividing decimals work? Essentially, we are using the steps we used in Section 0.1 to change a fraction to an equivalent fraction by multiplying both the numerator and denominator by the same number. Let's reexamine Example 12.

$$0.04288 \div 3.2 = \frac{0.04288}{3.2}$$

Write the original problem using fraction notation.

$$= \frac{0.04288 \times 10}{3.2 \quad \times 10}$$

Multiply the numerator and denominator by 10. Since this is the same as multiplying by 1, we are not changing the fraction.

$$= \frac{0.4288}{32}$$

Write the result of multiplication by 10.

$$= 0.4288 \div 32$$

Rewrite the fraction as an equivalent problem with division notation.

Notice that we have obtained a new problem that is the same as the problem in Example 12 when we moved the decimal one place to the right in the divisor and dividend. We see that the reason we can move the decimal point as many places as necessary to the right in divisor and dividend is that this is the same as multiplying the numerator and denominator of a fraction by a power of 10 to obtain an equivalent fraction.

Multiplying and Dividing a Decimal by a Multiple of 10

When multiplying by 10, 100, 1000, and so on, a simple rule may be used to obtain the answer. For every zero in the multiplier, move the decimal point one place to the right.

EXAMPLE 13 Multiply.

(a) 3.24×10 **(b)** 15.6×100 **(c)** 0.0026×1000

Solution

(a) $3.24 \times 10 = 32.4$ One zero—move decimal point one place to the right.
(b) $15.6 \times 100 = 1560$ Two zeros—move decimal point two places to the right.
(c) $0.0026 \times 1000 = 2.6$ Three zeros—move decimal point three places to the right.

Practice Problem 13 Multiply.

(a) 0.0016×100 **(b)** 2.34×1000 **(c)** $56.75 \times 10{,}000$

The reverse rule is true for division. When dividing by 10, 100, 1000, 10,000, and so on, move the decimal point one place to the left for every zero in the divisor.

EXAMPLE 14 Divide.

(a) $52.6 \div 10$ **(b)** $0.0038 \div 100$ **(c)** $5936.2 \div 1000$

Solution

(a) $\dfrac{52.6}{10} = 5.26$ Move one place to the left.

(b) $\dfrac{0.0038}{100} = 0.000038$ Move two places to the left.

(c) $\dfrac{5936.2}{1000} = 5.9362$ Move three places to the left.

Practice Problem 14 Divide.

(a) $\dfrac{5.82}{10}$ **(b)** $123.4 \div 1000$ **(c)** $\dfrac{0.00614}{10{,}000}$

Developing Your Study Skills

Making a Friend in the Class

Attempt to make a friend in your class. You may find that you enjoy sitting together and drawing support and encouragement from one another. Exchange phone numbers so you can call each other whenever you get stuck in your study. Set up convenient times to study together on a regular basis, to do homework, and to review for exams.

You must not depend on a friend or fellow student to tutor you, do your work for you, or in any way be responsible for your learning. However, you will learn from one another as you seek to master the course. Studying with a friend and comparing notes, methods, and solutions can be very helpful. And it can make learning mathematics a lot more fun!

Verbal and Writing Skills

1. A decimal is another way of writing a fraction whose denominator is _____.

2. We write 0.42 in words as _____.

3. When dividing 7432.9 by 1000 we move the decimal point _____ places to the _____.

4. When dividing 96.3 by 10,000 we move the decimal point _____ places to the _____.

Write each fraction as a decimal.

5. $\dfrac{5}{8}$ **6.** $\dfrac{6}{25}$ **7.** $\dfrac{3}{15}$ **8.** $\dfrac{12}{15}$ **9.** $\dfrac{7}{11}$ **10.** $\dfrac{2}{3}$

Write each decimal as a fraction in simplified form.

11. 0.8   **12.** 0.5   **13.** 0.25   **14.** 0.15   **15.** 0.625   **16.** 0.475

17. 0.06   **18.** 0.08   **19.** 3.4   **20.** 4.8   **21.** 5.5   **22.** 6.25

Add or subtract.

23. $1.71 + 0.38$ **24.** $3.42 + 0.38$ **25.** $2.5 + 3.42 + 4.9$ **26.** $6.31 + 4.2 + 8.5$

27. $46.03 + 215.1 + 0.078$ **28.** $33.01 + 0.38 + 175.401$ **29.** $147.18 - 15.39$ **30.** $131.43 - 86.95$

31. $53.783 - 2.34$ **32.** $48.575 - 5.44$ **33.** $125.43 - 2.8$ **34.** $30 - 0.82$

Multiply or divide.

35. 7.21×4.2 **36.** 7.12×2.6 **37.** 0.04×0.08 **38.** 5.23×1.41

39. 4.23×0.025 **40.** 3.84×0.0017 **41.** $58{,}200 \times 0.0015$ **42.** $23{,}000 \times 0.0042$

43. $3.616 \div 64$ **44.** $12.6672 \div 39$ **45.** $7.9728 \div 3.02$ **46.** $6.519 \div 2.05$

47. $0.5230 \div 0.002$ **48.** $0.031 \div 0.005$ **49.** $0.03048 \div 0.06$ **50.** $0.00855 \div 0.09$

Multiply or divide by moving the decimal point.

51. 3.45×1000 **52.** 1.36×1000 **53.** $0.76 \div 100$ **54.** $175{,}318 \div 1000$

55. $7.36 \times 10{,}000$ **56.** $0.00243 \times 100{,}000$ **57.** $73{,}892 \div 100{,}000$ **58.** $3.52 \div 1000$

59. 0.1498×100 **60.** $1.931 \div 100$ **61.** $85.54 \times 10{,}000$ **62.** $96.12 \div 10{,}000$

Mixed Practice *Perform the indicated calculations.*

63. 54.8×0.15

64. 8.252×0.005

65. $13.75 + 2.55 + 0.078$

66. $1.109 + 0.088 + 16.4$

67. $0.05724 \div 0.027$

68. $77.136 \div 0.003$

69. 0.7683×1000

70. $34.72 \times 10,000$

71. $25.98 - 2.33$

72. $12.1 - 0.23$

73. $153.7 \div 100$

74. $0.0388 \div 1000$

Applications

75. *Measurement* While mixing solutions in her chemistry lab, Mia needed to change the measured data from pints to liters. There are 0.4732 liters in one pint and the original measurement was 5.5 pints. What is the measured data in liters?

76. *Hybrid Cars* Chris Smith bought a Toyota Prius, a hybrid electric-gas-powered car, because of its great gas mileage. The car averages 51 miles per gallon in the city, and the tank holds 11.9 gallons of gas. How many miles can Chris drive in the city on a full tank of gas?

77. *Wages* Harry has a part-time job at Stop and Shop. He earns $9 an hour. He requested enough hours of work each week so that he could earn at least $185 a week. How many one-hour shifts will he have to work to achieve his goal? By how much will he exceed his earning goal of $185 per week?

78. *Drinking Water* The EPA standard for safe drinking water is a maximum of 1.3 milligrams of copper per liter of water. A water testing firm found 6.8 milligrams of copper in a 5-liter sample drawn from Jim and Sharon LeBlanc's house. Is the water safe or not? By how much does the amount of copper exceed or fall short of the maximum allowed?

Cumulative Review *Perform each operation. Simplify all answers.*

79. **[0.3.2]** $3\frac{1}{2} \div 5\frac{1}{4}$

80. **[0.3.1]** $\frac{3}{8} \cdot \frac{12}{27}$

81. **[0.2.3]** $\frac{12}{25} + \frac{9}{20}$

82. **[0.2.4]** $1\frac{3}{5} - \frac{1}{2}$

Quick Quiz 0.4 Perform the calculations indicated.

1. $8.0567 - 2.3489$

2. 58.7×0.06

3. $4.608 \div 0.16$

4. **Concept Check** Explain how you would place the decimal points when performing the calculation $0.252 \div 0.0035$.

0.5 PERCENTS, ROUNDING, AND ESTIMATING

Student Learning Objectives

After studying this section, you will be able to:

1 Change a decimal to a percent.

2 Change a percent to a decimal.

3 Find the percent of a given number.

4 Find the missing percent when given two numbers.

5 Use rounding to estimate.

 Changing a Decimal to a Percent

A **percent** is a fraction that has a denominator of 100. When you say "sixty-seven percent" or write 67%, you are just expressing the fraction $\frac{67}{100}$ in another way. The word *percent* is a shortened form of the Latin words *per centum,* which means "by the hundred." In everyday use, percent means per one hundred.

Russell Camp owns 100 acres of land in Montana. 49 of the acres are covered with trees. The rest of the land is open fields. We say that 49% of his land is covered with trees.

It is important to see that 49% means 49 parts out of 100 parts. It can also be written as a fraction, $\frac{49}{100}$, or as a decimal, 0.49. Understanding the meaning of the notation allows you to change from one form to another. For example,

$$49\% = 49 \text{ out of } 100 \text{ parts} = \frac{49}{100} = 0.49.$$

Similarly, you can express a fraction with denominator 100 as a percent or a decimal.

$$\frac{11}{100} \text{ means } 11 \text{ parts out of } 100 \text{ or } 11\%. \quad \text{So } \frac{11}{100} = 11\% = 0.11.$$

Now that we understand the concept, we can use some quick procedures to change from decimals to percent, and vice versa.

> **CHANGING A DECIMAL TO A PERCENT**
> 1. Move the decimal point two places to the right.
> 2. Add the % symbol.

Be sure to follow this procedure for percents that are less than 1%. Remember, 0.01 is 1%. Thus we would expect 0.001 to be less than 1%. 0.001 = 0.1% or one-tenth (0.1) of a percent.

EXAMPLE 1 Change to a percent.

(a) 0.0364   **(b)** 0.0008   **(c)** 0.4

Solution We move the decimal point two places to the right and add the % symbol.

(a) 0.0364 = 3.64%   **(b)** 0.0008 = 0.08%   **(c)** 0.4 = 0.40 = 40%

Practice Problem 1 Change to a percent.

(a) 0.92   **(b)** 0.0736   **(c)** 0.7   **(d)** 0.0003

Be sure to follow the same procedure for percents that are greater than 100%. Remember that 1 is 100%. Thus we would expect 1.5 to be greater than 100%. In fact, 1.5 = 150%.

EXAMPLE 2 Change to a percent.

(a) 1.48   **(b)** 2.938   **(c)** 4.5

Solution   We move the decimal point two places to the right and add the % symbol.

(a) 1.48 = 148%   **(b)** 2.938 = 293.8%   **(c)** 4.5 = 4.50 = 450%

Practice Problem 2 Change to a percent.

(a) 3.04   **(b)** 5.186   **(c)** 2.1

② Changing a Percent to a Decimal

In this procedure we move the decimal point to the left and remove the % symbol.

> **CHANGING A PERCENT TO A DECIMAL**
> **1.** Move the decimal point two places to the left.
> **2.** Remove the % symbol.

EXAMPLE 3 Change to a decimal.

(a) 4%   **(b)** 0.6%   **(c)** 254.8%

Solution   First we move the decimal point two places to the left. Then we remove the % symbol.

(a) 4% = 4.% = 0.04
           ↑
    The unwritten decimal point is understood to be here.

(b) 0.6% = 0.006
(c) 254.8% = 2.548

Practice Problem 3 Change to a decimal.

(a) 7%   **(b)** 9.3%   **(c)** 131%   **(d)** 0.04%

③ Finding the Percent of a Given Number

How do we find 60% of 20? Let us relate it to problems we did in Section 0.3 involving multiplication of fractions.

Consider the following problem.

$$\text{What} \quad \text{is} \quad \frac{3}{5} \quad \text{of} \quad 20?$$
$$\downarrow \qquad \downarrow \qquad \downarrow \qquad \downarrow \qquad \downarrow$$
$$\boxed{?} \quad = \quad \frac{3}{5} \quad \times \quad 20$$

$$\boxed{?} = \frac{3}{\cancel{5}} \times \overset{4}{\cancel{20}} = 12 \quad \text{The answer is 12.}$$

Calculator

 Percent to Decimal

You can use a calculator to change 52% to a decimal. If your calculator has a $\boxed{\%}$ key, do the following:

Enter: 52 $\boxed{\%}$

The display should read:

$$\boxed{0.52}$$

If your calculator does not have a $\boxed{\%}$ key, divide the number by 100.
Enter:

$$52 \boxed{\div} 100 \boxed{=}$$

Try the following:

(a) 46%   **(b)** 137%
(c) 9.3%   **(d)** 6%

Note: The calculator divides by 100 when the percent key is pressed. If you do not have a $\boxed{\%}$ key, then you can divide by 100 instead.

Since a percent is really a fraction, a percent problem is solved similarly to the way a fraction problem is solved. Since $\frac{3}{5} = \frac{3 \cdot 20}{5 \cdot 20} = \frac{60}{100} = 60\%$, we could write the problem as

What is 60% of 20?

$\downarrow$ $\quad$ $\downarrow$ $\quad$ $\downarrow$ $\quad$ $\downarrow$ $\quad$ $\downarrow$

$\boxed{?}$ $=$ $60\% \times 20$

$\boxed{?} = 0.60 \times 20$

$\boxed{?} = 12.0$ The answer is 12.

Thus we have developed the following rule.

FINDING THE PERCENT OF A NUMBER

To find the percent of a number, change the percent to a decimal and multiply the number by the decimal.

EXAMPLE 4 Find.

(a) 10% of 36 **(b)** 2% of 350 **(c)** 182% of 12 **(d)** 0.3% of 42

Solution

(a) 10% of 36 = 0.10 × 36 = 3.6 **(b)** 2% of 350 = 0.02 × 350 = 7
(c) 182% of 12 = 1.82 × 12 = 21.84 **(d)** 0.3% of 42 = 0.003 × 42 = 0.126

Practice Problem 4 Find.

(a) 18% of 50 **(b)** 4% of 64 **(c)** 156% of 35

There are many real-life applications for finding the percent of a number. When you go shopping in a store, you may find sale merchandise marked 35% off. This means that the sale price is 35% off the regular price. That is, 35% of the regular price is subtracted from the regular price to get the sale price.

EXAMPLE 5 A store is having a sale of 35% off the retail price of all sofas. Melissa wants to buy a particular sofa that normally sells for $595.

(a) How much will Melissa save if she buys the sofa on sale?
(b) What will the purchase price be if Melissa buys the sofa on sale?

Solution

(a) To find 35% of $595 we will need to multiply 0.35 × 595.

$$\begin{array}{r} 595 \\ \times\, 0.35 \\ \hline 2975 \\ 1785 \\ \hline 208.25 \end{array}$$

Thus Melissa will save $208.25 if she buys the sofa on sale.

(b) The purchase price is the difference between the original price and the amount saved.

$$\begin{array}{r} 595.00 \\ -\, 208.25 \\ \hline 386.75 \end{array}$$

If Melissa buys the sofa on sale, she will pay $386.75.

Practice Problem 5 John received a 4.2% pay raise at work this year. He had previously earned $38,000 per year.

(a) What was the amount of his pay raise in dollars?
(b) What is his new salary?

Finding the Missing Percent When Given Two Numbers

Recall that we can write $\frac{3}{4}$ as $\frac{75}{100}$ or 75%. If we were asked the question, "What percent is 3 of 4?" we would say 75%. This gives us a procedure for finding what percent one number is of a second number.

FINDING THE MISSING PERCENT

1. Write a fraction with the two numbers. The number *after* the word *of* is always the denominator, and the other number is the numerator.

2. Simplify the fraction (if possible).

3. Change the fraction to a decimal.

4. Express the decimal as a percent.

EXAMPLE 6   What percent of 24 is 15?

Solution   This can be solved quickly as follows.

Step 1 $\dfrac{15}{24}$ Write the relationship as a fraction. The number after "of" is 24, so 24 is the denominator.

Step 2 $\dfrac{15}{24} = \dfrac{5}{8}$ Simplify the fraction (when possible).

Step 3 $= 0.625$ Change the fraction to a decimal.

Step 4 $= 62.5\%$ Change the decimal to a percent.

Practice Problem 6   What percent of 148 is 37?

NOTE TO STUDENT: Fully worked-out solutions to all of the Practice Problems can be found at the back of the text starting at page SP-1

The question in Example 6 can also be written as "15 is what percent of 24?" To answer the question, we begin by writing the relationship as $\frac{15}{24}$. Remember that "of 24" means 24 will be the denominator.

EXAMPLE 7

(a) What percent of 16 is 3.8?   **(b)** $150 is what percent of $120?

Solution

(a) What percent of 16 is 3.8?

$\dfrac{3.8}{16}$ Write the relationship as a fraction.

You can divide to change the fraction to a decimal and then change the decimal to a percent.

$$\begin{array}{r} 0.2375 \to 23.75\% \\ 16\overline{)3.8000} \end{array}$$

(b) $150 is what percent of $120?

$\dfrac{150}{120} = \dfrac{5}{4}$ Reduce the fraction whenever possible to make the division easier.

$$\begin{array}{r} 1.25 \to 125\% \\ 4\overline{)5.00} \end{array}$$

Practice Problem 7

(a) What percent of 48 is 24?   **(b)** 4 is what percent of 25?

Calculator

Finding the Percent Given Two Numbers

You can use a calculator to find a missing percent. What percent of 95 is 19?

1. Enter as a fraction.

19 ÷ 95

2. Change to a percent.

19 ÷ 95 × 100 =

The display should read

20

This means 20%.
What percent of 625 is 250?

⑤ Using Rounding to Estimate

Before we proceed in this section, we will take some time to be sure you understand the idea of rounding a number. You will probably recall the following simple rule from your previous mathematics courses.

ROUNDING A NUMBER

If the first digit to the right of the round-off place is

1. less than 5, we make no change to the digit in the round-off place.

2. 5 or more, we increase the digit in the round-off place by 1.

To illustrate, 4689 rounded to the nearest hundred is 4700. Rounding 233,987 to the nearest ten thousand, we obtain 230,000. We will now use our experience in rounding as we discuss the general area of estimation.

Estimation is the process of finding an approximate answer. It is not designed to provide an exact answer. Estimation will give you a rough idea of what the answer might be. For any given problem, you may choose to estimate in many different ways.

ESTIMATION BY ROUNDING

1. Round each number so that there is one nonzero digit.

2. Perform the calculation with the rounded numbers.

EXAMPLE 8   Find an estimate of the product 5368×2864.

Solution

Step 1 Round 5368 to 5000.
        Round 2864 to 3000.

Step 2 Multiply.

$$5000 \times 3000 = 15,000,000$$

An estimate of the product is 15,000,000.

Practice Problem 8   Find an estimate of the product $128,621 \times 378$.

EXAMPLE 9   Won Lin has a small compact car. He drove 396.8 miles in his car and used 8.4 gallons of gas.

(a) Estimate the number of miles he gets per gallon.

(b) Estimate how much it will cost him for fuel to drive on a cross-country trip of 2764 miles if gasoline usually costs 3.59\frac{9}{10}$ per gallon.

Solution

(a) Round 396.8 miles to 400 miles. Round 8.4 gallons to 8 gallons. Now divide.

$$8\overline{)400} \quad \overset{50}{}$$

Won Lin's car gets about 50 miles per gallon.

(b) We will need to use the information we found in part **(a)** to determine how many gallons of gasoline Won Lin will use on his trip. Round 2764 miles to 3000 miles and divide 3000 miles by 50 miles per gallon.

$$\begin{array}{r} 60 \\ 50\overline{)3000} \end{array}$$

Won Lin will use about 60 gallons of gas for his cross-country trip.

To estimate the cost, we need to ask ourselves, "What kind of an estimate are we looking for?" It may be sufficient to round 3.59\frac{9}{10}$ to $4.00 and multiply.

$$60 \times \$4.00 = \$240.00$$

Keep in mind that this is a broad estimate. You may want an estimate that will be closer to the exact answer. In that case round 3.59\frac{9}{10}$ to $3.60 and multiply.

$$60 \times \$3.60 = \$216.00$$

CAUTION: An estimate is only a rough guess. If we are estimating the cost of something, we may want to round the unit cost to a value that is closer to the actual amount so that our estimate is more accurate. When we estimate costs, it is a good idea to round to a value above the actual unit cost to make sure that we will have enough money for the expenditure. The actual cost to the nearest penny of the cross-country trip in Example 9 is $210.59.

Practice Problem 9 Roberta drove 422.8 miles in her truck and used 19.3 gallons of gas. Assume that gasoline costs 3.69\frac{9}{10}$ per gallon.

(a) Estimate the number of miles she gets per gallon.

(b) Estimate how much it will cost her to drive to Chicago and back, a distance of 3862 miles.

NOTE TO STUDENT: Fully worked-out solutions to all of the Practice Problems can be found at the back of the text starting at page SP-1

Other words that indicate estimation in word problems are *about* and *approximate*.

EXAMPLE 10 Follow the principles of estimation to find an approximate value. 2.68% of $54,361.92

Solution   Begin by writing 2.68% as a decimal.

$$2.68\% = 0.0268$$

Next round 0.0268 to the nearest hundredth.

0.0268 is 0.03 rounded to the nearest hundredth.

Finally, round $54,361.92 to $50,000.
    Now multiply.

$0.03 \times 50,000 = 1500$ Remember that to find a percent of a number, you multiply.

Thus the approximate value is $1500.

To get a better approximation of $2.68\% \times \$54,361.92$, you may wish to round $54,361.92 to the nearest thousand. Try it and compare this approximation to the previous one. How close you should try to get to the exact answer depends on what you need the approximation for.

Practice Problem 10 Follow the principles of estimation to find an approximate value. 56.93% of $293,567.12.

PRACTICE   WATCH   DOWNLOAD   READ   REVIEW

Verbal and Writing Skills

1. When you write 19%, what do you really mean? Describe the meaning in your own words.

2. When you try to solve a problem like "What percent of 80 is 30?" how do you know if you should write the fraction as $\frac{80}{30}$ or as $\frac{30}{80}$?

Change to a percent.

3. 0.28

4. 0.61

5. 0.568

6. 0.089

7. 0.076

8. 0.325

9. 2.39

10. 7.4

Change to a decimal.

11. 3%

12. 66%

13. 0.4%

14. 0.62%

15. 250%

16. 175%

17. 7.4%

18. 87.59%

Find the following.

19. What is 8% of 65?

20. What is 7% of 69?

21. What is 10% of 130?

22. What is 25% of 600?

23. What is 112% of 65?

24. What is 154% of 270?

25. 36 is what percent of 24?

26. 49 is what percent of 28?

27. What percent of 340 is 17?

28. 48 is what percent of 600?

29. 75 is what percent of 30?

30. What percent of 35 is 28?

Applications

31. *Exam Grades* Dave took an exam with 80 questions. He answered 68 correctly. What was his grade for the exam? Write the grade as a percent.

32. *Hockey* For the 2006–07 season, Jordan Staal of the Pittsburgh Penguins led the National Hockey League in shooting percentage. Of his 131 shots on goal, 29 went in the net. What percent of the time did one of Staal's shots on goal result in a goal? (Round to the nearest whole percent.)

33. *Tipping* Diana and Russ ate a meal costing $32.80 when they went out to dinner. If they want to leave the standard 15% tip for their server, how much will they tip and what will their total bill be?

34. *CD Failure* Music CDs have a failure rate of 1.8%. (They skip or get stuck on a song.) If 36,000 CDs were manufactured last week, how many of them were defective?

35. *Food Budget* The Gonzalez family has a combined monthly income of $1850. Their food budget is $380/month. What percentage of their monthly income is budgeted for food? (Round your answer to the nearest whole percent.)

36. *Survey* In a local college survey, it was discovered that 137 out of 180 students had a grandparent not born in the United States. What percent of the students had a grandparent not born in the United States? (Round your answer to the nearest hundredth of a percent.)

37. *Gift Returns* Last Christmas season, the Jones Mill Outlet Store chain calculated that they sold 36,000 gift items. If they assume that there will be a 1.5% return rate after the holidays, how many gifts can they expect to be exchanged?

38. *Rent* The total cost of a downtown Boston apartment, including rent, heat, electricity, and water, is $3690 per month. Sarah is renting this apartment with four friends. Based on a complicated formula, which takes into account the size of her bedroom and the furniture she brought to the apartment, she must pay 17% of the apartment cost each month. What is her monthly cost?

39. *Salary* Abdul sells computers for a local computer outlet. He gets paid $450 per month plus a commission of 3.8% on all the computer hardware he sells. Last year he sold $780,000 worth of computer hardware. **(a)** What was his sales commission on the $780,000 worth of hardware he sold? **(b)** What was his annual salary (that is, his monthly pay and commission combined)?

40. *Business Travel* Bruce sells medical supplies on the road. He logged 18,600 miles in his car last year. He declared 65% of his mileage as business travel. **(a)** How many miles did he travel on business last year? **(b)** If his company reimburses him 31 cents for each mile traveled, how much should he be paid for travel expenses?

In exercises 41–52, follow the principles of estimation to find an approximate value. Round each number so that there is one nonzero digit. Do not find the exact value.

41. 693×307

42. 437×892

43. 2862×5986

44. 4893×6174

45. $14 + 73 + 80 + 21 + 56$

46. $318 + 494 + 613 + 243$

47. $41\overline{)829{,}346}$

48. $16\overline{)5{,}846{,}213}$

49. $\dfrac{2714}{31{,}500}$

50. $\dfrac{53{,}610}{786}$

51. Find 17% of $21,365.85.

52. Find 4.9% of $9321.88.

Applications

In exercises 53–58, determine an estimate of the exact answer. Use estimation by rounding. Do not find the exact value.

53. *Checkout Sales* A typical customer at the local Piggly Wiggly supermarket spends approximately $82 at the checkout register. The store keeps four registers open, and each handles 22 customers per hour. Estimate the amount of money the store receives in one hour.

54. *Weekend Spending Money* The Westerly Community Credit Union in Rhode Island has found that on Fridays, the average customer withdraws $85 from the ATM for weekend spending money. Each ATM averages 19 customers per hour. There are five machines. Estimate the amount of money withdrawn by customers in one hour.

55. *Gas Mileage* Rod's trip from Salt Lake City to his cabin in the mountains was 117.7 miles. If his car used 3.8 gallons of gas for the trip, estimate the number of miles his car gets to the gallon.

56. *Book Transport* The Paltrows are transferring their stock of books from one store to another. To save money, they are moving the books themselves. If their car can transport 430 books per trip, and they have 11,900 books to move, estimate the number of trips it will take.

57. *Revenue* Total revenue for May, June, and July of 2007 for Krispy Kreme stores was $104,100,000. If there were 411 stores systemwide, estimate the average revenue per store.

58. *Happy Meals Sales* The local McDonald's registered $437,690 in sales last month. They know from experience that 18% of their sales are Happy Meals. Estimate the amount of money spent on Happy Meals last month.

Income and Expenditures *The data for exercises 59–62 were obtained at www.bls.gov.*

59. In 2005 the average consumer aged 25–34 years had annual expenditures of $45,068 and spent 34% of this amount on housing. Estimate the annual expenditure for housing for a consumer in this age group.

60. In 2005 the average consumer aged 35–44 years had annual expenditures of $55,190 and spent 13% of this amount on food. Estimate the annual expenditure for food for a consumer in this age group.

61. In 2005, the average consumer aged 65–74 years had annual expenditures of $38,573. Of this amount, $4176 was spent on health care. Estimate the percent of total expenditures spent on health care for consumers in this age group.

62. In 2005, the average consumer aged 45–54 years had annual expenditures of $55,854. Of this amount, $3034 was spent on entertainment. Estimate the percent of total expenditures spent on entertainment for consumers in this age group.

Cumulative Review

63. **[0.4.4]** *Checking Account Balance* Janeen had an opening balance of $45.50 in her checking account this month. She deposited her paycheck of $1189 and a refund check of $33.90. She was charged by the bank $1.50 for ATM fees. Then she made out checks for $98.00, $128.00, $56.89, and $445.88. What will be her final balance at the end of the month?

64. **[0.4.5]** *Car Loan* Tally is buying a car and has a bank loan. He has to make 33 more monthly payments of $188.50. How much money will he pay to the bank over the next 33 months?

65. **[0.4.4]** *Gas Mileage* Dan took a trip in his Ford Taurus. At the start of the trip his car odometer read 68,459.5 miles. At the end of the trip his car odometer read 69,229.5 miles. He used 35 gallons of gas on the trip. How many miles per gallon did his car achieve?

66. **[0.4.6]** *Rainfall* In Hilo, Hawaii, Brad spent three months working in a restaurant for a summer job. In June he observed that 4.6 inches of rain fell. In July the rainfall was 4.5 inches and in August it was 2.9 inches. What was the average monthly rainfall that summer for those three months in Hilo?

Quick Quiz 0.5

1. 63 is what percent of 420?

2. What is 114% of 85?

Estimate the answer. First round each number to one nonzero digit. Then calculate.

3. $34{,}987\overline{)567{,}238}$

4. **Concept Check** Explain how you would change 0.0078 to a percent.

Putting Your Skills to Work: Use Math to Save Money

MANAGING DEBTS AND PAYMENTS

Can you imagine the joy of taking a great vacation without going into debt for it? Can you think about how great it would be to have all your debts paid off? It is a wonderful feeling! Paying off your debts and then being able to take a vacation is an excellent goal. But how is that done? Consider the story of Tracy and Max.

Facing Up to the Debt

Tracy and Max were overwhelmed with debt. Besides their ordinary living expenses, they had so much debt that they could barely make the money they earned last until the end of the month. They had little money left for extras and for having fun, and no money for a vacation. Each of their three credit cards was maxed out at $8000; they had a hospital debt of $12,000; they still owed $2000 on their car; and they had also borrowed money from friends in sums of $100 and $300.

Making a Plan

They decided to write down all of their debts, putting them in order from smallest to largest.

Then they made minimum payments on all the debts, but aimed at paying off the three smallest debts first.

1. Put the couple's debts in order from smallest to largest. Remember, there are three credit cards.

The minimum payment on each credit card averaged $25 per month. They had arranged with the hospital to pay the debt off at $50 per month. Their car payment was $200 per month, and they agreed to pay each of their friends $20 per month.

2. What is the total amount of their minimum monthly payment?

What Tracy and Max Accomplished

Tracy and Max decided they would eliminate any extras and not spend money on fun activities so they could use this money to pay off their three smallest debts. As a result they were able to pay off the first two smallest debts in just two months while making minimum payments on all other debts. Then each month they took the $40 they would have used to pay these two small debts and applied it towards the third smallest debt. Again, they made sure they made minimum payments on all other debts.

3. How many more months will it take Tracy and Max to pay off the third smallest debt if they follow the plan stated above?

After eliminating the smaller debts, they took the money they would have spent on those debts and used it on the principal of the remaining debts. In other words, they paid more than the minimum payment on the remaining debts. Because they hated the way they felt when they were in debt, they stopped using credit cards for new purchases. Max also took a temporary part-time job so they could pay off their debts faster. Finally they paid off all the debts!

Applying It to Your Life

Many debt counselors have a simple, practical suggestion for people in debt. Arrange debts in order, pay off the smallest first, and then let the consequences of that action help pay off the rest of the debts more quickly. Many people have been able to get out of debt in about two years by using this approach, other strategies for budgeting, and wise choices for living.

Topic	Procedure	Examples
Simplifying fractions, p. 2.	1. Write the **numerator** and **denominator** as a product of prime factors. 2. Use the basic rule of fractions that $$\frac{a \times c}{b \times c} = \frac{a}{b}$$ for any factor that appears in both the numerator and the denominator. 3. Multiply the remaining factors for the numerator and for the denominator separately.	$$\frac{15}{25} = \frac{\cancel{5} \cdot 3}{\cancel{5} \cdot 5} = \frac{3}{5}$$ $$\frac{36}{48} = \frac{\cancel{2} \cdot \cancel{2} \cdot 3 \cdot \cancel{3}}{\cancel{2} \cdot \cancel{2} \cdot 2 \cdot 2 \cdot \cancel{3}} = \frac{3}{4}$$ $$\frac{26}{39} = \frac{2 \cdot \cancel{13}}{3 \cdot \cancel{13}} = \frac{2}{3}$$
Changing improper fractions to mixed numbers, p. 4.	1. Divide the denominator into the numerator to obtain the whole-number part of the mixed number. 2. The remainder from the division will be the numerator of the fraction. 3. The denominator remains unchanged.	$\frac{14}{3} = 4\frac{2}{3}$ $\frac{19}{8} = 2\frac{3}{8}$ since $3\overline{)14}$ since $8\overline{)19}$ $\quad\frac{12}{2}$ $\frac{16}{3}$
Changing mixed numbers to improper fractions, p. 6.	1. Multiply the whole number by the denominator and add the result to the numerator. This will yield the new numerator. 2. The denominator does not change.	$$4\frac{5}{6} = \frac{(4 \times 6) + 5}{6} = \frac{24 + 5}{6} = \frac{29}{6}$$ $$3\frac{1}{7} = \frac{(3 \times 7) + 1}{7} = \frac{21 + 1}{7} = \frac{22}{7}$$
Changing fractions to equivalent fractions with a given denominator, p. 7.	1. Divide the original denominator into the new denominator. This result is the value that we use for multiplication. 2. Multiply the numerator and the denominator of the original fraction by that value.	$$\frac{4}{7} = \frac{?}{21}$$ $\underset{7\overline{)21}}{3} \leftarrow$ Use this to multiply $\frac{4 \times 3}{7 \times 3} = \frac{12}{21}$.
Finding the LCD (least common denominator) of two or more fractions, p. 12.	1. Write each denominator as the product of prime factors. 2. The LCD is a product containing each different factor. 3. If a factor occurs more than once in any one denominator, the LCD will contain that factor repeated the greatest number of times that it occurs in any one denominator.	Find the LCD of $\frac{4}{15}$ and $\frac{3}{35}$. $\qquad 15 = 5 \cdot 3$ $\qquad 35 = 5 \cdot 7$ $\qquad$ LCD $= 3 \cdot 5 \cdot 7 = 105$ Find the LCD of $\frac{11}{18}$ and $\frac{7}{45}$. $\quad 18 = 3 \cdot 3 \cdot 2$ (factor 3 appears twice) $\quad 45 = 3 \cdot 3 \cdot 5$ (factor 3 appears twice) $\quad$ LCD $= 2 \cdot 3 \cdot 3 \cdot 5 = 90$
Adding and subtracting fractions that do not have a common denominator, p. 13.	1. Find the LCD. 2. Change each fraction to an equivalent fraction with the LCD for a denominator. 3. Add or subtract the fractions and simplify the answer if possible.	$$\frac{3}{8} + \frac{1}{3} = \frac{3 \cdot 3}{8 \cdot 3} + \frac{1 \cdot 8}{3 \cdot 8} = \frac{9}{24} + \frac{8}{24} = \frac{17}{24}$$ $$\frac{11}{12} - \frac{1}{4} = \frac{11}{12} - \frac{1 \cdot 3}{4 \cdot 3} = \frac{11}{12} - \frac{3}{12} = \frac{8}{12} = \frac{2}{3}$$
Adding and subtracting mixed numbers, p. 16.	1. Change the mixed numbers to improper fractions. 2. Follow the rules for adding and subtracting fractions. 3. If necessary, change your answer to a mixed number.	$$1\frac{2}{3} + 1\frac{3}{4} = \frac{5}{3} + \frac{7}{4} = \frac{5 \cdot 4}{3 \cdot 4} + \frac{7 \cdot 3}{4 \cdot 3}$$ $$= \frac{20}{12} + \frac{21}{12} = \frac{41}{12} = 3\frac{5}{12}$$ $$2\frac{1}{4} - 1\frac{3}{4} = \frac{9}{4} - \frac{7}{4} = \frac{2}{4} = \frac{1}{2}$$
Multiplying fractions, p. 22.	1. If there are no common factors, multiply the numerators. Then multiply the denominators. 2. If possible, write the numerators and denominators as the product of prime factors. Use the basic rule of fractions to divide out any value that appears in both a numerator and a denominator. Multiply the remaining factors in the numerator. Multiply the remaining factors in the denominator.	$$\frac{3}{7} \times \frac{2}{13} = \frac{6}{91}$$ $$\frac{6}{15} \times \frac{35}{91} = \frac{2 \cdot \cancel{3}}{\cancel{3} \cdot \cancel{5}} \times \frac{\cancel{5} \cdot \cancel{7}}{\cancel{7} \cdot 13} = \frac{2}{13}$$ $$3 \times \frac{5}{8} = \frac{3}{1} \times \frac{5}{8} = \frac{15}{8} \text{ or } 1\frac{7}{8}$$

Topic	Procedure	Examples
Dividing fractions, p. 24.	**1.** Invert the second fraction. **2.** Multiply the fractions.	$\dfrac{4}{7} \div \dfrac{11}{3} = \dfrac{4}{7} \times \dfrac{3}{11} = \dfrac{12}{77}$ $\dfrac{5}{9} \div \dfrac{5}{7} = \dfrac{\cancel{5}}{9} \times \dfrac{7}{\cancel{5}} = \dfrac{7}{9}$
Multiplying and dividing mixed numbers, pp. 24, 26.	**1.** Change each mixed number to an improper fraction. **2.** Use the rules for multiplying or dividing fractions. **3.** If necessary, change your answer to a mixed number.	$2\dfrac{1}{4} \times 3\dfrac{3}{5} = \dfrac{9}{4} \times \dfrac{18}{5}$ $\quad = \dfrac{3 \cdot 3}{2 \cdot \cancel{2}} \times \dfrac{\cancel{2} \cdot 3 \cdot 3}{5} = \dfrac{81}{10} = 8\dfrac{1}{10}$ $1\dfrac{1}{4} \div 1\dfrac{1}{2} = \dfrac{5}{4} \div \dfrac{3}{2} = \dfrac{5}{2 \cdot \cancel{2}} \times \dfrac{\cancel{2}}{3} = \dfrac{5}{6}$
Changing fractional form to decimal form, p. 32.	Divide the denominator into the numerator.	$\dfrac{5}{8} = 0.625 \quad$ since $8\overline{)5.000}^{\,0.625}$
Changing decimal form to fractional form, p. 33.	**1.** Write the decimal as a fraction with a denominator of 10, 100, 1000, and so on. **2.** Simplify the fraction, if possible.	$0.37 = \dfrac{37}{100} \qquad 0.375 = \dfrac{375}{1000} = \dfrac{3}{8}$
Adding and subtracting decimals, p. 34.	**1.** Carefully line up the decimal points as indicated for addition and subtraction. (Extra zeros may be added to the right-hand side of the decimals if desired.) **2.** Add or subtract the appropriate digits.	Add. $\qquad\qquad$ Subtract. $1.236 + 7.825 \qquad 2 - 1.32$ $\quad 1.236 \qquad\qquad\; 2.00$ $\underline{+\ 7.825} \qquad\qquad \underline{-\ 1.32}$ $\quad 9.061 \qquad\qquad\; 0.68$
Multiplying decimals, p. 35.	**1.** First multiply the digits. **2.** Count the total number of decimal places in the numbers being multiplied. This number determines the number of decimal places in the answer.	0.9 (one place) $\qquad 0.009$ (three places) $\underline{\times\ 0.7}$ (one place) $\qquad \underline{\times\ 0.07}$ (two places) 0.63 (two places) $\qquad 0.00063$ (five places)
Dividing decimals, p. 36.	**1.** Count the number of decimal places in the divisor. **2.** Move the decimal point to the right the same number of places in both the divisor and dividend. **3.** Mark that position with a caret ($_\wedge$). **4.** Perform the division. Line up the decimal point in the quotient with the position indicated by the caret.	Divide. $7.5 \div 0.6$. Move decimal point one place to the right. $\qquad\quad 12.5$ $0.6_\wedge\overline{)7.5_\wedge 0}\qquad$ Therefore, $\qquad\ \underline{6}\qquad\qquad 7.5 \div 0.6 = 12.5$ $\qquad\ 1\,5$ $\qquad\ \underline{1\,2}$ $\qquad\quad 3\,0$ $\qquad\quad \underline{3\,0}$ $\qquad\qquad 0$
Changing a decimal to a percent, p. 42.	**1.** Move the decimal point two places to the right. **2.** Add the % symbol.	$0.46 = 46\%$ $0.002 = 0.2\% \qquad 1.59 = 159\%$ $0.013 = 1.3\% \qquad 0.0007 = 0.07\%$
Changing a percent to a decimal, p. 43.	**1.** Move the decimal point two places to the left. **2.** Remove the % symbol.	$49\% = 0.49 \qquad\qquad 180\% = 1.8$ $59.8\% = 0.598 \qquad 0.13\% = 0.0013$
Finding a percent of a number, p. 43.	**1.** Convert the percent to a decimal. **2.** Multiply the decimal by the number.	Find 12% of 86. $12\% = 0.12 \qquad 86 \times 0.12 = 10.32$ Therefore, 12% of 86 is 10.32.

(Continued on next page)

Topic	Procedure	Examples
Finding what percent one number is of another number, p. 45.	1. Place the number after the word *of* in the denominator. 2. Place the other number in the numerator. 3. If possible, simplify the fraction. 4. Change the fraction to a decimal. 5. Express the decimal as a percent.	What percent of 8 is 7? $$\frac{7}{8} = 0.875 = 87.5\%$$ 42 is what percent of 12? $$\frac{42}{12} = \frac{7}{2} = 3.5 = 350\%$$
Estimation, p. 46.	1. Round each number so that there is one nonzero digit. 2. Perform the calculation with the rounded numbers.	Estimate the number of square feet in a room that is 22 feet long and 13 feet wide. Assume that the room is rectangular. 1. We round 22 to 20. We round 13 to 10. 2. To find the area of a rectangle, we multiply length times width. $$20 \times 10 = 200.$$ We estimate that there are 200 square feet in the room.

Chapter 0 Review Problems

Section 0.1

In exercises 1–4, simplify.

1. $\dfrac{21}{28}$

2. $\dfrac{12}{40}$

3. $\dfrac{36}{82}$

4. $\dfrac{20}{35}$

5. Write $4\dfrac{3}{5}$ as an improper fraction.

6. Write $\dfrac{34}{5}$ as a mixed number.

7. Write $\dfrac{39}{6}$ as a mixed number.

Find the missing numerator.

8. $\dfrac{5}{8} = \dfrac{?}{24}$

9. $\dfrac{1}{7} = \dfrac{?}{35}$

10. $\dfrac{5}{9} = \dfrac{?}{72}$

11. $\dfrac{2}{5} = \dfrac{?}{55}$

Section 0.2

Combine.

12. $\dfrac{3}{5} + \dfrac{1}{4}$

13. $\dfrac{7}{12} + \dfrac{5}{8}$

14. $\dfrac{7}{20} - \dfrac{1}{12}$

15. $\dfrac{7}{10} - \dfrac{4}{15}$

16. $3\dfrac{1}{6} + 2\dfrac{3}{5}$

17. $1\dfrac{1}{4} + 2\dfrac{7}{10}$

18. $6\dfrac{2}{9} - 3\dfrac{5}{12}$

19. $3\dfrac{1}{15} - 1\dfrac{3}{20}$

Section 0.3

Multiply.

20. $6 \times \dfrac{5}{11}$

21. $2\dfrac{1}{3} \times 4\dfrac{1}{2}$

22. $1\dfrac{1}{8} \times 2\dfrac{1}{9}$

23. $\dfrac{4}{7} \times 5$

Divide.

24. $\dfrac{3}{8} \div 6$

25. $\dfrac{\dfrac{8}{3}}{\dfrac{5}{9}}$

26. $\dfrac{15}{16} \div 6\dfrac{1}{4}$

27. $2\dfrac{6}{7} \div \dfrac{10}{21}$

Section 0.4

Combine.

28. $1.634 + 3.007 + 2.560$ **29.** $24.831 - 17.094$ **30.** $47.251 - 17.69$ **31.** $1.9 + 2.53 + 0.006$

Multiply.

32. 0.007×5.35 **33.** 362.341×1000 **34.** $2.6 \times 0.03 \times 1.02$ **35.** $1.08 \times 0.06 \times 160$

Divide.

36. $0.186 \div 100$ **37.** $71.32 \div 1000$ **38.** $0.523 \div 0.4$ **39.** $1.35 \div 0.015$

40. $4.186 \div 2.3$ **41.** $0.19 \div 0.38$

42. Write as a decimal. $\frac{3}{8}$ **43.** Write as a fraction in simplified form. 0.36

Section 0.5

In exercises 44–47, write each percentage in decimal form.

44. 1.4%   **45.** 36.1%   **46.** 0.02%   **47.** 125.3%

In exercises 48–51, write each decimal as a percentage.

48. 0.0025   **49.** 0.325   **50.** 0.9   **51.** 0.1

52. What is 85% of 600?   **53.** Find 7.2% of 55.   **54.** 48 is what percent of 75?

55. What percent of 120 is 15?   **56.** What percent of 1250 is 750?

57. *Education Level* In 2005, 80% of California residents age 25 or older reported that they had at least a 4-year high school education. If the population of California age 25 or older was 22,299,041, how many people reported at least a high school education? (*Source:* www.census.gov)

58. *Math Deficiency* In a given university, 720 of the 960 freshmen had a math deficiency. What percentage of the class had a math deficiency?

In exercises 59–64, estimate. Do not find an exact value.

59. $234,897 \times 1,936,112$ **60.** $357 + 923 + 768 + 417$ **61.** $734,318 - 384,000$

62. $7\frac{1}{3} + 3\frac{5}{6} + 8\frac{3}{7}$ **63.** Find 18% of $56,297.   **64.** $12,482 \div 389$

65. *Salary* Estimate Sara's salary for the week if she makes $8.35 per hour, and worked 38.5 hours.

66. *Apartment Sharing* Estimate the monthly cost for each of three roommates who want to share an apartment that costs $923.50 per month.

How Am I Doing? Chapter 0 Test

Remember to use your Chapter Test Prep Video CD to see the worked-out solutions to the test problems you want to review.

In exercises 1 and 2, simplify.

1. $\dfrac{16}{18}$

2. $\dfrac{48}{36}$

3. Write as an improper fraction. $6\dfrac{3}{7}$

4. Write as a mixed number. $\dfrac{105}{9}$

In exercises 5–12, perform the operations indicated. Simplify answers whenever possible.

5. $\dfrac{2}{3} + \dfrac{5}{6} + \dfrac{3}{8}$

6. $1\dfrac{1}{8} + 3\dfrac{3}{4}$

7. $3\dfrac{2}{3} - 2\dfrac{5}{6}$

8. $\dfrac{5}{7} \times \dfrac{28}{15}$

9. $\dfrac{5}{18} \times \dfrac{3}{4}$

10. $5\dfrac{3}{8} \div 2\dfrac{3}{4}$

11. $2\dfrac{1}{2} \times 3\dfrac{1}{4}$

12. $\dfrac{\frac{7}{4}}{\frac{1}{2}}$

In exercises 13–18, perform the calculations indicated.

13. $1.6 + 3.24 + 9.8$

14. $7.0046 - 3.0149$

15. 32.8×0.04

16. 0.07385×1000

17. $12.88 \div 0.056$

18. $26{,}325.9 \div 100$

19. Write as a percent. 0.073

20. Write as a decimal. 196.5%

21. What is 3.5% of 180?

22. What is 2% of 16.8?

23. 39 is what percent of 650?

24. What percent of 460 is 138?

25. A 4-inch stack of computer chips is on the table. Each computer chip is $\frac{2}{9}$ of an inch thick. How many computer chips are in the stack?

In exercises 26–27, estimate. Round each number to one nonzero digit. Then calculate.

26. $52{,}344\overline{)4{,}678{,}987}$

27. $285.36 + 311.85 + 113.6$

1. _____

2. _____

3. _____

4. _____

5. _____

6. _____

7. _____

8. _____

9. _____

10. _____

11. _____

12. _____

13. _____

14. _____

15. _____

16. _____

17. _____

18. _____

19. _____

20. _____

21. _____

22. _____

23. _____

24. _____

25. _____

26. _____

27. _____

CHAPTER

1

Doctors and health fitness professionals are discovering that certain mathematical measurements of your body can be used in a formula to determine the level of fitness of your body. Suppose you would like to improve your fitness level. What can you do to change it? Doctors have found that through regular exercise, the level of fitness and overall health of almost any person can be significantly improved. Doctors and exercise physiologists use the mathematics that is covered in this chapter to study the progress that people can make through regular exercise sessions of walking or riding a bicycle.

Real Numbers and Variables

① Identifying Different Types of Numbers

Let's review some of the basic terms we use to talk about numbers.

Whole numbers are numbers such as $0, 1, 2, 3, 4, \ldots$

Integers are numbers such as $\ldots, -3, -2, -1, 0, 1, 2, 3, \ldots.$

Rational numbers are numbers such as $\frac{3}{2}, \frac{5}{7}, -\frac{3}{8}, -\frac{4}{13}, \frac{6}{1},$ and $-\frac{8}{2}.$

Rational numbers can be written as one integer divided by another integer (as long as the denominator is not zero!). Integers can be written as fractions ($3 = \frac{3}{1}$, for example), so we can see that all integers are rational numbers. Rational numbers can be expressed in decimal form. For example, $\frac{3}{2} = 1.5, -\frac{3}{8} = -0.375,$ and $\frac{1}{3} = 0.333\ldots$ or $0.\overline{3}$. It is important to note that rational numbers in decimal form are either terminating decimals or repeating decimals.

Irrational numbers are numbers that cannot be expressed as one integer divided by another integer. The numbers π, $\sqrt{2}$, and $\sqrt[3]{7}$ are irrational numbers.

Irrational numbers can be expressed in decimal form. The decimal form of an irrational number is a nonterminating, nonrepeating decimal. For example, $\sqrt{2} = 1.414213\ldots$ can be carried out to an infinite number of decimal places with no repeating pattern of digits.

Finally, **real numbers** are all the rational numbers and all the irrational numbers.

EXAMPLE 1 Classify as an integer, a rational number, an irrational number, and/or a real number.

(a) 5   **(b)** $-\dfrac{1}{3}$ **(c)** 2.85   **(d)** $\sqrt{2}$ **(e)** $0.777\ldots$

Solution   Make a table. Check off the description of the number that applies.

	Number	Integer	Rational Number	Irrational Number	Real Number
(a)	5	✓	✓		✓
(b)	$-\frac{1}{3}$		✓		✓
(c)	2.85		✓		✓
(d)	$\sqrt{2}$			✓	✓
(e)	$0.777\ldots$		✓		✓

Practice Problem 1   Classify.

(a) $-\dfrac{2}{5}$ **(b)** $1.515151\ldots$ **(c)** -8 **(d)** π

Any real number can be pictured on a **number line.**

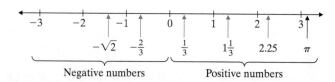

Positive numbers are to the right of 0 on a number line.

Negative numbers are to the left of 0 on a number line.

The **real numbers** include the positive numbers, the negative numbers, and zero.

Using Real Numbers in Real-Life Situations

We often encounter practical examples of number lines that include positive and negative rational numbers. For example, we can tell by reading the accompanying thermometer that the temperature is 20° below 0. From the stock market report, we see that the stock opened at 36 and closed at 34.5, so the net change for the day was −1.5.

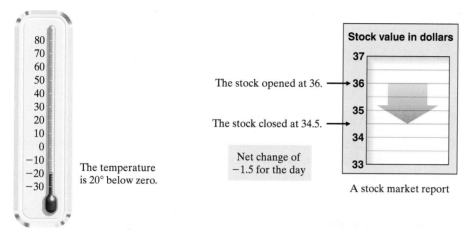

In the following example we use real numbers to represent real-life situations.

EXAMPLE 2   Use a real number to represent each situation.

(a) A temperature of 128.6°F below zero is recorded at Vostok, Antarctica.

(b) The Himalayan peak K2 rises 29,064 feet above sea level.

(c) The Dow gains 10.24 points.

(d) An oil drilling platform extends 328 feet below sea level.

Solution   A key word can help you to decide whether a number is positive or negative.

(a) 128.6°F *below* zero is −128.6.

(b) 29,064 feet *above* sea level is +29,064.

(c) A *gain* of 10.24 points is +10.24.

(d) 328 feet *below* sea level is −328.

Practice Problem 2   Use a real number to represent each situation.

(a) A population growth of 1259   **(b)** A depreciation of $763

(c) A wind-chill factor of minus 10°F

In everyday life we consider positive numbers the opposite of negative numbers. For example, a gain of 3 yards in a football game is the opposite of a loss of 3 yards; a check written for $2.16 on a checking account is the opposite of a deposit of $2.16.

Each positive number has an opposite negative number. Similarly, each negative number has an opposite positive number. **Opposite numbers,** also called **additive inverses,** have the same magnitude but different signs and can be represented on a number line.

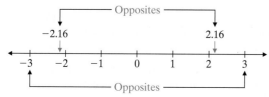

EXAMPLE 3 Find the additive inverse (that is, the opposite).

(a) -7 **(b)** $\dfrac{1}{4}$

Solution

(a) The opposite of -7 is $+7$.

(b) The opposite of $\dfrac{1}{4}$ is $-\dfrac{1}{4}$.

Practice Problem 3 Find the additive inverse (the opposite).

(a) $\dfrac{2}{5}$ **(b)** -1.92 **(c)** A loss of 12 yards on a football play

 Adding Real Numbers with the Same Sign

To use a real number, we need to be clear about its sign. When we write the number three as $+3$, the sign indicates that it is a positive number. The positive sign can be omitted. If someone writes three (3), it is understood that it is a positive three $(+3)$. To write a negative number such as negative three (-3), we must include the sign.

A concept that will help us add and subtract real numbers is the idea of absolute value. The **absolute value** of a number is the distance between that number and zero on a number line. The absolute value of 3 is written $|3|$.

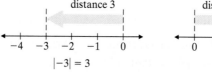

Distance is always a positive number regardless of the direction of travel. This means that the absolute value of any number will be a positive value or zero. We place the symbols $|$ and $|$ around a number to mean the absolute value of the number.

The distance from 0 to 3 is 3, so $|3| = 3$. This is read "the absolute value of 3 is 3."

The distance from 0 to -3 is 3, so $|-3| = 3$. This is read "the absolute value of -3 is 3."

Some other examples are

$$|-22| = 22, \qquad |5.6| = 5.6, \qquad \text{and} \qquad |0| = 0.$$

Thus, the absolute value of a number can be thought of as the magnitude of the number, without regard to its sign.

EXAMPLE 4 Find the absolute value.

(a) $|-4.62|$ **(b)** $\left|\dfrac{3}{7}\right|$ **(c)** $|0|$

Solution

(a) $|-4.62| = 4.62$ **(b)** $\left|\dfrac{3}{7}\right| = \dfrac{3}{7}$ **(c)** $|0| = 0$

Practice Problem 4 Find the absolute value.

(a) $|-7.34|$ **(b)** $\left|\dfrac{5}{8}\right|$ **(c)** $\left|\dfrac{0}{2}\right|$

Now let's look at addition of real numbers when the two numbers have the same sign. Suppose that you are keeping track of your checking account at a local

bank. When you make a deposit of 5 dollars, you record it as +5. When you write a check for 4 dollars, you record it as −4, as a debit. Consider two situations.

SITUATION 1: Total Deposit   You made a deposit of 20 dollars on one day and a deposit of 15 dollars the next day. You want to know the total value of your deposits.

Your record for situation 1.

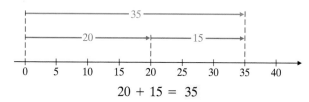

$$20 + 15 = 35$$

The amount of the deposit on the first day added to the amount of the deposit on the second day is the total of the deposits made over the two days.

SITUATION 2: Total Debit   You write a check for 25 dollars to pay one bill and two days later write a check for 5 dollars. You want to know the total value of the debits to your account for the two checks.

Your record for situation 2.

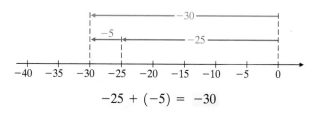

$$-25 + (-5) = -30$$

The value of the first check added to the value of the second check is the total debit to your account.

In each situation we found that we added the absolute value of each number. (That is, we added the numbers without regarding their sign.) The answer always contained the sign that was common to both numbers.

We will now state these results as a formal rule.

ADDITION RULE FOR TWO NUMBERS WITH THE SAME SIGN

To add two numbers with the same sign, add the absolute values of the numbers and use the common sign in the answer.

EXAMPLE 5   Add.

(a) $14 + 16$

(b) $-8 + (-7)$

Solution

(a) $14 + 16$

$14 + 16 = 30$ Add the absolute values of the numbers.

$14 + 16 = +30$ Use the common sign in the answer. Here the common sign is the + sign.

(b) $-8 + (-7)$

$8 + 7 = 15$ Add the absolute values of the numbers.

$-8 + (-7) = -15$ Use the common sign in the answer. Here the common sign is the − sign.

Practice Problem 5   Add.

(a) $37 + 19$

(b) $-23 + (-35)$

EXAMPLE 6 Add. $\frac{2}{3} + \frac{1}{7}$

Solution

$$\frac{2}{3} + \frac{1}{7}$$

$$\frac{14}{21} + \frac{3}{21}$$ Change each fraction to an equivalent fraction with a common denominator of 21.

$$\frac{14}{21} + \frac{3}{21} = +\frac{17}{21} \text{ or } \frac{17}{21}$$ Add the absolute values of the numbers. Use the common sign in the answer. Note that if no sign is written, the number is understood to be positive.

NOTE TO STUDENT: *Fully worked-out solutions to all of the Practice Problems can be found at the back of the text starting at page SP-1*

Practice Problem 6 Add.

$$-\frac{3}{5} + \left(-\frac{4}{7}\right)$$

EXAMPLE 7 Add. $-4.2 + (-3.94)$

Solution

$$-4.2 + (-3.94)$$
$$4.20 + 3.94 = 8.14 \qquad \text{Add the absolute values of the numbers.}$$
$$-4.20 + (-3.94) = -8.14 \qquad \text{Use the common sign in the answer.}$$

Practice Problem 7 Add. $-12.7 + (-9.38)$

The rule for adding two numbers with the same sign can be extended to more than two numbers. If we add more than two numbers with the same sign, the answer will have the sign common to all.

EXAMPLE 8 Add. $-7 + (-2) + (-5)$

Solution

$-7 + (-2) + (-5)$ We are adding three real numbers, all with the same sign. We begin by adding the first two numbers.

$= -9 + (-5)$ Add $-7 + (-2) = -9$.
$= -14$ Add $-9 + (-5) = -14$.

Of course, this can be shortened by adding the three numbers without regard to sign and then using the common sign for the answer.

Practice Problem 8 Add. $-7 + (-11) + (-33)$

Adding Real Numbers with Different Signs

What if the signs of the numbers you are adding are different? Let's consider our checking account again to see how such a situation might occur.

SITUATION 3: Net Increase You made a deposit of 30 dollars on one day. On the next day you write a check for 25 dollars. You want to know the result of your two transactions.

Your record for situation 3.

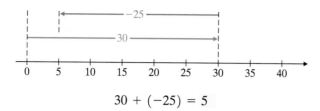

$$30 + (-25) = 5$$

A positive 30 for the deposit added to a negative 25 for the check, which is a debit, gives a net increase of 5 dollars in the account.

SITUATION 4: Net Decrease   You made a deposit of 10 dollars on one day. The next day you write a check for 40 dollars. You want to know the result of your two transactions.

Your record for situation 4.

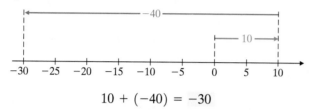

$$10 + (-40) = -30$$

A positive 10 for the deposit added to a negative 40 for the check, which is a debit, gives a net decrease of 30 dollars in the account.

The result is a negative thirty (-30), because the check was larger than the deposit. If you did not have at least 30 dollars in your account at the start of situation 4, you have overdrawn your account.

What do we observe from situations 3 and 4? In each case, first we found the difference of the absolute values of the two numbers. Then the sign of the result was always the sign of the number with the greater absolute value. Thus, in situation 3, 30 is larger than 25. The sign of 30 is positive. The sign of the answer (5) is positive. In situation 4, 40 is larger than 10. The sign of 40 is negative. The sign of the answer (-30) is negative.

We will now state these results as a formal rule.

ADDITION RULE FOR TWO NUMBERS WITH DIFFERENT SIGNS

1. Find the difference between the larger absolute value and the smaller one.
2. Give the answer the sign of the number having the larger absolute value.

EXAMPLE 9   Add. $8 + (-7)$

Solution

$8 + (-7)$	We are to add two numbers with different signs.
$8 - 7 = 1$	Find the difference between the two absolute values, which is 1.
$+8 + (-7) = +1$ or 1	The answer will have the sign of the number with the larger absolute value. That number is $+8$. Its sign is **positive,** so the answer will be $+1$.

Practice Problem 9   Add. $-9 + 15$

Using the Addition Properties for Real Numbers

It is useful to know the following three properties of real numbers.

1. *Addition is commutative.*
   This property states that if two numbers are added, the result is the same no matter which number is written first. The order of the numbers does not affect the result.

$$3 + 6 = 6 + 3 = 9$$
$$-7 + (-8) = (-8) + (-7) = -15$$
$$-15 + 3 = 3 + (-15) = -12$$

2. *Addition of zero to any given number will result in that given number again.*

$$0 + 5 = 5$$
$$-8 + 0 = -8$$

3. *Addition is associative.*
   This property states that if three numbers are added, it does not matter which two numbers are grouped by parentheses and added first.

$$3 + (5 + 7) = (3 + 5) + 7$$ First combine numbers inside parentheses; then
$$3 + (12) = (8) + 7$$ combine the remaining numbers. The results are
$$15 = 15$$ the same no matter which numbers are grouped
and added first.

We can use these properties along with the rules we have for adding real numbers to add three or more numbers. We go from left to right, adding two numbers at a time.

EXAMPLE 10 Add. $\dfrac{3}{17} + \left(-\dfrac{8}{17}\right) + \dfrac{4}{17}$

Solution

$$-\dfrac{5}{17} + \dfrac{4}{17}$$ Add $\frac{3}{17} + \left(-\frac{8}{17}\right) = -\frac{5}{17}$.
The answer is negative since the larger of the two absolute values is negative.

$$= -\dfrac{1}{17}$$ Add $-\frac{5}{17} + \frac{4}{17} = -\frac{1}{17}$.
The answer is negative since the larger of the two absolute values is negative.

Practice Problem 10 Add.

$$-\dfrac{5}{12} + \dfrac{7}{12} + \left(-\dfrac{11}{12}\right)$$

Sometimes the numbers being added have the same signs; sometimes the signs are different. When adding three or more numbers, you may encounter both situations.

EXAMPLE 11 Add. $-1.8 + 1.4 + (-2.6)$

Solution

$-0.4 + (-2.6)$ We take the difference of 1.8 and 1.4 and use the sign of the number with the larger absolute value.

$= -3.0$ Add $-0.4 + (-2.6) = -3.0$. The signs are the same; we add the absolute values of the numbers and use the common sign.

Practice Problem 11 Add. $-6.3 + (-8.0) + 3.5$

If many real numbers are added, it is often easier to add numbers with the same sign in a column format. Remember that addition is commutative and associative; therefore, real numbers can be added *in any order*. You do *not* need to combine the first two numbers as your first step.

EXAMPLE 12 Add. $-8 + 3 + (-5) + (-2) + 6 + 5$

Solution

-8		$+3$	
-5	All the signs are the same.	$+6$	All the signs are the same.
-2	Add the three negative	$+5$	Add the three positive
-15	numbers to obtain -15.	$+14$	numbers to obtain $+14$.

Add the two results.

$$-15 + 14 = -1$$

The answer is negative because the number with the larger absolute value is negative.

Practice Problem 12 Add. $-6 + 5 + (-7) + (-2) + 5 + 3$

A word about notation: The only time we really need to show the sign of a number is when the number is negative—for example, -3. The only time we need to show parentheses when we add real numbers is when we have two different signs preceding a number. For example, $-5 + (-6)$.

EXAMPLE 13 Add.

(a) $2.8 + (-1.3)$ **(b)** $-\dfrac{2}{5} + \left(-\dfrac{3}{4}\right)$

Solution

(a) $2.8 + (-1.3) = 1.5$

(b) $-\dfrac{2}{5} + \left(-\dfrac{3}{4}\right) = -\dfrac{8}{20} + \left(-\dfrac{15}{20}\right) = -\dfrac{23}{20}$ or $-1\dfrac{3}{20}$

Practice Problem 13 Add.

(a) $-2.9 + (-5.7)$ **(b)** $\dfrac{2}{3} + \left(-\dfrac{1}{4}\right)$

Verbal and Writing Skills

Check off any description of the number that applies.

	Number	Integer	Rational Number	Irrational Number	Real Number
1.	23				
2.	$-\frac{4}{5}$				
3.	π				
4.	2.34				
5.	$-6.666\ldots$				

	Number	Integer	Rational Number	Irrational Number	Real Number
6.	$-\frac{7}{9}$				
7.	$-2.3434\ldots$				
8.	14				
9.	$\sqrt{2}$				
10.	$3.232232223\ldots$				

Use a real number to represent each situation.

11. Jules Verne wrote a book with the title *20,000 Leagues Under the Sea.*

12. The value of the dollar is up $0.07 with respect to the yen.

13. Ramona lost $37\frac{1}{2}$ pounds on Weight Watchers.

14. The scouts hiked from sea level to the top of a 3642-foot-high mountain.

15. The temperature rises 7°F.

16. The lowest point in Australia is Lake Eyre at 52 feet below sea level.

Find the additive inverse (opposite).

17. 8

18. $-\frac{4}{5}$

19. -2.73

20. 85.4

Find the absolute value.

21. $|-1.3|$

22. $|-5.9|$

23. $\left|\frac{5}{6}\right|$

24. $\left|\frac{7}{12}\right|$

Add.

25. $-6 + (-5)$

26. $-13 + (-3)$

27. $-20 + (-30)$

28. $-15 + (-25)$

29. $-\frac{7}{20} + \frac{13}{20}$

30. $-\frac{2}{9} + \left(-\frac{4}{9}\right)$

31. $-\frac{2}{13} + \left(-\frac{5}{13}\right)$

32. $-\frac{5}{18} + \frac{11}{18}$

33. $-\frac{2}{5} + \frac{3}{7}$

34. $-\frac{2}{7} + \frac{3}{14}$

35. $-10.3 + (-8.9)$

36. $-5.4 + (-12.8)$

37. $0.6 + (-0.2)$

38. $-0.8 + 0.5$

39. $-5.26 + (-8.9)$

40. $-6.48 + (-3.7)$

41. $-8 + 5 + (-3)$

42. $7 + (-8) + (-4)$

43. $-2 + (-8) + 10$

44. $-9 + 6 + (-12)$

45. $-\dfrac{3}{10} + \dfrac{3}{4}$ **46.** $-\dfrac{3}{8} + \dfrac{11}{24}$ **47.** $-14 + 9 + (-3)$ **48.** $-18 + 10 + (-5)$

Mixed Practice *Add.*

49. $8 + (-11)$ **50.** $16 + (-24)$ **51.** $-83 + 142$ **52.** $-114 + 186$

53. $-\dfrac{4}{9} + \dfrac{5}{6}$ **54.** $-\dfrac{3}{5} + \dfrac{2}{3}$ **55.** $-\dfrac{1}{10} + \dfrac{1}{2}$ **56.** $-\dfrac{2}{3} + \left(-\dfrac{1}{4}\right)$

57. $5.18 + (-7.39)$ **58.** $8.33 + (-14.2)$ **59.** $4 + (-8) + 16$

60. $27 + (-11) + (-4)$ **61.** $34 + (-18) + 11 + (-27)$ **62.** $-23 + 4 + (-11) + 17$

63. $17.85 + (-2.06) + 0.15$ **64.** $23.17 + 5.03 + (-11.81)$

Applications

65. ***Profit/Loss*** Holly paid $47 for a vase at an estate auction. She resold it to an antiques dealer for $214. What was her profit or loss?

66. ***Temperature*** When we skied at Jackson Hole, Wyoming, yesterday, the temperature at the summit was −12°F. Today when we called the ski report, the temperature had risen 7°F. What is the temperature at the summit today?

67. ***Home Equity Line of Credit*** Ramon borrowed $2300 from his home equity line of credit to pay off his car loan. He then borrowed another $1500 to pay to have his kitchen repainted. Represent how much Ramon owed on his home equity line of credit as a real number.

68. ***Time Change*** During the winter, New York City is on Eastern Standard Time (EST). Melbourne, Australia, is 15 hours ahead of New York. If it is 11 P.M. in Melbourne, what time is it in New York?

69. ***Football*** During the Homecoming football game, Quentin lost 15 yards, gained 3 yards, and gained 21 yards in three successive running plays. What was his total gain or loss?

70. ***School Fees*** Wanda's financial aid account at school held $643.85. She withdrew $185.50 to buy books for the semester. Does she have enough left in her account to pay the $475.00 registration fee for the next semester? If so, how much extra money does she have? If not, how much is she short?

71. ***Butterfly Population*** The population of a particular butterfly species was 8000. Twenty years later there were 3000 fewer. Today, there are 1500 fewer. Study the graph to the right. What is the new population?

72. ***Credit Card Balance*** Aaron owes $258 to a credit card company. He makes a purchase of $32 with the card and then makes a payment of $150 on the account. How much does he still owe?

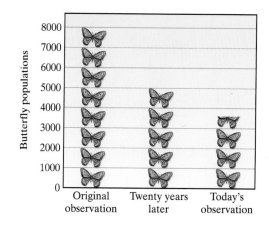

Profit/Loss *During the first five months of 2005, a regional Midwest airline posted profit and loss figures for each month of operation, as shown in the accompanying bar graph.*

73. For the first three months of 2005, what were the total earnings of the airline?

74. For the first five months of 2005, what were the total earnings for the airline?

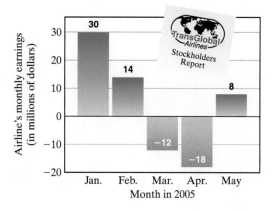

To Think About

75. What number must be added to −13 to get 5?

76. What number must be added to −18 to get 10?

Cumulative Review *Perform the indicated calculations.*

77. **[0.2.3]** $\dfrac{15}{16} + \dfrac{1}{4}$

78. **[0.3.1]** $\left(\dfrac{3}{7}\right)\left(\dfrac{14}{9}\right)$

79. **[0.2.3]** $\dfrac{2}{15} - \dfrac{1}{20}$

80. **[0.3.2]** $2\dfrac{1}{2} \div 3\dfrac{2}{5}$

81. **[0.4.4]** $0.72 + 0.8$

82. **[0.4.4]** $1.63 - 0.98$

83. **[0.4.5]** $(1.63)(0.7)$

84. **[0.4.6]** $0.208 \div 0.8$

Quick Quiz 1.1 Add.

1. $-18 + (-16)$

2. $-2.7 + 8.6 + (-5.4)$

3. $-\dfrac{5}{6} + \dfrac{7}{24}$

4. **Concept Check** Explain why when you add two negative numbers, you always obtain a negative number, but when you add one negative number and one positive number, you may obtain zero, a positive number, or a negative number.

1.2 SUBTRACTING REAL NUMBERS

 Subtracting Real Numbers with the Same or Different Signs

So far we have developed the rules for adding real numbers. We can use these rules to subtract real numbers. Let's look at a checkbook situation to see how.

SITUATION 5: Subtract a Deposit and Add a Debit You have a balance of 20 dollars in your checking account. The bank calls you and says that a deposit of 5 dollars that belongs to another account was erroneously added to your account. They say they will correct the account balance to 15 dollars. You want to keep track of what's happening to your account.

Your record for situation 5.

$$20 - (+5) = 15$$

From your present balance *subtract* the *deposit* to get the new balance.

This equation shows what needs to be done to your account. The bank tells you that because the error happened in the past, they cannot "take it away." However, they can add to your account a debit of 5 dollars. Here is the equivalent addition.

$$20 + (-5) = 15$$

To your present balance *add* a *debit* to get the new balance.

We see that subtracting a positive 5 has the same effect as adding a negative 5.

> ### SUBTRACTION OF REAL NUMBERS
> To subtract real numbers, add the opposite of the second number (that is, the number you are subtracting) to the first.

The rule tells us to do three things when we subtract real numbers. First, change subtraction to addition. Second, replace the second number by its opposite. Third, add the two numbers using the rules for addition of real numbers.

EXAMPLE 1 Subtract. $6 - (-2)$

Solution 6 $-$ (-2)

| Change subtraction to addition. | Write the opposite of the second number. |

$= 6 + (+2)$

| Add the two real numbers with the same sign. |

$= 8$

Practice Problem 1 Subtract. $9 - (-3)$

EXAMPLE 2 Subtract. $-8 - (-6)$

Solution

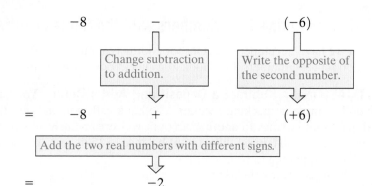

$$= \quad -8 \quad + \quad (+6)$$

Add the two real numbers with different signs.

$$= \quad -2$$

NOTE TO STUDENT: Fully worked-out solutions to all of the Practice Problems can be found at the back of the text starting at page SP-1

Practice Problem 2 Subtract. $-12 - (-5)$

EXAMPLE 3 Subtract.

(a) $\dfrac{3}{7} - \dfrac{6}{7}$ 　　　　　(b) $-\dfrac{7}{18} - \left(-\dfrac{1}{9}\right)$

Solution

(a) $\dfrac{3}{7} - \dfrac{6}{7} = \dfrac{3}{7} + \left(-\dfrac{6}{7}\right)$　Change the subtraction problem to one of adding the opposite of the second number. We note that the problem has two fractions with the same denominator.

$\quad = -\dfrac{3}{7}$　Add two numbers with different signs.

(b) $-\dfrac{7}{18} - \left(-\dfrac{1}{9}\right) = -\dfrac{7}{18} + \dfrac{1}{9}$　Change subtracting to adding the opposite.

$\quad = -\dfrac{7}{18} + \dfrac{2}{18}$　Change $\dfrac{1}{9}$ to $\dfrac{2}{18}$ since the LCD $= 18$.

$\quad = -\dfrac{5}{18}$　Add two numbers with different signs.

Calculator

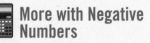

More with Negative Numbers

Subtract. Remember to use the $\boxed{+/-}$ key to change the sign of a number from $+$ to $-$ or from $-$ to $+$.

(a) $-18 - (-24)$

(b) $-6 + (-10) - (-15)$

Ans:

(a) 6 　　　(b) -1

Practice Problem 3 Subtract.

(a) $\dfrac{5}{9} - \dfrac{7}{9}$ 　　　　　(b) $-\dfrac{5}{21} - \left(-\dfrac{3}{7}\right)$

EXAMPLE 4 Subtract. $-5.2 - (-5.2)$

Solution

$-5.2 - (-5.2) = -5.2 + 5.2$　Change the subtraction problem to one of adding the opposite of the second number.

$\quad = 0$　Add two numbers with different signs.

Practice Problem 4 Subtract. $-17.3 - (-17.3)$

Example 4 illustrates what is sometimes called the **additive inverse property.** When you add two real numbers that are opposites of each other, you will obtain zero. Examples of this are the following:

$$5 + (-5) = 0 \qquad -186 + 186 = 0 \qquad -\dfrac{1}{8} + \dfrac{1}{8} = 0.$$

EXAMPLE 5 Subtract.

(a) $-8 - 2$ **(b)** $23 - 28$ **(c)** $5 - (-3)$ **(d)** $\dfrac{1}{4} - 8$

Solution

(a) $-8 - 2 = -8 + (-2)$ Notice that we are subtracting a positive 2. Change to addition.

             $= -10$ Add.

In a similar fashion we have

(b) $23 - 28 = 23 + (-28) = -5$

(c) $5 - (-3) = 5 + 3 = 8$

(d) $\dfrac{1}{4} - 8 = \dfrac{1}{4} + (-8) = \dfrac{1}{4} + \left(-\dfrac{32}{4}\right) = -\dfrac{31}{4}$ or $-7\dfrac{3}{4}$

Practice Problem 5 Subtract.

(a) $-21 - 9$ **(b)** $17 - 36$ **(c)** $12 - (-15)$ **(d)** $\dfrac{3}{5} - 2$

NOTE TO STUDENT: Fully worked-out solutions to all of the Practice Problems can be found at the back of the text starting at page SP-1

EXAMPLE 6 A satellite is recording radioactive emissions from nuclear waste buried 3 miles below sea level. The satellite orbits Earth at 98 miles above sea level. How far is the satellite from the nuclear waste?

Solution We want to find the difference between +98 miles and −3 miles. This means we must subtract −3 from 98.

$$98 - (-3) = 98 + 3$$
$$= 101$$

The satellite is 101 miles from the nuclear waste.

Practice Problem 6 A helicopter is directly over a sunken vessel. The helicopter is 350 feet above sea level. The vessel lies 186 feet below sea level. How far is the helicopter from the sunken vessel?

Developing Your Study Skills

Reading the Textbook

Begin reading your textbook with paper and pencil in hand. As you come across a new definition or concept, underline it in the text and/or write it down in your notebook. Whenever you encounter an unfamiliar term, look it up and make a note of it. When you come to an example, work through it step-by-step. Be sure to read each word and follow directions carefully.

Notice the helpful hints the author provides. They guide you to correct solutions and prevent you from making errors. Take advantage of these pieces of expert advice.

Be sure that you understand what you are reading. Make a note of any of those things that you do not understand and ask your instructor about them. Do not hurry through the material. Learning mathematics takes time.

Verbal and Writing Skills

1. Explain in your own words how you would perform the necessary steps to find $-8 - (-3)$.

2. Explain in your own words how you would perform the necessary steps to find $-10 - (-15)$.

Subtract by adding the opposite.

3. $18 - 35$

4. $16 - 48$

5. $19 - 23$

6. $10 - 20$

7. $-14 - (-3)$

8. $-24 - (-7)$

9. $-52 - (-60)$

10. $-48 - (-80)$

11. $0 - (-5)$

12. $0 - (-7)$

13. $-18 - (-18)$

14. $-24 - (-24)$

15. $-17 - (-20)$

16. $-11 - (-19)$

17. $\dfrac{2}{5} - \dfrac{4}{5}$

18. $\dfrac{2}{9} - \dfrac{7}{9}$

19. $\dfrac{3}{4} - \left(-\dfrac{3}{5}\right)$

20. $-\dfrac{2}{3} - \dfrac{1}{4}$

21. $-\dfrac{3}{4} - \dfrac{5}{6}$

22. $-\dfrac{7}{10} - \dfrac{10}{15}$

23. $-0.6 - 0.3$

24. $-0.9 - 0.5$

25. $2.64 - (-1.83)$

26. $-0.03 - 0.06$

Mixed Practice   *Calculate.*

27. $\dfrac{3}{5} - 4$

28. $\dfrac{5}{6} - 3$

29. $-\dfrac{2}{3} - 4$

30. $-\dfrac{1}{6} - 5$

31. $34 - 87$

32. $19 - 76$

33. $-25 - 48$

34. $-74 - 11$

35. $2.3 - (-4.8)$

36. $8.4 - (-2.7)$

37. $8 - \left(-\dfrac{3}{4}\right)$

38. $\dfrac{2}{3} - (-6)$

39. $\dfrac{5}{6} - 7$

40. $9 - \dfrac{2}{3}$

41. $-\dfrac{2}{7} - \dfrac{4}{5}$

42. $-\dfrac{5}{6} - \dfrac{1}{5}$

43. $-135 - (-126.5)$

44. $-97.6 - (-146)$

45. $\dfrac{1}{5} - 6$

46. $\dfrac{2}{7} - (-3)$

47. $4.5 - (-1.56)$

48. $5.2 - (-3.88)$

49. $-3 - 2.047$

50. $-1.043 - 4$

51. Subtract -9 from -2.

52. Subtract -12 from 20.

53. Subtract 13 from -35.

One Step Further *Change each subtraction operation to "adding the opposite." Then combine the numbers.*

54. $9 + 6 - (-5)$

55. $7 + (-6) - 3$

56. $11 + (-3) - 8$

57. $-13 + 12 - (-1)$

58. $18 - (-15) - 3$

59. $7 + (-42) - 27$

60. $-4.2 - (-3.8) + 1.5$

61. $-6.4 - (-2.7) + 5.3$

Applications

62. *Sea Rescue* A rescue helicopter is 300 feet above sea level. The captain has located an ailing submarine directly below it that is 126 feet below sea level. How far is the helicopter from the submarine?

63. *Checking Account Balance* Yesterday Jackie had $112 in her checking account. Today her account reads "balance $-$37." Find the difference in these two amounts.

64. *Federal Taxes* In 2007, Damian had $2435 withheld from his paychecks in federal taxes. When he filed his tax return, he received a $308 refund. How much did he actually pay in taxes?

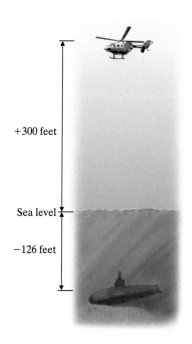

+300 feet

Sea level

−126 feet

Cumulative Review *In exercises 65–69, perform the indicated operations.*

65. **[1.1.4]** $-37 + 16$

66. **[1.1.3]** $-37 + (-14)$

67. **[1.1.3]** $-3 + (-6) + (-10)$

68. **[1.1.4]** *Temperature* On a winter morning, Alisa noticed that the outside temperature was $-5°$F. By the afternoon the temperature had risen $20°$F. What was the afternoon temperature?

69. **[0.3.1]** *Hiking* Sean and Khalid went hiking in the Blue Ridge Mountains. During their $8\frac{1}{3}$-mile hike, $\frac{4}{5}$ of the distance was covered with snow. How many miles were snow covered?

Quick Quiz 1.2 Subtract.

1. $-8 - (-15)$

2. $-1.3 - 0.6$

3. $\dfrac{5}{8} - \left(-\dfrac{2}{7}\right)$

4. **Concept Check** Explain the different results that are possible when you start with a negative number and then subtract a negative number.

Student Learning Objectives

After studying this section, you will be able to:

 1 Multiply real numbers.

2 Use the multiplication properties for real numbers.

3 Divide real numbers.

 1 Multiplying Real Numbers

We are familiar with the meaning of multiplication for positive numbers. For example, $5 \times 90 = 450$ might mean that you receive five weekly checks of 90 dollars each and you gain \$450. Let's look at a situation that corresponds to $5 \times (-90)$. What might that mean?

SITUATION 6: Checking an Account Balance   You write a check for five weeks in a row to pay your weekly room rent of 90 dollars. You want to know the total impact on your checking account balance.

   Your record for situation 6.

$$(+5) \qquad \times \qquad (-90) \qquad = \qquad -450$$

| The number of checks you have written | times | negative 90, the value of each check that was a debit to your account, | gives | negative 450 dollars, a net debit to your account. |

Note that a multiplication symbol is not needed between the $(+5)$ and the (-90) because the two sets of parentheses indicate multiplication. The multiplication $(5)(-90)$ is the same as repeated addition of five (-90)'s. Note that 5 multiplied by -90 can be written as $5(-90)$ or $(5)(-90)$.

$$(-90) + (-90) + (-90) + (-90) + (-90) = -450$$

This example seems to show that a positive number multiplied by a negative number is negative.

   What if the negative number is the one that is written first? If $(5)(-90) = -450$, then $(-90)(5) = -450$ by the commutative property of multiplication. This is an example showing that *when two numbers with different signs* (one positive, one negative) *are multiplied, the result is negative.*

   But what if both numbers are negative? Consider the following situation.

SITUATION 7: Renting a Room   Last year at college you rented a room at 90 dollars per week for 36 weeks, which included two semesters and summer school. This year you will not attend the summer session, so you will be renting the room for only 30 weeks. Thus the number of weekly rental checks will be six fewer than last year. You are making out your budget for this year. You want to know the financial impact of renting the room for six fewer weeks.

   Your record for situation 7.

$$(-6) \qquad \times \qquad (-90) \qquad = \qquad 540$$

| The difference in the number of checks this year compared to last year is -6, which is negative to show a decrease, | times | -90, the value of each check paid out, | gives | $+540$ dollars. The product is positive, because your financial situation will be 540 dollars better this year. |

You could check that the answer is positive by calculating the total rental expenses.

	Dollars in rent last year	$(36)(90) =$	3240
(subtract)	Dollars in rent this year	$-(30)(90) =$	-2700
	Extra dollars available this year	$=$	$+540$

This agrees with our previous answer: $(-6)(-90) = +540$.

   Note that negative 6 times -90 can be written as $-6(-90)$ or $(-6)(-90)$.

In this situation it seems reasonable that a negative number times a negative number yields a positive answer. We already know from arithmetic that a positive number times a positive number yields a positive answer. Thus we might see the general rule that *when two numbers with the same sign* (both positive or both negative) *are multiplied, the result is positive.*

We will now state our rule.

MULTIPLICATION OF REAL NUMBERS

To multiply two real numbers with **the same sign,** multiply the absolute values. The sign of the result is **positive.**

To multiply two real numbers with **different signs,** multiply the absolute values. The sign of the result is **negative.**

EXAMPLE 1 Multiply.

(a) $(3)(6)$ **(b)** $\left(-\dfrac{5}{7}\right)\left(-\dfrac{2}{9}\right)$ **(c)** $-4(8)$ **(d)** $\left(\dfrac{2}{7}\right)(-3)$

Solution

(a) $(3)(6) = 18$

> When multiplying two numbers with the same sign, the result is a positive number.

(b) $\left(-\dfrac{5}{7}\right)\left(-\dfrac{2}{9}\right) = \dfrac{10}{63}$

(c) $-4(8) = -32$

> When multiplying two numbers with different signs, the result is a negative number.

(d) $\left(\dfrac{2}{7}\right)(-3) = \left(\dfrac{2}{7}\right)\left(-\dfrac{3}{1}\right) = -\dfrac{6}{7}$

Practice Problem 1 Multiply.

(a) $(-6)(-2)$ **(b)** $(7)(9)$ **(c)** $\left(-\dfrac{3}{5}\right)\left(\dfrac{2}{7}\right)$ **(d)** $\left(\dfrac{5}{6}\right)(-7)$

NOTE TO STUDENT: Fully worked-out solutions to all of the Practice Problems can be found at the back of the text starting at page SP-1

To multiply more than two numbers, multiply two numbers at a time.

EXAMPLE 2 Multiply. $(-4)(-3)(-2)$

Solution

$(-4)(-3)(-2) = (+12)(-2)$ We begin by multiplying the first two numbers, (-4) and (-3). The signs are the same. The answer is positive 12.

$\qquad\qquad\qquad = -24$ Now we multiply $(+12)$ and (-2). The signs are different. The answer is negative 24.

Practice Problem 2 Multiply. $(-5)(-2)(-6)$

EXAMPLE 3 Multiply.

(a) $-3(-1.5)$ **(b)** $\left(-\dfrac{1}{2}\right)(-1)(-4)$ **(c)** $-2(-2)(-2)(-2)$

Solution Multiply two numbers at a time. See if you find a pattern.

(a) $-3(-1.5) = 4.5$ Be sure to place the decimal point in your answer.

(b) $\left(-\dfrac{1}{2}\right)(-1)(-4) = +\dfrac{1}{2}(-4) = -2$

(c) $-2(-2)(-2)(-2) = +4(-2)(-2) = -8(-2) = +16$ or 16

What kind of answer would we obtain if we multiplied five negative numbers? If you guessed "negative," you probably see the pattern.

NOTE TO STUDENT: *Fully worked-out solutions to all of the Practice Problems can be found at the back of the text starting at page SP-1*

Practice Problem 3 Determine the sign of the product. Then multiply to check.

(a) $-2(-3)$ **(b)** $(-1)(-3)(-2)$

(c) $-4\left(-\dfrac{1}{4}\right)(-2)(-6)$

When you multiply two or more nonzero real numbers:

1. The result is always **positive** if there is an **even** number of negative signs.
2. The result is always **negative** if there is an **odd** number of negative signs.

2 Using the Multiplication Properties for Real Numbers

For convenience, we will list the properties of multiplication.

1. *Multiplication is commutative.*

   This property states that if two real numbers are multiplied, the order of the numbers does not affect the result. The result is the same no matter which number is written first.

   $$(5)(7) = (7)(5) = 35, \quad \left(\dfrac{1}{3}\right)\left(\dfrac{2}{7}\right) = \left(\dfrac{2}{7}\right)\left(\dfrac{1}{3}\right) = \dfrac{2}{21}$$

2. *Multiplication of any real number by zero will result in zero.*

   $$(5)(0) = 0, \quad (-5)(0) = 0, \quad (0)\left(\dfrac{3}{8}\right) = 0, \quad (0)(0) = 0$$

3. *Multiplication of any real number by 1 will result in that same number.*

   $$(5)(1) = 5, \quad (1)(-7) = -7, \quad (1)\left(-\dfrac{5}{3}\right) = -\dfrac{5}{3}$$

4. *Multiplication is associative.*

   This property states that if three real numbers are multiplied, it does not matter which two numbers are grouped by parentheses and multiplied first.

   $2 \times (3 \times 4) = (2 \times 3) \times 4$ First multiply the numbers in parentheses. Then multiply the remaining numbers.

   $2 \times (12) = (6) \times 4$ The results are the same no matter which numbers are grouped and multiplied first.

   $24 = 24$

Dividing Real Numbers

What about division? Any division problem can be rewritten as a multiplication problem.

> We know that $20 \div 4 = 5$ because $4(5) = 20$.
> Similarly, $-20 \div (-4) = 5$ because $-4(5) = -20$.

In both division problems the answer is positive 5. Thus we see that *when you divide two numbers with the same sign* (both positive or both negative), *the answer is positive.* What if the signs are different?

> We know that $-20 \div 4 = -5$ because $4(-5) = -20$.
> Similarly, $20 \div (-4) = -5$ because $-4(-5) = 20$.

In these two problems the answer is negative 5. So we have reasonable evidence to see that *when you divide two numbers with different signs* (one positive and one negative), *the answer is negative.*

We will now state our rule for division.

DIVISION OF REAL NUMBERS

To divide two real numbers with **the same sign,** divide the absolute values. The sign of the result is **positive.**

To divide two real numbers with **different signs,** divide the absolute values. The sign of the result is **negative.**

EXAMPLE 4   Divide.

(a) $12 \div 4$ **(b)** $(-25) \div (-5)$ **(c)** $\dfrac{-36}{18}$ **(d)** $\dfrac{42}{-7}$

Solution

(a) $12 \div 4 = 3$ ⟵ ⎱ When dividing two numbers with the
(b) $(-25) \div (-5) = 5$ ⟵ ⎰ same sign, the result is a positive number.

(c) $\dfrac{-36}{18} = -2$ ⟵ ⎱ When dividing two numbers with differ-
(d) $\dfrac{42}{-7} = -6$ ⟵ ⎰ ent signs, the result is a negative number.

Practice Problem 4   Divide.

(a) $-36 \div (-2)$ **(b)** $-49 \div 7$ **(c)** $\dfrac{50}{-10}$ **(d)** $\dfrac{-39}{13}$

EXAMPLE 5   Divide.   **(a)** $-36 \div 0.12$ **(b)** $-2.4 \div (-0.6)$

Solution

(a) $-36 \div 0.12$ Look at the problem to determine the sign. When dividing two numbers with different signs, the result will be a negative number.

We then divide the absolute values.

$$
\begin{array}{r}
3\,00. \\
0.12_{\wedge}\overline{)36.00_{\wedge}} \\
\underline{36} \\
00
\end{array}
$$

Thus $-36 \div 0.12 = -300$. The answer is a negative number.

(b) $-2.4 \div (-0.6)$ Look at the problem to determine the sign. When dividing two numbers with the same sign, the result will be positive.

We then divide the absolute values.

$$0.6_\wedge \overline{)2.4_\wedge} \;\; \begin{array}{r} 4. \\ \hline \\ 2\,4 \end{array}$$

Thus $-2.4 \div (-0.6) = 4$. The answer is a positive number.

NOTE TO STUDENT: *Fully worked-out solutions to all of the Practice Problems can be found at the back of the text starting at page SP-1*

Practice Problem 5 Divide.

(a) $-12.6 \div (-1.8)$ **(b)** $0.45 \div (-0.9)$

Note that the rules for multiplication and division are the same. When you **multiply** or **divide** two numbers with the **same** sign, you obtain a **positive** number. When you **multiply** or **divide** two numbers with **different** signs, you obtain a **negative** number.

EXAMPLE 6 Divide. $-\dfrac{12}{5} \div \dfrac{2}{3}$

Solution

$$= \left(-\frac{12}{5}\right)\left(\frac{3}{2}\right)$$ Divide two fractions. We invert the second fraction and multiply by the first fraction.

$$= \left(-\frac{\overset{6}{\cancel{12}}}{5}\right)\left(\frac{3}{\underset{1}{\cancel{2}}}\right)$$

$$= -\frac{18}{5} \quad \text{or} \quad -3\frac{3}{5}$$ The answer is negative since the two numbers divided have different signs.

Practice Problem 6 Divide.

$$-\frac{5}{16} \div \left(-\frac{10}{13}\right)$$

Note that division can be indicated by the symbol $\div$ or by the fraction bar —. $\frac{2}{3}$ means $2 \div 3$.

EXAMPLE 7 Divide. **(a)** $\dfrac{\frac{7}{8}}{-21}$ **(b)** $\dfrac{-\frac{2}{3}}{-\frac{7}{13}}$

Solution

(a) $\dfrac{\frac{7}{8}}{-21}$

$$= \frac{7}{8} \div \left(-\frac{21}{1}\right)$$ Change -21 to a fraction. $-21 = -\frac{21}{1}$

$$= \frac{\overset{1}{\cancel{7}}}{8}\left(-\frac{1}{\underset{3}{\cancel{21}}}\right)$$ Change the division to multiplication. Cancel where possible.

$$= -\frac{1}{24}$$ Simplify.

(b) $\dfrac{-\dfrac{2}{3}}{-\dfrac{7}{13}} = -\dfrac{2}{3} \div \left(-\dfrac{7}{13}\right) = -\dfrac{2}{3}\left(-\dfrac{13}{7}\right) = \dfrac{26}{21}$ or $1\dfrac{5}{21}$

Practice Problem 7 Divide.

(a) $\dfrac{-12}{-\dfrac{4}{5}}$ **(b)** $\dfrac{-\dfrac{2}{9}}{\dfrac{8}{13}}$

---

1. *Division of 0 by any nonzero real number gives 0 as a result.*

$$0 \div 5 = 0, \qquad 0 \div \frac{2}{3} = 0, \qquad \frac{0}{5.6} = 0, \qquad \frac{0}{1000} = 0$$

You can divide zero by 5, $\frac{2}{3}$, 5.6, 1000, or any number (except 0).

2. *Division of any real number by 0 is* **undefined.**

$$7 \div 0 \qquad\qquad \frac{64}{0}$$
$$\uparrow \qquad\qquad\quad \uparrow$$

Neither of these operations is possible. **Division by zero is undefined.**

You may be wondering why division by zero is undefined. Let us think about it for a minute. We said that $7 \div 0$ is undefined. Suppose there were an answer. Let us call the answer a. So we assume for a minute that $7 \div 0 = a$. Then it would have to follow that $7 = 0(a)$. But this is impossible. Zero times any number is zero. So we see that if there were such a number, it would contradict known mathematical facts. Therefore there is no number a such that $7 \div 0 = a$. Thus we conclude that division by zero is undefined.

When combining two numbers, it is important to be sure you know which rule applies. Think about the concepts in the following chart. See if you agree with each example.

Operation	Two Real Numbers with the Same Sign	Two Real Numbers with Different Signs
Addition	Result may be positive or negative. $9 + 2 = 11$ $-5 + (-6) = -11$	Result may be positive or negative. $-3 + 7 = 4$ $4 + (-12) = -8$
Subtraction	Result may be positive or negative. $15 - 6 = 15 + (-6) = 9$ $-12 - (-3) = -12 + 3 = -9$	Result may be positive or negative. $-12 - 3 = -12 + (-3) = -15$ $5 - (-6) = 5 + 6 = 11$
Multiplication	Result is always positive. $9(3) = 27$ $-8(-5) = 40$	Result is always negative. $-6(12) = -72$ $8(-3) = -24$
Division	Result is always positive. $150 \div 6 = 25$ $-72 \div (-2) = 36$	Result is always negative. $-60 \div 10 = -6$ $30 \div (-6) = -5$

EXAMPLE 8   The Hamilton-Wenham Generals recently analyzed the 48 plays their team made while in possession of the football during their last game. The bar graph illustrates the number of plays made in each category. The team statistician prepared the following chart indicating the average number of yards gained or lost during each type of play.

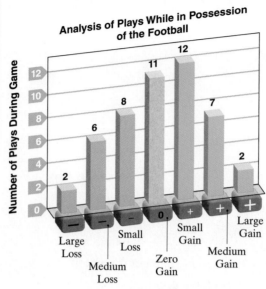

Type of Play	Average Yards Gained or Lost for Play
Large gain	+25
Medium gain	+15
Small gain	+5
Zero gain	0
Small loss	−5
Medium loss	−10
Large loss	−15

(a) How many yards were lost by the Generals in the plays that were considered small losses?

(b) How many yards were gained by the Generals in the plays that were considered small gains?

(c) If the total yards gained in small gains were combined with the total yards lost in small losses, what would be the result?

Solution

(a) We multiply the number of small losses by the average number of total yards lost on each small loss:

$$8(-5) = -40.$$

The team lost approximately 40 yards with plays that were considered small losses.

(b) We multiply the number of small gains by the average number of yards gained on each small gain:

$$12(5) = 60.$$

The team gained approximately 60 yards with plays that were considered small gains.

(c) We combine the results for **(a)** and **(b)**:

$$-40 + 60 = 20.$$

A total of 20 yards was gained during the plays that were small losses and small gains.

NOTE TO STUDENT: *Fully worked-out solutions to all of the Practice Problems can be found at the back of the text starting at page SP-1*

Practice Problem 8   Using the information provided in Example 8, answer the following:

(a) How many yards were lost by the Generals in the plays that were considered medium losses?

(b) How many yards were gained by the Generals in the plays that were considered medium gains?

(c) If the total yards gained in medium gains were combined with the total yards lost in medium losses, what would be the result?

Verbal and Writing Skills

1. Explain in your own words the rule for determining the correct sign when multiplying two real numbers.

2. Explain in your own words the rule for determining the correct sign when multiplying three or more real numbers.

Multiply. Be sure to write your answer in the simplest form.

3. $5(-4)$

4. $8(-2)$

5. $0(-12)$

6. $0(-150)$

7. $14(3.5)$

8. $8.5(6)$

9. $(-1.32)(-0.2)$

10. $(-2.3)(-0.11)$

11. $1.8(-2.5)$

12. $2.3(-4.4)$

13. $\left(\dfrac{3}{8}\right)(-4)$

14. $(5)\left(-\dfrac{7}{10}\right)$

15. $\left(-\dfrac{3}{5}\right)\left(-\dfrac{15}{11}\right)$

16. $\left(-\dfrac{4}{9}\right)\left(-\dfrac{3}{5}\right)$

17. $\left(\dfrac{12}{13}\right)\left(\dfrac{-5}{24}\right)$

18. $\left(\dfrac{14}{17}\right)\left(-\dfrac{3}{28}\right)$

Divide.

19. $-36 \div (-9)$

20. $0 \div (-15)$

21. $-48 \div (-8)$

22. $-45 \div (9)$

23. $-120 \div (-8)$

24. $-180 \div (-4)$

25. $156 \div (-13)$

26. $-0.6 \div 0.3$

27. $-9.1 \div (0.07)$

28. $8.1 \div (-0.03)$

29. $0.54 \div (-0.9)$

30. $-7.2 \div 8$

31. $-6.3 \div 7$

32. $\dfrac{2}{7} \div \left(-\dfrac{3}{5}\right)$

33. $\left(-\dfrac{1}{5}\right) \div \left(\dfrac{2}{3}\right)$

34. $\left(-\dfrac{5}{6}\right) \div \left(-\dfrac{7}{18}\right)$

35. $-\dfrac{5}{7} \div \left(-\dfrac{3}{28}\right)$

36. $\left(-\dfrac{4}{9}\right) \div \left(-\dfrac{8}{15}\right)$

37. $\left(-\dfrac{7}{12}\right) \div \left(-\dfrac{5}{6}\right)$

38. $\dfrac{12}{-\dfrac{2}{5}}$

39. $\dfrac{-6}{-\dfrac{3}{7}}$

40. $\dfrac{-\dfrac{3}{8}}{-\dfrac{2}{3}}$

41. $\dfrac{\dfrac{-2}{3}}{\dfrac{8}{15}}$

42. $\dfrac{\dfrac{9}{2}}{-3}$

43. $\dfrac{\dfrac{8}{3}}{-4}$

Multiply. You may want to determine the sign of the product before you multiply.

44. $-6(2)(-3)(4)$

45. $-1(-2)(-3)(4)$

46. $-3(2)(-1)(-2)(5)$

47. $-2(4)(3)(-1)(-3)$

48. $-3(2)(-4)(0)(-2)$

49. $-3(-2)\left(\dfrac{1}{3}\right)(-4)(2)$

50. $-4(0.5)(-0.02)(2)$

51. $25(-0.04)(-0.3)(-1)$

52. $\left(\dfrac{3}{8}\right)\left(\dfrac{1}{2}\right)\left(-\dfrac{5}{6}\right)$

53. $\left(-\dfrac{4}{5}\right)\left(-\dfrac{6}{7}\right)\left(-\dfrac{1}{3}\right)$

54. $\left(-\dfrac{1}{2}\right)\left(\dfrac{4}{5}\right)\left(-\dfrac{7}{8}\right)\left(-\dfrac{2}{3}\right)$

55. $\left(-\dfrac{3}{4}\right)\left(-\dfrac{7}{15}\right)\left(-\dfrac{8}{21}\right)\left(-\dfrac{5}{9}\right)$

Mixed Practice *Take a minute to review the chart before Example 8. Be sure that you can remember the sign rules for each operation. Then do exercises 56–65. Perform the indicated calculations.*

56. $-5 - (-2)$

57. $-36 \div (-4)$

58. $-3(-9)$

59. $12 + (-8)$

60. $(-30) \div 5$

61. $8 - (-9)$

62. $-6 + (-3)$

63. $6(-12)$

64. $18 \div (-18)$

65. $-37 \div 37$

Applications

66. *Stock Trading* In September 2007, the value of one share of NutriSystem stock opened at $56.95. At the end of the day, it closed at $50.08. If you owned 80 shares, what was your profit or loss that day?

67. *Equal Contributions* Ed, Ned, Ted, and Fred went camping. They each contributed an equal share of money toward food. Fred did the shopping. When he returned from the store, he had $17.60 left. How much money did Fred give back to each person?

68. *Student Loans* Ramon will pay $6480 on his student loan over the next 3 years. If $180 is automatically deducted from his bank account each month to pay the loan off, how much does he still owe after one year?

69. *Car Payments* Muriel will pay the Volkswagen dealer a total of $14,136 to be paid in 60 equal monthly installments. What is her monthly bill?

Football *The Beverly Panthers recently analyzed the 37 plays their team made while in possession of the football during their last game. The team statistician prepared the following chart indicating the number of plays in each category and the average number of yards gained or lost during each type of play. Use this chart to answer exercises 70–75.*

Type of Play	Number of Plays	Average Yards Gained or Lost per Play
Large gain	1	+25
Medium gain	6	+15
Small gain	4	+5
Zero gain	5	0
Small loss	10	−5
Medium loss	7	−10
Large loss	4	−15

70. How many yards were lost by the Panthers in the plays that were considered small losses?

71. How many yards were gained by the Panthers in the plays that were considered small gains?

72. If the total yards gained in small gains were combined with the total yards lost in small losses, what would be the result?

73. How many yards were lost by the Panthers in the plays that were considered medium losses?

74. How many yards were gained by the Panthers in the plays that were considered medium gains?

75. If the total yards gained in medium gains were combined with the total yards lost in medium losses, what would be the result?

Cumulative Review

76. **[1.1.5]** $-17.4 + 8.31 + 2.40$

77. **[1.1.5]** $-\dfrac{3}{4} + \left(-\dfrac{2}{3}\right) + \left(-\dfrac{5}{12}\right)$

78. **[1.2.1]** $-47 - (-32)$

79. **[1.2.1]** $-37 - 51$

Quick Quiz 1.3 Perform the indicated operations.

1. $\left(-\dfrac{3}{8}\right)(5)$

2. $-4(3)(-5)(-2)$

3. $-2.4 \div (-0.6)$

4. **Concept Check** Explain how you can determine the sign of the answer if you multiply several negative numbers.

Developing Your Study Skills

Steps Toward Success in Mathematics

Mathematics is a building process, mastered one step at a time. The foundation of this process consists of a few basic requirements. Those who are successful in mathematics realize the absolute necessity for building a study of mathematics on the firm foundation of these six minimum requirements.

1. Attend class every day.
2. Read the textbook.
3. Take notes in class.
4. Do assigned homework every day.
5. Get help immediately when needed.
6. Review regularly.

Writing Numbers in Exponent Form

In mathematics, we use exponents as a way to abbreviate repeated multiplication.

Long Notation		Exponent Form
$2 \cdot 2 \cdot 2 \cdot 2 \cdot 2 \cdot 2$	$=$	2^6

There are two parts to exponent notation: (1) the **base** and (2) the **exponent.** The **base** tells you what number is being multiplied and the **exponent** tells you how many times this number is used as a factor. (A *factor,* you recall, is a number being multiplied.)

$$2 \cdot 2 \cdot 2 \cdot 2 \cdot 2 \cdot 2 = 2^6$$

The *base* is 2
(the number being multiplied)

The *exponent* is 6
(the number of times 2 is used as a factor)

If the base is a *positive* real number, the exponent appears to the right and slightly above the level of the number as in, for example, 5^6 and 8^3. If the base is a *negative* real number, then parentheses are used around the number and the exponent appears outside the parentheses. For example, $(-2)(-2)(-2) = (-2)^3$.

In algebra, if we do not know the value of a number, we use a letter to represent the unknown number. We call the letter a **variable.** This is quite useful in the case of exponents. Suppose we do not know the value of a number, but we know the number is multiplied by itself several times. We can represent this with a variable base and a whole-number exponent. For example, when we have an unknown number represented by the variable x, and this number occurs as a factor four times, we have

$$(x)(x)(x)(x) = x^4.$$

Likewise, if an unknown number, represented by the variable w, occurs as a factor five times, we have

$$(w)(w)(w)(w)(w) = w^5.$$

EXAMPLE 1 Write in exponent form.

(a) $9(9)(9)$ **(b)** $13(13)(13)(13)$ **(c)** $-7(-7)(-7)(-7)(-7)$
(d) $-4(-4)(-4)(-4)(-4)(-4)$ **(e)** $(x)(x)$ **(f)** $(y)(y)(y)$

Solution

(a) $9(9)(9) = 9^3$ **(b)** $13(13)(13)(13) = 13^4$
(c) The -7 is used as a factor five times. The answer must contain parentheses. Thus $-7(-7)(-7)(-7)(-7) = (-7)^5$.
(d) $-4(-4)(-4)(-4)(-4)(-4) = (-4)^6$
(e) $(x)(x) = x^2$ **(f)** $(y)(y)(y) = y^3$

Practice Problem 1 Write in exponent form.

(a) $6(6)(6)(6)$ **(b)** $-2(-2)(-2)(-2)(-2)$
(c) $108(108)(108)$ **(d)** $-11(-11)(-11)(-11)(-11)(-11)$
(e) $(w)(w)(w)$ **(f)** $(z)(z)(z)(z)$

If the base has an exponent of 2, we say the base is **squared.**
If the base has an exponent of 3, we say the base is **cubed.**
If the base has an exponent greater than 3, we say the base is raised **to the (exponent)-th power.**
x^2 is read "x squared."
y^3 is read "y cubed."
3^6 is read "three to the sixth power" or simply "three to the sixth."

Evaluating Numerical Expressions That Contain Exponents

EXAMPLE 2 Evaluate.

(a) 2^5

(b) $2^3 + 4^4$

Solution

(a) $2^5 = (2)(2)(2)(2)(2) = 32$

(b) First we evaluate each power.

$2^3 = 8$ $4^4 = 256$

Then we add. $8 + 256 = 264$

Practice Problem 2 Evaluate.

(a) 3^5

(b) $2^2 + 3^3$

If the base is negative, be especially careful in determining the sign. Notice the following:

$$(-3)^2 = (-3)(-3) = +9 \qquad (-3)^3 = (-3)(-3)(-3) = -27$$

From Section 1.3 we know that when you multiply two or more real numbers, first you multiply their absolute values.

- The result is positive if there is an even number of negative signs.
- The result is negative if there is an odd number of negative signs.

SIGN RULE FOR EXPONENTS

Suppose that a number is written in exponent form and the base is negative. The result is **positive** if the exponent is **even.** The result is **negative** if the exponent is **odd.**

Be careful how you read expressions with exponents and negative signs.

$$(-3)^4 \text{ means } (-3)(-3)(-3)(-3) \text{ or } +81.$$
$$-3^4 \text{ means } -(3)(3)(3)(3) \text{ or } -81.$$

EXAMPLE 3 Evaluate.

(a) $(-2)^3$ **(b)** $(-4)^6$ **(c)** -3^6 **(d)** $-(5^4)$

Solution

(a) $(-2)^3 = -8$ The answer is negative since the base is negative and the exponent 3 is odd.

(b) $(-4)^6 = +4096$ The answer is positive since the exponent 6 is even.

(c) $-3^6 = -729$ The negative sign is not contained in parentheses. Thus we find 3 raised to the sixth power and then take the opposite of that value.

(d) $-(5^4) = -625$ The negative sign is outside the parentheses.

Practice Problem 3 Evaluate.

(a) $(-3)^3$ **(b)** $(-2)^6$ **(c)** -2^4 **(d)** $-(3^6)$

Calculator

Exponents

You can use a calculator to evaluate 3^5. Press the following keys:

$$\boxed{3}\ \boxed{y^x}\ \boxed{5}\ \boxed{=}$$

The display should read

$$\boxed{243}$$

Try the following.

(a) 4^6 **(b)** $(0.2)^5$
(c) 18^6 **(d)** 3^{12}

Ans:

(a) 4096   **(b)** 0.00032
(c) 34,012,224   **(d)** 531,441

The steps needed to raise a number to a power are slightly different on some calculators.

EXAMPLE 4 Evaluate.

(a) $\left(\dfrac{1}{2}\right)^4$ **(b)** $(0.2)^4$ **(c)** $\left(\dfrac{2}{5}\right)^3$

(d) $(3)^3(2)^5$ **(e)** $2^3 - 3^4$

Solution

(a) $\left(\dfrac{1}{2}\right)^4 = \left(\dfrac{1}{2}\right)\left(\dfrac{1}{2}\right)\left(\dfrac{1}{2}\right)\left(\dfrac{1}{2}\right) = \dfrac{1}{16}$

(b) $(0.2)^4 = (0.2)(0.2)(0.2)(0.2) = 0.0016$

(c) $\left(\dfrac{2}{5}\right)^3 = \left(\dfrac{2}{5}\right)\left(\dfrac{2}{5}\right)\left(\dfrac{2}{5}\right) = \dfrac{8}{125}$

(d) First we evaluate each power.
$$3^3 = 27 \qquad 2^5 = 32$$
Then we multiply. $(27)(32) = 864$

(e) $2^3 - 3^4 = 8 - 81 = -73$

NOTE TO STUDENT: *Fully worked-out solutions to all of the Practice Problems can be found at the back of the text starting at page SP-1*

Practice Problem 4 Evaluate.

(a) $\left(\dfrac{1}{3}\right)^3$ **(b)** $(0.3)^4$ **(c)** $\left(\dfrac{3}{2}\right)^4$

(d) $(3)^4(4)^2$ **(e)** $4^2 - 2^4$

Verbal and Writing Skills

1. Explain in your own words how to evaluate 4^4.

2. Explain in your own words how to evaluate 9^2.

3. Explain how you would determine whether $(-5)^3$ is negative or positive.

4. Explain how you would determine whether $(-2)^5$ is negative or positive.

5. Explain the difference between $(-2)^4$ and -2^4. What answers do you obtain when you evaluate the expressions?

6. Explain the difference between $(-3)^4$ and -3^4. What answers do you obtain when you evaluate the expressions?

Write in exponent form.

7. $(5)(5)(5)(5)(5)(5)(5)$

8. $(7)(7)(7)(7)(7)$

9. $(w)(w)$

10. $(z)(z)(z)$

11. $(p)(p)(p)(p)$

12. $(y)(y)(y)(y)(y)(y)$

13. $(3q)(3q)(3q)$

14. $(2w)(2w)(2w)(2w)(2w)$

Evaluate.

15. 3^3

16. 4^2

17. 3^4

18. 8^3

19. 6^3

20. 15^2

21. $(-3)^3$

22. $(-2)^3$

23. $(-4)^2$

24. $(-5)^4$

25. -5^2

26. -4^2

27. $\left(\dfrac{1}{4}\right)^2$

28. $\left(\dfrac{3}{4}\right)^2$

29. $\left(\dfrac{2}{5}\right)^3$

30. $\left(\dfrac{3}{2}\right)^3$

31. $(2.1)^2$

32. $(1.5)^2$

33. $(0.2)^4$

34. $(0.7)^3$

35. $(-16)^2$

36. $(-7)^4$

37. -16^2

38. -7^4

Evaluate.

39. $5^3 + 6^2$

40. $7^2 + 6^3$

41. $10^2 - 11^2$

42. $3^3 - 2^4$

43. $(-4)^2 - (12)^2$

44. $13^2 - (-7)^2$

45. $2^5 - (-3)^2$

46. $8^2 - (-2)^3$

47. $(-4)^3(-3)^2$

Cumulative Review *Evaluate.*

48. **[1.1.5]** $(-11) + (-13) + 6 + (-9) + 8$

49. **[0.3.2]** $\dfrac{3}{4} \div \left(-\dfrac{9}{20}\right)$

50. **[1.2.1]** $-17 - (-9)$

51. **[0.4.5]** $(-2.1)(-1.2)$

52. **[0.5.3]** Amanda decided to invest her summer job earnings of $1600. At the end of the year she earned 6% on her investment. How much money did Amanda have at the end of the year?

Quick Quiz 1.4 *Evaluate.*

1. $(-4)^4$

2. $(1.8)^2$

3. $\left(\dfrac{3}{4}\right)^3$

4. **Concept Check** Explain the difference between $(-2)^6$ and -2^6. How do you decide if the answers are positive or negative?

Using the Order of Operations to Simplify Numerical Expressions

It is important to know *when* to do certain operations as well as how to do them. For example, to simplify the expression $2 - 4 \cdot 3$, should we subtract first or multiply first?

Also remember that multiplication can be written several ways. Thus $4 \cdot 3$, 4×3, $4(3)$, and $(4)(3)$ all indicate that we are multiplying 4 times 3.

The following list will assist you. It tells which operations to do first: the correct **order of operations.** You might think of it as a *list of priorities*.

Student Learning Objective

After studying this section, you will be able to:

1 Use the order of operations to simplify numerical expressions.

> **ORDER OF OPERATIONS FOR NUMBERS**
>
> Follow this order of operations:
>
> Do first   **1.** Do all operations inside parentheses.
>
> **2.** Raise numbers to a power.
>
> **3.** Multiply and divide numbers from left to right.
>
> Do last   **4.** Add and subtract numbers from left to right.

Let's return to the problem $2 - 4 \cdot 3$. There are no parentheses or numbers raised to a power, so the first thing we do is multiply. Then we subtract since this comes last on our list.

$$2 - 4 \cdot 3 = 2 - 12 \quad \text{Follow the order of operations by first multiplying } 4 \cdot 3 = 12.$$
$$= -10 \quad \text{Combine } 2 - 12 = -10.$$

EXAMPLE 1 Evaluate. $8 \div 2 \cdot 3 + 4^2$

Solution

$$8 \div 2 \cdot 3 + 4^2 = 8 \div 2 \cdot 3 + 16 \quad \text{Evaluate } 4^2 = 16 \text{ because the highest priority in this problem is raising to a power.}$$
$$= 4 \cdot 3 + 16 \quad \text{Next multiply and divide from left to right. So } 8 \div 2 = 4.$$
$$= 12 + 16 \quad 4 \cdot 3 = 12.$$
$$= 28 \quad \text{Finally, add.}$$

Practice Problem 1 Evaluate. $25 \div 5 \cdot 6 + 2^3$

Note: Multiplication and division have equal priority. We do not do multiplication first. Rather, we work from left to right, doing any multiplication or division that we encounter. Similarly, addition and subtraction have equal priority.

EXAMPLE 2 Evaluate. $(-3)^3 - 2^4$

Solution The highest priority is to raise the expressions to the appropriate powers.

$$(-3)^3 - 2^4 = -27 - 16 \quad \text{In } (-3)^3 \text{ we are cubing the number } -3 \text{ to obtain } -27.$$
$$\text{Be careful; } -2^4 \text{ is not } (-2)^4!$$
$$\text{Raise 2 to the fourth power and subtract the result.}$$
$$= -43 \quad \text{The last step is to add and subtract from left to right.}$$

Practice Problem 2 Evaluate. $(-4)^3 - 2^6$

NOTE TO STUDENT: Fully worked-out solutions to all of the Practice Problems can be found at the back of the text starting at page SP-1

EXAMPLE 3 Evaluate. $2 \cdot (2 - 3)^3 + 6 \div 3 + (8 - 5)^2$

Solution

$$2 \cdot (2 - 3)^3 + 6 \div 3 + (8 - 5)^2$$
$$= 2 \cdot (-1)^3 + 6 \div 3 + 3^2 \qquad$$ Combine the numbers inside the parentheses. Note that we need parentheses for -1 because of the negative sign but that they are not needed for 3.
$$= 2 \cdot (-1) + 6 \div 3 + 9 \qquad$$ Next, raise to a power.
$$= -2 + 2 + 9 \qquad$$ Next, multiply and divide from left to right.
$$= 9 \qquad$$ Finally, add and subtract from left to right.

Practice Problem 3 Evaluate. $6 - (8 - 12)^2 + 8 \div 2$

EXAMPLE 4 Evaluate. $\left(-\dfrac{1}{5}\right)\left(\dfrac{1}{2}\right) - \left(\dfrac{3}{2}\right)^2$

Solution The highest priority is to raise $\frac{3}{2}$ to the second power.

$$\left(\frac{3}{2}\right)^2 = \left(\frac{3}{2}\right)\left(\frac{3}{2}\right) = \frac{9}{4}$$

$$\left(-\frac{1}{5}\right)\left(\frac{1}{2}\right) - \left(\frac{3}{2}\right)^2 = \left(-\frac{1}{5}\right)\left(\frac{1}{2}\right) - \frac{9}{4}$$

$$= -\frac{1}{10} - \frac{9}{4} \qquad$$ Next we multiply.

$$= -\frac{1 \cdot 2}{10 \cdot 2} - \frac{9 \cdot 5}{4 \cdot 5} \qquad$$ We need to write each fraction as an equivalent fraction with the LCD of 20.

$$= -\frac{2}{20} - \frac{45}{20}$$

$$= -\frac{47}{20} \text{ or } -2\frac{7}{20} \qquad$$ Add.

Practice Problem 4 Evaluate.

$$\left(-\frac{1}{7}\right)\left(-\frac{14}{5}\right) + \left(-\frac{1}{2}\right) \div \frac{3}{4}$$

Calculator

 Order of Operations

Use your calculator to evaluate $3 + 4 \cdot 5$. Enter

3 $+$ 4 $\times$ 5 $=$

If the display is $\boxed{23}$, the correct order of operations is built in. If the display is not 23, you will need to modify the way you enter the problem. You should use

4 $\times$ 5 $+$ 3 $=$

Try $6 + 3 \cdot 4 - 8 \div 2$.

Ans:
14

Developing Your Study Skills

Previewing New Material

Part of your study time each day should consist of looking over the sections in your text that are to be covered the following day. You do not necessarily need to study and learn the material on your own, but a survey of the concepts, terminology, diagrams, and examples will help the new ideas seem more familiar as the instructor presents them.

You can look for concepts that appear confusing or difficult and be ready to listen carefully for your instructor's explanations. You can be prepared to ask questions that will increase your understanding. Previewing new material enables you to see what is coming and prepares you to be ready to absorb it.

Verbal and Writing Skills

Game Points *You have lost a game of UNO and are counting the points left in your hand. You announce that you have three fours and six fives.*

1. Write this as a number expression.

2. How many points do you have in your hand?

3. What answer would you get for the number expression if you simplified it by
   (a) performing the operations from left to right?
   (b) following the order of operations?

4. Which procedure in exercise 3 gives the correct number of total points?

Evaluate.

5. $(2 - 5)^2 \div 3 \times 4$

6. $(3 - 7)^2 \div 2 \times 5$

7. $9 + 4(5 + 2 - 8)$

8. $12 + 5(3 - 9 + 4)$

9. $8 - 2^3 \cdot 5 + 3$

10. $6 - 3^2 \cdot 6 + 4$

11. $4 + 42 \div 3 \cdot 2 - 8$

12. $7 + 36 \div 12 \cdot 3 - 14$

13. $3 \cdot 5 + 7 \cdot 3 - 5 \cdot 3$

14. $2 \cdot 6 + 5 \cdot 3 - 7 \cdot 4$

15. $8 - 5(2)^3 \div (-8)$

16. $11 - 3(4)^2 \div (-6)$

17. $3(5 - 7)^2 - 6(3)$

18. $-2(3 - 6)^2 - (-2)$

19. $5 \cdot 6 - (3 - 5)^2 + 8 \cdot 2$

20. $(-3)^2 \cdot 6 \div 9 + 4 \cdot 2$

21. $\dfrac{1}{2} \div \dfrac{2}{3} + 6 \cdot \dfrac{1}{4}$

22. $\dfrac{5}{6} \div \dfrac{2}{3} - 6 \cdot \left(\dfrac{1}{2}\right)^2$

23. $0.8 + 0.3(0.6 - 0.2)^2$

24. $0.05 + 1.4 - (0.5 - 0.7)^3$

25. $\dfrac{3}{8}\left(-\dfrac{1}{6}\right) - \dfrac{7}{8} + \dfrac{1}{2}$

26. $\dfrac{1}{2} \div \dfrac{4}{5} - \dfrac{3}{4}\left(\dfrac{5}{6}\right)$

Mixed Practice

27. $(3 - 7)^2 \div 8 + 3$

28. $\left(\dfrac{3}{5}\right)\left(\dfrac{5}{6}\right) - \dfrac{3}{4} \div 6$

29. $\left(\dfrac{3}{4}\right)^2(-16) + \dfrac{4}{5} \div \dfrac{-8}{25}$

30. $\left(2\dfrac{4}{7}\right) \div \left(-1\dfrac{1}{5}\right)$

31. $-6.3 - (-2.7)(1.1) + (3.3)^2$

32. $4.35 + 8.06 \div (-2.6) - (2.1)^2$

33. $\left(\dfrac{1}{2}\right)^3 + \dfrac{1}{4} - \left(\dfrac{1}{6} - \dfrac{1}{12}\right) - \dfrac{2}{3} \cdot \left(\dfrac{1}{4}\right)^2$

34. $(2.4 \cdot 1.2)^2 - 1.6 \cdot 2.2 \div 4.0 - 3.6$

Applications *In January 2007, Tiger Woods won the Buick Invitational. The following scorecard shows the results of Round 1. There are 18 holes per round and the table shows the number of times Woods got each type of score.*

Score on a Hole	Number of Times the Score Occurred
Eagle (-2)	2
Birdie (-1)	4
Par (0)	10
Bogey $(+1)$	2

35. Write his score as the sum of eagles, birdies, pars, and bogeys.

36. What was his final score for the round, when compared to par?

37. What answer do you get if you do the arithmetic left to right?

38. Explain why the answers in exercises 36 and 37 do not match.

Cumulative Review *Simplify.*

39. **[1.4.2]** $(0.5)^3$

40. **[0.2.3]** $-\dfrac{3}{4} - \dfrac{5}{6}$

41. **[1.4.2]** -1^{20}

42. **[0.3.2]** $3\dfrac{3}{5} \div 6\dfrac{1}{4}$

Quick Quiz 1.5 Evaluate.

1. $7 - 3^4 + 2 - 5$

2. $(0.3)^2 - 4.2(-4) + 0.07$

3. $(7 - 9)^4 + 22 \div (-2) + 6$

4. **Concept Check** Explain in what order you would perform the calculations to evaluate the expression $4 - (-3 + 4)^3 + 12 \div (-3)$.

How are you doing with your homework assignments in Sections 1.1 to 1.5? Do you feel you have mastered the material so far? Do you understand the concepts you have covered? Before you go further in the textbook, take some time to do each of the following problems.

Simplify each of the following. If the answer is a fraction, be sure to leave it in reduced form.

1.1

1. $3 + (-12)$

2. $-\dfrac{5}{6} + \left(-\dfrac{7}{8}\right)$

3. $0.34 + 0.9$

4. $-3.5 + 9 + 2.3 - (-3)$

1.2

5. $-23 - (-34)$

6. $-\dfrac{1}{6} - \dfrac{4}{5}$

7. $4.5 - (-7.8)$

8. $-4 - (-5) + 9$

1.3

9. $(-3)(-8)(2)(-2)$

10. $\left(-\dfrac{6}{11}\right)\left(-\dfrac{5}{3}\right)$

11. $-0.072 \div 0.08$

12. $\dfrac{5}{8} \div \left(-\dfrac{17}{16}\right)$

1.4

Evaluate.

13. $(0.7)^3$

14. $(-4)^4$

15. -2^8

16. $\left(\dfrac{2}{3}\right)^3$

17. $-3^3 + 3^4$

1.5

18. $20 - 12 \div 3 - 8(-1)$

19. $15 + 3 - 2 + (-6)$

20. $(9 - 13)^2 + 15 \div (-3)$

21. $-0.12 \div 0.6 + (-3)(1.2) - (-0.5)$

22. $\left(\dfrac{3}{4}\right)\left(-\dfrac{2}{5}\right) + \left(-\dfrac{1}{2}\right)\left(\dfrac{4}{5}\right) + \left(\dfrac{1}{2}\right)^2$

1. _____

2. _____

3. _____

4. _____

5. _____

6. _____

7. _____

8. _____

9. _____

10. _____

11. _____

12. _____

13. _____

14. _____

15. _____

16. _____

17. _____

18. _____

19. _____

20. _____

21. _____

22. _____

Now turn to page SA-3 for the answer to each of these problems. Each answer also includes a reference to the objective in which the problem is first taught. If you missed any of these problems, you should stop and review the Examples and Practice Problems in the referenced objective. A little review now will help you master the material in the upcoming sections of the text.

Using the Distributive Property to Simplify Algebraic Expressions

As we learned previously, we use letters called *variables* to represent unknown numbers. If a number is multiplied by a variable, we do not need any symbol between the number and variable. Thus, to indicate $(2)(x)$, we write $2x$. To indicate $3 \cdot y$, we write $3y$. If one variable is multiplied by another variable, we place the variables next to each other. Thus, $(a)(b)$ is written ab. We use exponent form if an unknown number (a variable) is used several times as a factor. Thus, $x \cdot x \cdot x = x^3$. Similarly, $(y)(y)(y)(y) = y^4$.

In algebra, we need to be familiar with several definitions. We will use them throughout the remainder of this book. Take some time to think through how each of these definitions is used.

An **algebraic expression** is a quantity that contains numbers and variables, such as $a + b$, $2x - 3$, and $5ab^2$. In this chapter we will be learning rules about adding and multiplying algebraic expressions. A **term** is a number, a variable, or a product of numbers and variables. 17, x, $5xy$, and $22xy^3$ are all examples of terms. We will refer to terms when we discuss the distributive property.

An important property of algebra is the **distributive property.** We can state it in an equation as follows:

> **DISTRIBUTIVE PROPERTY**
>
> For all real numbers a, b, and c,
>
> $$a(b + c) = ab + ac.$$

A numerical example shows that it does seem reasonable.

$$5(3 + 6) = 5(3) + 5(6)$$
$$5(9) = 15 + 30$$
$$45 = 45$$

We can use the distributive property to multiply any term by the sum of two or more terms. In Section 0.1, we defined the word *factor*. Two or more algebraic expressions that are multiplied are called **factors.** Consider the following examples of multiplying algebraic expressions.

EXAMPLE 1 Multiply.

(a) $5(a + b)$　　　　　　　　**(b)** $-3(3x + 2y)$

Solution

(a) $5(a + b) = 5a + 5b$ 　　　Multiply the factor $(a + b)$ by the factor 5.

(b) $-3(3x + 2y) = -3(3x) + (-3)(2y)$ 　　Multiply the factor $(3x + 2y)$ by the factor (-3).

$$= -9x - 6y$$

Practice Problem 1 Multiply.

(a) $3(x + 2y)$　　　　　　　　**(b)** $-2(a - 3b)$

If the parentheses are preceded by a negative sign, we consider this to be the product of (-1) and the expression inside the parentheses.

EXAMPLE 2 Multiply. $-(a - 2b)$

Solution

$$-(a - 2b) = (-1)(a - 2b) = (-1)(a) + (-1)(-2b) = -a + 2b$$

Practice Problem 2 Multiply. $-(-3x + y)$

In general, we see that in all these examples we have multiplied each term of the expression in the parentheses by the expression in front of the parentheses.

EXAMPLE 3 Multiply.

(a) $\dfrac{2}{3}(x^2 - 6x + 8)$　　　　　　　**(b)** $1.4(a^2 + 2.5a + 1.8)$

Solution

(a) $\dfrac{2}{3}(x^2 - 6x + 8) = \left(\dfrac{2}{3}\right)(1x^2) + \left(\dfrac{2}{3}\right)(-6x) + \left(\dfrac{2}{3}\right)(8)$

$$= \dfrac{2}{3}x^2 + (-4x) + \dfrac{16}{3}$$

$$= \dfrac{2}{3}x^2 - 4x + \dfrac{16}{3}$$

(b) $1.4(a^2 + 2.5a + 1.8) = 1.4(1a^2) + (1.4)(2.5a) + (1.4)(1.8)$

$$= 1.4a^2 + 3.5a + 2.52$$

Practice Problem 3 Multiply.

(a) $\dfrac{3}{5}(a^2 - 5a + 25)$　　　　　　　**(b)** $2.5(x^2 - 3.5x + 1.2)$

NOTE TO STUDENT: Fully worked-out solutions to all of the Practice Problems can be found at the back of the text starting at page SP-1

There are times we multiply a variable by itself and use exponent notation. For example, $(x)(x) = x^2$ and $(x)(x)(x) = x^3$. In other cases there will be numbers and variables multiplied at the same time.

We will see problems like $(2x)(x) = (2)(x)(x) = 2x^2$. Some expressions will involve multiplication of more than one variable. We will see problems like $(3x)(xy) = (3)(x)(x)(y) = 3x^2y$. There will be times when we use the distributive property and all of these methods will be used. For example,

$$2x(x - 3y + 2) = 2x(x) + (2x)(-3y) + (2x)(2)$$
$$= 2x^2 + (-6)(xy) + 4(x)$$
$$= 2x^2 - 6xy + 4x.$$

We will discuss this type of multiplication of variables with exponents in more detail in Section 5.1. At that point we will expand these examples and other similar examples to develop the general rule for multiplication $(x^a)(x^b) = x^{a+b}$.

EXAMPLE 4 Multiply. $-2x(3x + y - 4)$

Solution

$$-2x(3x + y - 4) = -2(x)(3)(x) + (-2)(x)(y) + (-2)(x)(-4)$$
$$= -2(3)(x)(x) + (-2)(xy) + (-2)(-4)(x)$$
$$= -6x^2 - 2xy + 8x$$

Practice Problem 4 Multiply. $-4x(x - 2y + 3)$

The distributive property can also be presented with the *a* on the right.

$$(b + c)a = ba + ca$$

The *a* is "distributed" over the *b* and *c* inside the parentheses.

EXAMPLE 5   Multiply. $(2x^2 - x)(-3)$

Solution

$$(2x^2 - x)(-3) = 2x^2(-3) + (-x)(-3)$$
$$= -6x^2 + 3x$$

Practice Problem 5   Multiply.

$$(3x^2 - 2x)(-4)$$

EXAMPLE 6   A farmer has a rectangular field that is 300 feet wide. One portion of the field is $2x$ feet long. The other portion of the field is $3y$ feet long. Use the distributive property to find an expression for the area of this field.

Solution   First we draw a picture of a field that is 300 feet wide and $2x + 3y$ feet long.

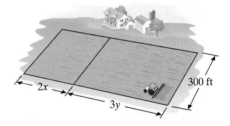

To find the area of the field, we multiply the width times the length.

$$300(2x + 3y) = 300(2x) + 300(3y) = 600x + 900y$$

Thus the area of the field in square feet is $600x + 900y$.

NOTE TO STUDENT: Fully worked-out solutions to all of the Practice Problems can be found at the back of the text starting at page SP-1

Practice Problem 6   A farmer has a rectangular field that is 400 feet wide. One portion of the field is $6x$ feet long. The other portion of the field is $9y$ feet long. Use the distributive property to find an expression for the area of this field.

1.6 EXERCISES

MyMathLab

PRACTICE

WATCH

DOWNLOAD

READ

REVIEW

Verbal and Writing Skills

In exercises 1 and 2, complete each sentence by filling in the blank.

1. A _____ is a symbol used to represent an unknown number.

2. When we write an expression with numbers and variables such as $7x$, it indicates that we are _____ 7 by x.

3. Explain in your own words how we multiply a problem like $(4x)(x)$.

4. Explain why you think the property $a(b + c) = ab + ac$ is called the distributive property. What does "distribute" mean?

5. Does the following distributive property work?
$$a(b - c) = ab - ac$$
Why or why not? Give an example.

6. Susan tried to use the distributive property and wrote
$$-5(x + 3y - 2) = -5x - 15y - 10.$$
What did she do wrong?

Multiply. Use the distributive property.

7. $3(x - 2y)$

8. $4(3x - y)$

9. $-2(4a - 3b)$

10. $-3(2a - 5b)$

11. $3(3x + y)$

12. $2(4x + y)$

13. $8(-m - 3n)$

14. $10(-2m - n)$

15. $-(x - 3y)$

16. $-(-4x + y)$

17. $-9(9x - 5y + 8)$

18. $-5(3x + 9 - 7y)$

19. $2(-5x + y - 6)$

20. $3(2x - 6y - 5)$

21. $\frac{5}{6}(12x^2 - 24x + 18)$

22. $\frac{2}{3}(-27a^4 + 9a^2 - 21)$

23. $\frac{x}{5}(x + 10y - 4)\left(Hint: \frac{x}{5} = \frac{1}{5}x\right)$

24. $\frac{y}{3}(3y - 4x - 6)\left(Hint: \frac{y}{3} = \frac{1}{3}y\right)$

25. $5x(x + 2y + z)$

26. $3a(2a + b - c)$

27. $(-4.5x + 5)(-3)$

28. $(-2.3x + 4)(-5)$

29. $(6x + y - 1)(3x)$

30. $(3x - 3y + 4)(2x)$

Mixed Practice *Multiply. Use the distributive property.*

31. $(3x + 2y - 1)(-xy)$

32. $(4a - 2b - 1)(-ab)$

33. $(-a - 2b + 4)5ab$

34. $(3a - b - 2)2ab$

35. $\frac{1}{3}(6a^2 - 12a + 8)$

36. $\frac{1}{2}(-4a^2 + 10a - 5)$

37. $-0.3x(-1.2x^2 - 0.3x + 0.5)$
Hint: $x(x^2) = x(x)(x) = x^3$

38. $-0.6q(1.2q^2 + 2.5r - 0.7s)$
Hint: $q(q^2) = q(q)(q) = q^3$

39. $0.4q(-3.3q^2 - 0.7r - 10)$
Hint: $q(q^2) = q(q)(q) = q^3$

Applications

▲ **40.** *Geometry* Gary Roswell owns a large rectangular field where he grows corn and wheat. The width of the field is 850 feet. The portion of the field where corn grows is $10x$ feet long and the portion where wheat grows is $7y$ feet long. Use the distributive property to find an expression for the area of this field.

▲ **41.** *Geometry* Kathy DesMaris has a rectangular field that is 800 feet wide. One portion of the field is $5x$ feet long. The other portion of the field is $14y$ feet long. Use the distributive property to find an expression for the area of this field.

To Think About

▲ **42.** *Athletic Field* The athletic field at Gordon College is $2x$ feet wide. It used to be 1800 feet long. An old, run-down building was torn down, making the field $3y$ feet longer. Use the distributive property to find an expression for the area of the new field.

▲ **43.** *Airport Runway* The Beverly Airport runway is $4x$ feet wide. The airport was supposed to have a 3000-foot-long runway. However, some of the land was wetland, so a runway could not be built on all of it. Therefore, the length of the runway was decreased by $2y$ feet. Use the distributive property to find an expression for the area of the final runway.

Cumulative Review *In exercises 44–48, evaluate.*

44. **[1.1.4]** $-18 + (-20) + 36 + (-14)$

45. **[1.4.2]** $(-2)^6$

46. **[1.2.1]** $-27 - (-41)$

47. **[1.5.3]** $25 \div 5(2) + (-6)$

48. **[1.5.3]** $(12 - 10)^2 + (-3)(-2)$

Quick Quiz 1.6 *Multiply. Use the distributive property.*

1. $5(-3a - 7b)$

2. $-2x(x - 4y + 8)$

3. $-3ab(4a - 5b - 9)$

4. **Concept Check** Explain how you would multiply to obtain the answer for $\left(-\frac{3}{7}\right)(21x^2 - 14x + 3)$.

1.7 COMBINING LIKE TERMS

Identifying Like Terms

We can add or subtract quantities that are *like quantities*. This is called **combining** like quantities.

$$5 \text{ inches} + 6 \text{ inches} = 11 \text{ inches}$$
$$20 \text{ square inches} - 16 \text{ square inches} = 4 \text{ square inches}$$

However, we cannot combine things that are not the same.

$$16 \text{ square inches} - 4 \text{ inches (Cannot be done!)}$$

Similarly, in algebra we can **combine like terms.** This means to add or subtract like terms. Remember, we cannot combine terms that are not the same. Recall that a *term* is a number, a variable, or a product of numbers and variables. **Like terms** are terms that have identical variables and exponents. In other words, like terms must have exactly the same letter parts.

Student Learning Objectives

After studying this section, you will be able to:

 Identify like terms.

 Combine like terms.

EXAMPLE 1 List the like terms of each expression.

(a) $5x - 2y + 6x$

(b) $2x^2 - 3x - 5x^2 - 8x$

Solution

(a) $5x$ and $6x$ are like terms. These are the only like terms in this expression.

(b) $2x^2$ and $-5x^2$ are like terms.

$-3x$ and $-8x$ are like terms.

Note that x^2 and x are not like terms.

Practice Problem 1 List the like terms of each expression.

(a) $5a + 2b + 8a - 4b$

(b) $x^2 + y^2 + 3x - 7y^2$

NOTE TO STUDENT: Fully worked-out solutions to all of the Practice Problems can be found at the back of the text starting at page SP-1

Do you really understand what a term is? A term is a number, a variable, or a product of numbers and variables. Terms are the parts of an algebraic expression separated by plus or minus signs. The sign in front of the term is considered part of the term.

Combining Like Terms

It is important to know how to combine like terms. Since

$$4 \text{ inches} + 5 \text{ inches} = 9 \text{ inches},$$

we would expect in algebra that $4x + 5x = 9x$.

Why is this true? Let's take a look at the distributive property.

> Like terms may be added or subtracted by using the distributive property:
> $$ab + ac = a(b + c) \quad \text{and} \quad ba + ca = (b + c)a.$$

For example,

$$-7x + 9x = (-7 + 9)x = 2x$$
$$5x^2 + 12x^2 = (5 + 12)x^2 = 17x^2.$$

EXAMPLE 2 Combine like terms.

(a) $-4x^2 + 8x^2$ **(b)** $5x + 3x + 2x$

Solution

(a) Notice that each term contains the factor x^2. Using the distributive property, we have
$$-4x^2 + 8x^2 = (-4 + 8)x^2 = 4x^2.$$

(b) Note that each term contains the factor x. Using the distributive property, we have
$$5x + 3x + 2x = (5 + 3 + 2)x = 10x.$$

Practice Problem 2 Combine like terms.

(a) $16y^3 + 9y^3$ **(b)** $5a + 7a + 4a$

In this section, the direction *simplify* means to remove parentheses and/or combine like terms.

EXAMPLE 3 Simplify. $5a^2 - 2a^2 + 6a^2$

Solution
$$5a^2 - 2a^2 + 6a^2 = (5 - 2 + 6)a^2 = 9a^2$$

Practice Problem 3 Simplify. $-8y^2 - 9y^2 + 4y^2$

After doing a few problems, you will find that it is not necessary to write out the step of using the distributive property. We will omit this step for the remaining examples in this section.

EXAMPLE 4 Simplify.

(a) $5.6a + 2b + 7.3a - 6b$
(b) $3x^2y - 2xy^2 + 6x^2y$
(c) $2a^2b + 3ab^2 - 6a^2b^2 - 8ab$

Solution

(a) $5.6a + 2b + 7.3a - 6b = 12.9a - 4b$ We combine the a terms and the b terms separately.

(b) $3x^2y - 2xy^2 + 6x^2y = 9x^2y - 2xy^2$ **Note:** x^2y and xy^2 are not like terms because of different powers.

(c) $2a^2b + 3ab^2 - 6a^2b^2 - 8ab$ These terms cannot be combined; there are no like terms in this expression.

Practice Problem 4 Simplify.

(a) $1.3x + 3a - 9.6x + 2a$
(b) $5ab - 2ab^2 - 3a^2b + 6ab$
(c) $7x^2y - 2xy^2 - 3x^2y^2 - 4xy$

The two skills in this section that a student must practice are identifying like terms and correctly adding and subtracting like terms. If a problem involves many terms, you may find it helpful to rearrange the terms so that like terms are together.

EXAMPLE 5 Simplify. $3a - 2b + 5a^2 + 6a - 8b - 12a^2$

Solution There are three pairs of like terms.

$$3a + 6a - 2b - 8b + 5a^2 - 12a^2$$

a terms b terms a^2 terms

$$= 9a - 10b - 7a^2$$

You can rearrange the terms so that like terms are together, making it easier to combine them. Combine like terms.

Because of the commutative property, the order of terms in an answer to this problem is not significant. These three terms can be rearranged in a different order. $-10b + 9a - 7a^2$ and $-7a^2 + 9a - 10b$ are also correct. Later, we will learn the preferred way to write the answer.

Practice Problem 5 Simplify. $5xy - 2x^2y + 6xy^2 - xy - 3xy^2 - 7x^2y$

NOTE TO STUDENT: Fully worked-out solutions to all of the Practice Problems can be found at the back of the text starting at page SP-1

Use extra care with fractional values.

EXAMPLE 6 Simplify. $\frac{3}{4}x^2 - 5y - \frac{1}{8}x^2 + \frac{1}{3}y$

Solution We need the least common denominator for the x^2 terms, which is 8. Change $\frac{3}{4}$ to eighths by multiplying the numerator and denominator by 2.

$$\frac{3}{4}x^2 - \frac{1}{8}x^2 = \frac{3 \cdot 2}{4 \cdot 2}x^2 - \frac{1}{8}x^2 = \frac{6}{8}x^2 - \frac{1}{8}x^2 = \frac{5}{8}x^2$$

The least common denominator for the y terms is 3. Change -5 to thirds.

$$-\frac{5}{1}y + \frac{1}{3}y = \frac{-5 \cdot 3}{1 \cdot 3}y + \frac{1}{3}y = \frac{-15}{3}y + \frac{1}{3}y = -\frac{14}{3}y$$

Thus, our solution is $\frac{5}{8}x^2 - \frac{14}{3}y$.

Practice Problem 6 Simplify.

$$\frac{1}{7}a^2 - \frac{5}{12}b + 2a^2 - \frac{1}{3}b$$

EXAMPLE 7 Simplify. $6(2x + 3xy) - 8x(3 - 4y)$

Solution First remove the parentheses; then combine like terms.

$$6(2x + 3xy) - 8x(3 - 4y) = 12x + 18xy - 24x + 32xy$$ Use the distributive property.

$$= -12x + 50xy$$ Combine like terms.

Practice Problem 7 Simplify. $5a(2 - 3b) - 4(6a + 2ab)$

Verbal and Writing Skills

1. Explain in your own words the mathematical meaning of the word *term*.

2. Explain in your own words the mathematical meaning of the phrase *like terms*.

3. Explain which terms are like terms in the expression $5x - 7y - 8x$.

4. Explain which terms are like terms in the expression $12a - 3b - 9a$.

5. Explain which terms are like terms in the expression $7xy - 9x^2y - 15xy^2 - 14xy$.

6. Explain which terms are like terms in the expression $-3a^2b - 12ab + 5ab^2 + 9ab$.

Combine like terms.

7. $-14b^2 - 11b^2$

8. $-17x^5 + 3x^5$

9. $5a^3 - 7a^2 + a^3$

10. $a^5 + 9a^3 - 3a^5$

11. $3x + 2y - 8x - 7y$

12. $4a - 3b - 2a - 8b$

13. $1.3x - 2.6y + 5.8x - 0.9y$

14. $3.1a - 0.2b - 0.8a + 5.3b$

15. $1.6x - 2.8y - 3.6x - 5.9y$

16. $1.9x - 2.4b - 3.8x - 8.2b$

17. $3p - 4q + 2p + 3 + 5q - 21$

18. $6x - 5y - 3y + 7 - 11x - 5$

19. $2ab + 5bc - 6ac - 2ab$

20. $7ab - 3bc - 12ac + 8ab$

21. $2x^2 - 3x - 5 - 7x + 8 - x^2$

22. $5x + 7 - 6x^2 + 6 - 11x + 4x^2$

23. $2y^2 - 8y + 9 - 12y^2 - 8y + 3$

24. $3y^2 + 9y - 12 - 4y^2 - 6y + 2$

25. $\frac{1}{3}x - \frac{2}{3}y - \frac{2}{5}x + \frac{4}{7}y$

26. $\frac{2}{5}s - \frac{3}{8}t - \frac{4}{15}s - \frac{5}{12}t$

27. $\frac{3}{4}a^2 - \frac{1}{3}b - \frac{1}{5}a^2 - \frac{1}{2}b$

28. $\frac{2}{5}y - \frac{3}{4}x^2 - \frac{1}{3}y + \frac{7}{8}x^2$

29. $3rs - 8r + s - 5rs + 10r - s$

30. $-rs + 10s + 5r - rs + 6s - 2r$

Simplify. Use the distributive property to remove parentheses; then combine like terms.

31. $5(2a - b) - 3(5b - 6a)$

32. $8(3x - 2y) + 4(3y - 5x)$

33. $-3b(5a - 3b) + 4(-3ab - 5b^2)$

34. $2x(x - 3y) - 4(-3x^2 - 2xy)$

35. $6(c - 2d^2) - 2(4c - d^2)$

36. $-5(2cd + c^2) + 3c(3d - 2c)$

37. $4(2 - x) - 3(-5 - 12x)$

38. $7(3 - x) - 6(8 - 13x)$

To Think About

▲ **39.** *Fencing in a Pool* Mr. Jimenez has a pool behind his house that needs to be fenced in. The backyard is an odd quadrilateral shape. The four sides are $3a$, $2b$, $4a$, and $7b$ in length. How much fencing (the length of the perimeter) would he need to enclose the pool?

▲ **40.** *Framing a Masterpiece* The new Degas masterpiece purchased by the Museum of Fine Arts in Boston needs to be reframed. If the rectangular picture measures $6x - 3$ wide by $8x - 7$ high, what is the perimeter of the painting?

▲ **41.** *Geometry* A rectangle is $5x - 10$ feet long and $2x + 6$ feet wide. What is the perimeter of the rectangle?

▲ **42.** *Geometry* A triangle has sides of length $3a + 8$ meters, $5a - b$ meters, and $12 - 2b$ meters. What is the perimeter of the triangle?

Cumulative Review *Evaluate.*

43. [0.2.3] $-\dfrac{3}{4} - \dfrac{1}{3}$

44. [0.3.1] $\left(\dfrac{2}{3}\right)\left(-\dfrac{9}{16}\right)$

45. [0.2.3] $\dfrac{4}{5} + \left(-\dfrac{1}{25}\right) + \left(-\dfrac{3}{10}\right)$

46. [0.3.2] $\left(\dfrac{5}{7}\right) \div \left(-\dfrac{14}{3}\right)$

Quick Quiz 1.7 Combine like terms.

1. $3xy - \dfrac{2}{3}x^2y - \dfrac{5}{6}xy + \dfrac{7}{3}x^2y$

2. $8.2a^2b + 5.5ab^2 - 7.6a^2b - 9.9ab^2$

3. $2(3x - 5y) - 2(-7x - 4y)$

4. **Concept Check** Explain how you would remove parentheses and then combine like terms to obtain the answer for $1.2(3.5x - 2.2y) - 4.5(2.0x + 1.5y)$.

1.8 USING SUBSTITUTION TO EVALUATE ALGEBRAIC EXPRESSIONS AND FORMULAS

Student Learning Objectives

After studying this section, you will be able to:

 1 Evaluate an algebraic expression for a specified value.

 2 Evaluate a formula by substituting values.

 1 Evaluating an Algebraic Expression for a Specified Value

You will use the order of operations to **evaluate** variable expressions. Suppose we are asked to evaluate

$$6 + 3x \text{ for } x = -4.$$

In general, x represents some unknown number. Here we are told x has the value -4. We can replace x with -4. Use parentheses around -4. Note that we always put replacement values in parentheses.

$$6 + 3(-4) = 6 + (-12) = -6$$

When we replace a variable by a particular value, we say we have **substituted** the value for the variable. We then evaluate the expression (that is, find a value for it).

> **EXAMPLE 1** Evaluate $\frac{2}{3}x - 5$ for $x = -6$.
>
> **Solution**
>
> $$\frac{2}{3}x - 5 = \frac{2}{3}(-6) - 5 \qquad \text{Substitute } -6 \text{ for } x. \text{ Be sure to enclose the } -6 \text{ in parentheses.}$$
>
> $$= -4 - 5 \qquad \text{Multiply } \left(\frac{2}{3}\right)\left(-\frac{6}{1}\right) = -4.$$
>
> $$= -9 \qquad \text{Combine.}$$

> **Practice Problem 1** Evaluate $4 - \frac{1}{2}x$ for $x = -8$.

NOTE TO STUDENT: Fully worked-out solutions to all of the Practice Problems can be found at the back of the text starting at page SP-1

Compare parts **(a)** and **(b)** in the next example. The two parts illustrate that you must be careful what value you raise to a power. *Note:* In part **(b)** we will need parentheses within parentheses. To avoid confusion, we use brackets [ ] to represent the outside parentheses.

> **EXAMPLE 2** Evaluate for $x = -3$.
>
> **(a)** $2x^2$ **(b)** $(2x)^2$
>
> **Solution**
>
> **(a)** Here the value x is squared.   **(b)** Here the value $(2x)$ is squared.
>
> $2x^2 = 2(-3)^2$ $(2x)^2 = [2(-3)]^2$
>
> $= 2(9)$ First square -3. $= (-6)^2$ First multiply the numbers inside the brackets.
>
> $= 18$ Then multiply. $= 36$ Then square -6.

> **Practice Problem 2** Evaluate for $x = -3$.
>
> **(a)** $4x^2$ **(b)** $(4x)^2$

Carefully study the solutions to Example 2**(a)** and Example 2**(b)**. You will find that taking the time to see *how* and *why* they are different is a good investment of time.

EXAMPLE 3 Evaluate $x^2 + 3x$ for $x = -4$.

Solution

$$
\begin{aligned}
x^2 + 3x &= (-4)^2 + 3(-4) &&\text{Replace each } x \text{ by } -4 \text{ in the original expression.}\\
&= 16 + (3)(-4) &&\text{Raise to a power.}\\
&= 16 - 12 &&\text{Multiply.}\\
&= 4 &&\text{Finally, add.}
\end{aligned}
$$

Practice Problem 3 Evaluate $2x^2 - 3x$ for $x = -2$.

2 Evaluating a Formula by Substituting Values

We can *evaluate a formula* by substituting values for the variables. For example, the area of a triangle can be found using the formula $A = \frac{1}{2}ab$, where b is the length of the base of the triangle and a is the altitude of the triangle (see figure). If we know values for a and b, we can substitute those values into the formula to find the area. The units for area are *square units*.

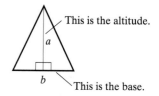
This is the altitude.
a
b
This is the base.

Because some of the examples and exercises in this section involve geometry, it may be helpful to review this topic.

The following information is very important. If you have forgotten some of this material (or if you have never learned it), please take the time to learn it completely now. Throughout the entire book we will be using this information in solving applied problems.

Perimeter is the distance around a plane figure. Perimeter is measured in linear units (inches (in.), feet (ft), centimeters (cm), miles (mi)). **Area** is a measure of the amount of surface in a region. Area is measured in square units (square inches (in.^{2}), square feet (ft^2), square centimeters (cm^2)).

In our sketches we will show angles of 90° by using a small square $\left(\raise1pt\hbox{$\llcorner$}\right)$. This indicates that the two lines are at right angles. All angles that measure 90° are called **right angles.** An **altitude** is perpendicular to the base of a figure. That is, the altitude forms right angles with the base. The small corner square in a sketch helps us identify the altitude of the figure.

The following box provides a handy guide to some facts and formulas you will need to know. Use it as a reference when solving word problems involving geometric figures.

GEOMETRIC FORMULAS: TWO-DIMENSIONAL FIGURES

A **parallelogram** is a four-sided figure with opposite sides parallel. In a parallelogram, opposite sides are equal and opposite angles are equal.

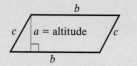
b
c a = altitude c
b

Perimeter = the sum of all four sides
Area = ab

A **rectangle** is a parallelogram with all interior angles measuring 90°.

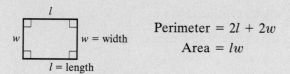
l
w w = width
l = length

Perimeter = $2l + 2w$
Area = lw

A **square** is a rectangle with all four sides equal.

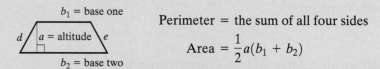

$$\text{Perimeter} = 4s$$
$$\text{Area} = s^2$$

A **trapezoid** is a four-sided figure with two sides parallel. The parallel sides are called the *bases* of the trapezoid.

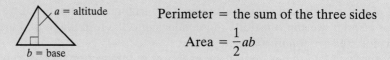

Perimeter = the sum of all four sides

$$\text{Area} = \frac{1}{2}a(b_1 + b_2)$$

A **triangle** is a closed plane figure with three sides.

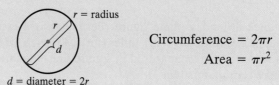

Perimeter = the sum of the three sides

$$\text{Area} = \frac{1}{2}ab$$

A **circle** is a plane curve consisting of all points at an equal distance from a given point called the center.

Circumference is the distance around a circle.

r = radius

d = diameter = $2r$

$$\text{Circumference} = 2\pi r$$
$$\text{Area} = \pi r^2$$

π (the number *pi*) is a constant associated with circles. It is an irrational number that is approximately 3.141592654. We usually use 3.14 as a sufficiently accurate approximation. Thus we write $\pi \approx 3.14$ for most of our calculations involving π.

▲ **EXAMPLE 4** Find the area of a triangle with a base of 16 centimeters (cm) and an altitude of 12 centimeters (cm).

Solution

Use the formula $A = \frac{1}{2}ab$.

Substitute 12 centimeters for a and 16 centimeters for b.

$$A = \frac{1}{2}(12 \text{ centimeters})(16 \text{ centimeters})$$

$$= \frac{1}{2}(12)(16)(\text{cm})(\text{cm}) \quad \text{If you take } \tfrac{1}{2} \text{ of 12 first, it will make your calculation easier.}$$

$$= (6)(16)(\text{cm})^2 = 96 \text{ square centimeters}$$

The area of the triangle is 96 square centimeters or 96 cm^2.

NOTE TO STUDENT: Fully worked-out solutions to all of the Practice Problems can be found at the back of the text starting at page SP-1

▲ **Practice Problem 4** Find the area of a triangle with an altitude of 3 meters and a base of 7 meters.

The area of a circle is given by

$$A = \pi r^2.$$

We will use 3.14 as an approximation for the *irrational number* π.

▲ **EXAMPLE 5**   Find the area of a circle if the radius is 2 inches.

Solution

$A = \pi r^2 \approx (3.14)(2 \text{ inches})^2$ Write the formula and substitute the given values for the letters.

    $= (3.14)(4)(\text{in.})^2$ Raise to a power. Then multiply.

    $= 12.56$ square inches or 12.56 in.^{2}

▲ **Practice Problem 5**   Find the area of a circle if the radius is 3 meters.

The formula $C = \frac{5}{9}(F - 32)$ allows us to find the Celsius temperature if we know the Fahrenheit temperature. That is, we can substitute a value for F in degrees Fahrenheit into the formula to obtain a temperature C in degrees Celsius.

EXAMPLE 6   What is the Celsius temperature when the Fahrenheit temperature is $F = -22°$?

Solution   Use the formula.

$C = \dfrac{5}{9}(F - 32)$

   $= \dfrac{5}{9}[(-22) - 32]$ Substitute -22 for F in the formula.

   $= \dfrac{5}{9}(-54)$ Combine the numbers inside the brackets.

   $= (5)(-6)$ Simplify.

   $= -30$ Multiply.

The temperature is $-30°$ Celsius or $-30°$C.

Practice Problem 6   What is the Celsius temperature when the Fahrenheit temperature is $F = 68°$? Use the formula $C = \frac{5}{9}(F - 32)$.

When driving in Canada or Mexico, we must observe speed limits posted in kilometers per hour. A formula that converts r (miles per hour) to k (kilometers per hour) is $k \approx 1.61r$. Note that this is an approximation.

EXAMPLE 7 You are driving on a highway in Mexico. It has a posted maximum speed of 100 kilometers per hour. You are driving at 61 miles per hour. Are you exceeding the speed limit?

Solution Use the formula.

$$k \approx 1.61r$$
$$= (1.61)(61) \quad \text{Replace } r \text{ by } 61.$$
$$= 98.21 \quad \text{Multiply the numbers.}$$

You are driving at approximately 98 kilometers per hour. You are not exceeding the speed limit.

Practice Problem 7 You are driving behind a heavily loaded truck on a Canadian highway. The highway has a posted minimum speed of 65 kilometers per hour. When you travel at exactly the same speed as the truck ahead of you, you observe that the speedometer reads 35 miles per hour. Assuming that your speedometer is accurate, determine whether the truck is violating the minimum speed law.

Developing Your Study Skills

Problems with Accuracy

Strive for accuracy. Mistakes are often made as a result of human error rather than from lack of understanding. Such mistakes are frustrating. A simple arithmetic or sign error can lead to an incorrect answer.

These five steps will help you cut down on errors.

1. Work carefully and take your time. Do not rush through a problem just to get it done.
2. Concentrate on one problem at a time. Sometimes working on problems becomes mechanical, and your mind begins to wander. You can become careless and make a mistake.
3. Check your problem. Be sure that you copied it correctly from the book.
4. Check your computations from step to step. Check the solution to the problem. Does it work? Does it make sense?
5. Keep practicing new skills. Remember the old saying, "Practice makes perfect." An increase in practice results in an increase in accuracy. Many errors are due simply to lack of practice.

There is no magic formula for eliminating all errors, but these five steps will be a tremendous help in reducing them.

1.8 EXERCISES

MyMathLab

 PRACTICE

 WATCH

 DOWNLOAD

 READ

REVIEW

Evaluate.

1. $-2x + 1$ for $x = 3$

2. $-4x - 2$ for $x = 5$

3. $\dfrac{2}{5}y - 8$ for $y = -10$

4. $\dfrac{3}{8}y - 9$ for $y = -8$

5. $5x + 10$ for $x = \dfrac{1}{2}$

6. $7x + 20$ for $x = -\dfrac{1}{2}$

7. $2 - 4x$ for $x = 7$

8. $3 - 5x$ for $x = 8$

9. $3.5 - 2x$ for $x = 2.4$

10. $6.3 - 3x$ for $x = 2.3$

11. $9x + 13$ for $x = -\dfrac{3}{4}$

12. $5x + 7$ for $x = -\dfrac{2}{3}$

13. $x^2 - 3x$ for $x = -2$

14. $x^2 + 3x$ for $x = 4$

15. $5y^2$ for $y = -1$

16. $8y^2$ for $y = -1$

17. $-3x^3$ for $x = 2$

18. $-7x^2$ for $x = 5$

19. $-5x^2$ for $x = -2$

20. $-2x^2$ for $x = -3$

21. $2x^2 + 3x$ for $x = -3$

22. $18 + 3x^2$ for $x = -3$

23. $(2x)^2 + x$ for $x = 3$

24. $2 - x^2$ for $x = -2$

25. $2 - (-x)^2$ for $x = -2$

26. $2x - 3x^2$ for $x = -4$

27. $10a + (4a)^2$ for $a = -2$

28. $9a - (2a)^2$ for $a = -3$

29. $4x^2 - 6x$ for $x = \dfrac{1}{2}$

30. $5 - 9x^2$ for $x = \dfrac{1}{3}$

31. $x^2 - 7x + 3$ for $x = 3$

32. $4x^2 - 3x + 9$ for $x = 2$

33. $\dfrac{1}{2}x^2 - 3x + 9$ for $x = -4$

34. $\dfrac{1}{3}x^2 + 2x - 5$ for $x = -3$

35. $2r^2 + 3s^2 - rs$ for $r = -1$ and $s = 3$

36. $-r^2 + 5rs + 4s^2$ for $r = -2$ and $s = 3$

37. $a^3 + 2abc - 3c^2$ for $a = 5, b = 9$, and $c = -1$

38. $a^2 - 2ab + 2c^2$ for $a = 3, b = 2$, and $c = -4$

39. $\dfrac{a^2 + ab}{3b}$ for $a = -1$ and $b = -2$

40. $\dfrac{x^2 - 2xy}{2y}$ for $x = -2$ and $y = -3$

Applications

▲ **41.** *Geometry* A sign is made in the shape of a parallelogram. The base measures 22 feet. The altitude measures 16 feet. What is the area of the sign?

▲ **42.** *Geometry* A field is shaped like a parallelogram. The base measures 92 feet. The altitude measures 54 feet. What is the area of the field?

▲ **43.** *TV Parts* A square support unit in a television is made with a side measuring 3 centimeters. A new model being designed for next year will have a larger square with a side measuring 3.2 centimeters. By how much will the area of the square be increased?

▲ **44.** *Computer Chips* A square computer chip for last year's computer had a side measuring 23 millimeters. This year the computer chip has been reduced in size. The new square chip has a side of 20 millimeters. By how much has the area of the chip decreased?

▲ **45.** *Carpentry* A carpenter cut out a small trapezoid as a wooden support for the front step. It has an altitude of 4 inches. One base of the trapezoid measures 9 inches and the other base measures 7 inches. What is the area of this support?

▲ **46.** *Signal Tower* The Comcast signal tower has a small trapezoid frame on the top of the tower. The frame has an altitude of 9 inches. One base of the trapezoid is 20 inches and the other base measures 17 inches. What is the area of this small trapezoidal frame?

▲ **47.** *Geometry* Bradley Palmer State Park has a triangular piece of land on the border. The altitude of the triangle is 400 feet. The base of the triangle is 280 feet. What is the area of this piece of land?

▲ **48.** *Roofing* The ceiling in the Madisons' house has a leak. The roofer exposed a triangular region that needs to be sealed and then reroofed. The region has an altitude of 14 feet. The base of the region is 19 feet. What is the area of the region that needs to be reroofed?

▲ **49.** *Geometry* The radius of a circular tablecloth is 3 feet. What is the area of the tablecloth?

▲ **50.** *Geometry* A new shot put ring is being constructed. The ring is required to have a concrete surface and a diameter of 8 feet. What will the area of the concrete ring be?

Temperature *For exercises 51 and 52, use the formula* $C = \dfrac{5}{9}(F - 32)$ *to find the Celsius temperature.*

51. Dry ice is solid carbon dioxide. Dry ice does not melt, but goes directly from the solid state to the gaseous state. Dry ice changes from a solid to a gas at $-109.3°F$. What is this temperature in Celsius?

52. Ivana was an exchange student from Russia, attending school in Montgomery, Alabama. Her host family told her to take a jacket to school because the temperature was not supposed to rise above 50°. Since Ivana was used to temperatures in Celsius, this made no sense at all, for 50°C is incredibly warm. The host family realized they needed to change the temperature into Celsius for Ivana. What was the temperature in Celsius?

Solve.

▲ **53.** *Sail Dimensions* Find the total cost of making a triangular sail that has a base dimension of 12 feet and a height of 20 feet if the price for making the sail is $19.50 per square foot.

▲ **54.** *Window Coating* A semicircular window of radius 15 inches is to be laminated with a sunblock coating that costs $0.85 per square inch to apply. What is the total cost of coating the window, to the nearest cent? (Use $\pi \approx 3.14$.)

55. *Temperature Extremes* The highest temperature ever recorded in Europe was 50° Celsius, and the lowest temperature was −55°C. What are the corresponding temperatures in degrees Fahrenheit? (Use the formula $F = \frac{9}{5}C + 32$.)

56. *Tour de France* The 2007 Tour de France was 3550 kilometers long and was completed in 20 stages. What was the average length of each stage in miles? Use the formula $r = 0.62k$, where r is the number of miles and k is the number of kilometers.

Cumulative Review *In exercises 57 and 58, simplify.*

57. **[1.5.1]** $(-2)^4 - 4 \div 2 - (-2)$

58. **[1.7.2]** $3(x - 2y) - (x^2 - y) - (x - y)$

Quick Quiz 1.8 Evaluate.

1. $2x^2 - 4x - 14$ for $x = -2$

2. $5a - 6b$ for $a = \frac{1}{2}$ and $b = -\frac{1}{3}$

3. $x^3 + 2x^2y + 5y + 2$ for $x = -2$ and $y = 3$

4. **Concept Check** Explain how you would find the area of a circle if you know its diameter is 12 meters.

1 Simplifying Algebraic Expressions by Removing Grouping Symbols

Many expressions in algebra use **grouping symbols** such as parentheses, brackets, and braces. Sometimes expressions are inside other expressions. Because it can be confusing to have more than one set of parentheses, brackets and braces are also used. How do we know what to do first when we see an expression like $2[5 - 4(a + b)]$?

To simplify the expression, we start with the innermost grouping symbols. Here it is a set of parentheses. We first use the distributive property to multiply.

$$2[5 - 4(a + b)] = 2[5 - 4a - 4b]$$

We use the distributive property again.

$$= 10 - 8a - 8b$$

There are no like terms, so this is our final answer.

Notice that we started with two sets of grouping symbols, but our final answer has none. So we can say we *removed* the grouping symbols. Of course, we didn't just take them away; we used the distributive property and the rules for real numbers to simplify as much as possible. Although simplifying expressions like this involves many steps, we sometimes say "remove parentheses" as a shorthand direction. Sometimes we say "simplify."

Remember to remove the innermost grouping symbols first. Keep working from the inside out.

EXAMPLE 1 Simplify. $3[6 - 2(x + y)]$

Solution We want to remove the innermost parentheses first. Therefore, we first use the distributive property to simplify $-2(x + y)$.

$$3[6 - 2(x + y)] = 3[6 - 2x - 2y] \quad \text{Use the distributive property.}$$
$$= 18 - 6x - 6y \quad \text{Use the distributive property again.}$$

Practice Problem 1 Simplify. $5[4x - 3(y - 2)]$

You recall that a negative sign in front of parentheses is equivalent to having a coefficient of negative 1. You can write the -1 and then multiply by -1 using the distributive property.

$$-(x + 2y) = -1(x + 2y) = -x - 2y$$

Notice that this has the effect of removing the parentheses. Each term in the result now has its sign changed.

Similarly, a positive sign in front of parentheses can be viewed as multiplication by $+1$.

$$+(5x - 6y) = +1(5x - 6y) = 5x - 6y$$

If a grouping symbol has a positive or negative sign in front, we mentally multiply by $+1$ or -1, respectively.

Fraction bars are also considered grouping symbols. In exercises 39 and 40 of Section 1.8, your first step was to simplify expressions above and below the fraction bars. Later in this book we will encounter further examples requiring the same first step. This type of operation will have some similarities to the operation of removing parentheses.

EXAMPLE 2 Simplify. $-2[3a - (b + 2c) + (d - 3e)]$

Solution

$= -2[3a - b - 2c + d - 3e]$ Remove the two innermost sets of parentheses. Since one is not inside the other, we remove both sets at once.

$= -6a + 2b + 4c - 2d + 6e$ Now we remove the brackets by multiplying each term by -2.

Practice Problem 2 Simplify. $-3[2a - (3b - c) + 4d]$

EXAMPLE 3 Simplify. $2[3x - (y + w)] - 3[2x + 2(3y - 2w)]$

Solution

$= 2[3x - y - w] - 3[2x + 6y - 4w]$ In each set of brackets, remove the inner parentheses.

$= 6x - 2y - 2w - 6x - 18y + 12w$ Remove each set of brackets by multiplying by the appropriate number.

$= -20y + 10w$ or $10w - 20y$ Combine like terms. (Note that $6x - 6x = 0x = 0$.)

Practice Problem 3 Simplify. $3[4x - 2(1 - x)] - [3x + (x - 2)]$

You can always simplify problems with many sets of grouping symbols by the method shown. Essentially, you just keep removing one level of grouping symbols at each step. Finally, at the end you combine the like terms if possible.

Sometimes it is possible to combine like terms at each step.

EXAMPLE 4 Simplify. $-3\{7x - 2[x - (2x - 1)]\}$

Solution

$= -3\{7x - 2[x - 2x + 1]\}$ Remove the inner parentheses by multiplying each term within the parentheses by -1.

$= -3\{7x - 2[-x + 1]\}$ Combine like terms by combining $+x - 2x$.

$= -3\{7x + 2x - 2\}$ Remove the brackets by multiplying each term within them by -2.

$= -3\{9x - 2\}$ Combine the x terms.

$= -27x + 6$ Remove the braces by multiplying each term by -3.

Practice Problem 4 Simplify. $-2\{5x - 3[2x - (3 - 4x)]\}$

Developing Your Study Skills

The Night Before an Exam

With adequate preparation, you can spend the night before an exam pulling together the final details.

1. Look over each section to be covered in the exam. Review the steps needed to solve each type of problem.

2. Review your list of terms, rules, and formulas that you are expected to know for the exam.

3. Take the Practice Test at the end of the chapter just as though you were taking the actual exam. Do not look in your text or get help in any way. Time yourself so that you know how long it takes you to complete the test.

4. Check the Practice Test. Redo the problems you missed.

5. Be sure you have ready the necessary supplies for taking your exam.

Verbal and Writing Skills

1. Rewrite the expression $-3x - 2y$ using a negative sign and parentheses.

2. Rewrite the expression $-x + 5y$ using a negative sign and parentheses.

3. To simplify expressions with grouping symbols, we use the _____ property.

4. When an expression contains many grouping symbols, remove the _____ grouping symbol first.

Simplify. Remove grouping symbols and combine like terms.

5. $6x - 3(x - 2y)$

6. $-4x - 2(y - 3x)$

7. $5(c - 3d) - (3c + d)$

8. $6(2d - c) - (9d + c)$

9. $-3(x + 3y) + 2(2x + y)$

10. $-4(a + 2b) + 5(2b - a)$

11. $2x[4x^2 - 2(x - 3)]$

12. $4y[-3y^2 + 2(4 - y)]$

13. $2[5(x + y) - 2(3x - 4y)]$

14. $-3[2(3a + b) - 5(a - 2b)]$

15. $[10 - 4(x - 2y)] + 3(2x + y)$

16. $[-5(-x + 3y) - 12] - 4(2x - 3)$

17. $5[3a - 2a(3a + 6b) + 6a^2]$

18. $3[x - y(3x + y) + y^2]$

19. $6a(2a^2 - 3a - 4) - a(a - 2)$

20. $7b(3b^2 - 2b - 5) - 2b(4 - b)$

21. $3a^2 - 4[2b - 3b(b + 2)]$

22. $2b^2 - 3[5b + 2b(2 - b)]$

23. $5b + \{-[3a + 2(5a - 2b)] - 1\}$

24. $8a - \{-2[a - 3(2a + b)] - 5\}$

25. $3\{3b^2 + 2[5b - (2 - b)]\}$

26. $2\{3x^2 + 4[2x - (3 - x)]\}$

27. $-4\{3a^2 - 2[4a^2 - (b + a^2)]\}$

28. $-2\{x^2 - 3[x - (x - 2x^2)]\}$

Cumulative Review

29. **[1.8.2]** *Melting Point of Gold* The melting point of pure gold is 1064.18°C. Use $F = 1.8C + 32$ to find the melting point of pure gold in degrees Fahrenheit. Round to the nearest hundredth of a degree.

▲ **30.** **[1.8.2]** *Geometry* Use 3.14 as an approximation for π to compute the area covered by a circular irrigation system with radial arm of length 380 feet. Use $A = \pi r^2$.

31. **[1.8.2]** *Dog Weight* An average Great Dane weighs between 120 and 150 pounds. Express the range of weight for a Great Dane in kilograms. Use the formula $k = 0.45p$ (where k = kilograms, p = pounds).

32. **[1.8.2]** *Dog Weight* An average Miniature Pinscher weighs between 9 and 14 pounds. Express the range of weight for a Miniature Pinscher in kilograms. Use the formula $k = 0.45p$ (where k = kilograms, p = pounds).

Quick Quiz 1.9 Simplify.

1. $2[3x - 2(5x + y)]$

2. $3[x - 3(x + 4) + 5y]$

3. $-4\{2a + 2[2ab - b(1 - a)]\}$

4. **Concept Check** Explain how you would simplify the following expression if you collect like terms whenever possible. $3\{2 - 3[4x - 2(x + 3) + 5x]\}$

Putting Your Skills to Work: Use Math to Save Money

TIME TO BUDGET

One way to improve your financial situation is to learn to manage the money you have with a budget. A budget can maximize your efforts to ensure you have enough money to cover your fixed expenses (including housing costs, insurance payments, taxes, credit or loan payments, and savings) as well as your variable expenses (including food, education expenses, clothing expenses, and entertainment). Consider the story of Michael.

Going Back to School

Michael is a teacher in Revere, Massachusetts. Many states, including Massachusetts, require teachers to earn an advanced degree to maintain their teaching certificates. One of Michael's goals is to go back to school to earn his Master's Degree in Education. He knows that this will not only help him further his career, but also provide better financial stability in the long run. His net monthly income, after deductions for items such as taxes and insurance, is currently $2500. So, Michael knows he'll need to put himself on a budget for a period of time in order to save money to go back to school. Michael's research shows that consumer credit counseling services recommend allocating the following percentages for each category of the monthly budget:

Housing	25%
Transportation	10%
Savings	5%
Utilities	5%
Medical	5%
Debt Payments	20%
Food	15%
Clothing	5%
Misc.	10%

1. If Michael follows these recommendations, how much will he have saved at the end of one year?

After investigating several schools in his area, Michael chooses to attend a state college that offers a one-year program for the degree he wishes to pursue. Michael will need $3000 for tuition and fees the first semester plus $450 for textbooks.

2. What is the total cost for each semester?

3. How much will Michael need to save for two semesters of tuition and fees and textbooks?

4. How many months will it take Michael to save the entire amount needed to complete his degree?

5. How much will Michael need to save per month to have the total tuition cost in two years?

6. Michael decided to cut back on some of his variable monthly expenses such as entertainment so that he could increase his savings to 10% of his net monthly income instead of 5%. How much will he now save per month? How many months will it take Michael to save the entire amount needed for college?

7. Michael's friend told him he could save about 50% on textbook costs if he bought eBooks. If he plans to buy eBooks each semester, how many fewer months would it take him to save the entire amount needed for college?

Once Michael earns his advanced degree, his salary will increase on the following schedule:

    Year 1 Michael will earn an additional $4200
    Year 2 Michael will earn an additional $4800
    Year 3 Michael will earn an additional $5200
    Year 4 Michael will earn an additional $5500
    Year 5 Michael will earn an additional $5800

8. Over 5 years, how much more money will Michael earn because he went back to school to get an advanced degree?

Making It Personal for You

9. Do you have a budget?

10. How would you adjust Michael's budget to fit your needs?

11. What advice would you give Michael for achieving his goal?

Topic	Procedure	Examples
Absolute value, p. 60.	The absolute value of a number is the distance between that number and zero on the number line. The absolute value of any number will be positive or zero.	$\|3\| = 3$ $\|-2\| = 2$ $\|0\| = 0$ $\left\|-\dfrac{5}{6}\right\| = \dfrac{5}{6}$ $\|-1.38\| = 1.38$
Adding real numbers with the same sign, p. 61.	If the signs are the same, add the absolute values of the numbers. Use the common sign in the answer.	$-3 + (-7) = -10$
Adding real numbers with different signs, p. 62.	If the signs are different: **1.** Find the difference between the larger and the smaller absolute values. **2.** Give the answer the sign of the number having the larger absolute value.	$(-7) + 13 = 6$ $7 + (-13) = -6$
Adding several real numbers, p. 64.	When adding several real numbers, separate them into two groups by sign. Find the sum of all the positive numbers and the sum of all the negative numbers. Combine these two subtotals by the method described above.	$-7 + 6 + 8 + (-11) + (-13) + 22$ $\begin{array}{rr} -7 & +6 \\ -11 & +8 \\ \underline{-13} & \underline{+22} \\ -31 & +36 \end{array}$ $-31 + 36 = 5$ The answer is positive since 36 is positive.
Subtracting real numbers, p. 69.	Add the opposite of the second number.	$-3 - (-13) = -3 + (+13) = 10$
Multiplying and dividing real numbers, pp. 74, 77.	**1.** If the two numbers have the same sign, multiply (or divide) the absolute values. The result is positive. **2.** If the two numbers have different signs, multiply (or divide) the absolute values. The result is negative.	$-5(-3) = +15$ $-36 \div (-4) = +9$ $28 \div (-7) = -4$ $-6(3) = -18$
Exponent form, p. 84.	The base tells you what number is being multiplied. The exponent tells you how many times this number is used as a factor.	$2^5 = 2 \cdot 2 \cdot 2 \cdot 2 \cdot 2 = 32$ $4^3 = 4 \cdot 4 \cdot 4 = 64$ $(-3)^4 = (-3)(-3)(-3)(-3) = 81$
Raising a negative number to a power, p. 85.	When the base is negative, the result is positive for even exponents and negative for odd exponents.	$(-3)^3 = -27$ but $(-2)^4 = 16$
Order of operations, p. 89.	Remember the proper order of operations: **1.** Perform operations inside parentheses. **2.** Raise to powers. **3.** Multiply and divide from left to right. **4.** Add and subtract from left to right.	$3(5 + 4)^2 - 2^2 \cdot 3 \div (9 - 2^3)$ $= 3 \cdot 9^2 - 4 \cdot 3 \div (9 - 8)$ $= 3 \cdot 81 - 12 \div 1$ $= 243 - 12 = 231$
Removing parentheses, p. 94.	Use the distributive property to remove parentheses. $a(b + c) = ab + ac$	$3(5x + 2) = 15x + 6$ $-4(x - 3y) = -4x + 12y$

(Continued on next page)

Topic	Procedure	Examples
Combining like terms, p. 99.	Combine terms that have identical variables and exponents.	$7x^2 - 3x + 4y + 2x^2 - 8x - 9y = 9x^2 - 11x - 5y$
Substituting into variable expressions, p. 104.	1. Replace each variable by the numerical value given for it. 2. Follow the order of operations in evaluating the expression.	Evaluate $2x^3 + 3xy + 4y^2$ for $x = -3$ and $y = 2$. $2(-3)^3 + 3(-3)(2) + 4(2)^2$ $\quad = 2(-27) + 3(-3)(2) + 4(4)$ $\quad = -54 - 18 + 16$ $\quad = -56$
Using formulas, p. 105.	1. Replace the variables in the formula by the given values. 2. Evaluate the expression. 3. Label units carefully.	Find the area of a circle with radius 4 feet. Use $A = \pi r^2$, with π as approximately 3.14. $A \approx (3.14)(4 \text{ ft})^2$ $\quad = (3.14)(16 \text{ ft}^2)$ $\quad = 50.24 \text{ ft}^2$ The area of the circle is approximately 50.24 square feet.
Removing grouping symbols, p. 112.	1. Remove innermost grouping symbols first. 2. Then remove remaining innermost grouping symbols. 3. Continue until all grouping symbols are removed. 4. Combine like terms.	$5\{3x - 2[4 + 3(x - 1)]\}$ $\quad = 5\{3x - 2[4 + 3x - 3]\}$ $\quad = 5\{3x - 8 - 6x + 6\}$ $\quad = 15x - 40 - 30x + 30$ $\quad = -15x - 10$

Chapter 1 Review Problems

Section 1.1

Add.

1. $-6 + (-2)$ **2.** $-12 + 7.8$ **3.** $5 + (-2) + (-12)$ **4.** $3.7 + (-1.8)$

5. $\dfrac{1}{2} + \left(-\dfrac{5}{6}\right)$ **6.** $-\dfrac{3}{11} + \left(-\dfrac{1}{22}\right)$ **7.** $\dfrac{3}{4} + \left(-\dfrac{1}{12}\right) + \left(-\dfrac{1}{2}\right)$ **8.** $\dfrac{2}{15} + \dfrac{1}{6} + \left(-\dfrac{4}{5}\right)$

Section 1.2

Add or subtract.

9. $5 - (-3)$ **10.** $-2 - (-15)$ **11.** $-30 - (+3)$ **12.** $8 - (-1.2)$

13. $-\dfrac{7}{8} + \left(-\dfrac{3}{4}\right)$ **14.** $-\dfrac{3}{8} + \dfrac{5}{6}$ **15.** $-20.8 - 1.9$ **16.** $-151 - (-63)$

Section 1.3

17. $87 \div (-29)$ **18.** $-10.4 \div (-0.8)$ **19.** $\dfrac{-24}{-\dfrac{3}{4}}$ **20.** $-\dfrac{2}{3} \div \left(-\dfrac{4}{5}\right)$

21. $\dfrac{5}{7} \div \left(-\dfrac{5}{25}\right)$ **22.** $-6(3)(4)$ **23.** $-1(-4)(-3)(-5)$ **24.** $(-5)\left(-\dfrac{1}{2}\right)(4)(-3)$

Mixed Practice

Sections 1.1–1.3

Perform the operations indicated. Simplify all answers.

25. $-5 + (-2) - (-3)$ 　　　**26.** $6 - (-4) + (-2) + 8$ 　　　**27.** $-16 + (-13)$ 　　　**28.** $-11 - (-12)$

29. $-\dfrac{4}{3} + \dfrac{2}{3} + \dfrac{1}{6}$ 　　　**30.** $-\dfrac{6}{7} + \dfrac{1}{2} + \left(-\dfrac{3}{14}\right)$ 　　　**31.** $-3(-2)(-5)$ 　　　**32.** $-6 + (-2) - (-3)$

33. $3.5(-2.6)$ 　　　**34.** $-5.4 \div (-6)$ 　　　**35.** $5 - (-3.5) + 1.6$ 　　　**36.** $-8 + 2 - (-4.8)$

37. $17 + 3.4 + (-16) + (-2.5)$ 　　　**38.** $37 + (-44) + 12.5 + (-6.8)$

Solve.

39. *Football* The Dallas Cowboys football team had three plays in which they lost 8 yards each time. What was the total yardage lost?

40. *Temperature Change* The low temperature in Anchorage, Alaska, last night was $-34°F$. During the day the temperature rose $12°F$. What was the temperature during the day?

41. *Elevation* The highest point in California is Mount Whitney, at 14,494 feet above sea level. The lowest point in California is in Death Valley, at 282 feet below sea level. What is the difference in height between these two elevations?

42. *Stock Prices* One day in 2007, the value of Stock A rose $2.28 on Monday, rose $2.45 on Tuesday, and fell $6.58 on Wednesday. The value of Stock B fell $3.35 on Monday, rose $2.75 on Tuesday, and fell $1.84 on Wednesday. Which stock had the greatest loss over the three days?

Section 1.4

Evaluate.

43. $(-3)^5$ 　　　**44.** $(-2)^6$ 　　　**45.** $(-5)^4$ 　　　**46.** $\left(-\dfrac{2}{3}\right)^3$

47. -9^2 　　　**48.** $(0.6)^2$ 　　　**49.** $\left(\dfrac{5}{6}\right)^2$ 　　　**50.** $\left(\dfrac{3}{4}\right)^3$

Section 1.5

Simplify using the order of operations.

51. $5(-4) + 3(-2)^3$ 　　　**52.** $20 - (-10) - (-6) + (-5) - 1$ 　　　**53.** $(3 - 6)^2 + (-12) + -3(-2)$

Section 1.6

Use the distributive property to multiply.

54. $7(-3x + y)$ 　　　**55.** $3x(6 - x + 3y)$ 　　　**56.** $-(7x^2 - 3x + 11)$ 　　　**57.** $(2xy + x - y)(-3y^2)$

Section 1.7

Combine like terms.

58. $3a^2b - 2bc + 6bc^2 - 8a^2b - 6bc^2 + 5bc$ 　　　**59.** $9x + 11y - 12x - 15y$

60. $4x^2 - 13x + 7 - 9x^2 - 22x - 16$

61. $-x + \dfrac{1}{2} + 14x^2 - 7x - 1 - 4x^2$

Section 1.8

Evaluate for the given value of the variable.

62. $7x - 6$ for $x = -7$

63. $7 - \dfrac{3}{4}x$ for $x = 8$

64. $x^2 + 3x - 4$ for $x = -3$

65. $-x^2 + 5x - 9$ for $x = 3$

66. $2x^3 - x^2 + 6x + 9$ for $x = -1$

67. $b^2 - 4ac$ for $a = -1, b = 5,$ and $c = -2$

68. $\dfrac{mMG}{r^2}$ for $m = -4, M = 15, G = -1,$ and $r = -2$

Solve.

69. *Simple Interest* Find the simple interest on a loan of $6000 at an annual interest rate of 18% per year for $\frac{3}{4}$ of a year. Use $I = prt$, where p = principal, r = rate per year, and t = time in years.

70. *Medication* The label of a medication warns the user that it must be stored at a temperature between 20°C and 25°C. What is this temperature range in degrees Fahrenheit? Use the formula $F = \dfrac{9C + 160}{5}$.

▲ **71.** *Sign Painting* How much will it cost to paint a circular sign with a radius of 4 feet if the painter charges $1.50 per square foot? Use $A = \pi r^2$, where π is approximately 3.14.

72. *Profit* Find the daily profit P at a furniture factory if the initial cost of setting up the factory $C = \$1200$, rent $R = \$300$, and sale price of furniture $S = \$56$. Use the profit formula $P = 180S - R - C$.

▲ **73.** *Parking Lot Sealer* A parking lot is in the shape of a trapezoid. The altitude of the trapezoid is 200 feet, and the bases of the trapezoid are 300 feet and 700 feet. What is the area of the parking lot? If the parking lot had a sealer applied that costs $2 per square foot, what was the cost of the amount of sealer needed for the entire parking lot?

▲ **74.** *Signal Paint* The Green Mountain Telephone Company has a triangular signal tester at the top of a communications tower. The altitude of the triangle is 3.8 feet and the base is 5.5 feet. What is the area of the triangular signal tester? If the signal tester was painted with a special metallic surface paint that costs $66 per square foot, what was the cost of the amount of paint needed to paint one side of the triangle?

Section 1.9

Simplify.

75. $5x - 7(x - 6)$

76. $3(x - 2) - 4(5x + 3)$

77. $2[3 - (4 - 5x)]$

78. $-3x[x + 3(x - 7)]$

79. $2xy^3 - 6x^3y - 4x^2y^2 + 3(xy^3 - 2x^2y - 3x^2y^2)$

80. $-5(x + 2y - 7) + 3x(2 - 5y)$

81. $-(a + 3b) + 5[2a - b - 2(4a - b)]$

82. $-5\{2a - [5a - b(3 + 2a)]\}$

83. $-3\{2x - [x - 3y(x - 2y)]\}$

84. $2\{3x + 2[x + 2y(x - 4)]\}$

Mixed Practice

Simplify the following.

85. $-6.3 + 4$

86. $4 + (-8) + 12$

87. $-\dfrac{2}{3} - \dfrac{4}{5}$

88. $-\dfrac{7}{8} - \left(-\dfrac{3}{4}\right)$

89. $3 - (-4) + (-8)$

90. $-1.1 - (-0.2) + 0.4$

91. $\left(-\dfrac{9}{10}\right)\left(-2\dfrac{1}{4}\right)$

92. $3.6 \div (-0.45)$

93. $-14.4 \div (-0.06)$

94. $(-8.2)(3.1)$

95. *Jeopardy* A Jeopardy quiz show contestant began the second round (Double Jeopardy) with $400. She buzzed in on the first two questions, answering a $1000 question correctly but then giving the incorrect answer to an $800 question. What was her score?

Simplify the following.

96. $(-0.3)^4$

97. -0.5^4

98. $9(5) - 5(2)^3 + 5$

99. $3.8x - 0.2y - 8.7x + 4.3y$

100. Evaluate $\dfrac{2p + q}{3q}$ for $p = -2$ and $q = 3$.

101. Evaluate $\dfrac{4s - 7t}{s}$ for $s = -3$ and $t = -2$.

102. *Dog Body Temperature* The normal body temperature of a dog is 38.6°C. Your dog has a temperature of 101.1°F. Does your dog have a fever? Use the formula $F = \dfrac{9}{5}C + 32$ to convert normal temperature in Fahrenheit.

103. $-7(x - 3y^2 + 4) + 3y(4 - 6y)$

104. $-2\{6x - 3[7y - 2y(3 - x)]\}$

How Am I Doing? Chapter 1 Test

Remember to use your Chapter Test Prep Video CD to see the worked-out solutions to the test problems you want to review

Simplify.

1. $-2.5 + 6.3 + (-4.1)$
2. $-5 - (-7)$
3. $\left(-\dfrac{2}{3}\right)(7)$

4. $-5(-2)(7)(-1)$
5. $-12 \div (-3)$
6. $-1.8 \div (0.6)$

7. $(-4)^3$
8. $(1.6)^2$
9. $\left(\dfrac{2}{3}\right)^4$

10. $(0.2)^2 - (2.1)(-3) + 0.46$
11. $3(4 - 6)^3 + 12 \div (-4) + 2$

12. $-5x(x + 2y - 7)$
13. $-2ab^2(-3a - 2b + 7ab)$

14. $6ab - \dfrac{1}{2}a^2b + \dfrac{3}{2}ab + \dfrac{5}{2}a^2b$
15. $2.3x^2y - 8.1xy^2 + 3.4xy^2 - 4.1x^2y$

16. $3(2 - a) - 4(-6 - 2a)$
17. $5(3x - 2y) - (x + 6y)$

In questions 18–20, evaluate for the value of the variable indicated.

18. $x^3 - 3x^2y + 2y - 5$ for $x = 3$ and $y = -4$

19. $3x^2 - 7x - 11$ for $x = -3$

20. $2a - 3b$ for $a = \dfrac{1}{3}$ and $b = -\dfrac{1}{2}$

21. If you are traveling 60 miles per hour on a highway in Canada, how fast are you traveling in kilometers per hour? (Use $k = 1.61r$, where r = rate in miles per hour and k = rate in kilometers per hour.)

▲22. A field is in the shape of a trapezoid. The altitude of the trapezoid is 120 feet and the bases of the trapezoid are 180 feet and 200 feet. What is the area of the field?

▲23. Jeff Slater's garage has a triangular roof support beam. The support beam is covered with a sheet of plywood. The altitude of the triangular region is 6.8 feet and the base is 8.5 feet. If the triangular piece of plywood was painted with paint that cost $0.80 per square foot, what was the cost of the amount of paint needed to coat one side of the triangle?

▲24. You wish to apply blacktop sealer to your driveway, but do not know how much to buy. If your rectangular driveway measures 60 feet long by 10 feet wide, and a can of blacktop sealer claims to cover 200 square feet, how many cans should you buy?

Simplify.

25. $3[x - 2y(x + 2y) - 3y^2]$
26. $-3\{a + b[3a - b(1 - a)]\}$

Answer lines:
1.
2.
3.
4.
5.
6.
7.
8.
9.
10.
11.
12.
13.
14.
15.
16.
17.
18.
19.
20.
21.
22.
23.
24.
25.
26.

CHAPTER

2

In states where there is significant snowfall in the winter, much of the snowplowing is done by four-wheel-drive pickup trucks with an adjustable snowplow attached to the front. With rising fuel prices, the owners of these trucks are concerned about gasoline consumption while operating the snowplows. Most of these trucks can operate in four-wheel drive or two-wheel drive. Those who operate snowplowing services can use the mathematics presented in this chapter to determine when it is more fuel efficient to operate snowplows using two-wheel drive and when it is actually more fuel efficient to operate them with four-wheel drive. In general, owners of small businesses will find the mathematics of Chapter 2 to be most useful in making wise business decisions.

Equations, Inequalities, and Applications

Student Learning Objective

After studying this section, you will be able to:

 Use the addition principle to solve equations of the form $x + b = c$.

 Using the Addition Principle to Solve Equations of the Form $x + b = c$

When we use an equals sign ($=$), we are indicating that two expressions are equal in value. Such a statement is called an **equation.** For example, $x + 5 = 23$ is an equation. A **solution** of an equation is a number that when substituted for the variable makes the equation true. Thus 18 is a solution of $x + 5 = 23$ because $18 + 5 = 23$. Equations that have exactly the same solutions are called **equivalent equations.** By following certain procedures, we can often transform an equation to a simpler, equivalent one that has the form $x = $ some number. Then this number is a solution of the equation. The process of finding all solutions of an equation is called **solving the equation.**

One of the first procedures used in solving equations has an application in our everyday world. Suppose that we place a 10-kilogram box on one side of a seesaw and a 10-kilogram stone on the other side. If the center of the box is the same distance from the balance point as the center of the stone, we would expect the seesaw to balance. The box and the stone do not look the same, but their weights are equal. If we add a 2-kilogram lead weight to the center of weight of each object at the same time, the seesaw should still balance. The weights are still equal.

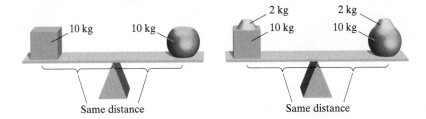

There is a similar principle in mathematics. We can state it in words as follows.

> **THE ADDITION PRINCIPLE**
>
> If the same number is added to both sides of an equation, the results on both sides are equal in value.

We can restate it in symbols this way.

> For real numbers a, b, and c, if $a = b$, then $a + c = b + c$.

Here is an example.

$$\text{If } 3 = \frac{6}{2}, \quad \text{then } 3 + 5 = \frac{6}{2} + 5.$$

Since we added the same amount, 5, to both sides, the sides remain equal to each other.

$$3 + 5 = \frac{6}{2} + 5$$

$$8 = \frac{6}{2} + \frac{10}{2}$$

$$8 = \frac{16}{2}$$

$$8 = 8$$

We can use the addition principle to solve certain equations.

EXAMPLE 1 Solve for x. $x + 16 = 20$

Solution $x + 16 + (-16) = 20 + (-16)$ Use the addition principle to add -16 to both sides.

$$x + 0 = 4$$ Simplify.

$$x = 4$$ The value of x is 4.

We have just found a solution of the equation. A **solution** is a value for the variable that makes the equation true. We then say that the value 4 in our example **satisfies** the equation. We can easily verify that 4 is a solution by substituting this value into the original equation. This step is called **checking** the solution.

Check.

$$x + 16 = 20$$
$$4 + 16 \overset{?}{=} 20$$
$$20 = 20 \checkmark$$

When the same value appears on both sides of the equals sign, we call the equation an **identity.** Because the two sides of the equation in our check have the same value, we know that the original equation has been solved correctly. We have found a solution, and since no other number makes the equation true, it is the only solution.

Practice Problem 1 Solve for x and check your solution. $x + 14 = 23$

NOTE TO STUDENT: Fully worked-out solutions to all of the Practice Problems can be found at the back of the text starting at page SP-1

Notice that when you are trying to solve these types of equations, you must add a particular number to both sides of the equation. What is the number to choose? Look at the number that is on the same side of the equation with x, that is, the number added to x. Then think of the number that is **opposite in sign.** This is called the **additive inverse** of the number. The additive inverse of 16 is -16. The additive inverse of -3 is 3. The number to add to both sides of the equation is precisely this additive inverse.

It does not matter which side of the equation contains the variable. The x-term may be on the right or left. In the next example the x-term will be on the right.

EXAMPLE 2 Solve for x. $14 = x - 3$

Solution $14 + 3 = x - 3 + 3$ Notice that -3 is being added to x in the original equation. Add 3 to both sides, since 3 is the additive inverse of -3. This will eliminate the -3 on the right and isolate x.

$$17 = x + 0$$ Simplify.

$$17 = x$$ The value of x is 17.

Check.

$$14 = x - 3$$
$$14 \overset{?}{=} 17 - 3$$ Replace x by 17.
$$14 = 14 \checkmark$$ Simplify. It checks. The solution is 17.

Practice Problem 2 Solve for x and check your solution. $17 = x - 5$

Before you add a number to both sides, you should always simplify the equation. The following example shows how combining numbers by addition—separately, on both sides of the equation—simplifies the equation.

EXAMPLE 3 Solve for x. $1.5 + 0.2 = 0.3 + x + 0.2$

Solution

$1.7 = x + 0.5$	Simplify by adding.
$1.7 + (-0.5) = x + 0.5 + (-0.5)$	Add the value -0.5 to both sides, since -0.5 is the additive inverse of 0.5.
$1.2 = x$	Simplify. The value of x is 1.2.

Check.

$1.5 + 0.2 = 0.3 + x + 0.2$	
$1.5 + 0.2 \stackrel{?}{=} 0.3 + 1.2 + 0.2$	Replace x by 1.2 in the original equation.
$1.7 = 1.7$ ✓	It checks.

Practice Problem 3 Solve for x and check your solution.

$$0.5 - 1.2 = x - 0.3$$

In Example 3 we added -0.5 to each side. You could subtract 0.5 from each side and get the same result. In Chapter 1 we discussed how subtracting a 0.5 is the same as adding a negative 0.5. Do you see why?

Just as it is possible to add the same number to both sides of an equation, it is also possible to subtract the same number from both sides of an equation. This is so because any subtraction problem can be rewritten as an addition problem. For example, $1.7 - 0.5 = 1.7 + (-0.5)$. Thus the addition principle tells us that we can subtract the same number from both sides of the equation.

We can determine whether a value is the solution to an equation by following the same steps used to check an answer. Substitute the value to be tested for the variable in the original equation. We will obtain an identity if the value is the solution.

EXAMPLE 4 Is 10 the solution to the equation $-15 + 2 = x - 3$? If it is not, find the solution.

Solution We substitute 10 for x in the equation and see if we obtain an identity.

$-15 + 2 = x - 3$	
$-15 + 2 \stackrel{?}{=} 10 - 3$	
$-13 \neq 7$	The values are not equal. The statement is not an identity.

Thus, 10 is not the solution. Now we take the original equation and solve to find the solution.

$-15 + 2 = x - 3$	
$-13 = x - 3$	Simplify by adding.
$-13 + 3 = x - 3 + 3$	Add 3 to both sides. 3 is the additive inverse of -3.
$-10 = x$	

Check to see if -10 is the solution. The value 10 was incorrect because of a sign error. We must be especially careful to write the correct sign for each number when solving equations.

Practice Problem 4 Is -2 the solution to the equation $x + 8 = -22 + 6$? If it is not, find the solution.

EXAMPLE 5   Find the value of x that satisfies the equation.

$$\frac{1}{5} + x = -\frac{1}{10} + \frac{1}{2}$$

Solution   To be combined, the fractions must have common denominators. The least common denominator (LCD) of the fractions is 10.

$$\frac{1 \cdot 2}{5 \cdot 2} + x = -\frac{1}{10} + \frac{1 \cdot 5}{2 \cdot 5} \qquad \text{Change each fraction to an equivalent fraction with a denominator of 10.}$$

$$\frac{2}{10} + x = -\frac{1}{10} + \frac{5}{10} \qquad \text{This is an equivalent equation.}$$

$$\frac{2}{10} + x = \frac{4}{10} \qquad \text{Simplify by adding.}$$

$$\frac{2}{10} + \left(-\frac{2}{10}\right) + x = \frac{4}{10} + \left(-\frac{2}{10}\right) \qquad \text{Add the additive inverse of } \frac{2}{10} \text{ to each side. You could also say that you are subtracting } \frac{2}{10} \text{ from each side.}$$

$$x = \frac{2}{10} \qquad \text{Add the fractions.}$$

$$x = \frac{1}{5} \qquad \text{Simplify the answer.}$$

Check.   We substitute $\frac{1}{5}$ for x in the original equation and see if we obtain an identity.

$$\frac{1}{5} + x = -\frac{1}{10} + \frac{1}{2}$$

$$\frac{1}{5} + \frac{1}{5} \overset{?}{=} -\frac{1}{10} + \frac{1}{2} \qquad \text{Substitute } \frac{1}{5} \text{ for } x.$$

$$\frac{2}{5} \overset{?}{=} -\frac{1}{10} + \frac{5}{10}$$

$$\frac{2}{5} \overset{?}{=} \frac{4}{10}$$

$$\frac{2}{5} = \frac{2}{5} \checkmark \qquad \text{It checks.}$$

Practice Problem 5   Find the value of x that satisfies the equation.

$$\frac{1}{20} - \frac{1}{2} = x + \frac{3}{5}$$

NOTE TO STUDENT: Fully worked-out solutions to all of the Practice Problems can be found at the back of the text starting at page SP-1

Developing Your Study Skills

Why Study Mathematics?

In our present-day, technological world, it is easy to see mathematics at work. Many vocational and professional areas—such as the fields of business, statistics, economics, psychology, finance, computer science, chemistry, physics, engineering, electronics, nuclear energy, banking, quality control, and teaching—require a certain level of expertise in mathematics. Those who want to work in these fields must be able to function at a given mathematical level. Those who cannot will not make it. So if your field of study requires you to take higher-level mathematics courses, be sure to master the topics of this course. Then you will be ready for the next one.

PRACTICE   WATCH   DOWNLOAD   READ   REVIEW

Verbal and Writing Skills *In exercises 1–3, fill in each blank with the appropriate word.*

1. When we use the _____ sign, we indicate two expressions are _____ in value.

2. If the _____ _____ is added to both sides of an equation, the results on each side are equal in value.

3. The _____ of an equation is a value of the variable that makes the equation true.

4. What is the additive inverse of -20?

5. Why do we add the additive inverse of a to each side of $x + a = b$ to solve for x?

6. What is the additive inverse of a?

Solve for x. Check your answers.

7. $x + 11 = 15$

8. $x + 12 = 18$

9. $20 = 9 + x$

10. $18 = 10 + x$

11. $x - 3 = 14$

12. $x - 11 = 5$

13. $0 = x + 5$

14. $0 = x - 7$

15. $x - 6 = -19$

16. $x - 11 = -13$

17. $-12 + x = 50$

18. $-18 + x = 48$

19. $3 + 5 = x - 7$

20. $8 - 2 = x + 5$

21. $32 - 17 = x - 6$

22. $27 - 12 = x - 9$

23. $4 + 8 + x = 6 + 6$

24. $19 - 3 + x = 10 + 6$

25. $18 - 7 + x = 7 + 9 - 5$

26. $3 - 17 + 8 = 8 + x - 3$

27. $-12 + x - 3 = 15 - 18 + 9$

28. $-19 + x - 7 = 20 - 42 + 10$

In exercises 29–36, determine whether the given solution is correct. If it is not, find the solution.

29. Is $x = 5$ the solution to $-7 + x = 2$?

30. Is $x = 7$ the solution to $-13 + x = 4$?

31. Is -6 the solution to $-11 + 5 = x + 8$?

32. Is -9 the solution to $-13 - 4 = x - 8$?

33. Is -33 the solution to $x - 23 = -56$?

34. Is -8 the solution to $-39 = x - 47$?

35. Is 35 the solution to $15 - 3 + 20 = x - 3$?

36. Is -12 the solution to $x + 8 = 12 - 19 + 3$?

Find the value of x that satisfies each equation.

37. $2.5 + x = 0.7$

38. $8.2 + x = 3.2$

39. $12.5 + x - 8.2 = 4.9$

40. $4.3 + x - 2.6 = 3.4$

41. $x - \dfrac{1}{4} = \dfrac{3}{4}$

42. $x + \dfrac{1}{3} = \dfrac{2}{3}$

43. $\dfrac{2}{3} + x = \dfrac{1}{6} + \dfrac{1}{4}$

44. $\dfrac{2}{5} + x = \dfrac{1}{2} - \dfrac{3}{10}$

Mixed Practice   *Solve for x.*

45. $3 + x = -12 + 8$

46. $12 + x = -7 + 20$

47. $5\dfrac{1}{6} + x = 8$

48. $7\dfrac{1}{8} = -20 + x$

49. $\dfrac{5}{12} - \dfrac{5}{6} = x - \dfrac{3}{2}$

50. $\dfrac{4}{15} - \dfrac{3}{5} = x - \dfrac{4}{3}$

51. $1.6 + x - 3.2 = -2 + 5.6$

52. $0.7 - 4.2 = 3.6 + x$

53. $x - 18.225 = 1.975$

54. $x - 10.012 = -16.835$

Cumulative Review   *Simplify by adding like terms.*

55. **[1.7.2]** $x + 3y - 5x - 7y + 2x$

56. **[1.7.2]** $y^2 + y - 12 - 3y^2 - 5y + 16$

Quick Quiz 2.1   Solve for the variable.

1. $x - 4.7 = 9.6$

2. $-8.6 + x = -12.1$

3. $3 - 12 + 7 = 8 + x - 2$

4. **Concept Check** Explain how you would check to verify whether $x = 3.8$ is the solution to $-1.3 + 1.6 + 3x = -6.7 + 4x + 3.2$.

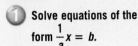

Student Learning Objectives

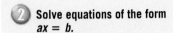

After studying this section, you will be able to:

1 Solve equations of the form $\frac{1}{a}x = b$.

2 Solve equations of the form $ax = b$.

 Solving Equations of the Form $\frac{1}{a}x = b$

The addition principle allows us to add the same number to both sides of an equation. What would happen if we multiplied each side of an equation by the same number? For example, what would happen if we multiplied each side of an equation by 3?

To answer this question, let's return to our simple example of the box and the stone on a balanced seesaw. If we triple the weight on each side (that is, multiply the weight on each side by 3), the seesaw should still balance. The weight values of both sides remain equal.

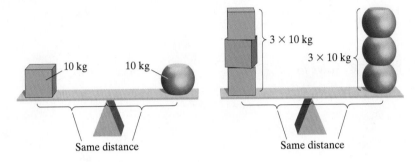

In words we can state this principle thus.

MULTIPLICATION PRINCIPLE

If both sides of an equation are multiplied by the same nonzero number, the results on both sides are equal in value.

In symbols we can restate the multiplication principle this way.

For real numbers a, b, and c with $c \neq 0$, if $a = b$, then $ca = cb$.

Let us look at an equation where it would be helpful to multiply each side by 3.

EXAMPLE 1 Solve for x. $\frac{1}{3}x = -15$

Solution We know that $(3)\left(\frac{1}{3}\right) = 1$. We will multiply each side of the equation by 3 because we want to isolate the variable x.

$$3\left(\frac{1}{3}x\right) = 3(-15) \qquad \text{Multiply each side of the equation by 3 since } (3)\left(\frac{1}{3}\right) = 1.$$

$$\left(\frac{3}{1}\right)\left(\frac{1}{3}\right)(x) = -45$$

$$1x = -45 \qquad \text{Simplify.}$$

$$x = -45 \qquad \text{The solution is } -45.$$

Check.

$$\frac{1}{3}(-45) \overset{?}{=} -15 \qquad \text{Substitute } -45 \text{ for } x \text{ in the original equation.}$$

$$-15 = -15 \quad \checkmark \quad \text{It checks.}$$

NOTE TO STUDENT: Fully worked-out solutions to all of the Practice Problems can be found at the back of the text starting at page SP-1

Practice Problem 1 Solve for x.

$$\frac{1}{8}x = -2$$

Note that $\frac{1}{5}x$ can be written as $\frac{x}{5}$. To solve the equation $\frac{x}{5} = 3$, we could multiply each side of the equation by 5. Try it. Then check your solution.

Solving Equations of the Form *ax* = *b*

We can see that using the multiplication principle to multiply each side of an equation by $\frac{1}{2}$ is the same as dividing each side of the equation by 2. Thus, it would seem that the multiplication principle would allow us to divide each side of the equation by any nonzero real number. Is there a real-life example of this idea?

Let's return to our simple example of the box and the stone on a balanced seesaw. Suppose that we were to cut the two objects in half (so that the amount of weight of each was divided by 2). We then return the objects to the same places on the seesaw. The seesaw would still balance. The weight values of both sides remain equal.

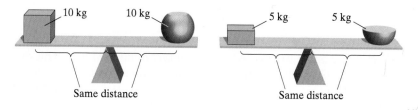

In words we can state this principle thus.

DIVISION PRINCIPLE

If both sides of an equation are divided by the same nonzero number, the results on both sides are equal in value.

Note: We put a restriction on the number by which we are dividing. We cannot divide by zero. We say that expressions like $\frac{2}{0}$ are not defined. Thus we restrict our divisor to *nonzero* numbers. We can restate the division principle this way.

For real numbers a, b, and c where $c \neq 0$, if $a = b$, then $\dfrac{a}{c} = \dfrac{b}{c}$.

EXAMPLE 2   Solve for x. $5x = 125$

Solution

$$\frac{5x}{5} = \frac{125}{5}$$ Divide both sides by 5.

$$x = 25$$ Simplify. The solution is 25.

Check.

$$5x = 125$$
$$5(25) \stackrel{?}{=} 125$$ Replace x by 25.
$$125 = 125 \ \checkmark$$ It checks.

Practice Problem 2   Solve for x. $9x = 72$

For equations of the form $ax = b$ (a number multiplied by x equals another number), we solve the equation by choosing to divide both sides by a particular number. What is the number to choose? We look at the side of the equation that contains x. We notice the number that is multiplied by x. We divide by that number. The division principle tells us that we can still have a true equation provided that we divide by that number *on both sides* of the equation.

The solution to an equation may be a proper fraction or an improper fraction.

EXAMPLE 3 Solve for x. $4x = 38$

Solution

$$\frac{4x}{4} = \frac{38}{4}$$ Divide both sides by 4.

$$x = \frac{19}{2}$$ Simplify. The solution is $\frac{19}{2}$.

Check.

$$4x = 38$$

$$\overset{2}{\cancel{4}}\left(\frac{19}{\cancel{2}}\right) \overset{?}{=} 38$$ Replace x by $\frac{19}{2}$.

$$38 = 38 \checkmark$$ It checks.

Practice Problem 3 Solve for x.

$$6x = 50$$

In Examples 2 and 3 we *divided by the number multiplied by x.* This procedure is followed regardless of whether the sign of that number is positive or negative. In equations of the form $ax = b$, a is a number multiplied by x. The **coefficient** of x is a. A coefficient is a multiplier.

Sidelight As you work through the exercises in this book, you will notice that the solutions of equations can be integers, fractions, or decimals. Recall from page 58 that a **terminating decimal** is one that has a definite number of digits. Unless directions state that the solution should be rounded, the decimal form of a solution should be given only if it is a terminating decimal.

EXAMPLE 4 Solve for x. $-3x = 48$

Solution

$$\frac{-3x}{-3} = \frac{48}{-3}$$ Divide both sides by -3.

$$x = -16$$ The solution is -16.

Check. Can you check this solution?

Practice Problem 4 Solve for x. $-27x = 54$

The coefficient of x may be 1 or -1. You may have to rewrite the equation so that the coefficient of 1 or -1 is obvious. With practice you may be able to recognize the coefficient without actually rewriting the equation.

EXAMPLE 5 Solve for x. $-x = -24$.

Solution $-1x = -24$ Rewrite the equation. $-1x$ is the same as $-x$. Now the coefficient of -1 is obvious.

$$\frac{-1x}{-1} = \frac{-24}{-1}$$ Divide both sides by -1.

$$x = 24$$ The solution is 24.

Check. Can you check this solution?

Practice Problem 5 Solve for x. $-x = 36$

The variable can be on either side of the equation. The equation $-78 = -3x$ can be solved in exactly the same way as $-3x = -78$.

EXAMPLE 6 Solve for x. $-78 = -3x$

Solution
$$\frac{-78}{-3} = \frac{-3x}{-3} \qquad \text{Divide both sides by } -3.$$
$$26 = x \qquad \text{The solution is 26.}$$

Check.
$$-78 = -3x$$
$$-78 \overset{?}{=} -3(26) \qquad \text{Replace } x \text{ by 26.}$$
$$-78 = -78 \quad \checkmark \quad \text{It checks.}$$

Practice Problem 6 Solve for x.
$$-51 = -6x$$

There is a mathematical concept that unites what we have learned in this section. The concept uses the idea of a multiplicative inverse. For any nonzero number a, the **multiplicative inverse** of a is $\frac{1}{a}$. Likewise, for any nonzero number a, the multiplicative inverse of $\frac{1}{a}$ is a. So to solve an equation of the form $ax = b$, we say that we need to multiply each side by the multiplicative inverse of a. Thus to solve $5x = 45$, we would multiply each side of the equation by the multiplicative inverse of 5, which is $\frac{1}{5}$. In similar fashion, if we wanted to solve the equation $\frac{1}{6}x = 4$, we would multiply each side of the equation by the multiplicative inverse of $\frac{1}{6}$, which is 6. In general, all the problems we have covered so far in this section can be solved by multiplying both sides of the equation by the multiplicative inverse of the coefficient of x.

EXAMPLE 7 Solve for x. $31.2 = 6.0x - 0.8x$

Solution
$$31.2 = 6.0x - 0.8x \qquad \text{There are like terms on the right side.}$$
$$31.2 = 5.2x \qquad \text{Combine like terms.}$$
$$\frac{31.2}{5.2} = \frac{5.2x}{5.2} \qquad \text{Divide both sides by 5.2 (which is the same as multiplying}$$
$$\qquad\qquad\qquad \text{both sides by the multiplicative inverse of 5.2).}$$
$$6 = x \qquad \text{The solution is 6.}$$

Note: Be sure to place the decimal point in the quotient directly above the caret ($_\wedge$) when performing the division.

$$5.2_\wedge\overline{)31.2_\wedge}$$
$$\begin{array}{r} 6. \\ \underline{31\ 2} \\ 0 \end{array}$$

Check. The check is left to you.

Practice Problem 7 Solve for x. $16.2 = 5.2x - 3.4x$

Developing Your Study Skills

Getting Help

Getting the right kind of help at the right time can be a key ingredient in being successful in mathematics. When you have gone to class on a regular basis, taken careful notes, methodically read your textbook, and diligently done your homework—all of which means making every effort possible to learn the mathematics—you may find that you are still having difficulty. If this is the case, then you need to seek help. Make an appointment with your instructor to find out what help is available to you. The instructor, tutoring services, a mathematics lab, videotapes, and computer software may be among the resources you can draw on.

Once you discover the resources available in your school, you need to take advantage of them. Do not put it off, or you will find yourself getting behind. You cannot afford that. When studying mathematics, you must keep up with your work.

Verbal and Writing Skills

1. To solve the equation $6x = -24$, divide each side of the equation by _____.

2. To solve the equation $-7x = 56$, divide each side of the equation by _____.

3. To solve the equation $\frac{1}{7}x = -2$, multiply each side of the equation by _____.

4. To solve the equation $\frac{1}{9}x = 5$, multiply each side of the equation by _____.

Solve for x. Be sure to reduce your answer. Check your solution.

5. $\frac{1}{9}x = 4$

6. $\frac{1}{7}x = 6$

7. $\frac{1}{2}x = -15$

8. $\frac{1}{8}x = -10$

9. $\frac{x}{5} = 16$

10. $\frac{x}{10} = 8$

11. $-3 = \frac{x}{5}$

12. $\frac{x}{3} = -12$

13. $13x = 52$

14. $15x = 60$

15. $56 = 7x$

16. $46 = 2x$

17. $-16 = 6x$

18. $-35 = 21x$

19. $1.5x = 75$

20. $2x = 0.36$

21. $-15 = -x$

22. $32 = -x$

23. $-112 = 16x$

24. $-108 = -18x$

25. $0.4x = 0.08$

26. $2.5x = 0.5$

27. $-3.9x = -15.6$

28. $-4.7x = -14.1$

Determine whether the given solution is correct. If it is not, find the correct solution.

29. Is 7 the solution for $-3x = 21$?

30. Is 8 the solution for $5x = -40$?

31. Is -15 the solution for $-x = 15$?

32. Is -8 the solution for $-11x = 88$?

Mixed Practice *Find the value of the variable that satisfies the equation.*

33. $7y = -0.21$

34. $-3y = 0.42$

35. $-56 = -21t$

36. $34 = -51q$

37. $4.6y = -3.22$

38. $-2.8y = -3.08$

39. $4x + 3x = 21$

40. $5x + 4x = 36$

41. $2x - 7x = 20$

42. $3x - 9x = 18$

43. $-8x - 4x = -5$

44. $y - 10y = 6$

45. $12 - 19 = -7x$

46. $31 - 16 = -5m$

47. $8m = -14 + 30$

48. $11x = -20 + 42$

49. $\frac{2}{3}x = 18$

50. $\frac{3}{5}x = 39$

51. $-2.5133x = 26.38965$

52. $-5.42102x = -45.536568$

Cumulative Review *Evaluate using the correct order of operations. (Be careful to avoid sign errors.)*

53. **[1.5.1]** $(-6)(-8) + (-3)(2)$

54. **[1.5.1]** $(-3)^3 + (-20) \div 2$

55. **[1.5.1]** $5 + (2 - 6)^2$

56. **[1.1.4]** *Humpback Whales* During the summer months, a group of humpback whales gather at Stellwagen Bank, near Gloucester, Massachusetts, to feed. When they return to the Caribbean for winter, they will lose up to 25% of their body weight in blubber. If a humpback whale weighs 30 tons after feeding at Stellwagen Bank, how much will it weigh after losing 25% of its body weight?

57. **[0.5.3]** *Earthquakes* In an average year, worldwide, there are 20 earthquakes of magnitude 7 on the Richter scale. If next year is predicted to be an exceptional year, and the number of earthquakes of magnitude 7 is expected to increase by 35%, about how many earthquakes of magnitude 7 can be expected?

Quick Quiz 2.2 Solve for the variable.

1. $2.5x = -95$

2. $-3.9x = -54.6$

3. $7x - 12x = 60$

4. **Concept Check** Explain how you would check to verify whether $x = 36\frac{2}{3}$ is the solution to $-22 = -\frac{3}{5}x$.

Student Learning Objectives

After studying this section, you will be able to:

1. Solve equations of the form $ax + b = c$.

2. Solve equations with the variable on both sides of the equation.

3. Solve equations with parentheses.

1 Solving Equations of the Form $ax + b = c$

Jenny Crawford scored several goals in field hockey during April. Her teammates scored three more than five times the number she scored. Her teammates scored 18 goals. How many did Jenny score? To solve this problem, we need to solve the equation $5x + 3 = 18$.

To solve an equation of the form $ax + b = c$, we must use both the addition principle and the multiplication principle.

EXAMPLE 1 Solve for x to determine how many goals Jenny scored and check your solution.

$$5x + 3 = 18$$

Solution We first want to isolate the variable term.

$5x + 3 + (-3) = 18 + (-3)$	Use the addition principle to add -3 to both sides.
$5x = 15$	Simplify.
$\dfrac{5x}{5} = \dfrac{15}{5}$	Use the division principle to divide both sides by 5.
$x = 3$	The solution is 3. Thus Jenny scored 3 goals.

Check.

$$5(3) + 3 \overset{?}{=} 18$$
$$15 + 3 \overset{?}{=} 18$$
$$18 = 18 \checkmark \qquad \text{It checks.}$$

Practice Problem 1 Solve for x and check your solution. $9x + 2 = 38$

2 Solving Equations with the Variable on Both Sides of the Equation

In some cases the variable appears on both sides of the equation. We would like to rewrite the equation so that all the terms containing the variable appear on one side. To do this, we apply the addition principle to the variable term.

EXAMPLE 2 Solve for x. $9x = 6x + 15$

Solution

$9x + (-6x) = 6x + (-6x) + 15$	Add $-6x$ to both sides. Notice $6x + (-6x)$ eliminates the variable on the right side.
$3x = 15$	Combine like terms.
$\dfrac{3x}{3} = \dfrac{15}{3}$	Divide both sides by 3.
$x = 5$	The solution is 5.

Check. The check is left to the student.

Practice Problem 2 Solve for x. $13x = 2x - 66$

In many problems the variable terms and constant terms appear on both sides of the equation. You will want to get all the variable terms on one side and all the constant terms on the other side.

EXAMPLE 3 Solve for x and check your solution. $9x + 3 = 7x - 2$

Solution First we want to isolate the variable term.

$$9x + (-7x) + 3 = 7x + (-7x) - 2 \qquad \text{Add } -7x \text{ to both sides of the equation.}$$
$$2x + 3 = -2 \qquad \text{Combine like terms.}$$
$$2x + 3 + (-3) = -2 + (-3) \qquad \text{Add } -3 \text{ to both sides.}$$
$$2x = -5 \qquad \text{Simplify.}$$
$$\frac{2x}{2} = \frac{-5}{2} \qquad \text{Divide both sides by 2.}$$
$$x = -\frac{5}{2} \qquad \text{The solution is } -\frac{5}{2}.$$

Check. $\qquad 9x + 3 = 7x - 2$

$$9\left(-\frac{5}{2}\right) + 3 \overset{?}{=} 7\left(-\frac{5}{2}\right) - 2 \qquad \text{Replace } x \text{ by } -\frac{5}{2}.$$
$$-\frac{45}{2} + 3 \overset{?}{=} -\frac{35}{2} - 2 \qquad \text{Simplify.}$$
$$-\frac{45}{2} + \frac{6}{2} \overset{?}{=} -\frac{35}{2} - \frac{4}{2} \qquad \text{Change to equivalent fractions with a common denominator.}$$
$$-\frac{39}{2} = -\frac{39}{2} \checkmark \qquad \text{It checks. The solution is } -\frac{5}{2}.$$

Practice Problem 3 Solve for x and check your solution.

$$3x + 2 = 5x + 2$$

In our next example we will study equations that need simplifying before any other steps are taken. Where it is possible, you should first collect like terms on one or both sides of the equation. The variable terms can be collected on the right side or the left side. In this example we will collect all the x-terms on the right side.

EXAMPLE 4 Solve for x. $5x + 26 - 6 = 9x + 12x$

Solution

$$5x + 20 = 21x \qquad \text{Combine like terms.}$$
$$5x + (-5x) + 20 = 21x + (-5x) \qquad \text{Add } -5x \text{ to both sides.}$$
$$20 = 16x \qquad \text{Combine like terms.}$$
$$\frac{20}{16} = \frac{16x}{16} \qquad \text{Divide both sides by 16.}$$
$$\frac{5}{4} = x \qquad \text{Don't forget to reduce the resulting fraction.}$$

Check. The check is left to the student.

Practice Problem 4 Solve for z.

$$-z + 8 - z = 3z + 10 - 3$$

Do you really need all these steps? No. As you become more proficient you will be able to combine or eliminate some of these steps. However, it is best to write each step in its entirety until you are consistently obtaining the correct solution.

It is much better to show every step than to take a lot of shortcuts and possibly obtain a wrong answer. This is a section of the algebra course where working neatly and accurately will help you—both now and as you progress through the course.

 ## Solving Equations with Parentheses

The equations that you just solved are simpler versions of equations that we will now discuss. These equations contain parentheses. If the parentheses are first removed, the problems then become just like those encountered previously. We use the distributive property to remove the parentheses.

EXAMPLE 5 Solve for x and check your solution.

$$4(x + 1) - 3(x - 3) = 25$$

Solution

$$4(x + 1) - 3(x - 3) = 25$$
$$4x + 4 - 3x + 9 = 25$$

Multiply by 4 and -3 to remove parentheses. Be careful of the signs. Remember that $(-3)(-3) = 9$.

After removing the parentheses, it is important to combine like terms on each side of the equation. Do this before going on to isolate the variable.

$$x + 13 = 25$$ Combine like terms.

$$x + 13 - 13 = 25 - 13$$ Subtract 13 from both sides to isolate the variable.

$$x = 12$$ The solution is 12.

Check. $4(12 + 1) - 3(12 - 3) \overset{?}{=} 25$ Replace x by 12.

$$4(13) - 3(9) \overset{?}{=} 25$$ Combine numbers inside parentheses.

$$52 - 27 \overset{?}{=} 25$$ Multiply.

$$25 = 25 \ \checkmark$$ Simplify. It checks.

Practice Problem 5 Solve for x and check your solution.

$$4x - (x + 3) = 12 - 3(x - 2)$$

NOTE TO STUDENT: *Fully worked-out solutions to all of the Practice Problems can be found at the back of the text starting at page SP-1*

EXAMPLE 6 Solve for x. $3(-x - 7) = -2(2x + 5)$

Solution $-3x - 21 = -4x - 10$ Remove parentheses. Watch the signs carefully.

$$-3x + 4x - 21 = -4x + 4x - 10$$ Add $4x$ to both sides.

$$x - 21 = -10$$ Simplify.

$$x - 21 + 21 = -10 + 21$$ Add 21 to both sides.

$$x = 11$$ The solution is 11.

Check.   The check is left to the student.

Practice Problem 6 Solve for x. $4(-2x - 3) = -5(x - 2) + 2$

In problems that involve decimals, great care should be taken. In some steps you will be multiplying decimal quantities, and in other steps you will be adding them.

EXAMPLE 7 Solve for x. $0.3(1.2x - 3.6) = 4.2x - 16.44$

Solution

$0.36x - 1.08 = 4.2x - 16.44$	Remove parentheses.
$0.36x - 0.36x - 1.08 = 4.2x - 0.36x - 16.44$	Subtract $0.36x$ from both sides.
$-1.08 = 3.84x - 16.44$	Combine like terms.
$-1.08 + 16.44 = 3.84x - 16.44 + 16.44$	Add 16.44 to both sides.
$15.36 = 3.84x$	Simplify.
$\dfrac{15.36}{3.84} = \dfrac{3.84x}{3.84}$	Divide both sides by 3.84.
$4 = x$	The solution is 4.

Check. The check is left to the student.

Practice Problem 7 Solve for x.

$$0.3x - 2(x + 0.1) = 0.4(x - 3) - 1.1$$

EXAMPLE 8 Solve for z and check. $2(3z - 5) + 2 = 4z - 3(2z + 8)$

Solution

$6z - 10 + 2 = 4z - 6z - 24$	Remove parentheses.
$6z - 8 = -2z - 24$	Combine like terms.
$6z + 2z - 8 = -2z + 2z - 24$	Add $2z$ to both sides.
$8z - 8 = -24$	Simplify.
$8z - 8 + 8 = -24 + 8$	Add 8 to both sides.
$8z = -16$	Simplify.
$\dfrac{8z}{8} = \dfrac{-16}{8}$	Divide both sides by 8.
$z = -2$	Simplify. The solution is -2.

Check.

$2[3(-2) - 5] + 2 \stackrel{?}{=} 4(-2) - 3[2(-2) + 8]$	Replace z by -2.
$2[-6 - 5] + 2 \stackrel{?}{=} -8 - 3[-4 + 8]$	Multiply.
$2[-11] + 2 \stackrel{?}{=} -8 - 3[4]$	Simplify.
$-22 + 2 \stackrel{?}{=} -8 - 12$	
$-20 = -20$ ✓	It checks.

Practice Problem 8 Solve for z and check.

$$5(2z - 1) + 7 = 7z - 4(z + 3)$$

Find the value of the variable that satisfies the equation in exercises 1–22. Check your solution. Answers that are not integers may be left in fractional form or decimal form.

1. $4x + 13 = 21$ **2.** $7x + 4 = 53$ **3.** $4x - 11 = 13$ **4.** $5x - 11 = 39$

5. $7x - 18 = -46$ **6.** $6x - 23 = -71$ **7.** $-4x + 17 = -35$ **8.** $-6x + 25 = -83$

9. $2x + 3.2 = 9.4$ **10.** $4x + 4.6 = 9.2$ **11.** $\frac{1}{4}x + 6 = 13$ **12.** $\frac{1}{5}x + 2 = 9$

13. $\frac{1}{3}x + 5 = -4$ **14.** $\frac{1}{8}x - 3 = -9$ **15.** $8x = 48 + 2x$ **16.** $5x = 22 + 3x$

17. $-6x = -27 + 3x$ **18.** $-7x = -26 + 6x$ **19.** $44 - 2x = 6x$ **20.** $21 - 5x = 7x$

21. $54 - 2x = -8x$ **22.** $72 - 4x = -12x$

In exercises 23–26, determine whether the given solution is correct. If it is not, find the solution.

23. Is 2 the solution for $2y + 3y = 12 - y$?   **24.** Is 4 the solution for $5y + 2 = 6y - 6 + y$?

25. Is 11 a solution for $7x + 6 - 3x = 2x - 5 + x$?   **26.** Is -12 a solution for $9x + 2 - 5x = -8 + 5x - 2$?

Solve for the variable. You may move the variable terms to the right or to the left.

27. $14 - 2x = -5x + 11$ **28.** $8 - 3x = 7x + 8$ **29.** $x - 6 = 8 - x$

30. $-x + 12 = -4 + x$ **31.** $0.6y + 0.8 = 0.1 - 0.1y$ **32.** $1.1y + 0.3 = -1.3 + 0.3y$

33. $5x - 9 = 3x + 23$ **34.** $9x - 5 = 7x + 43$

To Think About *For exercises 35 and 36, first combine like terms on each side of the equation. Then solve for y by getting all the y-terms on the left. Then solve for y by getting all the y-terms on the right. Which approach is better?*

35. $-3 + 10y + 6 = 15 + 12y - 18$

36. $7y + 21 - 5y = 5y - 7 + y$

Remove the parentheses and solve for the variable. Check your solution. Answers that are not integers may be left in fractional form or decimal form.

37. $5(x + 3) = 35$

38. $7(x + 3) = 28$

39. $5(4x - 3) + 8 = -2$

40. $4(2x + 1) - 7 = 6 - 5$

41. $7x - 3(5 - x) = 10$

42. $8x - 2(4 - x) = 14$

43. $0.5x - 0.3(2 - x) = 4.6$

44. $0.4x - 0.2(3 - x) = 1.8$

45. $4(a - 3) + 2 = 2(a - 5)$

46. $6(a + 3) - 2 = -4(a - 4)$

47. $-2(x + 3) + 4 = 3(x + 4) + 2$

48. $-3(x + 5) + 2 = 4(x + 6) - 9$

49. $-3(y - 3y) + 4 = -4(3y - y) + 6 + 13y$

50. $2(4x - x) + 6 = 2(2x + x) + 8 - x$

Mixed Practice *Solve for the variable.*

51. $5.7x + 3 = 4.2x - 3$

52. $4x - 3.1 = 5.3 - 3x$

53. $5z + 7 - 2z = 32 - 2z$

54. $8 - 7z + 2z = 20 + 5z$

55. $-0.3a + 1.4 = -1.2 - 0.7a$

56. $-0.7b + 1.6 = -1.7 - 1.5b$

57. $6x + 8 - 3x = 11 - 12x - 13$ **58.** $4 - 7x - 13 = 8x - 3 - 5x$ **59.** $-3.5x + 1.3 = -2.7x + 1.5$

60. $2.8x - 0.9 = 5.2x - 3.3$ **61.** $5(4 + x) = 3(3x - 1) - 9$ **62.** $6(3x - 1) = 4(2x + 5) - 6$

63. $4x + 3.2 - 1.9x = 0.3x - 4.9$ **64.** $3x + 2 - 1.7x = 0.6x + 31.4$

Cumulative Review *Simplify.*

65. **[1.7.2]** $-3y(2x + y) + 5(3xy - y^2)$ **66.** **[1.9.1]** $-\{2(x - 3) + 3[x - (2x - 5)]\}$

67. **[0.7.1]** *Investments* On October 5, 2007, Marcella owned three different stocks: Motorola, General Mills, and CVS. Her portfolio contained the following:

   18.5 shares of Motorola valued at $18.66 per share,
   9.25 shares of General Mills valued at $57.56 per share, and
   5.0 shares of CVS valued at $29.26 per share.

Find the market value of Marcella's stock holdings on October 5, 2007.

68. **[0.7.1]** *Employee Discount* Marvin works at Best Buy and gets a 10% discount on anything he buys from the store. A GPS navigation system that Marvin wishes to purchase costs $899 and is on sale at a 20% discount.

(a) What is the sale price if Marvin has a total discount of 30%? (Disregard sales tax.)

(b) What is the price if Marvin gets a 10% discount on the 20% sale price? (Disregard sales tax.)

Quick Quiz 2.3 Solve for the variable.

1. $7x - 6 = -4x - 10$ **2.** $-3x + 6.2 = -5.8$

3. $2(3x - 2) = 4(5x + 3)$ **4.** **Concept Check** Explain how you would solve the equation $3(x - 2) + 2 = 2(x - 4)$.

Solving Equations with Fractions

Equations with fractions can be rather difficult to solve. This difficulty is simply due to the extra care we usually have to use when computing with fractions. The actual equation-solving procedures are the same, with fractions or without. To avoid unnecessary work, we transform the given equation with fractions to an equivalent equation that does not contain fractions. How do we do this? We multiply each side of the equation by the least common denominator of all the fractions contained in the equation. We then use the distributive property so that the LCD is multiplied by each term of the equation.

Student Learning Objective

After studying this section, you will be able to:

1 Solve equations with fractions.

EXAMPLE 1 Solve for x. $\dfrac{1}{4}x - \dfrac{2}{3} = \dfrac{5}{12}x$

Solution   First we find that the LCD = 12.

$$12\left(\frac{1}{4}x - \frac{2}{3}\right) = 12\left(\frac{5}{12}x\right) \qquad \text{Multiply both sides by 12.}$$

$$\left(\frac{12}{1}\right)\left(\frac{1}{4}\right)(x) - \left(\frac{12}{1}\right)\left(\frac{2}{3}\right) = \left(\frac{12}{1}\right)\left(\frac{5}{12}\right)(x) \qquad \text{Use the distributive property.}$$

$$3x - 8 = 5x \qquad \text{Simplify.}$$

$$3x + (-3x) - 8 = 5x + (-3x) \qquad \text{Add } -3x \text{ to both sides.}$$

$$-8 = 2x \qquad \text{Simplify.}$$

$$\frac{-8}{2} = \frac{2x}{2} \qquad \text{Divide both sides by 2.}$$

$$-4 = x \qquad \text{Simplify.}$$

Check.

$$\frac{1}{4}(-4) - \frac{2}{3} \overset{?}{=} \frac{5}{12}(-4)$$

$$-1 - \frac{2}{3} \overset{?}{=} -\frac{5}{3}$$

$$-\frac{3}{3} - \frac{2}{3} \overset{?}{=} -\frac{5}{3}$$

$$-\frac{5}{3} = -\frac{5}{3} \quad ✓ \qquad \text{It checks.}$$

Practice Problem 1 Solve for x.

$$\frac{3}{8}x - \frac{3}{2} = \frac{1}{4}x$$

NOTE TO STUDENT: Fully worked-out solutions to all of the Practice Problems can be found at the back of the text starting at page SP-1

In Example 1 we multiplied both sides of the equation by the LCD. However, most students prefer to go immediately to the second step and multiply each term by the LCD. This avoids having to write out a separate step using the distributive property.

EXAMPLE 2 Solve for x and check your solution. $\dfrac{x}{3} + 3 = \dfrac{x}{5} - \dfrac{1}{3}$

Solution

$$15\left(\dfrac{x}{3}\right) + 15(3) = 15\left(\dfrac{x}{5}\right) - 15\left(\dfrac{1}{3}\right)$$
The LCD is 15. Use the multiplication principle to multiply each term by 15.

$$5x + 45 = 3x - 5$$ Simplify.

$$5x - 3x + 45 = 3x - 3x - 5$$ Subtract $3x$ from both sides.

$$2x + 45 = -5$$ Combine like terms.

$$2x + 45 - 45 = -5 - 45$$ Subtract 45 from both sides.

$$2x = -50$$ Simplify.

$$\dfrac{2x}{2} = \dfrac{-50}{2}$$ Divide both sides by 2.

$$x = -25$$ The solution is -25.

Check.

$$\dfrac{-25}{3} + 3 \stackrel{?}{=} \dfrac{-25}{5} - \dfrac{1}{3}$$

$$-\dfrac{25}{3} + \dfrac{9}{3} \stackrel{?}{=} -\dfrac{5}{1} - \dfrac{1}{3}$$

$$-\dfrac{16}{3} \stackrel{?}{=} -\dfrac{15}{3} - \dfrac{1}{3}$$

$$-\dfrac{16}{3} = -\dfrac{16}{3} \checkmark$$

Practice Problem 2 Solve for x and check your solution.

$$\dfrac{5x}{4} - 1 = \dfrac{3x}{4} + \dfrac{1}{2}$$

EXAMPLE 3 Solve for x. $\dfrac{x+5}{7} = \dfrac{x}{4} + \dfrac{1}{2}$

Solution

$$\dfrac{x}{7} + \dfrac{5}{7} = \dfrac{x}{4} + \dfrac{1}{2}$$
First we rewrite the left side as two fractions. This is actually multiplying $\dfrac{1}{7}(x + 5) = \dfrac{x}{7} + \dfrac{5}{7}$.

$$28\left(\dfrac{x}{7}\right) + 28\left(\dfrac{5}{7}\right) = 28\left(\dfrac{x}{4}\right) + 28\left(\dfrac{1}{2}\right)$$
We observe that the LCD is 28, so we multiply each term by 28.

$$4x + 20 = 7x + 14$$ Simplify.

$$4x - 4x + 20 = 7x - 4x + 14$$ Subtract $4x$ from both sides.

$$20 = 3x + 14$$ Combine like terms.

$$20 - 14 = 3x + 14 - 14$$ Subtract 14 from both sides.

$$6 = 3x$$ Combine like terms.

$$\dfrac{6}{3} = \dfrac{3x}{3}$$ Divide both sides by 3.

$$2 = x$$ The solution is 2.

Check. The check is left to the student.

Practice Problem 3 Solve for x.

$$\dfrac{x+6}{9} = \dfrac{x}{6} + \dfrac{1}{2}$$

If a problem contains both parentheses and fractions, it is best to remove the parentheses first. Many students find it is helpful to have a written procedure to follow in solving these more involved equations.

> **PROCEDURE TO SOLVE EQUATIONS**
> 1. Remove any parentheses.
> 2. If fractions exist, multiply all terms on both sides by the least common denominator of all the fractions.
> 3. Combine like terms if possible.
> 4. Add or subtract terms on both sides of the equation to get all terms with the variable on one side of the equation.
> 5. Add or subtract a constant value on both sides of the equation to get all terms not containing the variable on the other side of the equation.
> 6. Divide both sides of the equation by the coefficient of the variable.
> 7. Simplify the solution (if possible).
> 8. Check your solution.

Let's use each step in solving the next example.

EXAMPLE 4   Solve for x and check your solution.

$$\frac{1}{3}(x - 2) = \frac{1}{5}(x + 4) + 2$$

Solution

Step 1 $\dfrac{x}{3} - \dfrac{2}{3} = \dfrac{x}{5} + \dfrac{4}{5} + 2$ Remove parentheses.

Step 2 $15\left(\dfrac{x}{3}\right) - 15\left(\dfrac{2}{3}\right) = 15\left(\dfrac{x}{5}\right) + 15\left(\dfrac{4}{5}\right) + 15(2)$ Multiply by the LCD, 15.

$5x - 10 = 3x + 12 + 30$ Simplify.

Step 3 $5x - 10 = 3x + 42$ Combine like terms.

Step 4 $5x - 3x - 10 = 3x - 3x + 42$ Subtract $3x$ from both sides.

$2x - 10 = 42$ Simplify.

Step 5 $2x - 10 + 10 = 42 + 10$ Add 10 to both sides.

$2x = 52$ Simplify.

Step 6 $\dfrac{2x}{2} = \dfrac{52}{2}$ Divide both sides by 2.

Step 7 $x = 26$ Simplify the solution.

Step 8 *Check.* $\dfrac{1}{3}(26 - 2) \overset{?}{=} \dfrac{1}{5}(26 + 4) + 2$ Replace x by 26.

$\dfrac{1}{3}(24) \overset{?}{=} \dfrac{1}{5}(30) + 2$ Combine values within parentheses.

$8 \overset{?}{=} 6 + 2$ Simplify.

$8 = 8 \ \checkmark$ The solution is 26.

NOTE TO STUDENT: Fully worked-out solutions to all of the Practice Problems can be found at the back of the text starting at page SP-1

Practice Problem 4 Solve for x and check your solution.

$$\frac{1}{2}(x + 5) = \frac{1}{5}(x - 2) + \frac{1}{2}$$

Remember that not every step will be needed in each problem. You can combine some steps as well, *as long as you are consistently obtaining the correct solution.* However, you are encouraged to write out every step as a way of helping you to avoid careless errors.

It is important to remember that when we write decimals, these numbers are really fractions written in a special way. Thus, $0.3 = \frac{3}{10}$ and $0.07 = \frac{7}{100}$. It is possible to take an equation containing decimals and to multiply each term by the appropriate value to obtain integer coefficients.

EXAMPLE 5 Solve for x. $0.2(1 - 8x) + 1.1 = -5(0.4x - 0.3)$

Solution

$0.2 - 1.6x + 1.1 = -2.0x + 1.5$	Remove parentheses.
$10(0.2) - 10(1.6x) + 10(1.1) = 10(-2.0x) + 10(1.5)$	Multiply each term by 10.
$2 - 16x + 11 = -20x + 15$	Multiplying by 10 moves the decimal point one place to the right.
$-16x + 13 = -20x + 15$	Simplify.
$-16x + 20x + 13 = -20x + 20x + 15$	Add $20x$ to both sides.
$4x + 13 = 15$	Simplify.
$4x + 13 - 13 = 15 - 13$	Subtract 13 from both sides.
$4x = 2$	Simplify.
$\dfrac{4x}{4} = \dfrac{2}{4}$	Divide both sides by 4.
$x = \dfrac{1}{2} \quad \text{or} \quad 0.5$	Simplify.

Check.

$$0.2[1 - 8(0.5)] + 1.1 \overset{?}{=} -5[0.4(0.5) - 0.3]$$
$$0.2[1 - 4] + 1.1 \overset{?}{=} -5[0.2 - 0.3]$$
$$0.2[-3] + 1.1 \overset{?}{=} -5[-0.1]$$
$$-0.6 + 1.1 \overset{?}{=} 0.5$$
$$0.5 = 0.5 \quad \checkmark$$

Practice Problem 5 Solve for x.

$$2.8 = 0.3(x - 2) + 2(0.1x - 0.3)$$

TO THINK ABOUT: Does Every Equation Have One Solution? Actually, no. There are some rare cases where an equation has no solution at all. Suppose we try to solve the equation

$$5(x + 3) = 2x - 8 + 3x.$$

If we remove the parentheses and combine like terms, we have

$$5x + 15 = 5x - 8.$$

If we add $-5x$ to each side, we obtain

$$15 = -8.$$

Clearly this is impossible. There is no value of x for which these two numbers are equal. We would say this equation has **no solution.**

One additional surprise may happen. An equation may have an infinite number of solutions. Suppose we try to solve the equation

$$9x - 8x - 7 = 3 + x - 10.$$

If we combine like terms on each side, we have the equation

$$x - 7 = x - 7.$$

If we add $-x$ to each side, we obtain

$$-7 = -7.$$

Now this statement is always true, no matter what the value of x. We would say this equation has **an infinite number of solutions.**

In the To Think About exercises in this section, we will encounter some equations that have no solution or an infinite number of solutions.

Developing Your Study Skills

Taking Notes in Class

An important part of studying mathematics is taking notes. To take meaningful notes, you must be an active listener. Keep your mind on what the instructor is saying, and be ready with questions whenever you do not understand something.

If you have previewed the lesson material, you will be prepared to take good notes. The important concepts will seem somewhat familiar. If you frantically try to write all that the instructor says or copy all the examples done in class, you may find your notes nearly worthless when you are home alone. Write down *important* ideas and examples as the instructor lectures, making sure that you are listening and following the logic. Include any helpful hints or suggestions that your instructor gives you or refers to in your text.

In exercises 1–16, solve for the variable and check your answer. Noninteger answers may be left in fractional form or decimal form.

1. $\dfrac{1}{2}x + \dfrac{2}{3} = \dfrac{1}{6}$

2. $\dfrac{1}{3} + \dfrac{5}{12}x = \dfrac{3}{4}$

3. $\dfrac{2}{3}x = \dfrac{1}{15}x + \dfrac{3}{5}$

4. $\dfrac{5}{21}x = \dfrac{2}{3}x - \dfrac{1}{7}$

5. $\dfrac{x}{2} + \dfrac{x}{5} = \dfrac{7}{10}$

6. $\dfrac{x}{4} - \dfrac{x}{16} = \dfrac{3}{8}$

7. $5 - \dfrac{1}{3}x = \dfrac{1}{12}x$

8. $15 - \dfrac{1}{2}x = \dfrac{1}{4}x$

9. $2 + \dfrac{y}{2} = \dfrac{3y}{4} - 3$

10. $\dfrac{x}{3} - 1 = -\dfrac{1}{2} - x$

11. $\dfrac{x - 3}{5} = 1 - \dfrac{x}{3}$

12. $\dfrac{y - 5}{4} = 1 - \dfrac{y}{5}$

13. $\dfrac{x + 3}{4} = \dfrac{x}{2} + \dfrac{1}{6}$

14. $\dfrac{x - 2}{3} = \dfrac{x}{12} + \dfrac{5}{4}$

15. $0.6x + 5.9 = 3.8$

16. $-1.2x - 4.3 = 1.1$

17. Is 4 a solution to $\dfrac{1}{2}(y - 2) + 2 = \dfrac{3}{8}(3y - 4)$?

18. Is 2 a solution to $\dfrac{1}{5}(y + 2) = \dfrac{1}{10}y + \dfrac{3}{5}$?

19. Is $\dfrac{5}{8}$ a solution to $\dfrac{1}{2}\left(y - \dfrac{1}{5}\right) = \dfrac{1}{5}(y + 2)$?

20. Is $\dfrac{13}{3}$ a solution to $\dfrac{y}{2} - \dfrac{7}{9} = \dfrac{y}{6} + \dfrac{2}{3}$?

Remove parentheses first. Then combine like terms. Solve for the variable. Noninteger answers may be left in fractional form or decimal form.

21. $\dfrac{3}{4}(3x + 1) = 2(3 - 2x) + 1$

22. $\dfrac{1}{4}(3x + 1) = 2(2x - 4) - 8$

23. $2(x - 2) = \dfrac{2}{5}(3x + 1) + 2$

24. $2(x - 4) = \dfrac{5}{6}(x + 6) - 6$

25. $0.3x - 0.2(3 - 5x) = -0.5(x - 6)$

26. $0.2(x + 1) + 0.5x = -0.3(x - 4)$

27. $-8(0.1x + 0.4) - 0.9 = -0.1$

28. $0.6x + 1.5 = 0.3x - 0.6(2x + 5)$

Mixed Practice *Solve. Noninteger answers may be left in fractional form or decimal form.*

29. $\frac{1}{3}(y + 2) = 3y - 5(y - 2)$

30. $\frac{1}{4}(y + 6) = 2y - 3(y - 3)$

31. $\frac{1 + 2x}{5} + \frac{4 - x}{3} = \frac{1}{15}$

32. $\frac{1 + 3x}{2} + \frac{2 - x}{3} = \frac{5}{6}$

33. $\frac{1}{5}(x + 3) = 2x - 3(2 - x) - 3$

34. $\frac{2}{3}(x + 4) = 6 - \frac{1}{4}(3x - 2) - 1$

35. $\frac{1}{3}(x - 2) = 3x - 2(x - 1) + \frac{16}{3}$

36. $\frac{3}{4}(x - 2) + \frac{3}{5} = \frac{1}{5}(x + 1)$

37. $\frac{4}{5}x - \frac{2}{3} = \frac{3x + 1}{2}$

38. $\frac{4}{7}x + \frac{1}{3} = \frac{3x - 2}{14}$

39. $0.4x - 0.5(2x + 3) = -0.7(x + 3)$

40. $0.6(x + 0.1) = 2(0.4x - 0.2)$

To Think About *Solve. Be careful to examine your work to see if the equation may have no solution or an infinite number of solutions.*

41. $-1 + 5(x - 2) = 12x + 3 - 7x$

42. $x + 3x - 2 + 3x = -11 + 7(x + 2)$

43. $9(x + 3) - 6 = 24 - 2x - 3 + 11x$

44. $7(x + 4) - 10 = 3x + 20 + 4x - 2$

45. $-3(4x - 1) = 5(2x - 1) + 8$

46. $11x - 8 = -4(x + 3) + 4$

47. $3(4x + 1) - 2x = 2(5x - 3)$

48. $5(-3 + 4x) = 4(2x + 4) + 12x$

Cumulative Review

49. **[0.3.1]** Multiply. $\left(-3\frac{1}{4}\right)\left(5\frac{1}{3}\right)$

50. **[0.3.2]** Divide. $5\frac{1}{2} \div 1\frac{1}{4}$

51. **[0.7.5]** *Peregrine Falcons* Peregrine falcons are known for being the fastest birds on record, reaching horizontal speeds of 40–55 miles per hour. They also are one of the few animals in which the females are larger than the males. If female peregrine falcons are 30% larger than the males, and males measure 440–750 grams, what is the weight range for females?

52. **[1.8.2]** *Auditorium Seating* The seating area of an auditorium is shaped like a trapezoid, with front and back sides parallel. The front of the auditorium measures 88 feet across, the back of the auditorium measures 150 feet across, and the auditorium is 200 feet from front to back. If each seat requires a space that is 2.5 feet wide by 3 feet deep, how many seats will the auditorium hold? (This will only be an approximation because of the angled side walls. Round off to the nearest whole number.)

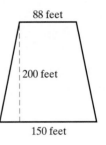

88 feet

200 feet

150 feet

Quick Quiz 2.4 Solve for the variable. When needed, leave your answer in fractional form. Reduce all fractions.

1. $\frac{3}{4}x + \frac{5}{12} = \frac{1}{3}x - \frac{1}{6}$

2. $\frac{2}{3}x - \frac{3}{5} + \frac{7}{5}x + \frac{1}{3} = 1$

3. $\frac{2}{3}(x + 2) + \frac{1}{4} = \frac{1}{2}(5 - 3x)$

4. **Concept Check** Explain how you would solve the equation $\frac{x + 5}{6} = \frac{x}{2} + \frac{3}{4}$.

How are you doing with your homework assignments in Sections 2.1 to 2.4? Do you feel you have mastered the material so far? Do you understand the concepts you have covered? Before you go further in the textbook, take some time to do each of the following problems.

Solve for x. If the solution is not an integer, you may express your answer as a fraction or as a decimal.

2.1

1. $5 - 8 + x = -12$

2. $-2.8 + x = 4.7$

2.2

3. $-45 = -5x$

4. $12x - 6x = -48$

2.3

5. $-1.2x + 3.5 = 2.7$

6. $-14x + 9 = 2x + 7$

7. $14x + 2(7 - 2x) = 20$

8. $0.5(1.2x - 3.4) = -1.4x + 5.8$

9. $3(x + 6) = -2(4x - 1) + x$

2.4

10. $\dfrac{x}{3} + \dfrac{x}{4} = \dfrac{5}{6}$

11. $\dfrac{1}{4}(x + 3) = 4x - 2(x - 3)$

12. $\dfrac{1}{2}(x - 1) + 2 = 3(2x - 1)$

13. $\dfrac{1}{7}(7x - 14) - 2 = \dfrac{1}{3}(x - 2)$

14. $0.2(x - 3) = 4(0.2x - 0.1)$

1. _____

2. _____

3. _____

4. _____

5. _____

6. _____

7. _____

8. _____

9. _____

10. _____

11. _____

12. _____

13. _____

14. _____

Now turn to page SA-5 for the answer to each of these problems. Each answer also includes a reference to the objective in which the problem is first taught. If you missed any of these problems, you should stop and review the Examples and Practice Problems in the referenced objective. A little review now will help you master the material in the upcoming sections of the text.

Student Learning Objectives

After studying this section, you will be able to:

 Translate English phrases into algebraic expressions.

 Write an algebraic expression to compare two or more quantities.

 Translating English Phrases into Algebraic Expressions

One of the most useful applications of algebra is solving word problems. One of the first steps in solving a word problem is translating the conditions of the problem into algebra. In this section we show you how to translate common English phrases into algebraic symbols. This process is similar to translating between languages like Spanish and French.

Several English phrases describe the operation of addition. If we represent an unknown number by the variable x, all of the following phrases can be translated into algebra as $x + 3$.

English Phrases Describing Addition	Algebraic Expression	Diagram
Three *more than* a number		
The *sum of* a number and three		
A number *increased by* three	$x + 3$	
Three is *added to* a number.		
Three *greater than* a number		
A number *plus* three		

In a similar way we can use algebra to express English phrases that describe the operations of subtraction, multiplication, and division.

CAUTION: Since subtraction is not commutative, the order is essential. A number decreased by five is $x - 5$. It is not correct to say $5 - x$. Use extra care as you study each example. Make sure you understand the proper order.

English Phrases Describing Subtraction	Algebraic Expression	Diagram
A number *decreased by* four		
Four *less than* a number		
Four is *subtracted from* a number.		
Four *smaller than* a number	$x - 4$	
Four *fewer than* a number		
A number *diminished by* four		
A number *minus* four		
The *difference between* a number and four		

English Phrases Describing Multiplication	Algebraic Expression	Diagram
Double a number		
Twice a number		
The *product* of two and a number	$2x$	
Two *of* a number		
Two *times* a number		

Since division is not commutative, the order is essential. A number divided by 3 is $\frac{x}{3}$. It is not correct to say $\frac{3}{x}$. Use extra care as you study each example.

English Phrases Describing Division	Algebraic Expression	Diagram
A number *divided by* five		
One-*fifth* of a number	$\dfrac{x}{5}$	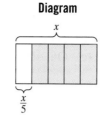
The *quotient* of a number and five		

Often other words are used in English instead of the word *number*. We can use a variable, such as x, here also.

EXAMPLE 1 Write each English phrase as an algebraic expression.

English Phrase	Algebraic Expression
(a) A *quantity* is increased by five.	$x + 5$
(b) Double the *value*	$2x$
(c) One-third of the *weight*	$\dfrac{x}{3}$ or $\dfrac{1}{3}x$
(d) Twelve more than an *unknown number*	$x + 12$
(e) Seven less than a *number*	$x - 7$

Note that the algebraic expression for "seven less than a number" does not follow the order of the words in the English phrase. The variable or expression that follows the words *less than* always comes first.

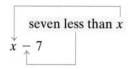

seven less than x

$x - 7$

The variable or expression that follows the words *more than* technically comes before the plus sign. However, since addition is commutative, it also can be written after the plus sign.

Practice Problem 1 Write each English phrase as an algebraic expression.

(a) Four more than a number   **(b)** Triple a value

(c) Eight less than a number   **(d)** One-fourth of a height

More than one operation can be described in an English phrase. Sometimes parentheses must be used to make clear which operation is done first.

EXAMPLE 2 Write each English phrase as an algebraic expression.

English Phrase	Algebraic Expression
(a) Seven more than double a number	$2x + 7$ *Note that these are **not** the same.*
(b) The value of the number is increased by seven and then doubled.	$2(x + 7)$ *Note that the word **then** tells us to add x and 7 before doubling.*
(c) One-half of the sum of a number and 3	$\dfrac{1}{2}(x + 3)$

Practice Problem 2   Write each English phrase as an algebraic expression.

(a) Eight more than triple a number

(b) A number is increased by eight and then it is tripled.

(c) One-third of the sum of a number and 4

 ## Writing an Algebraic Expression to Compare Two or More Quantities

Often in a word problem two or more quantities are described in terms of another. We will want to use a variable to represent one quantity and then write an algebraic expression using *the same variable* to represent the other quantity. Which quantity should we let the variable represent? We usually let the variable represent the quantity that is the basis of comparison: the quantity that the others are being *compared to*.

EXAMPLE 3   Use a variable and an algebraic expression to describe the two quantities in the English sentence "Mike's salary is $2000 more than Fred's salary."

Solution   The two quantities that are being compared are Mike's salary and Fred's salary. Since Mike's salary is being *compared to* Fred's salary, we let the variable represent Fred's salary. The choice of the letter f helps us to remember that the variable represents Fred's salary.

$$\text{Let } f = \text{Fred's salary.}$$

Then $f + \$2000 = $ Mike's salary, since Mike's salary is $2000 *more than* Fred's.

Practice Problem 3   Use a variable and an algebraic expression to describe the two quantities in the English sentence "Marie works 17 hours per week less than Ann."

EXAMPLE 4   The length of a rectangle is 3 meters shorter than twice the width. Use a variable and an algebraic expression to describe the length and the width. Draw a picture of the rectangle and label the length and width.

Solution   The length of the rectangle is being *compared to* the width. Use the letter w for width.

$$\text{Let } w = \text{the width.}$$

$$\overbrace{\text{3 meters shorter than twice the width}}$$

Then $2w - 3 = $ the length.

A picture of the rectangle is shown.

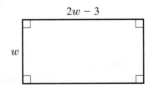

Practice Problem 4   The length of a rectangle is 5 meters longer than double the width. Use a variable and an algebraic expression to describe the length and the width. Draw a picture of the rectangle and label the length and width.

EXAMPLE 5 The first angle of a triangle is triple the second angle. The third angle of the triangle is 12° more than the second angle. Describe each angle algebraically. Draw a diagram of the triangle and label its parts.

Solution Since the first and third angles are described in terms of the second angle, we let the variable represent the number of degrees in the second angle.

Let s = the number of degrees in the second angle.

Then $3s$ = the number of degrees in the first angle.

And $s + 12$ = the number of degrees in the third angle.

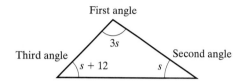

Practice Problem 5 The first angle of a triangle is 16° less than the second angle. The third angle is double the second angle. Describe each angle algebraically. Draw a diagram of the triangle and label its parts.

Some comparisons will involve fractions.

EXAMPLE 6 A theater manager was examining the records of attendance for last year. The number of people attending the theater in January was one-half of the number of people attending the theater in February. The number of people attending the theater in March was three-fifths of the number of people attending the theater in February. Use algebra to describe the attendance each month.

Solution What are we looking for? The *number of people* who attended the theater *each month*. The basis of comparison is February. That is where we begin.

Let f = the number of people who attended in February.

Then $\dfrac{1}{2}f$ = the number of people who attended in January.

And $\dfrac{3}{5}f$ = the number of people who attended in March.

Practice Problem 6 The college dean noticed that in the spring the number of students on campus was two-thirds of the number of students on campus in the fall. She also noticed that in the summer the number of students on campus was one-fifth the number of students on campus in the fall. Use algebra to describe the number of students on campus in each of these three time periods.

Verbal and Writing Skills *Write an algebraic expression for each quantity. Let x represent the unknown value.*

1. a quantity increased by 5

2. nine greater than a number

3. twelve less than a number

4. a value decreased by seven

5. one-eighth of a quantity

6. one-half of a quantity

7. twice a quantity

8. triple a number

9. three more than half of a number

10. five more than one-third of a number

11. double a quantity increased by nine

12. ten times a number increased by 1

13. one-third of the sum of a number and seven

14. one-fourth of the sum of a number and 5

15. one-third of a number reduced by twice the same number

16. one-fifth of a number reduced by double the same number

17. five times a quantity decreased by eleven

18. four less than seven times a number

Write an algebraic expression for each of the quantities being compared.

19. *Stock Value* The value of a share of IBM stock yesterday was $74.50 more than the value of a share of AT&T stock.

20. *Investments* One day in September 2007, the value of one share of Xerox stock was $49.10 less than the value of a share of Target stock.

▲ **21.** *Geometry* The length of the rectangle is 7 inches more than double the width.

▲ **22.** *Geometry* The length of the rectangle is 3 meters more than triple the width.

23. *Cookie Sales* The number of boxes of cookies sold by Sarah was 43 fewer than the number of boxes of cookies sold by Keiko. The number of boxes of cookies sold by Imelda was 53 more than the number sold by Keiko.

24. *April Rainfall* The average April rainfall in Savannah, Georgia, is about 13 inches more than that of Burlington, Vermont. The average rainfall in Phoenix, Arizona, is about 28 inches less than that of Burlington.

▲ **25.** *Geometry* The first angle of a triangle is 16 degrees less than the second angle. The third angle of the triangle is double the second angle.

▲ **26.** *Geometry* The first angle of a triangle is 19 degrees more than the third angle. The second angle is triple the third angle.

27. _Exports_ The value of the exports of Japan was twice the value of the exports of Canada.

28. _Olympic Medals_ The total number of medals won by Canada in all Winter Olympic Games through 2004 was 2 more than Switzerland.

▲ **29. _Geometry_** The first angle of a triangle is triple the second angle. The third angle of the triangle is 14 degrees less than the second angle.

30. _Book Cost_ The cost of Hiro's biology book was $13 more than the cost of his history book. The cost of his English book was $27 less than the cost of his history book.

To Think About

Kayak Rentals _The following bar graph depicts the number of people who rented sea kayaks at Essex Boat Rental during the summer of 2007. Use the bar graph to answer exercises 31 and 32._

31. Write an expression for the number of men who rented sea kayaks at Essex Boat Rental in each age category. Start by using x for the number of men aged 16 to 24 who rented kayaks.

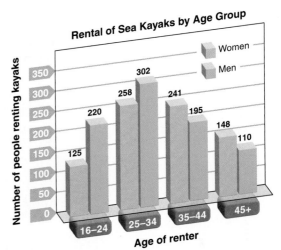

32. Write an expression for the number of women who rented sea kayaks at Essex Boat Rental. Start by using y for the number of women aged 25 to 34 who rented kayaks.

Cumulative Review _Solve for the variable._

33. [2.4.1] $x + \dfrac{1}{2}(x - 3) = 9$

34. [2.4.1] $\dfrac{3}{5}x - 3(x - 1) = 9$

Quick Quiz 2.5 Write an algebraic expression for each quantity. Let x represent the unknown value.

1. Ten greater than a number

2. Five less than double a number

3. The first angle of a triangle is 15 degrees more than the second angle. The third angle of the triangle is double the second angle. Write an algebraic expression for the measure of each of the three angles.

4. Concept Check Explain how you would decide whether to use $\frac{1}{3}(x + 7)$ or $\frac{1}{3}x + 7$ as the algebraic expression for the phrase "one-third of the sum of a number and seven."

2.6 USING EQUATIONS TO SOLVE WORD PROBLEMS

Student Learning Objectives

After studying this section, you will be able to:

1. Solve number problems.

2. Use the Mathematics Blueprint to solve applied word problems.

3. Use formulas to solve word problems.

In Chapters 0 and 1, you solved word problems involving percent and geometric formulas. If you found these word problems challenging or would like to review solving word problems, refer to Appendix B. In Appendix B we introduce a Mathematics Blueprint for Problem Solving that will help you organize information and plan an approach to solve a word problem.

In this section we are going to focus our attention on solving applied problems that require the use of variables, translating English phrases into algebraic expressions, and setting up equations. The process is a little more involved. Some students find the following outline a helpful way to keep organized while solving such problems.

1. *Understand the problem.*
   (a) Read the word problem carefully to get an overview.
   (b) Determine what information you will need to solve the problem.
   (c) Draw a sketch. Label it with the known information. Determine what needs to be found.
   (d) Choose a variable to represent one unknown quantity.
   (e) If necessary, represent other unknown quantities in terms of that very same variable.

2. *Write an equation.*
   (a) Look for key words to help you translate the words into algebraic symbols and expressions.
   (b) Use a given relationship in the problem or an appropriate formula to write an equation.

3. *Solve and state the answer.*

4. *Check.*
   (a) Check the solution in the original equation. Is the answer reasonable?
   (b) Be sure the solution to the equation answers the question in the word problem. You may need to do some additional calculations if it does not.

1 Solving Number Problems

EXAMPLE 1 Two-thirds of a number is eighty-four. What is the number?

Solution

1. *Understand the problem.* Draw a sketch.

   Let x = the unknown number.

2. *Write an equation.*

3. Solve and state the answer.

$$\frac{2}{3}x = 84$$

$$3\left(\frac{2}{3}x\right) = 3(84) \quad \text{Multiply both sides of the equation by 3.}$$

$$2x = 252 \quad \text{Simplify.}$$

$$\frac{2x}{2} = \frac{252}{2} \quad \text{Divide both sides by 2.}$$

$$x = 126$$

The number is 126.

4. Check. Is two-thirds of 126 eighty-four?

$$\frac{2}{3}(126) \overset{?}{=} 84$$

$$84 = 84 \quad \checkmark$$

Practice Problem 1 Three-fourths of a number is negative eighty-one. What is the number?

NOTE TO STUDENT: *Fully worked-out solutions to all of the Practice Problems can be found at the back of the text starting at page SP-1*

Learning to solve problems like Examples 1 and 2 is a very useful skill. You will find that learning the material in the rest of the text will be much easier if you can master the procedure used in these two examples.

EXAMPLE 2 Five more than six times a quantity is three hundred five. Find the number.

Solution

1. Understand the problem. Read the problem carefully. You may not need to draw a sketch.

Let x = the unknown quantity.

2. Write an equation.

Five more than	six times a quantity	is	three hundred five.
5 +	6x	=	305

3. Solve and state the answer. You may want to rewrite the equation to make it easier to solve.

$$6x + 5 = 305$$

$$6x + 5 - 5 = 305 - 5 \quad \text{Subtract 5 from both sides.}$$

$$6x = 300 \quad \text{Simplify.}$$

$$\frac{6x}{6} = \frac{300}{6} \quad \text{Divide both sides by 6.}$$

$$x = 50$$

The quantity, or number, is 50.

4. Check. Is five more than six times 50 three hundred five?

$$6(50) + 5 \overset{?}{=} 305$$

$$300 + 5 \overset{?}{=} 305$$

$$305 = 305 \quad \checkmark$$

Practice Problem 2 Two less than triple a number is forty-nine. Find the number.

EXAMPLE 3   The larger of two numbers is three more than twice the smaller. The sum of the numbers is thirty-nine. Find each number.

Solution

1. *Understand the problem.* Read the problem carefully. The problem refers to *two* numbers. We must write an algebraic expression for *each number* before writing the equation. The larger number is being compared to the smaller number. We want to use *one variable* to describe each number.

$$\text{Let } s = \text{the smaller number.}$$
$$\text{Then } 2s + 3 = \text{the larger number.}$$
$$\underbrace{\phantom{2s + 3 = \text{the larger number.}}}$$
three more than twice the smaller number

2. *Write an equation.*   The sum of the numbers   is   thirty-nine.
$$s + (2s + 3) \qquad\qquad = \qquad 39$$

3. *Solve.* $s + (2s + 3) = 39$
$$3s + 3 = 39 \quad \text{Combine like terms.}$$
$$3s = 36 \quad \text{Subtract 3 from each side.}$$
$$s = 12 \quad \text{Divide both sides by 3.}$$

4. *Check.* $12 + [2(12) + 3] \stackrel{?}{=} 39$
$$39 = 39 \quad \checkmark$$

The solution checks, but have we solved the word problem? We need to find *each* number. 12 is the smaller number. Substitute 12 into the expression $2s + 3$ to find the larger number.

$$2s + 3 = 2(12) + 3 = 27$$

The smaller number is 12. The larger number is 27.

Practice Problem 3   Consider two numbers. The second number is twelve less than triple the first number. The sum of the two numbers is twenty-four. Find each number.

NOTE TO STUDENT: Fully worked-out solutions to all of the Practice Problems can be found at the back of the text starting at page SP-1

 Using the Mathematics Blueprint to Solve Applied Word Problems

To facilitate understanding more involved word problems, we will use a Mathematics Blueprint for Problem Solving similar to the one we use in Appendix B. This format is a simple way to organize facts, determine what to set variables equal to, and select a method or approach that will assist you in finding the desired quantity. You will find using this form helpful, particularly in those cases when you read through a word problem and mentally say to yourself, "Now where do I begin?" You begin by responding to the headings of the blueprint. Soon a procedure for solving the problem will emerge.

Mathematics Blueprint for Problem Solving			
Gather the Facts	Assign the Variable	Basic Formula or Equation	Key Points to Remember

EXAMPLE 4 The mean annual snowfall in Juneau, Alaska, is 105.8 inches. This is 20.2 inches less than three times the annual snowfall in Boston. What is the annual snowfall in Boston?

Solution

Understand the problem and write an equation.

Mathematics Blueprint for Problem Solving

Gather the Facts	Assign the Variable	Basic Formula or Equation	Key Points to Remember
Snowfall in Juneau is 105.8 inches. This is 20.2 inches less than three times the snowfall in Boston.	We do not know the snowfall in Boston. Let b = annual snowfall in Boston. Then $3b - 20.2$ = annual snowfall in Juneau.	Set $3b - 20.2$ equal to 105.8, which is the snowfall in Juneau.	All measurements of snowfall are recorded in inches.

Juneau's snowfall is 20.2 less than three times Boston's snowfall.

$$105.8 = 3b - 20.2$$

Solve and state the answer. You may want to rewrite the equation to make it easier to solve.

$$3b - 20.2 = 105.8$$
$$3b = 126 \qquad \text{Add 20.2 to both sides.}$$
$$b = 42 \qquad \text{Divide both sides by 3.}$$

The annual snowfall in Boston is 42 inches.

Check. Reread the word problem. Work backward.

Three times 42 is 126.
126 less 20.2 is 105.8.

Is this the annual snowfall in Juneau? Yes. ✓

Practice Problem 4 The maximum recorded rainfall for a 24-hour period in the United States occurred in Alvin, Texas, on July 25–26, 1979. The maximum recorded rainfall for a 24-hour period in Canada occurred in Ucluelet Brynnor Mines, British Columbia, on October 6, 1977. Alvin, Texas, received 43 inches of rain in that period. The amount recorded in Texas was 14 inches less than three times the amount recorded in Canada. How much rainfall was recorded at Ucluelet Brynnor Mines in Canada? (*Source:* National Oceanic and Atmospheric Administration.)

Some word problems require a simple translation of the facts. Others require a little more detective work. You will not need to use the Mathematics Blueprint to solve every word problem. As you gain confidence in problem solving, you will no doubt leave out some of the steps. We suggest that you use the procedure when you find yourself on unfamiliar ground. It is a powerful organizational tool.

Using Formulas to Solve Word Problems

Sometimes the relationship between two quantities is so well understood that we have developed a formula to describe that relationship. The following examples show how you can use a formula to solve a word problem.

EXAMPLE 5 Two people traveled in separate cars. They each traveled a distance of 330 miles on an interstate highway. To maximize fuel economy, Fred traveled at exactly 50 mph. Sam traveled at exactly 55 mph. How much time did the trip take each person? (Use the formula distance = rate · time or $d = rt$.)

Solution

Mathematics Blueprint for Problem Solving

Gather the Facts	Assign the Variable	Basic Formula or Equation	Key Points to Remember
Each person drove 330 miles. Fred drove at 50 mph. Sam drove at 55 mph.	Time is the unknown quantity for each driver. Use subscripts to denote different values of t. t_f = Fred's time t_s = Sam's time	distance = (rate)(time) or $d = rt$	The time is expressed in hours.

We must find t in the formula $d = rt$. To simplify calculations we can solve for t before we substitute values:

$$\frac{d}{r} = \frac{\cancel{r}t}{\cancel{r}} \rightarrow \frac{d}{r} = t$$

Substitute the known values into the formula and solve for t.

$$\frac{d}{r} = t \qquad\qquad \frac{d}{r} = t$$

$$\frac{330}{50} = t_f \qquad\qquad \frac{330}{55} = t_s$$

$$6.6 = t_f \qquad\qquad 6 = t_s$$

It took Fred 6.6 hours to drive 330 miles.   It took Sam 6 hours to drive 330 miles.

Check. Is this reasonable? Yes, you would expect Fred to take longer to drive the same distance because Fred drove at a lower speed.

Note: You may wish to express 6.6 hours in hours and minutes. To change 0.6 hour to minutes, proceed as follows:

$$0.6 \ \cancel{\text{hour}} \cdot \frac{60 \text{ minutes}}{1 \ \cancel{\text{hour}}} = (0.6)(60) \text{ minutes} = 36 \text{ minutes}$$

Thus, Fred drove for 6 hours and 36 minutes.

NOTE TO STUDENT: *Fully worked-out solutions to all of the Practice Problems can be found at the back of the text starting at page SP-1*

Practice Problem 5 Sarah left the city to visit her aunt and uncle, who live in a rural area north of the city. She traveled the 220-mile trip in 4 hours. On her way home she took a slightly longer route, which measured 225 miles on the car odometer. The return trip took 4.5 hours.

(a) What was her average speed on the trip leaving the city?

(b) What was her average speed on the return trip?

(c) On which trip did she travel faster and by how much?

EXAMPLE 6 A teacher told Melinda that she has a course average of 78 based on her six math tests. When she got home, Melinda found five of her tests. She had scores of 87, 63, 79, 71, and 96 on the five tests. She could not find her sixth test. What score did she obtain on that test? (Use the formula that an average = the sum of scores ÷ the number of scores.)

Solution

Mathematics Blueprint for Problem Solving

Gather the Facts	Assign the Variable	Basic Formula or Equation	Key Points to Remember
Her five known test scores are 87, 63, 79, 71, and 96. Her course average is 78.	We do not know the score Melinda received on her sixth test. Let x = the score on the sixth test.	average = $\dfrac{\text{sum of scores}}{\text{number of scores}}$	Since there are six test scores, we will need to divide the sum by 6.

When you average anything, you total up the sum of all the values and then divide it by the number of values.

We now write the equation for the average of six items. This involves adding the scores for all the tests and dividing by 6.

$$\frac{87 + 63 + 79 + 71 + 96 + x}{6} = 78$$

$$\frac{396 + x}{6} = 78 \qquad \text{Add the numbers in the numerator.}$$

$$\frac{396}{6} + \frac{x}{6} = 78 \qquad \begin{array}{l}\text{First we rewrite the left side as two}\\ \text{fractions. This is actually multiplying}\\ \frac{1}{6}(396 + x) = \frac{396}{6} + \frac{x}{6}\end{array}$$

$$6 \cdot \frac{396}{6} + 6 \cdot \frac{x}{6} = 6(78) \qquad \begin{array}{l}\text{Multiply both sides of the equation by}\\ \text{6 to remove the fraction.}\end{array}$$

$$396 + x = 468 \qquad \text{Simplify.}$$

$$x = 72 \qquad \text{Subtract 396 from both sides to find } x.$$

Melinda's score on the sixth test was 72.

Check. To verify that this is correct, we check that the average of the 6 tests is 78.

$$\frac{87 + 63 + 79 + 71 + 96 + 72}{6} \stackrel{?}{=} 78$$

$$\frac{468}{6} \stackrel{?}{=} 78$$

$$78 = 78 \checkmark$$

The problem checks. We know that the score on the sixth test was 72.

Practice Problem 6 Barbara's math course has four tests and one final exam. The final exam counts as much as two tests. Barbara has test scores of 78, 80, 100, and 96 on her four tests. What grade does she need on the final exam if she wants to have a 90 average for the course?

Solve. Check your solution.

1. What number minus 543 gives 718?

2. What number added to 74 gives 265?

3. A number divided by eight is 296. What is the number?

4. Eighteen less than a number is 23. What is the number?

5. Seventeen greater than a number is 199. Find the number.

6. Three times a number is one. What is the number?

7. A number is doubled and then increased by seven. The result is ninety-three. What is the original number?

8. Four less than nine times a number is one hundred twenty-two. Find the original number.

9. When eighteen is reduced by two-thirds of a number, the result is 12. Find the number.

10. Twice a number is increased by one-third the same number. The result is 42. Find the number.

11. Eight less than triple a number is the same as five times the same number. Find the number.

12. Ten less than double a number is the same as seven times the number. Find the number.

13. A number, half of that number, and one-third of that number are added. The result is 22. What is the original number?

14. A number, twice that number, and one-third of that number are added. The result is 20. What is the original number?

Applications *Solve. Check to see if your answer is reasonable.*

15. *Motorcycle Inventory* A Harley-Davidson motorcycle shop maintains an inventory of four times as many new bikes as used bikes. If there are 60 new bikes, how many used bikes are now in stock?

16. *Shirt Sizes* The Enro Shirt Outlet store maintains an inventory of five times as many shirts in regular men's sizes (size XL and smaller) than it does in large men's sizes (size 2X and larger). There are 430 shirts in inventory in the store that are regular men's sizes. How many shirts in large men's sizes are in inventory?

17. *Wildfires* Between 1/1/06 and 9/25/06 there were 82,955 wildfires in the United States. This was 14,725 less than twice the number of wildfires between 1/1/03 and 9/25/03. How many wildfires were there during the period between 1/1/03 and 9/25/03? (*Source*: www .nifc.gov)

18. *On-Demand Movies* Joseph just changed his cable television plan to include on-demand movies for $2.95 each. The charge for the basic cable package is $24.95 per month. If his bill one month was $39.70, how many on-demand movies did he watch?

19. *CD Purchase* The sale price of a new Panasonic compact disc player is $218 at a local discount store. At the store where this sale is going on, each new CD is on sale for $11 each. If Kyle purchases a player and some CDs for $284, how many CDs did he purchase?

20. *iPod Nano* Kerry just bought a new Apple iPod Nano for $129. On the same day, she bought several albums from iTunes for $11.99 each. If she spent $224.92 total, how many albums did she buy?

21. ***One-Day Sale*** Raquelle went to Weller's Department Store's one-day sale. She bought two blouses for $38 each and a pair of shoes for $49. She also wanted to buy some jewelry. Each item of jewelry was bargain priced at $11.50 each. If she brought $171 with her, how many pieces of jewelry could she buy?

22. ***Waiting Tables*** Brad is a waiter at the "Steaks-Are-Us" restaurant. He gets paid $5.75 per hour, and he can keep his tips. He knows that his tips average $8.80 per table. If he worked an eight-hour shift and took home $169.20, how many tables did he serve?

23. ***TGV*** On May 18, 1990, the fastest speed of any national railroad was achieved by the French high-speed train *Train à Grande Vitesse* (TGV) as it traveled over a distance from Cortalain to Tours, France. A commentator said that this speed was so fast that if it continued at that rate, the train would travel 6404 miles in 20 hours. How fast did the train travel on that date? (*Source:* www.guinnessworldrecords.com)

24. ***Submarine Speed*** In 1958, the nuclear-powered submarine *Nautilus* took 6 days 12 hours to travel submerged 5068 km across the Atlantic Ocean from Portsmouth, England, to New York City. What was its average speed, in kilometers per hour, for this trip? (Round to the nearest whole number.)

25. ***Gravity*** It has been shown that the force of gravity on a planet varies with the mass of the planet. The force of gravity on Jupiter, for example, is about two and a half times that on Earth. Using this information, approximately how much would a 220-lb astronaut weigh on Jupiter?

26. ***Sunday Comics*** Charles Schulz, the creator of *Peanuts*, once estimated that he had drawn close to 2600 Sunday comics over his career. At that time, about how many years had he been drawing *Peanuts*?

27. ***In-Line Skating*** Two in-line skaters, Nell and Kristin, start from the same point and skate in the same direction. Nell skates at 12 miles per hour and Kristin skates at 14 miles per hour. If they can keep up that pace for 2.5 hours, how far apart will they be at the end of that time?

28. ***Bus Travel*** Two Greyhound buses leave the Jackson City station at the same time. One travels north at 55 miles per hour. The other travels south at 60 miles per hour. In how many hours will the two buses be 460 miles apart?

29. ***Travel Routes*** Nella drove from Albuquerque, New Mexico, to the Garden of the Gods rock formation in Colorado Springs. It took her six hours to travel 312 miles over the mountain road. She came home on the highway. On the highway it took five hours to travel 320 miles. How fast did she travel using the mountain route? How much faster (in miles per hour) did she travel using the highway route?

30. ***Travel Speeds*** Allison drives 30 miles per hour through the city and 55 miles per hour on the New Jersey Turnpike. She drove 90 miles from Battery Park to the Jersey Shore. How much of the time was city driving if she spent 1.2 hours on the turnpike?

31. *Grading System* Danielle's chemistry professor has a complicated grading system. Each lab counts once and each test counts as two labs. At the end of the semester there is a final lab that counts as three regular labs. Danielle received scores of 84, 87, and 93 on her labs, and scores of 89 and 94 on her tests. What score must she get on the final lab to receive an A (an A is 90)?

32. *Politicians' Salaries* In 2007, Speaker of the House Nancy Pelosi and Vice President Dick Cheney both earned the same annual salary. President George W. Bush earned $27,400 more than twice what Pelosi and Cheney each earned. The average of the three salaries was $257,533.33. What did each earn in 2007? Round your answers to the nearest whole dollar. (*Source:* www.govspot.com)

To Think About

33. *Cricket Chirps* In warmer climates, approximate temperature predictions can be made by counting the number of chirps of a cricket during a minute. The Fahrenheit temperature decreased by forty is equivalent to one-fourth of the number of cricket chirps.

(a) Write an equation for this relationship.
(b) Approximately how many chirps per minute should be recorded if the temperature is 90°F?
(c) If a person recorded 148 cricket chirps in a minute, what would be the Fahrenheit temperature according to this formula?

Cumulative Review *Simplify.*

34. **[1.7.2]** $5x(2x^2 - 6x - 3)$

35. **[1.7.2]** $-2a(ab - 3b + 5a)$

36. **[1.7.2]** $7x - 3y - 12x - 8y + 5y$

37. **[1.7.2]** $5x^2y - 7xy^2 - 8xy - 9x^2y$

Quick Quiz 2.6

1. A number is tripled and then decreased by 15. The result is 36. What is the original number?

2. The sum of one-half of a number, one-sixth of the number, and one-eighth of the number is thirty-eight. Find the original number.

3. James scored 84, 89, 73, and 80 on four tests. What must he score on the next test to have an 80 average in the course?

4. **Concept Check** Explain how you would set up an equation to solve the following problem.

Phil purchased two shirts for $23 each and then purchased several pairs of socks. The socks were priced at $0.75 per pair. How many pairs of socks did he purchase if the total cost was $60.25?

1 Finding Area, Perimeter, and Missing Angles

In Section 1.8 we reviewed a number of area and perimeter formulas. We repeat this list of formulas here for convenience.

AREA AND PERIMETER FORMULAS

A **parallelogram** is a four-sided figure with opposite sides parallel. In a parallelogram, opposite sides are equal and opposite angles are equal.

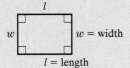

Perimeter = the sum of all four sides

Area = ab

A **rectangle** is a parallelogram with all interior angles measuring 90°.

Perimeter = $2l + 2w$

Area = lw

A **square** is a rectangle with all four sides equal.

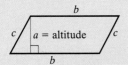

Perimeter = $4s$

Area = s^2

A **trapezoid** is a four-sided figure with two sides parallel. The parallel sides are called the bases of the trapezoid.

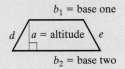

Perimeter = the sum of all four sides

Area = $\frac{1}{2}a(b_1 + b_2)$

A **triangle** is a closed plane figure with three sides.

Perimeter = the sum of the three sides

Area = $\frac{1}{2}ab$

A **circle** is a plane curve consisting of all points at an equal distance from a given point called the center.

Circumference is the distance around a circle. A **radius** is a line segment from the center of the circle to a point on the circle. A **diameter** is a line segment across the circle that passes through the center with endpoints on the circle.

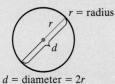

Circumference = $2\pi r$

Area = πr^2

π is a constant associated with circles. It is an irrational number that is approximately 3.141592654. We usually use 3.14 as a sufficiently accurate approximation. Thus we write $\pi \approx 3.14$ for most of our calculations involving π.

We frequently encounter triangles in word problems. There are four important facts about triangles, which we list for convenient reference.

TRIANGLE FACTS

1. The sum of the interior angles of any triangle is 180°. That is,

   measure of ∠A + measure of ∠B + measure of ∠C = 180°.

2. An **equilateral** triangle is a triangle with three sides equal in length and three angles that measure 60° each.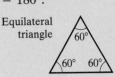

3. An **isosceles** triangle is a triangle with exactly two equal sides. The two angles opposite the equal sides are also equal.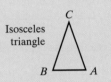

   measure ∠A = measure ∠B

4. A **right** triangle is a triangle with one angle that measures 90°.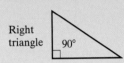

EXAMPLE 1   The area of an NBA basketball court is 4700 square feet. Find the width of the court if the length is 94 feet.

Solution   Draw a diagram.

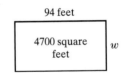

94 feet

4700 square feet w

The formula for the area is $A = lw$. Since we must find the width, we solve the formula for w, then substitute values. Do you see why this simplifies calculations?

$$A = lw \quad \text{Write the formula.}$$

$$\frac{A}{l} = \frac{lw}{l} \quad \text{Solve for } w.$$

$$\frac{A}{l} = w \quad \text{Simplify.}$$

$$\frac{4700\ (\text{ft})(\text{ft})}{94\ \text{ft}} = w \quad \text{Substitute known values. } A = 4700\ \text{ft}^2; l = 94\ \text{ft.}$$

$$50\ \text{ft} = w$$

The width of the basketball court is 50 feet.

▲ **Practice Problem 1**   The area of a rectangular field is 120 square yards. If the width of the field is 8 yards, what is the length?

▲ **EXAMPLE 2**   The area of a trapezoid is 400 square inches. The altitude is 20 inches and one of the bases is 15 inches. Find the length of the other base.

Solution

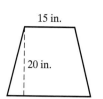

$$A = \frac{1}{2}a\,(b_1 + b_2)$$ Write the formula for the area of a trapezoid.

$$400(\text{in.})^2 = \frac{1}{2}(20 \text{ in.})(15 \text{ in.} + b_2)$$ Substitute the known values.

$400(\text{in.})(\text{in.}) = 10(\text{in.})(15 \text{ in.} + b_2)$ Simplify.

$400(\text{in.})(\text{in.}) = 150(\text{in.})(\text{in.}) + (10 \text{ in.})b_2$ Remove parentheses.

$250(\text{in.})(\text{in.}) = (10 \text{ in.})b_2$ Subtract 150 in.^{2} from both sides.

$25 \text{ in.} = b_2$ Divide both sides by 10 in.

The other base is 25 inches long.

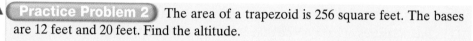

▲ **Practice Problem 2**   The area of a trapezoid is 256 square feet. The bases are 12 feet and 20 feet. Find the altitude.

▲ **EXAMPLE 3**   Find the area of a circular sign whose diameter is 14 inches. (Use 3.14 for π.) Round to the nearest square inch.

Solution $A = \pi r^2$ Write the formula for the area of a circle.

Note that the length of the diameter is given in the word problem. We need to know the length of the radius to use the formula. Since $d = 2r$, we must solve for r: $\frac{d}{2} = \frac{2r}{2}$ so $\frac{d}{2} = r$. Now we replace d with 14 in. to find r: $\frac{14 \text{ in.}}{2} = r$. 7 in. $= r$.

$A = \pi(7 \text{ in.})(7 \text{ in.})$ Substitute known values into the formula.

$= (3.14)(7 \text{ in.})(7 \text{ in.})$

$= (3.14)(49 \text{ in.}^2)$

$= 153.86 \text{ in.}^2$

Rounded to the nearest square inch, the area of the circle is approximately 154 square inches.

In this example, you were asked to find the area of the circle. Do not confuse the formula for area with the formula for circumference. $A = \pi r^2$, while $C = 2\pi r$. Remember that area involves square units, so it is only natural that in the area formula, you would square the radius.

▲ **Practice Problem 3**   Find the circumference of a circle whose radius is 15 meters. (Use 3.14 for π.) Round your answer to the nearest meter.

▲ **EXAMPLE 4**   Find the perimeter of a parallelogram whose longer sides are 4 feet and whose shorter sides are 2.6 feet.

Solution   Draw a picture.

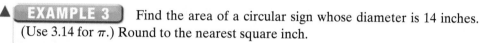

The perimeter is the distance around a figure. To find the perimeter, add the lengths of the sides. Since the opposite sides of a parallelogram are equal, we can write the following.

$P = 2(4 \text{ feet}) + 2(2.6 \text{ feet})$

$= 8 \text{ feet} + 5.2 \text{ feet}$

$= 13.2 \text{ feet}$

The perimeter of the parallelogram is 13.2 feet.

▲ **Practice Problem 4**   Find the perimeter of an equilateral triangle with a side that measures 15 centimeters.

EXAMPLE 5 The smallest angle of an isosceles triangle measures 24°. The other two angles are larger. What are the measurements of the other two angles?

Solution We know that in an isosceles triangle the measures of two angles are equal. We also know that the sum of the measures of all three angles is 180°. Both of the larger angles must be different from 24°, therefore these two larger angles must be equal.

Let x = the measure in degrees of each of the larger angles.

Then we can write

$$24° + x + x = 180°$$
$$24° + 2x = 180° \quad \text{Add like terms.}$$
$$2x = 156° \quad \text{Subtract 24° from each side.}$$
$$x = 78° \quad \text{Divide each side by 2.}$$

Thus the measures of each of the other two angles of the triangle must be 78°.

Practice Problem 5 The largest angle of an isosceles triangle measures 132°. The other two angles are smaller. What are the measurements of the other two angles?

EXAMPLE 6 In a triangle, the measure of the first angle is 25° more than the measure of the second angle. The measure of the third angle is three times the measure of the second angle. What are the measurements of all three angles?

Solution The sum of the measures of all three angles of a triangle is 180°.

Let x = the measure of the second angle. Then $x + 25°$ = the measure of the first angle, and $3x$ = the measure of the third angle.

Then we can write

$$x + 25° + x + 3x = 180°$$
$$5x + 25° = 180° \quad \text{Add like terms.}$$
$$5x = 155° \quad \text{Subtract 25° from each side.}$$
$$x = 31° \quad \text{Divide each side by 5.}$$

Thus, the measure of the second angle is 31°, the measure of the first angle is $31° + 25° = 56°$, and the measure of the third angle is $3(31°) = 93°$.

Practice Problem 6 In a triangle, the measure of the second angle is five degrees more than twice the measure of the first angle. The measure of the third angle is half the measure of the first angle. What are the measurements of all three angles?

2 Finding Volume and Surface Area

Now let's examine three-dimensional figures. **Surface area** is the total area of the faces of a figure. You can find surface area by calculating the area of each face and then finding the sum. **Volume** is the measure of the amount of space inside a figure. Some formulas for the surface area and the volume of regular figures can be found in the following table.

GEOMETRIC FORMULAS: THREE-DIMENSIONAL FIGURES

Rectangular prism

Note that all of the faces are rectangles.

h = height

w = width

l = length

Surface area = $2lw + 2wh + 2lh$

Volume = lwh

GEOMETRIC FORMULAS: THREE-DIMENSIONAL FIGURES (continued)

Sphere

r = radius

Surface area $= 4\pi r^2$

Volume $= \dfrac{4}{3}\pi r^3$

Right circular cylinder

r = radius
h = height

Surface area $= 2\pi rh + 2\pi r^2$

Volume $= \pi r^2 h$

▲ **EXAMPLE 7**   Find the volume of a sphere with radius 4 centimeters. (Use 3.14 for π.) Round your answer to the nearest cubic centimeter.

Solution

$$V = \frac{4}{3}\pi r^3 \qquad \text{Write the formula for the volume of a sphere.}$$

$$= \frac{4}{3}(3.14)(4 \text{ cm})^3 \quad \text{Substitute the known values into the formula.}$$

$$= \frac{4}{3}(3.14)(64) \text{ cm}^3 \approx 267.946667 \text{ cm}^3$$

Rounded to the nearest whole number, the volume of the sphere is 268 cubic centimeters.

▲ **Practice Problem 7**   Find the surface area of a sphere with radius 5 meters. Round your answer to the nearest square meter.

NOTE TO STUDENT: Fully worked-out solutions to all of the Practice Problems can be found at the back of the text starting at page SP-1

▲ **EXAMPLE 8**   A can is made of aluminum. It has a flat top and a flat bottom. The height is 5 inches and the radius is 2 inches. How much aluminum is needed to make the can? How much aluminum is needed to make 10,000 cans?

Solution

Understand the problem. What do we need to find? Reword the problem.

We need to find the total surface area of a cylinder.

Side of the cylinder	5
Circumference	
$2\pi r = 4\pi$	

Top — 2   Bottom — 2
$A = \pi r^2$ $A = \pi r^2$
$A = 4\pi$ $A = 4\pi$

You may calculate the area of each piece and find the sum or you may use the formula.

We will use 3.14 to approximate π.

$$\text{Surface area} = 2\pi rh + 2\pi r^2$$
$$= 2(3.14)(2 \text{ in.})(5 \text{ in.}) + 2(3.14)(2 \text{ in.})^2$$
$$= 62.8 \text{ in.}^2 + 25.12 \text{ in.}^2 = 87.92 \text{ in.}^2$$

87.92 square inches of aluminum are needed to make each can.

Have we answered all the questions in the word problem? Reread the problem. How much aluminum is needed to make 10,000 cans?

$$(\text{aluminum for 1 can})(10{,}000) = (87.92 \text{ in.}^2)(10{,}000) = 879{,}200 \text{ square inches}$$

It would take 879,200 square inches of aluminum to make 10,000 cans.

▲ **Practice Problem 8** Sand is stored in a cylindrical drum that is 4 feet high and has a radius of 3 feet. How much sand can be stored in the drum? Round your answer to the nearest cubic foot.

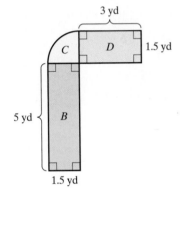

③ Solving More Complicated Geometric Problems

▲ **EXAMPLE 9** A quarter-circle (of radius 1.5 yards) is connected to two rectangles with dimensions as labeled on the sketch in the margin. You need to lay a strip of carpet in your house according to this sketch. How many square yards of carpeting will be needed on the floor? (Use 3.14 for π. Round your final answer to the nearest tenth.)

Solution   The desired area is the sum of three areas, which we will call B, C, and D. Area B and area D are rectangular in shape. It is relatively easy to find the areas of these shapes.

$$A_B = (5 \text{ yd})(1.5 \text{ yd}) = 7.5 \text{ yd}^2 \qquad A_D = (3 \text{ yd})(1.5 \text{ yd}) = 4.5 \text{ yd}^2$$

Area C is one-fourth of a circle. The radius of the circle is 1.5 yd.

$$A_C = \frac{\pi r^2}{4} = \frac{(3.14)(1.5 \text{ yd})^2}{4} = \frac{(3.14)(2.25) \text{ yd}^2}{4} = 1.76625 \text{ yd}^2 \approx 1.8 \text{ yd}^2$$

The total area is $A_B + A_D + A_C \approx 7.5 \text{ yd}^2 + 4.5 \text{ yd}^2 + 1.8 \text{ yd}^2 \approx 13.8 \text{ yd}^2$.
13.8 square yards of carpeting will be needed to cover the floor.

▲ **Practice Problem 9** John has a swimming pool that measures 8 feet by 12 feet. He plans to make a concrete walkway around the pool that is 3 feet wide. What will be the total cost of the walkway at $12 per square foot?

EXAMPLE 10 Find the weight of the water in a full cylinder if the height of the cylinder is 5 feet and the radius is 2 feet. 1 cubic foot of water weighs 62.4 pounds. Round your answer to the nearest ten pounds. Use 3.14 for π.

Solution

Understand the problem.
Draw a diagram.

Write the formula for the volume of a cylinder.

$$\begin{aligned} V &= \pi r^2 h \\ &= (3.14)(2 \text{ ft})^2(5 \text{ ft}) \\ &= (3.14)(4 \text{ ft}^2)(5 \text{ ft}) \\ &= 62.8 \text{ ft}^3 \end{aligned}$$

The volume of the cylinder is 62.8 cubic feet.
Each cubic foot of water weighs 62.4 pounds. This can be written as 62.4 lb/ft^3. Since the volume of the cylinder is 62.8 ft^3, we multiply.

$$\text{Weight of the water} = (62.8 \text{ ft}^3)\left(\frac{62.4 \text{ lb}}{1 \text{ ft}^3}\right)$$
$$= (62.8)(62.4) \text{ lb} = 3918.72 \text{ lb}$$

The weight of the water in the cylinder is 3920 rounded to the nearest ten pounds.

Practice Problem 10 Find the weight of the water in a full rectangular container that is 5 feet wide, 6 feet long, and 8 feet high. Remember, 1 cubic foot of water weighs 62.4 pounds. Round your answer to the nearest 100 pounds.

NOTE TO STUDENT: Fully worked-out solutions to all of the Practice Problems can be found at the back of the text starting at page SP-1

Verbal and Writing Skills

Geometry *Fill in the blank to complete each sentence.*

▲ **1.** Perimeter is the _____ a plane figure.

▲ **2.** _____ is the distance around a circle.

▲ **3.** Area is a measure of the amount of _____ in a region.

▲ **4.** A _____ is a four-sided figure with exactly two sides parallel.

▲ **5.** The sum of the interior angles of any triangle is _____.

▲ **6.** An equilateral triangle is a triangle in which each angle measures _____.

▲ **7.** ***Geometry*** Find the area of a triangle whose altitude is 24 inches and whose base is 13 inches.

▲ **8.** ***Geometry*** A triangle has an altitude of 9 m and a base of 21 m. Find the area of the triangle.

▲ **9.** ***Geometry*** Find the area of a parallelogram whose altitude is 14 inches and whose base is 7 inches.

▲ **10.** ***Geometry*** Find the area of a parallelogram whose altitude is 15 meters and whose base is 10 meters.

▲ **11.** ***Geometry*** The area of a triangle is 90 square feet. Find the altitude if the base is 30 feet.

▲ **12.** ***Geometry*** The area of a triangle is 120 square inches. Find the altitude if the base is 24 inches.

▲ **13.** ***Geometry*** The perimeter of a parallelogram is 46 inches. If the length of one side is 14 inches, what is the length of a side adjacent to it?

▲ **14.** ***Geometry*** If one side of a parallelogram measures 18.5 cm and an adjacent side measures 17.5 cm, what is the perimeter of the parallelogram?

▲ **15.** ***Geometry*** Find the area of a circular sign whose radius is 7.00 feet. (Use 3.14 to approximate π.)

▲ **16.** ***Flower Bed*** Find the area of a circular flower bed whose diameter is 6 meters. (Use 3.14 to approximate π.)

▲ **17.** ***Mercury*** The diameter of the planet Mercury is approximately 3032 miles. Find the distance around its equator. (Use 3.14 to approximate π.)

▲ **18.** ***Olympic Medals*** The medals made for the 2008 Beijing Olympics measure 70 mm, or about 2.76 inches, in diameter. Find the area of the medals in square inches. (Use 3.14 to approximate π.) Round your answer to the nearest hundredth.

▲ **19.** ***Geometry*** The area of a trapezoid is 600 square inches and the bases are 20 inches and 30 inches. Find the altitude.

▲ **20.** ***Geometry*** The area of a trapezoid is 900 square inches. The bases are 40 inches and 50 inches. Find the altitude.

Applications

▲ **21.** *Streets of Manhattan* In midtown Manhattan, the street blocks have a uniform size of 80 meters north-south by 280 meters east-west. If a typical New York City neighborhood is two blocks east-west by three blocks north-south, how much area does it cover?

▲ **22.** *Computer Cases* A computer manufacturing plant makes its computer cases from plastic. Two sides of a triangular scrap of plastic measure 7 cm and 25 cm.

   (a) If the perimeter of the plastic piece is 56 cm, how long is the third side?

   (b) If each side of the triangular piece is decreased by 0.1 cm, what would the new perimeter be?

▲ **23.** *Geometry* The circumference of a circle is 31.4 centimeters. Find the radius. (Use 3.14 to approximate π.)

▲ **24.** *Geometry* The perimeter of an equilateral triangle is 27 inches. Find the length of each side of the triangle.

▲ **25.** *Tires* An automobile tire has a diameter of 64 cm. What is the circumference of the tire? (Use 3.14 to approximate π.)

▲ **26.** *Golden Ratio* The ancient Greeks discovered that rectangles where the length-to-width ratio is 8 to 5 (known as the golden ratio) are most pleasing to the eye. Kayla wants to create a painting that will be framed by some antique molding that she found. If she calculates that the molding can be used to make a frame of 104 inches in perimeter, what dimensions should her painting be so that it is visually appealing?

▲ **27.** *Geometry* Two angles of a triangle measure 47 degrees and 59 degrees. What is the measure of the third angle?

▲ **28.** *Geometry* A right triangle has one angle that measures 77 degrees. What does the other acute angle measure?

▲ **29.** *Geometry* Each of the equal angles of an isosceles triangle is twice as large as the third angle. What is the measure of each angle?

▲ **30.** *Geometry* The measure of the first angle in a triangle is triple the measure of the second angle. The measure of the third angle is 10 degrees more than the second angle. What is the measure of each angle?

▲ **31.** *Playhouse Construction* Roger constructs children's playhouses. At the top of the houses he places an isosceles triangle to form the roof. The largest angle measures 120°. What are the measurements of the other two angles in this triangle?

▲ **32.** *TV Antenna* The largest angle of an isosceles triangle used in holding a cable TV antenna measures 146°. The other two angles are smaller. What are the measurements of the other two angles in the triangular piece of the cable TV antenna?

▲ **33.** *Parking Lot* A triangular region of the community college parking lot was measured. The measure of the first angle of this triangle is double the measure of the second angle. The measure of the third angle is 19° greater than double the measure of the first angle. What are the measurements of all three angles?

▲ **34.** *Lobster Pot* A lobster pot used in Gloucester, Massachusetts, has a small triangular piece of wood at the point where the lobster enters. The measure of the first angle of this triangle is triple the measure of the second angle. The third angle of the triangle is 30° less than double the measure of the first angle. What are the measurements of all three angles?

▲ **35.** *Garden Border* Kim wants to put a small fence in the shape of a circle around a tree in her backyard. She wants the radius of the circle to be 3 feet, leaving room to plant flowers around the tree. How many feet of fencing does she need to buy if it is sold by the foot? (Use 3.14 to approximate π.)

▲ **36.** *Rock Garden* Nate and Jenny want to order large rocks to use as a border around three rectangular gardens. Each garden measures 4 feet by 9 feet, and each rock will measure about 6 inches across. How many rocks do they need to order?

▲ **37.** *Painting a Shed* The cost for paint to cover the exterior of the four sides of a shed was $40. A 5-gallon can of paint costs $20 and will cover 102 square feet. The shed consists of four sides, each shaped like a trapezoid. They each have one base of 8 feet and one base of 9 feet. What is the altitude of each trapezoid?

▲ **38.** *Painting a Sign* The cost for paint to cover both sides of a large road-side sign (front side and back side) was $150. A 5-gallon can of paint costs $25 and will cover 250 square feet. The sign is constructed in the shape of a trapezoid. It has one base measuring 35 feet and one base measuring 40 feet. What is the altitude of this trapezoid?

▲ **39.** *Geometry* What is the volume of a cylinder whose height is 8 inches and whose radius is 10 inches? (Use 3.14 to approximate π.)

▲ **40.** *Gas Cylinder* A cylinder holds propane gas. The volume of the cylinder is 235.5 cubic feet. Find the height if the radius is 5 feet. (Use 3.14 to approximate π.)

▲ **41.** *Weather Balloon* A spherical weather balloon needs to hold at least 175 cubic feet of helium to be buoyant enough to lift the meteorological instruments. Will a helium-filled balloon with a diameter of 8 feet stay aloft?

▲ **42.** *Storage Freezer* George is considering purchasing a new freezer. He has space for a freezer that measures 5 feet long and 2.5 feet high. He also needs to have 25 cubic feet of storage space. How wide can the freezer be?

▲ **43.** *Milk Container* A plastic cylinder made to hold milk is constructed with a solid top and bottom. The radius is 6 centimeters and the height is 4 centimeters. **(a)** Find the volume of the cylinder. **(b)** Find the total surface area of the cylinder. (Use 3.14 to approximate π.)

▲ **44.** *Pyrex Sphere* A Pyrex glass sphere is made to hold liquids in a science lab. The radius of the sphere is 3 centimeters. **(a)** Find the volume of the sphere. **(b)** Find the total surface area of the sphere. (Use 3.14 to approximate π.)

▲ **45. *Goat Tether*** Dan Perkin's goat is attached to the corner of the 8-foot-by-4-foot pump house by an 8-foot length of rope. The goat eats the grass around the pump house as shown in the diagram. This pattern consists of three-fourths of a circle of radius 8 feet and one-fourth of a circle of radius 4 feet. What area of grass can Dan Perkin's goat eat? (Use 3.14 to approximate π.)

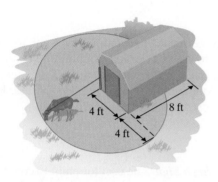

▲ **46. *Seven-Layer Torte*** For a party, Julian is making a seven-layer torte, which consists of seven 9-inch-diameter cake layers with a cream topping on each. He knows that one recipe of cream topping will cover 65 square inches of cake. How many batches of the topping must he make to complete his torte? (Use 3.14 to approximate π.)

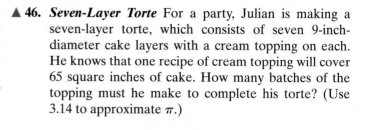

▲ **47. *Walkway*** A cement walkway is to be poured. It consists of the two rectangles with dimensions as shown in the diagram and a quarter of a circle with radius 1.5 yards. (Use 3.14 to approximate π.) **(a)** How many square yards will the walkway be? **(b)** If a painter paints it for $2.50 per square yard, how much will the painting cost?

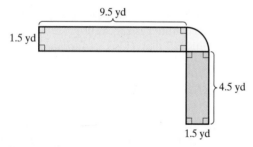

▲ **48. *Aluminum Plate*** An aluminum plate is made according to the following dimensions. The outer radius is 10 inches and the inner radius is 3 inches. (Use 3.14 to approximate π.) **(a)** How many square inches in area is the aluminum plate (the shaded region in the sketch)? **(b)** If the cost to construct the plate is $7.50 per square inch, what is the construction cost?

49. *Toy Car Track* New Toy Company is producing a racing car set, complete with plastic track. The track is a solid surface and has a straight edge with circular ends. See the following sketch for dimensions. What is the distance around the outside of the track? (Use 3.14 to approximate π.)

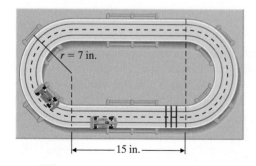

50. *Uniform Design* The Viceroy soccer team has designed new uniforms. The front of their jerseys will have a blue rectangle, with a red V inside. (See the diagram.) The plain jersey shirt costs $19.00. The red color costs $0.05 per square inch and the blue color costs $0.02 per square inch.

(a) What is the area of the red V?

(b) What is the area of the blue color?

(c) What is the cost per jersey?

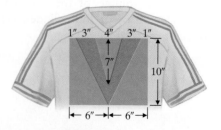

▲ **51.** *Painting Frame* It cost $120.93 to frame a painting with a flat aluminum strip. This was calculated at $1.39 per foot for materials and labor. Find the dimensions of the painting if its length is 3 feet less than four times its width.

▲ **52.** *Driveway* A concrete driveway measures 18 yards by 4 yards. The driveway will be 0.5 yards thick. **(a)** How many cubic yards of concrete will be needed? **(b)** How much will it cost if concrete is $3.50 per cubic yard?

To Think About

▲ **53.** *Moon* Assume that a very long rope supported by poles that are 3 feet tall is stretched around the moon. (Neglect gravitational pull and assume that the rope takes the shape of a large circle.) The radius of the moon is approximately 1080 miles. How much longer would you need to make the rope if you wanted the rope to be supported by poles that are 4 feet tall? (Use 3.14 to approximate π.)

Cumulative Review *Simplify.*

54. **[1.9.1]** $2(x - 6) - 3[4 - x(x + 2)]$

55. **[1.9.1]** $-5(x + 3) - 2[4 + 3(x - 1)]$

56. **[0.5.5]** In 2002, 203 billion pieces of mail were delivered in the United States. By 2006, that number had increased by 5%. How many pieces of mail were delivered in 2006? (*Source:* www.usps.com) Round your answer to the nearest billion.

57. **[0.5.5]** During the first six months of 2005, 92,558 hybrid cars were sold in the United States. During the same period of 2004, 36,276 hybrids were sold. (*Source:* www.allabouthybridcars.com)

(a) How many more hybrid cars were sold in 2005 than in 2004?

(b) This increase in 2005 sales is what percent of the number of hybrid cars sold in the first six months of 2004? Round to the nearest percent.

Quick Quiz 2.7

1. Find the area of a trapezoid if the two bases are 12 feet and 17 feet and the altitude is 8 feet.

2. Find the volume of a sphere with a radius of 3 inches. Use 3.14 as an approximation for π. Round your answer to the nearest cubic inch.

3. The hallway at the gym is a square that measures 5 yards on a side. The square is connected to a rectangle that measures 5 yards by 14 yards. How much would it cost to carpet this hallway with all-weather carpeting that costs $24 per square yard?

4. **Concept Check** Explain how you would set up an equation to solve the following problem.

Two angles of a triangle measure 135° and 11°. What is the measure of the third angle?

Student Learning Objectives

After studying this section, you will be able to:

1. Interpret inequality statements.

2. Graph an inequality on a number line.

3. Translate English phrases into algebraic statements.

4. Solve and graph an inequality.

1 Interpreting Inequality Statements

We frequently speak of one value being greater than or less than another value. We say that "5 is less than 7" or "9 is greater than 4." These relationships are called **inequalities.** We can write inequalities in mathematics by using symbols. We use the symbol $<$ to represent the words "**is less than.**" We use the symbol $>$ to represent the words "**is greater than.**"

Statement in Words	Statement in Algebra
5 is less than 7.	$5 < 7$
9 is greater than 4.	$9 > 4$

Note: "5 is less than 7" and "7 is greater than 5" have the same meaning. Similarly, $5 < 7$ and $7 > 5$ have the same meaning. They represent two equivalent ways of describing the same relationship between the two numbers 5 and 7.

We can better understand the concept of inequality if we examine a number line.

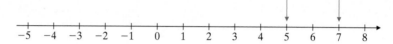

We say that one number is greater than another if it is to the right of the other on the number line. Thus $7 > 5$, since 7 is to the right of 5.

What about negative numbers? We can say "-1 is greater than -3" and write it in symbols as $-1 > -3$ because we know that -1 lies to the right of -3 on the number line.

EXAMPLE 1 In each statement, replace the question mark with the symbol $<$ or $>$.

(a) $3 ? -1$ **(b)** $-2 ? 1$ **(c)** $-3 ? -4$ **(d)** $0 ? 3$ **(e)** $-3 ? 0$

Solution

(a) $3 > -1$ Use $>$, since 3 is to the right of -1 on the number line.

(b) $-2 < 1$ Use $<$, since -2 is to the left of 1. (Or equivalently, we could say that 1 is to the right of -2.)

(c) $-3 > -4$ Note that -3 is to the right of -4.

(d) $0 < 3$

(e) $-3 < 0$

Practice Problem 1 In each statement, replace the question mark with the symbol $<$ or $>$.

(a) $7 ? 2$ **(b)** $-2 ? -4$ **(c)** $-1 ? 2$ **(d)** $-8 ? -5$ **(e)** $0 ? -2$ **(f)** $5 ? -3$

2 Graphing an Inequality on a Number Line

Sometimes we will use an inequality to express the relationship between a variable and a number. $x > 3$ means that x could have the value of *any number* greater than 3.

Any number that makes an inequality true is called a **solution** of the inequality. The set of all numbers that make the inequality true is called the **solution set.** A picture that represents all of the solutions of an inequality is called a **graph** of the inequality.

The inequality $x > 3$ can be graphed on a number line as follows:

Case 1

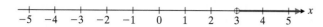

Note that all of the points to the right of 3 are shaded. The open circle at 3 indicates that we do not include the point for the number 3.

Sometimes a variable will be either less than or equal to a certain number. In the statement "x is less than or equal to -2," we are implying that x could have the value of -2 or any number less than -2. We write this as $x \le -2$. We graph it as follows:

Case 2

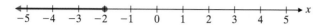

The closed circle at -2 indicates that we *do* include the point for the number -2. *Note:* Be careful that you do not confuse $\circ\!\!\longrightarrow$ with $\bullet\!\!\longrightarrow$. It is important to decide if you need an open circle or a closed one. Case 1 uses an open circle and Case 2 uses a closed circle.

EXAMPLE 2   State each mathematical relationship in words and then graph it.

(a) $x < -2$ **(b)** $x \ge -2$ **(c)** $-3 < x$ **(d)** $x \le -\dfrac{1}{2}$

Solution

(a) We state that "x is less than -2."

$x < -2$

(b) We state that "x is greater than or equal to -2."

$x \ge -2$

(c) We can state that "-3 is less than x" or, equivalently, that "x is greater than -3." Be sure you see that $-3 < x$ is equivalent to $x > -3$. Although both statements are correct, we *usually write the variable first* in a simple inequality containing a variable and a numerical value.

$x > -3$

(d) We state that "x is less than or equal to $-\frac{1}{2}$."

$x \le -\frac{1}{2}$

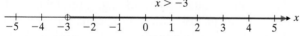

Practice Problem 2   State each mathematical relationship in words and then graph it on the given number line.

(a) $x > 5$

(b) $x \le -2$

(c) $3 > x$

(d) $x \ge -\dfrac{3}{2}$

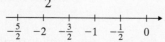

TO THINK ABOUT: Comparing Results   What is the difference between the graphs in Example 2(a), 2(b), and Case 2 above? Why are these graphs different?

Translating English Phrases into Algebraic Statements

We can translate many everyday situations into algebraic statements with an unknown value and an inequality symbol. This is the first step in solving word problems using inequalities.

EXAMPLE 3 Translate each English statement into an algebraic statement.

(a) The police on the scene said that the car was traveling more than 80 miles per hour. (Use the variable *s* for speed.)

(b) The owner of the trucking company said that the payload of a truck must never exceed 4500 pounds. (Use the variable *p* for payload.)

Solution

(a) Since the speed must be greater than 80, we have $s > 80$.

(b) If the payload of the truck can never exceed 4500 pounds, then the payload must be always less than or equal to 4500 pounds. Thus we write $p \leq 4500$.

Practice Problem 3 Translate each English statement into an algebraic statement.

(a) During the drying cycle, the temperature inside the clothes dryer must never exceed 180 degrees Fahrenheit. (Use the variable *t* for temperature.)

(b) The bank loan officer said that the total consumer debt incurred by Wally and Mary must be less than $15,000 if they want to qualify for a mortgage to buy their first home. (Use the variable *d* for debt.)

NOTE TO STUDENT: Fully worked-out solutions to all of the Practice Problems can be found at the back of the text starting at page SP-1

 Solving and Graphing Inequalities on a Number Line

When we **solve an inequality,** we are finding *all* the values that make it true. To solve an inequality, we simplify it to the point where we can clearly see all possible values for the variable. We've solved equations by adding, subtracting, multiplying by, and dividing by a particular value on both sides of the equation. Here we perform similar operations with inequalities with one important exception. We'll show some examples so that you can see how these operations can be used with inequalities just as with equations.

We will first examine the pattern that occurs when we perform these operations *with a positive value* on both sides of an inequality.

Original Inequality	Operations with a Positive Number	New Inequality
$4 < 6$	Add 2 to both sides.	$6 < 8$
	Subtract 2 from both sides.	$2 < 4$
	Multiply both sides by 2.	$8 < 12$
	Divide both sides by 2.	$2 < 3$

Notice that the inequality symbol remains the same when these operations are performed with a positive value.

Now let us examine what happens when we perform these operations *with a negative value.*

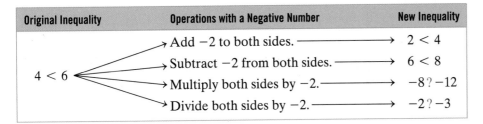

Original Inequality	Operations with a Negative Number	New Inequality
4 < 6	Add −2 to both sides.	2 < 4
	Subtract −2 from both sides.	6 < 8
	Multiply both sides by −2.	−8 ? −12
	Divide both sides by −2.	−2 ? −3

What happens to the inequality sign when we multiply both sides by a negative number? Since −8 is to the right of −12 on a number line, we know that the new inequality should be −8 > −12 if we want the statement to remain true. Notice how we reverse the direction of the inequality from < (less than) to > (greater than) when we multiply by a negative value. Thus we have the following.

$$4 < 6 \longrightarrow \text{Multiply both sides by } -2. \longrightarrow -8 > -12$$

The same thing happens when we divide by a negative number. The inequality is reversed from < to >. We know this since −2 is to the right of −3 on a number line.

$$4 < 6 \longrightarrow \text{Divide both sides by } -2. \longrightarrow -2 > -3$$

Similar reversals take place in the next example.

EXAMPLE 4   Perform the given operations and write the new inequalities.

Original Inequality			New Inequality
(a) −2 < −1	⟶	Multiply both sides by −3. ⟶	6 > 3
(b) 0 > −4	⟶	Divide both sides by −2. ⟶	0 < 2
(c) 8 ≥ 4	⟶	Divide both sides by −4. ⟶	−2 ≤ −1

Notice that we perform the arithmetic with signed numbers just as we always do. But the new inequality signs are reversed (from those of the original inequalities). *Whenever both sides of an inequality are multiplied or divided by a negative quantity, the direction of the inequality is reversed.*

Practice Problem 4   Perform the given operations and write the new inequalities.

(a) 7 > 2	Multiply each side by −2.
(b) −3 < −1	Multiply each side by −1.
(c) −10 ≥ −20	Divide each side by −10.
(d) −15 ≤ −5	Divide each side by −5.

NOTE TO STUDENT: Fully worked-out solutions to all of the Practice Problems can be found at the back of the text starting at page SP-1

PROCEDURE FOR SOLVING INEQUALITIES

You may use the same procedures to solve inequalities that you used to solve equations *except* that the direction of an inequality is *reversed* if you *multiply* or *divide* both sides *by a negative number.*

It may be helpful to think over quickly what we have discussed here. The inequalities remain the same when we add a number to both sides or subtract a number from both sides of the inequality. The inequalities remain the same when we multiply both sides by a positive number or divide both sides by a positive number.

However, if we *multiply* both sides of an inequality by a *negative number* or if we *divide* both sides of an inequality by a *negative number*, then *the inequality is reversed.*

EXAMPLE 5 Solve and graph $3x + 7 \geq 13$.

Solution

$$3x + 7 - 7 \geq 13 - 7 \qquad \text{Subtract 7 from both sides.}$$
$$3x \geq 6 \qquad \text{Simplify.}$$
$$\frac{3x}{3} \geq \frac{6}{3} \qquad \text{Divide both sides by 3.}$$
$$x \geq 2 \qquad \text{Simplify. Note that the direction of the inequality is not changed, since we have divided by a positive number.}$$

The graph is as follows:

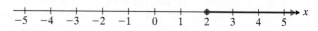

Practice Problem 5 Solve and graph $8x - 2 < 3$.

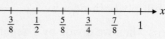

EXAMPLE 6 Solve and graph $5 - 3x > 7$.

Solution

$$5 - 5 - 3x > 7 - 5 \qquad \text{Subtract 5 from both sides.}$$
$$-3x > 2 \qquad \text{Simplify.}$$
$$\frac{-3x}{-3} < \frac{2}{-3} \qquad \text{Divide by } -3 \text{ and \textbf{reverse the inequality} since we are dividing by a negative number.}$$
$$x < -\frac{2}{3} \qquad \text{Note the direction of the inequality.}$$

The graph is as follows:

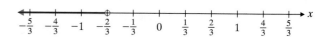

Practice Problem 6 Solve and graph $4 - 5x > 7$.

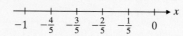

Just like equations, some inequalities contain parentheses and fractions. The initial steps to solve these inequalities will be the same as those used to solve equations with parentheses and fractions. When the variable appears on both sides of the inequality, it is advisable to combine the x-terms on the left side of the inequality symbol.

EXAMPLE 7 Solve and graph $-\frac{13x}{2} \leq \frac{x}{2} - \frac{15}{8}$.

Solution

$$8\left(\frac{-13x}{2}\right) \leq 8\left(\frac{x}{2}\right) - 8\left(\frac{15}{8}\right) \qquad \text{Multiply all terms by LCD} = 8. \text{ We do \textbf{not} reverse the direction of the inequality symbol since we are multiplying by a positive number.}$$

$$-52x \leq 4x - 15 \qquad \text{Simplify.}$$
$$-52x - 4x \leq 4x - 15 - 4x \qquad \text{Subtract } 4x \text{ from both sides.}$$
$$-56x \leq -15 \qquad \text{Combine like terms.}$$
$$\frac{-56x}{-56} \geq \frac{-15}{-56} \qquad \text{Divide both sides by } -56. \text{ We \textbf{reverse} the direction of the inequality when we divide both sides by a negative number.}$$
$$x \geq \frac{15}{56}$$

The graph is as follows: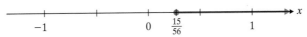

Practice Problem 7 Solve and graph

$$\frac{1}{2}x + 3 < \frac{2}{3}x.$$

16 17 18 19 20 21

EXAMPLE 8 Solve and graph $\frac{1}{3}(3 - 2x) \le -4(x + 1)$.

Solution

$$1 - \frac{2x}{3} \le -4x - 4 \qquad \text{Remove parentheses.}$$

$$3(1) - 3\left(\frac{2x}{3}\right) \le 3(-4x) - 3(4) \qquad \text{Multiply all terms by LCD} = 3.$$

$$3 - 2x \le -12x - 12 \qquad \text{Simplify.}$$

$$3 - 2x + 12x \le -12x + 12x - 12 \qquad \text{Add } 12x \text{ to both sides.}$$

$$3 + 10x \le -12 \qquad \text{Combine like terms.}$$

$$3 - 3 + 10x \le -12 - 3 \qquad \text{Subtract 3 from both sides.}$$

$$10x \le -15 \qquad \text{Simplify.}$$

$$\frac{10x}{10} \le \frac{-15}{10} \qquad \text{Divide both sides by 10. Since we are dividing by a \textbf{positive} number, the inequality is \textbf{not} reversed.}$$

$$x \le -\frac{3}{2}$$

The graph is as follows: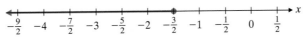

Practice Problem 8 Solve and graph

$$\frac{1}{2}(3 - x) \le 2x + 5.$$

-2 $-\frac{9}{5}$ $-\frac{8}{5}$ $-\frac{7}{5}$ $-\frac{6}{5}$ -1

CAUTION: The most common error students make when solving inequalities is forgetting to reverse the direction of the inequality symbol when multiplying or dividing both sides of the inequality by a negative number.

Normally when you solve inequalities you solve for x by putting the variables on the left side. If you solve by placing the variables on the right side, you will end up with statements like $3 > x$. This is equivalent to $x < 3$. It is wise to express your answer with the variables on the left side.

EXAMPLE 9 A hospital director has determined that the costs of operating one floor of the hospital for an eight-hour shift must never exceed $2370. An expression for the cost of operating one floor of the hospital is $130n + 1200$, where n is the number of nurses. This expression is based on an estimate of $1200 in fixed costs and a cost of $130 per nurse for an eight-hour shift. Solve the inequality $130n + 1200 \le 2370$ to determine the number of nurses that may be on duty on this floor during an eight-hour shift if the director's cost control measure is to be followed.

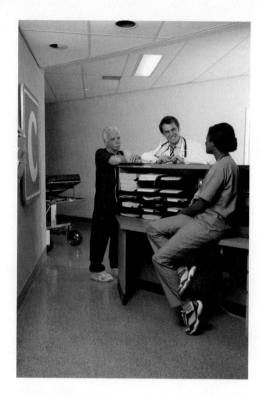

Solution

$$130n + 1200 \leq 2370 \qquad \text{The inequality we must solve.}$$

$$130n + 1200 - 1200 \leq 2370 - 1200 \qquad \text{Subtract 1200 from each side.}$$

$$130n \leq 1170 \qquad \text{Simplify.}$$

$$\frac{130n}{130} \leq \frac{1170}{130} \qquad \text{Divide each side by 130.}$$

$$n \leq 9$$

The number of nurses on duty on this floor during an eight-hour shift must always be less than or equal to nine.

Practice Problem 9 The company president of Staywell, Inc., wants the monthly profits never to be less than $2,500,000. He has determined that an expression for monthly profit for the company is $2000n - 700,000$. In the expression, n is the number of exercise machines manufactured each month. The profit on each machine is $2000, and the $-\$700,000$ in the expression represents the fixed costs of running the company.

Solve the inequality $2000n - 700,000 \geq 2,500,000$ to find how many machines must be made and sold each month to satisfy these financial goals.

Developing Your Study Skills

Getting Organized for an Exam

Studying adequately for an exam requires careful preparation. Begin early so that you will be able to spread your review over several days. Even though you may still be learning new material at this time, you can be reviewing concepts previously learned in the chapter. Giving yourself plenty of time for review will take the pressure off. You need this time to process what you have learned and to tie concepts together.

Adequate preparation enables you to feel confident and to think clearly with less tension and anxiety.

Verbal and Writing Skills

1. Is the statement $5 > -6$ equivalent to the statement $-6 < 5$? Why?

2. Is the statement $-8 < -3$ equivalent to the statement $-3 > -8$? Why?

Replace the ? by $<$ or $>$.

3. $9 \; ? \; -3$ 4. $-2 \; ? \; 5$ 5. $0 \; ? \; -8$ 6. $-9 \; ? \; 0$ 7. $-4 \; ? \; -2$ 8. $-3 \; ? \; -6$

9. (a) $-7 \; ? \; 2$
   (b) $2 \; ? \; -7$

10. (a) $-5 \; ? \; 11$
    (b) $11 \; ? \; -5$

11. (a) $15 \; ? \; -15$
    (b) $-15 \; ? \; 15$

12. (a) $-17 \; ? \; 17$
    (b) $17 \; ? \; -17$

13. $\dfrac{3}{5} \; ? \; \dfrac{4}{7}$

14. $\dfrac{4}{6} \; ? \; \dfrac{7}{9}$

15. $\dfrac{7}{8} \; ? \; \dfrac{25}{31}$

16. $\dfrac{9}{11} \; ? \; \dfrac{41}{53}$

17. $-6.6 \; ? \; -8.9$

18. $-4.2 \; ? \; -7.3$

19. $-4.2 \; ? \; 3.5$

20. $-2.6 \; ? \; 7.5$

21. $-\dfrac{10}{3} \; ? \; -3$

22. $-6 \; ? \; -\dfrac{17}{3}$

23. $-\dfrac{5}{8} \; ? \; -\dfrac{3}{5}$

24. $-\dfrac{2}{3} \; ? \; -\dfrac{3}{4}$

Graph each inequality on the number line.

25. $x > 7$

26. $x < 1$

27. $x \geq -6$

28. $x \leq -2$

29. $x < -\dfrac{1}{4}$

30. $x \leq -\dfrac{3}{2}$

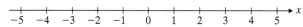

31. $x \leq -5.3$

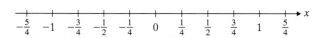

32. $x > -3.5$

33. $25 < x$

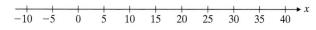

34. $35 \geq x$

Translate each graph to an inequality using the variable x.

35.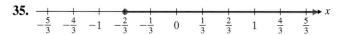

36.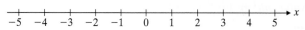

37.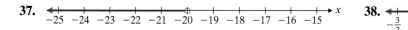

38.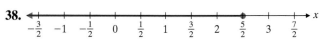

39.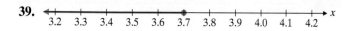

40.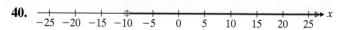

Translate each English statement into an inequality.

41. *Full-Time Student* At Normandale Community College the number of credits a student takes per semester must not be less than 12 to be considered full time. (Use the variable c for credits.)

42. *BMI Category* A person is considered underweight if his or her BMI (body mass index) measurement is smaller than 18.5. (Use the variable B for BMI.)

43. *Height Limit* In order for you to be allowed to ride the roller coaster at the theme park, your height must be at least 48 inches. (Use h for height.)

44. *Boxing Category* To box in the featherweight category, your weight must not exceed 126 pounds. (Use w for weight.)

To Think About

45. Suppose that the variable x must satisfy *all* of these conditions.

$$x \le 2, \quad x > -3, \quad x < \frac{5}{2}, \quad x \ge -\frac{5}{2}$$

Graph on a number line the region that satisfies all of the conditions.

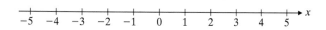

46. Suppose that the variable x must satisfy *all* of these conditions.

$$x < 4, \quad x > -4, \quad x \le \frac{7}{2}, \quad x \ge -\frac{9}{2}$$

Graph on a number line the region that satisfies all of the conditions.

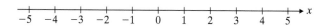

Solve and graph the result.

47. $x + 7 \le 4$

48. $x - 5 < -3$

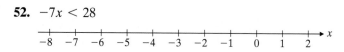

49. $5x \le 25$

50. $6x \ge -42$

51. $-2x < 18$

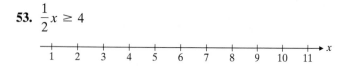

52. $-7x < 28$

53. $\frac{1}{2}x \ge 4$

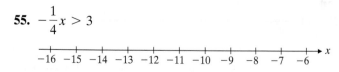

54. $\frac{1}{3}x \le 2$

55. $-\frac{1}{4}x > 3$

56. $-\frac{1}{5}x < 10$

57. $8 - 5x > 13$

58. $9 - 4x \le 21$

59. $-4 + 5x < -3x + 8$

60. $-6 - 4x < 1 - 6x$

61. $\dfrac{5x}{6} - 5 > \dfrac{x}{6} - 9$

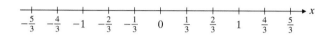

62. $\dfrac{x}{4} - 2 < \dfrac{3x}{4} + 5$

63. $2(3x + 4) > 3(x + 3)$

64. $5(x - 3) \le 2(x - 3)$

Verbal and Writing Skills

65. Add -2 to both sides of the inequality $5 > 3$. What is the result? Why is the direction of the inequality not reversed?

66. Divide -3 into both sides of the inequality $-21 > -29$. What is the result? Why is the direction of the inequality reversed?

Mixed Practice *Solve. Collect the variable terms on the left side of the inequality.*

67. $3x + 8 < 7x - 4$

68. $7x + 3 > 9x - 5$

69. $6x - 2 \ge 4x + 6$

70. $5x - 5 \le 2x + 10$

71. $0.3(x - 1) < 0.1x - 0.5$

72. $0.2(3 - x) + 0.1 > 0.1(x - 2)$

73. $3 + 5(2 - x) \ge -3(x + 5)$

74. $9 - 3(2x - 1) \le 4(x + 2)$

75. $\dfrac{x + 6}{7} - \dfrac{6}{14} > \dfrac{x + 3}{2}$

76. $\dfrac{3x + 5}{4} - \dfrac{7}{12} > -\dfrac{x}{6}$

Applications

77. *Course Average* To pass a course with a B grade, a student must have an average of 80 or greater. A student's grades on three tests are 75, 83, and 86. Solve the inequality $\dfrac{75 + 83 + 86 + x}{4} \ge 80$ to find what score the student must get on the next test to get a B average or better.

78. *Payment Options* Sharon sells very expensive European sports cars. She may choose to receive $10,000.00 per month or 8% of her sales as payment for her work. Solve the inequality $0.08x > 10,000$ to find how much she needs to sell to make the 8% offer a better deal.

79. *Elephant Weight* The average African elephant weighs 268 pounds at birth. During the first three weeks of life a baby elephant will usually gain about 4 pounds per day. Assuming that growth rate, solve the inequality $268 + 4x \geq 300$ to find how many days it will be until a baby elephant weighs at least 300 pounds.

80. *Car Loan* Rennie is buying a used car that costs $4500. The deal called for a $600 down payment, and payments of $260 monthly. He wants to know whether he can pay off the car within a year. Solve the inequality $600 + 260x \geq 4500$ to find out the minimum number of months it will take to pay off the car.

Cumulative Review

81. **[0.5.3]** Find 16% of 38.

82. **[0.5.3]** 18 is what percent of 120?

83. **[0.5.2]** *Percent Accepted* For the most coveted graduate study positions, only 16 out of 800 students are accepted. What percent are accepted?

84. **[0.5.1]** Write the fraction $\frac{3}{8}$ as a percent.

Quick Quiz 2.8

1. Graph $x \leq -3.5$ on the given number line.

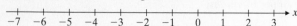

Solve and graph the result for each of the following.

2. $-12 + 4x \leq 2x$

3. $\dfrac{x}{2} - 1 < \dfrac{3}{2}x + 4$

4. **Concept Check** Explain the difference between $12 < x$ and $x > 12$. Would the graphs of these inequalities be the same or different?

Putting Your Skills to Work: Use Math to Save Money

GAS PRICES

It's July 2008 in Stockton, California, and Sam needs to put gas in his car. He is on a street that has an ARCO gas station and a SHELL station. Sam will use his debit/credit card to pay for the gas. The ARCO station is charging $4.43 per gallon of gas while the SHELL station is charging $4.55 per gallon.

If Sam's goal is to save money it would seem obvious that he should go to ARCO, right? But Sam knows from experience it's not that simple. He knows

that ARCO will charge an extra $0.45 as an "ATM Transaction Fee" in addition to the gas he buys:

1. If Sam plans on buying just **one gallon** of gas, which gas station should he choose?

2. If Sam plans on buying **three gallons** of gas, which gas station should he choose?

3. If Sam plans on buying **four gallons** of gas, which gas station should he choose?

4. If Sam plans on buying **ten gallons** of gas, which gas station should he choose?

5. How many gallons of gas would Sam need to buy for the cost to be **exactly the same** at the two gas stations? Consider the results of Question 2 and Question 3 when formulating your answer.

6. Does the station where you normally get gas charge the same for cash or credit?

7. Do you know if the station charges an "ATM transaction fee"?

8. Has the increase in gas prices caused you to change your driving habits? If so, please explain.

Topic	Procedure	Examples
Solving equations without parentheses or fractions, p. 136.	1. On both sides of the equation, combine like terms if possible. 2. Add or subtract terms on both sides of the equation in order to get all terms with the variable on one side of the equation. 3. Add or subtract a value on both sides of the equation to get all terms not containing the variable on the other side of the equation. 4. Divide both sides of the equation by the coefficient of the variable. 5. If possible, simplify the solution. 6. Check your solution by substituting the obtained value into the original equation.	Solve for x. $$5x + 2 + 2x = -10 + 4x + 3$$ $$7x + 2 = -7 + 4x$$ $$7x - 4x + 2 = -7 + 4x - 4x$$ $$3x + 2 = -7$$ $$3x + 2 - 2 = -7 - 2$$ $$3x = -9$$ $$\frac{3x}{3} = \frac{-9}{3}$$ $$x = -3$$ *Check:* Is -3 the solution of $$5x + 2 + 2x = -10 + 4x + 3?$$ $$5(-3) + 2 + 2(-3) \overset{?}{=} -10 + 4(-3) + 3$$ $$-15 + 2 + (-6) \overset{?}{=} -10 + (-12) + 3$$ $$-13 - 6 \overset{?}{=} -22 + 3$$ $$-19 = -19 \checkmark$$
Solving equations with parentheses and/or fractions, p. 138.	1. Remove any parentheses. 2. Simplify, if possible. 3. If fractions exist, multiply all terms on both sides by the least common denominator of all the fractions. 4. Now follow the remaining steps for solving an equation without parentheses or fractions.	Solve for y. $$5(3y - 4) = \frac{1}{4}(6y + 4) - 48$$ $$15y - 20 = \frac{3}{2}y + 1 - 48$$ $$15y - 20 = \frac{3}{2}y - 47$$ $$2(15y) - 2(20) = 2\left(\frac{3}{2}y\right) - 2(47)$$ $$30y - 40 = 3y - 94$$ $$30y - 3y - 40 = 3y - 3y - 94$$ $$27y - 40 = -94$$ $$27y - 40 + 40 = -94 + 40$$ $$27y = -54$$ $$\frac{27y}{27} = \frac{-54}{27}$$ $$y = -2$$
Geometric formulas, p. 167.	Know the definitions of and formulas related to basic geometric shapes such as the triangle, rectangle, parallelogram, trapezoid, and circle. Review these formulas from the beginning of Section 2.7 as needed. Formulas for the volume and total surface area of three solids are also given there.	The volume of a right circular cylinder found in a four-cylinder car engine is determined by the formula $V = \pi r^2 h$. Find V if $r = 6$ centimeters and $h = 5$ centimeters. Use 3.14 for π. $$V = 3.14(6 \text{ cm})^2(5 \text{ cm})$$ $$= 3.14(36)(5) \text{ cm}^3$$ $$= 565.2 \text{ cc (or cm}^3)$$ Four such cylinders would make this a 2260.8-cc engine.

Procedure for Solving Applied Problems, p. 152 and p. 158

1. **Understand the problem.**
   (a) Read the word problem carefully to get an overview.
   (b) Determine what information you will need to solve the problem.
   (c) Draw a sketch. Label it with the known information. Determine what needs to be found.
   (d) Choose a variable to represent one unknown quantity.
   (e) If necessary, represent other unknown quantities in terms of that same variable.
2. **Write an equation.**
   (a) Look for key words to help you to translate the words into algebraic symbols.
   (b) Use a given relationship in the problem or an appropriate formula in order to write an equation.
3. **Solve and state the answer.**
4. **Check.**
   (a) Check the solution in the original equation. Is the answer reasonable?
   (b) Be sure the solution to the equation answers the question in the word problem.

EXAMPLE

The perimeter of a rectangle is 126 meters. The length of the rectangle is 6 meters less than double the width. Find the dimensions of the rectangle.

1. **Understand the problem.**
   We want to find the length and the width of a rectangle whose perimeter is 126 meters.

   The length is *compared to* the width, so we start with the width.   Let w = width.

   The length is 6 meters less than double the width.   Then $l = 2w - 6$.

2. **Write an equation.**
   The perimeter of a rectangle is $P = 2w + 2l$.

   $$126 = 2w + 2(2w - 6)$$

3. **Solve.**

   $$126 = 2w + 4w - 12$$
   $$126 = 6w - 12$$
   $$138 = 6w$$
   $$23 = w \quad \text{The width is 23 meters.}$$
   $$2w - 6 = 2(23) - 6 =$$
   $$46 - 6 = 40 \quad \text{The length is 40 meters.}$$

4. **Check:**
   Is this reasonable? Yes. A rectangle 23 meters wide and 40 meters long seems to be about right for the perimeter to be 126 meters. Is the perimeter exactly 126 meters? Is the length exactly 6 meters less than double the width?

   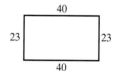

   $$2(23) + 2(40) \overset{?}{=} 126$$
   $$46 + 80 \overset{?}{=} 126$$
   $$126 = 126 \quad ✓$$
   $$40 \overset{?}{=} 2(23) - 6$$
   $$40 = 40 \quad ✓$$

Topic	Procedure	Examples
Solving inequalities, p. 178.	1. Follow the steps for solving an equation up to the multiplication or division step. 2. If you divide or multiply both sides of the inequality by a *positive number*, the **direction** of the inequality **is not reversed.** 3. If you divide or multiply both sides of the inequality by a *negative number*, the **direction** of the inequality **is reversed.**	Solve for x and graph your solution. $\frac{1}{2}(3x - 2) \leq -5 + 5x - 3$ First remove parentheses and simplify. $\frac{3}{2}x - 1 \leq -8 + 5x$ $2\left(\frac{3}{2}x\right) - 2(1) \leq 2(-8) + 2(5x)$ Now multiply each term by 2. $3x - 2 \leq -16 + 10x$ $3x - 10x - 2 \leq -16 + 10x - 10x$ $-7x - 2 \leq -16$ $-7x - 2 + 2 \leq -16 + 2$ $-7x \leq -14$ $\frac{-7x}{-7} \geq \frac{-14}{-7}$ When we divide both sides by a negative number, the inequality is reversed. $x \geq 2$ Graph of the solution: 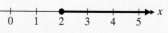

Chapter 2 Review Problems

Sections 2.1–2.3

Solve for the variable. Noninteger answers may be left in fractional form or decimal form.

1. $5x = -35$

2. $x - 19 = -22$

3. $6 - 18x = 4 - 17x$

4. $18 - 10x = 63 + 5x$

5. $x - (0.5x + 2.6) = 17.6$

6. $-0.2(x + 1) = 0.3(x + 11)$

7. $3(x - 2) = -4(5 + x)$

8. $\dfrac{2}{3}x = -18$

9. $\dfrac{3}{4}x = 15$

10. $4(2x + 3) = 5(x - 3)$

11. $3(x - 3) = 13x + 21$

12. $1.2x - 0.8 = 0.8x + 0.4$

13. $3.2 - 0.6x = 0.4(x - 2)$

14. $12 - 5x = -7x - 2$

15. $2(3 - x) = 1 - (x - 2)$

16. $4(x + 5) - 7 = 2(x + 3)$

17. $3 = 2x + 5 - 3(x - 1)$

18. $2(5x - 1) - 7 = 3(x - 1) + 5 - 4x$

Section 2.4

Solve for the variable. Noninteger answers may be left in fractional form or decimal form.

19. $\dfrac{3}{4}x - 3 = \dfrac{1}{2}x + 2$

20. $1 = \dfrac{5x}{6} + \dfrac{2x}{3}$

21. $\dfrac{7x}{5} = 5 + \dfrac{2x}{5}$

22. $\dfrac{7x - 3}{2} - 4 = \dfrac{5x + 1}{3}$

23. $\dfrac{3x - 2}{2} + \dfrac{x}{4} = 2 + x$

24. $\dfrac{-3}{2}(x + 5) = 1 - x$

25. $\dfrac{-4}{3}(2x + 1) = -x - 2$

26. $\dfrac{1}{3}(x - 2) = \dfrac{x}{4} + 2$

27. $\frac{1}{5}(x - 3) = 2 - \frac{x}{2}$

28. $\frac{3}{4} - \frac{2}{3}x = \frac{1}{3}x + \frac{3}{4}$

29. $x + \frac{1}{8} + \frac{1}{2}x = \frac{1}{4}\left(-3x + \frac{1}{2}\right)$

30. $\frac{3}{2}x - \frac{5}{6} + x = \frac{1}{2}x + \frac{2}{3}$

31. $-\frac{8}{3}x - 8 + 2x - 5 = -\frac{5}{3}$

32. $\frac{1}{6} + \frac{1}{3}(x - 3) = \frac{1}{2}(x + 9)$

33. $\frac{1}{7}(x + 5) - \frac{6}{14} = \frac{1}{2}(x + 3)$

34. $\frac{1}{6}(8x + 3) = \frac{1}{2}(2x + 7)$

Section 2.5

Write an algebraic expression. Use the variable x to represent the unknown value.

35. 19 more than a number

36. two-thirds of a number

37. triple the sum of a number and 4

38. twice a number decreased by three

Write an expression for each of the quantities compared. Use the letter specified.

39. *Workforce* The number of working people is four times the number of retired people. The number of unemployed people is one-half the number of retired people. (Use the letter *r*.)

▲ **40.** *Geometry* The length of a rectangle is 5 meters more than triple the width. (Use the letter *w*.)

▲ **41.** *Geometry* A triangle has three angles *A*, *B*, and *C*. The number of degrees in angle *A* is double the number of degrees in angle *B*. The number of degrees in angle *C* is 17 degrees less than the number in angle *B*. (Use the letter *b*.)

42. *Class Size* There are 29 more students in biology class than in algebra. There are one-half as many students in geology class as in algebra. (Use the letter *a*.)

Section 2.6

Solve.

43. Fourteen is taken away from triple a number. The result is −5. What is the number?

44. When twice a number is reduced by seven, the result is −21. What is the number?

45. *Table Set* Six chairs and a table cost $450. If the table costs $210, how much does one chair cost?

46. *Age* Jon is twice as old as his cousin David. If Jon is 32, how old is David?

47. *Speed Rates* Two cars left San Francisco and drove to San Diego, 800 miles away. One car was driven at 60 miles an hour. The other car was driven at 65 miles an hour. How long did it take each car to make the trip? Round your answer to the nearest tenth.

48. *Biology* Zach took four biology tests and got scores of 83, 86, 91, and 77. He needs to complete one more test. If he wants to get an average of 85 on all five tests, what score does he need on his last test?

Section 2.7

Solve.

▲ **49.** *Geometry* Find the surface area of a tennis ball that has a diameter of 2.6 inches. (Use $\pi \approx 3.14$.) Round your answer to the nearest hundredth.

▲ **50.** *Trench Length* Wilfredo and Tina bought a new screen house for camping. It is hexagonal (6 sides) in shape and 4 ft long on a side. If they want to dig a trench around their house so that rainwater will flow away, how long will the trench be? (You are looking for the perimeter of the screen house.)

▲ **51.** *Geometry* Find the third angle of a triangle if one angle measures 62 degrees and the second angle measures 47 degrees.

▲ **52.** *Geometry* Find the area of a triangle whose base is 8 miles and whose altitude is 10.5 miles.

▲ **53.** *Geometry* Find the volume of a storage room that has a width of 6 feet, a length of 11 feet, and a height of 7 feet.

▲ **54.** *Geometry* Find the volume of a basketball whose radius is 4.5 inches. (Use $\pi \approx 3.14$.)

▲ **55.** *Geometry* The Clock Tower (also known as Big Ben) in London has four clock faces measuring 23 feet in diameter. Find the area of each clock face. (Use $\pi \approx 3.14$.) Round your answer to the nearest hundredth.

▲ **56.** *Geometry* The perimeter of an isosceles triangle is 46 inches. The two equal sides are each 17 inches long. How long is the third side?

▲ **57.** *House Painting* Jim wants to have the exterior siding on his house painted. Two sides of the house have a height of 18 feet and a length of 22 feet, and the other two sides of the house have a height of 18 feet and a length of 19 feet. The painter has quoted a price of $2.50 per square foot. The windows and doors of the house take up 100 square feet. How much will the painter charge to paint the siding?

▲ **58.** *Aluminum Cylinder* Find the cost to build a cylinder out of aluminum if the radius is 5 meters, the height is 14 meters, and the cylinder will have a flat top and flat bottom. The cost factor has been determined to be $40.00 per square meter. (*Hint:* First find the total surface area.) (Use $\pi \approx 3.14$.)

Section 2.8

Solve each inequality and graph the result.

59. $9 + 2x \leq 6 - x$

60. $4x - 7 \geq -x + 8$

61. $2x - 3 + x > 5(x + 1)$

62. $-x + 4 < 3x + 16$

63. $8 - \dfrac{1}{3}x \le x$

```
  +---+---+---+---+---+---+---+---+---+---+--> x
 -1   0   1   2   3   4   5   6   7   8   9
```

64. $7 - \dfrac{3}{5}x > 4$

```
  +---+---+---+---+---+---+---+---+---+---+--> x
 -3  -2  -1   0   1   2   3   4   5   6   7
```

65. $-4x - 14 < 4 - 2(3x - 1)$

```
      +---+---+---+---+---+---+---+---+---+---+--> x
      7   8   9   10  11  12  13  14  15  16  17
```

66. $3(x - 2) + 8 < 7x + 14$

```
   +---+---+---+---+---+---+---+---+---+---+--> x
  -5  -4  -3  -2  -1   0   1   2   3   4   5
```

Use an inequality to solve.

67. *Wages* Julian earns $15 per hour as a plasterer's assistant. His employer determines that the current job allows him to pay $480 in wages to Julian. What are the maximum number of hours that Julian can work on this job? (*Hint:* Use $15h \le 480$.)

68. *Hiring a Substitute* The cost of hiring a substitute elementary teacher for a day is $110. Let $n =$ the number of substitute teachers. Set up an inequality to determine how many times a substitute teacher may be hired if the school's budget for substitute teachers is $2420 per month. What is the maximum number of times a substitute teacher may be hired during a month? (*Hint:* Use $110n \le 2420$.)

Mixed Practice

Solve for the variable. Noninteger answers may be left in fractional form or decimal form.

69. $10(2x + 4) - 13 = 8(x + 7) - 3$

70. $-9x + 15 - 2x = 4 - 3x$

71. $-2(x - 3) = -4x + 3(3x + 2)$

72. $\dfrac{1}{2} + \dfrac{5}{4}x = \dfrac{2}{5}x - \dfrac{1}{10} + 4$

73. $\dfrac{1}{6}x - \dfrac{2}{3} = \dfrac{1}{3}(x - 4)$

74. $\dfrac{1}{2}(x - 3) = \dfrac{1}{4}(3x - 1)$

Solve each inequality and graph the result.

75. $5 - \dfrac{1}{2}x > 4$

```
   +---+---+---+---+---+---+---+---+---+---+--> x
  -5  -4  -3  -2  -1   0   1   2   3   4   5
```

76. $2(x - 1) \ge 3(2 + x)$

```
   +---+---+---+---+---+---+---+---+---+---+--> x
 -10 -9  -8  -7  -6  -5  -4  -3  -2  -1   0
```

77. $\dfrac{1}{3}(x + 2) \le \dfrac{1}{2}(3x - 5)$

```
   +----+----+---+----+----+----+----+----+---+----+--> x
   12   13   2  15   16   17   18   19   20   3  22
   --   --      --   --   --   --   --   --      --
    7    7       7    7    7    7    7    7       7
```

78. $4(2 - x) - (-5x + 1) \ge -8$

```
   +----+----+----+----+----+----+----+----+----+----+--> x
  -18  -17  -16  -15  -14  -13  -12  -11  -10  -9   -8
```

How Am I Doing? Chapter 2 Test

Remember to use your Chapter Test Prep Video CD to see the worked-out solutions to the test problems you want to review.

Solve for the variable. Noninteger answers may be left in fractional form or decimal form.

1. $3x + 5.6 = 11.6$

2. $9x - 8 = -6x - 3$

3. $2(2y - 3) = 4(2y + 2)$

4. $\frac{1}{7}y + 3 = \frac{1}{2}y$

5. $4(7 - 4x) = 3(6 - 2x)$

6. $0.8x + 0.18 - 0.4x = 0.3(x + 0.2)$

7. $\frac{2y}{3} + \frac{1}{5} - \frac{3y}{5} + \frac{1}{3} = 1$

8. $3 - 2y = 2(3y - 2) - 5y$

9. $5(20 - x) + 10x = 165$

10. $5(x + 40) - 6x = 9x$

11. $-2(2 - 3x) = 76 - 2x$

12. $20 - (2x + 6) = 5(2 - x) + 2x$

In questions 13–17, solve for x.

13. $2x - 3 = 12 - 6x + 3(2x + 3)$

14. $\frac{1}{3}x - \frac{3}{4}x = \frac{1}{12}$

15. $\frac{3}{5}x + \frac{7}{10} = \frac{1}{3}x + \frac{3}{2}$

16. $\frac{15x - 2}{28} = \frac{5x - 3}{7}$

17. $\frac{2}{3}(x + 8) + \frac{3}{5} = \frac{1}{5}(11 - 6x)$

Solve and graph the inequality.

18. $3(x - 2) \geq 5x$

19. $2 - 7(x + 1) - 5(x + 2) < 0$

20. $5 + 8x - 4 < 2x + 13$

21. $\frac{1}{4}x + \frac{1}{16} \leq \frac{1}{8}(7x - 2)$

1. _____

2. _____

3. _____

4. _____

5. _____

6. _____

7. _____

8. _____

9. _____

10. _____

11. _____

12. _____

13. _____

14. _____

15. _____

16. _____

17. _____

18. _____

19. _____

20. _____

21. _____

Solve.

22. A number is doubled and then decreased by 11. The result is 59. What is the original number?

23. The sum of one-half of a number, one-ninth of the number, and one-twelfth of the number is twenty-five. Find the original number.

24. Double the sum of a number and 5 is the same as fourteen more than triple the same number. Find the number.

▲ 25. A triangular region has a perimeter of 66 meters. The first side is two-thirds of the second side. The third side is 14 meters shorter than the second side. What are the lengths of the three sides of the triangular region?

▲ 26. A rectangle has a length 7 meters longer than double the width. The perimeter is 134 meters. Find the dimensions of the rectangle.

▲ 27. Find the circumference of a circle with radius 34 inches. Use $\pi \approx 3.14$ as an approximation.

▲ 28. Find the area of a trapezoid if the two bases are 10 inches and 14 inches and the altitude is 16 inches.

▲ 29. Find the volume of a sphere with radius 10 inches. Use $\pi \approx 3.14$ as an approximation and round your answer to the nearest cubic inch.

▲ 30. Find the area of a parallelogram with a base of 12 centimeters and an altitude of 8 centimeters.

▲ 31. How much would it cost to carpet the area shown in the figure if carpeting costs $12 per square yard?

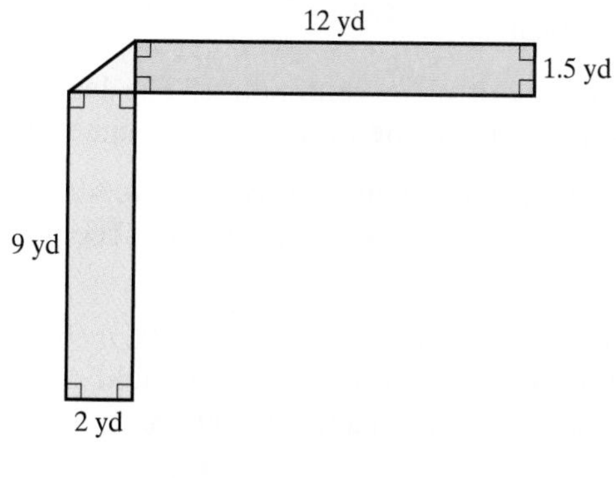

22. _____

23. _____

24. _____

25. _____

26. _____

27. _____

28. _____

29. _____

30. _____

31. _____

Cumulative Test for Chapters 0–2

1. _____

2. _____

3. _____

4. _____

5. _____

6. _____

7. _____

8. _____

9. _____

10. _____

11. _____

12. _____

13. _____

14. _____

15. _____

16. _____

17. _____

18. _____

19. _____

20. _____

21. _____

22. _____

198

Approximately one-half of this test covers the content of Chapters 0–1. The remainder covers the content of Chapter 2.

1. Divide. $2.4\overline{)8.856}$

2. Add. $\dfrac{3}{8} + \dfrac{5}{12} + \dfrac{1}{2}$

3. Simplify.
$-4(4x - y + 5) - 3(6x - 2y)$

4. Evaluate.
$12 - 3(2 - 4) + 12 \div 4$

5. Multiply. $(-3)(-5)(-1)(2)(-1)$

6. Combine like terms. $3st - 8s^2t + 12st^2 + s^2t - 5st + 2st^2$

7. Simplify. $(5x)^2$

8. Simplify. $2\{3x - 4[5 - 3y(2 - x)]\}$

9. Solve for x. $-2(x + 5) = 4x - 15$

10. Solve for x. $\dfrac{1}{3}(x + 5) = 2x - 5$

11. Solve for y. $\dfrac{2y}{3} - \dfrac{1}{4} = \dfrac{1}{6} + \dfrac{y}{4}$

In questions 12–16, solve and graph the inequality.

12. $7x - 2 \le 3x + 10 - 2x$

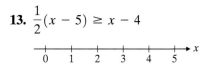

13. $\dfrac{1}{2}(x - 5) \ge x - 4$

14. $4(2 - x) > 1 - 5x - 8$

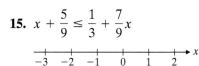

15. $x + \dfrac{5}{9} \le \dfrac{1}{3} + \dfrac{7}{9}x$

16. $7x - 13 \le 3(4 - 5x) - 3$

17. The football team will not let Chuck play unless he passes biology with a C (70 or better) average. There are five tests in the semester, and he failed (0) the first one. However, he found a tutor and received an 82, an 89, and an 87 on the next three tests. Solve the inequality $\dfrac{0 + 82 + 89 + 87 + x}{5} \ge 70$ to find what his minimum score must be on the last test in order for him to pass the course and play football.

18. When a number is doubled and then increased by 15, the result is 1. Find the number.

19. The number of students in a literature class is 12 fewer than the number of students in sociology. The total enrollment for the two classes is 96 students. How many students are in each class?

▲ **20.** A rectangle has a perimeter of 78 centimeters. The length of the rectangle is 11 centimeters longer than triple the width. Find the dimensions of the rectangle.

▲ **21.** Find the area of a triangle with an altitude of 13 meters and a base of 25 meters. What is the cost to construct such a triangle out of sheet metal that costs $4.50 per square meter?

▲ **22.** Find the volume of a sphere that has a radius of 3.00 inches. How much will the contents of the sphere weigh if it is filled with a liquid that weighs 1.50 pounds per cubic inch? Use $\pi \approx 3.14$ as an approximation.

CHAPTER

3

Vegetable gardens both small and large are present in every state of the United States. Many Americans enjoy the produce from their gardens. Large vegetable farms produce millions of dollars worth of fresh vegetables that are exported daily to countries all around the world. The graphing techniques covered in this chapter can be used to predict how much a vegetable crop is worth.

Graphing and Functions

Student Learning Objectives

After studying this section, you will be able to:

1. Plot a point, given the coordinates.

2. Determine the coordinates of a plotted point.

3. Find ordered pairs for a given linear equation.

1 Plotting a Point, Given the Coordinates

Often we can better understand an idea if we see a picture. This is the case with many mathematical concepts, including those relating to algebra. We can illustrate algebraic relationships with drawings called **graphs.** Before we can draw a graph, however, we need a frame of reference.

In Chapter 1 we showed that any real number can be represented on a number line. Look at the following number line. The arrow indicates the positive direction.

To form a **rectangular coordinate system,** we draw a second number line vertically. We construct it so that the 0 point on each number line is exactly at the same place. We refer to this location as the **origin.** The horizontal number line is often called the **x-axis.** The vertical number line is often called the **y-axis.** Arrows show the positive direction for each axis.

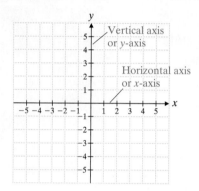

We can represent a point in this rectangular coordinate system by using an **ordered pair** of numbers. For example, $(5, 2)$ is an ordered pair that represents a point in the rectangular coordinate system. The numbers in an ordered pair are often referred to as the **coordinates** of the point. The first number is called the **x-coordinate** and it represents the distance from the origin measured along the horizontal or x-axis. If the x-coordinate is positive, we count the proper number of squares to the right (that is, in the positive direction). If the x-coordinate is negative, we count to the left. The second number in the pair is called the **y-coordinate** and it represents the distance from the origin measured along the y-axis. If the y-coordinate is positive, we count the proper number of squares upward (that is, in the positive direction). If the y-coordinate is negative, we count downward.

$$(5, 2)$$
x-coordinate ⌐ ⌐ y-coordinate

Suppose the directory for the map on the left indicated that you would find a certain street in the region C2. To find the street you would first scan across the horizontal scale until you found section C; from there you would scan up the map until you hit section 2 along the vertical scale. As we will see in the next example, plotting a point in the rectangular coordinate system is much like finding a street on a map with grids.

EXAMPLE 1 Plot the point $(5, 2)$ on a rectangular coordinate system. Label this point as A.

Solution Since the x-coordinate is 5, we first count 5 units to the right on the x-axis. Then, because the y-coordinate is 2, we count 2 units up from the point

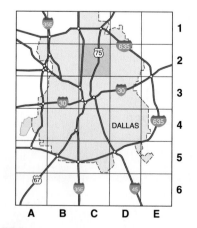

where we stopped on the *x*-axis. This locates the point corresponding to (5, 2). We mark this point with a dot and label it *A*.

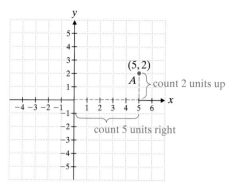

Practice Problem 1 Plot the point (3, 4) on the preceding coordinate system. Label this point as *B*.

NOTE TO STUDENT: Fully worked-out solutions to all of the Practice Problems can be found at the back of the text starting at page SP-1

It is important to remember that the first number in an ordered pair is the *x*-coordinate and the second number is the *y*-coordinate. The ordered pairs (5, 2) and (2, 5) represent different points.

EXAMPLE 2 Plot each point on the following coordinate system. Label the points *F*, *G*, and *H*, respectively.

(a) (−5, 3)   **(b)** (2, −6)   **(c)** (−4, −5)

Solution

(a) Notice that the *x*-coordinate, −5, is negative. On the coordinate grid, negative *x*-values appear to the left of the origin. Thus, we will begin by counting 5 squares to the left, starting at the origin. Since the *y*-coordinate, 3, is positive, we will count 3 units up from the point where we stopped on the *x*-axis.

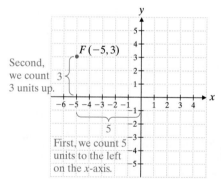

(b) The *x*-coordinate is positive. Begin by counting 2 squares to the right of the origin. Then count down because the *y*-coordinate is negative.

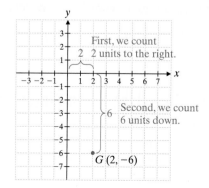

(c) The x-coordinate is negative. Begin by counting 4 squares to the left of the origin. Then count down because the y-coordinate is negative.

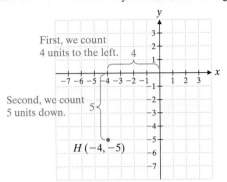

PRACTICE PROBLEM 2

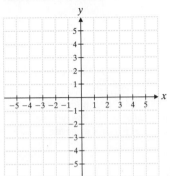

Practice Problem 2 Use the coordinate system in the margin to plot each point. Label the points I, J, and K, respectively.

(a) $(-2, -4)$ **(b)** $(-4, 5)$ **(c)** $(4, -2)$

EXAMPLE 3 Plot the following points.

$F: (0, 5)$ $\quad$ $G: \left(3, \frac{3}{2}\right)$ $\quad$ $H: (-6, 4)$ $\quad$ $I: (-3, -4)$

$J: (-4, 0)$ $\quad$ $K: (2, -3)$ $\quad$ $L: (6.5, -7.2)$

Solution These points are plotted in the figure.

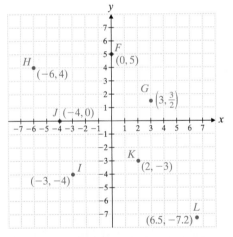

Note: When you are plotting decimal values like $(6.5, -7.2)$, plot the point halfway between 6 and 7 in the x-direction (for the 6.5) and at your best approximation of -7.2 in the y-direction.

PRACTICE PROBLEM 3

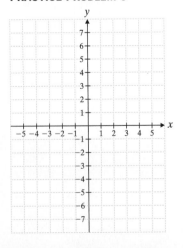

Practice Problem 3 Plot the following points. Label each point with both the letter and the ordered pair. Use the coordinate system provided in the margin.

$$A: (3, 7); B: (0, -6); C: (3, -4.5); D: \left(-\frac{7}{2}, 2\right)$$

② Determining the Coordinates of a Plotted Point

Sometimes we need to find the coordinates of a point that has been plotted. First, we count the units we need on the x-axis to get as close as possible to the point. Next we count the units up or down that we need to go from the x-axis to reach the point.

EXAMPLE 4 What ordered pair of numbers represents point A on the graph below?

Solution If we move along the x-axis until we get as close as possible to A, we end up at the number 5. Thus we obtain 5 as the first number of the ordered pair. Then we count 4 units upward on a line parallel to the y-axis to reach A. So we obtain 4 as the second number of the ordered pair. Thus point A is represented by the ordered pair $(5, 4)$.

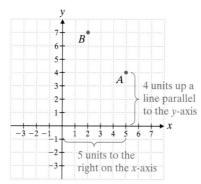

Practice Problem 4 What ordered pair of numbers represents point B on the graph above?

EXAMPLE 5 Write the coordinates of each point plotted on the following graph.

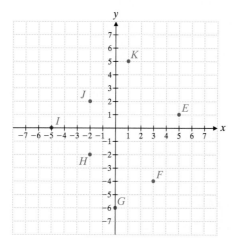

Solution The coordinates of each point are as follows.

$E = (5, 1)$ $I = (-5, 0)$

$F = (3, -4)$ $J = (-2, 2)$ Be very careful that you put the x-coordinate

$G = (0, -6)$ $K = (1, 5)$ first and the y-coordinate second. Be careful

$H = (-2, -2)$ that each sign is correct.

Practice Problem 5 Give the coordinates of each point plotted on the graph in the margin.

PRACTICE PROBLEM 5

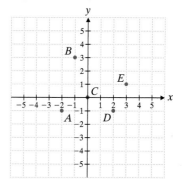

In examining data from real-world situations, we often find that plotting data points shows useful trends. In such cases, it is often necessary to use a different scale, one that displays only positive values.

EXAMPLE 6 The number of motor vehicle accidents in millions is recorded in the following table for the years 1980 to 2000.

(a) Plot points that represent this data on the given coordinate system.

(b) What trends are apparent from the plotted data?

Number of Years Since 1980	Number of Motor Vehicle Accidents (in Millions)
0	18
5	19
10	12
15	11
20	15

Source: U.S. National Highway Traffic Safety Administration

Solution

(a)

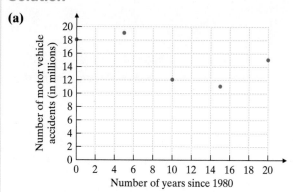

(b) From 1980 to 1985, there was a slight increase in the number of accidents. From 1985 to 1995, there was a significant decrease in the number of accidents. From 1995 to 2000, there was a moderate increase in the number of accidents.

Practice Problem 6 The number of motor vehicle deaths in thousands is recorded in the following table for the years 1980 to 2000.

(a) Plot points that represent this data on the given coordinate system.

(b) What trends are apparent from the plotted data?

Number of Years Since 1980	Number of Motor Vehicle Deaths (in Thousands)
0	51
5	44
10	45
15	42
20	43

Source: U.S. National Highway Traffic Safety Administration

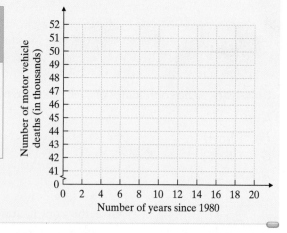

③ Finding Ordered Pairs for a Given Linear Equation

Equations such as $3x + 2y = 5$ and $6x + y = 3$ are called linear equations in two variables.

A **linear equation in two variables** is an equation that can be written in the form $Ax + By = C$ where A, B, and C are real numbers but A and B are not *both* zero.

Replacement values for x and y that make *true mathematical statements* of the equation are called *truth values,* and an ordered pair of these truth values is called a **solution.**

EXAMPLE 7 Are the following ordered-pair solutions to the equation $3x + 2y = 5$?

(a) $(-1, 4)$ **(b)** $(2, 2)$

Solution We replace values for x and y in the equation to verify that we obtain a true statement.

(a)
$$3x + 2y = 5$$
$$3(-1) + 2(4) \stackrel{?}{=} 5 \qquad \text{Replace } x \text{ with } -1 \text{ and } y \text{ with } 4.$$
$$-3 + 8 \stackrel{?}{=} 5$$
$$5 = 5 \checkmark \qquad \text{True statement}$$

The ordered pair $(-1, 4)$ is a solution to $3x + 2y = 5$ because when we replace x by -1 and y by 4, we obtain a true statement.

(b)
$$3x + 2y = 5$$
$$3(2) + 2(2) \stackrel{?}{=} 5 \qquad \text{Replace } x \text{ with } 2 \text{ and } y \text{ with } 2.$$
$$6 + 4 \stackrel{?}{=} 5$$
$$10 = 5 \qquad \text{False statement}$$

The ordered pair $(2, 2)$ is *not* a solution to $3x + 2y = 5$ because when we replace x by 2 and y by 2, we obtain a *false* statement.

Practice Problem 7 Are the following ordered-pair solutions to the equation $3x + 2y = 5$?

(a) $(3, -1)$ **(b)** $\left(2, -\dfrac{1}{2}\right)$

If one value of an ordered-pair solution to a linear equation is known, the other can be quickly obtained. To do so, we replace the proper variable in the equation by the known value. Then, using the methods learned in Chapter 2, we solve the resulting equation for the other variable.

EXAMPLE 8 Find the missing coordinate to complete the following ordered-pair solutions to the equation $2x + 3y = 15$.

(a) $(0, ?)$ **(b)** $(?, 1)$

Solution

(a) For the ordered pair $(0, ?)$, we know that $x = 0$. Replace x by 0 in the equation and solve for y.

$$2x + 3y = 15$$
$$2(0) + 3y = 15 \qquad \text{Replace } x \text{ with } 0.$$
$$0 + 3y = 15 \qquad \text{Simplify.}$$
$$y = 5 \qquad \text{Divide both sides by 3.}$$

Thus we have the ordered pair $(0, 5)$.

(b) For the ordered pair $(?, 1)$, we *do not know* the value of x. However, we do know that $y = 1$. So we start by replacing the variable y by 1. We will end up with an equation with one variable, x. We can then solve for x.

$$2x + 3y = 15$$
$$2x + 3(1) = 15 \quad \text{Replace } y \text{ with 1.}$$
$$2x + 3 = 15 \quad \text{Simplify.}$$
$$2x = 12$$
$$x = 6 \quad \text{Solve for } x.$$

Thus we have the ordered pair $(6, 1)$.

NOTE TO STUDENT: Fully worked-out solutions to all of the Practice Problems can be found at the back of the text starting at page SP-1

Practice Problem 8 Find the missing coordinate to complete the following ordered-pair solutions to the equation $3x - 4y = 12$.

(a) $(0, ?)$ **(b)** $(?, 3)$ **(c)** $(?, -6)$

Sometimes it is convenient to first solve for y, then find the missing coordinates of the ordered-pair solutions.

EXAMPLE 9 Find the missing coordinates to complete the ordered-pair solutions to $3x - 2y = 6$.

(a) $(2, ?)$ **(b)** $(4, ?)$

Solution Since we are given the values for x and must find the values for y in each ordered pair, solving the equation for y first simplifies the calculations.

$$-2y = 6 - 3x \quad \text{We want to isolate the term containing } y,$$
$$\text{so we subtract } 3x \text{ from both sides.}$$

$$\frac{-2y}{-2} = \frac{6 - 3x}{-2} \quad \text{Divide both sides by the coefficient of } y.$$

$$y = \frac{6}{-2} + \frac{-3x}{-2} \quad \begin{array}{l}\text{Change subtracting to adding the opposite:} \\ 6 - 3x = 6 + (-3x). \text{ Then rewrite the fraction} \\ \text{on the right side as two fractions.}\end{array}$$

$$y = \frac{3}{2}x - 3 \quad \text{Simplify and reorder the terms on the right.}$$

Now we find the values of y by replacing x with the given values.

(a) $y = \dfrac{3}{2}(2) - 3$ **(b)** $y = \dfrac{3}{2}(\overset{2}{4}) - 3$
$\quad\quad y = 3 - 3$ $y = 6 - 3$
$\quad\quad y = 0$ $y = 3$

Thus we have the ordered pairs $(2, 0)$ and $(4, 3)$.

NOTE TO STUDENT: Fully worked-out solutions to all of the Practice Problems can be found at the back of the text starting at page SP-1

Practice Problem 9 Find the missing coordinates to complete the ordered-pair solutions to $8 - 2y + 3x = 0$.

(a) $(0, ?)$ **(b)** $(2, ?)$

TO THINK ABOUT: An Alternate Method Find the missing coordinates in Example 9 using the equation in the form $3x - 2y = 6$. Which form do you prefer? Why?

Verbal and Writing Skills

Unless otherwise indicated, assume each grid line represents one unit.

1. What is the *x*-coordinate of the origin?

2. What is the *y*-coordinate of the origin?

3. Explain why (5, 1) is referred to as an *ordered* pair of numbers.

4. Explain how you would locate the point (4, 3) on graph paper.

5. Plot the following points.

$J: (-4, 3.5)$ $K: (6, 0)$
$L: (5, -6)$ $M: (0, -4)$

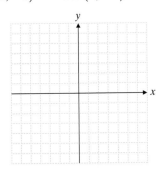

6. Plot the following points.

$R: (-3, 0)$ $S: (3.5, 4)$
$T: (-2, -2.5)$ $V: (0, 5)$

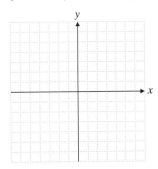

Consider the points plotted on the graph at right.

7. Give the coordinates for points *R*, *S*, *X*, and *Y*.

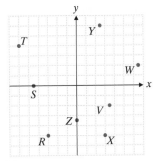

8. Give the coordinates for points *T*, *V*, *W*, and *Z*.

In exercises 9 and 10, six points are plotted in each figure. List all the ordered pairs needed to represent the points.

9.

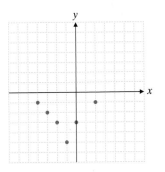

10.

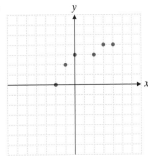

Are the following ordered pairs solutions to the equation $2x - y = 6$?

11. $(1, 0)$ **12.** $(6, 2)$ **13.** $(2, -2)$ **14.** $(0, -6)$

In each equation, solve for y.

15. $y - 4 = -\dfrac{2}{3}x.$

16. $y + 9 = \dfrac{3}{8}x.$

17. $4x + y = 11.$

18. $5x + y = 3.$

19. $8x - 12y = 24.$

20. $3y - 5x = 9.$

To Think About

21. (a) Find the missing coordinate $(3, ?)$ to complete the ordered-pair solution to the equation $3y - 5x = 9$.
   (b) Solve the equation $3y - 5x = 9$ for y, then find the missing coordinate $(6, ?)$ to complete the ordered-pair solution.
   (c) Which form did you find easier to use to complete the ordered-pair solution? Why?

22. (a) Find the missing coordinate $(2, ?)$ to complete the ordered-pair solution to the equation $2y - 7x = 8$.
   (b) Solve the equation $2y - 7x = 8$ for y, then find the missing coordinate $(4, ?)$ to complete the ordered-pair solution.
   (c) Which form did you find easier to use to complete the ordered-pair solution? Why?

Find the missing coordinate to complete the ordered-pair solution to the given linear equation.

23. $y = 4x + 7$
   (a) $(0, \ )$
   (b) $(2, \ )$

24. $y = 6x + 5$
   (a) $(0, \ )$
   (b) $(3, \ )$

25. $y + 6x = 5$
   (a) $(-1, \ )$
   (b) $(3, \ )$

26. $y + 4x = 9$
   (a) $(-2, \ )$
   (b) $(5, \ )$

27. $3x - 4y = 11$
   (a) $(-3, \ )$
   (b) $(\ , 1)$

28. $5x - 2y = 9$
   (a) $(7, \ )$
   (b) $(\ , -7)$

29. $3x + 2y = -6$
   (a) $(-2, \ )$
   (b) $(\ , 3)$

30. $-4x + 5y = -20$
   (a) $(10, \ )$
   (b) $(\ , -8)$

31. $y - 1 = \dfrac{2}{7}x$
   (a) $(7, \ )$
   (b) $\left(\ , \dfrac{5}{7}\right)$

32. $y - 5 = \dfrac{3}{5}x$
   (a) $(5, \ )$
   (b) $\left(\ , \dfrac{4}{5}\right)$

33. $2x + \dfrac{1}{3}y = 5$
   (a) $(\ , 18)$
   (b) $\left(\dfrac{8}{3}, \ \right)$

34. $3x + \dfrac{1}{4}y = 11$
   (a) $(\ , 8)$
   (b) $\left(\dfrac{13}{4}, \ \right)$

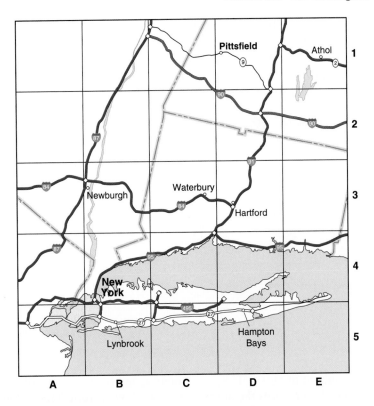

Using Road Maps *The preceding map shows a portion of New York, Connecticut, and Massachusetts. Like many maps used in driving or flying, it has horizontal and vertical grid markers for ease of use. For example, Newburgh, New York, is located in grid B3. Use the grid labels to indicate the locations of the following cities.*

35. Lynbrook, New York

36. Hampton Bays, New York

37. Athol, Massachusetts

38. Pittsfield, Massachusetts

39. Hartford, Connecticut

40. Waterbury, Connecticut

41. ***CD Shipments*** According to music CD manufacturers, the number of CDs shipped decreased significantly from 2001 to 2006. The number of CDs shipped during these years is recorded in the following table, and is measured in millions. For example, 882 in the second column means 882 million, or 882,000,000.

(a) Plot points that represent this data on the given rectangular coordinate system.

(b) What trends are apparent from the plotted data?

Number of Years Since 2001	Number of CDs Shipped (in Millions)
0	882
1	803
2	746
3	767
4	705
5	615

Source: www.riaa.com

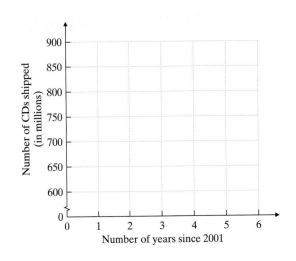

42. ***U.S. Oil Consumption*** The number of barrels of oil used per day in the United States for selected years starting in 1995 is recorded in the following table. The number of barrels is measured in millions. For example, 18 in the second column means 18 million, or 18,000,000 barrels.

(a) Plot points that represent the data on the given rectangular coordinate system.
(b) What trends are apparent from the plotted data?

Number of Years Since 1995	Oil Consumption per Day (in Millions of Barrels)
0	18
5	20
10	22
15	23.5*
20	24.5*
25	26.5*

*Estimate
Source: www.energy.senate.gov

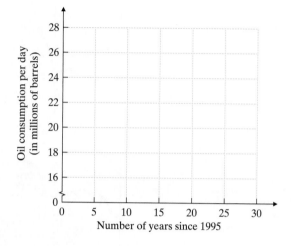

43. ***Buying Books Online*** The amount of money U.S. consumers spend buying books online continues to increase at a significant rate. The following chart records the amount spent each year from 2003 to 2006.

Year	Total Amount Spent (in Billions of Dollars)
2003	3.2
2004	3.6
2005	3.9
2006*	4.2

*Estimate
Source: Statistical Abstract of the United States: 2007.

(a) Plot points that represent the data on the given rectangular coordinate system.
(b) Based on your graph, estimate the amount consumers spent on buying books online in 2007.

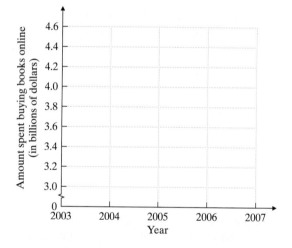

44. ***Buying Personal Computers Online*** The amount of money U.S. consumers spend buying personal computers online continues to increase at a significant rate. The following chart records the amount spent each year from 2003 to 2006.

Year	Total Amount Spent (in Billions of Dollars)
2003	53.9
2004	67.3
2005	80.9
2006*	95.3

*Estimate
Source: Statistical Abstract of the United States: 2007

(a) Plot points that represent the data on the given rectangular coordinate system.
(b) Based on your graph, estimate the amount consumers spent buying personal computers online in 2007.

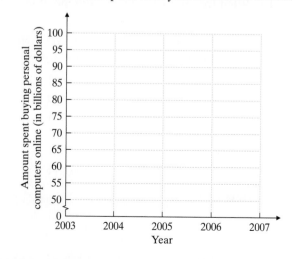

Cumulative Review

▲ **45.** **[1.8.2]** *Circular Swimming Pool* The circular pool at the hotel where Bob and Linda stayed in Orlando, Florida, has a radius of 19 yards. What is the area of the pool? (Use $\pi \approx 3.14$.)

46. **[2.6.1]** A number is doubled and then decreased by three. The result is twenty-one. What is the original number?

Quick Quiz 3.1

1. Plot and label the following points.

$A: (3, -4)$
$B: (-6, -2)$
$C: (0, 5)$
$D: (-3, 6)$

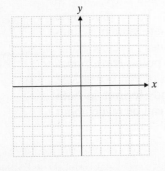

Find the missing coordinate to complete the ordered-pair solution to the given linear equation.

2. $y = -5x - 7$
  (a) $(-2, \ \ )$
  (b) $(3, \ \ )$
  (c) $(0, \ \ )$

3. $4x - 3y = -12$
  (a) $(3, \ \ )$
  (b) $(\ \ , -8)$
  (c) $(\ \ , 10)$

4. **Concept Check** Explain how you would find the missing coordinate to complete the ordered-pair solution to the equation $2.5x + 3y = 12$ if the ordered pair was of the form $(\ \ , -6)$.

Student Learning Objectives

After studying this section, you will be able to:

  Graph a linear equation by plotting three ordered pairs.

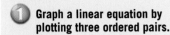 **Graph a straight line by plotting its intercepts.**

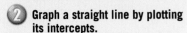 **Graph horizontal and vertical lines.**

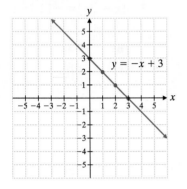

Graphing a Linear Equation by Plotting Three Ordered Pairs

We have seen that a solution to a linear equation in two variables is an ordered pair. The graph of an ordered pair is a point. Thus we can graph an equation by graphing the points corresponding to its ordered-pair solutions.

A linear equation in two variables has an infinite number of ordered-pair solutions. We can see that this is true by noting that we can substitute any number for x in the equation and solve it to obtain a y-value. For example, if we substitute $x = 0, 1, 2, 3, \ldots$ into the equation $y = -x + 3$ and solve for y, we obtain the ordered-pair solutions $(0, 3)$, $(1, 2)$, $(2, 1)$, ... Likewise, we can choose values for y, say $y = 0$, and solve for x to obtain the ordered pair $(3, 0)$. If we plot these points on a rectangular coordinate system, we notice that they fall on a *straight line*.

It turns out that all of the points corresponding to the ordered-pair solutions of $y = -x + 3$ lie on this line, and the line extends forever in both directions. A similar statement can be made about any linear equation in two variables.

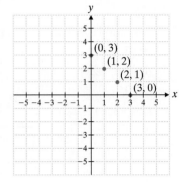

> The graph of any linear equation in two variables is a straight line.

From geometry, we know that two points determine a line. Thus to graph a linear equation in two variables, we need graph only two ordered-pair solutions of the equation and then draw the line that passes through them. Having said this, we recommend that you use three points to graph a line. Two points will determine where the line is. The third point verifies that you have drawn the line correctly. For ease in plotting, it is better if the ordered pairs contain integers.

> ### TO GRAPH A LINEAR EQUATION
> 1. Look for three ordered pairs that are solutions to the equation.
> 2. Plot the points.
> 3. Draw a line through the points.

EXAMPLE 1 Find three ordered pairs that satisfy $y = -2x + 4$. Then graph the resulting straight line.

Solution Since we can choose any value for x, we choose numbers that are convenient. To organize the results, we will make a table of values. We will let $x = 0$, $x = 1$, and $x = 3$, respectively. We write these numbers under x in our table of values. For each of these x-values, we find the corresponding y-value in the equation $y = -2x + 4$.

Table of Values

x	y
0	
1	
3	

$y = -2x + 4$ $y = -2x + 4$ $y = -2x + 4$

$y = -2(0) + 4$ $y = -2(1) + 4$ $y = -2(3) + 4$

$y = 0 + 4$ $y = -2 + 4$ $y = -6 + 4$

$y = 4$ $y = 2$ $y = -2$

We record these results by placing each y-value in the table next to its corresponding x-value. Keep in mind that these values represent ordered pairs, each of which is a solution to the equation. To make calculating and graphing easier, we choose integer values whenever possible. If we plot these ordered pairs and connect the three points, we get a straight line that is the graph of the equation $y = -2x + 4$. The graph of the equation is shown in the figure at the right.

Table of Values

x	y
0	4
1	2
3	-2

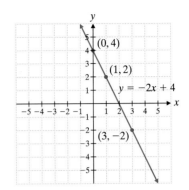

Practice Problem 1 Find three ordered pairs that satisfy $x + y = 10$. Then graph the resulting straight line. Use the given coordinate system.

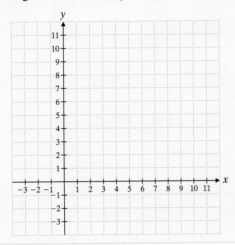

SIDELIGHT Determining Values for Ordered Pairs Why can we choose any value for either x or y when we are finding ordered pairs that are solutions to an equation?

The numbers for x and y that are solutions to the equation come in pairs. When we try to find a pair that fits, we have to start with some number. Usually, it is an x-value that is small and easy to work with. Then we must find the value for y so that the pair of numbers (x, y) is a solution to the given equation.

EXAMPLE 2 Graph $5x - 4y + 2 = 2$.

Solution First, we simplify the equation $5x - 4y + 2 = 2$ by subtracting 2 from each side.

$$5x - 4y + 2 - 2 = 2 - 2$$
$$5x - 4y = 0$$

Table of Values

x	y
0	0

Since we are free to choose any value of x, $x = 0$ is a natural choice. Calculate the value of y when $x = 0$ and place the results in the table of values.

$$5(0) - 4y = 0$$
$$-4y = 0 \quad \text{Remember: Any number times 0 is 0.}$$
$$y = 0 \quad \text{Since } -4y = 0, y \text{ must equal 0.}$$

Now let's see what happens when $x = 1$.

$$5(1) - 4y = 0$$
$$5 - 4y = 0$$
$$-4y = -5$$
$$y = \frac{-5}{-4} \quad \text{or} \quad \frac{5}{4} \quad \text{This is not an easy number to graph.}$$

A better choice for a replacement of x is a number that is divisible by 4. Let's see why. Let $x = 4$ and let $x = -4$.

$$5(4) - 4y = 0 \qquad\qquad 5(-4) - 4y = 0$$
$$20 - 4y = 0 \qquad\qquad -20 - 4y = 0$$
$$-4y = -20 \qquad\qquad -4y = 20$$
$$y = \frac{-20}{-4} \quad \text{or} \quad 5 \qquad\qquad y = \frac{20}{-4} \quad \text{or} \quad -5$$

Now we can put these numbers into our table of values and graph the line.

PRACTICE PROBLEM 2

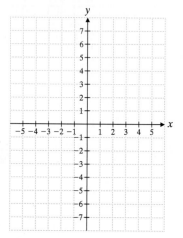

Table of Values

x	y
0	0
4	5
-4	-5

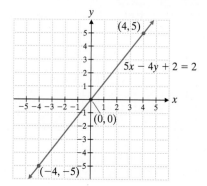

NOTE TO STUDENT: Fully worked-out solutions to all of the Practice Problems can be found at the back of the text starting at page SP-1

Practice Problem 2 Graph $7x + 3 = -2y + 3$ on the coordinate system in the margin.

TO THINK ABOUT: An Alternative Approach In Example 2, when we picked the value of 1 for x, we found that the corresponding value for y was a fraction. To avoid fractions, we can solve the equation for the variable y *first*, then choose the values for x.

$$5x - 4y = 0 \qquad \text{We must isolate } y.$$
$$-4y = -5x \qquad \text{Subtract } 5x \text{ from each side.}$$
$$\frac{-4y}{-4} = \frac{-5x}{-4} \qquad \text{Divide each side by } -4.$$
$$y = \frac{5}{4}x$$

Now let $x = -4$, $x = 0$, and $x = 4$, and find the corresponding values of y. Explain why you would choose multiples of 4 as replacements for x in this equation. Graph the equation and compare it to the graph in Example 2.

In the previous two examples we began by picking values for x. We could just as easily have chosen values for y.

② Graphing a Straight Line by Plotting Its Intercepts

What values should we pick for x and y? Which points should we use for plotting? For many straight lines it is easiest to pick the two *intercepts*. Some lines have only one intercept. We will discuss these separately.

> The **x-intercept** of a line is the point where the line crosses the x-axis; it has the form $(a, 0)$. The **y-intercept** of a line is the point where the line crosses the y-axis; it has the form $(0, b)$.

INTERCEPT METHOD OF GRAPHING

To graph an equation using intercepts, we:

1. Find the x-intercept by letting $y = 0$ and solving for x.

2. Find the y-intercept by letting $x = 0$ and solving for y.

3. Find one additional ordered pair so that we have three points with which to plot the line.

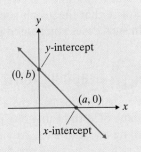

EXAMPLE 3 Complete **(a)** and **(b)** for the equation $5y - 3x = 15$.

(a) State the x- and y-intercepts.

(b) Use the intercept method to graph.

Solution Substitute values in the equation $5y - 3x = 15$.

(a) Let $y = 0$.

$\quad 5(0) - 3x = 15$ Replace y by 0.

$\quad\quad\quad -3x = 15$ Divide both sides by -3.

$\quad\quad\quad\quad x = -5$

x	y
−5	0

x-intercept

The ordered pair $(-5, 0)$ is the x-intercept.

Let $x = 0$.

$\quad 5y - 3(0) = 15$ Replace x by 0.

$\quad\quad\quad 5y = 15$ Divide both sides by 5.

$\quad\quad\quad\ y = 3$

x	y
0	3

y-intercept

The ordered pair $(0, 3)$ is the y-intercept.

(b) We find another ordered pair to have a third point and then graph.

Let $y = 6$.

$\quad 5(6) - 3x = 15$ Replace y by 6.

$\quad 30 - 3x = 15$ Simplify.

$\quad\quad -3x = -15$ Subtract 30 from both sides.

$\quad\quad\quad x = \dfrac{-15}{-3}$ or 5

The ordered pair is $(5, 6)$.
Our table of values is

x	y
−5	0
0	3
5	6

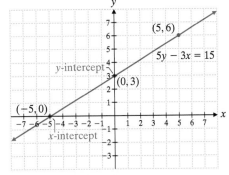

PRACTICE PROBLEM 3

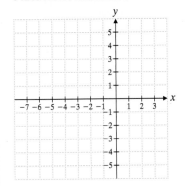

CAUTION: The three points on the graph must form a straight line. If the three points do not form a straight line, you made a calculation error.

Practice Problem 3 Use the intercept method to graph $2y - x = 6$. Use the given coordinate system.

TO THINK ABOUT: Lines That Go Through the Origin Can you draw all straight lines by the intercept method? Not really. Some straight lines may go through the origin and have only one intercept. If a line goes through the origin, it will have an equation of the form $Ax + By = 0$, where $A \neq 0$ or $B \neq 0$ or both. Refer to Example 2. When we simplified the equation, we obtained $5x - 4y = 0$. Notice that the graph goes through the origin, and thus there is only one intercept. In such cases you should plot two additional points besides the origin.

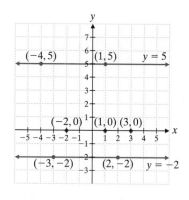

3 Graphing Horizontal and Vertical Lines

You will notice that the x-axis is a horizontal line. It is the line $y = 0$, since for any value of x, the value of y is 0. Try a few points. The points $(1, 0)$, $(3, 0)$, and $(-2, 0)$ all lie on the x-axis. Any horizontal line will be parallel to the x-axis. Lines such as $y = 5$ and $y = -2$ are horizontal lines. What does $y = 5$ mean? It means that for any value of x, y is 5. Likewise $y = -2$ means that for any value of x, $y = -2$.

How can we recognize the equation of a line that is horizontal, that is, parallel to the x-axis?

> If the graph of an equation is a straight line that is parallel to the x-axis (that is, a horizontal line), the equation will be of the form $y = b$, where b is some real number.

EXAMPLE 4 Graph $y = -3$.

Solution You could write the equation as $0x + y = -3$. Then it is clear that for any value of x that you substitute, you will always obtain $y = -3$. Thus, as shown in the figure, $(4, -3)$, $(0, -3)$, and $(-3, -3)$ are all ordered pairs that satisfy the equation $y = -3$. Since the y-coordinate of every point on this line is -3, it is easy to see that the horizontal line will be 3 units below the x-axis.

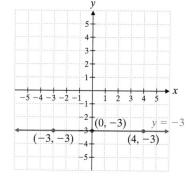

x	y
4	−3
0	−3
−3	−3

Practice Problem 4 Graph $2y - 3 = 0$ on the given coordinate system.

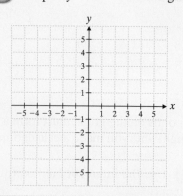

Notice that the y-axis is a vertical line. This is the line $x = 0$, since for any y, x is 0. Try a few points. The points $(0, 2)$, $(0, -3)$, and $\left(0, \frac{1}{2}\right)$ all lie on the y-axis. Any vertical line will be parallel to the y-axis. Lines such as $x = 2$ and $x = -3$ are vertical lines. Think of what $x = 2$ means. It means that for any value of y, x is 2. The graph of $x = 2$ is a vertical line two units to the right of the y-axis.

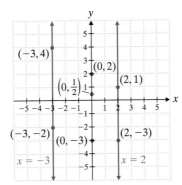

How can we recognize the equation of a line that is vertical, that is, parallel to the y-axis?

> If the graph of an equation is a straight line that is parallel to the y-axis (that is, a vertical line), the equation will be of the form $x = a$, where a is some real number.

EXAMPLE 5   Graph $2x + 1 = 11$.

Solution   Notice that there is only one variable, x, in the equation. This is an indication that we can simplify the equation to the form $x = a$.

$$2x + 1 = 11 \quad \text{x is the only variable in the equation.}$$
$$2x = 10 \quad \text{Solve for x.}$$
$$x = 5$$

Since the x-coordinate of every point on this line is 5, we can see that the vertical line will be 5 units to the right of the y-axis.

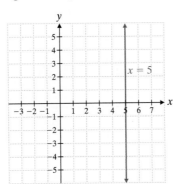

Practice Problem 5   Graph $x + 3 = 0$ on the following coordinate system.

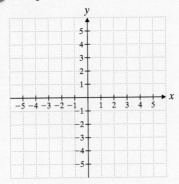

PRACTICE   WATCH   DOWNLOAD   READ   REVIEW

Verbal and Writing Skills

Unless otherwise indicated, assume each grid line represents one unit.

1. Is the point $(-2, 5)$ a solution to the equation $2x + 5y = 0$? Why or why not?

2. The graph of a linear equation in two variables is a _____ _____.

3. The x-intercept of a line is the point where the line crosses the _____.

4. The graph of the equation $y = b$ is a _____ line.

Complete the ordered pairs so that each is a solution of the given linear equation. Then plot each solution and graph the equation by connecting the points by a straight line.

5. $y = -2x + 1$
(0,)
(−2,)
(1,)

6. $y = -3x - 4$
(−2,)
(−1,)
(0,)

7. $y = x - 4$
(0,)
(2,)
(4,)

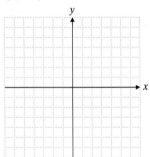 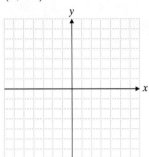 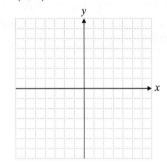

8. $y = 3x - 1$
(0,)
(2,)
(−1,)

9. $y = -2x + 3$
(0,)
(2,)
(4,)

10. $y = 2x - 5$
(0,)
(2,)
(4,)

11. $y = 3x + 2$
(−1,)
(0,)
(1,)

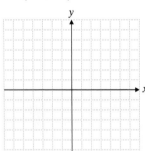 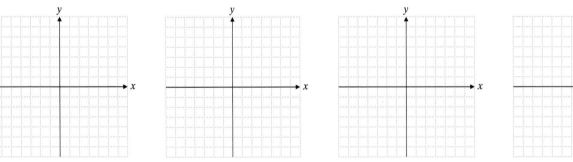 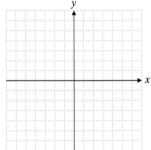

Graph each equation by plotting three points and connecting them.

12. $y = -3x + 2$

13. $3x - 2y = 0$

14. $2y - 5x = 0$

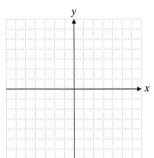 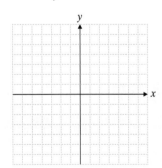 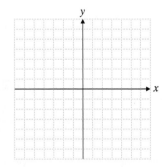

15. $y = -\dfrac{3}{4}x + 3$ **16.** $y = \dfrac{2}{3}x + 2$ **17.** $4x + 3y = 12$ **18.** $-2x + 4y = 12$

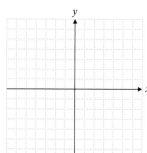   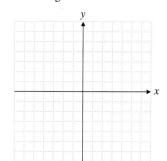   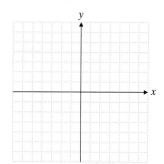   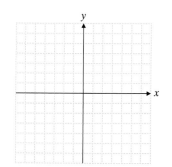

Graph each equation by plotting the intercepts and one other point.

19. $y = 6 - 2x$ **20.** $y = 4 - 2x$

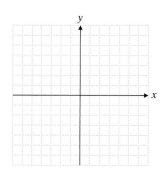   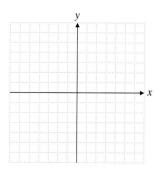

21. $x + 3 = 6y$ **22.** $x - 6 = 2y$

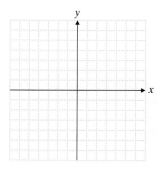   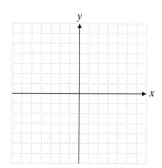

Mixed Practice *Graph the equation. Be sure to simplify the equation before graphing it.*

23. $y - 2 = 3y$ **24.** $3y + 1 = 7$ **25.** $2x + 9 = 5x$

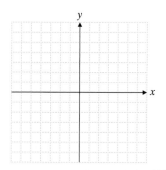   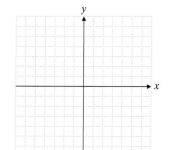   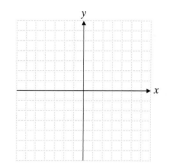

26. $3x - 4 = -13$

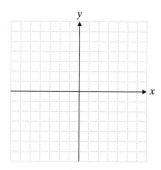

27. $2x + 5y - 2 = -12$

28. $3x - 4y - 5 = -17$

Applications

29. ***Cross-Country Skiing*** The number of calories burned by an average person while cross-country skiing is given by the equation $C = 8m$, where m is the number of minutes. (*Source:* National Center for Health Statistics.) Graph the equation for $m = 0, 15, 30, 45, 60,$ and 75.

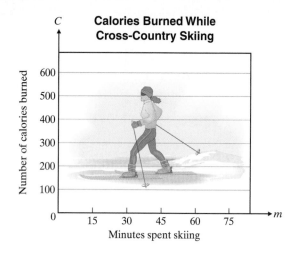

30. ***Calories Burned While Jogging*** The number of calories burned by an average person while jogging is given by the equation $C = \dfrac{28}{3}m$, where m is the number of minutes. (*Source:* National Center for Health Statistics.) Graph the equation for $m = 0, 15, 30, 45, 60,$ and 75.

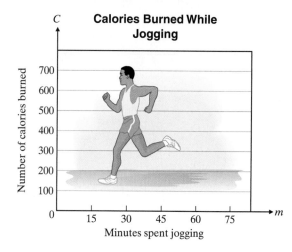

31. ***Foreign Students in the United States*** The number of foreign students enrolled in college in the United States can be approximated by the equation $S = 11t + 395$, where t stands for the number of years since 1990, and S is the number of foreign students (in thousands). (*Source:* www.opendoors.iienetwork.org.) Graph the equation for $t = 0, 4, 8, 16$.

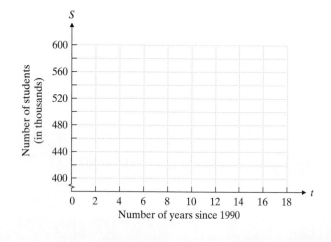

32. ***Restaurant Sales*** The amount of money U.S. consumers spend on food and beverages in restaurants can be approximated by the equation $S = 16.8t + 223$, where S is the sales in billions of dollars and t is the number of years since 1990. (*Source: Statistical Abstract of the United States: 2007.*) Graph the equation for $t = 0, 6, 10, 14, 18$.

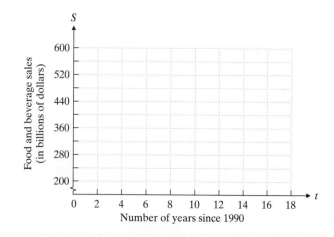

Cumulative Review

33. **[2.3.3]** Solve. $2(x + 3) + 5x = 3x - 2$

34. **[2.8.4]** Solve and graph on a number line. $4 - 3x \leq 18$

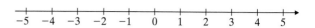

Quick Quiz 3.2 Graph each equation.

1. $y = -2x + 4$

2. $y = \dfrac{2}{5}x - 3$

3. $4y + 1 = x + 9$

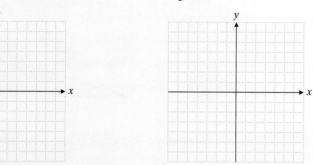

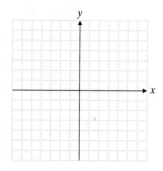

4. Concept Check In graphing the equation $3y - 7x = 0$, what is the most important ordered pair to obtain before drawing a graph of the line? Why is that ordered pair so essential to drawing the graph?

Finding the Slope of a Line Given Two Points on the Line

We often use the word *slope* to describe the incline (the steepness) of a hill. A carpenter or a builder will refer to the *pitch* or *slope* of a roof. The slope is the change in the vertical distance (the rise) compared to the change in the horizontal distance (the run) as you go from one point to another point along the roof. If the change in the vertical distance is greater than the change in the horizontal distance, the slope will be steep. If the change in the horizontal distance is greater than the change in the vertical distance, the slope will be gentle.

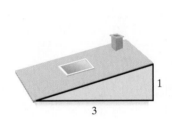

In a coordinate plane, the **slope** of a straight line is defined by the change in *y* divided by the change in *x*.

$$\text{slope} = \frac{\text{change in } y}{\text{change in } x}$$

Consider the line drawn through points A and B in the figure. If we measure the change from point A to point B in the x-direction and the y-direction, we will have an idea of the steepness (or the slope) of the line. From point A to point B, the change in y-values is from 2 to 4, a *change of* 2. From point A to point B, the change in x-values is from 1 to 5, a *change of* 4. Thus

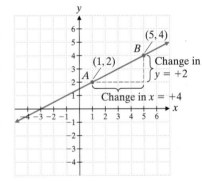

$$\text{slope} = \frac{\text{change in } y}{\text{change in } x} = \frac{2}{4} = \frac{1}{2}.$$

Informally, we can describe this move as the rise over the run: $\text{slope} = \dfrac{\text{rise}}{\text{run}}$.

We now state a more formal (and more frequently used) definition.

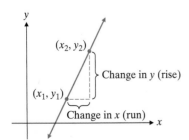

DEFINITION OF SLOPE OF A LINE

The **slope** of any *nonvertical* straight line that contains the points with coordinates (x_1, y_1) and (x_2, y_2) is defined by the difference ratio

$$\text{slope} = m = \frac{y_2 - y_1}{x_2 - x_1} \qquad \text{where } x_2 \neq x_1.$$

The use of subscripted terms such as x_1, x_2, and so on, is just a way of indicating that the first x-value is x_1 and the second x-value is x_2. Thus (x_1, y_1) are the coordinates of the first point and (x_2, y_2) are the coordinates of the second point. The letter m is commonly used for the slope.

EXAMPLE 1 Find the slope of the line that passes through $(2, 0)$ and $(4, 2)$.

Solution Let $(2, 0)$ be the first point (x_1, y_1) and $(4, 2)$ be the second point (x_2, y_2).

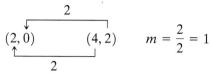

$$(2, 0) \qquad (4, 2) \qquad m = \frac{2}{2} = 1$$

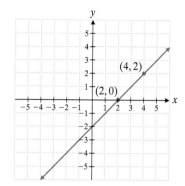

By the formula, $\quad \text{slope} = m = \dfrac{y_2 - y_1}{x_2 - x_1} = \dfrac{2 - 0}{4 - 2} = \dfrac{2}{2} = 1$.

A sketch of the line is shown in the figure at the right.

Note that the slope of the line will be the same if we let $(4, 2)$ be the first point (x_1, y_1) and $(2, 0)$ be the second point (x_2, y_2).

$$(4, 2) \qquad (2, 0)$$

$$m = \frac{y_2 - y_1}{x_2 - x_1} = \frac{0 - 2}{2 - 4} = \frac{-2}{-2} = 1$$

Thus, given two points, it does not matter which you call (x_1, y_1) and which you call (x_2, y_2).

CAUTION: Be careful, however, not to put the x's in one order and the y's in another order when finding the slope from two points on a line.

Practice Problem 1 Find the slope of the line that passes through $(6, 1)$ and $(-4, -1)$.

It is a good idea to have some concept of what different slopes mean. In downhill skiing, a very gentle slope used for teaching beginning skiers might drop 1 foot vertically for each 10 feet horizontally. The slope would be $\frac{1}{10}$. The speed of a skier on a hill with such a gentle slope would be only about 6 miles per hour.

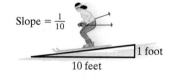

A triple diamond slope for experts might drop 11 feet vertically for each 10 feet horizontally. The slope would be $\frac{11}{10}$. The speed of a skier on such an expert trail would be in the range of 60 miles per hour.

It is important to see how positive and negative slopes affect the graphs of lines.

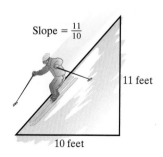

POSITIVE SLOPE

$$m = \frac{5 - 2}{4 - 1} = \frac{3}{3} = 1$$

$(4, 5)$

y is increasing from 2 to 5 as you go from left to right

$(1, 2)$

NEGATIVE SLOPE

$$m = \frac{5 - 1}{1 - 4} = \frac{4}{-3} = -\frac{4}{3}$$

$(1, 5)$

y is decreasing from 5 to 1 as you go from left to right

$(4, 1)$

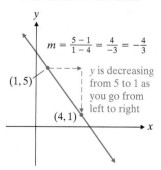

1. If the y-values increase as you go from left to right, the slope of the line is positive.
2. If the y-values decrease as you go from left to right, the slope of the line is negative.

EXAMPLE 2 Find the slope of the line that passes through $(-3, 2)$ and $(2, -4)$.

Solution Let $(-3, 2)$ be (x_1, y_1) and $(2, -4)$ be (x_2, y_2).

$$m = \frac{y_2 - y_1}{x_2 - x_1} = \frac{-4 - 2}{2 - (-3)} = \frac{-4 - 2}{2 + 3} = \frac{-6}{5} = -\frac{6}{5}$$

The slope of this line is negative. We would expect this, since the y-value decreased from 2 to -4 as the x-value increased. What does the graph of this line look like? Plot the points and draw the line to verify.

Practice Problem 2 Find the slope of the line that passes through $(2, 0)$ and $(-1, 1)$.

NOTE TO STUDENT: *Fully worked-out solutions to all of the Practice Problems can be found at the back of the text starting at page SP-1*

TO THINK ABOUT: Using the Slope to Describe a Line Describe the line in Practice Problem 2 by looking at its slope. Then verify by drawing the graph.

EXAMPLE 3 Find the slope of the line that passes through the given points.

(a) $(0, 2)$ and $(5, 2)$ **(b)** $(-4, 0)$ and $(-4, -4)$

Solution

(a) Take a moment to look at the y-values. What do you notice? What does this tell you about the line? Now calculate the slope.

$$m = \frac{2 - 2}{5 - 0} = \frac{0}{5} = 0$$

Since any two points on a horizontal line will have the same y-value, the slope of a horizontal line is 0.

(b) Take a moment to look at the x-values. What do you notice? What does this tell you about the line? Now calculate the slope.

$$m = \frac{-4 - 0}{-4 - (-4)} = \frac{-4}{0}$$

Recall that division by 0 is undefined.

The slope of a vertical line is undefined. We say that a vertical line has **no slope.**

Notice in our definition of slope that $x_2 \neq x_1$. Thus it is not appropriate to use the formula for slope for the points in **(b)**. We did so to illustrate what would happen if $x_2 = x_1$. We get an impossible situation, $\frac{y_2 - y_1}{0}$. Now you can see why we include the restriction $x_2 \neq x_1$ in our definition.

Example 3(a)

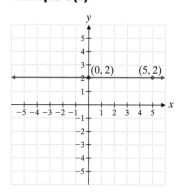

Example 3(b)

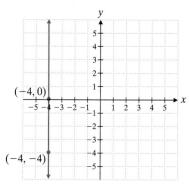

Practice Problem 3 Find the slope of the line that passes through the given points.

(a) $(-5, 6)$ and $(-5, 3)$ **(b)** $(-7, -11)$ and $(3, -11)$

SLOPE OF A STRAIGHT LINE

Positive slope
Line goes upward to the right

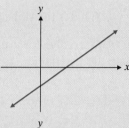

1. *Lines with positive slopes go upward* as you go from left to right.

Negative slope
Line goes downward to the right

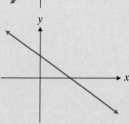

2. *Lines with negative slopes go downward* as you go from left to right.

Zero slope
Horizontal line

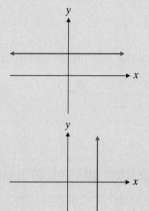

3. Horizontal lines have a slope of 0.

Undefined slope
Vertical line

4. A vertical line is said to have undefined slope. The slope of a vertical line is not defined. In other words, a vertical line has no slope.

② Finding the Slope and *y*-Intercept of a Line Given Its Equation

Recall that the equation of a line is a linear equation in two variables. This equation can be written in several different ways. A very useful form of the equation of a straight line is the slope–intercept form. This form can be derived in the following way. Suppose that a straight line with slope m crosses the y-axis at a point $(0, b)$. Consider any other point on the line and label the point (x, y). Then we have the following.

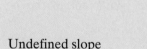

$$\frac{y_2 - y_1}{x_2 - x_1} = m \qquad \text{Definition of slope.}$$

$$\frac{y - b}{x - 0} = m \qquad \text{Substitute } (0, b) \text{ for } (x_1, y_1) \text{ and } (x, y) \text{ for } (x_2, y_2).$$

$$\frac{y - b}{x} = m \qquad \text{Simplify.}$$

$$y - b = mx \qquad \text{Multiply both sides by } x.$$

$$y = mx + b \qquad \text{Add } b \text{ to both sides.}$$

This form of a linear equation immediately reveals the slope of the line, m, and the y-coordinate of the point where the line intercepts (crosses) the y-axis, b.

SLOPE–INTERCEPT FORM OF A LINE

The slope–intercept form of the equation of the line that has slope m and y-intercept $(0, b)$ is given by

$$y = mx + b.$$

By using algebraic operations, we can write any linear equation in slope–intercept form and use this form to identify the slope and the y-intercept of the line.

EXAMPLE 4   What is the slope and the y-intercept of the line $5x + 3y = 2$?

Solution   We want to solve for y and get the equation in the form $y = mx + b$. We need to isolate the y-variable.

$$5x + 3y = 2$$
$$3y = -5x + 2 \qquad \text{Subtract } 5x \text{ from both sides.}$$
$$y = \frac{-5x + 2}{3} \qquad \text{Divide both sides by 3.}$$
$$y = -\frac{5}{3}x + \frac{2}{3} \qquad \text{Using the property } \frac{a + b}{c} = \frac{a}{c} + \frac{b}{c}, \text{ write the right-hand side as two fractions.}$$
$$m = -\frac{5}{3} \text{ and } b = \frac{2}{3}$$

The *slope* is $-\dfrac{5}{3}$.   The *y-intercept* is $\left(0, \dfrac{2}{3}\right)$.

Note: We write the y-intercept as an ordered pair of the form $(0, b)$.

Practice Problem 4   What is the slope and the y-intercept of the line $4x - 2y = -5$?

NOTE TO STUDENT: *Fully worked-out solutions to all of the Practice Problems can be found at the back of the text starting at page SP-1*

③ Writing the Equation of a Line Given the Slope and y-Intercept

If we know the slope of a line and the y-intercept, we can write the equation of the line, $y = mx + b$.

EXAMPLE 5   Find an equation of the line with slope $\frac{2}{5}$ and y-intercept $(0, -3)$.

(a) Write the equation in slope–intercept form, $y = mx + b$.
(b) Write the equation in the form $Ax + By = C$.

Solution

(a) We are given that $m = \frac{2}{5}$ and $b = -3$. Thus we have the following.

$$y = mx + b$$
$$y = \frac{2}{5}x + (-3)$$
$$y = \frac{2}{5}x - 3$$

(b) Recall, for the form $Ax + By = C$, that A, B, and C are integers. We first clear the equation of fractions. Then we move the x-term to the left side.

$$5y = 5\left(\frac{2x}{5}\right) - 5(3) \qquad \text{Multiply each term by 5.}$$
$$5y = 2x - 15 \qquad \text{Simplify.}$$
$$-2x + 5y = -15 \qquad \text{Subtract } 2x \text{ from each side.}$$
$$2x - 5y = 15 \qquad \text{Multiply each term by } -1. \text{ The form } Ax + By = C \text{ is usually written with } A \text{ as a positive integer.}$$

Practice Problem 5   Find an equation of the line with slope $-\frac{3}{7}$ and y-intercept $\left(0, \frac{2}{7}\right)$.

(a) Write the equation in slope–intercept form.
(b) Write the equation in the form $Ax + By = C$.

4 Graphing a Line Using the Slope and *y*-Intercept

If we know the slope of a line and the *y*-intercept, we can draw the graph of the line.

EXAMPLE 6 Graph the line with slope $m = \frac{2}{3}$ and *y*-intercept $(0, -3)$. Use the given coordinate system.

Solution Recall that the *y*-intercept is the point where the line crosses the *y*-axis. We need a point to start the graph with. So we plot the point $(0, -3)$ on the *y*-axis.

Recall that slope $= \dfrac{\text{rise}}{\text{run}}$. Since the slope of this line is $\frac{2}{3}$, we will go up (rise) 2 units and go over (run) to the right 3 units from the point $(0, -3)$. Look at the figure to the right. This is the point $(3, -1)$. Plot the point. Draw a line that connects the two points $(0, -3)$ and $(3, -1)$.

This is the graph of the line with slope $\frac{2}{3}$ and *y*-intercept $(0, -3)$.

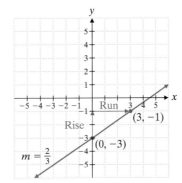

Practice Problem 6 Graph the line with slope $= \frac{3}{4}$ and *y*-intercept $(0, -1)$. Use the coordinate system in the margin.

PRACTICE PROBLEM 6

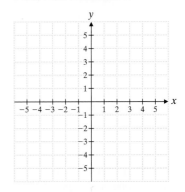

Let's summarize the process used in Example 6. When graphing we use the fact that slope $= \frac{\text{rise}}{\text{run}}$. We rise in the *y*-direction. We move up if the number is positive and down if it is negative. We run in the *x*-direction. We move right if the number is positive and left if the number is negative. Complete the following three steps to graph a line using the slope and *y*-intercept:

Step 1. Plot *y*-intercept point on graph.

Step 2. Begin at the *y*-intercept point. Move up or down the number of units represented by the numerator (rise). Then move left or right the number of units represented by the denominator (run).

Step 3. Plot this point and draw a line through the 2 points.

EXAMPLE 7 Graph the equation $y = -\frac{1}{2}x + 4$. Use the following coordinate system.

Solution First we identify the *y*-intercept, $(0, 4)$, and the slope, $m = -\frac{1}{2}$. Then we begin with the *y*-intercept. Since $b = 4$, plot the point $(0, 4)$. Now look at the slope, $-\frac{1}{2}$. This can be written as $\frac{-1}{2}$. Begin at $(0, 4)$ and go *down* 1 unit and to the right 2 units. This is the point $(2, 3)$. Plot the point. Draw the line that connects the points $(0, 4)$ and $(2, 3)$.

This is the graph of the equation $y = -\frac{1}{2}x + 4$.

Note: The slope $-\frac{1}{2} = \frac{-1}{2} = \frac{1}{-2}$. Therefore we get the same line if we write the slope as $\frac{1}{-2}$. Try it.

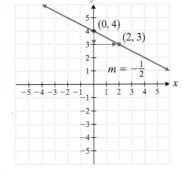

PRACTICE PROBLEM 7

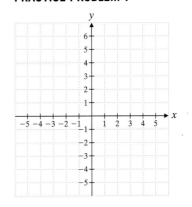

Practice Problem 7 Graph the equation $y = -\frac{2}{3}x + 5$. Use the coordinate system in the margin.

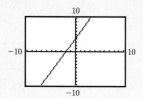
TO THINK ABOUT: Graphing Using Different Methods In this section and in Section 3.2 we learned the following three ways to graph a linear equation:

1. Plot any 3 points.
2. Find the intercepts.
3. Find the slope and y-intercept.

For the equation $y = -\frac{1}{2}x + 4$ in Example 7, we used the slope–intercept method because the equation was in the form $y = mx + b$. We can also write this equation as $y + \frac{1}{2}x = 4$ or $2y + x = 8$. Which one of these forms would you prefer to use if you were graphing the equation using the intercept method? Try it and verify that the points lie on the same line.

5 **Finding the Slopes of Parallel and Perpendicular Lines**

Parallel lines are two straight lines that never touch. Look at the parallel lines in the figure. Notice that the slope of line a is -3 and the slope of line b is also -3. Why do you think the slopes must be equal? What would happen if the slope of line b were -1? Graph it and see.

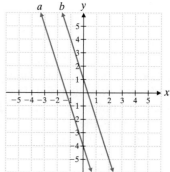

PARALLEL LINES

Parallel lines are two straight lines that never touch.
Parallel lines have the same slope but different y-intercepts.

$$m_1 = m_2$$

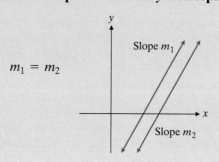

Perpendicular lines are two lines that meet at a 90° angle. Look at the perpendicular lines in the figure at the left. The slope of line c is -3. The slope of line d is $\frac{1}{3}$. Notice that

$$(-3)\left(\frac{1}{3}\right) = \left(-\frac{3}{1}\right)\left(\frac{1}{3}\right) = -1.$$

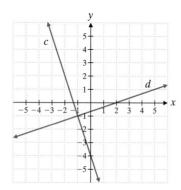

PERPENDICULAR LINES

Perpendicular lines are two lines that meet at a 90° angle.
Perpendicular lines have slopes whose product is -1. If m_1 and m_2 are slopes of perpendicular lines, then

$$m_1 m_2 = -1 \quad \text{or} \quad m_1 = -\frac{1}{m_2}.$$

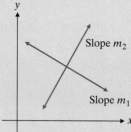

EXAMPLE 8 Line h has a slope of $-\frac{2}{3}$.

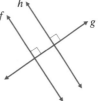

(a) If line f is parallel to line h, what is its slope?

(b) If line g is perpendicular to line h, what is its slope?

Solution

(a) Parallel lines have the same slope. Line f has a slope of $-\frac{2}{3}$.

(b) Perpendicular lines have slopes whose product is -1.

$$m_1 m_2 = -1$$

$$-\frac{2}{3}m_2 = -1 \qquad \text{Substitute } -\frac{2}{3} \text{ for } m_1.$$

$$\left(-\frac{3}{2}\right)\left(-\frac{2}{3}\right)m_2 = -1\left(-\frac{3}{2}\right) \quad \text{Multiply both sides by } -\frac{3}{2}.$$

$$m_2 = \frac{3}{2}$$

Thus line g has a slope of $\frac{3}{2}$.

Practice Problem 8 Line h has a slope of $\frac{1}{4}$.

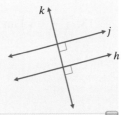

(a) If line j is parallel to line h, what is its slope?

(b) If line k is perpendicular to line h, what is its slope?

EXAMPLE 9 The equation of line l is $y = -2x + 3$.

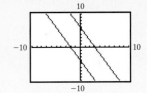

(a) What is the slope of line f, which is parallel to line l?

(b) What is the slope of line s, which is perpendicular to line l?

Solution

(a) Looking at the equation, we can see that the slope of line l is -2.
Since the slope of a line that is parallel to line l is -2, the slope of line f is -2.

(b) Perpendicular lines have slopes whose product is -1.

$$m_1 m_2 = -1$$

$$(-2)m_2 = -1 \quad \text{Substitute } -2 \text{ for } m_1.$$

$$m_2 = \frac{1}{2} \quad \text{Because } (-2)\left(\frac{1}{2}\right) = -1.$$

The slope of a line s which is perpendicular to line l is $\frac{1}{2}$.

Practice Problem 9 The equation of line n is $y = \frac{2}{3}x - 4$.

(a) What is the slope of line p, which is parallel to line n?

(b) What is the slope of line t, which is perpendicular to line n?

Graphing Calculator

Graphing Parallel Lines

If two equations are in the form $y = mx + b$, then it will be obvious that they are parallel because the slope will be the same. On a graphing calculator, graph both of these equations:

$$y = -2x + 6$$
$$y = -2x - 4$$

Use a window of -10 to 10 for both x and y.
Display:

Verbal and Writing Skills

1. Can you find the slope of the line passing through $(5, -12)$ and $(5, -6)$? Why or why not?

2. Can you find the slope of the line passing through $(6, -2)$ and $(-8, -2)$? Why or why not?

Find the slope of the straight line that passes through the given pair of points.

3. $(4, 1)$ and $(6, 7)$

4. $(7, 5)$ and $(12, 10)$

5. $(5, 10)$ and $(6, 5)$

6. $(4, 7)$ and $(1, 13)$

7. $(-2, 1)$ and $(3, 4)$

8. $(5, 6)$ and $(-3, 1)$

9. $(-6, -5)$ and $(2, -7)$

10. $(-8, -3)$ and $(4, -9)$

11. $(-3, 0)$ and $(0, -4)$

12. $(0, 5)$ and $(5, 3)$

13. $\left(\frac{3}{4}, -4\right)$ and $(2, -8)$

14. $\left(\frac{5}{3}, -2\right)$ and $(3, 6)$

Find the slope and the y-intercept.

15. $y = 8x + 9$

16. $y = 2x + 10$

17. $3x + y - 4 = 0$

18. $8x + y + 7 = 0$

19. $y = -\frac{8}{7}x + \frac{3}{4}$

20. $y = \frac{5}{3}x - \frac{4}{5}$

21. $y = -6x$

22. $y = -2$

23. $6x + y = \frac{4}{5}$

24. $2x + y = -\frac{3}{4}$

25. $3x + 4y - 6 = 0$

26. $5x + 3y - 1 = 0$

27. $7x - 3y = 4$

28. $9x - 4y = 18$

Write the equation of the line in slope–intercept form.

29. $m = \dfrac{3}{5}$, *y*-intercept $(0, 3)$

30. $m = \dfrac{2}{3}$, *y*-intercept $(0, 5)$

31. $m = 4$, *y*-intercept $(0, -5)$

32. $m = 2$, *y*-intercept $(0, -1)$

33. $m = -\dfrac{5}{4}$, *y*-intercept $\left(0, -\dfrac{3}{4}\right)$

34. $m = -4$, *y*-intercept $\left(0, \dfrac{1}{2}\right)$

Graph the line $y = mx + b$ for the given values.

35. $m = \dfrac{3}{4}$, $b = -4$

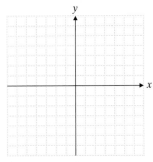

36. $m = \dfrac{1}{3}$, $b = -2$

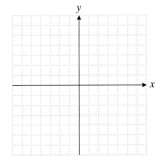

37. $m = -\dfrac{5}{3}$, $b = 2$

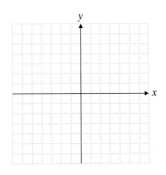

38. $m = -\dfrac{3}{2}$, $b = 4$

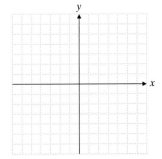

In exercises 39–44, graph the line.

39. $y = \dfrac{2}{3}x + 2$

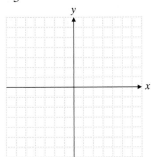

40. $y = \dfrac{3}{4}x + 1$

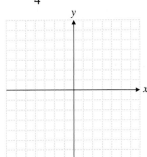

41. $y + 2x = 3$

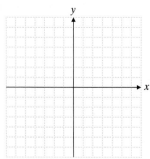

42. $y + 4x = 5$

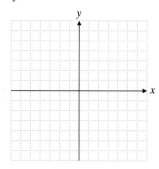

43. $y = 2x$

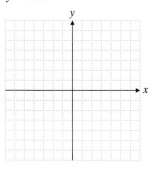

44. $y = 3x$

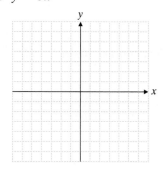

45. Line t has a slope of $\dfrac{5}{6}$.

    (a) What is the slope of the line parallel to it?
    (b) What is the slope of the line perpendicular to it?

46. Line k has a slope of $\dfrac{11}{4}$.

    (a) What is the slope of the line parallel to it?
    (b) What is the slope of the line perpendicular to it?

47. Line s has a slope of 6.

    (a) What is the slope of the line parallel to it?
    (b) What is the slope of the line perpendicular to it?

48. Line l has a slope of $-\dfrac{5}{8}$.

    (a) What is the slope of the line parallel to it?
    (b) What is the slope of the line perpendicular to it?

49. The equation of a line is $y = \dfrac{2}{3}x + 6$.

    (a) What is the slope of a line parallel to it?
    (b) What is the slope of a line perpendicular to it?

50. The equation of a line is $y = 3x - 8$.

    (a) What is the slope of a line parallel to it?
    (b) What is the slope of a line perpendicular to it?

To Think About

51. Do the points $(3, -4)$, $(18, 6)$, and $(9, 0)$ all lie on the same line? If so, what is the equation of the line?

52. Do the points $(2, 1)$, $(-3, -2)$, and $(7, 4)$ lie on the same line? If so, what is the equation of the line?

53. *Home Health Aides* It is projected that one of the fastest-growing occupations in the United States between 2004 and 2014 will be home health aides. During this 10-year period, the number of home health aides can be approximated by the equation $y = 5(7x + 125)$, where x is the number of years since 2004 and y is the number of employees in thousands. (*Source:* www.bls.gov.)

(a) Write the equation in slope–intercept form.

(b) Find the slope and the y-intercept.

(c) In this specific equation, what is the meaning of the slope? What does it indicate?

54. *Physician Assistants* It is projected that one of the fastest-growing occupations in the United States between 2004 and 2014 will be physician assistants. During this 10-year period, the number of physician assistants can be approximated by the equation $y = \dfrac{1}{10}(31x + 620)$, where x is the number of years since 2004 and y is the number of employees in thousands. (*Source:* www.bls.gov.)

(a) Write the equation in slope–intercept form.

(b) Find the slope and the y-intercept.

(c) In this specific equation, what is the meaning of the slope? What does it indicate?

Cumulative Review

55. [0.2.3] Add. $\dfrac{5}{12} + \dfrac{1}{3} + \dfrac{3}{5}$

56. [0.2.3] Multiply. $\dfrac{25}{36} \times \dfrac{54}{45}$

Solve for x and graph the solution.

57. [2.8.4] $3x + 8 > 2x + 12$

58. [2.8.4] $\dfrac{1}{4}x + 3 > \dfrac{2}{3}x + 2$

59. [2.8.4] $\dfrac{1}{2}(x + 2) \le \dfrac{1}{3}x + 5$

60. [2.8.4] $7x - 2(x + 3) \le 4(x - 7)$

Quick Quiz 3.3

1. Find the slope of the straight line that passes through the points $(-2, 5)$ and $(-6, 3)$.

2. Find the slope and the y-intercept of the line $6x - 2y - 3 = 0$.

3. Write the equation of the line in slope–intercept form that passes through $(0, -5)$ and has a slope of $-\frac{5}{7}$.

4. Concept Check The equation of a given line is $y = \frac{7}{11}x - \frac{3}{4}$. Explain how you would find the slope of a line that is perpendicular to that line.

1. _____

How are you doing with your homework assignments in Sections 3.1 to 3.3? Do you feel you have mastered the material so far? Do you understand the concepts you have covered? Before you go further in the textbook, take some time to do each of the following problems.

3.1

1. Plot the following points.

 E: $(-6, 3)$

 F: $(5, -4)$

 G: $(-2, -8)$

 H: $(6.5, 3.5)$

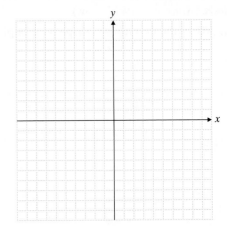

2. _____

2. Give the coordinates for points *A, B, C, D*.

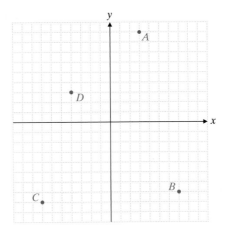

3. _____

3. Complete the ordered pairs for the equation $y = -7x + 3$.

 (4,)

 (0,)

 (−2,)

4. _____

3.2

Find three points and then graph.

4. $4x + y = -3$

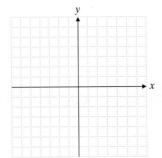

5. $y = \dfrac{3}{4}x - 1$

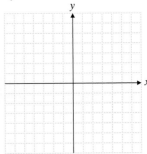

5. _____

6. Graph $6x - 5y = -30$ by plotting the intercepts and one other point.

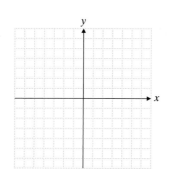

7. Graph $3x - 5y = 0$.

8. Graph $-5x + 2 = -13$.

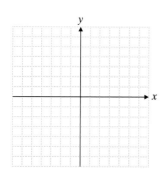

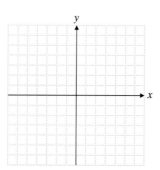

3.3

9. Find the slope of the straight line that passes through $(-2, 5)$ and $(0, 1)$.

10. Find the slope and y-intercept of $-12x + 6y + 8 = 0$.

11. A line has a slope of $\frac{3}{4}$. What is the slope of a line perpendicular to it?

12. An equation of a line is $3y - 5x + 10 = 0$. What is the slope of a line parallel to it?

Now turn to page SA-9 for the answer to each of these problems. Each answer also includes a reference to the objective in which the problem is first taught. If you missed any of these problems, you should stop and review the Examples and Practice Problems in the referenced objective. A little review now will help you master the material in the upcoming sections of the text.

6. _____

7. _____

8. _____

9. _____

10. _____

11. _____

12. _____

3.4 WRITING THE EQUATION OF A LINE

Student Learning Objectives

After studying this section, you will be able to:

 Write an equation of a line given a point and the slope.

2 Write an equation of a line given two points.

3 Write an equation of a line given a graph of the line.

1 Writing an Equation of a Line Given a Point and the Slope

If we know the slope of a line and the y-intercept, we can write the equation of the line in slope–intercept form. Sometimes we are given the slope and a point on the line. We use this information to find the y-intercept. Then we can write the equation of the line.

It may be helpful to summarize our approach.

> **TO FIND AN EQUATION OF A LINE GIVEN A POINT AND THE SLOPE**
> 1. Write the slope-intercept form of the equation of a line: $y = mx + b$.
> 2. Substitute the given values of x, y, and m into the equation.
> 3. Solve for b.
> 4. Use the values of b and m to write the equation in the form $y = mx + b$.

EXAMPLE 1 Find an equation of the line that passes through $(-3, 6)$ with slope $-\frac{2}{3}$.

Solution We are given the values $m = -\frac{2}{3}$, $x = -3$, $y = 6$, and we must find b.

$$y = mx + b$$ 1. Write the equation in slope–intercept form.

$$6 = \left(-\frac{2}{3}\right)(-3) + b$$ 2. Substitute known values.

$$6 = 2 + b$$ 3. Solve for b.

$$4 = b$$ 4. Now use the values for b and m to write an equation in slope–intercept form.

An equation of the line is $y = -\frac{2}{3}x + 4$.

Practice Problem 1 Find an equation of the line that passes through $(-8, 12)$ with slope $-\frac{3}{4}$.

NOTE TO STUDENT: Fully worked-out solutions to all of the Practice Problems can be found at the back of the text starting at page SP-1

2 Writing an Equation of a Line Given Two Points

Our procedure can be extended to the case for which two points are given. Recall from Section 3.3 that if we are given 2 points on a line, we can find the slope m using the formula $m = \dfrac{y_2 - y_1}{x_2 - x_1}$. In the next example we will use this information to find the values of m and b so that we can write an equation in the form $y = mx + b$.

EXAMPLE 2 Find an equation of the line that passes through $(2, 5)$ and $(6, 3)$.

Solution We must find both m and b. We first find m using the formula for slope. Then we proceed as in Example 1 to find b.

$$m = \frac{y_2 - y_1}{x_2 - x_1}$$

$$m = \frac{3 - 5}{6 - 2}$$ Substitute $(x_1, y_1) = (2, 5)$ and $(x_2, y_2) = (6, 3)$ into the formula.

$$= \frac{-2}{4} = -\frac{1}{2}$$

Now that we know m, we can find b.

Choose either point, say $(2, 5)$, to substitute into $y = mx + b$ as in Example 1.

$y = mx + b$ 1. Write the equation in slope–intercept form.

$5 = -\dfrac{1}{2}(2) + b$ 2. Substitute known values.

$5 = -1 + b$ 3. Solve for b.

$6 = b$ 4. Use the values for b and m to write the equation.

An equation of the line is $y = -\dfrac{1}{2}x + 6$.

    Note: We could have substituted the slope and the other point, $(6, 3)$, into the slope–intercept form and arrived at the same answer. Try it.

Practice Problem 2   Find an equation of the line that passes through $(3, 5)$ and $(-1, 1)$.

NOTE TO STUDENT: Fully worked-out solutions to all of the Practice Problems can be found at the back of the text starting at page SP-1

 Writing an Equation of a Line Given a Graph of the Line

As we saw in Examples 1 and 2, once we find m and b we can write the equation of a line. Now, if we are given the graph of a line, can we write the equation of this line? That is, from the graph can we find m and b? We will look at this in Example 3.

EXAMPLE 3   What is the equation of the line in the figure at right?

Solution   First, look for the y-intercept. The line crosses the y-axis at $(0, 4)$. Thus $b = 4$.
    Second, find the slope.

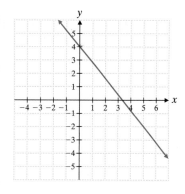

$$m = \frac{\text{change in } y}{\text{change in } x} = \frac{\text{rise}}{\text{run}}$$

Look for another point on the line. We choose $(5, -2)$. Count the number of vertical units from 4 to -2 (rise). Count the number of horizontal units from 0 to 5 (run).

$$m = \frac{-6}{5}$$

    Now, using $m = -\frac{6}{5}$ and $b = 4$, we can write an equation of the line.

$$y = mx + b$$

$$y = -\frac{6}{5}x + 4$$

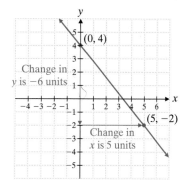

PRACTICE PROBLEM 3

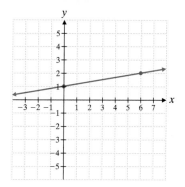

Practice Problem 3   What is the equation of the line in the figure at the right?

Find an equation of the line that has the given slope and passes through the given point.

1. $m = 4, (-3, 0)$

2. $m = 3, (2, -2)$

3. $m = -2, (3, 5)$

4. $m = -1, (5, 0)$

5. $m = -3, \left(\dfrac{1}{2}, 2\right)$

6. $m = -2, \left(3, \dfrac{1}{3}\right)$

7. $m = \dfrac{1}{4}, (4, 5)$

8. $m = \dfrac{2}{3}, (3, -2)$

Write an equation of the line passing through the given points.

9. $(3, -12)$ and $(-4, 2)$

10. $(-1, 8)$ and $(0, 5)$

11. $(2, -6)$ and $(-1, 6)$

12. $(-1, 1)$ and $(5, 7)$

13. $(3, 5)$ and $(-1, -15)$

14. $(-1, -19)$ and $(2, 2)$

15. $\left(1, \dfrac{5}{6}\right)$ and $\left(3, \dfrac{3}{2}\right)$

16. $(2, 0)$ and $\left(\dfrac{3}{2}, \dfrac{1}{2}\right)$

Mixed Practice

17. Find an equation of the line with a slope of -3 that passes through the point $(-1, 3)$.

18. Find an equation of the line with a slope of $-\frac{1}{2}$ that passes through the point $(6, -2)$.

19. Find an equation of the line that passes through $(2, -3)$ and $(-1, 6)$.

20. Find an equation of the line that passes through $(1, -8)$ and $(2, -14)$.

Write an equation of each line.

21.

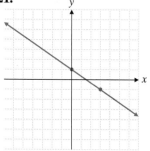

22.

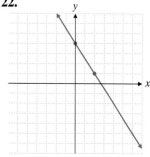

23.

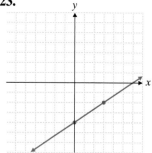

24.

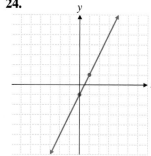

25.

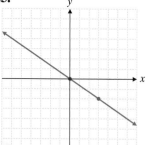

26.

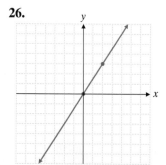

27.

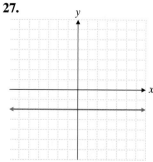

28.

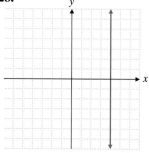

To Think About *Find an equation of the line that fits each description.*

29. Passes through $(7, -2)$ and has zero slope

30. Passes through $(4, 6)$ and has undefined slope

31. Passes through $(4, -6)$ and is perpendicular to the x-axis

32. Passes through $(-3, 5)$ and is perpendicular to the y-axis

33. Passes through $(0, 5)$ and is parallel to $y = \frac{1}{3}x + 4$

34. Passes through $(0, 5)$ and is perpendicular to $y = \frac{1}{3}x + 4$

35. Passes through $(2, 3)$ and is perpendicular to $y = 2x - 9$

36. Passes through $(2, 9)$ and is parallel to $y = 5x - 3$

37. *Population Growth* The growth of the population of the United States during the period from 1980 to 2008 can be approximated by an equation of the form $y = mx + b$, where x is the number of years since 1980 and y is the population measured in millions. (*Source:* U.S. Census Bureau.) Find the equation if two ordered pairs that satisfy it are $(0, 227)$ and $(10, 251)$.

38. *Home Equity Loans* The amount of debt outstanding on home equity loans in the United States during the period from 1993 to 2008 can be approximated by an equation of the form $y = mx + b$, where x is the number of years since 1993 and y is the debt measured in billions of dollars. (*Source:* Board of Governors of the Federal Reserve System.) Find the equation if two ordered pairs that satisfy it are $(1, 280)$ and $(6, 500)$.

Cumulative Review *Solve.*

39. **[2.8.4]** $10 - 3x > 14 - 2x$

40. **[2.8.4]** $2x - 3 \geq 7x - 18$

41. **[2.8.4]** $30 - 2(x + 1) \leq 4x$

42. **[2.8.4]** $2(x + 3) - 22 < 4(x - 2)$

43. **[2.6.2]** *Leaky Faucet* Frank has a leaky faucet in his bathtub. He closed the drain and measured that the faucet drips 8 gallons of water in one 24-hour period.

    (a) If the faucet isn't fixed, how many gallons of water will be wasted in one year?

    (b) If the faucet isn't fixed and water costs $8.50 per 1000 gallons, how much extra money will Frank have to pay on his water bill for the year?

▲ **44.** **[2.7.1]** *Archeology* Archeologists are searching a rectangular region in Mexico for evidence of a primitive civilization over 3000 years old. The perimeter of the region is 3440 feet. The length of the region is 70 feet longer than twice the width. Find the dimensions of the region.

Quick Quiz 3.4

1. Write the equation in slope–intercept form of the line that passes through the point $(3, -5)$ and has a slope of $\frac{2}{3}$.

2. Write the equation in slope–intercept form of the line that passes through the points $(-2, 7)$ and $(-4, -5)$.

3. Write the equation of the line that passes through $(4, 5)$ and $(4, -2)$. What is the slope of this line?

4. **Concept Check** How would you find the equation of the line that passes through $(-2, -3)$ and has zero slope?

Student Learning Objective

After studying this section, you will be able to:

 Graph linear inequalities in two variables.

In Section 2.8 we discussed inequalities in one variable. Look at the inequality $x < -2$ (x is less than -2). Some of the solutions to the inequality are -3, -5, and $-5\frac{1}{2}$. In fact all numbers to the left of -2 on a number line are solutions. The graph of the inequality is given in the following figure. Notice that the open circle at -2 indicates that -2 is *not* a solution.

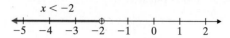

Now we will extend our discussion to consider linear inequalities in two variables.

Graphing Linear Inequalities in Two Variables

Consider the inequality $y \geq x$. The solution of the inequality is the set of all possible ordered pairs that when substituted into the inequality will yield a true statement. Which ordered pairs will make the statement $y \geq x$ true? Let's try some.

$(0, 6)$	$(-2, 1)$	$(1, -2)$	$(3, 5)$	$(4, 4)$
$6 \geq 0$, true	$1 \geq -2$, true	$-2 \geq 1$, false	$5 \geq 3$, true	$4 \geq 4$, true

$(0, 6)$, $(-2, 1)$, $(3, 5)$, and $(4, 4)$ are solutions to the inequality $y \geq x$. In fact, every point at which the y-coordinate is greater than or equal to the x-coordinate is a solution to the inequality. This is shown by the solid line and the shaded region in the graph at the left.

Is there an easier way to graph a linear inequality in two variables? It turns out that we can graph such an inequality by first graphing the associated linear equation and then testing one point that is not on that line. That is, we can change the inequality symbol to an equals sign and graph the equation. If the inequality symbol is $\geq$ or $\leq$, we use a solid line to indicate that the points on the line are included in the solution of the inequality. If the inequality symbol is $>$ or $<$, we use a dashed line to indicate that the points on the line are not included in the solution of the inequality. Then we test one point that is not on the line. If the point is a solution to the inequality, we shade the region on the side of the line that includes the point. If the point is not a solution, we shade the region on the other side of the line.

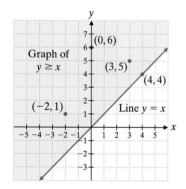

Graph of $y \geq x$

PRACTICE PROBLEM 1

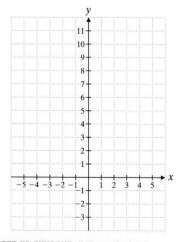

NOTE TO STUDENT: Fully worked-out solutions to all of the Practice Problems can be found at the back of the text starting at page SP-1

EXAMPLE 1 Graph $5x + 3y > 15$. Use the given coordinate system.

Solution We begin by graphing the line $5x + 3y = 15$. You may use any method discussed previously to graph the line. Since there is no equals sign in the inequality, we will draw a dashed line to indicate that the line is *not* part of the solution set.

Look for a test point. The easiest point to test is $(0, 0)$. Substitute $(0, 0)$ for (x, y) in the inequality.

$$5x + 3y > 15$$
$$5(0) + 3(0) > 15$$
$$0 > 15 \quad \text{false}$$

$(0, 0)$ is not a solution. Shade the side of the region on the line that does *not* include $(0, 0)$.

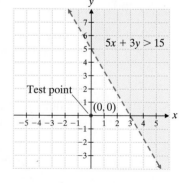

Practice Problem 1 Graph $x - y \geq -10$. Use the coordinate system in the margin.

GRAPHING LINEAR INEQUALITIES

1. Replace the inequality symbol by an equality symbol. Graph the line.
   (a) The line will be solid if the inequality is $\geq$ or $\leq$.
   (b) The line will be dashed if the inequality is $>$ or $<$.

2. Test the point $(0, 0)$ in the inequality if $(0, 0)$ does not lie on the graphed line in step 1.
   (a) If the inequality is true, shade the side of the line that includes $(0, 0)$.
   (b) If the inequality is false, shade the side of the line that does not include $(0, 0)$.

3. If the point $(0, 0)$ is a point on the line, choose another test point and proceed accordingly.

EXAMPLE 2   Graph $2y \leq -3x$.

PRACTICE PROBLEM 2

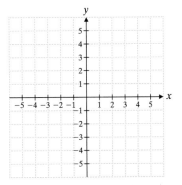

Solution

Step 1   Graph $2y = -3x$. Since $\leq$ is used, the line should be a solid line.

Step 2   We see that the line passes through $(0, 0)$.

Step 3   Choose another test point. We will choose $(-3, -3)$.

$$2y \leq -3x$$
$$2(-3) \leq -3(-3)$$
$$-6 \leq 9 \quad \text{true}$$

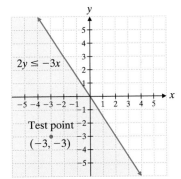

Shade the region that includes $(-3, -3)$, that is, the region below the line.

Practice Problem 2   Graph $2y > x$ on the coordinate system in the margin.

If we are graphing the inequality $x < -2$ on a coordinate plane, the solution will be a region. Notice that this is very different from the solution of $x < -2$ on a number line discussed earlier.

EXAMPLE 3   Graph $x < -2$.

Solution

Step 1   Graph $x = -2$. Since $<$ is used, the line should be dashed.

Step 2   Test $(0, 0)$ in the inequality.

$$x < -2$$
$$0 < -2 \quad \text{false}$$

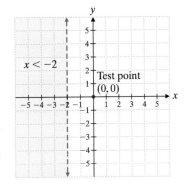

Shade the region that does not include $(0, 0)$, that is, the region to the left of the line $x = -2$. Observe that every point in the shaded region has an x-value that is less than -2.

PRACTICE PROBLEM 3

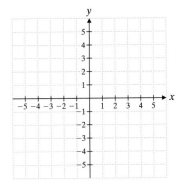

Practice Problem 3   Graph $y \geq -3$ on the coordinate system in the margin.

PRACTICE   WATCH   DOWNLOAD   READ   REVIEW

Verbal and Writing Skills

1. Does it matter what point you use as your test point? Justify your response.

2. Explain when to use a solid line or a dashed line when graphing a linear inequality in two variables.

Graph the region described by the inequality.

3. $y < 2x - 4$

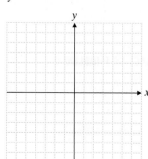

4. $y > 1 - 3x$

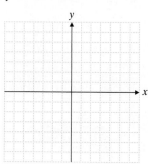

5. $2x - 3y < 6$

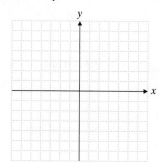

6. $3x + 2y < -6$

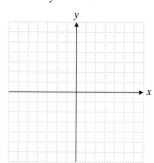

7. $2x - y \geq 3$

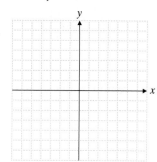

8. $3x - y \geq 4$

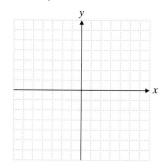

9. $y \geq 4x$

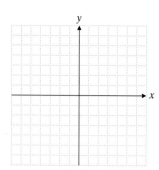

10. $y \leq -2x$

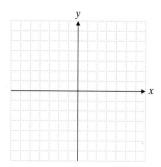

11. $y < -\frac{1}{2}x$

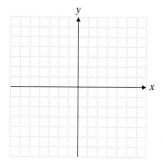

12. $y > \frac{1}{5}x$

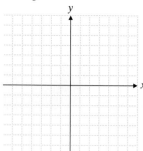

13. $x \geq 2$

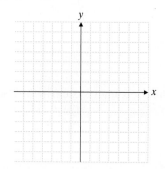

14. $y \leq -2$

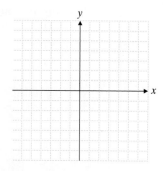

15. $2x - 3y + 6 \geq 0$

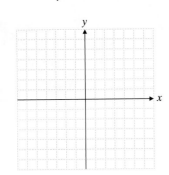

16. $3x + 4y - 8 \leq 0$

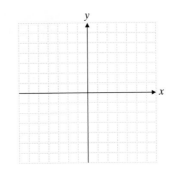

17. $2x > -3y$

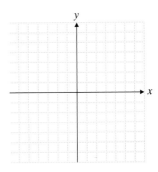

18. $3x \leq -2y$

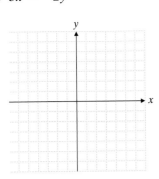

19. $2x > 3 - y$

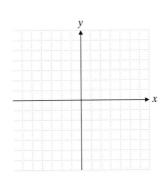

20. $x > 4 - y$

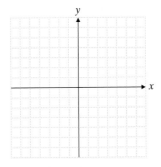

21. $x > -2y$

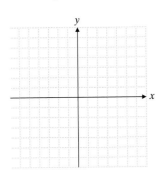

22. $x < -3y$

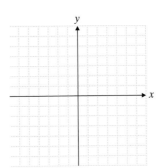

Cumulative Review *Perform the operations in the proper order.*

23. [1.5.1] $5(3) + 6 \div (-2)$

24. [1.5.1] $5(-3) - 2(12 - 15)^2 \div 9$

Evaluate for $x = -2$ and $y = 3$.

25. [1.8.1] $2x^2 + 3xy - 2y^2$

26. [1.8.1] $x^3 - 5x^2 + 3y - 6$

Satellite Parts **[0.5.4]** *Brian sells high-tech parts to satellite communications companies. In his negotiations he originally offers to sell one company 200 parts for a total of $22,400. However, after negotiations, he offers to sell that company the same parts at a 15% discount if the company agrees to sign a purchasing contract for 200 additional parts at some future date.*

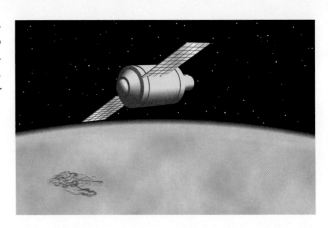

27. What is the average cost per part if the parts are sold at the discounted price?

28. How much will the total bill be for the 200 parts at the discounted price?

Quick Quiz 3.5

1. When you graph the inequality $y > -3x + 1$, should you use a solid line for a boundary or a dashed line for a boundary? Why?

2. Graph the region described by $3y \leq -7x$.

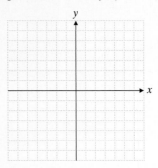

3. Graph the region described by $-5x + 2y > -3$.

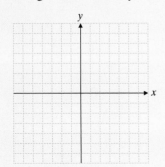

4. **Concept Check** Explain how you would determine if you should shade above the line or below the line if you were to graph the inequality $y > -3x + 4$ using $(0, 0)$ as a test point.

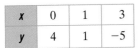 **Understanding the Meanings of a Relation and a Function**

Student Learning Objectives

After studying this section, you will be able to:

1. Understand the meanings of a relation and a function.

2. Graph simple nonlinear equations.

3. Determine whether a graph represents a function.

4. Use function notation.

Thus far you have studied linear equations in two variables. You have seen that such an equation can be represented by a table of values, by the algebraic equation itself, or by a graph.

x	0	1	3
y	4	1	−5

$$y = -3x + 4$$

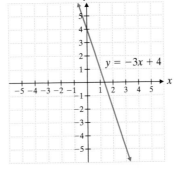

The solutions to the linear equation are all the ordered pairs that satisfy the equation (make the equation true). They are all the points that lie on the graph of the line. These ordered pairs can be represented in a table of values. Notice the relationship between the ordered pairs. We can choose any value for x. But once we have chosen a value for x, the value of y is determined. For the preceding equation, if x is 0, then y must be 4. We say that x is the **independent variable** and that y is the **dependent variable.**

Mathematicians call such a pairing of two values a *relation*.

DEFINITION OF A RELATION

A **relation** is any set of ordered pairs.

All the first coordinates in all of the ordered pairs of the relation make up the **domain** of the relation. All the second coordinates in all of the ordered pairs make up the **range** of the relation. Notice that the definition of a relation is very broad. Some relations cannot be described by an equation. These relations may simply be a set of discrete ordered pairs.

EXAMPLE 1 State the domain and range of the relation.

$$\{(5, 7), (9, 11), (10, 7), (12, 14)\}$$

Solution The domain consists of all the first coordinates in the ordered pairs.

Domain
$$\{(5, 7), (9, 11), (10, 7), (12, 14)\}$$
Range

The range consists of all the second coordinates in the ordered pairs. We usually list the values of a domain or range from smallest to largest.

The domain is $\{5, 9, 10, 12\}$.
The range is $\{7, 11, 14\}$.   We list 7 only once.

Practice Problem 1 State the domain and range of the relation.

$$\{(-3, -5), (3, 5), (0, -5), (20, 5)\}$$

NOTE TO STUDENT: *Fully worked-out solutions to all of the Practice Problems can be found at the back of the text starting at page SP-1*

Some relations have the special property that no two different ordered pairs have the same first coordinate. Such relations are called **functions.** The relation $y = -3x + 4$ is a function. If we substitute a value for x, we get just one value for y. Thus no two ordered pairs will have the same x-coordinate and different y-coordinates.

DEFINITION OF A FUNCTION

A **function** is a relation in which no two different ordered pairs have the same first coordinate.

EXAMPLE 2 Determine whether the relation is a function.

(a) $\{(3, 9), (4, 16), (5, 9), (6, 36)\}$

(b) $\{(7, 8), (9, 10), (12, 13), (7, 14)\}$

Solution

(a) Look at the ordered pairs. No two ordered pairs have the same first coordinate. Thus this set of ordered pairs defines a function. Note that the ordered pairs $(3, 9)$ and $(5, 9)$ have the same second coordinate, but the relation is still a function. It is the first coordinates that cannot be the same.

(b) Look at the ordered pairs. Two different ordered pairs, $(7, 8)$ and $(7, 14)$, have the same first coordinate. Thus this relation is *not* a function.

Practice Problem 2 Determine whether the relation is a function.

(a) $\{(-5, -6), (9, 30), (-3, -3), (8, 30)\}$

(b) $\{(60, 30), (40, 20), (20, 10), (60, 120)\}$

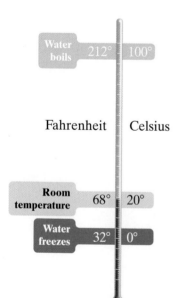

A functional relationship is often what we find when we analyze two sets of data. Look at the following table of values, which compares Celsius temperature with Fahrenheit temperature. Is there a relationship between degrees Fahrenheit and degrees Celsius? Is the relation a function?

Temperature

°F	23	32	41	50
°C	−5	0	5	10

Since every Fahrenheit temperature produces a unique Celsius temperature, we would expect this to be a function. We can verify our assumption by looking at the formula $C = \frac{5}{9}(F - 32)$ and its graph. The formula is a linear equation, and its graph is a line with slope $\frac{5}{9}$ and y-intercept at about -17.8. The relation is a function. In the equation given here, notice that the *dependent variable* is C, since the value of C *depends* on the value of F. We say that F is the *independent variable*. The *domain* can be described as the set of possible values of the independent variable. The *range* is the set of corresponding values of the dependent variable. Scientists believe that the coldest temperature possible is approximately $-273°C$. They call this temperature **absolute zero.** Thus,

Domain = {all possible Fahrenheit temperatures from absolute zero to infinity}

Range = {all corresponding Celsius temperatures from $-273°C$ to infinity}.

EXAMPLE 3 Each of the following tables contains some data pertaining to a relation. Determine whether the relation suggested by the table is a function. If it is a function, identify the domain and range.

(a) Circle

Radius	1	2	3	4	5
Area	3.14	12.56	28.26	50.24	78.5

(b) $4000 Loan at 8% for a Minimum of One Year

Time (yr)	1	2	3	4	5
Interest	$320	$665.60	$1038.85	$1441.96	$1877.31

Solution

(a) Looking at the table, we see that no two different ordered pairs have the same first coordinate. The area of a circle is a function of the length of the radius.

Next we need to identify the independent variable to determine the domain. Sometimes it is easier to identify the dependent variable. Here we notice that the area of the circle depends on the length of the radius. Thus radius is the independent variable. Since a negative length does not make sense, the radius cannot be a negative number. Although only integer radius values are listed in the table, the radius of a circle can be any nonnegative real number.

$$\text{Domain} = \{\text{all nonnegative real numbers}\}$$
$$\text{Range} = \{\text{all nonnegative real numbers}\}$$

(b) No two different ordered pairs have the same first coordinate. Interest is a function of time.

Since the amount of interest paid on a loan depends on the number of years (term of the loan), interest is the dependent variable and time is the independent variable. Negative numbers do not apply in this situation. Although the table includes only integer values for the time, the length of a loan in years can be any real number that is greater than or equal to 1.

$$\text{Domain} = \{\text{all real numbers greater than or equal to 1}\}$$
$$\text{Range} = \{\text{all positive real numbers greater than or equal to \$320}\}$$

Practice Problem 3 Determine whether the relation suggested by the table is a function. If it is a function, identify the domain and the range.

NOTE TO STUDENT: Fully worked-out solutions to all of the Practice Problems can be found at the back of the text starting at page SP-1

(a) 28 Mpg at $4.16 per Gallon

Distance	0	28	42	56	70
Cost	$0	$4.16	$6.24	$8.32	$10.40

(b) Store's Inventory of Shirts

Number of Shirts	5	10	5	2	8
Price of Shirt	$20	$25	$30	$45	$50

TO THINK ABOUT: Is It a Function? Look at the following bus schedule. Determine whether the relation is a function. Which is the independent variable? Explain your choice.

Bus Schedule

Bus Stop	Main St.	8th Ave.	42nd St.	Sunset Blvd.	Cedar Lane
Time	7:00	7:10	7:15	7:30	7:39

Graphing Simple Nonlinear Equations

Thus far in this chapter we have graphed linear equations in two variables. We now turn to graphing a few nonlinear equations. We will need to plot more than three points to get a good idea of what the graph of a nonlinear equation will look like.

EXAMPLE 4   Graph $y = x^2$.

Solution   Begin by constructing a table of values. We select values for x and then determine by the equation the corresponding values of y. We will include negative values for x as well as positive values. We then plot the ordered pairs and connect the points with a smooth curve.

x	$y = x^2$	y
-2	$y = (-2)^2 = 4$	4
-1	$y = (-1)^2 = 1$	1
0	$y = (0)^2 = 0$	0
1	$y = (1)^2 = 1$	1
2	$y = (2)^2 = 4$	4

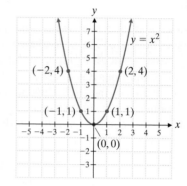

This type of curve is called a *parabola*.

Practice Problem 4   Graph $y = x^2 - 2$ on the coordinate system in the margin.

Some equations are solved for x. Usually, in those cases we pick values of y and then obtain the corresponding values of x from the equation.

EXAMPLE 5   Graph $x = y^2 + 2$.

Solution   Since the equation is solved for x, we start by picking a value of y. We will find the value of x from the equation in each case. We will select $y = -2$ first. Then we substitute it into the equation to obtain x. For convenience in graphing, we will repeat the y column at the end so that it is easy to write the ordered pairs (x, y).

y	$x = y^2 + 2$	x	y
-2	$x = (-2)^2 + 2 = 4 + 2 = 6$	6	-2
-1	$x = (-1)^2 + 2 = 1 + 2 = 3$	3	-1
0	$x = (0)^2 + 2 = 0 + 2 = 2$	2	0
1	$x = (1)^2 + 2 = 1 + 2 = 3$	3	1
2	$x = (2)^2 + 2 = 4 + 2 = 6$	6	2

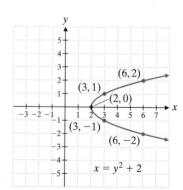

Practice Problem 5   Graph $x = y^2 - 1$ on the coordinate system in the margin.

 Graphing Calculator

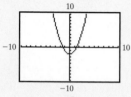 **Graphing Nonlinear Equations**

You can graph nonlinear equations solved for y using a graphing calculator. For example, graph $y = x^2 - 2$ on a graphing calculator using an appropriate window. Display:

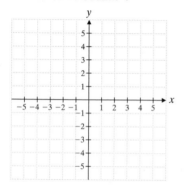

Try graphing $y = \dfrac{5}{x}$.

PRACTICE PROBLEM 4

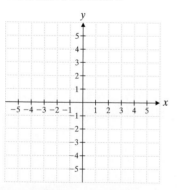

PRACTICE PROBLEM 5

If the equation involves fractions with variables in the denominator, we must use extra caution. Remember that you may never divide by zero.

EXAMPLE 6 Graph $y = \dfrac{4}{x}$.

Solution   It is important to note that x cannot be zero because division by zero is not defined. $y = \frac{4}{0}$ is not allowed! Observe that when we draw the graph we get two separate branches that do not touch.

x	$y = \dfrac{4}{x}$	y
-4	$y = \dfrac{4}{-4} = -1$	-1
-2	$y = \dfrac{4}{-2} = -2$	-2
-1	$y = \dfrac{4}{-1} = -4$	-4
0	We cannot divide by zero.	There is no value.
1	$y = \dfrac{4}{1} = 4$	4
2	$y = \dfrac{4}{2} = 2$	2
4	$y = \dfrac{4}{4} = 1$	1

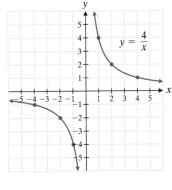

PRACTICE PROBLEM 6

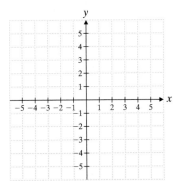

Practice Problem 6   Graph $y = \dfrac{6}{x}$ on the coordinate system in the margin.

③ Determining Whether a Graph Represents a Function

Can we tell whether a graph represents a function? Recall that a function cannot have two different ordered pairs with the same first coordinate. That is, each value of x must have a separate, unique value of y. Look at the graph below of the function $y = x^2$. Each x-value has a unique y-value. Now look at the graph of $x = y^2 + 2$ below. At $x = 6$ there are two y-values, 2 and -2. In fact, for every x-value greater than 2, there are two y-values. $x = y^2 + 2$ is not a function.

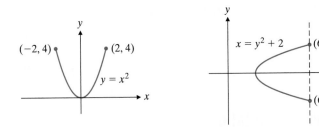

Observe that we can draw a vertical line through $(6, 2)$ and $(6, -2)$. Any graph that is not a function will have at least one region in which a vertical line will cross the graph more than once.

VERTICAL LINE TEST

If a vertical line can intersect the graph of a relation more than once, the relation is not a function. If no such line can be drawn, then the relation is a function.

EXAMPLE 7 Determine whether each of the following is the graph of a function.

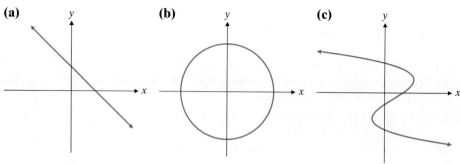

(a)   (b)   (c)

Solution

(a) The graph of the straight line is a function. Any vertical line will cross this straight line in only one location.

(b) and **(c)** Each of these graphs is not the graph of a function. In each case there exists a vertical line that will cross the curve in more than one place.

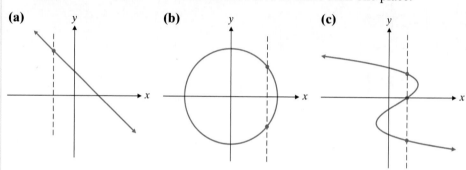

(a)   (b)   (c)

Practice Problem 7 Determine whether each of the following is the graph of a function.

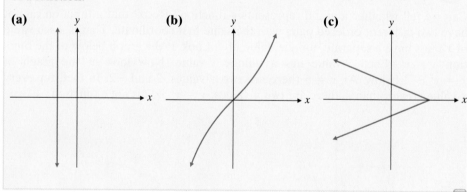

(a)   (b)   (c)

4 Using Function Notation

We have seen that an equation like $y = 2x + 7$ is a function. For each value of x, the equation assigns a unique value to y. We could say, "y is a function of x." If we name the function f, this statement can be symbolized by using the **function notation** $y = f(x)$. Many times we avoid using the y-variable completely and write the function as $f(x) = 2x + 7$.

CAUTION: Be careful. The notation $f(x)$ does not mean f multiplied by x.

EXAMPLE 8   If $f(x) = 3x^2 - 4x + 5$, find each of the following.

(a) $f(-2)$ **(b)** $f(4)$ **(c)** $f(0)$

Solution

(a) $f(-2) = 3(-2)^2 - 4(-2) + 5 = 3(4) - 4(-2) + 5 = 12 + 8 + 5 = 25$

(b) $f(4) = 3(4)^2 - 4(4) + 5 = 3(16) - 4(4) + 5 = 48 - 16 + 5 = 37$

(c) $f(0) = 3(0)^2 - 4(0) + 5 = 3(0) - 4(0) + 5 = 0 - 0 + 5 = 5$

Practice Problem 8   If $f(x) = -2x^2 + 3x - 8$, find each of the following.

(a) $f(2)$ **(b)** $f(-3)$ **(c)** $f(0)$

NOTE TO STUDENT: Fully worked-out solutions to all of the Practice Problems can be found at the back of the text starting at page SP-1

When evaluating a function, it is helpful to place parentheses around the value that is being substituted for x. Taking the time to do this will minimize sign errors in your work.

Some functions are useful in medicine, anthropology, and forensic science. For example, the approximate length of a man's femur (thigh bone) is given by the function $f(x) = 0.53x - 17.03$, where x is the height of the man in inches. If a man is 68 inches tall, $f(68) = 0.53(68) - 17.03 = 19.01$. A man 68 inches tall would have a femur length of approximately 19.01 inches.

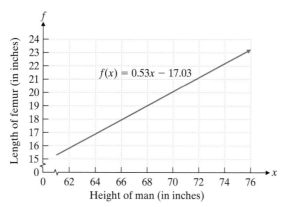

Source: www.nsbri.org

SIDELIGHT: Linear versus Nonlinear

When graphing, keep the following facts in mind.

The graph of a linear equation will be a straight line.

The graph of a nonlinear equation will *not* be a straight line.

The graphs in Examples 4, 5, and 6 are *not* straight lines because the equations are *not* linear equations. The graph shown above *is* a straight line because its equation *is* a linear equation.

Verbal and Writing Skills

1. What are the three ways you can describe a function?

2. What is the difference between a function and a relation?

3. The domain of a function is the set of _____ _____ of the _____ variable.

4. The range of a function is the set of _____ _____ of the _____ variable.

5. How can you tell whether a graph is the graph of a function?

6. Without drawing a graph, how could you tell if the equation $x = y^2$ is a function or not?

(a) *Find the domain and range of the relation.* **(b)** *Determine whether the relation is a function.*

7. $\left\{ \left(\frac{2}{5}, 4 \right), \left(3, \frac{2}{5} \right), \left(-3, \frac{2}{5} \right), \left(\frac{2}{5}, -1 \right) \right\}$

8. $\left\{ \left(\frac{3}{8}, -6 \right), (-6, 5), \left(\frac{3}{8}, 2 \right), (5, -6) \right\}$

9. $\{(6, 2.5), (3, 1.5), (0, 0.5)\}$

10. $\{(7.3, 1), (0, 8), (2, 1)\}$

11. $\{(12, 1), (14, 3), (1, 12), (9, 12)\}$

12. $\{(5.6, 8), (5.8, 6), (6, 5.8), (5, 6)\}$

13. $\{(3, 75), (5, 95), (3, 85), (7, 100)\}$

14. $\{(85, 3), (95, 11), (110, 15), (110, 20)\}$

Graph the equation.

15. $y = x^2 + 3$

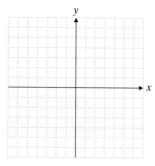

16. $y = x^2 - 1$

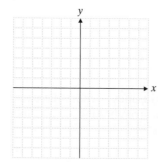

17. $y = 2x^2$

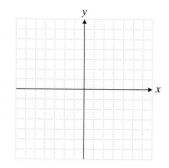

18. $y = \dfrac{1}{2}x^2$

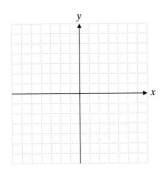

19. $x = -2y^2$

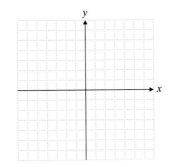

20. $x = \dfrac{1}{2}y^2$

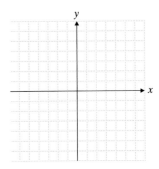

21. $x = y^2 - 4$

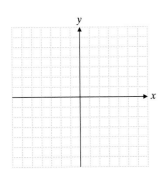

22. $x = 2y^2$

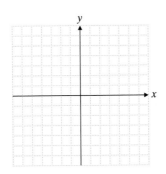

23. $y = \dfrac{2}{x}$

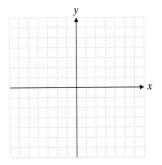

24. $y = -\dfrac{2}{x}$

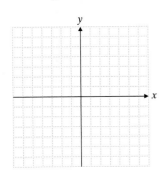

25. $y = \dfrac{4}{x^2}$

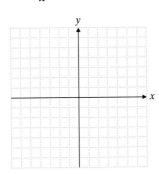

26. $y = -\dfrac{6}{x^2}$

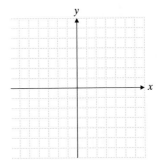

27. $x = (y + 1)^2$

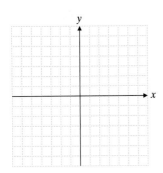

28. $y = (x - 3)^2$

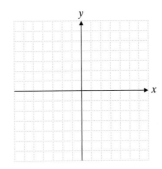

29. $y = \dfrac{4}{x - 2}$

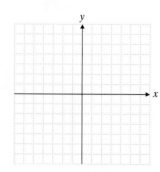

30. $x = \dfrac{2}{y + 1}$

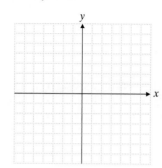

Determine whether each relation is a function.

31.

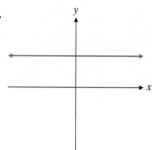

32.

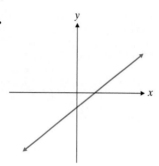

33.

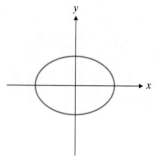

34.

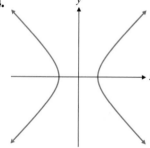

35.

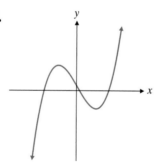

36.

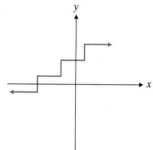

37.

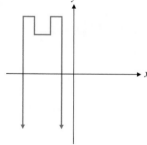

38.

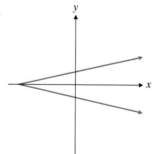

Given the following functions, find the indicated values.

39. $f(x) = 2 - 3x$
    (a) $f(-8)$ **(b)** $f(0)$ **(c)** $f(2)$

40. $f(x) = 7 - 4x$
    (a) $f(-2)$ **(b)** $f(1)$ **(c)** $f(3)$

41. $f(x) = 2x^2 - x + 3$
    (a) $f(0)$ **(b)** $f(-3)$ **(c)** $f(2)$

42. $f(x) = -x^2 - x + 5$
    (a) $f(0)$ **(b)** $f(-1)$ **(c)** $f(2)$

Applications

43. California Population During a recent population growth period in California, from 1995 to 2005, the approximate population of the state measured in millions could be predicted by the function $f(x) = 0.02x^2 + 0.08x + 31.6$, where x is the number of years since 1995. (*Source:* www.census.gov.) Find $f(0)$, $f(4)$, and $f(10)$. Graph the function. What pattern do you observe?

44. Kentucky Population During a recent population growth period in Kentucky, from 1995 to 2005, the approximate population of the state measured in thousands could be predicted by the function $f(x) = -0.64x^2 + 30.2x + 3860$, where x is the number of years since 1995. (*Source:* www.census.gov.) Find $f(0)$, $f(5)$, and $f(10)$. Use the function to predict the population in 2011 and 2015. Graph the function. What pattern do you observe?

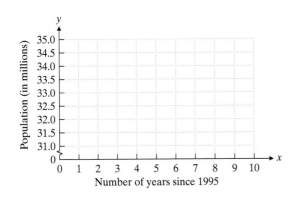

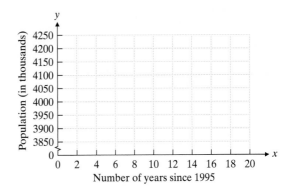

Cumulative Review *Simplify.*

45. [1.9.1] $-4x(2x^2 - 3x + 8)$

46. [1.9.1] $5a(ab + 6b - 2a)$

47. [1.7.2] $-7x + 10y - 12x - 8y - 2$

48. [1.7.2] $3x^2y - 6xy^2 + 7xy + 6x^2y$

Quick Quiz 3.6

1. Is this relation a function? $\{(5, 7), (7, 5), (5, 5)\}$ Why?

2. For $f(x) = 3x^2 - 4x + 2$:
   (a) Find $f(-3)$.   **(b)** Find $f(4)$.

3. For $g(x) = \dfrac{7}{x - 3}$:
   (a) Find $g(2)$.   **(b)** Find $g(-5)$.

4. Concept Check In the relation $\{(3, 4), (5, 6), (3, 8), (2, 9)\}$, why is there a different number of elements in the domain than in the range?

Putting Your Skills to Work: Use Math to Save Money

RETIREMENT PLANNING

It is never too early to begin thinking about what you will be doing twenty, thirty, or more years from now. For example, how do you envision your retirement? One thing you will need when you retire is income. Social Security is projected to run out of funds, so we cannot be sure that we will have this cushion to fall back on. It is therefore important to consider investing money in a retirement account. Consider the story of Louis.

Louis accepts a job that pays $42,000 per year. He begins investing money in his retirement account at the age of 25 with a goal of retiring at the age of 65. His company has a retirement option that deducts approximately 5% of his salary and invests it in a retirement account that pays a yearly average of 8%.

To calculate Louis's projected earnings at retirement, we use the retirement formula

$$FV = PMT \times \frac{(1 + I)^N - 1}{I},$$

which has the following variables:

- *FV* is the future value of the retirement account.
- *PMT* is the monthly payment contributed to the retirement account.
- *I* is the interest rate in decimal notation divided by 12, which is the amount of interest being earned each month.
- *N* is 12 times the number of years of the account, which is the number of payments made into the account.

Here, *FV*, the future value, is the variable we will be solving for.

Louis's *PMT* is $175, since 5% of Louis's annual salary is $2100, and that amount divided by 12 equals $175.

Also, $I = \frac{0.08}{12} \approx 0.0067$, and $N = 12 \times 40 = 480$, since Louis will be working for 40 years.

1. Substitute these values into the retirement formula to find the future value of Louis's retirement account when he turns 65. Round your answer to the nearest ten thousand dollars.

2. How much money will Louis personally contribute into his retirement account?

3. Approximately how much interest is the account projected to earn?

Making It Personal for You

4. Look in the classified ads or contact the human resources department of a company whose jobs interest you. Find out about the starting salary of a job, as well as the contributions that are made to a retirement account. Using your age and the age at which you wish to retire, calculate the amount of money you could have in a retirement account when you retire.

5. How much money would you have contributed by the time you retire?

6. How much interest would you have earned by the time you retire?

Topic	Procedure	Examples
Graphing straight lines, *p. 212.*	An equation of the form $$Ax + By = C$$ has a graph that is a straight line. To graph such an equation, plot any three points; two give the line and the third checks it. (Where possible, use the *x*- and *y*-intercepts.)	Graph $3x + 2y = 6$. <table><tr><td>**x**</td><td>**y**</td></tr><tr><td>0</td><td>3</td></tr><tr><td>4</td><td>−3</td></tr><tr><td>2</td><td>0</td></tr></table>
Finding the slope given two points, *p. 222.*	Nonvertical lines passing through distinct points (x_1, y_1) and (x_2, y_2) have slope $$m = \frac{y_2 - y_1}{x_2 - x_1}.$$ The slope of a horizontal line is 0. The slope of a vertical line is undefined.	What is the slope of the line through $(2, 8)$ and $(5, 1)$? $$m = \frac{1 - 8}{5 - 2} = -\frac{7}{3}$$
Finding the slope and y-intercept of a line given the equation, *p. 225.*	1. Rewrite the equation in the form $y = mx + b$. 2. The slope is m. 3. The *y*-intercept is $(0, b)$.	Find the slope and *y*-intercept. $$3x - 4y = 8$$ $$-4y = -3x + 8$$ $$y = \tfrac{3}{4}x - 2$$ The slope is $\frac{3}{4}$. The *y*-intercept is $(0, -2)$.
Finding an equation of a line given the slope and y-intercept, *p. 226.*	The slope–intercept form of the equation of a line is $$y = mx + b.$$ The slope is m and the *y*-intercept is $(0, b)$.	Find an equation of the line with *y*-intercept $(0, 7)$ and with slope $m = 3$. $$y = 3x + 7$$
Graphing a line using slope and y-intercept, *p. 227.*	1. Plot the *y*-intercept. 2. Starting from $(0, b)$, plot a second point using the slope. $$\text{slope} = \frac{\text{rise}}{\text{run}}$$ 3. Draw a line that connects the two points.	Graph $y = -4x + 1$. First plot the *y*-intercept at $(0, 1)$. Slope $= -4$ or $\frac{-4}{1}$.
Finding the slopes of parallel and perpendicular lines, *p. 228.*	Parallel lines have the same slope. Perpendicular lines have slopes whose product is -1.	Line q has a slope of 2. The slope of a line parallel to q is 2. The slope of a line perpendicular to q is $-\frac{1}{2}$.
Finding the equation of a line through a point with a given slope, *p. 236.*	1. Write the slope–intercept form of the equation of a line: $y = mx + b$. 2. Substitute the known values into the equation $y = mx + b$. 3. Solve for b. 4. Use the values of m and b to write the slope–intercept form of the equation.	Find an equation of the line through $(3, 2)$ with slope $m = \frac{4}{5}$. $$y = mx + b \qquad 2 = \frac{4}{5}(3) + b$$ $$2 = \frac{12}{5} + b$$ $$-\frac{2}{5} = b$$ An equation is $y = \frac{4}{5}x - \frac{2}{5}$.

(Continued on next page)

Topic	Procedure	Examples
Finding an equation of the line through two points, p. 236.	1. Find the slope. 2. Use the procedure when given a point and the slope.	Find the equation of the line through $(3, 2)$ and $(13, 10)$. $$m = \frac{y_2 - y_1}{x_2 - x_1} = \frac{10 - 2}{13 - 3} = \frac{8}{10} = \frac{4}{5}$$ We choose the point $(3, 2)$. $$y = mx + b \qquad 2 = \frac{4}{5}(3) + b$$ $$2 = \frac{12}{5} + b$$ $$-\frac{2}{5} = b$$ The equation is $y = \frac{4}{5}x - \frac{2}{5}$.
Graphing linear inequalities, p. 240.	1. Graph as if it were an equation. If the inequality symbol is $>$ or $<$, use a dashed line. If the inequality symbol is $\geq$ or $\leq$, use a solid line. 2. Look for a test point. The easiest test point is $(0, 0)$, unless the line passes through $(0, 0)$. In that case, choose another test point. 3. Substitute the coordinates of the test point into the inequality. 4. If it is a true statement, shade the side of the line containing the test point. If it is a false statement, shade the side of the line that does *not* contain the test point.	Graph $y \geq 3x + 2$. Graph the line $y = 3x + 2$. Use a solid line. $\qquad$ Test $(0, 0)$. $\qquad 0 \geq 3(0) + 2$ $\qquad\qquad\qquad\qquad\quad 0 \geq 2 \quad$ false Shade the side of the line that does not contain $(0, 0)$.
Determining whether a relation is a function, p. 246.	A function is a relation in which no two different ordered pairs have the same first coordinate.	Is this relation a function? $$\{(5, 7), (3, 8), (5, 10)\}$$ It is *not* a function since $(5, 7)$ and $(5, 10)$ are two different ordered pairs with the same x-coordinate, 5.
Determining whether a graph represents a function, p. 250.	If a vertical line can intersect the graph of a relation more than once, the relation is not a function. If no such line exists, the relation is a function.	Does this graph represent a function? Yes. Any vertical line will intersect it at most once.

Chapter 3 Review Problems

Section 3.1

1. Plot and label the following points.
 $A: (2, -3)$ $B: (-1, 0)$ $C: (3, 2)$ $D: (-2, -3)$

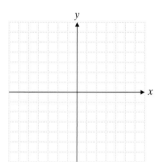

2. Give the coordinates of each point.

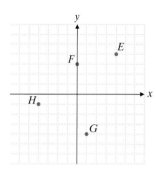

Complete the ordered pairs so that each is a solution to the given equation.

3. $y = 7 - 3x$
  (a) $(0, \ )$ **(b)** $(\ , 10)$

4. $2x + 5y = 12$
  (a) $(1, \ )$ **(b)** $(\ , 4)$

5. $x = 6$
  (a) $(\ , -1)$ **(b)** $(\ , 3)$

Section 3.2

6. Graph $5y + x = -15$.

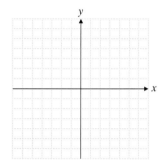

7. Graph $2y + 4x = -8 + 2y$.

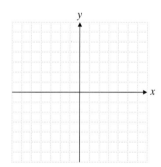

8. Graph $3y = 2x + 6$ and label the intercepts.

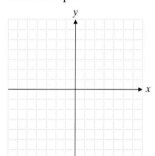

Section 3.3

9. Find the slope of the line passing through $(5, -3)$ and $\left(2, -\frac{1}{2}\right)$.

10. Find the slope and y-intercept of the line $9x - 11y + 15 = 0$.

11. Write an equation of the line with slope $-\frac{1}{2}$ and y-intercept $(0, 3)$.

12. The equation of a line is $y = \frac{3}{5}x - 2$. What is the slope of a line perpendicular to that line?

13. Graph $y = -\dfrac{1}{2}x + 3$.

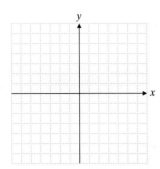

14. Graph $2x - 3y = -12$.

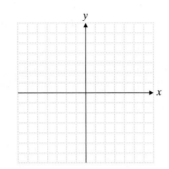

15. Graph $6x - 2y = 6 + 6x$.

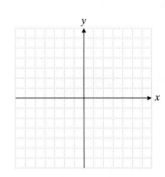

Section 3.4

16. Write an equation of the line passing through $(2, 5)$ having a slope of 1.

17. Write an equation of the line passing through $(3, -4)$ having a slope of -6.

18. Write an equation of the line passing through $(-1, 4)$ having a slope of $-\frac{1}{3}$.

19. Write an equation of the line passing through $(3, 7)$ and $(-6, 7)$.

20. What is the slope of a line parallel to a line whose equation is $y = -\frac{2}{3}x + 4$?

21. What is the slope of a line perpendicular to the line whose equation is $-2x + 6y = 3$?

Write an equation of the graph.

22.

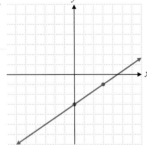

23.

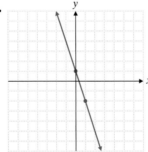

24.

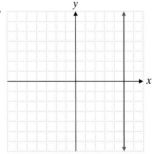

Section 3.5

25. Graph $y < \frac{1}{3}x + 2$.

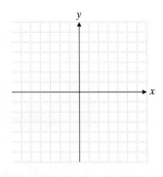

26. Graph $3y + 2x \geq 12$.

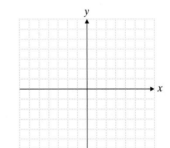

27. Graph $x \leq 2$.

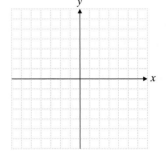

Section 3.6

Determine the domain and range of the relation. Determine whether the relation is a function.

28. $\{(5, -6), (-6, 5), (-5, 5), (-6, -6)\}$

29. $\{(2, -3)(5, -3)(6, 4)(-2, 4)\}$

In exercises 30–32, determine whether the graphs represent a function.

30.

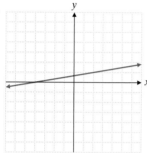

31.

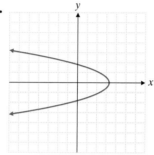

32.

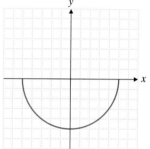

33. Graph $y = x^2 - 5$.

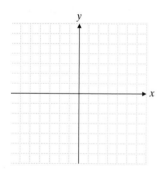

34. Graph $x = y^2 + 3$.

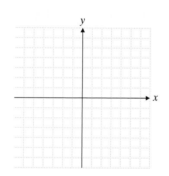

35. Graph $y = (x - 3)^2$.

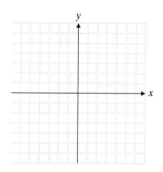

Given the following functions, find the indicated values.

36. $f(x) = 7 - 6x$ **(a)** $f(0)$ **(b)** $f(-4)$

37. $g(x) = -2x^2 + 3x + 4$ **(a)** $g(-1)$ **(b)** $g(3)$

38. $f(x) = 8x - 3$ **(a)** $f(-1)$ **(b)** $f\left(\dfrac{1}{2}\right)$

39. $f(x) = \dfrac{2}{x + 4}$ **(a)** $f(-2)$ **(b)** $f(6)$

40. $f(x) = x^2 - 2x + \dfrac{3}{x}$ **(a)** $f(-1)$ **(b)** $f(3)$

Applications

Monthly Electric Bill *Russ and Norma Camp found that their monthly electric bill could be calculated by the equation* $y = 30 + 0.09x$. *In this equation y represents the amount of the monthly bill in dollars and x represents the number of kilowatt-hours used during the month.*

41. What would be their monthly bill if they used 2000 kilowatt-hours of electricity?

42. What would be their monthly bill if they used 1600 kilowatt-hours of electricity?

43. Write the equation in the form $y = mx + b$, and determine the numerical value of the y-intercept. What is the significance of this y-intercept? What does it tell us?

44. If the equation is placed in the form $y = mx + b$, what is the numerical value of the slope? What is the significance of this slope? What does it tell us?

45. If Russ and Norma have a monthly bill of $147, how many kilowatt-hours of electricity did they use?

46. If Russ and Norma have a monthly bill of $246, how many kilowatt-hours of electricity did they use?

Petroleum Consumption *Scientists have found that the amount of petroleum consumed worldwide for a given year can be predicted by the equation* $y = 1.15x + 65.6$, *where x is the number of years since 1990 and y is the number of barrels of petroleum consumed per day measured in millions. Use this equation to answer the following questions. (Source: www.eia.doe.gov.)*

47. How many barrels of petroleum were consumed in 2000?

48. How many barrels of petroleum will be consumed in 2015?

49. The given equation is in the form $y = mx + b$. What is the slope and what is its significance? What does it tell us?

50. The given equation is in the form $y = mx + b$. What is the y-intercept? What is the significance of this y-intercept? What does it tell us?

51. In what year will the consumption of petroleum reach 96.65 million barrels?

52. In what year will the consumption of petroleum reach 100.1 million barrels?

Mixed Practice

53. Graph $5x + 3y = -15$.

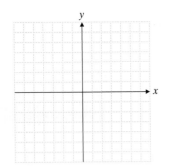

54. Graph $y = \frac{3}{4}x - 3$.

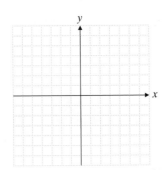

55. Find the slope of the line through $(2, -7)$ and $(-3, -5)$.

56. What is the slope and the y-intercept of the line $7x + 6y - 10 = 0$?

57. Write an equation of the line that passes through $(3, -5)$ and has a slope of $\frac{2}{3}$.

58. Graph $y < -2x + 1$.

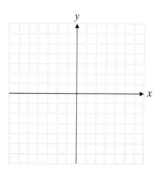

*Job Losses in Manufacturing Economists in the Labor Department are concerned about the continued job loss in the manufacturing industry. The number of people employed in manufacturing jobs in the United States can be predicted by the equation $y = -269x + 17{,}020$, where x is the number of years since 1994 and y is the number of employees in **thousands** in the manufacturing industry. Use this data to answer the following questions. (Source: www.bls.gov.)*

59. How many people were employed in manufacturing in 1994? In 2000? In 2008?

60. Use your answers for 59 to draw a graph of the equation $y = -269x + 17{,}020$.

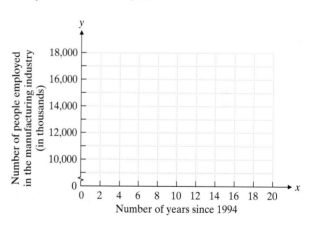

61. What is the slope of this equation? What is its significance?

62. What is the y-intercept of this equation? What is its significance?

63. Use the equation to predict in what year the number of manufacturing jobs will be 11,640,000.

64. Use the equation to predict in what year the number of manufacturing jobs will be 10,295,000.

How Am I Doing? Chapter 3 Test

Remember to use your Chapter Test Prep Video CD to see the worked-out solutions to the test problems you want to review.

1. Plot and label the following points.

    B: $(6, 1)$ C: $(-4, -3)$
    D: $(-3, 0)$ E: $(5, -2)$

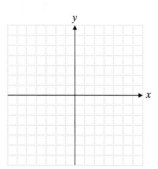

2. Graph the line $6x - 3 = 5x - 2y$.

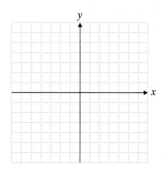

3. Graph the line $8x + 2y = -4$.

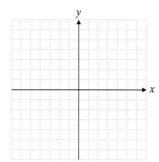

4. Graph $y = \frac{2}{3}x - 4$.

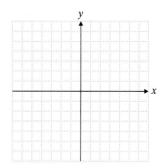

5. What is the slope and the y-intercept of the line $3x + 2y - 5 = 0$?

6. Find the slope of the line that passes through $(8, 6)$ and $(-3, -5)$.

7. Write an equation for the line that passes through $(4, -2)$ and has a slope of $\frac{1}{2}$.

8. Find the slope of the line through $(-3, 11)$ and $(6, 11)$.

9. Find an equation for the line passing through $(5, -4)$ and $(-3, 8)$.

10. Write an equation for the line through $(2, 7)$ and $(2, -2)$. What is the slope of this line?

1. _____

2. _____

3. _____

4. _____

5. _____

6. _____

7. _____

8. _____

9. _____

10. _____

11. Graph the region described by $4y \leq 3x$.

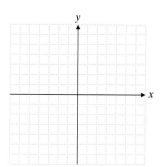

12. Graph the region described by $-3x - 2y > 10$.

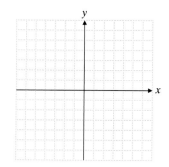

11. _____

12. _____

13. Is this relation a function? $\{(2, -8), (3, -7), (2, 5)\}$ Why?

13. _____

14. Look at the relation graphed below. Is this relation a function? Why?

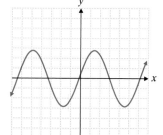

14. _____

15. Graph $y = 2x^2 - 3$.

x	y
-2	
-1	
0	
1	
2	

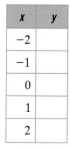

15. _____

16. For $f(x) = -x^2 - 2x - 3$:
  (a) Find $f(0)$.
  (b) Find $f(-2)$.

16. (a) _____

  (b) _____

17. For $g(x) = \dfrac{3}{x - 4}$:
  (a) Find $g(3)$.
  (b) Find $g(-11)$.

17. (a) _____

  (b) _____

1. _____

2. _____

3. _____

4. _____

5. _____

6. _____

7. _____

8. _____

9. _____

10. _____

11. _____

12. _____

13. _____

14. _____

Approximately one-half of this test covers the content of Chapters 0–2. The remainder covers the content of Chapter 3.

1. Evaluate. $\left(-\dfrac{1}{3}\right)\left(\dfrac{2}{5}\right) - \left(\dfrac{2}{5}\right)^2$

2. Simplify. $-5x[3x + 2(x - 6)]$

Solve.

3. $2 - 3(4 - x) = x - (3 - x)$

4. $x = 8 - 2^4 \div 8 - 10$

▲ **5.** Ricardo was hired by a new fitness center to paint a mural on one of the gym walls. The length of the mural is to be 3 feet less than triple the height. The perimeter of the mural will be 90 feet. What is the length and the height of the mural?

6. Add. $1.386 + (-2.9)$

7. Subtract. $9 - 0.48$

8. Multiply. -12.04×0.72

9. Solve $2x + 3y = 8$ for y.

10. Find an equation of the line through $(5, 3)$ and $(2, -3)$.

In questions 11 and 12, give the equation of the line that fits each description.

11. Passes through $(7, -4)$ and is vertical

12. Passes through $(3, -5)$ with slope $-\dfrac{2}{3}$

13. Find the slope of the line through $(-8, -3)$ and $(11, -3)$.

14. What is the slope of the line $4x - 3y = 6$?

15. Graph $y = \frac{2}{3}x - 4$.

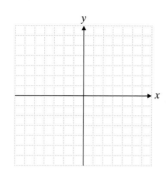

16. Graph $3x + 8 = 5x$.

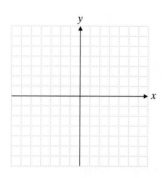

17. Graph the region $2x + 5y \leq -10$.

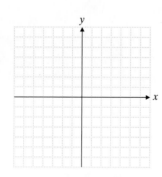

18. Graph $x = y^2 + 3$. Is this a function?

x	y
	−2
	−1
	0
	1
	2

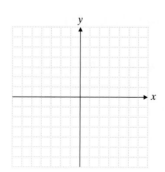

19. Is this relation a function? $\{(3, 10), (10, -3), (0, -3), (5, 10)\}$

20. For $f(x) = 3x^2 - 2x + \dfrac{4}{x}$:

    (a) Find $f(-1)$.   **(b)** Find $f(2)$.

15. _____

16. _____

17. _____

18. _____

19. _____

20. (a) _____

(b) _____

CHAPTER 4

Systems of Equations

The Great Wall of China is a sight that many people come to see firsthand. Tourists, travelers to the 2008 Beijing Olympics, business officials, and government leaders come each day to see the remains of this remarkable historic structure. But how many soldiers manned this ancient structure in historic times? How many tons of stone were used to construct this wall? These and many other historical questions are being answered today by mathematicians working with archaeologists and historians. They use the mathematics of this chapter to help answer these interesting questions.

In this chapter we will examine systems of linear equations. In Chapter 3 we examined linear equations and inequalities in two variables. Two or more equations or inequalities in several variables that are considered simultaneously are called a **system of equations** or a **system of inequalities.** Recall that the graph of a linear equation is a straight line. If we have the graphs of two linear equations on one coordinate system, how might they relate to one another?

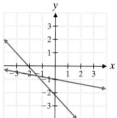

The lines may intersect.

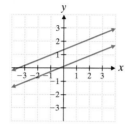

The lines may be parallel.

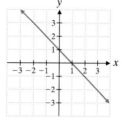

The lines may coincide.

The solutions to a system of linear equations are those points that both lines have in common. Note that parallel lines have no points in common. Thus the second system has no solution. Intersecting lines intersect at one point. Thus the first system has one solution. Lines that coincide have an infinite number of points in common. Thus the third system has an infinite number of solutions. You can determine the solution to a system of equations by graphing.

Graphing Calculator

Graphing a System of Linear Equations

To solve the system

$$3x + 4y = -28$$
$$x - 4y = 12,$$

we first need to solve each equation for y. This will give us the equivalent system

$$y = -0.75x - 7$$
$$y = 0.25x - 3.$$

We graph this system with a scale of -10 to 10 for both x and y.

Display.

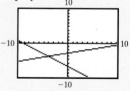

If we use the Zoom function, we will locate the intersection at $x = -4$, $y = -4$.

① Using Graphing to Solve a System of Linear Equations with a Unique Solution

To solve a system of equations by graphing, graph both equations on the same coordinate system. If the lines intersect, the system has a unique solution. The ordered pair of coordinates of the point of intersection is the solution to the system of equations.

EXAMPLE 1 Solve by graphing.

$$-2x + 3y = 6$$
$$2x + 3y = 18$$

Solution Graph both equations on the same coordinate system and determine the point of intersection.

We will graph the system by first obtaining a table of values for each equation. First we solve the first equation for y.

$$-2x + 3y = 6$$
$$3y = 2x + 6$$

$$y = \frac{2}{3}x + 2$$

Let $x = 0$. $y = \frac{2}{3}(0) + 2 = 0 + 2 = 2$

Let $x = -3$. $y = \frac{2}{3}(-3) + 2 = -2 + 2 = 0$

Let $x = 6$. $y = \frac{2}{3}(6) + 2 = 4 + 2 = 6$

x	y
0	2
−3	0
6	6

This allows us to use the three sets of ordered pairs to graph $-2x + 3y = 6$.

Next we solve the second equation for y.

$$2x + 3y = 18$$
$$3y = -2x + 18$$

$$y = -\frac{2}{3}x + 6$$

x	y
0	6
−3	8
6	2

Let $x = 0$. $y = -\frac{2}{3}(0) + 6 = 0 + 6 = 6$

Let $x = -3$. $y = -\frac{2}{3}(-3) + 6 = 2 + 6 = 8$

Let $x = 6$. $y = -\frac{2}{3}(6) + 6 = -4 + 6 = 2$

PRACTICE PROBLEM 1

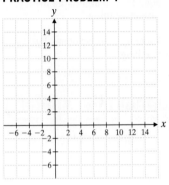

This allows us to use these three ordered pairs to graph $2x + 3y = 18$. (Of course, each equation could be graphed by any method we have discussed in the text. It is not necessary to use this method.) The graph of the system is shown in the figure at the right. The lines intersect at the point for which $x = 3$ and $y = 4$. Thus the unique solution to the system of equations is $(3, 4)$. Always check your answer.

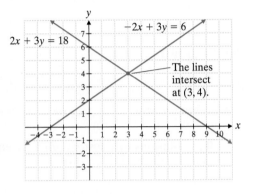

NOTE TO STUDENT: Fully worked-out solutions to all of the Practice Problems can be found at the back of the text starting at page SP-1

$$-2x + 3y \overset{?}{=} 6 \qquad 2x + 3y \overset{?}{=} 18$$
$$-2(3) + 3(4) \overset{?}{=} 6 \qquad 2(3) + 3(4) \overset{?}{=} 18$$
$$6 = 6 \qquad\qquad 18 = 18 \quad \checkmark$$

A linear system of equations that has one solution is said to be **consistent.**

Practice Problem 1 Solve by graphing. $x + y = 12$
$$-x + y = 4$$

② Graphing a System of Linear Equations to Determine the Type of Solution

Two lines in a plane may intersect, may never intersect (be parallel lines), or may be the same line. The corresponding system of equations will have one solution, no solution, or an infinite number of solutions, respectively. Thus far we have focused on systems of linear equations that have a unique solution. We now draw your attention to the other two cases.

EXAMPLE 2 Solve by graphing.
$$3x - y = 1$$
$$3x - y = -7$$

PRACTICE PROBLEM 2

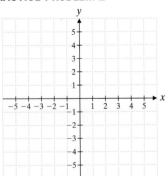

Solution We graph both equations on the same coordinate system. The graph is at the right. Notice that the lines are parallel. They do not intersect. Hence there is no solution to this system of equations. A system of linear equations that has no solution is called **inconsistent.**

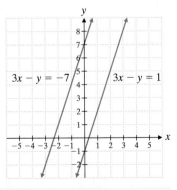

Practice Problem 2 Solve by graphing. $4x + 2y = 8$
$$-6x - 3y = 6$$

EXAMPLE 3 Solve by graphing.

$$x + y = 4$$
$$3x + 3y = 12$$

Solution We graph both equations on the same coordinate system at the right. Notice that the equations represent the same line (coincide). The coordinates of every point on the line will satisfy both equations. That is, every point on the line represents a solution to $x + y = 4$ and to $3x + 3y = 12$. Thus there is *an infinite number of solutions* to this system. Such equations are said to be **dependent.**

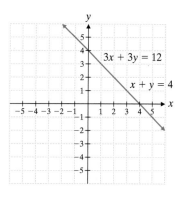

Practice Problem 3 Solve by graphing.

$$3x - 9y = 18$$
$$-4x + 12y = -24$$

PRACTICE PROBLEM 3

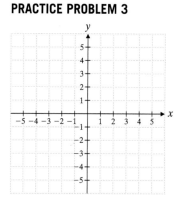

TO THINK ABOUT: Determining the Type of Solution Without Graphing
Without graphing you can tell that the system in Example 2 represents parallel lines because both lines have the *same slope*. Verify this fact by writing each equation in slope–intercept form. What can you say about the *y*-intercepts for each of these lines? Now look at the system in Example 3 and describe how, without graphing, you could determine that this system is dependent (lines coincide) rather than inconsistent (lines don't intersect).

③ Solving Applied Problems by Graphing

EXAMPLE 4 Walter and Barbara need some plumbing repairs done at their house. They called two plumbing companies for estimates of the work that needed to be done. Roberts Plumbing and Heating charges $40 for a house call and then $35 per hour for labor. Instant Plumbing Repairs charges $70 for a house call and then $25 per hour for labor.

(a) Create a cost equation for each company, where *y* is the total cost of plumbing repairs and *x* is the number of hours of labor. Write a system of equations.

(b) Graph the two equations using the values $x = 0, 3,$ and 6.

(c) Determine from your graph how many hours of plumbing repairs would be required for the two companies to charge the same.

(d) Determine from your graph which company charges less if the estimated amount of time to complete the plumbing repairs is 4 hours.

Solution

(a) For each company we will obtain an equation of the form

$$y = (\text{cost of house call}) + (\text{cost per hour})x.$$

Roberts Plumbing and Heating charges $40 for a house call and $35 per hour. Thus we obtain the first equation, $y = 40 + 35x$.

Instant Plumbing Repairs charges $70 for a house call and $25 per hour. Thus we obtain the second equation, $y = 70 + 25x$.

This yields the following system of equations:

$$y = 40 + 35x$$
$$y = 70 + 25x.$$

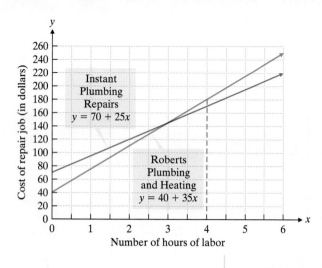

(b) We will graph the system by first obtaining a table of values for each equation.

Roberts Plumbing and Heating $y = 40 + 35x$

Let $x = 0$.	$y = 40 + 35(0) = 40$
Let $x = 3$.	$y = 40 + 35(3) = 145$
Let $x = 6$.	$y = 40 + 35(6) = 250$

x	y
0	40
3	145
6	250

Instant Plumbing Repairs $y = 70 + 25x$

Let $x = 0$.	$y = 70 + 25(0) = 70$
Let $x = 3$.	$y = 70 + 25(3) = 145$
Let $x = 6$.	$y = 70 + 25(6) = 220$

x	y
0	70
3	145
6	220

The graph of the system is to the left.

(c) We see that the graphs of the two lines intersect at (3, 145). Thus the two companies will charge the same if 3 hours of plumbing repairs are required.

(d) We draw a dashed line at $x = 4$. We see that the blue line representing Roberts Plumbing and Heating is higher than the red line representing Instant Plumbing Repairs after 3 hours. Thus the cost would be less if Walter and Barbara use Instant Plumbing Repairs for 4 hours of work.

Practice Problem 4   Jonathan and Joanne Wells need some electrical work done on their new house. They obtained estimates from two companies. Bill Tupper's Electrical Service charges $100 for a house call and $30 per hour for labor. Wire for Hire charges $50 for a house call and $40 per hour for labor.

(a) Create a cost equation for each company where y is the total cost of the electrical work and x is the number of hours of labor. Write a system of equations.

(b) Graph the two equations using the values $x = 0, 4$, and 8.

(c) Determine from your graph how many hours of electrical repairs would be required for the two companies to charge the same.

(d) Determine from your graph which company charges less if the estimated amount of time to complete the electrical repairs to the house is 6 hours.

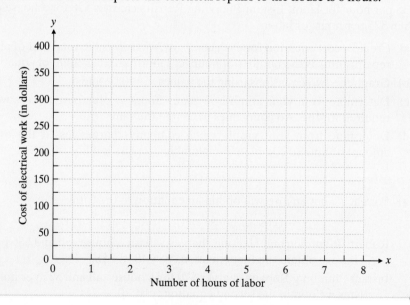

Verbal and Writing Skills

1. In the system $y = 2x - 5$ and $y = 2x + 6$, the lines have the same slope but different y-intercepts. What does this tell you about the graph of this system?

2. In the system $y = -3x + 4$ and $y = -3x + 4$, the lines have the same slope and the same y-intercepts. What does this tell you about the graph of the system?

3. Before you graph a system of equations, if you notice that the lines have different slopes, you can conclude that they _____ and that the system has _____ solution.

4. If lines have the same slope, you cannot conclude that the lines are parallel. Why? How else might the lines relate to each other?

5. Two lines have different slopes but the same y-intercept. What does this tell you about the graphs of the lines and about the solution of the system?

6. In the system $y = -3x + 5$ and $y = -3x - 2$, what can you say about the number of solutions?

Solve by graphing. If there isn't a unique solution to the system, state the reason.

7. $x - y = 3$
$x + y = 5$

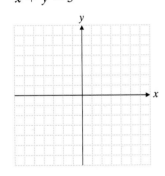

8. $x - y = 4$
$-x - y = 6$

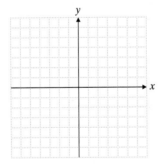

9. $y = -3x$
$y = 2x + 5$

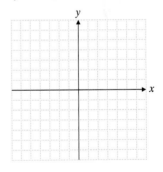

10. $x + 4y = 6$
$-x + 2y = 12$

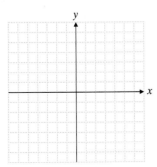

11. $4x + y = 5$
$3x - 2y = 12$

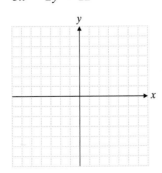

12. $x + 2y = 10$
$3x - 2y = 6$

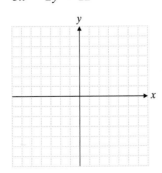

13. $-2x + y - 3 = 0$
$4x + y + 3 = 0$

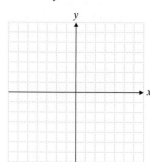

14. $x - 2y - 10 = 0$
$2x + 3y - 6 = 0$

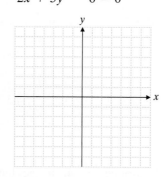

15. $3x - 2y = -18$
$2x + 3y = 14$

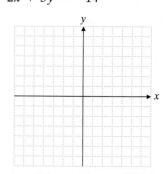

Mixed Practice *Solve by graphing. If there isn't a unique solution to the system, state the reason.*

16. $3x + 2y = -10$
$-2x + 3y = 24$

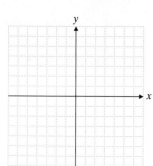

17. $y = \dfrac{3}{4}x + 7$

$y = -\dfrac{1}{2}x + 2$

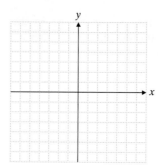

18. $y = \dfrac{5}{7}x - 2$

$y = \dfrac{1}{3}x + \dfrac{2}{3}$

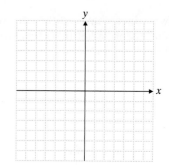

19. $3x - 2y = -4$
$-9x + 6y = -9$

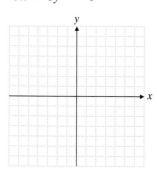

20. $4x - 6y = 8$
$-2x + 3y = -4$

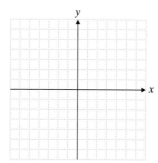

21. $y - 2x - 6 = 0$

$\dfrac{1}{2}y - 3 = x$

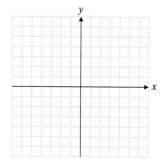

22. $2y + x - 6 = 0$

$y + \dfrac{1}{2}x = 4$

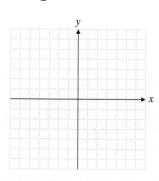

23. $y = \dfrac{1}{2}x - 2$

$y = \dfrac{2}{3}x - 1$

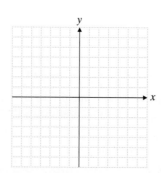

24. $2y + 6x - 8 = 0$

$y = \dfrac{3}{2}x + 4$

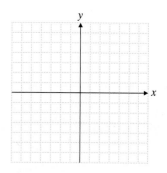

25. *Telephone Charges* The Newtown Telephone Company advertises that if you use their 10-10 service, you will always pay only 7¢ per minute, no matter where you call within the continental United States. Its major rival, the Tryus Telephone Company, charges only 3¢ per minute, but you must pay a 36¢ "connection fee" per call.

(a) Create a cost equation for each telephone company, where y is the total cost for each phone call and x is the number of minutes the phone call lasts. Write a system of equations.

(b) Graph the two equations using the values $x = 1$, 4, 8, and 12. Fill in a table for those values of x and y.

(c) Determine from your graph how many minutes long your phone call must be for the cost of using each telephone company to be the same.

(d) Determine from your graph which company would cost you less if you typically make phone calls of 10 minutes or more.

26. *Landscaping Costs* Fred and Amy want to have shrubs planted in front of their new house. They have obtained estimates from two landscaping companies. Camp Property Care charges $100 for an initial consultation and $60 an hour for labor. Manchester Landscape Designs charges $200 for an initial consultation and $40 an hour for labor.

(a) Create a cost equation for each company where y is the total cost of the landscaping work and x is the number of hours of labor. Write a system of equations.

(b) Graph the two equations using the values $x = 0, 5$, and 10. Fill in a table for those values of x and y.

(c) Determine from your graph how many hours of labor would be required for the cost of landscaping from each of the two companies to be the same.

(d) Determine from your graph which company would charge less if 7 hours of landscaping are required.

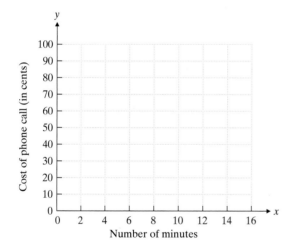

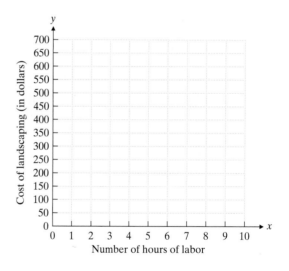

If you have a graphing calculator, solve each system by graphing the equations and estimating the point of intersection to the nearest hundredth. You will need to adjust the window for the appropriate x- and y-values in order to see the region where the straight lines intersect.

 27. $88x + 57y = 683.10$
$95x - 48y = 7460.64$

28. $64x + 99y = -8975.52$
$58x - 73y = 3503.58$

29. $y = 56x + 1808$
$y = -62x - 2086$

30. $y = 47x - 960$
$y = -36x + 783$

Cumulative Review

31. **[2.8.2]** Solve. $7 - 3(4 - x) \le 11x - 6$

32. **[2.3.3]** Simplify. $2(x + 3) - 3(x - 1)$

33. **[3.1.3]** Solve for x. $3x - 2y = 6$

34. **[3.1.3]** Solve for y. $4x + 2y = 16$

Quick Quiz 4.1

1. Solve the following system by graphing.

$x - 3y = 6$
$4x + 3y = 9$

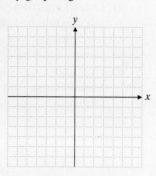

2. Find the slope and the y-intercept of each of the following lines.

$$y = \frac{3}{5}x + 2$$

$$5y - 3x = 35$$

What does this tell you if you want to find the solution to the system of equations by the graphing method?

3. What will happen when you try to find the solution of the following system of equations by graphing?

$$y = 3x - 2$$
$$-6x + 2y = -4$$

4. **Concept Check** You are attempting to find the solution to this system of equations by graphing.

$$-3x + 4y = -16$$
$$0 = 6x - 8y + 16$$

What will you discover?

Solving a System of Two Linear Equations with Integer Coefficients by the Substitution Method

Since the solution to a system of two linear equations may be a point where the two lines intersect, we would expect the solution to be an ordered pair. Since the point must lie on both lines, the ordered pair (x, y) must satisfy both equations. For example, the solution to the system $3x + 2y = 6$ and $x + y = 3$ is $(0, 3)$. Let's check.

$$3x + 2y = 6 \qquad\qquad x + y = 3$$
$$3(0) + 2(3) \overset{?}{=} 6 \qquad\qquad 0 + 3 \overset{?}{=} 3$$
$$6 = 6 \;\checkmark \qquad\qquad 3 = 3 \;\checkmark$$

Given a system of linear equations, we can develop algebraic methods for finding a solution. These methods reduce the system to one equation in one variable, which we already know how to solve. Once we have solved for this variable, we can substitute this known value into any one of the original equations and solve for the second variable.

The first method we will discuss is the **substitution method.**

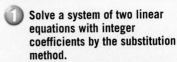

Student Learning Objectives

After studying this section, you will be able to:

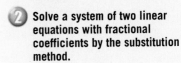

① Solve a system of two linear equations with integer coefficients by the substitution method.

② Solve a system of two linear equations with fractional coefficients by the substitution method.

PROCEDURE FOR SOLVING A SYSTEM OF EQUATIONS BY THE SUBSTITUTION METHOD

1. Solve one of the two equations for one variable. If possible, solve for a variable with a coefficient of 1 or −1.

2. Substitute the expression from step 1 into the *other* equation.

3. You now have one equation with one variable. Solve this equation to find the value for that one variable.

4. Substitute this value for the variable into one of the original equations to obtain a value for the other variable.

5. Check the solution in each equation to verify your results.

EXAMPLE 1 Find the solution.
$$x - 2y = 7 \quad \textbf{(1)}$$
$$-5x + 4y = -5 \quad \textbf{(2)}$$

Solution

Step 1 Solve one equation for one variable.

$x - 2y = 7$ Equation **(1)** is the easiest one in which to isolate a variable.

$ x = 7 + 2y$ Add $2y$ to both sides to solve for x.

Step 2 Now substitute this expression into the other equation.

$-5x + 4y = -5$ Write equation **(2)**.

$-5(\,7 + 2y\,) + 4y = -5$ Substitute the value $7 + 2y$ for x in this equation.

Now we have one equation in one variable, y.

Step 3 Solve this equation.

$-35 - 10y + 4y = -5$ Remove the parentheses.

$ -35 - 6y = -5$ Combine like terms.

$ -6y = 30$ Add 35 to both sides.

$ y = -5$ Divide both sides by −6.

Step 4 We will now use the value $y = -5$ in one of the original equations to find the value for x.

$$x - 2y = 7 \quad \textbf{(1)}$$ The easiest equation to use is the one that we solved for x.

$$x - 2(-5) = 7$$ Replace y by -5.

$$x + 10 = 7$$ Simplify.

$$x = -3$$ Subtract 10 from both sides.

The solution is $(-3, -5)$.

Step 5 *Check.* To be sure that we have the correct solution, we will need to check that the values obtained for x and y can be substituted into *both original* equations to obtain true mathematical statements.

$$x - 2y = 7 \quad \textbf{(1)} \qquad\qquad -5x + 4y = -5 \quad \textbf{(2)}$$
$$-3 - 2(-5) \overset{?}{=} 7 \qquad\qquad -5(-3) + 4(-5) \overset{?}{=} -5$$
$$-3 + 10 \overset{?}{=} 7 \qquad\qquad 15 - 20 \overset{?}{=} -5$$
$$7 = 7 \ \checkmark \qquad\qquad -5 = -5 \ \checkmark$$

NOTE TO STUDENT: Fully worked-out solutions to all of the Practice Problems can be found at the back of the text starting at page SP-1

Practice Problem 1 Find the solution.

$$5x + 3y = 19$$
$$2x - y = 12$$

② Solving a System of Two Linear Equations with Fractional Coefficients by the Substitution Method

If a system of equations contains fractions, clear the equations of fractions *before* performing any other steps.

EXAMPLE 2 Find the solution.

$$\frac{3}{2}x + y = \frac{5}{2} \quad \textbf{(1)}$$
$$-y + 2x = -1 \quad \textbf{(2)}$$

Solution We want to clear the first equation of fractions. We observe that the LCD of the fractions is 2.

$$2\left(\frac{3}{2}x\right) + 2(y) = 2\left(\frac{5}{2}\right)$$ Multiply each term of equation **(1)** by the LCD.

$$3x + 2y = 5$$ This is equivalent to $\frac{3}{2}x + y = \frac{5}{2}$; and we label it **(3)**.

The new system is as follows.

$$3x + 2y = 5 \quad \textbf{(3)}$$
$$-y + 2x = -1 \quad \textbf{(2)}$$

Now follow the five-step procedure.

Step 1 Solve for one variable.

$$-y + 2x = -1$$ Equation **(2)** is the easiest one in which to isolate a variable.

$$-y = -1 - 2x$$ Add $-2x$ to both sides.

$$y = 1 + 2x$$ Multiply each term by -1.

Step 2 Substitute the resulting expression into equation **(3)**.

$$3x + 2(1 + 2x) = 5 \quad \textbf{(3)}$$

Step 3 Solve this equation for the variable.

$$3x + 2 + 4x = 5 \quad \text{Remove parentheses.}$$
$$7x + 2 = 5 \quad \text{Simplify.}$$
$$7x = 3$$
$$x = \frac{3}{7} \quad \text{Solve for } x.$$

Step 4 Find the value of the second variable, y.

$$-y + 2x = -1 \quad \text{We will use equation (2).}$$
$$-y + 2\left(\frac{3}{7}\right) = -1 \quad \text{Replace } x \text{ with } \frac{3}{7}.$$
$$-y + \frac{6}{7} = -1 \quad \text{Simplify.}$$
$$y - \frac{6}{7} = 1 \quad \text{Multiply by } -1.$$
$$y = 1 + \frac{6}{7} = \frac{13}{7} \quad 1 = \frac{7}{7}; \frac{7}{7} + \frac{6}{7} = \frac{13}{7}.$$

The solution to the system is $\left(\frac{3}{7}, \frac{13}{7}\right)$.

Step 5 *Check*.

$$\frac{3}{2}x + y = \frac{5}{2} \quad \textbf{(1)}$$
$$\frac{3}{2}\left(\frac{3}{7}\right) + \frac{13}{7} \overset{?}{=} \frac{5}{2}$$
$$\frac{9}{14} + \frac{13}{7} \overset{?}{=} \frac{5}{2}$$
$$\frac{9}{14} + \frac{26}{14} \overset{?}{=} \frac{35}{14}$$
$$\frac{35}{14} = \frac{35}{14} \quad \checkmark$$

$$-y + 2x = -1 \quad \textbf{(2)}$$
$$-\frac{13}{7} + 2\left(\frac{3}{7}\right) \overset{?}{=} -1$$
$$-\frac{13}{7} + \frac{6}{7} \overset{?}{=} -\frac{7}{7}$$
$$-\frac{7}{7} = -\frac{7}{7} \quad \checkmark$$

Practice Problem 2 Find the solution. Be sure to check your answer.

$$\frac{1}{3}x - \frac{1}{2}y = 1$$
$$x + 4y = -8$$

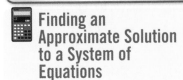

Graphing Calculator

Finding an Approximate Solution to a System of Equations

We can solve systems of equations graphically by using a graphing calculator. For example, to solve the system of equations in Example 2, first rewrite each equation in slope–intercept form.

$$y = -\frac{3}{2}x + \frac{5}{2}$$
$$y = 2x + 1$$

Then graph $y_1 = -\frac{3}{2}x + \frac{5}{2}$ and $y_2 = 2x + 1$ on the same screen. Display:

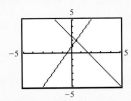

Next you can use the Trace and Zoom features to find the intersection of the two lines. Some graphing calculators have a command to find and calculate the intersection. Display:

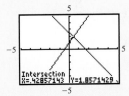

Rounded to four decimal places, the solution is (0.4286, 1.8571). Observe that

$$\frac{3}{7} = 0.4285714\ldots$$

and

$$\frac{13}{7} = 1.8571428\ldots,$$

so the answer agrees with the answer found in Example 2.

Find the solution to each system of equations by the substitution method. Check your answers. Write your solution in the form $(x, y), (a, b), (p, q),$ *or* (s, t).

1. $4x + 3y = 9$
$\quad x = 3y + 6$

2. $\quad x = 4 - 4y$
$\quad -x + 2y = 2$

3. $2x + y = 4$
$\quad 2x - y = 0$

4. $7x - 3y = -10$
$\quad x + 3y = \quad 2$

5. $5x + 2y = 5$
$\quad 3x + \ y = 4$

6. $4x + 2y = 4$
$\quad 3x + \ y = 4$

7. $4x - 3y = -9$
$\quad x - \ y = -3$

8. $3x - \ y = 3$
$\quad x + 2y = -6$

9. $2a - 3b = \ 0$
$\quad 3a + \ b = 22$

10. $\ a - 4b = 21$
$\quad 5a + \ b = \ 0$

11. $3x - \ y = \ 3$
$\quad x + 3y = 11$

12. $4x - \ y = \ 9$
$\quad 3x - 5y = 11$

13. $\ p + 2q - 4 = 0$
$\quad 7p - \ q - 3 = 0$

14. $7s + 2t - 10 = 0$
$\quad 5s - \ t + \ 5 = 0$

15. $3x - \ y - 9 = 0$
$\quad 8x + 5y - 1 = 0$

16. $\ 8x + 2y - 7 = 0$
$\quad -2x - \ y + 2 = 0$

Mixed Practice *Find the solution to each system of equations. Write your answer in the form* (x, y).

17. $\dfrac{5}{3}x + \dfrac{1}{3}y = -3$
$\quad -2x + 3y = 24$

18. $-x + \ y = -4$
$\quad \dfrac{3}{7}x + \dfrac{2}{3}y = \ 5$

19. $4x + 5y = \ 2$
$\quad \dfrac{1}{5}x + \ y = -\dfrac{7}{5}$

20. $\ x + 2y = -6$
$\quad \dfrac{2}{3}x + \dfrac{1}{3}y = -3$

21. $\dfrac{4}{7}x + \dfrac{2}{7}y = \ 2$
$\quad 3x + \ y = 13$

22. $\ x - 4y = 11$
$\quad \dfrac{2}{3}x + \dfrac{1}{2}y = \ 1$

23. $\ 3x + y = -1$
$\quad -2x - y = -1$

24. $\ x - 2y = -4$
$\quad 5x - 4y = \ 4$

25. $3x - 4y = 36$
$\quad y + 3 = 3(x + 1)$

26. $2x + 3y = -15$
$\quad y + 15 = 2(x + 1)$

27. $2x = 4(2y + 2)$
$\quad 3(x - 3y) + 2 = 17$

28. $2(2x - y - 11) = 1 - y$
$\quad 3(3x + y) - 17 = 2(4x + y)$

To Think About

29. How many equations do you think you would need to solve a system with three unknowns? With seven unknowns? Explain your reasoning.

30. The point where the graphs of two linear equations intersect is the _____ to the system of linear equations.

31. The solution to a system of two equations must satisfy _____ equations.

32. How many solutions will a system of equations have if the graphs of the lines intersect? Justify your answer.

33. How many solution(s) will a system of equations have if the graphs of the equations are parallel lines? Why?

Applications

34. *Construction Costs* Tim Martinez is involved in a large construction project in the city. He needs to rent a heavy construction crane for several weeks. He is considering renting from one of two companies. Boston Construction will rent a crane for an initial delivery charge of $1500 and a rental fee of $900 per week. North End Contractors will rent a crane for an initial delivery charge of $500 and a rental fee of $1000 per week.

  (a) Create a cost equation for each company where y is the total cost of renting a crane and x is the number of weeks the crane is rented. Write a system of equations.

  (b) Solve the system of equations by the substitution method to find out how many weeks would be required for the two companies to charge the same. What would the cost be for each company?

  (c) Tim remembers that last year he considered renting from the same two companies. For the number of weeks he needed the crane, the cost of renting from one company was $4,000 more than the other. How many weeks was the rental last year? Which company was less expensive for that period last year?

35. *Snow Removal Costs* Fred Driscoll has just moved from the sunny Southwest to the wintry Northeast, outside Buffalo, New York. His farmhouse has a very long driveway and Fred has little experience with snow. He decides to hire a snow removal company for the winter. Lake Erie Plowing requires a $50 nonrefundable reservation fee at the start of the winter and charges $90 every time they come out to plow. Adirondack Plowing requires a reservation fee of $100 but only charges $80 each time they come out to plow.

  (a) Create a cost equation for each plowing company where y is the total cost of plowing and x is the number of times Fred needs to be plowed out. Write a system of equations.

  (b) Solve the system of equations by the substitution method to find out how many times Fred would have to be plowed out for the cost of the plowing companies to be the same. How many times would this be? What would be the cost for each company?

  (c) Fred calculated that if he chose one company over the other, based on the average number of times that his neighbors claimed he would need to be plowed, he would save himself $80. How many times do his neighbors expect Fred to need plowing? Which company should he hire to save that money?

Cumulative Review

36. **[3.3.2]** Find the slope and y-intercept of $7x + 11y = 19$.   **37.** **[3.5.1]** Graph $6x + 3y \geq 9$.

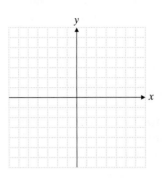

38. **[0.5.5]** *Gas Mileage* Darrin purchased a 2008 Infiniti G37 Coupe as soon as the car was available. It has a 20-gallon gasoline tank and gets 24 miles per gallon on the highway. The average price for premium gasoline (recommended for this car) was $4.20 per gallon in August 2008. Darrin travels 60 miles roundtrip on the highway each day to work and back home. How much does gasoline cost per day for Darrin to drive to work?

39. **[0.5.3]** *Holiday Spending* In 2007, the average consumer spent $64.82 on Halloween candy, decorations, and costumes. The average amount spent on costumes was $23.33. What percent of the total was spent on decorations and candy? If the average 18–24-year-old spent 10% more on Halloween expenses than the average consumer, how much did the average 18–24-year-old consumer spend on Halloween?

Quick Quiz 4.2 Find the solution to each system of equations by the substitution method. Write your solution in the form (x, y).

1. $10x + 3y = 8$
$2x + y = 2$

2. $-3x - y = -9$
$-x + 2y = -10$

3. $-14x - 4y + 20 = 0$
$5x - y + 5 = 0$

4. **Concept Check** Explain how you would solve the following system by the substitution method.

$$3x - y = -4$$
$$-2x + 4y = 36$$

Using the Addition Method to Solve a System of Two Linear Equations with Integer Coefficients

The substitution method is useful for solving a system of equations when the coefficient of one variable is 1. In the following system, the variable x in the first equation has a coefficient of 1, and we can easily solve for x.

$$x - 2y = 7 \Rightarrow x = 7 + 2y$$
$$-5x + 4y = -5$$

This makes the substitution method a natural choice for solving this system. But we may not always have 1 as a coefficient of a variable. We need a method of solving systems of equations that will work for integer coefficients, fractional coefficients, and decimal coefficients. One such method is the **addition method.**

EXAMPLE 1 Solve by addition.

$$5x + 2y = 7 \quad (1)$$
$$3x - y = 13 \quad (2)$$

Solution We would like the coefficients of the y-terms to be opposites. One should be $+2y$ and the other $-2y$. Thus, we multiply each term of equation **(2)** by 2.

$$2(3x) - 2(y) = 2(13) \qquad \text{Multiply each term of equation (2) by 2.}$$
$$5x + 2y = 7 \quad (1)$$
$$\underline{6x - 2y = 26} \quad (3) \qquad \text{We label this equation (3).}$$
$$11x \qquad = 33 \qquad \text{Add the two equations. This will eliminate the } y\text{-variable.}$$
$$x = 3 \qquad \text{Divide both sides by 11.}$$

Substitute $x = 3$ into one of the *original* equations to find y.

$$5(x) + 2y = 7 \qquad \text{Arbitrarily, we pick equation (1).}$$
$$5(3) + 2y = 7 \qquad \text{Substitute for } x.$$
$$15 + 2y = 7 \qquad \text{Remove parentheses.}$$
$$2y = -8 \qquad \text{Subtract 15 from both sides.}$$
$$y = -4 \qquad \text{Divide both sides by 2.}$$

The solution is $(3, -4)$.

Check. Replace x by 3 and y by -4 in *both* original equations.

$$5x + 2y = 7 \quad (1) \qquad\qquad 3x - y = 13 \quad (2)$$
$$5(3) + 2(-4) \stackrel{?}{=} 7 \qquad\qquad 3(3) - (-4) \stackrel{?}{=} 13$$
$$15 - 8 \stackrel{?}{=} 7 \qquad\qquad 9 + 4 \stackrel{?}{=} 13$$
$$7 = 7 \ \checkmark \qquad\qquad 13 = 13 \ \checkmark$$

Practice Problem 1 Solve by addition.

$$3x + y = 7$$
$$5x - 2y = 8$$

Notice that when we added the two equations together, one variable was eliminated. The addition method is therefore often called the **elimination method.**

For convenience, we will make a list of the steps we use to solve a system of equations by the addition method.

Student Learning Objectives

After studying this section, you will be able to:

1 Use the addition method to solve a system of two linear equations with integer coefficients.

2 Use the addition method to solve a system of two linear equations with fractional coefficients.

3 Use the addition method to solve a system of two linear equations with decimal coefficients.

NOTE TO STUDENT: Fully worked-out solutions to all of the Practice Problems can be found at the back of the text starting at page SP-1

PROCEDURE FOR SOLVING A SYSTEM OF EQUATIONS BY THE ADDITION METHOD

1. Multiply each term of one or both equations by some nonzero integer so that the coefficients of one of the variables are opposites.
2. Add the equations of this new system so that one variable is eliminated.
3. Solve the resulting equation for the remaining variable.
4. Substitute the value found in step 3 into either of the original equations to find the value of the other variable.
5. Check your solution in both of the original equations.

Care should be used when the solution of a system contains fractions.

EXAMPLE 2 Solve by addition.

$$3x + 4y = 17 \quad \textbf{(1)}$$
$$2x + 7y = 19 \quad \textbf{(2)}$$

Solution Suppose we want the x-terms to have opposite coefficients. One equation could have $+6x$ and the other $-6x$. This would happen if we multiply equation **(1)** by $+2$ and equation **(2)** by -3.

$(2)(3x) + (2)(4y) = (2)(17)$	Multiply each term of equation **(1)** by 2.
$(-3)(2x) + (-3)(7y) = (-3)(19)$	Multiply each term of equation **(2)** by -3. Often these multiplication steps can be done mentally.

We now have an equivalent system of equations labeled **(3)** and **(4)**.

$6x + 8y = 34$	**(3)**	
$-6x - 21y = -57$	**(4)**	The coefficients of the x-terms are opposites.
$-13y = -23$		Add the two equations to eliminate the variable x.
$y = \dfrac{-23}{-13} = \dfrac{23}{13}$		Divide both sides by -13.

Substitute $y = \dfrac{23}{13}$ into one of the *original* equations to find x.

$3x + 4\left(\dfrac{23}{13}\right) = 17 \quad \textbf{(1)}$	We pick equation **(1)**.
$3x + \dfrac{92}{13} = 17$	Solve for x.
$13(3x) + 13\left(\dfrac{92}{13}\right) = 13(17)$	Multiply both sides by 13 to clear the fraction.
$39x + 92 = 221$	Remove the parentheses.
$39x = 129$	Subtract 92 from both sides.
$x = \dfrac{43}{13}$	Divide both sides by 39. The solution is $\left(\dfrac{43}{13}, \dfrac{23}{13}\right)$.

Alternative solution: You could also have eliminated the y-variable. For example, if you multiply equation **(1)** by 7 and equation **(2)** by -4, you obtain the equivalent system shown here.

$$21x + 28y = 119$$
$$-8x - 28y = -76$$

You can add these equations to eliminate the y-variable. Since the numbers involved in this approach are somewhat larger, it is probably wiser to eliminate the x-variable in this example.

CAUTION: A common error occurs when students forget that we want to obtain two terms with *opposite signs* that when added will equal zero. If all the coefficients in the equations are positive, such as in Example 2, it will be necessary to multiply *one* of the equations by a negative number.

Practice Problem 2 Solve by addition.

$$4x + 5y = 17$$
$$3x + 7y = 12$$

NOTE TO STUDENT: *Fully worked-out solutions to all of the Practice Problems can be found at the back of the text starting at page SP-1*

Using the Addition Method to Solve a System of Two Linear Equations with Fractional Coefficients

If the system of equations has fractional coefficients, you should first clear each equation of the fractions. To do so, you will need to multiply each term in the equation by the LCD of the fractions.

EXAMPLE 3 Solve.

$$x - \frac{5}{2}y = \frac{5}{2} \quad \textbf{(1)}$$
$$\frac{4}{3}x + y = \frac{23}{3} \quad \textbf{(2)}$$

Solution

$$2(x) - 2\left(\frac{5}{2}y\right) = 2\left(\frac{5}{2}\right) \qquad \text{Multiply each term of equation \textbf{(1)} by 2.}$$
$$2x - 5y = 5$$

$$3\left(\frac{4}{3}x\right) + 3(y) = 3\left(\frac{23}{3}\right) \qquad \text{Multiply each term of equation \textbf{(2)} by 3.}$$
$$4x + 3y = 23$$

We now have an equivalent system of equations that does not contain fractions.

$$2x - 5y = 5 \quad \textbf{(3)}$$
$$4x + 3y = 23 \quad \textbf{(4)}$$

Let us eliminate the *x*-variable. We want the coefficients of *x* to be opposites.

$$(-2)(2x) - (-2)5y = (-2)(5) \qquad \text{Multiply each term of equation \textbf{(3)} by } -2.$$

$$\begin{array}{rl}
-4x + 10y = -10 & \textbf{(5)} \quad \text{We now have an equivalent system of equations.} \\
\underline{4x + 3y = 23} & \textbf{(4)} \quad \text{The coefficients of the } x\text{-terms are opposites.} \\
13y = 13 & \phantom{\textbf{(4)}} \quad \text{Add the two equations to eliminate the variable } x. \\
y = 1 & \phantom{\textbf{(4)}} \quad \text{Divide both sides by 13.}
\end{array}$$

$$\begin{array}{rl}
4x + 3(1) = 23 & \text{Substitute } y = 1 \text{ into equation \textbf{(4)}.} \\
4x + 3 = 23 & \\
4x = 20 & \\
x = 5 &
\end{array}$$

The solution is $(5, 1)$.

Check. Check the solution in both of the *original* equations.

Practice Problem 3 Solve.

$$\frac{2}{3}x - \frac{3}{4}y = 3$$
$$-2x + y = 6$$

③ Using the Addition Method to Solve a System of Two Linear Equations with Decimal Coefficients

Some linear equations will have decimal coefficients. It will be easier to work with the equations if we change the decimal coefficients to integer coefficients. To do so, we will multiply each term of the equation by a power of 10.

EXAMPLE 4 Solve.

$$0.12x + 0.05y = -0.02 \quad \textbf{(1)}$$
$$0.08x - 0.03y = -0.14 \quad \textbf{(2)}$$

Solution Since the decimals are hundredths, we will multiply each term of both equations by 100.

$$100(0.12x) + 100(0.05y) = 100(-0.02)$$
$$100(0.08x) + 100(-0.03y) = 100(-0.14)$$

$$12x + 5y = -2 \quad \textbf{(3)} \qquad \text{We now have an equivalent system of equations that has integer coefficients.}$$

$$8x - 3y = -14 \quad \textbf{(4)}$$

We will eliminate the variable y. We want the coefficients of the y-terms to be opposites.

$$\begin{aligned} 36x + 15y &= -6 \\ \underline{40x - 15y} &= \underline{-70} \end{aligned} \qquad \text{Multiply equation \textbf{(3)} by 3 and equation \textbf{(4)} by 5.}$$

$$76x \qquad\quad = -76 \qquad \text{Add the equations.}$$
$$x = -1 \qquad\qquad \text{Solve for } x.$$
$$12(-1) + 5y = -2 \qquad \text{Replace } x \text{ by } -1 \text{ in equation \textbf{(3)}, and solve for } y.$$
$$-12 + 5y = -2$$
$$5y = -2 + 12$$
$$5y = 10$$
$$y = 2$$

The solution is $(-1, 2)$.

Check. We substitute $x = -1$ and $y = 2$ into the original equations. Most students probably would rather not use equation **(1)** and equation **(2)**! However, *we must check the solutions in the original equations* if we want to be sure of our answers. It is possible to make an error going from equations **(1)** and **(2)** to equations **(3)** and **(4)**. The solution could satisfy equations **(3)** and **(4)**, but might not satisfy equations **(1)** and **(2)**.

$$0.12x + 0.05y = -0.02 \quad \textbf{(1)} \qquad\qquad 0.08x - 0.03y = -0.14 \quad \textbf{(2)}$$
$$0.12(-1) + 0.05(2) \overset{?}{=} -0.02 \qquad\qquad 0.08(-1) - 0.03(2) \overset{?}{=} -0.14$$
$$-0.12 + 0.10 \overset{?}{=} -0.02 \qquad\qquad -0.08 - 0.06 \overset{?}{=} -0.14$$
$$-0.02 = -0.02 \ \checkmark \qquad\qquad -0.14 = -0.14 \ \checkmark$$

Practice Problem 4 Solve.

$$0.2x + 0.3y = -0.1$$
$$0.5x - 0.1y = -1.1$$

CAUTION: A common error when solving a system is to find the value for x and then stop. You have not solved a system in x and y until you find both the x- and the y-values that make both equations true.

PRACTICE   WATCH   DOWNLOAD   READ   REVIEW

Verbal and Writing Skills *Look at the following systems of equations. Decide which variable to eliminate in each system, and explain how you would eliminate that variable.*

1. $3x + 2y = 5$
$5x - y = 3$

2. $2x - 9y = 1$
$2x + 3y = 2$

3. $4x - 3y = 10$
$5x + 4y = 0$

4. $7x + 6y = 13$
$-2x + 5y = 3$

Find the solution by the addition method. Check your answers.

5. $-x + y = -3$
$-2x - y = 6$

6. $-x + y = 2$
$x + y = 4$

7. $2x + 3y = 1$
$x - 2y = 4$

8. $2x + y = 4$
$3x - 2y = -1$

9. $6x - y = -7$
$6x + 2y = 5$

10. $3x - 2y = -11$
$3x + y = 1$

11. $5x - 15y = 9$
$-x + 10y = 1$

12. $-4x + 5y = -16$
$8x + y = -1$

13. $2x + 5y = 2$
$3x + y = 3$

14. $5x - 3y = 14$
$2x - y = 6$

15. $8x + 6y = -2$
$10x - 9y = -8$

16. $4x + 9y = 0$
$8x - 5y = -23$

17. $2x + 3y = -8$
$5x + 4y = -34$

18. $3x - 5y = 11$
$4x + 3y = 5$

19. $4x + 3y = 1$
$2x - 5y = -19$

20. $-4x - 5y = 2$
$6x + 9y = -6$

21. $12x - 6y = -2$
$-9x - 7y = -10$

22. $2x - 4y = -22$
$-6x + 3y = 3$

Verbal and Writing Skills *Before using the addition method to solve each system, you will need to simplify the equation(s). Explain how you would change the fractional coefficients to integer coefficients in each system.*

23. $\frac{1}{4}x - \frac{3}{4}y = 2$
$2x + 3y = 1$

24. $5x + 2y = 9$
$\frac{4}{9}x - \frac{2}{3}y = 4$

25. $\dfrac{1}{2}x + \dfrac{2}{3}y = \dfrac{1}{3}$

$\dfrac{3}{4}x - \dfrac{4}{5}y = 2$

26. $\dfrac{1}{4}x + \dfrac{1}{3}y = 15$

$\dfrac{1}{5}x + \dfrac{3}{10}y = 13$

Find the solution. Check your answers.

27. $x + \dfrac{5}{4}y = \dfrac{9}{4}$

$\dfrac{2}{5}x - \ \ y = \dfrac{3}{5}$

28. $\dfrac{2}{3}x + \ \ y = 2$

$x + \dfrac{1}{2}y = 7$

29. $\dfrac{x}{6} + \dfrac{y}{2} = -\dfrac{1}{2}$

$x - 9y = \ \ 21$

30. $\dfrac{3}{2}x - \dfrac{y}{8} = \ -1$

$16x + 3y = -28$

31. $\dfrac{5}{6}x + \ \ y = -\dfrac{1}{3}$

$-8x + 9y = 28$

32. $\dfrac{2}{9}x - \dfrac{1}{3}y = \ \ 1$

$4x + 9y = -2$

33. $\dfrac{2}{3}x + \dfrac{3}{5}y = -\dfrac{1}{5}$

$\dfrac{1}{4}x + \dfrac{1}{3}y = \ \ \dfrac{1}{4}$

34. $\dfrac{x}{5} + \dfrac{1}{2}y = \dfrac{9}{10}$

$\dfrac{x}{3} + \dfrac{3}{4}y = \dfrac{5}{4}$

Verbal and Writing Skills *Explain how you would change each system to an equivalent system with integer coefficients. Write the equivalent system.*

35. $0.5x - 0.3y = 0.1$

$5x + \ \ 3y = 6$

36. $0.08x + \ \ \ \ y = 0.05$

$2x - 0.1y = 3$

37. $4x + \ 0.5y = 9$

$0.2x - 0.05y = 1$

Solve for x and y using the addition method.

38. $0.5x - 0.2y = 0.4$

$-0.3x + 0.4y = 0.6$

39. $0.2x + 0.3y = 0.4$

$0.5x + 0.4y = 0.3$

40. $0.02x - 0.04y = 0.26$

$0.07x - 0.09y = 0.66$

41. $0.04x - 0.03y = \ \ \ 0.05$

$0.05x + 0.08y = -0.76$

42. $0.4x - \ \ 5y = -1.2$

$-0.03x + 0.5y = 0.14$

43. $-0.6x - 0.08y = -4$

$3x + \ \ \ 2y = \ \ \ 4$

Mixed Practice *Solve for x and y using the addition method.*

44.
$$x + 6y = -1$$
$$-2x - 8y = 4$$

45.
$$3x - y = -8$$
$$-9x + 2y = 18$$

46.
$$\frac{5}{4}x + y = 16$$
$$x - \frac{5}{3}y = -2$$

47.
$$\frac{1}{3}x - \frac{1}{4}y = 0$$
$$\frac{1}{6}x - \frac{1}{2}y = -6$$

48.
$$0.2x + 0.1y = 1.1$$
$$0.01x + 0.03y = 0.18$$

49.
$$0.05x - 0.04y = -0.08$$
$$0.2x + 0.3y = 0.6$$

To Think About *Find the solution.*

50.
$$4(x - 2y) = 5 - (y - 3x)$$
$$-5(x + 1) = y - 6x$$

51.
$$3(3x + 2) = y + 5x$$
$$4 = -2(x - y)$$

Cumulative Review

52. [0.5.4] *Air Traffic* At any given moment of the day, between 5000 and 7000 commercial aircraft are flying in U.S. airspace. If 89% of the air traffic is flying over the contiguous states, how many commercial airplanes are flying over Alaska and Hawaii?

53. [2.6.2] *Used Car Purchase* A used car is priced at $5800, if you pay cash. An installment plan requires $1000 down, plus $230 a month for 24 months. How much more would you pay for the car under the installment plan?

Solve for the variable.

54. [2.3.3] $\frac{1}{3}(4 - 2x) = \frac{1}{2}x$

55. [2.3.3] $2(y - 3) - (2y + 4) = -6y$

Quick Quiz 4.3 Find the solution to the system by the addition method. Express your answer in the form (x, y).

1.
$$3x + 5y = -1$$
$$-5x + 4y = -23$$

2.
$$-7x + 3y = -31$$
$$4x + 6y = 10$$

3.
$$\frac{1}{3}x + y = \frac{8}{3}$$
$$\frac{4}{5}x - \frac{2}{5}y = \frac{18}{5}$$

4. Concept Check Explain how you would obtain an equivalent system that does not have decimals if you wanted to solve the following system.
$$0.5x + 0.4y = -3.4$$
$$0.02x + 0.03y = -0.08$$

1. _____

2. _____

3. _____

4. _____

5. _____

6. _____

7. _____

8. _____

9. _____

10. _____

How are you doing with your homework assignments in Sections 4.1 to 4.3? Do you feel you have mastered the material so far? Do you understand the concepts you have covered? Before you go further in the textbook, take some time to do each of the following problems.

4.1

Solve by graphing.

1. $2x - y = -8$
$\quad\ -x + 2y = 10$

2. $-2x + y = 4$
$\quad\ x - 3y = 3$

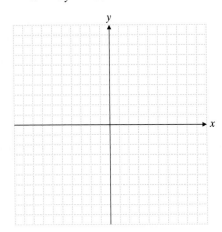
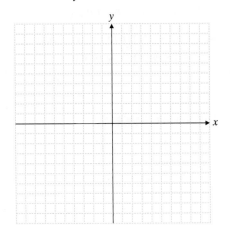

4.2

Solve by the substitution method.

3. $5x + y = 26$
$\quad 3x + 2y = 10$

4. $3x - 5y = 10$
$\quad x + 4y = -8$

5. $2x + 3y = 5$
$\quad \dfrac{1}{2}x + \dfrac{1}{6}y = -\dfrac{1}{2}$

6. $7x - 5y - 17 = 0$
$\quad x + 8y + 15 = 0$

4.3

Solve by the addition method.

7. $3x - 2y = 19$
$\quad x + y = 8$

8. $3x + 10y = -4$
$\quad 2x - 8y = -10$

9. $\dfrac{2}{3}x - \dfrac{3}{4}y = 3$
$\quad \dfrac{1}{9}x + \dfrac{1}{12}y = -2$

10. $0.6x - 0.5y = 2.1$
$\quad 0.4x - 0.3y = 1.3$

Now turn to page SA-13 for the answer to each of these problems. Each answer also includes a reference to the objective in which the problem is first taught. If you missed any of these problems, you should stop and review the Examples and Practice Problems in the referenced objective. A little review now will help you master the material in the upcoming sections of the text.

Choosing an Appropriate Method to Solve a System of Equations Algebraically

At this point we will review the algebraic methods for solving systems of linear equations and discuss the advantages and disadvantages of each method.

Method	Advantage	Disadvantage
Substitution	Works well if one or more variables have a coefficient of 1 or −1.	Often becomes difficult to use if no variable has a coefficient of 1 or −1.
Addition	Works well if equations have fractional or decimal coefficients. Works well if no variable has a coefficient of 1 or −1.	None

EXAMPLE 1 Select a method and solve each system of equations.

(a) $x + y = 3080$
$2x + 3y = 8740$

(b) $5x - 2y = 19$
$-3x + 7y = 35$

Solution

(a) Since there are x- and y-values that have coefficients of 1, we will select the substitution method.

$$y = 3080 - x \qquad \text{Solve the first equation for } y.$$
$$2x + 3(3080 - x) = 8740 \qquad \text{Substitute the expression into the second equation.}$$
$$2x + 9240 - 3x = 8740 \qquad \text{Remove parentheses.}$$
$$-1x = -500 \qquad \text{Simplify.}$$
$$x = 500 \qquad \text{Divide each side by } -1.$$
$$y = 3080 - x$$
$$= 3080 - 500 \qquad \text{Substitute 500 for } x.$$
$$= 2580 \qquad \text{Simplify.}$$

The solution is $(500, 2580)$.

(b) Because none of the x- and y-variables has a coefficient of 1 or −1, we select the addition method. We choose to eliminate the y-variable. Thus, we would like the coefficients of y to be −14 and 14.

$$7(5x) - 7(2y) = 7(19) \qquad \text{Multiply each term of the first equation by 7.}$$
$$2(-3x) + 2(7y) = 2(35) \qquad \text{Multiply each term of the second equation by 2.}$$
$$35x - 14y = 133 \qquad \text{We now have an equivalent system of equations.}$$
$$\underline{-6x + 14y = 70}$$
$$29x = 203 \qquad \text{Add the two equations.}$$
$$x = 7 \qquad \text{Divide each side by 29.}$$

Substitute $x = 7$ into one of the original equations.

$$5(7) - 2y = 19$$
$$35 - 2y = 19 \qquad \text{Solve for } y.$$
$$-2y = -16$$
$$y = 8$$

The solution is $(7, 8)$.

Student Learning Objectives

After studying this section, you will be able to:

1 Choose an appropriate method to solve a system of equations algebraically.

2 Use algebraic methods to identify inconsistent and dependent systems.

NOTE TO STUDENT: Fully worked-out solutions to all of the Practice Problems can be found at the back of the text starting at page SP-1

Practice Problem 1 Select a method and solve each system of equations.

(a) $3x + 5y = 1485$
$x + 2y = 564$

(b) $7x + 6y = 45$
$6x - 5y = -2$

② Using Algebraic Methods to Identify Inconsistent and Dependent Systems

Recall that an inconsistent system of linear equations is a system of parallel lines. Since parallel lines never intersect, the system has no solution. Can we determine this algebraically?

EXAMPLE 2 Solve algebraically.

$$3x - y = -1$$
$$3x - y = -7$$

Solution Clearly, the addition method would be very convenient in this case.

$3x - y = -1$ Keep the first equation unchanged.
$\underline{-3x + y = +7}$ Multiply each term in the second equation by -1.
$0 = 6$ Add the two equations.

Notice we have $0 = 6$, which we know is not true. The statement $0 = 6$ is inconsistent with known mathematical facts. No possible x- and y-values can make this equation true. Thus there is **no solution to this system of equations.**

This example shows us that, if we obtain a mathematical statement that is not true (inconsistent with known facts), we can identify the entire system as inconsistent. If graphed, these lines would be parallel.

Practice Problem 2 Solve algebraically.

$$4x + 2y = 2$$
$$-6x - 3y = 6$$

What happens if we try to solve a dependent system of equations algebraically? Recall that a dependent system consists of two lines that coincide (are the same line).

EXAMPLE 3 Solve algebraically.

$$x + y = 4$$
$$3x + 3y = 12$$

Solution Let us use the substitution method.

$y = 4 - x$ Solve the first equation for y.
$3x + 3(4 - x) = 12$ Substitute $4 - x$ for y in the second equation.
$3x + 12 - 3x = 12$ Remove parentheses.
$12 = 12$

Notice we have $12 = 12$, which is always true. Thus we obtain an equation that is true for any value of x. An equation that is always true is called an **identity.** All the solutions of one equation of the system are also solutions of the other equation.

Thus the lines coincide. There is an **infinite number of solutions to this system of equations.**

This example shows us that, if we obtain a mathematical statement that is always true (an identity), we can identify the equations as dependent. There is an unlimited number of solutions to a system that has dependent equations.

Practice Problem 3 Solve algebraically.

$$3x - 9y = 18$$
$$-4x + 12y = -24$$

TO THINK ABOUT: Two Linear Equations with Two Variables Now is a good time to look back over what we have learned. When you graph a system of two linear equations, what possible kinds of graphs will you obtain?

What will happen when you try to solve a system of two linear equations using algebraic methods? How many solutions are possible in each case? The following chart may help you to organize your answers to these questions.

Graph	Number of Solutions	Algebraic Interpretation
Two lines intersect at one point (6, −3)	**One unique solution**	You obtain one value for x and one value for y. For example, $x = 6$, $y = -3$.
Parallel lines	**No solution**	You obtain an equation that is inconsistent with known facts. For example, $0 = 6$. The system of equations is inconsistent.
Lines coincide	**Infinite number of solutions**	You obtain an equation that is always true. For example, $8 = 8$. The equations are dependent.

Think about each of the three possibilities in this chart. You need to understand each one. As you do each of exercises 5–30, you will need to identify which of the three possibilities is involved.

Verbal and Writing Skills

1. If there is no solution to a system of linear equations, the graphs of the equations are _____. Solving the system algebraically, you will obtain an equation that is _____ with known facts.

2. If an algebraic attempt at solving a system of linear equations results in an identity, the system is said to be _____. There are an _____ number of solutions, and the graphs of the lines _____.

3. If there is exactly one solution, the graphs of the equations _____. This system is said to be _____ and _____.

4. A student solves a system of equations and obtains the equation $-36x = 0$. He is unsure of what to do next. What should he do next? What should he conclude about the system of equations?

If possible, solve by an algebraic method (substitution or addition method, without the use of graphing). Otherwise, state that the problem has no solution or an infinite number of solutions.

5. $-2x - 3y = 15$
$5x + 2y = 1$

6. $3x + 4y = 8$
$-5x - 3y = 5$

7. $3x - 4y = 2$
$-5x + 6y = -12$

8. $4x + 3y = 7$
$6x + 5y = 15$

9. $2x - 4y = 5$
$-4x + 8y = 9$

10. $3x + 6y = 12$
$x + 2y = 7$

11. $-5x + 2y = 2$
$15x - 6y = -6$

12. $6x - 4y = 8$
$-9x + 6y = -12$

13. $5x - 3y = 13$
$7x + 2y = 43$

14. $-4x + 5y = 10$
$6x - 7y = 8$

15. $3x - 2y = 70$
$0.6x + 0.5y = 50$

16. $5x - 3y = 10$
$0.9x + 0.4y = 30$

17. $0.2x - 0.3y = 0.1$
$-0.5x + 0.8y = 0$

18. $0.05x + 0.04y = 0.12$
$0.03x - 0.08y = 0.28$

19. $\dfrac{4}{3}x + \dfrac{1}{2}y = 1$
$\dfrac{1}{3}x - y = -\dfrac{1}{2}$

20. $\dfrac{3}{5}x - \dfrac{3}{4}y = \dfrac{1}{5}$
$2x + \dfrac{3}{2}y = 2$

21. $\dfrac{2}{3}x + \dfrac{1}{6}y = 2$
$\dfrac{1}{4}x - \dfrac{1}{2}y = -\dfrac{3}{4}$

22. $\dfrac{1}{2}x + \dfrac{2}{5}y = \dfrac{9}{10}$
$\dfrac{2}{3}x + \dfrac{2}{15}y = \dfrac{1}{2}$

Mixed Practice *Solve by any algebraic method. If there is not one solution to a system of equations, state the reason.*

23. $4x + 3y = -2$
$5x - 2y = 9$

24. $3x = 4(y + 9)$
$y + 3 = 3(x + 1)$

25. $\dfrac{2}{3}x + y = 1$
$\dfrac{3}{4}x + \dfrac{1}{2}y = \dfrac{3}{4}$

26. $0.8x - 0.3y = 0.7$
$1.2x + 0.6y = 4.2$

27. $x - 2y + 2 = 0$
$3(x - 2y + 1) = 15$

28. $3(x - 2y + 2) = -3$
$-\dfrac{1}{3}x + \dfrac{2}{3}y = 1$

To Think About *Remove parentheses and solve the system for (a, b).*

29. $2(a + 3) = b + 1$
$3(a - b) = a + 1$

30. $3(a + b) - 2 = b + 2$
$a - 2(b - 1) = 3a - 4$

Cumulative Review *Solve for x.*

31. [2.3.3] $8x + 15\left(\dfrac{2}{3}x - \dfrac{2}{5}\right) = -10$

32. [2.3.3] $0.2(x + 0.3) + x = 0.3(x + 0.5)$

In 2007, federal income taxes for single filers could be estimated using the following chart.

If taxable income is over . . .	But not over . . .	Estimated income tax is . . .
$0	$7825	10% of amount over $0
$7825	$31,850	$782.50 + 15% of amount over $7825
$31,850	$77,100	$4386.25 + 25% of amount over $31,850
$77,100	$160,850	$15,698.75 + 28% of amount over $77,100

Source: www.irs.gov

33. [0.5.5] In 2007, Margot earned $27,600. Find the estimated federal income tax she had to pay.

34. [0.5.5] In 2007, Dean earned $32,900. During the year, a total of $5033 was deducted from his paychecks for federal income tax. Was this more or less than what he should have paid? Estimate the amount he still owed or his federal income tax refund.

Quick Quiz 4.4 Solve if possible. Otherwise, state that the problem has no solution or has an infinite number of solutions.

1. $12 - x = 3(x + y)$
$3x + 14 = 4x + 5(x + y)$

2. $-4x + 6y = 2$
$12y = 8x + 12$

3. $\dfrac{1}{2}x - \dfrac{1}{4}y = 1$
$\dfrac{4}{3}x + 4y = 12$

4. Concept Check Explain how you would remove the fractions in order to solve the following system.

$$\dfrac{3}{10}x + \dfrac{2}{5}y = \dfrac{1}{2}$$
$$\dfrac{1}{8}x - \dfrac{3}{16}y = -\dfrac{1}{2}$$

Student Learning Objective

After studying this section, you will be able to:

 Use a system of equations to solve word problems.

Using a System of Equations to Solve Word Problems

Word problems can be solved in a variety of ways. In Chapter 2 and throughout the text, you have used one equation with one unknown to describe situations found in real-life applications. Some of these same problems can be solved using two equations in two unknowns. The method you use will depend on what you find works best for you. Become familiar with both methods before you decide.

EXAMPLE 1 A worker in a large post office is trying to verify the rate at which two electronic card-sorting machines operate. Yesterday the first machine sorted for 3 minutes and the second machine sorted for 4 minutes. The total workload both machines processed during that time period was 10,300 cards. Two days ago the first machine sorted for 2 minutes and the second machine for 3 minutes. The total workload both machines processed during that time period was 7400 cards. Can you determine the number of cards per minute sorted by each machine?

Solution

1. **Understand the problem.** The number of cards processed by two different machines on two different days provides the basis for two linear equations with two unknowns.

   The two equations will represent what occurred on two different days:

   (1) yesterday and

   (2) two days ago.

   The number of cards sorted by each machine is what we need to find.

   Let x = the number of cards per minute sorted by the first machine and
   y = the number of cards per minute sorted by the second machine.

2. **Write the equations.** Yesterday the first machine sorted for 3 minutes and the second machine sorted for 4 minutes, processing a total of 10,300 cards.

first machine		*second machine*	*total no. of cards*
3(no. of cards per minute)		4(no. of cards per minute)	
$3x$	$+$	$4y$	$= \quad 10{,}300$ **(1)**

   Two days ago the first machine sorted for 2 minutes and the second machine sorted for 3 minutes, processing a total of 7400 cards.

first machine		*second machine*	*total no. of cards*
2(no. of cards per minute)		3(no. of cards per minute)	
$2x$	$+$	$3y$	$= \quad 7400$ **(2)**

3. **Solve and state the answer.** We will use the addition method to solve this system.

   $$3x + 4y = 10{,}300 \quad \textbf{(1)}$$
   $$2x + 3y = 7400 \quad \textbf{(2)}$$

   $$6x + 8y = 20{,}600 \qquad \text{Multiply equation \textbf{(1)} by 2.}$$
   $$\underline{-6x - 9y = -22{,}200} \qquad \text{Multiply equation \textbf{(2)} by } -3.$$
   $$-1y = -1600 \qquad \text{Add the two equations to eliminate } x \text{ and solve for } y.$$
   $$y = 1600$$

   $$2x + 3(1600) = 7400 \qquad \text{Substitute the value of } y \text{ into equation \textbf{(2)}.}$$
   $$2x + 4800 = 7400 \qquad \text{Simplify and solve for } x.$$
   $$2x = 2600$$
   $$x = 1300$$

Thus, the first machine sorts 1300 cards per minute and the second machine sorts 1600 cards per minute.

4. **Check.** We can verify each statement in the problem.

Yesterday: Were 10,300 cards sorted?

$$3(1300) + 4(1600) \overset{?}{=} 10{,}300$$
$$3900 + 6400 \overset{?}{=} 10{,}300$$
$$10{,}300 = 10{,}300 \checkmark$$

Two days ago: Were 7400 cards sorted?

$$2(1300) + 3(1600) \overset{?}{=} 7400$$
$$2600 + 4800 \overset{?}{=} 7400$$
$$7400 = 7400 \checkmark$$

Practice Problem 1 After a recent disaster, the Red Cross brought in a pump to evacuate water from a basement. After 3 hours a larger pump was brought in and the two ran for an additional 5 hours, removing 49,000 gallons of water. The next day, after running both pumps for 3 hours, the large one was taken to another site. The small pump finished the job after another 2 hours. If on the second day 30,000 gallons were pumped, how much water does each pump remove per hour?

NOTE TO STUDENT: Fully worked-out solutions to all of the Practice Problems can be found at the back of the text starting at page SP-1

EXAMPLE 2 Fred is considering working for a company that sells encyclopedias. Fred found out that all starting representatives receive the same annual base salary and a standard commission of a certain percentage of the sales they make during the first year. He has been told that one representative sold $50,000 worth of encyclopedias her first year and that she earned $14,000. He was able to find out that another representative sold $80,000 worth of encyclopedias and that he earned $17,600. Determine the base salary and the commission rate of a beginning sales representative.

Solution The earnings of the two different sales representatives will represent the two equations, each of the form

$$\text{base salary} + \text{commission} = \text{total earnings}.$$

What is unknown? The base salary and the commission rate are what we need to find.

Let b = the base salary and
c = the commission rate.

$b + 50{,}000c = 14{,}000$ **(1)** Earnings for the first sales representative
$b + 80{,}000c = 17{,}600$ **(2)** Earnings for the second sales representative

We will use the addition method to solve these equations.

$$-b - 50{,}000c = -14{,}000 \quad \text{Multiply equation (1) by } -1.$$
$$\underline{b + 80{,}000c = 17{,}600} \quad \text{Leave equation (2) unchanged.}$$
$$30{,}000c = 3600 \quad \text{Add the two equations and solve for } c.$$
$$c = \frac{3600}{30{,}000}$$
$$c = 0.12 \quad \text{The commission rate is 12\%.}$$
$$b + (50{,}000)(0.12) = 14{,}000 \quad \text{Substitute this value into equation (1).}$$
$$b + 6000 = 14{,}000 \quad \text{Simplify and solve for } b.$$
$$b = 8000 \quad \text{The base salary is \$8000 per year.}$$

Thus the base salary is $8000 per year, and the commission rate is 12%.

NOTE TO STUDENT: Fully worked-out solutions to all of the Practice Problems can be found at the back of the text starting at page SP-1

Practice Problem 2   Rick has two cars, one using premium gas and the other using regular gas. Last week Rick bought 7 gallons of unleaded premium and 8 gallons of unleaded regular gasoline and paid $47.15. This week he purchased 8 gallons of unleaded premium and 4 gallons of unleaded regular and paid $38.20. He forgot to record how much each type of fuel cost per gallon. Can you determine these values?

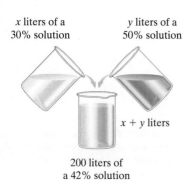

x liters of a 30% solution   *y* liters of a 50% solution

x + *y* liters

200 liters of a 42% solution

EXAMPLE 3   A lab technician is required to prepare 200 liters of a solution. The prepared solution must contain 42% fungicide. The technician wishes to combine a solution that contains 30% fungicide with a solution that contains 50% fungicide. How much of each solution should he use?

Solution   *Understand the problem.* The technician needs to make one solution from two different solutions. We need to find the amount (number of liters) of each solution that will give us 200 liters of a 42% solution. The two unknowns are easy to identify.

Let x = the number of liters of the 30% solution needed and
y = the number of liters of the 50% solution needed.

Now, what are the two equations? We know that the 200 liters of the final solution will be made by combining x and y.

$$x + y = 200 \quad \textbf{(1)}$$

The second piece of information concerns the percent of fungicide in each solution (30% of x liters, 50% of y liters, and 42% of 200 liters). See the picture at left.

$$0.3x + 0.5y = 0.42(200) \quad \textbf{(2)}$$

Thus our system of equations is

$$x + y = 200 \quad \textbf{(1)}$$
$$0.3x + 0.5y = 84. \quad \textbf{(2)}$$

We will solve this system by the addition method.

$$
\begin{aligned}
1.0x + 1.0y &= 200 \qquad &\text{Rewrite equation } \textbf{(1)} \text{ in an equivalent form.}\\
-0.6x - 1.0y &= -168 \qquad &\text{Multiply equation } \textbf{(2)} \text{ by } -2.\\
\hline
0.4x &= 32 \qquad &\text{Solve for } x.\\
\frac{0.4x}{0.4} &= \frac{32}{0.4}\\
x &= 80\\
80 + y &= 200 \qquad &\text{Substitute the value } x = 80 \text{ into one of the}\\
y &= 120 \qquad &\text{original equations and solve for } y.
\end{aligned}
$$

The technician should use 80 liters of the 30% fungicide and 120 liters of the 50% fungicide to obtain the required solution.

Check. Do the amounts total 200 liters?

$$80 + 120 = 200 \quad \checkmark$$

Do the amounts yield a 42% strength mixture?

$$0.30x + 0.50y = 0.42(200)$$
$$0.30(80) + 0.50(120) \stackrel{?}{=} 84$$
$$24 + 60 = 84 \quad \checkmark$$

Practice Problem 3   Dwayne Laboratories needs to ship 4000 liters of H_2SO_4 (sulfuric acid). Dwayne keeps in stock solutions that are 20% H_2SO_4 and 80% H_2SO_4. If the strength of the solution shipped is to be 65%, how should Dwayne's stock be mixed?

EXAMPLE 4 Mike recently rode his boat on Lazy River. He took a 48-mile trip up the river traveling against the current in exactly 3 hours. He refueled and made the return trip in exactly 2 hours. What was the speed of his boat in still water and the speed of the current in the river?

Solution   Let $x =$ the speed of the boat in still water in miles/hour and

$\quad\quad\quad\quad y =$ the speed of the river current in miles/hour.

You may want to draw a picture to show the boat traveling in each direction.

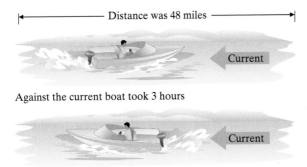

Against the current boat took 3 hours

With the current boat took 2 hours

Traveling *against* the current, the boat took 3 hours to travel 48 miles. We can calculate the rate (speed) the boat traveled upstream. Use distance = rate · time or $d = r \cdot t$.

$$48 = 3 \cdot r \quad\quad r = 16 \text{ miles/hour}$$

The boat was traveling *against* the current. Thus the speed of the boat in still water minus the speed of the current is 16.

$$x - y = 16 \quad \textbf{(1)}$$

Traveling *with* the current, the boat took 2 hours to travel 48 miles. We can calculate the rate (speed) the boat traveled downstream. Use $d = r \cdot t$.

$$48 = 2 \cdot r \quad\quad r = 24 \text{ miles/hour}$$

The boat was traveling *with* the current. Thus the speed of the boat in still water plus the speed of the current is 24.

$$x + y = 24 \quad \textbf{(2)}$$

This system is solved most easily by the addition method.

$$
\begin{array}{ll}
x - y = 16 & \textbf{(1)} \\
\underline{x + y = 24} & \textbf{(2)} \\
2x \quad\quad = 40 & \text{Add the two equations.} \\
\quad\quad x = 20 & \text{Solve for } x \text{ by dividing both sides by 2.}
\end{array}
$$

$\quad x + y = 24 \quad \textbf{(2)}$ Substitute the value of $x = 20$ into one of the original

$\quad\quad\quad\quad\quad\quad\quad\quad$ equations and solve for y. We'll use equation **(2)**.

$$20 + y = 24$$
$$\quad\quad y = 4$$

Thus the speed of Mike's boat in still water was 20 miles/hour and the speed of the current in Lazy River was 4 miles/hour.

Check. Can you verify these answers?

Practice Problem 4 Mr. Caminetti traveled downstream in his boat a distance of 72 miles in a time of 3 hours. His return trip up the river against the current took 4 hours. Find the speed of the boat in still water. Find the speed of the current in the river.

Applications *Solve using two equations with two variables.*

1. **Airline Travel** Sarah took the advertising department from her company on a round trip to Chicago to meet with a potential client. Including Sarah, a total of 14 people took the trip. She was able to purchase coach tickets for $250 and first-class tickets for $1150. She used her total budget for airfare for the trip, which was $10,700. How many first-class tickets did she buy? How many coach tickets did she buy?

2. **Baseball** Walter had to purchase tickets for 11 people in his office to go to a Red Sox game. He paid $449 to buy the 11 tickets. Infield Grandstand seats cost $44. Right Field Roof seats cost $37. How many of each kind of ticket did he purchase?

3. **Art Dealer** An art dealer sold 12 etchings for $455. He sold some of them at $35 each and the rest at $40 each. How many etchings did he sell at each price?

4. **High School Musical** El Segundo High School put on their annual musical. The students sold 650 tickets for a value of $4375. If orchestra seats cost $7.50 and balcony seats cost $3.50, how many of each kind of seat were sold?

5. **Income of a Nurse** Sharyn worked as a Registered Nurse in London last summer. The first week she was there she worked a total of 23 hours at a city hospital. She was paid $13.50 an hour for day shift and $16.50 an hour for night shift work. She earned $352.50 for her first week of work. How many hours did she spend on the day shift? How many hours did she spend on the night shift?

6. **Income of a Substitute Teacher** Tony worked as a substitute teacher last month. He substituted a total of 18 days. The first part of the month he worked in Tucson, Arizona, where he was paid $75 a day. Then he moved to Oregon, where he was paid $135 a day to substitute. During the month he earned $2010. How many days did he substitute in Tucson? How many days did he substitute in Oregon?

▲ 7. **Home Office Construction** The basement of your house is only partially finished. Currently, there is a room that you use for storage. You want to convert this space to a home office, but you want more space than is available in the room now. You decide to knock down two walls and enlarge the room. The existing room has a perimeter of 38 feet. If you extend the width of the room by 5 feet and make the new length 6 feet less than double the original length, the new perimeter will be 56 feet. What are the dimensions of the existing storage room? What are the dimensions of the new room?

existing room

enlarged room

8. ***Engineering Club*** The engineering club at MCTC ran a coffee and donut sale to raise money for the club's activities. The first week, they made a profit of $200. The second week, they sold three times as many donuts, but half the number of coffees, and made a profit of $225. If the profit for each donut was $0.50 and the profit for each coffee was $0.75, how many donuts and coffees did they sell the first week?

9. ***Automobile Radiators*** During her beginning mechanics course, Rose learned that her leaking radiator holds 16 quarts of water. She temporarily patched the leak when the radiator was partially full. At that time it was 50% antifreeze. After she filled the remaining space in the radiator with an 80% antifreeze solution and waited a little while, she found that it was 65% antifreeze. How many quarts of solution were in the radiator just before Rose added antifreeze? How many quarts of 80% antifreeze solution did she add?

10. ***Manufacturing Chocolate*** A candy company wants to produce 20 kilograms of a special 45% fat content chocolate to conform with customer dietary demands. To obtain this, a 50% fat content Hawaiian chocolate is combined with a 30% fat content domestic chocolate in a special melting process. How many kilograms of the 50% fat content chocolate and how many kilograms of the 30% fat content chocolate are used to make the required 20 kilograms?

11. ***Hiking Club*** The hiking club wants to sell 50 pounds of an energy snack food that will cost them only $1.80 per pound. The mixture will be made up of nuts that cost $2.00 per pound and raisins that cost $1.50 per pound. How much of each should they buy to obtain 50 pounds of the desired mixture?

12. ***Flower Costs*** How many bunches of local daisies that cost $2.50 per bunch should be mixed with imported daisies that cost $3.50 per bunch to obtain 100 bunches of daisies that will cost $2.90 per bunch?

13. ***Airline Travel*** An airplane traveled between two cities that are 3000 kilometers apart. The trip against the wind took 6 hours. The return trip with the benefit of the wind was 5 hours. What was the wind speed in kilometers per hour? What was the speed of the plane in still air?

14. ***Airline Travel*** An airplane flies between Boston and Cleveland. During a roundtrip flight, the plane flew a distance of 630 miles each way. The trip with a tailwind (the wind traveling the same direction as the plane) took 3 hours. The return trip traveling against the wind took 3.5 hours. Find the speed of the wind. Find the speed of the airplane in still air.

Mixed Practice *Solve the following applied problems using a system of equations.*

15. *Department Store Operations* Soledad Rivera's job was to ready the cash registers for the day's business at Foley's Department Store, in Houston. Each cash register began the day with the same amount of cash, $320. Each cash register received $25 in $1 bills and $55 in $5 bills. But from there each department had its own requirements. The Cosmetics department wanted two less than double the number of $10 bills and four fewer $20 bills than Jewelry wanted. How many $10 bills and $20 bills did these departments want?

16. *Canoe Trip* The Outdoor Adventures Club sponsored a canoe trip. The participants paddled 4 miles upstream from the canoe rental headquarters to a popular swimming and picnicking spot. They ate lunch, rested, and headed back down to where they'd started. The trip upstream took 2 hours and the trip back took only 1 hour. How fast were they paddling on average? How fast was the current?

17. *Fruit Purchase* Hank purchased 7 pounds of oranges and 5 pounds of apples and it cost him $8.78. His wife went to the store the next day and purchased for the same prices per pound a total of 3 pounds of oranges and 8 pounds of apples, for a total of $8.39. What is the price per pound of oranges? What is the price per pound of apples?

18. *Work Rates* Tom and Dave worked sorting discarded bottles for a recycle center. Tom worked for 3 hours and Dave worked for 2 hours on Saturday. They were able to sort through 2250 bottles. On Monday Tom worked for 2 hours and Dave worked for 3 hours. On Monday they sorted through 2650 bottles. How many bottles per hour can Tom sort? How many bottles per hour can Dave sort?

19. *Theater Tickets* The North Shore Community College theater department sold tickets for their annual production. General admission tickets were $8 and student admission tickets were $5.50. On opening night, 342 people attended the show. The receipts for the night totaled $2211. How many general admission tickets were sold? How many student admission tickets were sold?

20. *Airplane Flight* Walter and Mary Jones flew from Boston to Denver to visit their daughter in college. The flight of the plane was a distance of 2000 miles. It took 5 hours to fly west against the jet stream. Coming back home with the jet stream the trip only took 4 hours. What was the speed of the jet stream? What was the speed of the plane in still air?

21. *Population Growth* In 2005 the estimated population of Maryland was 5,467,000 and was projected to grow moderately. In the same year, the population of Missouri was 5,718,000 and was projected to grow at a slower rate. The relationship between the population of Maryland in thousands (y) and the number of years (x) since 2005 can be approximated by $40x - y = -5467$. A similar relationship for Missouri can be approximated by $53x - 2y = -11,447$. (*Source:* www.census.gov.) Solve the system of equations to determine the approximate year that the two states will have the same population. What will be the population of each state during that year?

22. *Population Growth* In 2005 the estimated population of Iowa was 3,040,000 and was projected to grow at a slow rate. In the same year, the population of Kansas was 3,108,000 and was projected to grow more rapidly. The relationship between the population of Iowa in thousands (y) and the number of years (x) since 2005 can be approximated by $50x - 10y = -29,410$. A similar relationship for Kansas can be approximated by $175x - 10y = -27,610$. (*Source:* www.census.gov.) Solve the system of equations to determine the approximate year that the two states will have the same population. What will be the population of each state during that year?

Cumulative Review

23. **[3.3.1]** Find the slope of the line passing through $(3, -4)$ and $(-1, -2)$.

24. **[3.3.2]** What is the slope and the y-intercept of the line defined by the equation $3x + 4y = -8$?

25. **[3.4.2]** Find an equation of the line passing through $(2, 6)$ and $(-2, 1)$. Write the equation in slope–intercept form.

26. **[3.3.3]** Find an equation of the line with slope $= -2$ and y-intercept $= (0, 10)$.

Quick Quiz 4.5

1. A college football game at Hampton University is played to a sell-out crowd of 18,500 people. Student tickets are $9 while general admission tickets are $15. A total of $217,500 was collected from ticket sales. How many people purchased student tickets? How many people purchased general admission tickets?

2. Last year the computer lab purchased 8 Dell PCs and 12 Apple iMacs. The cost of these computers was $20,800. This year the computer company said it would sell computers to the lab at the same cost per computer. This year the lab purchased 12 Dell PCs and 7 Apple iMacs for $18,000. How much did each Dell PC cost? How much did each Apple iMac cost?

3. The LaBells chartered a jet from Denver to Richmond, Virginia. The flight plan required that the plane travel exactly 1500 miles. The trip was aided by the jet stream and took only 3 hours. The return trip of 1500 miles was traveling against the jet stream. That trip took 5 hours. What is the speed of the plane in still air? What is the speed of the jet stream?

4. **Concept Check** A new company is making a drink that is 45% pure fruit juice. They make a test batch of 500 gallons of the new drink. They are using some juice that is 50% pure fruit juice and some juice that is 30% pure fruit juice. They want to find out how many gallons of each of these two types they will need. Explain how you would set up a system of two equations using the variables x and y to solve this problem.

Putting Your Skills to Work: Use Math to Save Money

ESTIMATING DISCOUNTS

There is always a sale somewhere claiming we can "save" 25%. The only true way to save money is not to spend it at all. However, when we need to purchase an item, looking for a bargain is a good idea. When we shop it helps if we can figure discounts quickly in our heads. Consider the story of Kate.

Kate needs to purchase two easy chairs. She has found the chairs at two different stores. Store A regularly sells each chair for $400, but now has the chairs on sale for 20% off. Store B sells each chair for $500, but if you buy one at full price, you get the second on sale for 20% off. In order to help her decide which store

was offering the best deal, Kate estimated the total price using the following values:

Approximate Values of Percents of Discounts

$10\% = \frac{1}{10}$	$20\% = \frac{1}{5}$	$25\% = \frac{1}{4}$	$30\% = \sim\frac{1}{3}$	$50\% = \frac{1}{2}$

1. Use the estimates to determine which store is offering the better buy.

2. Use the table to estimate how much you would save on a $28 shirt if it were reduced by each of the following percentages.
   (a) 10%
   (b) 20%
   (c) 30%

3. Use the table to find approximately how much you would save on a $62 pair of pants if they are on sale at
   (a) 20% off.
   (b) 25% off.
   (c) 30% off.

4. Amanda was shopping at a department store offering 20% off everything in the store. Use the table to estimate the amount Amanda would save if she purchased a lamp for $52, a rug for $36, and two pillows for $15 each.

5. Henry was shopping for shoes at the same department store as Amanda. He found a pair that he liked for $95. He had found a coupon in the newspaper for an additional 10% off one item in the store. The cashier informs Henry that first 20% will be taken off the original price. Then, an additional 10% will be taken off at the cash register to get the final sale price. Estimate Henry's savings.

6. Murray needs to buy six drapes for his living room. Store A sells the drapes for $40 each and is having a 25%-off sale. Store B sells the drapes for $30 and for every two drapes bought at full price, the third is 10% off. Which store is offering the better buy?

7. What shortcuts do you use to estimate without using a calculator?

Topic	Procedure	Examples
Solving a system of equations by graphing, pp. 269–271.	Graph both equations on the same coordinate axis. One of the following will be true. **1.** The lines intersect. The point of intersection is the solution to the system. Such a system is consistent.	**1.**
	2. The lines are parallel. There is no point of intersection. The system has no solution. Such a system is inconsistent.	**2.**
	3. The lines coincide. Every point on the line represents a solution. There is an infinite number of solutions. Such a system has dependent equations.	**3.**
Solving a system of equations: substitution method, p. 277.	**1.** Solve for one variable in terms of the other variable in one of the two equations. **2.** Substitute the expression you obtain for this variable into the other equation. **3.** Solve this equation to find the value of the variable. **4.** Substitute this value for the variable into one of the equations to obtain a value for the other variable. **5.** Check the solution in both original equations.	Solve. $$2x + 3y = -6$$ $$3x - y = 13$$ Solve the second equation for y. $$-y = -3x + 13$$ $$y = 3x - 13$$ Substitute this expression into the first equation. $$2x + 3(3x - 13) = -6$$ $$2x + 9x - 39 = -6$$ $$11x - 39 = -6$$ $$11x = 33$$ $$x = 3$$ Substitute $x = 3$ into $y = 3x - 13$. $$y = 3(3) - 13 = 9 - 13 = -4$$ *Check.* $$2(3) + 3(-4) \stackrel{?}{=} -6 \qquad 3(3) - (-4) \stackrel{?}{=} 13$$ $$6 - 12 \stackrel{?}{=} -6 \qquad 9 - (-4) \stackrel{?}{=} 13$$ $$-6 = -6 \ \checkmark \qquad 9 + 4 = 13$$ $$13 = 13 \ \checkmark$$ The solution is $(3, -4)$.

(Continued on next page)

Topic	Procedure	Examples
Solving a system of equations: addition method, p. 283.	1. Multiply each term of one or both equations by some nonzero integer so that the coefficients of one of the variables are opposites. 2. Add the equations of this new system so that one variable is eliminated. 3. Solve the resulting equation for the remaining variable. 4. Substitute the value found into either of the original equations to find the value of the other variable. 5. Check your solution in both of the original equations.	Solve. $$3x - 2y = 0 \quad \textbf{(1)}$$ $$-4x + 3y = 1 \quad \textbf{(2)}$$ Eliminate the x's: a common multiple of 3 and 4 is 12. Multiply equation **(1)** by 4 and equation **(2)** by 3; then add. $$\begin{aligned} 12x - 8y &= 0 \\ \underline{-12x + 9y} &= \underline{3} \\ y &= 3 \end{aligned}$$ Substitute $y = 3$ into either equation and solve for x. Let us pick $3x - 2y = 0$. $$3x - 2(3) = 0 \qquad 3x - 6 = 0 \qquad x = 2$$ *Check.* $$3(2) - 2(3) \stackrel{?}{=} 0 \qquad\qquad -4(2) + 3(3) \stackrel{?}{=} 1$$ $$6 - 6 \stackrel{?}{=} 0 \qquad\qquad\qquad -8 + 9 \stackrel{?}{=} 1$$ $$0 = 0 \ \checkmark \qquad\qquad\qquad\qquad 1 = 1 \ \checkmark$$ The solution is $(2, 3)$.
Choosing a method to solve a system of equations, p. 291.	1. Substitution works well if at least one variable has a coefficient of 1 or −1. 2. Addition works well for integer, fractional, or decimal coefficients.	Use substitution. $$\begin{aligned} x + y &= 8 \\ 2x - 3y &= 9 \end{aligned}$$ Use addition. $$\frac{3}{5}x - \frac{1}{4}y = 4$$ $$\frac{1}{5}x + \frac{3}{4}y = 8$$
Identifying inconsistent and dependent systems algebraically, p. 292.	1. If you obtain an equation that is inconsistent with known facts, such as $0 = 2$, the system is inconsistent. There is no solution. 2. If you obtain an equation that is always true, such as $0 = 0$, the equations are dependent. There is an infinite number of solutions.	Solve. $$\begin{aligned} x + 2y &= 1 \\ 3x + 6y &= 12 \end{aligned}$$ Multiply the first equation by −3. $$\begin{aligned} -3x - 6y &= -3 \\ \underline{3x + 6y} &= \underline{12} \\ 0 &= 9 \end{aligned}$$ 0 is not equal to 9, so there is *no solution*. The system is inconsistent. Solve. $$\begin{aligned} 2x + y &= 1 \\ 6x + 3y &= 3 \end{aligned}$$ Multiply the first equation by −3. $$\begin{aligned} -6x - 3y &= -3 \\ \underline{6x + 3y} &= \underline{3} \\ 0 &= 0 \end{aligned}$$ Since 0 is always equal to 0, the equations are dependent. There is an infinite number of solutions.

Topic	Procedure	Examples
Solving word problems using systems of equations, p. 296.	1. Understand the problem. Choose a variable to represent each unknown quantity. 2. Write a system of equations in two variables. 3. Solve the system of equations and state the answer. 4. Check the answer.	Apples sell for $0.35 per pound. Oranges sell for $0.40 per pound. Nancy bought 12 pounds of apples and oranges for $4.45. How many pounds of each fruit did she buy? Let x = number of pounds of apples and y = number of pounds of oranges. 12 pounds in all were purchased. $x + y = 12$ The purchase cost $4.45. $0.35x + 0.40y = 4.45$ Multiply the second equation by 100 to obtain the following system. $$\begin{aligned} x + y &= 12 \\ 35x + 40y &= 445 \end{aligned}$$ Multiply the first equation by -35 and add the equations. $$\begin{aligned} -35x - 35y &= -420 \\ \underline{35x + 40y} &= \underline{445} \\ 5y &= 25 \\ y &= 5 \end{aligned}$$ Substitute $y = 5$ into $x + y = 12$. $$\begin{aligned} x + 5 &= 12 \\ x &= 7 \end{aligned}$$ Nancy purchased 7 pounds of apples and 5 pounds of oranges. *Check.* Are there 12 pounds of apples and oranges? $7 + 5 = 12$ ✓ Would this purchase cost $4.45? $$0.35(7) + 0.40(5) \stackrel{?}{=} 4.45$$ $$2.45 + 2.00 \stackrel{?}{=} 4.45$$ $$4.45 = 4.45 \text{ ✓}$$

Chapter 4 Review Problems

Section 4.1

Solve by graphing the lines and finding the point of intersection.

1. $2x + 3y = 0$
$-x + 3y = 9$

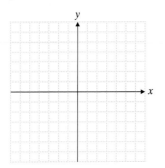

2. $-3x + y = -2$
$-2x - y = -8$

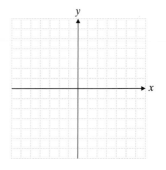

3. $2x - y = 6$
$6x + 3y = 6$

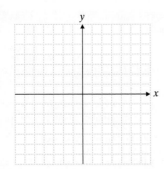

4. $2x - y = 1$
$3x + y = -6$

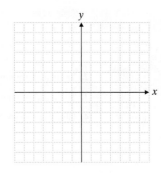

Section 4.2

Solve by the substitution method. Check your solution.

5. $x + y = 6$
$-2x + y = -3$

6. $x + 3y = 18$
$2x + y = 11$

7. $3x - 2y = 3$
$x - \dfrac{1}{3}y = 8$

8. $0.5x + y = 16$
$4x - 2y = 8$

Section 4.3

Solve by the addition method. Check your solution.

9. $6x - 2y = 10$
$2x + 3y = 7$

10. $4x - 5y = -4$
$-3x + 2y = 3$

11. $5x - 4y = 2$
$-3x + 10y = -5$

12. $9x + 2y = -5$
$12x - y = 8$

Section 4.4

Solve by any appropriate method. If it is not possible to obtain one solution, state the reason. Write the solution in the form (x, y).

13. $6x - y = 33$
$6x + 7y = 9$

14. $2x + 5y = 29$
$-3x + 10y = -26$

15. $7x + 3y = 2$
$-8x - 7y = 2$

16. $7x + 5y = 7$
$-3x - 4y = 1$

17. $4x + 5y = 4$
$-5x - 8y = 2$

18. $7x - 2y = 1$
$-5x + 3y = -7$

19. $2x - 4y = 6$
$-3x + 6y = 7$

20. $4x - 7y = 8$
$5x + 9y = 81$

21. $2x - 9y = 0$
$3x + 5 = 6y$

22. $2x + 10y = 1$
$-4x - 20y = -2$

23. $5x - 8y = 3x + 12$
$7x + y = 6y - 4$

24. $1 + x - y = y + 4$
$4(x - y) = 3 - x$

25. $3x + y = 9$
$x - 2y = 10$

26. $x + 12y = 0$
$3x - 5y = -2$

27. $2(x + 3) = y + 4$
$4x - 2y = -4$

28. $x + y = 3000$
$x - 2y = -120$

Mixed Practice

Solve by any appropriate method. If it is not possible to obtain the solution, state the reason.

29. $5x + 4y + 3 = 23$
$8x - 3y - 4 = 75$

30. $4x - 3y + 1 = 6$
$5x + 8y + 2 = -74$

31. $\dfrac{2x}{3} - \dfrac{3y}{4} = \dfrac{7}{12}$
$8x + 5y = 9$

32. $\dfrac{x}{2} + \dfrac{y}{5} = 4$
$\dfrac{x}{3} + \dfrac{y}{5} = \dfrac{10}{3}$

33. $\dfrac{1}{5}a + \dfrac{1}{2}b = 6$
$\dfrac{3}{5}a - \dfrac{1}{2}b = 2$

34. $\dfrac{2}{3}a + \dfrac{3}{5}b = -17$
$\dfrac{1}{2}a - \dfrac{1}{3}b = -1$

35. $4.8 + 0.6m = 0.9n$
$0.2m - 0.3n = 1.6$

36. $8.4 - 0.8m = 0.4n$
$0.2m + 0.1n = 2.1$

37. $6s - 4t = 5$
$4s - 5t = -2$

38. $10s + 3t = 4$
$4s - 2t = -5$

39. $3(x + 2) = -2 - (x + 3y)$
$3(x + y) = 3 - 2(y - 1)$

40. $13 - x = 3(x + y) + 1$
$14 + 2x = 5(x + y) + 3x$

41. $0.2b = 1.4 - 0.3a$
$0.1b + 0.6 = 0.5a$

42. $0.3a = 1.1 - 0.2b$
$0.3b = 0.4a - 0.9$

43. $\dfrac{b}{5} = \dfrac{2}{5} - \dfrac{a-3}{2}$
$4(a - b) = 3b - 2(a - 2)$

44. $9(b + 4) = 2(2a + 5b)$
$\dfrac{b}{5} + \dfrac{a}{2} = \dfrac{18}{5}$

Section 4.5

Solve.

45. *Museum of Modern Art* Admission to the Museum of Modern Art in New York City is $20 for adults and $12 for students. One Sunday, a total of 186 adults and students visited the museum. The receipts for that day totaled $3240. How many adults visited the museum that day? How many students?

46. *Fruit Packaging* A farmer near Napa has 252 pounds of apples to take to the farmer's market. He packages them into 2-lb and 5-lb bags to sell. If the number of 5-lb bags he used was double the number of 2-lb bags, how many of each type of bags did he prepare to bring to market?

47. *Airplane Travel* A plane travels 1500 miles in 5 hours with the benefit of a tailwind. On the return trip it requires 6 hours to fly against the wind. Can you find the speed of the plane in still air and the wind speed?

48. *Chemical Solutions* A chemist has 20% and 30% acid solutions and needs 40 liters of a 25% acid solution. How much of each solution should he mix to get the 25% solution?

49. *Scout Tree Sale* A boy scout troop is selling Christmas trees to raise money for a trip. On the first day they sold 79 trees. The balsam firs sold for $23 and the Norwegian pines were $28. The receipts for the day totaled $1942. How many trees of each type were sold?

50. *Sanding Roads in Maine* Carl manages a highway department garage in Maine. A mixture of salt and sand is stored in the garage for use during the winter. Carl needs 24 tons of a salt/sand mixture that is 25% salt. He will combine shipments of 15% salt and shipments of 30% salt to achieve the desired 24 tons. How much of each type should he use?

51. *Water Ski Boat* A water ski boat traveled 23 kilometers/hour going with the current. It went back in the opposite direction against the current and traveled only 15 kilometers/hour. What was the speed of the boat? What was the speed of the current?

52. *Manufacturing Costs* Fred is analyzing the cost of producing two different items at an electronics company. An electrical sensing device uses 5 grams of copper and requires 3 hours to assemble. A smaller sensing device made by the same company uses 4 grams of copper but requires 5 hours to assemble. The first device has a production cost of $27. The second device has a production cost of $32. How much does it cost the company for a gram of this type of copper? What is the hourly labor cost at this company? (Assume that production cost is obtained by adding the copper cost and the labor cost.)

Use the following information for exercises 53 and 54.

Toronto Blue Jays 2007 Season Ticket Prices

In-the-Action Seats	$195	Field Level Bases	$41
Premium Dugout Level	$59	Level Outfield	$23
Field Level Infield	$57	Upper Skydeck	$9

53. Connie and Dan Lacorazza took the youth group from their church to a Toronto Blue Jays game. They purchased 25 tickets. Some of the tickets were Level Outfield seats, while the rest were Upper Skydeck seats. The cost of all the tickets was $379. How many of each type of ticket did they buy?

54. Russ and Norma Camp took a group of senior citizens to a Toronto Blue Jays game. They purchased 22 tickets. Some of the tickets were Field Level Bases seats, while the rest were Field Level Infield seats. The cost of all the tickets was $1062. How many of each type of ticket did they buy?

How Am I Doing? Chapter 4 Test

Remember to use your Chapter Test Prep Video CD to see the worked-out solutions to the test problems you want to review.

Solve by the method specified.

1. Substitution method

$$3x - y = -5$$
$$-2x + 5y = -14$$

2. Addition method

$$3x + 4y = 7$$
$$2x + 3y = 6$$

3. Graphing method

$$2x - y = 4$$
$$4x + y = 2$$

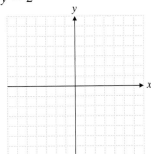

4. Any method

$$x + 3y = 12$$
$$2x - 4y = 4$$

Solve by any method. If there is not one solution to a system, state why.

5. $2x - y = 5$
   $-x + 3y = 5$

6. $2x + 3y = 13$
   $3x - 5y = 10$

7. $\frac{2}{3}x - \frac{1}{5}y = 2$
   $\frac{4}{3}x + 4y = 4$

8. $3x - 6y = 5$
   $-\frac{1}{2}x + y = \frac{7}{2}$

9. $5x - 2 = y$
   $10x = 4 + 2y$

10. $0.3x + 0.2y = 0$
   $1.0x + 0.5y = -0.5$

11. $2(x + y) = 2(1 - y)$
   $5(-x + y) = 2(23 - y)$

12. $3(x - y) = 12 + 2y$
   $8(y + 1) = 6x - 7$

Solve.

13. Twice one number plus three times a second is one. Twice the second plus three times the first is nine. Find the numbers.

14. Both $8 and $12 tickets were sold for a basketball game. In all, 30,500 people paid for admission, resulting in a "gate" of $308,000. How many of each type of ticket were sold?

15. Five shirts and three pairs of slacks cost $172. Three of the same shirts and four pairs of the same slacks cost $156. How much does each shirt cost? How much is a pair of slacks?

16. You are producing a commemorative booklet for the 25th anniversary of your church. Ace Printers charges a $200 typesetting fee and $0.25 per booklet. Wong Printers charges a $250 typesetting fee but only $0.20 per booklet. How many booklets must you order for the cost to be the same? What is the cost?

17. A jet plane flew 2000 kilometers against the wind in a time of 5 hours. It re-fueled and flew back the same distance with the wind in 4 hours. Find the wind speed in kilometers per hour. Find the speed of the jet in still air in kilometers per hour.

1. _____

2. _____

3. _____

4. _____

5. _____

6. _____

7. _____

8. _____

9. _____

10. _____

11. _____

12. _____

13. _____

14. _____

15. _____

16. _____

17. _____

311

1. _____

2. _____

3. _____

4. _____

5. _____

6. _____

7. _____

8. _____

9. _____

10. _____

11. _____

12. _____

13. _____

14. _____

15. _____

16. _____

17. _____

18. _____

Approximately one-half of this test covers the content of Chapters 0–3. The remainder covers the content of Chapter 4.

1. Subtract. $5\frac{1}{2} - 3\frac{7}{8}$

2. Evaluate.

$$(2 - 3)^3 + 20 \div (-10)\left(\frac{1}{5}\right)$$

3. Evaluate. $(-2)^2 + (-3)^3$

4. Simplify. $2x - 4[x - 3(2x + 1)]$

5. Solve for y: $2x + 4y = 10$

6. Solve for x:

$$\frac{1}{4}x + 5 = \frac{1}{3}(x - 2)$$

7. Graph the line $4x - 8y = 10$. Plot at least three points.

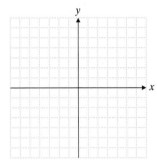

8. Find the slope of the line passing through $(6, -1)$ and $(-4, -2)$.

Solve the linear inequality and graph the solution on a number line.

9. $4x + 3 - 13x - 7 < 2(3 - 4x)$

10. Solve for x. $\dfrac{2x - 1}{3} \leq 7$

Solve by any method. If there is not one solution to a system, state why.

11. $2x + y = 8$
$3x + 4y = -8$

12. $\dfrac{1}{3}x - \dfrac{1}{2}y = 4$

$\dfrac{3}{8}x - \dfrac{1}{4}y = 7$

13. $1.3x - 0.7y = 0.4$
$-3.9x + 2.1y = -1.2$

14. $10x - 5y = 45$
$3x - 8y = 7$

Solve.

15. The equations in a system are graphed and parallel lines are obtained. What is this system called? What does that mean?

16. How much water must be added to 60 kilograms of an 80% acid solution to produce a 50% solution?

17. A boat trip 6 miles upstream takes 3 hours. The return takes 90 minutes. How fast is the stream?

18. Two printers were used for a 5-hour job of printing 15,000 labels. The next day, a second run of the same labels took 7 hours because one printer broke down after 2 hours. How many labels can each printer process per hour?

CHAPTER

5

In a fruit market, have you ever noticed a stack of oranges in a pyramid shape where each layer is square? Suppose you wanted to know how many oranges are in a pyramid with seven layers of oranges. After studying the mathematics in this chapter, you will have the knowledge to find out.

Exponents and Polynomials

Student Learning Objectives

After studying this section, you will be able to:

1 Use the product rule to multiply exponential expressions with like bases.

2 Use the quotient rule to divide exponential expressions with like bases.

3 Raise exponential expressions to a power.

1 Using the Product Rule to Multiply Exponential Expressions with Like Bases

Recall that x^2 means $x \cdot x$. That is, x appears as a factor two times. The 2 is called the **exponent.** The **base** is the variable x. The expression x^2 is called an **exponential expression.** What happens when we multiply $x^2 \cdot x^2$? Is there a pattern that will help us form a general rule?

$$(2^2)(2^3) = \overbrace{(2 \cdot 2)(2 \cdot 2 \cdot 2)}^{5 \text{ twos}} = 2^5$$

The exponent means 2 occurs 5 times as a factor.

$$(3^3)(3^4) = \overbrace{(3 \cdot 3 \cdot 3)(3 \cdot 3 \cdot 3 \cdot 3)}^{7 \text{ threes}} = 3^7$$

Notice that $3 + 4 = 7$.

$$(x^3)(x^5) = \overbrace{(x \cdot x \cdot x)(x \cdot x \cdot x \cdot x \cdot x)}^{8 \text{ } x\text{'s}} = x^8$$

The sum of the exponents is $3 + 5 = 8$.

$$(y^4)(y^2) = \overbrace{(y \cdot y \cdot y \cdot y)(y \cdot y)}^{6 \text{ } y\text{'s}} = y^6$$

The sum of the exponents is $4 + 2 = 6$.

We can state the pattern in words and then use variables.

> **THE PRODUCT RULE**
>
> To multiply two exponential expressions that have the same base, keep the base and *add the exponents.*
>
> $$x^a \cdot x^b = x^{a+b}$$

Be sure to notice that this rule applies only to expressions that have the *same base.* Here x represents the base, while the letters a and b represent the exponents that are added.

It is important that you apply this rule even when an exponent is 1. Every variable that does not have a written exponent is understood to have an exponent of 1. Thus $x^1 = x$, $y^1 = y$, and so on.

EXAMPLE 1 Multiply. **(a)** $x^3 \cdot x^6$ **(b)** $x \cdot x^5$

Solution

(a) $x^3 \cdot x^6 = x^{3+6} = x^9$

(b) $x \cdot x^5 = x^{1+5} = x^6$ Note that the exponent of the first x is 1.

Practice Problem 1 Multiply. **(a)** $a^7 \cdot a^5$ **(b)** $w^{10} \cdot w$

EXAMPLE 2 Simplify. **(a)** $y^5 \cdot y^{11}$ **(b)** $2^3 \cdot 2^5$ **(c)** $x^6 \cdot y^8$

Solution

(a) $y^5 \cdot y^{11} = y^{5+11} = y^{16}$

(b) $2^3 \cdot 2^5 = 2^{3+5} = 2^8$ Note that the base does not change! Only the exponent changes.

(c) $x^6 \cdot y^8$ The rule for multiplying exponential expressions does not apply since the bases are not the same. This cannot be simplified.

Practice Problem 2 Simplify, if possible.

(a) $x^3 \cdot x^9$ **(b)** $3^7 \cdot 3^4$ **(c)** $a^3 \cdot b^2$

We can now look at multiplying expressions such as $(2x^5)(3x^6)$.

The number 2 in $2x^5$ is called the **numerical coefficient.** Recall that a numerical coefficient is a number that is multiplied by a variable. When we multiply two expressions such as $2x^5$ and $3x^6$, we first multiply the numerical coefficients; we multiply the variables with exponents separately.

As you do the following problems, keep in mind the rule for multiplying expressions with exponents and the rules for multiplying signed numbers.

EXAMPLE 3   Multiply.

(a) $(2x^5)(3x^6)$ **(b)** $(5x^3)(x^6)$ **(c)** $(-6x)(-4x^5)$

Solution

(a) $(2x^5)(3x^6) = (2 \cdot 3)(x^5 \cdot x^6)$ Multiply the numerical coefficients.

$\qquad\qquad = 6(x^5 \cdot x^6)$ Use the rule for multiplying expressions with exponents. Add the exponents.

$\qquad\qquad = 6x^{11}$

(b) Every variable that does not have a visible numerical coefficient is understood to have a numerical coefficient of 1. Thus x^6 has a numerical coefficient of 1.

$$(5x^3)(x^6) = (5 \cdot 1)(x^3 \cdot x^6) = 5x^9$$

(c) $(-6x)(-4x^5) = (-6)(-4)(x^1 \cdot x^5) = 24x^6$ Remember that x has an exponent of 1.

Practice Problem 3   Multiply.

(a) $(-a^8)(a^4)$ **(b)** $(3y^2)(-2y^3)$ **(c)** $(-4x^3)(-5x^2)$

Problems of this type may involve more than one variable or more than two factors.

EXAMPLE 4   Multiply. $(5ab)\left(-\frac{1}{3}a\right)(9b^2)$

Solution $(5ab)\left(-\frac{1}{3}a\right)(9b^2) = (5)\left(-\frac{1}{3}\right)(9)(a \cdot a)(b \cdot b^2)$

$\qquad\qquad\qquad\qquad\qquad = -15a^2b^3$

Practice Problem 4   Multiply.

$$(2xy)\left(-\frac{1}{4}x^2y\right)(6xy^3)$$

Using the Quotient Rule to Divide Exponential Expressions with Like Bases

Frequently, we must divide exponential expressions. Because division by zero is undefined, in all problems in this chapter we assume that the denominator of any variable expression is not zero. We'll look at division in three separate parts.

Suppose that we want to simplify $x^5 \div x^2$. We could do the division the long way.

$$\frac{x^5}{x^2} = \frac{(x)(x)(x)\cancel{(x)}\cancel{(x)}}{\cancel{(x)}\cancel{(x)}} = x^3$$

Here we are using the arithmetical property of reducing fractions (see Section 0.1). When the same factor appears in both numerator and denominator, that factor can be removed.

A simpler way is to *subtract the exponents*. Notice that the base remains the same.

THE QUOTIENT RULE (PART 1)

$\dfrac{x^a}{x^b} = x^{a-b}$ Use this form if the larger exponent is in the numerator and $x \neq 0$.

EXAMPLE 5 Divide.

(a) $\dfrac{2^{16}}{2^{11}}$ (b) $\dfrac{x^5}{x^3}$ (c) $\dfrac{y^{16}}{y^7}$

Solution

(a) $\dfrac{2^{16}}{2^{11}} = 2^{16-11} = 2^5$ Note that the base does *not* change.

(b) $\dfrac{x^5}{x^3} = x^{5-3} = x^2$ (c) $\dfrac{y^{16}}{y^7} = y^{16-7} = y^9$

Practice Problem 5 Divide.

(a) $\dfrac{10^{13}}{10^7}$ (b) $\dfrac{x^{11}}{x}$ (c) $\dfrac{y^{18}}{y^8}$

NOTE TO STUDENT: *Fully worked-out solutions to all of the Practice Problems can be found at the back of the text starting at page SP-1*

Now we consider the situation where the larger exponent is in the denominator. Suppose that we want to simplify $x^2 \div x^5$.

$$\frac{x^2}{x^5} = \frac{\cancel{(x)}\cancel{(x)}}{\cancel{(x)}\cancel{(x)}(x)(x)(x)} = \frac{1}{x^3}$$

THE QUOTIENT RULE (PART 2)

$\dfrac{x^a}{x^b} = \dfrac{1}{x^{b-a}}$ Use this form if the larger exponent is in the denominator and $x \neq 0$.

EXAMPLE 6 Divide.

(a) $\dfrac{12^{17}}{12^{20}}$ (b) $\dfrac{b^7}{b^9}$ (c) $\dfrac{x^{20}}{x^{24}}$

Solution

(a) $\dfrac{12^{17}}{12^{20}} = \dfrac{1}{12^{20-17}} = \dfrac{1}{12^3}$ Note that the base does *not* change.

(b) $\dfrac{b^7}{b^9} = \dfrac{1}{b^{9-7}} = \dfrac{1}{b^2}$ (c) $\dfrac{x^{20}}{x^{24}} = \dfrac{1}{x^{24-20}} = \dfrac{1}{x^4}$

Practice Problem 6 Divide.

(a) $\dfrac{c^3}{c^4}$ (b) $\dfrac{10^{31}}{10^{56}}$ (c) $\dfrac{z^{15}}{z^{21}}$

When there are numerical coefficients, use the rules for dividing signed numbers to reduce fractions to lowest terms.

EXAMPLE 7 Divide.

(a) $\dfrac{5x^5}{25x^7}$ (b) $\dfrac{-12x^8}{4x^3}$ (c) $\dfrac{-16x^7}{-24x^8}$

Solution

(a) $\dfrac{5x^5}{25x^7} = \dfrac{1}{5x^{7-5}} = \dfrac{1}{5x^2}$ (b) $\dfrac{-12x^8}{4x^3} = -3x^{8-3} = -3x^5$ (c) $\dfrac{-16x^7}{-24x^8} = \dfrac{2}{3x^{8-7}} = \dfrac{2}{3x}$

Practice Problem 7 Divide.

(a) $\dfrac{-7x^7}{-21x^9}$ (b) $\dfrac{15x^{11}}{-3x^4}$ (c) $\dfrac{23x^8}{46x^9}$

You have to work very carefully if two or more variables are involved. Treat the coefficients and each variable separately.

EXAMPLE 8 Divide.

(a) $\dfrac{x^3 y^2}{5x y^6}$ (b) $\dfrac{-3x^2 y^5}{12x^6 y^8}$

Solution

(a) $\dfrac{x^3 y^2}{5x y^6} = \dfrac{x^2}{5y^4}$ (b) $\dfrac{-3x^2 y^5}{12x^6 y^8} = -\dfrac{1}{4x^4 y^3}$

Practice Problem 8 Divide.

(a) $\dfrac{x^7 y^9}{y^{10}}$ (b) $\dfrac{12x^5 y^6}{-24x^3 y^8}$

Suppose that a given base appears with the same exponent in the numerator and denominator of a fraction. In this case we can use the fact that *any nonzero number divided by itself is* 1.

EXAMPLE 9 Divide.

(a) $\dfrac{x^6}{x^6}$ (b) $\dfrac{3x^5}{x^5}$

Solution

(a) $\dfrac{x^6}{x^6} = 1$ (b) $\dfrac{3x^5}{x^5} = 3\left(\dfrac{x^5}{x^5}\right) = 3(1) = 3$

Practice Problem 9 Divide.

(a) $\dfrac{10^7}{10^7}$ (b) $\dfrac{12a^4}{15a^4}$

Do you see that if we had subtracted exponents when simplifying $\dfrac{x^6}{x^6}$ we would have obtained x^0 in Example 9? So we can surmise that any number (except 0) to the 0 power equals 1. We can write this fact as a separate rule.

THE QUOTIENT RULE (PART 3)

$$\frac{x^a}{x^a} = x^0 = 1 \qquad \text{if } x \neq 0 \qquad (0^0 \text{ remains undefined}).$$

TO THINK ABOUT: What Is 0 to the 0 Power? What about 0^0? Why is it undefined? $0^0 = 0^{1-1}$. If we use the quotient rule, $0^{1-1} = \dfrac{0}{0}$. Since division by zero is undefined, we must agree that 0^0 is undefined.

EXAMPLE 10 Divide.

(a) $\dfrac{4x^0y^2}{8^0y^5z^3}$

(b) $\dfrac{5x^2y}{10x^2y^3}$

Solution

(a) $\dfrac{4x^0y^2}{8^0y^5z^3} = \dfrac{4(1)y^2}{(1)y^5z^3} = \dfrac{4y^2}{y^5z^3} = \dfrac{4}{y^3z^3}$

(b) $\dfrac{5x^2y}{10x^2y^3} = \dfrac{1x^0}{2y^2} = \dfrac{(1)(1)}{2y^2} = \dfrac{1}{2y^2}$

Practice Problem 10 Divide.

(a) $\dfrac{-20a^3b^8c^4}{28a^3b^7c^5}$

(b) $\dfrac{5x^0y^6}{10x^4y^8}$

We can combine all three parts of the quotient rule we have developed.

THE QUOTIENT RULE

$\dfrac{x^a}{x^b} = x^{a-b}$ Use this form if the larger exponent is in the numerator and $x \neq 0$.

$\dfrac{x^a}{x^b} = \dfrac{1}{x^{b-a}}$ Use this form if the larger exponent is in the denominator and $x \neq 0$.

$\dfrac{x^a}{x^a} = x^0 = 1$ if $x \neq 0$.

We can combine the product rule and the quotient rule to simplify algebraic expressions that involve both multiplication and division.

EXAMPLE 11 Simplify. $\dfrac{(8x^2y)(-3x^3y^2)}{-6x^4y^3}$

Solution $\dfrac{(8x^2y)(-3x^3y^2)}{-6x^4y^3} = \dfrac{-24x^5y^3}{-6x^4y^3} = 4x$

Practice Problem 11   Simplify. $\dfrac{(-6ab^5)(3a^2b^4)}{16a^5b^7}$

Raising Exponential Expressions to a Power

How do we simplify an expression such as $(x^4)^3$? $(x^4)^3$ is x^4 raised to the third power. For this type of problem we say that we are raising a power to a power. A problem such as $(x^4)^3$ could be done by writing the following.

$$(x^4)^3 = x^4 \cdot x^4 \cdot x^4 \quad \text{By definition}$$
$$= x^{12} \qquad\qquad \text{By adding exponents}$$

Notice that when we add the exponents we get $4 + 4 + 4 = 12$. This is the same as multiplying 4 by 3. That is, $4 \cdot 3 = 12$. This process can be summarized by the following rule.

RAISING A POWER TO A POWER

To raise a power to a power, keep the same base and multiply the exponents.

$$(x^a)^b = x^{ab}$$

Recall what happens when you raise a negative number to a power. $(-1)^2 = 1$. $(-1)^3 = -1$. In general,

$$(-1)^n = \begin{cases} +1 & \text{if } n \text{ is even} \\ -1 & \text{if } n \text{ is odd.} \end{cases}$$

EXAMPLE 12 Simplify.

(a) $(x^3)^5$ 　　　　　　**(b)** $(2^7)^3$ 　　　　　　**(c)** $(-1)^8$

Solution

(a) $(x^3)^5 = x^{3 \cdot 5} = x^{15}$ 　　**(b)** $(2^7)^3 = 2^{7 \cdot 3} = 2^{21}$ 　　**(c)** $(-1)^8 = +1$

Note that in both parts (a) and (b) the base does not change.

Practice Problem 12 Simplify.

(a) $(a^4)^3$ 　　　　　　**(b)** $(10^5)^2$ 　　　　　　**(c)** $(-1)^{15}$

Here are two rules involving products and quotients that are very useful: the product raised to a power rule and the quotient raised to a power rule. We'll illustrate each with an example.

If a product in parentheses is raised to a power, the parentheses indicate that *each factor* must be raised to that power.

$$(xy)^2 = x^2y^2 \qquad (xy)^3 = x^3y^3$$

PRODUCT RAISED TO A POWER

$$(xy)^a = x^ay^a$$

EXAMPLE 13 Simplify.

(a) $(ab)^8$ **(b)** $(3x)^4$ **(c)** $(-2x^2)^3$

Solution

(a) $(ab)^8 = a^8 b^8$ **(b)** $(3x)^4 = (3)^4 x^4 = 81x^4$

(c) $(-2x^2)^3 = (-2)^3 \cdot (x^2)^3 = -8x^6$

Practice Problem 13 Simplify.

(a) $(3xy)^3$ **(b)** $(yz)^{37}$ **(c)** $(-3x^3)^2$

If a fractional expression within parentheses is raised to a power, the parentheses indicate that both numerator and denominator must be raised to that power.

$$\left(\frac{x}{y}\right)^5 = \frac{x^5}{y^5} \qquad \left(\frac{x}{y}\right)^2 = \frac{x^2}{y^2} \qquad \text{if } y \neq 0$$

QUOTIENT RAISED TO A POWER

$$\left(\frac{x}{y}\right)^a = \frac{x^a}{y^a} \qquad \text{if } y \neq 0$$

EXAMPLE 14 Simplify.

(a) $\left(\dfrac{x}{y}\right)^5$ **(b)** $\left(\dfrac{7}{w}\right)^4$

Solution

(a) $\left(\dfrac{x}{y}\right)^5 = \dfrac{x^5}{y^5}$ **(b)** $\left(\dfrac{7}{w}\right)^4 = \dfrac{7^4}{w^4} = \dfrac{2401}{w^4}$

Practice Problem 14 Simplify.

(a) $\left(\dfrac{x}{5}\right)^3$ **(b)** $\left(\dfrac{4a}{b}\right)^6$

Many expressions can be simplified by using the previous rules involving exponents. Be sure to take particular care to determine the correct sign, especially if there is a negative numerical coefficient.

EXAMPLE 15 Simplify. $\left(\dfrac{-3x^2 z^0}{y^3}\right)^4$

Solution $\left(\dfrac{-3x^2 z^0}{y^3}\right)^4 = \left(\dfrac{-3x^2}{y^3}\right)^4$ Simplify inside the parentheses first. Note that $z^0 = 1$.

$= \dfrac{(-3)^4 x^8}{y^{12}}$ Apply the rules for raising a power to a power. Notice that we wrote $(-3)^4$ and not -3^4. We are raising -3 to the fourth power.

$= \dfrac{81x^8}{y^{12}}$ Simplify the coefficient: $(-3)^4 = +81$.

NOTE TO STUDENT: *Fully worked-out solutions to all of the Practice Problems can be found at the back of the text starting at page SP-1*

Practice Problem 15

Simplify. $\left(\dfrac{-2x^3 y^0 z}{4xz^2}\right)^5$

We list here the rules of exponents we have discussed in Section 5.1.

$$x^a \cdot x^b = x^{a+b}$$

$$\frac{x^a}{x^b} = \begin{cases} x^{a-b} & \text{if} \quad a > b \\ \dfrac{1}{x^{b-a}} & \text{if} \quad b > a \\ x^0 = 1 & \text{if} \quad a = b \end{cases} \qquad x \neq 0$$

$$(x^a)^b = x^{ab}$$

$$(xy)^a = x^a y^a$$

$$\left(\frac{x}{y}\right)^a = \frac{x^a}{y^a} \qquad y \neq 0$$

Developing Your Study Skills

Why Is Review Necessary?

You master a course in mathematics by learning the concepts one step at a time. Thus understanding of mathematics is built step-by-step, with each step a supporting foundation for the next. The process is a carefully designed procedure, so no steps can be skipped. A student of mathematics needs to realize the importance of this building process to succeed.

Because new concepts depend on those previously learned, students often need to take time to review. The reviewing process will strengthen understanding and skills, which may be weak due to a lack of mastery or the passage of time. Review at the right time on the right concepts can strengthen previously learned skills and make progress possible.

Timely, periodic review of previously learned mathematical concepts is absolutely necessary for mastery of new concepts. You may have forgotten a concept or grown a bit rusty in applying it. Reviewing is the answer. Make use of the cumulative review problems in your textbook, whether they are assigned or not. Look back to previous chapters whenever you have forgotten how to do something. Review the chapter organizers from previous chapters. Study the examples and practice some exercises to refresh your understanding.

Be sure that you understand and can perform the computations of each new concept. This will enable you to move successfully on to the next one.

Remember, mathematics is a step-by-step building process. Learn each concept and reinforce and strengthen with review whenever necessary.

Verbal and Writing Skills

1. Write in your own words the product rule for exponents.

2. To be able to use the rules of exponents, what must be true of the bases?

3. If the larger exponent is in the denominator, the quotient rule states that $\dfrac{x^a}{x^b} = \dfrac{1}{x^{b-a}}$. Provide an example to show why this is true.

In exercises 4 and 5, identify the numerical coefficient, the base(s), and the exponent(s).

4. $-8x^5y^2$

5. $6x^{11}y$

6. Evaluate **(a)** $3x^0$ and **(b)** $(3x)^0$. **(c)** Why are the results different?

Write in simplest exponent form.

7. $2 \cdot 2 \cdot a \cdot a \cdot a \cdot b$

8. $5 \cdot x \cdot x \cdot x \cdot y \cdot y$

9. $(-5)(x)(y)(z)(y)(x)(x)(z)$

10. $(-8)(a)(b)(a)(a)(c)(b)$

Multiply. Leave your answer in exponent form.

11. $(7^4)(7^6)$

12. $(4^3)(4^5)$

13. $(5^{10})(5^{16})$

14. $(6^5)(6^8)$

15. $x^4 \cdot x^8$

16. $a^8 \cdot a^{12}$

17. $t^{15} \cdot t$

18. $z^{16} \cdot z^{10}$

Multiply.

19. $-5x^4(4x^2)$

20. $6x^2(-9x^3)$

21. $(5x)(10x^2)$

22. $(-4x^2)(-3x)$

23. $(2xy^3)(9x^2y^5)$

24. $(4ab)(8ab^5)$

25. $\left(\dfrac{2}{5}xy^3\right)\left(\dfrac{1}{3}x^2y^2\right)$

26. $\left(\dfrac{4}{5}x^5y\right)\left(\dfrac{15}{16}x^2y^4\right)$

27. $(1.1x^2z)(-2.5xy)$

28. $(2.3wx^4)(-3.5xy^4)$

29. $(8a)(2a^3b)(0)$

30. $(5ab)(2a^2)(0)$

31. $(-16x^2y^4)(-5xy^3)$

32. $(-12x^4y)(-7x^5y^3)$

33. $(-8x^3y^2)(3xy^5)$

34. $(9x^2y^6)(-11x^3y^3)$

35. $(-2x^3y^2)(0)(-3x^4y)$

36. $(-4x^8y^2)(13y^3)(0)$

37. $(8a^4b^3)(-3x^2y^5)$

38. $(5x^3y)(-2w^4z)$

39. $(2x^2y)(-3y^3z^2)(5xz^4)$

40. $(3ab)(5a^2c)(-2b^2c^3)$

Divide. Leave your answer in exponent form. Assume that all variables in any denominator are nonzero.

41. $\dfrac{y^{12}}{y^5}$

42. $\dfrac{x^{13}}{x^3}$

43. $\dfrac{y^5}{y^8}$

44. $\dfrac{b^{20}}{b^{23}}$

45. $\dfrac{11^{18}}{11^{30}}$

46. $\dfrac{8^9}{8^{12}}$

47. $\dfrac{2^{17}}{2^{10}}$

48. $\dfrac{7^{18}}{7^9}$

49. $\dfrac{a^{13}}{4a^5}$

50. $\dfrac{b^{16}}{5b^{13}}$

51. $\dfrac{x^7}{y^9}$

52. $\dfrac{x^{20}}{y^3}$

53. $\dfrac{48x^5y^3}{24xy^3}$

54. $\dfrac{45a^4b^3}{15a^4b^2}$

55. $\dfrac{16x^5y}{-32x^2y^3}$

56. $\dfrac{-36x^3y^7}{72x^5y}$

57. $\dfrac{1.8f^4g^3}{54f^2g^8}$

58. $\dfrac{3.1s^5t^3}{62s^8t}$

59. $\dfrac{-85x^5y^{10}}{-5xy^7}$

60. $\dfrac{-60x^6y^3}{-4x^6y}$

61. $\dfrac{8^0x^2y^3}{16x^5y}$

62. $\dfrac{3^2x^3y^7}{3^0x^5y^2}$

63. $\dfrac{18a^6b^3c^0}{24a^5b^3}$

64. $\dfrac{12a^7b^8}{16a^3b^8c^0}$

65. $\dfrac{12c^2}{42ab}$

66. $\dfrac{70a^2c^3}{40b}$

Simplify.

67. $\left(x^2\right)^6$

68. $\left(w^5\right)^8$

69. $\left(x^3y\right)^5$

70. $\left(ab^4\right)^2$

71. $\left(rs^2\right)^6$

72. $\left(m^3n^2\right)^5$

73. $\left(3a^3b^2c\right)^3$

74. $\left(2x^4yz^3\right)^2$

75. $\left(-3a^4\right)^2$

76. $\left(-2a^5\right)^4$

77. $\left(\dfrac{x}{2m^4}\right)^7$

78. $\left(\dfrac{p^5}{6x}\right)^5$

79. $\left(\dfrac{5x}{7y^2}\right)^2$

80. $\left(\dfrac{2a^4}{3b^3}\right)^4$

81. $\left(-3a^2b^3c^0\right)^4$

82. $\left(-2a^5b^2c^0\right)^5$

83. $\left(-2x^3y^0z\right)^3$

84. $\left(-4xy^0z^4\right)^3$

85. $\dfrac{(3x)^5}{(3x^2)^3}$

86. $\dfrac{(6y^3)^4}{(6y)^2}$

Mixed Practice

87. $(-5a^2b^3)^2(ab)$

88. $(-3a^3b)^3(a^2b)$

89. $\left(\dfrac{8}{y^5}\right)^2$

90. $\left(\dfrac{4}{x^6}\right)^3$

91. $\left(\dfrac{2x}{y^3}\right)^4$

Cumulative Review *Simplify.*

92. **[1.2.1]** $-3 - 8$

93. **[1.1.5]** $-17 + (-32) + (-24) + 27$

94. **[1.3.1]** $\left(-\dfrac{3}{5}\right)\left(-\dfrac{2}{15}\right)$

95. **[1.3.3]** $-\dfrac{5}{4} \div \dfrac{5}{16}$

Rain Forests *Brazil, the largest country in South America, boasts a land area of approximately 8,511,960 sq km. In 1970, the Amazon rain forest covered 4,100,000 sq km of Brazil's land area. By 2006, the rain forest in Brazil only covered 3,400,254 sq km. (Source: en.wikipedia.org.) Round your answers to the nearest tenth of a percent.*

96. **[0.5.3]** What percent of Brazil's land area was covered by rain forest in 2006?

97. **[0.5.3]** If Brazil is said to lose 2.3% of its rain forest per year, how much rain forest was lost during 2007?

Quick Quiz 5.1 Simplify the following.

1. $(2x^2y^3)(-5xy^4)$

2. $\dfrac{-28x^6y^6}{35x^3y^8}$

3. $(-3x^3y^5)^4$

4. **Concept Check** Explain the steps you would need to follow to simplify the expression.
$$\frac{(4x^3)^2}{(2x^4)^3}$$

① Using Negative Exponents

If n is an integer and $x \neq 0$, then x^{-n} is defined as follows:

Student Learning Objectives

After studying this section, you will be able to:

 Use negative exponents.

 Use scientific notation.

DEFINITION OF A NEGATIVE EXPONENT

$$x^{-n} = \frac{1}{x^n}, \quad x \neq 0$$

EXAMPLE 1 Write with positive exponents.

(a) y^{-3} **(b)** z^{-6} **(c)** w^{-1}

Solution

(a) $y^{-3} = \dfrac{1}{y^3}$ **(b)** $z^{-6} = \dfrac{1}{z^6}$ **(c)** $w^{-1} = \dfrac{1}{w^1} = \dfrac{1}{w}$

Practice Problem 1 Write with positive exponents.

(a) x^{-12} **(b)** w^{-5} **(c)** z^{-2}

NOTE TO STUDENT: Fully worked-out solutions to all of the Practice Problems can be found at the back of the text starting at page SP-1

To evaluate a numerical expression with a negative exponent, first write the expression with a positive exponent. Then simplify.

EXAMPLE 2 Evaluate.

(a) 3^{-2} **(b)** 2^{-5}

Solution

(a) $3^{-2} = \dfrac{1}{3^2} = \dfrac{1}{9}$ **(b)** $2^{-5} = \dfrac{1}{2^5} = \dfrac{1}{32}$

Practice Problem 2 Evaluate.

(a) 4^{-3} **(b)** 2^{-4}

All the previously studied laws of exponents are true for any integer exponent. These laws are summarized in the following box. Assume that $x, y \neq 0$.

LAWS OF EXPONENTS

The Product Rule

$$x^a \cdot x^b = x^{a+b}$$

The Quotient Rule

$$\frac{x^a}{x^b} = x^{a-b} \quad \text{Use if } a > b. \qquad \frac{x^a}{x^b} = \frac{1}{x^{b-a}} \quad \text{Use if } a < b.$$

Power Rules

$$(xy)^a = x^a y^a, \qquad (x^a)^b = x^{ab}, \qquad \left(\frac{x}{y}\right)^a = \frac{x^a}{y^a}$$

By using the definition of a negative exponent and the properties of fractions, we can derive two more helpful properties of exponents. Assume that $x, y \neq 0$.

PROPERTIES OF NEGATIVE EXPONENTS

$$\frac{1}{x^{-n}} = x^n \qquad \frac{x^{-m}}{y^{-n}} = \frac{y^n}{x^m}$$

EXAMPLE 3 Simplify. Write the expression with no negative exponents.

(a) $\dfrac{1}{x^{-6}}$ 　　　　(b) $\dfrac{x^{-3}y^{-2}}{z^{-4}}$ 　　　　(c) $x^{-2}y^3$

Solution

(a) $\dfrac{1}{x^{-6}} = x^6$ 　　(b) $\dfrac{x^{-3}y^{-2}}{z^{-4}} = \dfrac{z^4}{x^3y^2}$ 　　(c) $x^{-2}y^3 = \dfrac{y^3}{x^2}$

NOTE TO STUDENT: *Fully worked-out solutions to all of the Practice Problems can be found at the back of the text starting at page SP-1*

Practice Problem 3 Simplify. Write the expression with no negative exponents.

(a) $\dfrac{3}{w^{-4}}$ 　　　　(b) $\dfrac{x^{-6}y^4}{z^{-2}}$ 　　　　(c) $x^{-6}y^{-5}$

EXAMPLE 4 Simplify. Write the expression with no negative exponents.

(a) $(3x^{-4}y^2)^{-3}$ 　　　　　　　　　(b) $\dfrac{x^2y^{-4}}{x^{-5}y^3}$

Solution

(a) $(3x^{-4}y^2)^{-3} = 3^{-3}x^{12}y^{-6} = \dfrac{x^{12}}{3^3y^6} = \dfrac{x^{12}}{27y^6}$

(b) $\dfrac{x^2y^{-4}}{x^{-5}y^3} = \dfrac{x^2x^5}{y^4y^3} = \dfrac{x^7}{y^7}$ 　First rewrite the expression so that only positive exponents appear. Then simplify using the product rule.

Practice Problem 4 Simplify. Write the expression with no negative exponents.

(a) $(2x^4y^{-5})^{-2}$ 　　　　　　　　　(b) $\dfrac{y^{-3}z^{-4}}{y^2z^{-6}}$

---

2 Using Scientific Notation

One common use of negative exponents is in writing numbers in scientific notation. Scientific notation is most useful in expressing very large and very small numbers.

SCIENTIFIC NOTATION

A positive number is written in **scientific notation** if it is in the form $a \times 10^n$, where $1 \le a < 10$ and n is an integer.

EXAMPLE 5   Write in scientific notation.

(a) 4567   **(b)** 157,000,000

Solution

(a) $4567 = 4.567 \times 1000$ To change 4567 to a number that is greater than 1 but less than 10, we move the decimal point *three* places to the *left*. We must then multiply the number by a power of 10 so that $= 4.567 \times 10^3$ we do not change the value of the number. Use 1000.

Notice we moved the decimal point 3 places to the left, so we must multiply by 10^3.

(b) $157{,}000{,}000 = \underset{\text{8 places}}{1.57000000} \times \underset{\text{8 zeros}}{100000000}$

$= 1.57 \times 10^8$

Practice Problem 5   Write in scientific notation.

(a) 78,200   **(b)** 4,786,000

Numbers that are smaller than 1 will have a *negative power* of 10 if they are written in scientific notation.

EXAMPLE 6   Write in scientific notation.

(a) 0.061   **(b)** 0.000052

Solution

(a) We need to write 0.061 as a number that is greater than 1 but less than 10. In which direction do we move the decimal point?

$0.061 = 6.1 \times 10^{-2}$ Move the decimal point 2 places to the *right*.

(b) $0.000052 = 5.2 \times 10^{-5}$ Why?

Practice Problem 6   Write in scientific notation.

(a) 0.98   **(b)** 0.000092

The reverse procedure transforms scientific notation into ordinary decimal notation.

EXAMPLE 7   Write in decimal notation.   **(a)** 1.568×10^2 **(b)** 7.432×10^{-3}

Solution

(a) $1.568 \times 10^2 = 1.568 \times 100 = 156.8$

Alternative Method

$1.568 \times 10^2 = 156.8$ The exponent 2 tells us to move the decimal point 2 places to the right.

(b) $7.432 \times 10^{-3} = 7.432 \times \dfrac{1}{1000}$

$= 0.007432$

Alternative Method

$7.432 \times 10^{-3} = 0.007432$ The exponent -3 tells us to move the decimal point 3 places to the left.

Practice Problem 7   Write in decimal notation.

(a) 1.93×10^6 **(b)** 8.562×10^{-5}

Calculator

Scientific Notation

Most scientific calculators can display only eight digits at one time. Numbers with more than eight digits are usually shown in scientific notation. 1.12 E 08 or 1.12 8 means 1.12×10^8. You can use a calculator to compute with large numbers by entering the numbers using scientific notation. For example,

$$(7.48 \times 10^{24}) \times (3.5 \times 10^8)$$

is entered as follows.

7.48 | EXP | 24 | × |

3.5 | EXP | 8 | = |

Display: | 2.618 E 33 |

or | 2.618   33 |

Note: Some calculators have an | EE | key instead of | EXP |.

The distance light travels in one year is called a *light-year*. A light-year is a convenient unit of measure to use when investigating the distances between stars.

EXAMPLE 8 A light-year is a distance of 9,460,000,000,000,000 meters. Write this in scientific notation.

Solution   9,460,000,000,000,000 meters = 9.46×10^{15} meters

Practice Problem 8 Astronomers measure distances to faraway galaxies in parsecs. A parsec is a distance of 30,900,000,000,000,000 meters. Write this in scientific notation.

To perform a calculation involving very large or very small numbers, it is usually helpful to write the numbers in scientific notation and then use the laws of exponents to do the calculation.

EXAMPLE 9 Use scientific notation and the laws of exponents to find the following. Leave your answer in scientific notation.

(a) $(32,000,000)(1,500,000,000,000)$ **(b)** $\dfrac{0.00063}{0.021}$

Solution

(a) $(32,000,000)(1,500,000,000,000)$

$= (3.2 \times 10^7)(1.5 \times 10^{12})$ Write each number in scientific notation.

$= 3.2 \times 1.5 \times 10^7 \times 10^{12}$ Rearrange the order. Remember that multiplication is commutative.

$= 4.8 \times 10^{19}$ Multiply 3.2×1.5. Multiply $10^7 \times 10^{12}$.

(b) $\dfrac{0.00063}{0.021} = \dfrac{6.3 \times 10^{-4}}{2.1 \times 10^{-2}}$ Write each number in scientific notation.

$= \dfrac{6.3}{2.1} \times \dfrac{10^{-4}}{10^{-2}}$ Rearrange the order. We are actually using the definition of multiplication of fractions.

$= \dfrac{6.3}{2.1} \times \dfrac{10^2}{10^4}$ Rewrite with positive exponents.

$= 3.0 \times 10^{-2}$

Practice Problem 9 Use scientific notation and the laws of exponents to find the following. Leave your answer in scientific notation.

(a) $(56,000)(1,400,000,000)$ **(b)** $\dfrac{0.000111}{0.00000037}$

NOTE TO STUDENT: Fully worked-out solutions to all of the Practice Problems can be found at the back of the text starting at page SP-1

When we use scientific notation, we are often writing approximate numbers. We must include some zeros so that the decimal point can be properly located. However, all other digits except for these zeros are considered **significant digits.** The number 34.56 has four significant digits. The number 0.0049 has two significant digits. The zeros are considered placeholders. The number 634,000 has three significant digits (unless we have specific knowledge to the contrary). The zeros are considered placeholders. We sometimes round numbers to a specific number of significant digits. For example, 0.08746 rounded to two significant digits is 0.087. When we round 1,348,593 to three significant digits, we obtain 1,350,000.

EXAMPLE 10 The approximate distance from Earth to the star Polaris is 208 parsecs. A parsec is a distance of approximately 3.09×10^{13} kilometers. How long would it take a space probe traveling at 40,000 kilometers per hour to reach the star? Round to three significant digits.

Solution

1. *Understand the problem.* Recall that the distance formula is

$$\text{distance} = \text{rate} \times \text{time}.$$

We are given the distance and the rate. We need to find the time.
Let's take a look at the distance. The distance is given in parsecs, but the rate is given in kilometers per hour. We need to change the distance to kilometers. We are told that a parsec is approximately 3.09×10^{13} kilometers. That is, there are 3.09×10^{13} kilometers per parsec. We use this information to change 208 parsecs to kilometers.

$$208 \text{ parsecs} = \frac{(208 \text{ parsecs})(3.09 \times 10^{13} \text{ kilometers})}{1 \text{ parsec}} = 642.72 \times 10^{13} \text{ kilometers}$$

2. *Write an equation.* Use the distance formula.

$$d = r \times t$$

3. *Solve the equation and state the answer.* Substitute the known values into the formula and solve for the unknown, time.

$$642.72 \times 10^{13} \text{ km} = \frac{40,000 \text{ km}}{1 \text{ hr}} \times t$$

$$6.4272 \times 10^{15} \text{ km} = \frac{4 \times 10^4 \text{ km}}{1 \text{ hr}} \times t \qquad \text{Change the numbers to scientific notation.}$$

$$\frac{6.4272 \times 10^{15} \text{ km}}{\dfrac{4 \times 10^4 \text{ km}}{1 \text{ hr}}} = t \qquad \text{Divide both sides by } \frac{4 \times 10^4 \text{ km}}{1 \text{ hr}}.$$

$$\frac{(6.4272 \times 10^{15} \text{ km})(1 \text{ hr})}{4 \times 10^4 \text{ km}} = t$$

$$1.6068 \times 10^{11} \text{ hr} = t$$

1.6068×10^{11} is 160.68×10^9 or 160.68 billion hours. The space probe will take approximately 160.68 billion hours to reach the star.

Reread the problem. Are we finished? What is left to do? We need to round the answer to three significant digits. Rounding to three significant digits, we have

$$160.68 \times 10^9 \approx 161 \times 10^9.$$

This is approximately 161 billion hours or a little more than 18 million years.

4. *Check.* Unless you have had a great deal of experience working in astronomy, it would be difficult to determine whether this is a reasonable answer. You may wish to reread your analysis and redo your calculations as a check.

Practice Problem 10 The average distance from Earth to the distant star Betelgeuse is 159 parsecs. How many hours would it take a space probe to travel from Earth to Betelgeuse at a speed of 50,000 kilometers per hour? Round to three significant digits.

Simplify. Express your answer with positive exponents. Assume that all variables are nonzero.

1. x^{-4}

2. y^{-5}

3. 3^{-4}

4. 5^{-3}

5. $\dfrac{1}{y^{-8}}$

6. $\dfrac{1}{z^{-10}}$

7. $\dfrac{x^{-4}y^{-5}}{z^{-6}}$

8. $\dfrac{x^{-6}y^{-2}}{z^{-5}}$

9. $a^3 b^{-2}$

10. $a^5 b^{-8}$

11. $(2x^{-3})^{-3}$

12. $(4x^{-4})^{-2}$

13. $3x^{-2}$

14. $4y^{-4}$

15. $(3xy^2)^{-2}$

16. $(4x^2y^4)^{-3}$

Mixed Practice

17. $\dfrac{3xy^{-2}}{z^{-3}}$

18. $\dfrac{4x^{-2}y^{-3}}{y^4}$

19. $\dfrac{(3x)^{-2}}{(3x)^{-3}}$

20. $\dfrac{(2ab^2)^{-3}}{(2ab^2)^{-4}}$

21. $a^{-1}b^3c^{-4}d$

22. $x^{-5}y^{-2}z^3$

23. $(8^{-2})(2^3)$

24. $(9^2)(3^{-3})$

25. $\left(\dfrac{3x^0y^2}{z^4}\right)^{-2}$

26. $\left(\dfrac{2a^3b^0}{c^2}\right)^{-3}$

27. $\dfrac{x^{-2}y^{-3}}{x^4y^{-2}}$

28. $\dfrac{a^{-6}b^3}{a^{-2}b^{-5}}$

Write in scientific notation.

29. 123,780

30. 5,786,100

31. 0.063

32. 0.000742

33. 889,610,000,000

34. 7,652,000,000

35. 0.00000342

36. 0.0000000058

In exercises 37–42, write in decimal notation.

37. 3.02×10^5

38. 8.137×10^7

39. 4.7×10^{-4}

40. 6.53×10^{-3}

41. 9.83×10^5

42. 3.5×10^{-8}

43. *Bamboo Growth* The growth rate of some species of bamboo is 0.0000237 miles per hour. Write this in scientific notation.

44. *Neptune* Neptune is 2.793×10^9 miles from the sun. Write this in decimal notation.

45. *Astronomical Unit* The astronomical unit (AU) is a unit of length approximately equal to 1.496×10^8 km. Write this in decimal notation.

46. *Gold Atom* The average volume of an atom of gold is 0.0000000000000000000001695 cubic centimeters. Write this in scientific notation.

Evaluate by using scientific notation and the laws of exponents. Leave your answer in scientific notation.

47. $(42,000,000)(150,000,000)$

48. $(55,000,000,000)(16,000,000)$

49. $\dfrac{(5,000,000)(16,000)}{8,000,000,000}$

50. $(0.0075)(0.0000002)(0.001)$

51. $(0.003)^4$

52. $(500,000)^4$

53. $(150,000,000)(0.00005)(0.002)(30,000)$

54. $\dfrac{(1,600,000)(0.00003)}{2400}$

Applications

National Debt *In July 2006, the national debt was about* 8.42×10^{12} *dollars.* (*Source:* www.factfinder.census.gov.)

55. The Census Bureau estimates that in July 2006, the population of the United States was 2.99×10^8 people. If the national debt were evenly divided among every person in the country, how much debt would be assigned to each individual? Round to three significant digits.

56. The Census Bureau estimates that in July 2006, the number of people in the United States who were 18 years or older was approximately 2.26×10^8. If the national debt were evenly divided among every person 18 years or older in the country, how much would be assigned to each individual? Round to three significant digits.

57. ***Watch Hand*** The tip of a $\frac{1}{3}$-inch-long hour hand on a watch travels at a speed of 0.00000275 miles per hour. How far has it traveled in a day?

58. ***Neutron Mass*** The mass of a neutron is approximately 1.675×10^{-27} kilogram. Find the mass of 180,000 neutrons.

59. ***Mission to Pluto*** In January 2006, a spacecraft called *New Horizons* began its journey to Pluto. The trip is 3.5×10^9 miles long and will take $9\frac{1}{2}$ years. How many miles will *New Horizons* travel per year? (*Source:* www.pluto.jhuapl.edu.)

60. ***Moles per Molecule*** Avogadro's number says that there are approximately 6.02×10^{23} molecules/mole. How many molecules can one expect in 0.00483 mole?

61. ***Construction Cost*** In 1990 the cost for construction of new private buildings was estimated at $\$3.61 \times 10^{11}$. By 2010 the estimated cost for construction of new private buildings will be $\$9.36 \times 10^{11}$. What is the percent of increase from 1990 to 2010? Round to the nearest tenth of a percent. (*Source:* U.S. Census Bureau.)

62. ***Construction Cost*** In 1990 the cost for construction of new public buildings was estimated at $\$1.07 \times 10^{11}$. By 2010 the estimated cost for construction of new public buildings will be $\$3.06 \times 10^{11}$. What is the percent of increase from 1990 to 2010? Round to the nearest tenth of a percent. (*Source:* U.S. Census Bureau.)

Cumulative Review *Simplify.*

63. [1.2.1] $-2.7 - (-1.9)$

64. [1.4.2] $(-1)^{33}$

65. [1.1.4] $-\dfrac{3}{4} + \dfrac{5}{7}$

Quick Quiz 5.2 Simplify and write your answer with only positive exponents.

1. $3x^{-3}y^2z^{-4}$

2. $\dfrac{4a^3b^{-4}}{8a^{-5}b^{-3}}$

3. Write in scientific notation. 0.00876

4. **Concept Check** Explain how you would simplify a problem like the following so that your answer had only positive exponents.

$$(4x^{-3}y^4)^{-3}$$

1 Recognizing Polynomials and Determining Their Degrees

A **polynomial** in x is the sum of a finite number of terms of the form ax^n, where a is any real number and n is a whole number. Usually these polynomials are written in descending powers of the variable, as in

$$5x^3 + 3x^2 - 2x - 5 \quad \text{and} \quad 3.2x^2 - 1.4x + 5.6.$$

A **multivariable polynomial** is a polynomial with more than one variable. The following are multivariable polynomials:

$$5xy + 8, \qquad 2x^2 - 7xy + 9y^2, \qquad 17x^3y^9$$

The **degree of a term** is the sum of the exponents of all of the variables in the term. For example, the degree of $7x^3$ is three. The degree of $4xy$ is two. The degree of $10x^4y^2$ is six.

The **degree of a polynomial** is the highest degree of all of the terms in the polynomial. For example, the degree of $5x^3 + 8x^2 - 20x - 2$ is three. The degree of $6xy - 4x^2y + 2xy^3$ is four.

The polynomial 0 is said to have **no degree.** A polynomial consisting of a constant only is said to have degree 0.

There are special names for polynomials with one, two, or three terms.

A **monomial** has *one* term:

$$5a, \qquad 3x^3yz^4, \qquad 12xy$$

A **binomial** has *two* terms:

$$7x + 9y, \qquad -6x - 4, \qquad 5x^4 + 2xy^2$$

A **trinomial** has *three* terms:

$$8x^2 - 7x + 4, \qquad 2ab^3 - 6ab^2 - 15ab, \qquad 2 + 5y + y^4$$

EXAMPLE 1 State the degree of the polynomial, and whether it is a monomial, a binomial, or a trinomial.

(a) $5xy + 3x^3$ **(b)** $-7a^5b^2$ **(c)** $8x^4 - 9x - 15$

Solution

(a) This polynomial is of degree 3. It has two terms, so it is a binomial.

(b) The sum of the exponents is $5 + 2 = 7$. Therefore this polynomial is of degree 7. It has one term, so it is a monomial.

(c) This polynomial is of degree 4. It has three terms, so it is a trinomial.

Practice Problem 1 State the degree of the polynomial, and whether it is a monomial, a binomial, or a trinomial.

(a) $-7x^5 - 3xy$ **(b)** $22a^3b^4$ **(c)** $-3x^3 + 3x^2 - 6x$

2 Adding Polynomials

We usually write a polynomial in x so that the exponents on x decrease from left to right. For example, the polynomial

$$5x^2 - 6x + 2$$

is said to be written in **decreasing order** since each exponent is decreasing as we move from left to right.

You can add, subtract, multiply, and divide polynomials. Let us take a look at addition. To add two polynomials, we add their like terms.

EXAMPLE 2 Add. $(5x^2 - 6x - 12) + (-3x^2 - 9x + 5)$

Solution

$$
\begin{aligned}
(5x^2 - 6x - 12) + (-3x^2 - 9x + 5) &= [5x^2 + (-3x^2)] + [-6x + (-9x)] + [-12 + 5] \\
&= [(5 - 3)x^2] + [(-6 - 9)x] + [-12 + 5] \\
&= 2x^2 + (-15x) + (-7) \\
&= 2x^2 - 15x - 7
\end{aligned}
$$

Practice Problem 2 Add. $(-8x^3 + 3x^2 + 6) + (2x^3 - 7x^2 - 3)$

The numerical coefficients of polynomials may be any real number. Thus polynomials may have numerical coefficients that are decimals or fractions.

EXAMPLE 3 Add. $\left(\frac{1}{2}x^2 - 6x + \frac{1}{3}\right) + \left(\frac{1}{5}x^2 - 2x - \frac{1}{2}\right)$

Solution

$$
\begin{aligned}
\left(\frac{1}{2}x^2 - 6x + \frac{1}{3}\right) + \left(\frac{1}{5}x^2 - 2x - \frac{1}{2}\right) &= \left[\frac{1}{2}x^2 + \frac{1}{5}x^2\right] + [-6x + (-2x)] + \left[\frac{1}{3} + \left(-\frac{1}{2}\right)\right] \\
&= \left[\left(\frac{1}{2} + \frac{1}{5}\right)x^2\right] + [(-6 - 2)x] + \left[\frac{1}{3} + \left(-\frac{1}{2}\right)\right] \\
&= \left[\left(\frac{5}{10} + \frac{2}{10}\right)x^2\right] + [-8x] + \left[\frac{2}{6} - \frac{3}{6}\right] \\
&= \frac{7}{10}x^2 - 8x - \frac{1}{6}
\end{aligned}
$$

Practice Problem 3 Add.

$$
\left(-\frac{1}{3}x^2 - 6x - \frac{1}{12}\right) + \left(\frac{1}{4}x^2 + 5x - \frac{1}{3}\right)
$$

EXAMPLE 4 Add. $(1.2x^3 - 5.6x^2 + 5) + (-3.4x^3 - 1.2x^2 + 4.5x - 7)$

Solution Group like terms.

$$
\begin{aligned}
(1.2x^3 - 5.6x^2 + 5) + (-3.4x^3 - 1.2x^2 + 4.5x - 7) &= (1.2 - 3.4)x^3 + (-5.6 - 1.2)x^2 + 4.5x + (5 - 7) \\
&= -2.2x^3 - 6.8x^2 + 4.5x - 2
\end{aligned}
$$

Practice Problem 4 Add.

$$
(3.5x^3 - 0.02x^2 + 1.56x - 3.5) + (-0.08x^2 - 1.98x + 4)
$$

3 Subtracting Polynomials

Recall that subtraction of real numbers can be defined as adding the opposite of the second number. Thus $a - b = a + (-b)$. That is, $3 - 5 = 3 + (-5)$. A similar method is used to subtract two polynomials.

To subtract two polynomials, change the sign of each term in the second polynomial and then add.

EXAMPLE 5 Subtract. $(7x^2 - 6x + 3) - (5x^2 - 8x - 12)$

Solution We change the sign of each term in the second polynomial and then add.

$$(7x^2 - 6x + 3) - (5x^2 - 8x - 12) = (7x^2 - 6x + 3) + (-5x^2 + 8x + 12)$$
$$= (7 - 5)x^2 + (-6 + 8)x + (3 + 12)$$
$$= 2x^2 + 2x + 15$$

Practice Problem 5 Subtract.

$$(5x^3 - 15x^2 + 6x - 3) - (-4x^3 - 10x^2 + 5x + 13)$$

NOTE TO STUDENT: Fully worked-out solutions to all of the Practice Problems can be found at the back of the text starting at page SP-1

As mentioned previously, polynomials may involve more than one variable. When subtracting polynomials in two variables, you will need to use extra care in determining which terms are like terms. For example $6x^2y$ and $5x^2y$ are like terms. In a similar fashion, $3xy$ and $8xy$ are like terms. However, $7xy^2$ and $15x^2y^2$ are not like terms. Every exponent of every variable in the two terms must be the same if the terms are to be like terms. You will use this concept in Example 6.

EXAMPLE 6 Subtract.

$$(-6x^2y - 3xy + 7xy^2) - (5x^2y - 8xy - 15x^2y^2)$$

Solution Change the sign of each term in the second polynomial and add. Look for like terms.

$$(-6x^2y - 3xy + 7xy^2) + (-5x^2y + 8xy + 15x^2y^2)$$
$$= (-6 - 5)x^2y + (-3 + 8)xy + 7xy^2 + 15x^2y^2$$
$$= -11x^2y + 5xy + 7xy^2 + 15x^2y^2$$

Nothing further can be done to combine these four terms.

Practice Problem 6 Subtract.

$$(x^3 - 7x^2y + 3xy^2 - 2y^3) - (2x^3 + 4xy - 6y^3)$$

Evaluating Polynomials to Predict a Value

Sometimes polynomials are used to predict values. In such cases we need to **evaluate** the polynomial. We do this by substituting a known value for the variable and determining the value of the polynomial.

EXAMPLE 7 Automobiles sold in the United States have become more fuel efficient over the years due to regulations from Congress. The number of miles per gallon obtained by the average automobile in the United States can be described by the polynomial

$$0.3x + 12.9,$$

where x is the number of years since 1970. (*Source:* U.S. Federal Highway Administration.) Use this polynomial to estimate the number of miles per gallon obtained by the average automobile in

(a) 1972   **(b)** 2012

Solution

(a) The year 1972 is two years later than 1970, so $x = 2$.

Thus the number of miles per gallon obtained by the average automobile in 1972 can be estimated by evaluating $0.3x + 12.9$ when $x = 2$.

$$0.3(2) + 12.9 = 0.6 + 12.9$$
$$= 13.5$$

We estimate that the average car in 1972 obtained 13.5 miles per gallon.

(b) The year 2012 will be 42 years after 1970, so $x = 42$.

Thus the estimated number of miles per gallon obtained by the average automobile in 2012 can be predicted by evaluating $0.3x + 12.9$ when $x = 42$.

$$0.3(42) + 12.9 = 12.6 + 12.9$$
$$= 25.5$$

We therefore predict that the average car in 2012 will obtain 25.5 miles per gallon.

Practice Problem 7 The number of miles per gallon obtained by the average truck in the United States can be described by the polynomial $0.03x + 5.4$, where x is the number of years since 1970. (*Source:* U.S. Federal Highway Administration.) Use this polynomial to estimate the number of miles per gallon obtained by the average truck in

(a) 1974   **(b)** 2011

Verbal and Writing Skills

1. State in your own words a definition for a polynomial in x and give an example.

2. State in your own words a definition for a multivariable polynomial and give an example.

3. State in your own words how to determine the degree of a polynomial in x.

4. State in your own words how to determine the degree of a multivariable polynomial.

State the degree of the polynomial and whether it is a monomial, a binomial, or a trinomial.

5. $6x^3y$

6. $5xy^6$

7. $20x^5 + 6x^3 - 7x$

8. $13x^4 - 12x + 20$

9. $4x^2y^3 - 7x^3y^3$

10. $9x^4y^3 - 6xy^5$

Add.

11. $(-3x + 15) + (8x - 43)$

12. $(5x - 11) + (-7x + 34)$

13. $(6x^2 + 5x - 6) + (-8x^2 - 3x + 5)$

14. $(3x^2 - 2x - 5) + (-7x^2 + 8x - 2)$

15. $\left(\frac{1}{2}x^2 + \frac{1}{3}x - 4\right) + \left(\frac{1}{3}x^2 + \frac{1}{6}x - 5\right)$

16. $\left(\frac{1}{4}x^2 - \frac{2}{3}x - 10\right) + \left(-\frac{1}{3}x^2 + \frac{1}{9}x + 2\right)$

17. $(3.4x^3 - 7.1x + 3.4) + (2.2x^2 - 6.1x - 8.8)$

18. $(-4.6x^3 + 5.6x - 0.3) + (9.8x^2 + 4.5x - 1.7)$

Subtract.

19. $(2x - 19) - (-3x + 5)$

20. $(5x - 5) - (6x - 3)$

21. $\left(\frac{2}{5}x^2 - \frac{1}{2}x + 5\right) - \left(\frac{1}{3}x^2 - \frac{3}{7}x - 6\right)$

22. $\left(\frac{3}{8}x^2 - \frac{2}{3}x - 7\right) - \left(\frac{2}{3}x^2 - \frac{1}{2}x + 2\right)$

23. $(4x^3 + 3x) - (x^3 + x^2 - 5x)$

24. $(6x^3 + x^2 - x + 1) - (2x^3 - x^2 + 5)$

25. $(0.5x^4 - 0.7x^2 + 8.3) - (5.2x^4 + 1.6x + 7.9)$

26. $(1.3x^4 - 3.1x^3 + 6.3x) - (x^4 - 5.2x^2 + 6.5x)$

Perform the indicated operations.

27. $(8x + 2) + (x - 7) - (3x + 1)$

28. $(x - 5) - (3x + 8) - (5x - 2)$

29. $(5x^2y - 6xy^2 + 2) + (-8x^2y + 12xy^2 - 6)$

30. $(7x^2y^2 - 6xy + 5) + (-15x^2y^2 - 6xy + 18)$

31. $(3x^4 - 4x^2 - 18) - (2x^4 + 3x^3 + 6)$

32. $(2b^3 + 3b - 5) - (-3b^3 + 5b^2 + 7b)$

Prisons The number of prisoners held in federal and state prisons, measured in thousands, can be described by the polynomial $-2.06x^2 + 77.82x + 743$. The variable x represents the number of years since 1990. (Source: www.ojp.usdoj.gov.)

33. Estimate the number of prisoners in 1990.

34. Estimate the number of prisoners in 2000.

35. According to the polynomial, by how much did the prison population increase from 2002 to 2007?

36. According to the polynomial, by how much will the prison population decrease from 2008 to 2012?

Applications

▲ **37.** *Geometry* The lengths and the widths of the following three rectangles are labeled. Create a polynomial that describes the sum of the *area* of these three rectangles.

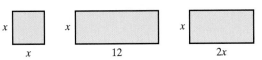

▲ **38.** *Geometry* The dimensions of the sides of the following figure are labeled. Create a polynomial that describes the *perimeter* of this figure.

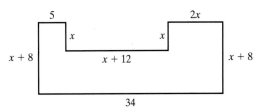

Cumulative Review

39. **[3.1.3]** Solve for y. $3y - 8x = 2$

40. **[2.7.1]** Solve for B. $3A = 2BCD$

41. **[2.3.3]** Solve for x. $-2(x - 5) + 6 = 2^2 - 9 + x$

42. **[2.4.1]** Solve for x. $\dfrac{x}{6} + \dfrac{x}{2} = \dfrac{4}{3}$

Quick Quiz 5.3 Combine.

1. $(3x^2 - 5x + 8) + (-7x^2 - 6x - 3)$

2. $(2x^2 - 3x - 7) - (-4x^2 + 6x + 9)$

3. $(5x - 3) - (2x - 4) + (-6x + 7)$

4. **Concept Check** Explain how you would determine the degree of the following polynomial and how you would decide if it is a monomial, a binomial, or a trinomial.

$$2xy^2 - 5x^3y^4$$

1. _____

2. _____

3. _____

4. _____

5. _____

6. _____

7. _____

8. _____

9. _____

10. _____

11. _____

12. _____

How are you doing with your homework assignments in Sections 5.1 to 5.3? Do you feel you have mastered the material so far? Do you understand the concepts you have covered? Before you go further in the textbook, take some time to do each of the following problems.

Simplify all answers. Reduce all fractions.

5.1

Multiply.

1. $(8x^2y^3)(-3xy^2)$

Divide. Assume that all variables are nonzero. Leave your answer with only positive exponents.

2. $-\dfrac{35xy^6}{25x^8y^3}$

3. $\dfrac{60x^7y^0}{15x^2y^9}$

Simplify.

4. $(-3x^5y)^4$

5.2

Simplify. Express your answer with positive exponents. Assume that all variables are nonzero.

5. $(4x^{-3}y^4)^{-2}$

6. $\dfrac{4x^4y^{-3}}{12x^{-1}y^2}$

7. Write in scientific notation. 58,740

8. Write in scientific notation. 0.00009362

5.3

Perform the indicated operations.

9. $(2x^2 + 0.5x - 2) + (0.3x^2 - 0.9x - 3.4)$

10. $(3x^2 + 7x - 10) - (-x^2 + 5x - 2)$

11. $\left(\dfrac{1}{2}x^3 + \dfrac{1}{4}x^2 - 2x\right) - \left(\dfrac{1}{3}x^3 - \dfrac{1}{8}x^2 - 5x\right)$

12. $\left(\dfrac{1}{16}x^2 + \dfrac{1}{8}\right) + \left(\dfrac{1}{4}x^2 - \dfrac{3}{10}x - \dfrac{1}{2}\right)$

Now turn to page SA-15 for the answer to each of these problems. Each answer also includes a reference to the objective in which the problem is first taught. If you missed any of these problems, you should stop and review the Examples and Practice Problems in the referenced objective. A little review now will help you master the material in the upcoming sections of the text.

Multiplying a Monomial by a Polynomial

We use the distributive property to multiply a monomial by a polynomial. Remember, the distributive property states that for real numbers a, b, and c,

$$a(b + c) = ab + ac.$$

EXAMPLE 1 Multiply. $3x^2(5x - 2)$

Solution

$$3x^2(5x - 2) = 3x^2(5x) + 3x^2(-2) \qquad \text{Use the distributive property.}$$
$$= (3 \cdot 5)(x^2 \cdot x) + (3)(-2)x^2$$
$$= 15x^3 - 6x^2$$

Practice Problem 1 Multiply. $4x^3(-2x^2 + 3x)$

Try to do as much of the multiplication as you can mentally.

EXAMPLE 2 Multiply.

(a) $2x(x^2 + 3x - 1)$ **(b)** $-2xy^2(x^2 - 2xy - 3y^2)$

Solution

(a) $2x(x^2 + 3x - 1) = 2x^3 + 6x^2 - 2x$

(b) $-2xy^2(x^2 - 2xy - 3y^2) = -2x^3y^2 + 4x^2y^3 + 6xy^4$

Notice in part (b) that you are multiplying each term by the negative expression $-2xy^2$. This will change the sign of each term in the product.

Practice Problem 2 Multiply.

(a) $-3x(x^2 + 2x - 4)$ **(b)** $6xy(x^3 + 2x^2y - y^2)$

When we multiply by a monomial, the monomial may be on the right side.

EXAMPLE 3 Multiply. $(x^2 - 2x + 6)(-2xy)$

Solution $(x^2 - 2x + 6)(-2xy) = -2x^3y + 4x^2y - 12xy$

Practice Problem 3 Multiply.

$$(-6x^3 + 4x^2 - 2x)(-3xy)$$

 Multiplying Two Binomials

We can build on our knowledge of the distributive property and our experience with multiplying monomials to learn how to multiply two binomials. Let's suppose that we want to multiply $(x + 2)(3x + 1)$. We can use the distributive property. Since a can represent any quantity, let $a = x + 2$. Then let $b = 3x$ and $c = 1$. We now have the following.

$$a(b + c) = ab + ac$$
$$(x + 2)(3x + 1) = (x + 2)(3x) + (x + 2)(1)$$
$$= 3x^2 + 6x + x + 2$$
$$= 3x^2 + 7x + 2$$

Let's take another look at the original problem, $(x + 2)(3x + 1)$. This time we will assign a letter to each term in the binomials. That is, let $a = x$, $b = 2$, $c = 3x$, and $d = 1$. Using substitution, we have the following.

$$
\begin{aligned}
(x + 2)(3x + 1) &= (a + b)(c + d) \\
&= (a + b)c + (a + b)d \\
&= ac + bc + ad + bd \\
&= (x)(3x) + (2)(3x) + (x)(1) + (2)(1) \quad \text{By substitution} \\
&= 3x^2 + 6x + x + 2 \\
&= 3x^2 + 7x + 2
\end{aligned}
$$

How does this compare with the preceding result?

The distributive property shows us *how* the problem can be done and *why* it can be done. In actual practice there is a memory device to help students remember the steps involved. It is often referred to as FOIL. The letters FOIL stand for the following.

> F multiply the *First* terms
>
> O multiply the *Outer* terms
>
> I multiply the *Inner* terms
>
> L multiply the *Last* terms

The FOIL letters are simply a way to remember the four terms in the final product and how they are obtained. Let's return to our original problem.

$(x + 2)(3x + 1)$ F Multiply the *first* terms to obtain $3x^2$.

$(x + 2)(3x + 1)$ O Multiply the *outer* terms to obtain x.

$(x + 2)(3x + 1)$ I Multiply the *inner* terms to obtain $6x$.

$(x + 2)(3x + 1)$ L Multiply the *last* terms to obtain 2.

The result so far is $3x^2 + x + 6x + 2$. These four terms are the same four terms that we obtained when we multiplied using the distributive property. We can combine the like terms to obtain the final answer: $3x^2 + 7x + 2$. Now let's study the FOIL method in a few examples.

EXAMPLE 4 Multiply. $(2x - 1)(3x + 2)$

Solution

First   Last   First + Outer + Inner + Last

$(2x - 1)(3x + 2)$ F   O   I   L

$$
\begin{aligned}
&= 6x^2 + \quad 4x \quad - \quad 3x \quad - \quad 2 \\
&= 6x^2 + x - 2 \quad \text{Combine like terms.}
\end{aligned}
$$

−3*x* Inner

+4*x* Outer

NOTE TO STUDENT: *Fully worked-out solutions to all of the Practice Problems can be found at the back of the text starting at page SP-1*

Notice that we combine the inner and outer terms to obtain the middle term.

Practice Problem 4 Multiply. $(5x - 1)(x - 2)$

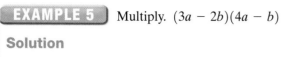 Multiply. $(3a - 2b)(4a - b)$

Solution

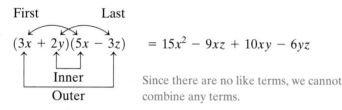

$$= 12a^2 - 3ab - 8ab + 2b^2$$
$$= 12a^2 - 11ab + 2b^2$$

Practice Problem 5 Multiply. $(8a - 5b)(3a - b)$

After you have done several problems, you may be able to combine the outer and inner products mentally.

In some problems the inner and outer products cannot be combined.

EXAMPLE 6 Multiply. $(3x + 2y)(5x - 3z)$

Solution

First   Last

$(3x + 2y)(5x - 3z)$ $= 15x^2 - 9xz + 10xy - 6yz$

  Inner   Since there are no like terms, we cannot
  Outer   combine any terms.

Practice Problem 6 Multiply. $(3a + 2b)(2a - 3c)$

EXAMPLE 7 Multiply. $(7x - 2y)^2$

Solution

$(7x - 2y)(7x - 2y)$ When we square a binomial, it is the same as multiplying
                       the binomial by itself.

First   Last

$(7x - 2y)(7x - 2y)$ $= 49x^2 - 14xy - 14xy + 4y^2$

  Inner $= 49x^2 - 28xy + 4y^2$
  Outer

Practice Problem 7 Multiply. $(3x - 2y)^2$

We can multiply binomials containing exponents that are greater than 1. That is, we can multiply binomials containing x^2 or y^3, and so on.

EXAMPLE 8 Multiply. $(3x^2 + 4y^3)(2x^2 + 5y^3)$

Solution $(3x^2 + 4y^3)(2x^2 + 5y^3) = 6x^4 + 15x^2y^3 + 8x^2y^3 + 20y^6$

$$= 6x^4 + 23x^2y^3 + 20y^6$$

Practice Problem 8 Multiply. $(2x^2 + 3y^2)(5x^2 + 6y^2)$

▲ **EXAMPLE 9** The width of a living room is $(x + 4)$ feet. The length of the room is $(3x + 5)$ feet. What is the area of the room in square feet?

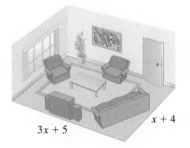

3x + 5

x + 4

Solution $A = (\text{length})(\text{width}) = (3x + 5)(x + 4)$

$$= 3x^2 + 12x + 5x + 20$$

$$= 3x^2 + 17x + 20$$

There are $(3x^2 + 17x + 20)$ square feet in the room.

NOTE TO STUDENT: Fully worked-out solutions to all of the Practice Problems can be found at the back of the text starting at page SP-1

▲ **Practice Problem 9** What is the area in square feet of a room that is $(2x - 1)$ feet wide and $(7x + 3)$ feet long?

PRACTICE   WATCH   DOWNLOAD   READ   REVIEW

Multiply.

1. $-2x(6x^3 - x)$

2. $5x(-3x^4 + 4x)$

3. $3x^2(7x - 3)$

4. $2x^2(8x - 5)$

5. $2x^3(-2x^3 + 5x - 1)$

6. $5x^2(-3x^3 + 6x - 2)$

7. $\frac{1}{2}(2x + 3x^2 + 5x^3)$

8. $\frac{2}{3}(4x + 6x^2 - 2x^3)$

9. $(2x^3 - 4x^2 + 5x)(-x^2y)$

10. $(3b^3 + 3b^2 - ab)(-5a^2)$

11. $(3x^3 + x^2 - 8x)(3xy)$

12. $(2x^3 + x^2 - 6x)(2xy)$

13. $(x^3 - 3x^2 + 5x - 2)(3x)$

14. $(-2x^3 + 4x^2 - 7x + 3)(2x)$

15. $(x^2y^2 - 6xy + 8)(-2xy)$

16. $(x^2y^2 + 5xy - 9)(-3xy)$

17. $(-7x^3 + 3x^2 + 2x - 1)(4x^2y)$

18. $(-5x^3 - 6x^2 + x - 1)(5xy^2)$

19. $(3d^4 - 4d^2 + 6)(-2c^2d)$

20. $(-4x^3 + 6x^2 - 5x)(-7xy^2)$

21. $6x^3(2x^4 - x^2 + 3x + 9)$

22. $8x^3(-2x^4 + 3x^2 - 5x - 14)$

23. $-2x^3(8x^3 - 5x^2 + 6x)$

24. $-6x^4(-3x^3 + 8x^2 - 4x)$

Multiply. Try to do most of the exercises mentally without writing down intermediate steps.

25. $(x + 5)(x + 7)$

26. $(x + 6)(x + 3)$

27. $(x + 6)(x + 2)$

28. $(x + 9)(x + 3)$

29. $(x - 8)(x + 2)$

30. $(x + 3)(x - 6)$

31. $(x - 5)(x - 4)$

32. $(x - 6)(x - 5)$

33. $(5x - 2)(-4x - 3)$

34. $(7x + 1)(-2x - 3)$

35. $(2x - 5)(x + 3y)$

36. $(x - 6)(3x + 2y)$

37. $(5x + 2)(3x - y)$

38. $(3x + 4)(5x - y)$

39. $(4y + 1)(5y - 3)$

40. $(5y + 1)(6y - 5)$

To Think About

41. What is wrong with this multiplication?
$(x - 2)(-3) = 3x - 6$

42. What is wrong with this answer?
$-(3x - 7) = -3x - 7$

43. What is the missing term?
$(5x + 2)(5x + 2) = 25x^2 +$ _____ $+ 4$

44. Multiply the binomials and write a brief description of what is special about the result. $(5x - 1)(5x + 1)$

Mixed Practice *Multiply.*

45. $(4x - 3y)(5x - 2y)$

46. $(3b - 5c)(2b - 7c)$

47. $(7x - 2)^2$

48. $(3x - 7)^2$

49. $(4a + 2b)^2$

50. $(5a + 3b)^2$

51. $(0.2x + 3)(4x - 0.3)$

52. $(0.5x - 2)(6x - 0.2)$

53. $\left(\dfrac{1}{2}x + \dfrac{1}{3}\right)\left(\dfrac{1}{2}x - \dfrac{1}{4}\right)$

54. $\left(\dfrac{1}{3}x + \dfrac{1}{5}\right)\left(\dfrac{1}{3}x - \dfrac{1}{2}\right)$

55. $(2a - 3b)(x - 5b)$

56. $(4b - 5c)(a - 3b)$

Find the area of the rectangle.

▲ **57.**

5x + 2

2x − 3

▲ **58.**

7x + 3

4x − 6

Cumulative Review

59. **[2.3.3]** Solve for x. $3(x - 6) = -2(x + 4) + 6x$

60. **[2.3.3]** Solve for w. $3(w - 7) - (4 - w) = 11w$

61. **[2.6.1]** *Paper Currency* Heather returned from the bank with $375. She had one more $20 bill than she had $10 bills. The number of $5 bills she had was one less than triple the number of $10 bills. How many of each denomination did she have?

Social Security Every year, the U.S. Department of the Treasury disburses billions of dollars in social security checks. The polynomial $17.7x + 240$ can be used to estimate the total value of social security checks sent in a year in billions of dollars, where x is the number of years since 1990. (Source: Statistical Abstract of the United States: 2007.)

62. **[5.3.4]** Estimate the total value of social security checks sent in 1990.

63. **[5.3.4]** Estimate the total value of social security checks sent in 2005.

64. **[5.3.4]** Predict the total value of social security checks that will be sent in 2010.

65. **[5.3.4]** Predict the total value of social security checks that will be sent in 2014.

Quick Quiz 5.4 *Multiply.*

1. $(2x^2y^2 - 3xy + 4)(4xy^2)$

2. $(2x + 3)(3x - 5)$

3. $(6a - 4b)(2a - 3b)$

4. **Concept Check** Explain how you would multiply $(7x - 3)^2$.

5.5 MULTIPLICATION: SPECIAL CASES

1 Multiplying Binomials of the Type $(a + b)(a - b)$

The case when you multiply $(x + y)(x - y)$ is interesting and deserves special consideration. Using the FOIL method, we find

$$(x + y)(x - y) = x^2 - xy + xy - y^2 = x^2 - y^2.$$

Notice that the sum of the inner product and the outer product is zero. We see that

$$(x + y)(x - y) = x^2 - y^2.$$

This works in all cases when the binomials are the sum and difference of the same two terms. That is, in one factor the terms are added, while in the other factor the same two terms are subtracted.

$$(5a + 2b)(5a - 2b) = 25a^2 - 10ab + 10ab - 4b^2$$
$$= 25a^2 - 4b^2$$

The product is the difference of the squares of the terms. That is, $(5a)^2 - (2b)^2$ or $25a^2 - 4b^2$.

Many students find it helpful to memorize this equation.

Student Learning Objectives

After studying this section, you will be able to:

1 **Multiply binomials of the type $(a + b)(a - b)$.**

2 **Multiply binomials of the type $(a + b)^2$ and $(a - b)^2$.**

3 **Multiply polynomials with more than two terms.**

> **MULTIPLYING BINOMIALS: A SUM AND A DIFFERENCE**
>
> $$(a + b)(a - b) = a^2 - b^2$$

You may use this relationship to find the product quickly in cases where it applies. The terms must be the same and there must be a sum and a difference.

EXAMPLE 1 Multiply. $(7x + 2)(7x - 2)$

Solution

$$(7x + 2)(7x - 2) = (7x)^2 - (2)^2 = 49x^2 - 4$$

Check: Multiply the binomials using FOIL to verify that the sum of the inner and outer products is zero.

Practice Problem 1 Multiply. $(6x + 7)(6x - 7)$

EXAMPLE 2 Multiply. $(5x - 8y)(5x + 8y)$

Solution

$$(5x - 8y)(5x + 8y) = (5x)^2 - (8y)^2 = 25x^2 - 64y^2$$

Practice Problem 2 Multiply. $(3x - 5y)(3x + 5y)$

2 Multiplying Binomials of the Type $(a + b)^2$ and $(a - b)^2$

A second case that is worth special consideration is a binomial that is squared. Consider the following problem.

$$(3x + 2)^2 = (3x + 2)(3x + 2)$$
$$= 9x^2 + 6x + 6x + 4$$
$$= 9x^2 + 12x + 4$$

If you complete enough problems of this type, you will notice a pattern. The answer always contains the square of the first term added to double the product of the first and last terms added to the square of the last term.

$3x$ is the first term	2 is the last term	Square the first term: $(3x)^2$	Double the product of the first and last terms: $2(3x)(2)$	Square the last term: $(2)^2$
↓	↓	↓	↓	↓
$(3x$ +	$2)^2$ =	$9x^2$ +	$12x$ +	4

We can show the same steps using variables instead of words.

$$(a + b)^2 = a^2 + 2ab + b^2$$

There is a similar formula for the square of a difference:

$$(a - b)^2 = a^2 - 2ab + b^2$$

We can use this formula to simplify $(2x - 3)^2$.

$$(2x - 3)^2 = (2x)^2 - 2(2x)(3) + (3)^2$$
$$= 4x^2 - 12x + 9$$

You may wish to multiply this product using FOIL to verify.

These two types of products, the square of a sum and the square of a difference, can be summarized as follows.

A BINOMIAL SQUARED

$$(a + b)^2 = a^2 + 2ab + b^2$$
$$(a - b)^2 = a^2 - 2ab + b^2$$

EXAMPLE 3 Multiply.

(a) $(5y - 2)^2$ **(b)** $(8x + 9y)^2$

Solution

(a) $(5y - 2)^2 = (5y)^2 - (2)(5y)(2) + (2)^2$
$$= 25y^2 - 20y + 4$$

(b) $(8x + 9y)^2 = (8x)^2 + (2)(8x)(9y) + (9y)^2$
$$= 64x^2 + 144xy + 81y^2$$

Practice Problem 3 Multiply.

(a) $(4a - 9b)^2$ **(b)** $(5x + 4)^2$

NOTE TO STUDENT: Fully worked-out solutions to all of the Practice Problems can be found at the back of the text starting at page SP-1

CAUTION: $(a + b)^2 \neq a^2 + b^2$! The two sides are not equal! Squaring the sum $(a + b)$ does not give $a^2 + b^2$! Beginning algebra students often make this error. Make sure you remember that when you square a binomial, there is always a *middle term*.

$$(a + b)^2 = a^2 + 2ab + b^2$$

Sometimes a numerical example helps you to see this.

$$(3 + 4)^2 \neq 3^2 + 4^2$$
$$7^2 \neq 9 + 16$$
$$49 \neq 25$$

Notice that what is missing on the right is $2ab = 2 \cdot 3 \cdot 4 = 24$.

 Multiplying Polynomials with More Than Two Terms

We used the distributive property to multiply two binomials $(a + b)(c + d)$, and we obtained $ac + ad + bc + bd$. We could also use the distributive property to multiply the polynomials $(a + b)$ and $(c + d + e)$, and we would obtain $ac + ad + ae + bc + bd + be$. Let us see if we can find a direct way to multiply products such as $(3x - 2)(x^2 - 2x + 3)$. It can be done quickly using an approach similar to that used in arithmetic for multiplying whole numbers. Consider the following arithmetic problem.

$$
\begin{array}{r}
128 \\
\times \ \ 43 \\
\hline
384 \\
512 \ \ \\
\hline
5504
\end{array}
$$

$\leftarrow$ The product of 128 and 3

$\leftarrow$ The product of 128 and 4 moved one space to the left

$\leftarrow$ The sum of the two partial products

Let us use a similar format to multiply the two polynomials. For example, multiply $(x^2 - 2x + 3)$ and $(3x - 2)$.

$$
\begin{array}{r}
x^2 - \ 2x + 3 \\
3x - 2 \\
\hline
-2x^2 + \ 4x - 6 \\
3x^3 - 6x^2 + \ 9x \ \ \ \ \ \\
\hline
3x^3 - 8x^2 + 13x - 6
\end{array}
$$

This is often called **vertical multiplication.**

$\leftarrow$ The product $(x^2 - 2x + 3)(-2)$

$\leftarrow$ The product $(x^2 - 2x + 3)(3x)$ moved one space to the left so that like terms are underneath each other

$\leftarrow$ The sum of the two partial products

EXAMPLE 4 Multiply vertically. $(3x^3 + 2x^2 + x)(x^2 - 2x - 4)$

Solution

$$
\begin{array}{r}
3x^3 + \ 2x^2 + \ \ x \\
x^2 - \ 2x - \ \ 4 \\
\hline
-12x^3 - \ 8x^2 - 4x \\
-6x^4 - \ 4x^3 - 2x^2 \ \ \ \ \ \ \ \\
3x^5 + 2x^4 + \ \ x^3 \ \ \ \ \ \ \ \ \ \ \ \ \ \ \ \ \\
\hline
3x^5 - 4x^4 - 15x^3 - 10x^2 - 4x
\end{array}
$$

We place one polynomial over the other.

$\leftarrow$ The product $(3x^3 + 2x^2 + x)(-4)$

$\leftarrow$ The product $(3x^3 + 2x^2 + x)(-2x)$

$\leftarrow$ The product $(3x^3 + 2x^2 + x)(x^2)$

$\leftarrow$ The sum of the three partial products

Note that the answers for each partial product are placed so that like terms are underneath each other.

Practice Problem 4 Multiply vertically. $(4x^3 - 2x^2 + x)(x^2 + 3x - 2)$

ALTERNATIVE METHOD: FOIL Horizontal Multiplication Some students prefer to do this type of multiplication using a horizontal format similar to the FOIL method. The following example illustrates this approach.

EXAMPLE 5 Multiply horizontally. $(x^2 + 3x + 5)(x^2 - 2x - 6)$

Solution We will use the distributive property repeatedly.

$$(x^2 + 3x + 5)(x^2 - 2x - 6) = x^2(x^2 - 2x - 6) + 3x(x^2 - 2x - 6) + 5(x^2 - 2x - 6)$$

$$= \boxed{x^4 - 2x^3 - 6x^2} + \boxed{3x^3 - 6x^2 - 18x} + \boxed{5x^2 - 10x - 30}$$

$$= x^4 + x^3 - 7x^2 - 28x - 30$$

Practice Problem 5 Multiply horizontally. $(2x^2 + 5x + 3)(x^2 - 3x - 4)$

Some problems may need to be done in two or more separate steps.

EXAMPLE 6 Multiply. $(2x - 3)(x + 2)(x + 1)$

Solution We first need to multiply any two of the binomials. Let us select the first pair.

$$\underbrace{(2x - 3)(x + 2)}(x + 1)$$
Find this product first.

$$(2x - 3)(x + 2) = 2x^2 + 4x - 3x - 6$$
$$= 2x^2 + x - 6$$

Now we replace the first two factors with their resulting product.

$$\underbrace{(2x^2 + x - 6)}(x + 1)$$
First product

We then multiply again.

$$(2x^2 + x - 6)(x + 1) = (2x^2 + x - 6)x + (2x^2 + x - 6)1$$
$$= 2x^3 + x^2 - 6x + 2x^2 + x - 6$$
$$= 2x^3 + 3x^2 - 5x - 6$$

The vertical format of Example 4 is an alternative method for this type of problem.

$$
\begin{array}{r}
2x^2 + x - 6 \\
x + 1 \\
\hline
2x^2 + x - 6 \\
2x^3 + x^2 - 6x \\
\hline
2x^3 + 3x^2 - 5x - 6
\end{array}
$$

$\left.\begin{array}{c} \\ \\ \end{array}\right\}$ Be sure to use special care in writing the exponents correctly if problems have more than one variable.

Thus we have

$$(2x - 3)(x + 2)(x + 1) = 2x^3 + 3x^2 - 5x - 6.$$

Note that it does not matter which two binomials are multiplied first. For example, you could first multiply $(2x - 3)(x + 1)$ to obtain $2x^2 - x - 3$ and then multiply that product by $(x + 2)$ to obtain the same result.

NOTE TO STUDENT: Fully worked-out solutions to all of the Practice Problems can be found at the back of the text starting at page SP-1

Practice Problem 6 Multiply. $(3x - 2)(2x + 3)(3x + 2)$
(*Hint:* Rearrange the factors.)

Sometimes we encounter a binomial raised to a third power. In such cases we would write out the binomial three times as a product and then multiply. So to evaluate $(3x + 4)^3$ we would first write $(3x + 4)(3x + 4)(3x + 4)$ and then follow the method of Example 6.

Verbal and Writing Skills

1. In the special case of $(a + b)(a - b)$, a binomial times a binomial is a _____.

2. Identify which of the following could be the answer to a problem using the formula for $(a + b)(a - b)$. Why?

    (a) $9x^2 - 16$

    (b) $4x^2 + 25$

    (c) $9x^2 + 12x + 4$

    (d) $x^4 - 1$

3. A student evaluated $(4x - 7)^2$ as $16x^2 + 49$. What is missing? State the correct answer.

4. The square of a binomial, $(a - b)^2$, always produces which of the following?

    (a) binomial

    (b) trinomial

    (c) four-term polynomial

Use the formula $(a + b)(a - b) = a^2 - b^2$ to multiply.

5. $(y - 7)(y + 7)$

6. $(x + 5)(x - 5)$

7. $(x - 8)(x + 8)$

8. $(x + 10)(x - 10)$

9. $(6x - 5)(6x + 5)$

10. $(4x - 9)(4x + 9)$

11. $(2x - 7)(2x + 7)$

12. $(3x - 10)(3x + 10)$

13. $(5x - 3y)(5x + 3y)$

14. $(8a - 3b)(8a + 3b)$

15. $(0.6x + 3)(0.6x - 3)$

16. $(5x - 0.2)(5x + 0.2)$

Use the formula for a binomial squared to multiply.

17. $(2y + 5)^2$

18. $(3x - 1)^2$

19. $(5x - 4)^2$

20. $(6x + 5)^2$

21. $(7x + 3)^2$

22. $(8x - 3)^2$

23. $(3x - 7)^2$

24. $(2x + 3y)^2$

25. $\left(\dfrac{2}{3}x + \dfrac{1}{4}\right)^2$

26. $\left(\dfrac{3}{4}x + \dfrac{1}{2}\right)^2$

27. $(9xy + 4z)^2$

28. $(7y - 3xz)^2$

Mixed Practice Exercises 29–36 *Multiply. Use the special formula that applies.*

29. $(7x + 3y)(7x - 3y)$

30. $(12x - 5y)(12x + 5y)$

31. $(3c - 5d)^2$

32. $(6c - d)^2$

33. $(9a - 10b)(9a + 10b)$ **34.** $(11a + 6b)(11a - 6b)$ **35.** $(5x + 9y)^2$ **36.** $(4x + 8y)^2$

Use the distributive property to multiply.

37. $(x^2 - x + 5)(x - 3)$ **38.** $(x^2 + 2x - 6)(x + 7)$ **39.** $(4x + 1)(x^3 - 2x^2 + x - 1)$

40. $(3x - 1)(x^3 + x^2 - 4x - 2)$ **41.** $(a^2 - 3a + 2)(a^2 + 4a - 3)$ **42.** $(x^2 + 4x - 5)(x^2 - 3x + 4)$

43. $(x + 3)(x - 1)(3x - 8)$ **44.** $(x - 7)(x + 4)(2x - 5)$ **45.** $(2x - 5)(x - 1)(x + 3)$

46. $(3x - 2)(x + 6)(x - 3)$ **47.** $(a - 5)(2a + 3)(a + 5)$ **48.** $(b + 7)(3b - 2)(b - 7)$

To Think About

▲ **49.** Find the volume of this object.

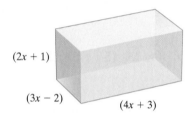

$(2x + 1)$
$(3x - 2)$
$(4x + 3)$

▲ **50.** *Geometry* The volume of a pyramid is $V = \dfrac{1}{3}Bh$, where B is the area of the base and h is the height. Find the volume of the following pyramid.

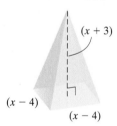

$(x + 3)$
$(x - 4)$
$(x - 4)$

Cumulative Review

51. **[2.6.1]** One number is three more than twice a second number. The sum of the two numbers is 60. Find both numbers.

▲ **52.** **[2.7.1]** *Room Dimensions* The perimeter of a rectangular room measures 34 meters. The width is 2 meters more than half the length. Find the dimensions of the room.

Quick Quiz 5.5 Multiply.

1. $(7x - 12y)(7x + 12y)$

2. $(2x + 3)(x - 2)(3x + 1)$

3. $(3x - 2)(5x^3 - 2x^2 - 4x + 3)$

4. **Concept Check** Using the formula $(a + b)^2 = a^2 + 2ab + b^2$, explain how to multiply $(6x - 9y)^2$.

Dividing a Polynomial by a Monomial

To divide a polynomial by a monomial, divide each term of the numerator by the denominator; then write the sum of the results. We are using the property of fractions that states that

$$\frac{a + b}{c} = \frac{a}{c} + \frac{b}{c}$$

Student Learning Objectives

After studying this section, you will be able to:

 Divide a polynomial by a monomial.

 Divide a polynomial by a binomial.

DIVIDING A POLYNOMIAL BY A MONOMIAL

1. Divide each term of the polynomial by the monomial.

2. When dividing variables, use the property $\frac{x^a}{x^b} = x^{a-b}$.

EXAMPLE 1 Divide. $\dfrac{8y^6 - 8y^4 + 24y^2}{8y^2}$

Solution $\dfrac{8y^6 - 8y^4 + 24y^2}{8y^2} = \dfrac{8y^6}{8y^2} - \dfrac{8y^4}{8y^2} + \dfrac{24y^2}{8y^2} = y^4 - y^2 + 3$

Practice Problem 1 Divide. $\dfrac{15y^4 - 27y^3 - 21y^2}{3y^2}$

NOTE TO STUDENT: Fully worked-out solutions to all of the Practice Problems can be found at the back of the text starting at page SP-1

 Dividing a Polynomial by a Binomial

Division of a polynomial by a binomial is similar to long division in arithmetic. Notice the similarity in the following division problems.

Division of a three-digit number by a two-digit number	Division of a polynomial by a binomial

$$\begin{array}{r} 32 \\ 21{\overline{\smash{\big)}\,672}} \\ \underline{63} \\ 42 \\ \underline{42} \\ 0 \end{array}$$

$$\begin{array}{r} 3x + 2 \\ 2x + 1{\overline{\smash{\big)}\,6x^2 + 7x + 2}} \\ \underline{6x^2 + 3x} \\ 4x + 2 \\ \underline{4x + 2} \\ 0 \end{array}$$

DIVIDING A POLYNOMIAL BY A BINOMIAL

1. Place the terms of the polynomial and binomial in descending order. Insert a 0 for any missing term.

2. Divide the first term of the polynomial by the first term of the binomial. The result is the first term of the answer.

3. Multiply the first term of the answer by the binomial and subtract the result from the first two terms of the polynomial. Bring down the next term to obtain a new polynomial.

4. Divide the new polynomial by the binomial using the process described in step 2.

5. Continue dividing, multiplying, and subtracting until the degree of the remainder is less than the degree of the binomial divisor.

6. Write the remainder as the numerator of a fraction that has the binomial divisor as its denominator.

EXAMPLE 2 Divide. $(x^3 + 5x^2 + 11x + 4) \div (x + 2)$

Solution

Step 1 The terms are arranged in descending order. No terms are missing.

Step 2 Divide the first term of the polynomial by the first term of the binomial. In this case, divide x^3 by x to get x^2.

$$
\begin{array}{r}
x^2 \\
x + 2 \overline{) x^3 + 5x^2 + 11x + 4}
\end{array}
$$

Step 3 Multiply x^2 by $x + 2$ and subtract the result from the first two terms of the polynomial, $x^3 + 5x^2$ in this case.

$$
\begin{array}{r}
x^2 \\
x + 2 \overline{) x^3 + 5x^2 + 11x + 4} \\
\underline{x^3 + 2x^2} \downarrow \\
3x^2 + 11x
\end{array}
$$

Subtract: $5x^2 - 2x^2 = 3x^2$
Bring down the next term.

Step 4 Continue to use the step 2 process. Divide $3x^2$ by x. Write the resulting $3x$ as the next term of the answer.

$$
\begin{array}{r}
x^2 + 3x \\
x + 2 \overline{) x^3 + 5x^2 + 11x + 4} \\
\underline{x^3 + 2x^2} \\
3x^2 + 11x
\end{array}
$$

Step 5 Continue multiplying, dividing, and subtracting until the degree of the remainder is less than the degree of the divisor. In this case, we stop when the remainder does not have an x.

$$
\begin{array}{r}
x^2 + 3x + 5 \\
x + 2 \overline{) x^3 + 5x^2 + 11x + 4} \\
\underline{x^3 + 2x^2} \\
3x^2 + 11x \\
\underline{3x^2 + 6x} \\
5x + 4 \\
\underline{5x + 10} \\
-6
\end{array}
$$

$3x(x + 2) = 3x^2 + 6x$
Bring down 4.
Subtract $4 - 10 = -6$.
The remainder is -6.

Step 6 The answer is $x^2 + 3x + 5 + \dfrac{-6}{x + 2}$.

To check the answer, we multiply $(x + 2)(x^2 + 3x + 5)$ and add the remainder, -6.

$$
\begin{aligned}
(x + 2)(x^2 + 3x + 5) + (-6) &= x^3 + 5x^2 + 11x + 10 - 6 \\
&= x^3 + 5x^2 + 11x + 4
\end{aligned}
$$

This is the original polynomial. It checks.

Practice Problem 2 Divide.

$$
(x^3 + 10x^2 + 31x + 35) \div (x + 4)
$$

Take great care with the subtraction step when negative numbers are involved.

EXAMPLE 3 Divide. $(5x^3 - 24x^2 + 9) \div (5x + 1)$

Solution We must first insert $0x$ to represent the missing x-term. Then we divide $5x^3$ by $5x$.

$$
\begin{array}{r}
x^2 \\
5x + 1 \overline{)5x^3 - 24x^2 + 0x + 9} \\
\underline{5x^3 + x^2 } \\
-25x^2
\end{array}
$$

Note that we are subtracting:
$$-24x^2 - (+1x^2) = -24x^2 - 1x^2$$
$$= -25x^2$$

Next we divide $-25x^2$ by $5x$.

$$
\begin{array}{r}
x^2 - 5x \\
5x + 1 \overline{)5x^3 - 24x^2 + 0x + 9} \\
\underline{5x^3 + x^2 } \\
-25x^2 + 0x \\
\underline{-25x^2 - 5x } \\
5x
\end{array}
$$

Note that we are subtracting:
$$0x - (-5x) = 0x + 5x = 5x$$

Finally, we divide $5x$ by $5x$.

$$
\begin{array}{r}
x^2 - 5x + 1 \\
5x + 1 \overline{)5x^3 - 24x^2 + 0x + 9} \\
\underline{5x^3 + x^2 } \\
-25x^2 + 0x \\
\underline{-25x^2 - 5x } \\
5x + 9 \\
\underline{5x + 1} \\
\boxed{8}
\end{array}
$$

$\longleftarrow$ The remainder is 8.

The answer is $x^2 - 5x + 1 + \dfrac{8}{5x + 1}$.

To check, multiply $(5x + 1)(x^2 - 5x + 1)$ and add the remainder, 8.

$$(5x + 1)(x^2 - 5x + 1) + 8 = 5x^3 - 24x^2 + 1 + 8$$
$$= 5x^3 - 24x^2 + 9$$

This is the original polynomial. Our answer is correct.

Practice Problem 3 Divide.

$$(2x^3 - x^2 + 1) \div (x - 1)$$

NOTE TO STUDENT: Fully worked-out solutions to all of the Practice Problems can be found at the back of the text starting at page SP-1

Now we will perform the division by writing a minimum of steps. See if you can follow each step.

EXAMPLE 4 Divide and check. $(12x^3 - 11x^2 + 8x - 4) \div (3x - 2)$

Solution

$$
\begin{array}{r}
4x^2 - x + 2 \\
3x - 2\overline{)12x^3 - 11x^2 + 8x - 4} \\
\underline{12x^3 - 8x^2} \\
-3x^2 + 8x \\
\underline{-3x^2 + 2x} \\
6x - 4 \\
\underline{6x - 4} \\
0
\end{array}
$$

Check. $(3x - 2)(4x^2 - x + 2) = 12x^3 - 3x^2 + 6x - 8x^2 + 2x - 4$

$ = 12x^3 - 11x^2 + 8x - 4$ Our answer is correct.

Practice Problem 4 Divide and check.

$$(20x^3 - 11x^2 - 11x + 6) \div (4x - 3)$$

Developing Your Study Skills

Taking an Exam

Allow yourself plenty of time to get to your exam. You may even find it helpful to arrive a little early in order to collect your thoughts and ready yourself. This will help you feel more relaxed.

After you get your exam, you will find it helpful to do the following.

1. Take two or three moderately deep breaths. Inhale, then exhale slowly. You will feel your entire body begin to relax.

2. Write down on the back of the exam any formulas or ideas that you need to remember.

3. Look over the entire test quickly in order to pace yourself and use your time wisely. Notice how many points each question is worth. Spend more time on items of greater worth.

4. Read directions carefully and be sure to answer all questions clearly. Keep your work neat and easy to read.

5. Ask your instructor about anything that is not clear to you.

6. Answer the questions that are easiest for you first. Then come back to the more difficult ones.

7. Do not get bogged down on one question for too long because it may jeopardize your chances of finishing other problems. Leave a tough question and come back to it when you have time later.

8. Check your work. This will help you to catch minor errors.

9. Stay calm if others leave before you do. You are entitled to use the full amount of allotted time. You will do better on the exam if you take your time and work carefully.

Divide.

1. $\dfrac{25x^4 - 15x^2 + 20x}{5x}$

2. $\dfrac{18b^5 - 12b^3 + 6b^2}{3b}$

3. $\dfrac{8y^4 - 12y^3 - 4y^2}{4y^2}$

4. $\dfrac{10y^4 - 35y^3 + 5y^2}{5y^2}$

5. $\dfrac{81x^7 - 36x^5 - 63x^3}{9x^3}$

6. $\dfrac{49x^8 - 35x^6 - 56x^3}{7x^3}$

7. $(48x^7 - 54x^4 + 36x^3) \div 6x^3$

8. $(72x^8 - 56x^5 - 40x^3) \div 8x^3$

Divide. Check your answers for exercises 9–16 by multiplication.

9. $\dfrac{6x^2 + 13x + 5}{2x + 1}$

10. $\dfrac{12x^2 + 19x + 5}{3x + 1}$

11. $\dfrac{x^2 - 8x - 17}{x - 5}$

12. $\dfrac{x^2 - 9x - 5}{x - 3}$

13. $\dfrac{3x^3 - x^2 + 4x - 2}{x + 1}$

14. $\dfrac{2x^3 - 3x^2 - 3x + 6}{x - 1}$

15. $\dfrac{4x^3 + 4x^2 - 19x - 15}{2x + 5}$

16. $\dfrac{6x^3 + 11x^2 - 8x + 5}{2x + 5}$

17. $\dfrac{10x^3 + 11x^2 - 11x + 2}{5x - 2}$

18. $\dfrac{5x^3 - 28x^2 - 20x + 21}{5x - 3}$

19. $\dfrac{12y^3 - 22y^2 + 23y - 20}{3y - 4}$

20. $\dfrac{8y^3 + 10y^2 - 49y + 30}{4y - 5}$

21. $(y^3 - y^2 - 13y - 12) \div (y + 3)$

22. $(y^3 - 2y^2 - 26y - 4) \div (y + 4)$

23. $(y^4 - 9y^2 - 5) \div (y - 2)$

24. $(2y^4 + 3y^2 - 5) \div (y - 2)$

Cumulative Review

25. **[0.5.3]** *Milk Prices* In 1995, the average price for a gallon of reduced-fat 2% milk was $2.36. By 2000, the price had increased by 20%. In 2005, the price increased 13% from the 2000 price. What was the average price of a gallon of reduced-fat 2% milk in 2000? What was the price in 2005? (*Source:* www.usda.gov.)

26. **[2.6.1]** *Page Numbers* Thomas was assigned to read a special two-page case study in his psychology book. He can't remember the page numbers, but does remember that the two numbers add to 341. What are the page numbers?

27. **[2.6.3]** *Hurricane Pattern* The National Hurricane Center has noticed an interesting pattern in the number of Atlantic hurricanes that begin during the period August 1 to October 31 each year. Examine the bar graph, then answer the following questions. Round all answers to the nearest tenth.

  (a) What was the mean number of hurricanes per year during the years 1999 to 2002?

  (b) What was the mean number of hurricanes per year during the years 2003 to 2006?

  (c) What was the percent of increase from the four-year period 1999 to 2002 to the four-year period 2003 to 2006?

  (d) If the same percent of increase for a four-year period continued during the years 2007 to 2010, what would be the mean number of hurricanes during the period 2007 to 2010? Round to the nearest tenth.

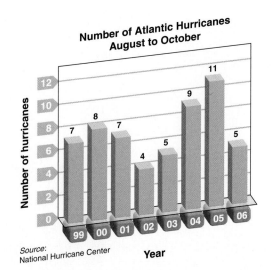

Quick Quiz 5.6 Divide.

1. $\dfrac{20x^5 - 64x^4 - 8x^3}{4x^2}$

2. $(8x^3 + 2x^2 - 19x - 6) \div (2x + 3)$

3. $(x^3 + 4x - 3) \div (x - 2)$

4. **Concept Check** Explain how you would check your answer to problem 3. Perform the check. Does your answer check?

Putting Your Skills to Work: Use Math to Save Money

TRIP TO THE GROCERY STORE

Is your dollar buying less each month at the grocery store? The rise in the cost of gas not only directly affects our budget, but also affects it indirectly by causing the price of groceries to increase. By using coupons, buying the store brand items, and purchasing staples in bulk, it is possible to get more for your money at the grocery store. Consider the story of Jenny.

Jenny is a single mom with two kids. Concerned with rising costs, she decided to track their grocery expenses for three months. In the first month they spent $450,

in the second month they spent $425, and in the third month they spent $460.

1. What are Jenny's average grocery expenses each month?

2. To help bring their monthly costs down, she and the kids started clipping coupons and cutting back on expensive treats. They saved 6% of their average cost by using the coupons and $30 by cutting out the treats. How much was their grocery bill the fourth month?

3. Jenny and the kids were pleased with the results, but knew they could do more. For the next month, she and the kids decided to choose store brand products when possible. They were happy to see that it saved them 20% off what they spent in the fourth month. How much was their grocery bill the fifth month?

4. Compare their original average costs and their costs in the fifth month and calculate the savings to the nearest tenth of a percent.

5. Can you think of other ways that Jenny's family could lower their monthly grocery bill?

Topic	Procedure	Examples
Multiplying monomials, p. 314.	$x^a \cdot x^b = x^{a+b}$ 1. Multiply the numerical coefficients. 2. Add the exponents of a given base.	$3^{12} \cdot 3^{15} = 3^{27}$ $(-3x^2)(6x^3) = -18x^5$ $(2ab)(4a^2b^3) = 8a^3b^4$
Dividing monomials, p. 315.	$\dfrac{x^a}{x^b} = \begin{cases} x^{a-b} & \text{Use if } a \text{ is greater than } b. \\ \dfrac{1}{x^{b-a}} & \text{Use if } b \text{ is greater than } a. \end{cases}$ 1. Divide or reduce the fraction created by the quotient of the numerical coefficients. 2. Subtract the exponents of a given base.	$\dfrac{16x^7}{8x^3} = 2x^4$ $\dfrac{5x^3}{25x^5} = \dfrac{1}{5x^2}$ $\dfrac{-12x^5y^7}{18x^3y^{10}} = -\dfrac{2x^2}{3y^3}$
Exponent of zero, pp. 317–318.	$x^0 = 1$ if $x \neq 0$	$5^0 = 1 \qquad \dfrac{x^6}{x^6} = 1$ $w^0 = 1 \qquad 3x^0y = 3y$
Raising a power, product, or quotient to a power, p. 319.	$(x^a)^b = x^{ab}$ $(xy)^a = x^a y^a$ $\left(\dfrac{x}{y}\right)^a = \dfrac{x^a}{y^a} \quad (y \neq 0)$ 1. Raise the numerical coefficient to the power outside the parentheses. 2. Multiply the exponent outside the parentheses times the exponent inside the parentheses.	$(x^9)^3 = x^{27}$ $(3x^2)^3 = 27x^6$ $\left(\dfrac{2x^2}{y^3}\right)^3 = \dfrac{8x^6}{y^9}$ $(-3x^4y^5)^4 = 81x^{16}y^{20}$ $(-5ab)^3 = -125a^3b^3$
Negative exponents, p. 325.	If $x \neq 0$ and $y \neq 0$, then $x^{-n} = \dfrac{1}{x^n}, \quad \dfrac{1}{x^{-n}} = x^n, \quad \dfrac{x^{-m}}{y^{-n}} = \dfrac{y^n}{x^m}$	Write with positive exponents. $x^{-6} = \dfrac{1}{x^6}, \quad \dfrac{1}{w^{-3}} = w^3, \quad \dfrac{w^{-12}}{z^{-5}} = \dfrac{z^5}{w^{12}}$
Scientific notation, pp. 326–327.	A positive number is written in scientific notation if it is in the form $a \times 10^n$, where $1 \leq a < 10$ and n is an integer.	$2{,}568{,}000 = 2.568 \times 10^6$ $0.0000034 = 3.4 \times 10^{-6}$
Add polynomials, pp. 332–333.	To add two polynomials, we add their like terms.	$(-7x^3 + 2x^2 + 5) + (x^3 + 3x^2 + x)$ $= -6x^3 + 5x^2 + x + 5$
Subtracting polynomials, p. 334.	To subtract the polynomials, change all signs of the second polynomial and add the result to the first polynomial. $a - b = a + (-b)$	$(5x^2 - 6) - (-3x^2 + 2) = (5x^2 - 6) + (3x^2 - 2)$ $= 8x^2 - 8$
Multiplying a monomial by a polynomial, p. 339.	Use the distributive property. $a(b + c) = ab + ac$ $(b + c)a = ba + ca$	Multiply. $-5x(2x^2 + 3x - 4) = -10x^3 - 15x^2 + 20x$ $(6x^3 - 5xy - 2y^2)(3xy) = 18x^4y - 15x^2y^2 - 6xy^3$
Multiplying two binomials, pp. 340, 345.	1. The product of the sum and difference of the same two terms: $\quad (a + b)(a - b) = a^2 - b^2$ 2. The square of a binomial: $\quad (a + b)^2 = a^2 + 2ab + b^2$ $\quad (a - b)^2 = a^2 - 2ab + b^2$ 3. Use FOIL for other binomial multiplication. The middle terms can often be combined, giving a trinomial.	$(3x + 7y)(3x - 7y) = 9x^2 - 49y^2$ $(3x + 7y)^2 = 9x^2 + 42xy + 49y^2$ $(3x - 7y)^2 = 9x^2 - 42xy + 49y^2$ $(3x - 5)(2x + 7) = 6x^2 + 21x - 10x - 35$ $= 6x^2 + 11x - 35$

Topic	Procedure	Examples
Multiplying two polynomials, p. 347.	To multiply two polynomials, multiply each term of one by each term of the other. This method is similar to the multiplication of many-digit numbers.	Vertical method: $$\begin{array}{r} 3x^2 - 7x + 4 \\ \times \quad\quad 3x - 1 \\ \hline -3x^2 + 7x - 4 \\ 9x^3 - 21x^2 + 12x \quad\quad \\ \hline 9x^3 - 24x^2 + 19x - 4 \end{array}$$ Horizontal method: $$(5x + 2)(2x^2 - x + 3)$$ $$= 10x^3 - 5x^2 + 15x + 4x^2 - 2x + 6$$ $$= 10x^3 - x^2 + 13x + 6$$
Multiplying three or more polynomials, p. 348.	1. Multiply any two polynomials. 2. Multiply the result by any remaining polynomials.	$$(2x + 1)(x - 3)(x + 4) = (2x^2 - 5x - 3)(x + 4)$$ $$= 2x^3 + 3x^2 - 23x - 12$$
Dividing a polynomial by a monomial, p. 351.	1. Divide each term of the polynomial by the monomial. 2. When dividing variables, use the property $$\frac{x^a}{x^b} = x^{a-b}.$$	Divide. $(15x^3 + 20x^2 - 30x) \div (5x)$ $$= \frac{15x^3}{5x} + \frac{20x^2}{5x} - \frac{30x}{5x}$$ $$= 3x^2 + 4x - 6$$
Dividing a polynomial by a binomial, p. 351.	1. Place the terms of the polynomial and binomial in descending order. Insert a 0 for any missing term. 2. Divide the first term of the polynomial by the first term of the binomial. 3. Multiply the first term of the answer by the binomial, and subtract the result from the first two terms of the polynomial. Bring down the next term to obtain a new polynomial. 4. Divide the new polynomial by the binomial using the process described in step 2. 5. Continue dividing, multiplying, and subtracting until the degree of the remainder is less than the degree of the binomial divisor. 6. Write the remainder as the numerator of a fraction that has the binomial divisor as its denominator.	Divide. $$(8x^3 + 2x^2 - 13x + 7) \div (4x - 1)$$ $$\begin{array}{r} 2x^2 + x - 3 \\ 4x - 1\overline{)8x^3 + 2x^2 - 13x + 7} \\ \underline{8x^3 - 2x^2} \\ 4x^2 - 13x \\ \underline{4x^2 - x} \\ -12x + 7 \\ \underline{-12x + 3} \\ 4 \end{array}$$ The answer is $$2x^2 + x - 3 + \frac{4}{4x - 1}.$$

Chapter 5 Review Problems

Section 5.1

Simplify. In problems 1–12, leave your answer in exponent form.

1. $(-6a^2)(3a^5)$

2. $(5^{10})(5^{13})$

3. $(3xy^2)(2x^3y^4)$

4. $(2x^3y^4)(-7xy^5)$

5. $\dfrac{7^{15}}{7^{27}}$

6. $\dfrac{x^{12}}{x^{17}}$

7. $\dfrac{y^{30}}{y^{16}}$

8. $\dfrac{9^{13}}{9^{24}}$

9. $\dfrac{-15xy^2}{25x^6y^6}$

10. $\dfrac{-12a^3b^6}{18a^2b^{12}}$

11. $(x^3)^8$

12. $(b^5)^6$

13. $(-3a^3b^2)^2$

14. $(3x^3y)^4$

15. $\left(\dfrac{5ab^2}{c^3}\right)^2$

16. $\left(\dfrac{x^0y^3}{4w^5z^2}\right)^3$

Section 5.2

Simplify. Write with positive exponents.

17. $a^{-3}b^5$

18. $m^8 p^{-5}$

19. $\dfrac{2x^{-6}}{y^{-3}}$

20. $(2x^{-5}y)^{-3}$

21. $(6a^4b^5)^{-2}$

22. $\dfrac{3x^{-3}}{y^{-2}}$

23. $\dfrac{4x^{-5}y^{-6}}{w^{-2}z^8}$

24. $\dfrac{3^{-3}a^{-2}b^5}{c^{-3}d^{-4}}$

Write in scientific notation.

25. 156,340,200,000

26. 179,632

27. 0.00092

28. 0.00000174

Write in decimal notation.

29. 1.2×10^5

30. 6.034×10^6

31. 3×10^6

32. 2.5×10^{-1}

33. 4.32×10^{-5}

34. 6×10^{-9}

Perform the indicated calculation. Leave your answer in scientific notation.

35. $\dfrac{(28,000,000)(5,000,000,000)}{7000}$

36. $(3.12 \times 10^5)(2.0 \times 10^6)(1.5 \times 10^8)$

37. $(1.6 \times 10^{-3})(3.0 \times 10^{-5})(2.0 \times 10^{-2})$

38. $\dfrac{(0.00078)(0.000005)(0.00004)}{0.002}$

39. ***Mission to Pluto*** The *New Horizons* spacecraft began its 3.5×10^9-mile journey to Pluto in January 2006. If the cost of this mission is $0.20 per mile, find the total cost. (*Source*: www.pluto.jhuapl.edu.)

40. ***Atomic Clock*** An atomic clock is based on the fact that cesium emits 9,192,631,770 cycles of radiation in one second. How many of these cycles occur in one day? Round to three significant digits.

41. ***Computer Speed*** Today's fastest modern computers can perform one operation in 1×10^{-8} second. How many operations can such a computer perform in 1 minute?

Section 5.3

Combine.

42. $(2.8x^2 - 1.5x + 3.4) + (2.7x^2 + 0.5x - 5.7)$

43. $(4.1x^2 + 1.8x - 3.3) + (1.5x^2 - 2.9x - 0.7)$

44. $(x^3 + x^2 - 6x + 2) - (2x^3 - x^2 - 5x - 6)$

45. $(4x^3 - x^2 - x + 3) - (-3x^3 + 2x^2 + 5x - 1)$

46. $\left(\dfrac{3}{5}x^2y - \dfrac{1}{3}x + \dfrac{3}{4}\right) - \left(\dfrac{1}{2}x^2y + \dfrac{2}{7}x + \dfrac{1}{3}\right)$

47. $\dfrac{1}{2}x^2 - \dfrac{3}{4}x + \dfrac{1}{5} - \left(\dfrac{1}{4}x^2 - \dfrac{1}{2}x + \dfrac{1}{10}\right)$

48. $(3x^2 - 5x) + (8x - 3) - (x^2 + 2)$

49. $(x^2 - 9) - (4x^2 + 5x) + (5x - 6)$

Section 5.4

Multiply.

50. $(3x + 1)(5x - 1)$

51. $(7x - 2)(4x - 3)$

52. $(2x + 3)(10x + 9)$

53. $5x(2x^2 - 6x + 3)$

54. $(3x^3 + 4x^2 - 5x + 8)(-5x)$

55. $(xy^2 + 5xy - 6)(-4xy^2)$

56. $(5a + 7b)(a - 3b)$

57. $(2x^2 - 3)(4x^2 - 5y)$

58. $-3x^2y(5x^4y + 3x^2 - 2)$

Section 5.5

Multiply.

59. $(4x + 3)^2$

60. $(a + 5b)(a - 5b)$

61. $(7x + 6y)(7x - 6y)$

62. $(5a - 2b)^2$

63. $(8x + 9y)^2$

64. $(x^2 + 7x + 3)(4x - 1)$

65. $(x - 6)(2x - 3)(x + 4)$

Section 5.6

Divide.

66. $(12y^3 + 18y^2 + 24y) \div (6y)$

67. $(30x^5 + 35x^4 - 90x^3) \div (5x^2)$

68. $(16x^3 - 24x^2 + 32x) \div (4x)$

69. $(106x^6 - 24x^5 + 38x^4 + 26x^3) \div (2x^3)$

70. $(15x^2 + 11x - 14) \div (5x + 7)$

71. $(12x^2 - x - 63) \div (4x + 9)$

72. $(6x^3 + x^2 + 6x + 5) \div (2x - 1)$

73. $(2x^3 - x^2 + 3x - 1) \div (x + 2)$

74. $(20x^2 + 3x - 2) \div (4x - 1)$

75. $(6x^2 + x - 9) \div (2x + 3)$

76. $(x^3 - x - 24) \div (x - 3)$

77. $(2x^3 - 3x + 1) \div (x - 2)$

Mixed Practice

Simplify. Leave your answer in exponent form.

78. $(4^6)(4^{23})$

79. $\dfrac{6^{50}}{6^{35}}$

Simplify. Write your answer with only positive exponents.

80. $(2x^3y^{-2})^{-2}$

81. $\dfrac{-3ab^{-1}}{bc^{-2}}$

82. $\left(\dfrac{3x^0y^{-2}}{z^3}\right)^{-2}$

Write in scientific notation.

83. 0.000922

Remove parentheses and simplify.

84. $(-6x^3 - 7x + 8) - (-2x^3 + 5x - 9)$

85. $-6x^2(2x^3 + 5x^2 - 2)$

86. Multiply. $(7x - 3)(2x - 1)(x + 2)$

87. Divide. $(6x^3 + 11x^2 - x - 2) \div (3x - 2)$

Applications

Solve. Express your answer in scientific notation.

88. *Fighting Disease* President Bush signed a foreign spending bill for fiscal year 2006 that included 2.8×10^9 dollars to fight HIV/AIDS, malaria, and tuberculosis worldwide. If the population of the United States in 2006 was 3×10^8 people, how much per person did the United States spend to fight these diseases? (*Source*: www.info.state.gov.) Write your answer in dollars and cents.

89. *Population* The population of India in 2010 is projected to be 1.18×10^9. The projected population of Bangladesh in 2010 is 1.6×10^8. (*Source: Statistical Abstract of the United States: 2007.*) What is the total population of the two countries projected to be? (*Hint:* First write 1.6×10^8 as 0.16×10^9 before you do the calculation.)

90. *Electron Mass* The mass of an electron is approximately 9.11×10^{-28} gram. Find the mass of 30,000 electrons.

91. *Gray Whales* During feeding season, gray whales eat 3.4×10^5 pounds of food per day. If the feeding period lasts for 140 days, how many pounds of food total will a gray whale consume?

To Think About

Find a polynomial that describes the shaded area.

▲ **92.**

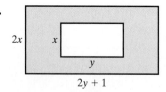

▲ **93.**

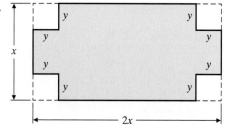

Remember to use your Chapter Test Prep Video CD to see the worked-out solutions to the test problems you want to review.

Simplify. Leave your answer in exponent form.

1. $(3^{10})(3^{24})$

2. $\dfrac{25^{18}}{25^{34}}$

3. $(8^4)^6$

In questions 4–8, simplify.

4. $(-3xy^4)(-4x^3y^6)$

5. $\dfrac{-35x^8y^{10}}{25x^5y^{10}}$

6. $(-5xy^6)^3$

7. $\left(\dfrac{7a^7b^2}{3c^0}\right)^2$

8. $\dfrac{(3x^2)^3}{(6x)^2}$

9. Evaluate. 4^{-3}

In questions 10 and 11, simplify and write with only positive exponents.

10. $6a^{-4}b^{-3}c^5$

11. $\dfrac{3x^{-3}y^2}{x^{-4}y^{-5}}$

12. Write in scientific notation. 0.0005482

13. Write in decimal notation. 5.82×10^8

14. Multiply. Leave your answer in scientific notation.
$(4.0 \times 10^{-3})(3.0 \times 10^{-8})(2.0 \times 10^4)$

Combine.

15. $(2x^2 - 3x - 6) + (-4x^2 + 8x + 6)$

16. $(3x^3 - 4x^2 + 3) - (14x^3 - 7x + 11)$

Multiply.

17. $-7x^2(3x^3 - 4x^2 + 6x - 2)$

18. $(5x^2y^2 - 6xy + 2)(3x^2y)$

19. $(5a - 4b)(2a + 3b)$

20. $(3x + 2)(2x + 1)(x - 3)$

21. $(7x^2 + 2y^2)^2$

22. $(5s - 11t)(5s + 11t)$

23. $(3x - 2)(4x^3 - 2x^2 + 7x - 5)$

24. $(3x^2 - 5xy)(x^2 + 3xy)$

Divide.

25. $(15x^6 - 5x^4 + 25x^3) \div 5x^3$

26. $(8x^3 - 22x^2 - 5x + 12) \div (4x + 3)$

27. $(2x^3 - 6x - 36) \div (x - 3)$

Solve. Express your answer in scientific notation. Round to the nearest hundredth.

28. At the end of 2002, Saudi Arabia was estimated to have 2.618×10^{11} barrels of oil reserves. By one estimate, they have 86 years of reserves remaining. If they disbursed the oil in an equal amount each year, how many barrels of oil would they pump each year? (*Source:* United Nations Review of World Energy, 2003.)

29. A space probe is traveling from Earth to Pluto at a speed of 2.49×10^4 miles per hour. How far would this space probe travel in one week?

1. _____

2. _____

3. _____

4. _____

5. _____

6. _____

7. _____

8. _____

9. _____

10. _____

11. _____

12. _____

13. _____

14. _____

15. _____

16. _____

17. _____

18. _____

19. _____

20. _____

21. _____

22. _____

23. _____

24. _____

25. _____

26. _____

27. _____

28. _____

29. _____

1. _____

2. _____

3. _____

4. _____

5. _____

6. _____

7. _____

8. _____

9. _____

10. _____

11. _____

12. _____

13. _____

14. _____

15. _____

16. _____

17. _____

18. _____

19. _____

20. _____

21. _____

22. _____

23. _____

24. _____

25. _____

Approximately one-half of this test covers the content of Chapters 0 through 4. The remainder covers the content of Chapter 5.

In questions 1–4, simplify.

1. $\dfrac{11}{15} - \dfrac{7}{10}$

2. $(-3.7) \times (0.2)$

3. $\left(-4\dfrac{1}{2}\right) \div \left(5\dfrac{1}{4}\right)$

4. $7x(3x - 4) - 5x(2x - 3) - (3x)^2$

5. Find 6% of 1842.5

6. Evaluate $3a^2 + ab - 4b^2$ for $a = 5$ and $b = -1$.

In questions 7–10, solve.

7. $7x - 3(4 - 2x) = 14x - (3 - x)$

8. $\dfrac{2}{3}x + 6 = 4(x - 11)$

9. $2 - 5x > 17$

10. $2 + 3x \geq x + 5$

11. A national walkout of 11,904 employees of the VBM Corp. occurred last month. This was 96% of the total number of employees. How many employees does VBM have?

12. Multiply. $(4x - 5)(5x + 1)$

13. Multiply. $(3x - 5)^2$

14. Multiply. $(3x + 2)(2x + 1)(x - 4)$

In questions 16–18, simplify.

15. $(-4x^4y^5)(5xy^3)$

16. $\dfrac{14x^8y^3}{-21x^5y^{12}}$

17. $(-2x^3y^2z^0)^4$

18. Write with only positive exponents. $\dfrac{9x^{-3}y^{-4}}{w^2z^{-8}}$

19. Write in scientific notation. 42,500,000,000

20. Write in scientific notation. 0.00056

21. Calculate. Leave your answer in scientific notation. $\dfrac{(2.0 \times 10^{-12})(8.0 \times 10^{-20})}{4.0 \times 10^3}$

22. Subtract. $(x^3 - 3x^2 - 5x + 20) - (-4x^3 - 10x^2 + x - 30)$

Multiply.

23. $-6xy^2(6x^2 - 3xy + 8y^2)$

24. $(2x^2 - 3x + 1)(3x - 5)$

25. Divide. $(x^2 + 2x - 12) \div (x - 3)$

CHAPTER

6

Niagara Falls is an amazing wonder. Every year hundreds of thousands of tourists flock to see this spectacular waterfall. Besides its natural beauty, it also powers three hydroelectric stations that produce 4.4 gigawatts of power. Many of the equations related to hydroelectric power are quadratic equations similar to the ones studied in this chapter.

Factoring

Student Learning Objective

After studying this section, you will be able to:

 Factor polynomials whose terms contain a common factor.

1 Factoring Polynomials Whose Terms Contain a Common Factor

Recall that when two or more numbers, variables, or algebraic expressions are multiplied, each is called a **factor**.

$$\underbrace{3}_{\text{factor}} \cdot \underbrace{2}_{\text{factor}} \qquad \underbrace{3x^2}_{\text{factor}} \cdot \underbrace{5x^3}_{\text{factor}} \qquad \underbrace{(2x - 3)}_{\text{factor}}\underbrace{(x + 4)}_{\text{factor}}$$

When you are asked **to factor** a number or an algebraic expression, you are being asked, "What factors, when multiplied, will give that number or expression?"

For example, you can factor 6 as $3 \cdot 2$ since $3 \cdot 2 = 6$. You can factor $15x^5$ as $3x^2 \cdot 5x^3$ since $3x^2 \cdot 5x^3 = 15x^5$. Factoring is simply the reverse of multiplying. 6 and $15x^5$ are simple expressions to factor and can be factored in different ways.

The factors of the polynomial $2x^2 + x - 12$ are not so easy to recognize. Factoring a polynomial changes an addition and/or subtraction problem into a multiplication problem. In this chapter we will be learning techniques for finding the factors of a polynomial. We will begin with **common factors.**

> **EXAMPLE 1** Factor. **(a)** $3x - 6y$ **(b)** $9x + 2xy$
>
> **Solution** Begin by looking for a common factor, a factor that both terms have in common. Then rewrite the expression as a product.
>
> **(a)** $3x - 6y = 3(x - 2y)$ This is true because $3(x - 2y) = 3x - 6y$.
> **(b)** $9x + 2xy = x(9 + 2y)$ This is true because $x(9 + 2y) = 9x + 2xy$.

Some people find it helpful to think of factoring as the distributive property in reverse. When we write $3x - 6y = 3(x - 2y)$, we are doing the reverse of distributing the 3. This is the main point of this section. We are doing problems of the form $ca + cb = c(a + b)$. The common factor c becomes the first factor of our answer.

NOTE TO STUDENT: Fully worked-out solutions to all of the Practice Problems can be found at the back of the text starting at page SP-1

> **Practice Problem 1** Factor.
>
> **(a)** $21a - 7b$ **(b)** $5xy + 8x$

When we factor, we begin by looking for the **greatest common factor.** For example, in the polynomial $48x - 16y$, a common factor is 2. We could factor $48x - 16y$ as $2(24x - 8y)$. However, this is not complete. To factor $48x - 16y$ completely, we look for the greatest common factor of 48 and of 16.

$$48x - 16y = 16(3x - y)$$

> **EXAMPLE 2** Factor $24xy + 12x^2 + 36x^3$. Remember to remove the greatest common factor.
>
> **Solution** Find the greatest common factor of 24, 12, and 36. You may want to factor each number, or you may notice that 12 is a common factor. 12 is the greatest numerical common factor.
>
> Notice also that x is a factor of each term. Thus, $12x$ is the greatest common factor.
>
> $$24xy + 12x^2 + 36x^3 = 12x(2y + x + 3x^2)$$

> **Practice Problem 2** Factor $12a^2 + 16ab^2 - 12a^2b$. Be careful to remove the greatest common factor.

FACTORING A POLYNOMIAL WITH COMMON FACTORS

1. Determine the greatest common numerical factor by asking, "What is the largest integer that will divide into the coefficients of all the terms?"

2. Determine the greatest common variable factor by first asking, "What variables are common to all the terms?" Then, for each variable that is common to all the terms ask, "What is the largest exponent of the variable that is common to all the terms?"

3. The common factors found in steps 1 and 2 are the first part of the answer.

4. After removing the common factors, what remains is placed in parentheses as the second factor.

EXAMPLE 3 Factor. **(a)** $12x^2 - 18y^2$ **(b)** $x^2y^2 + 3xy^2 + y^3$

Solution

(a) Note that the largest integer that is common to both terms is 6 (not 3 or 2).

$$12x^2 - 18y^2 = 6(2x^2 - 3y^2)$$

(b) Although y is common to all of the terms, we factor out y^2 since 2 is the largest exponent of y that is common to all terms. We do not factor out x, since x is not common to all of the terms.

$$x^2y^2 + 3xy^2 + y^3 = y^2(x^2 + 3x + y)$$

Practice Problem 3 Factor.

(a) $16a^3 - 24b^3$ **(b)** $r^3s^2 - 4r^4s + 7r^5$

Checking Is Very Important! You can check any factoring problem by multiplying the factors you obtain. The result should be the same as the original polynomial.

EXAMPLE 4 Factor. $8x^3y + 16x^2y^2 - 24x^3y^3$

Solution We see that 8 is the largest integer that will divide evenly into the three numerical coefficients. We can factor an x^2 out of each term. We can also factor y out of each term.

$$8x^3y + 16x^2y^2 - 24x^3y^3 = 8x^2y(x + 2y - 3xy^2)$$

Check.

$$8x^2y(x + 2y - 3xy^2) = 8x^3y + 16x^2y^2 - 24x^3y^3 \checkmark$$

Practice Problem 4 Factor. $18a^3b^2c - 27ab^3c^2 - 45a^2b^2c^2$

EXAMPLE 5 Factor. $9a^3b^2 + 9a^2b^2$

Solution We observe that both terms contain a common factor of 9. We can factor a^2 and b^2 out of each term. Thus the common factor is $9a^2b^2$.

$$9a^3b^2 + 9a^2b^2 = 9a^2b^2(a + 1)$$

CAUTION: Don't forget to include the 1 inside the parentheses in Example 5. The solution is wrong without it. You will see why if you try to check a result written without the 1.

Practice Problem 5 Factor and check. $30x^3y^2 - 24x^2y^2 + 6xy^2$

EXAMPLE 6 Factor. $3x(x - 4y) + 2(x - 4y)$

Solution Be sure you understand what are terms and what are factors of the polynomial in this example. There are two terms. The expression $3x(x - 4y)$ is one term. The expression $2(x - 4y)$ is the second term. Each term is made up of two factors. Observe that the binomial $(x - 4y)$ is a common factor of the terms. A common factor may be any type of polynomial. Thus we can factor out the common factor $(x - 4y)$.

$$3x(x - 4y) + 2(x - 4y) = (x - 4y)(3x + 2)$$

Practice Problem 6 Factor. $3(a + 5b) + x(a + 5b)$

EXAMPLE 7 Factor. $7x^2(2x - 3y) - (2x - 3y)$

Solution The common factor of the terms is $(2x - 3y)$. What happens when we factor out $(2x - 3y)$? What are we left with in the second term?
Recall that $(2x - 3y) = 1(2x - 3y)$. Thus

$$7x^2(2x - 3y) - (2x - 3y) = 7x^2(2x - 3y) - 1(2x - 3y)$$ Rewrite the original expression.

$$= (2x - 3y)(7x^2 - 1)$$ Factor out $(2x - 3y)$.

Practice Problem 7 Factor. $8y(9y^2 - 2) - (9y^2 - 2)$

NOTE TO STUDENT: Fully worked-out solutions to all of the Practice Problems can be found at the back of the text starting at page SP-1

▲ **EXAMPLE 8** A computer programmer is writing a program to find the total area of 4 circles. She uses the formula $A = \pi r^2$. The radii of the circles are a, b, c, and d, respectively. She wants the final answer to be in factored form with the value of π occurring only once, in order to minimize the rounding error. Write the total area with a formula that has π occurring only once.

Solution

For each circle, $A = \pi r^2$, where $r = a, b, c,$ or d.

The total area is $\pi a^2 + \pi b^2 + \pi c^2 + \pi d^2$.

In factored form the total area $= \pi(a^2 + b^2 + c^2 + d^2)$.

▲ **Practice Problem 8** Use $A = \pi r^2$ to find the shaded area. The radius of the larger circle is b. The radius of the smaller circle is a. Write the shaded-area formula in factored form so that π appears only once.

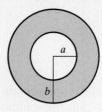

Verbal and Writing Skills *In exercises 1 and 2, write a word or words to complete each sentence.*

1. In the expression $3x^2 \cdot 5x^3$, $3x^2$ and $5x^3$ are called _____.

2. In the expression $3x^2 + 5x^3$, $3x^2$ and $5x^3$ are called _____.

3. We can factor $30a^4 + 15a^3 - 45a^2$ as $5a(6a^3 + 3a^2 - 9a)$. Is the factoring complete? Why or why not?

4. We can factor $4x^3 - 8x^2 + 20x$ as $4(x^3 - 2x^2 + 10x)$. Is the factoring complete?

Remove the largest possible common factor. Check your answers for exercises 5–28 by multiplication.

5. $3a^2 + 3a$

6. $2c^2 + 2c$

7. $21ab - 14ab^2$

8. $18wz - 27w^2z$

9. $2\pi rh + 2\pi r^2$

10. $9a^2b^2 - 36ab$

11. $5x^3 + 25x^2 - 15x$

12. $8x^3 - 10x^2 - 14x$

13. $12ab - 28bc + 20ac$

14. $12xy - 18yz - 36xz$

15. $16x^5 + 24x^3 - 32x^2$

16. $36x^6 + 45x^4 - 18x^2$

17. $14x^2y - 35xy - 63x$

18. $40a^2 - 16ab - 24a$

19. $54x^2 - 45xy + 18x$

20. $48xy - 24y^2 + 40y$

21. $3xy^2 - 2ay + 5xy - 2y$

22. $2ab^3 + 3xb^2 - 5b^4 + 2b^2$

23. $24x^2y - 40xy^2$

24. $35abc^2 - 49ab^2c$

25. $7x^3y^2 + 21x^2y^2$

26. $8x^3y^2 + 32xy^2$

27. $8x^4y + 24x^2y^2 - 32x^2y$

28. $30x^2y^2 - 24xy^3 + 12x^2y$

Hint: In exercises 29–42, refer to Examples 6 and 7.

29. $7a(x + 2y) - b(x + 2y)$

30. $6(3a + b) - z(3a + b)$

31. $3x(x - 4) - 2(x - 4)$

32. $5x(x - 7) + 3(x - 7)$

33. $6b(2a - 3c) - 5d(2a - 3c)$

34. $7x(3y + 5z) - 6t(3y + 5z)$

35. $7c(b - a^2) - 5d(b - a^2) + 2f(b - a^2)$

36. $2a(x^2 - y) - 5b(x^2 - y) + c(x^2 - y)$

37. $2a(xy - 3) - 4(xy - 3) - z(xy - 3)$

38. $3c(bc - 3a) - 2(bc - 3a) - 6b(bc - 3a)$

39. $4a^3(a - 3b) + (a - 3b)$

40. $(c + 4d) + 7c^2(c + 4d)$

41. $(a + 2) - x(a + 2)$

42. $y(5x - 1) - (5x - 1)$

Cumulative Review

43. **[2.6.1]** *Integers* The sum of three consecutive odd integers is 40 more than the smallest of these integers. Find each of these three consecutive odd integers. (*Hint:* Consecutive odd integers are numbers included in the following pattern: $-3, -1, 1, 3, 5, 7, 9, \ldots, x, x + 2, \ldots$).

Coffee Production   *In 2006, the top six green (unroasted beans) coffee-producing countries of the world produced 5,350,000 metric tons of coffee beans. The percent of this 5.35 million metric tons produced by each of these countries is shown in the following graph. Use the graph to answer the following questions. (Source: en.wikipedia.org)*

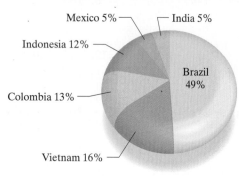

Percent of Coffee Produced by the Top Six Coffee-Producing Countries of the World in 2006

Mexico 5% — India 5%
Indonesia 12% —
Colombia 13% —
Brazil 49%
Vietnam 16% —

44. **[0.5.3]** How many metric tons of coffee were produced in Vietnam?

45. **[0.5.3]** How many metric tons of coffee were produced in Brazil?

46. **[2.6.2]** The population of Vietnam in 2006 was approximately 85,000,000 people. How many pounds of coffee were produced in Vietnam for each person? (Round your answer to the nearest whole number. One metric ton is about 2205 pounds.)

47. **[2.6.2]** The population of Brazil in 2006 was approximately 190,000,000 people. How many pounds of coffee were produced in Brazil for each person? (Round your answer to the nearest whole number. One metric ton is about 2205 pounds.)

Quick Quiz 6.1   Remove the largest possible common factor.

1. $3x - 4x^2 + 2xy$

2. $20x^3 - 25x^2 - 5x$

3. $8a(a + 3b) - 7b(a + 3b)$

4. **Concept Check** Explain how you would remove the greatest common factor from the following polynomial.
$$36a^3b^2 - 72a^2b^3$$

Factoring Expressions with Four Terms by Grouping

A common factor of a polynomial can be a number, a variable, or an algebraic expression. Sometimes the polynomial is written so that it is easy to recognize the common factor. This is especially true when the common factor is enclosed by parentheses.

Student Learning Objective

After studying this section, you will be able to:

1 Factor expressions with four terms by grouping.

EXAMPLE 1 Factor. $x(x - 3) + 2(x - 3)$

Solution Observe each term:

$$\underbrace{x(x - 3)}_{\substack{\text{first} \\ \text{term}}} + \underbrace{2(x - 3)}_{\substack{\text{second} \\ \text{term}}}$$

The common factor of the first and second terms is the quantity $(x - 3)$, so we have

$$x(x - 3) + 2(x - 3) = (x - 3)(x + 2).$$

Practice Problem 1 Factor. $3y(2x - 7) - 8(2x - 7)$

NOTE TO STUDENT: Fully worked-out solutions to all of the Practice Problems can be found at the back of the text starting at page SP-1

Now let us face a new challenge. Think carefully. Try to follow this new idea. Suppose the polynomial in Example 1, $x(x - 3) + 2(x - 3)$, were written in the form $x^2 - 3x + 2x - 6$. (Note that this form is obtained by multiplying the factors of the first and second terms.) How would we factor a four-term polynomial like this?

In such cases we remove a common factor from the first two terms and a different common factor from the second two terms. That is, we would factor x from $x^2 - 3x$ and 2 from $2x - 6$.

$$x^2 - 3x + 2x - 6 = x(x - 3) + 2(x - 3)$$

Because the resulting terms have a common factor (the binomial enclosed by the parentheses), we would then proceed as we did in Example 1 to obtain the factored form $(x - 3)(x + 2)$. This procedure for factoring is often called **factoring by grouping.**

EXAMPLE 2 Factor. $2x^2 + 3x + 6x + 9$

Solution
$$2x^2 + 3x \qquad + \qquad 6x + 9$$

Factor out a common factor of x from the first two terms.
Factor out a common factor of 3 from the second two terms.

$$\underbrace{x(2x + 3)} \qquad\qquad \underbrace{3(2x + 3)}$$

Note that the sets of parentheses in the two terms contain the same expression at this step.

The expression in parentheses is now a common factor of the terms. Now we finish the factoring.

$$2x^2 + 3x + 6x + 9 = x(2x + 3) + 3(2x + 3)$$
$$= (2x + 3)(x + 3)$$

Practice Problem 2 Factor. $6x^2 - 15x + 4x - 10$

EXAMPLE 3 Factor. $4x + 8y + ax + 2ay$

Solution

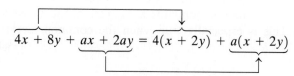

Factor out a common
factor of 4 from
the first two terms.

$$4x + 8y + ax + 2ay = 4(x + 2y) + a(x + 2y)$$

Factor out a common
factor of a from
the second two terms.

$4(x + 2y) + a(x + 2y) = (x + 2y)(4 + a)$ The common factor of the terms
is the expression in parentheses,
$x + 2y$.

Practice Problem 3 Factor by grouping. $ax + 2a + 4bx + 8b$

NOTE TO STUDENT: *Fully worked-out solutions to all of the Practice Problems can be found at the back of the text starting at page SP-1*

In some problems the terms are out of order. In this case, we have to rearrange the order of the terms first so that the first two terms have a common factor.

EXAMPLE 4 Factor. $bx + 4y + 4b + xy$

Solution

$bx + 4y + 4b + xy = bx + 4b + xy + 4y$ Rearrange the terms so that the
first terms have a common term.

$\qquad\qquad\qquad = b(x + 4) + y(x + 4)$ Factor out the common factor
of b from the first two terms
and the common factor of y
from the second two terms.

$\qquad\qquad\qquad = (x + 4)(b + y)$

Practice Problem 4 Factor. $6a^2 + 5bc + 10ab + 3ac$

Sometimes you will need to factor out a *negative common factor* from the second two terms to obtain two terms that contain the same parenthetical expression.

EXAMPLE 5 Factor. $2x^2 + 5x - 4x - 10$

Solution

$2x^2 + 5x - 4x - 10 = x(2x + 5) - 4x - 10$ Factor out the common factor
of x from the first two terms.

$\qquad\qquad\qquad = x(2x + 5) - 2(2x + 5)$ Factor out the common factor of -2
from the second two terms.

$\qquad\qquad\qquad = (2x + 5)(x - 2)$

CAUTION: If you factored out a common factor of $+2$ in the second step, the two resulting terms would not contain the same parenthetical expression. If the expressions inside the two sets of parentheses are *not exactly the same,* you cannot express the polynomial as a product of two factors!

Practice Problem 5 Factor. $6xy + 14x - 15y - 35$

EXAMPLE 6 Factor. $2ax - a - 2bx + b$

Solution

$$2ax - a - 2bx + b = a(2x - 1) - b(2x - 1)$$ Factor out the common factor of a from the first two terms. Factor out the common factor of $-b$ from the second two terms.

$$= (2x - 1)(a - b)$$ Since the two resulting terms contain the same parenthetical expression, we can complete the factoring.

Practice Problem 6 Factor. $3x + 6y - 5ax - 10ay$

CAUTION: Many students find that they make a factoring error in the first step of problems like Example 6. When factoring out $-b$, be sure to check your signs carefully: $-2bx + b = -b(2x - 1)$.

EXAMPLE 7 Factor and check your answer. $8ad + 21bc - 6bd - 28ac$

Solution We observe that the first two terms do not have a common factor.

$$8ad + 21bc - 6bd - 28ac = 8ad - 6bd - 28ac + 21bc$$ Rearrange the order using the commutative property of addition.

$$= 2d(4a - 3b) - 7c(4a - 3b)$$ Factor out the common factor of $2d$ from the first two terms and the common factor of $-7c$ from the last two terms.

$$= (4a - 3b)(2d - 7c)$$ Factor out the common factor of $(4a - 3b)$.

To check, we multiply the two binomials using the FOIL procedure.

$$(4a - 3b)(2d - 7c) = 8ad - 28ac - 6bd + 21bc$$
$$= 8ad + 21bc - 6bd - 28ac \quad \checkmark$$ Rearrange the order of the terms. This is the original problem. Thus it checks.

Practice Problem 7 Factor and check your answer.

$$10ad + 27bc - 6bd - 45ac$$

PRACTICE   WATCH   DOWNLOAD   READ   REVIEW

Verbal and Writing Skills

1. To factor $3x^2 - 6xy + 5x - 10y$, we must first remove a common factor of $3x$ from the first two terms. What do we do with the last two terms? What should we get for the answer?

2. To factor $5x^2 + 15xy - 2x - 6y$, we must first remove a common factor of 5 from the first two terms. What do we do with the last two terms? What should we get for the answer?

Factor by grouping. Check your answers for exercises 3–26.

3. $ab - 3a + 4b - 12$

4. $xy - x + 4y - 4$

5. $x^3 - 4x^2 + 3x - 12$

6. $x^3 - 6x^2 + 2x - 12$

7. $2ax + 6bx - ay - 3by$

8. $4x + 8y - 3wx - 6wy$

9. $3ax + bx - 6a - 2b$

10. $ad + 3a - d^2 - 3d$

11. $5a + 12bc + 10b + 6ac$

12. $4u^2 + v + 4uv + u$

13. $6c - 12d + cx - 2dx$

14. $xy + 5x - 5y - 25$

15. $y^2 - 2y - 3y + 6$

16. $x^2 - 4x + 3x - 12$

17. $54 - 6y + 9y - y^2$

18. $ax - 3a - 2bx + 6b$

19. $6ax - y + 2ay - 3x$

20. $6tx + r - 3t - 2rx$

21. $2x^2 + 8x - 3x - 12$

22. $3y^2 - y + 9y - 3$

23. $t^3 - t^2 + t - 1$

24. $x^2 - 2x - xy + 2y$

25. $28x^2 + 8xy^2 + 21xw + 6y^2w$

26. $15wx + 18x^2 + 12xy^2 + 10wy^2$

To Think About

27. Although $6a^2 - 12bd - 8ad + 9ab = 6(a^2 - 2bd) - a(8d - 9b)$ is true, it is not the correct solution to the problem "Factor $6a^2 - 12bd - 8ad + 9ab$." Explain. Can this expression be factored?

28. Tim was trying to factor $5x^2 - 3xy - 10x + 6y$. In his first step he wrote down $x(5x - 3y) + 2(-5x + 3y)$. Was he doing the problem correctly? What is the answer?

Cumulative Review

29. [0.5.4] **Online Sales** In 2010, it is estimated that online retail sales in the United States will total $328.6 billion. About $119.1 billion of this will be from sales related to travel. What percent of the total will be travel sales? Round your answer to the nearest tenth of a percent. (*Source:* www.census.gov.)

30. [2.6.1] **Professor Salaries** In 2006, the average salary for a professor at a public 2-year or 4-year institution was $11,200 more than twice the average salary for an instructor. The two salaries total $131,500. Find the average salaries of a professor and an instructor. (*Source:* www.census.gov.)

31. [2.6.2] **Wheat Production** During the years 2001–2002, 581 million metric tons of wheat were produced worldwide. By 2003–2004, total wheat production had decreased by 4.6%. During the years 2005–2006, worldwide wheat production had increased by 12.3% from the 2003–2004 amount. How many metric tons of wheat were produced in 2005–2006? (*Source:* www.census.gov.)

Quick Quiz 6.2 Factor by grouping.

1. $7ax + 12a - 14x - 24$

2. $2xy^2 - 15 + 6x - 5y^2$

3. $10xy - 3x + 40by - 12b$

4. **Concept Check** Explain how you would factor the following polynomial.
$$10ax + b^2 + 2bx + 5ab$$

Student Learning Objectives

After studying this section, you will be able to:

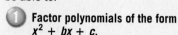 Factor polynomials of the form $x^2 + bx + c$.

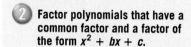 Factor polynomials that have a common factor and a factor of the form $x^2 + bx + c$.

 ## Factoring Polynomials of the Form $x^2 + bx + c$

Suppose that you wanted to factor $x^2 + 5x + 6$. After some trial and error you *might* obtain $(x + 2)(x + 3)$, or you might get discouraged and not get an answer. If you did get these factors, you could check this answer by the FOIL method.

$$(x + 2)(x + 3) = x^2 + 3x + 2x + 6$$
$$= x^2 + 5x + 6$$

But trial and error can be a long process. There is another way. Let's look at the preceding equation again.

$$
\begin{array}{cccc}
& \text{F} & \text{O} & \text{I} & \text{L} \\
(x + 2)(x + 3) = & x^2 & + 3x & + 2x & + 6 \\
& \downarrow & \downarrow & & \downarrow \\
= & x^2 & + 5x & & + 6
\end{array}
$$

The first thing to notice is that the product of the first terms in the factors gives the first term of the polynomial. That is, $x \cdot x = x^2$.

> The first term is the product of these terms.
> $x^2 + 5x + 6 \qquad = \qquad (x + 2)(x + 3)$

The next thing to notice is that the sum of the products of the outer and inner terms in the factors produces the middle term of the polynomial. That is, $(x \cdot 3) + (2 \cdot x) = 3x + 2x = 5x$. Thus we see that the sum of the second terms in the factors, $2 + 3$, gives the coefficient of the middle term, 5.

Finally, note that the product of the last terms of the factors gives the last term of the polynomial. That is, $2 \cdot 3 = 6$.

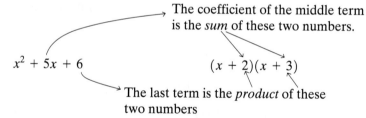

The coefficient of the middle term is the *sum* of these two numbers.

$x^2 + 5x + 6 \qquad\qquad (x + 2)(x + 3)$

The last term is the *product* of these two numbers

Let's summarize our observations in general terms and then try a few examples.

FACTORING TRINOMIALS OF THE FORM $x^2 + bx + c$

1. The answer will be of the form $(x + m)(x + n)$.

2. m and n are numbers such that:
   (a) When you multiply them, you get the last term, which is c.
   (b) When you add them, you get the coefficient of the middle term, which is b.

EXAMPLE 1 Factor. $x^2 + 7x + 12$

Solution The answer is of the form $(x + m)(x + n)$. We want to find the two numbers, m and n, that you can multiply to get 12 and add to get 7. The numbers are 3 and 4.

$$x^2 + 7x + 12 = (x + 3)(x + 4)$$

Practice Problem 1 Factor. $x^2 + 8x + 12$

NOTE TO STUDENT: Fully worked-out solutions to all of the Practice Problems can be found at the back of the text starting at page SP-1

EXAMPLE 2   Factor. $x^2 + 12x + 20$

Solution   We want two numbers that have a product of 20 and a sum of 12. The numbers are 10 and 2.

$$x^2 + 12x + 20 = (x + \underline{10})(x + \underline{2})$$

Note: If you cannot think of the numbers in your head, write down the possible factors whose product is 20.

Product	Sum
$1 \cdot 20 = 20$	$1 + 20 = 21$
$2 \cdot 10 = 20$	$2 + 10 = 12 \leftarrow$
$4 \cdot 5 = 20$	$4 + 5 = 9$

Then select the pair whose sum is 12.   Select this pair.

Practice Problem 2   Factor. $x^2 + 17x + 30$
You may find that it is helpful to list all the factors whose product is 30 first.

So far we have factored only trinomials of the form $x^2 + bx + c$, where b and c are positive numbers. The same procedure applies if b is a negative number and c is positive. Because m and n have a positive product and a negative sum, they must both be negative.

EXAMPLE 3   Factor. $x^2 - 8x + 15$

Solution   We want two numbers that have a product of $+15$ and a sum of -8. They must be negative numbers since the sign of the middle term is negative and the sign of the last term is positive.

$$\overset{\text{the sum } -5 + (-3)}{x^2 - 8x + 15 = (x - 5)(x - 3)}$$
the product $(-5)(-3)$

Think: $(-5)(-3) = +15$
and $-5 + (-3) = -8.$

Multiply using FOIL to check.

Practice Problem 3   Factor. $x^2 - 11x + 18$

EXAMPLE 4   Factor. $x^2 - 9x + 14$

Solution   We want two numbers whose product is 14 and whose sum is -9. The numbers are -7 and -2. So

$$x^2 - 9x + 14 = (x - 7)(x - 2) \text{ or } (x - 2)(x - 7).$$

Practice Problem 4   Factor. $x^2 - 11x + 24$

All the examples so far have had a positive last term. What happens when the last term is negative? If the last term is a negative number, one of the numbers m or n must be a positive number and the other must be a negative number. Why? The product of a positive number and a negative number is negative.

EXAMPLE 5 Factor. $x^2 - 3x - 10$

Solution We want two numbers whose product is -10 and whose sum is -3. The two numbers are -5 and $+2$.

$$x^2 - 3x - 10 = (x - 5)(x + 2)$$

Practice Problem 5 Factor. $x^2 - 5x - 24$

What if we made a sign error and *incorrectly* factored the trinomial $x^2 - 3x - 10$ as $(x + 5)(x - 2)$? We could detect the error immediately since the sum of $+5$ and -2 is 3. We need a sum of -3!

EXAMPLE 6 Factor $y^2 + 10y - 24$. Check your answer.

Solution The two numbers whose product is -24 and whose sum is $+10$ are $+12$ and -2.

$$y^2 + 10y - 24 = (y + 12)(y - 2)$$

CAUTION: It is very easy to make a sign error in these problems. Make sure that you mentally multiply your answer by FOIL to obtain the original expression. Check each sign carefully.

Check. $(y + 12)(y - 2) = y^2 - 2y + 12y - 24 = y^2 + 10y - 24$ ✓

Practice Problem 6 Factor $y^2 + 17y - 60$.
Multiply your answer to check.

EXAMPLE 7 Factor. $x^2 - 16x - 36$

Solution We want two numbers whose product is -36 and whose sum is -16.
   List all the possible factors of 36 (without regard to sign). Find the pair that has a difference of 16. We are looking for a difference because the signs of the factors are different.

Factors of 36	The Difference Between the Factors
36 and 1	35
18 and 2	16 ← This is the value we want.
12 and 3	9
9 and 4	5
6 and 6	0

Once we have picked the pair of numbers (18 and 2), it is not difficult to find the signs. For the coefficient of the middle term to be -16, we will have to add the numbers -18 and $+2$.

$$x^2 - 16x - 36 = (x - 18)(x + 2)$$

Practice Problem 7 Factor. $x^2 - 7x - 60$
You may find it helpful to list the pairs of numbers whose product is 60.

At this point you should work several problems to develop your factoring skill. This is one section where you really need to drill by doing many problems.

Feel a little confused about the signs? If you do, you may find these facts helpful.

FACTS ABOUT FACTORING TRINOMIALS OF THE FORM $x^2 + bx + c$

The *two numbers m and n* will have the *same sign* if the last term of the polynomial is *positive*.

$$x^2 + bx + c = (x \quad m)(x \quad n)$$

1. They will both be *positive* if the *coefficient* of the *middle* term is *positive*.

$$x^2 + 5x + 6 = (x + 2)(x + 3)$$

2. They will both be *negative* if the *coefficient* of the *middle* term is *negative*.

$$x^2 - 5x + 6 = (x - 2)(x - 3)$$

The two numbers *m* and *n* will have *opposite signs* if the last term is *negative*.

1. The *larger* of the absolute values of the two numbers will be given a plus sign if the coefficient of the *middle term* is *positive*.

$$x^2 + 6x - 7 = (x + 7)(x - 1)$$

2. The larger of the absolute values of the two numbers will be given a negative sign if the coefficient of the *middle term* is *negative*.

$$x^2 - 6x - 7 = (x - 7)(x + 1)$$

Do not memorize these facts; rather, try to understand the pattern.

Sometimes the exponent of the first term of the polynomial will be greater than 2. If the exponent is an even power, it is a square. For example, $x^4 = (x^2)(x^2)$. Likewise, $x^6 = (x^3)(x^3)$.

EXAMPLE 8   Factor. $y^4 - 2y^2 - 35$

Solution   Think: $y^4 = (y^2)(y^2)$ This will be the first term of each set of parentheses.

$(y^2 \quad)(y^2 \quad)$

$(y^2 + \quad)(y^2 - \quad)$ The last term of the polynomial is negative. Thus the signs of *m* and *n* will be different.

$(y^2 + 5)(y^2 - 7)$ Now think of factors of 35 whose difference is 2.

Practice Problem 8   Factor. $a^4 + a^2 - 42$

Factoring Polynomials That Have a Common Factor and a Factor of the Form $x^2 + bx + c$

Some factoring problems require two steps. Often we must first factor out a common factor from each term of the polynomial. Once this is done, we may find that the other factor is a trinomial that can be factored using the methods previously discussed in this section.

EXAMPLE 9   Factor. $2x^2 + 36x + 160$

Solution

$2x^2 + 36x + 160 = 2(x^2 + 18x + 80)$ First factor out the common factor of 2 from each term of the polynomial.

$= 2(x + 8)(x + 10)$ Then factor the remaining polynomial.

The final answer is $2(x + 8)(x + 10)$. *Be sure to list all parts of the answer.*

Check. $2(x + 8)(x + 10) = 2(x^2 + 18x + 80) = 2x^2 + 36x + 160$ ✓

Thus we are sure that the answer is $2(x + 8)(x + 10)$.

Practice Problem 9   Factor. $3x^2 + 45x + 150$

EXAMPLE 10 Factor. $3x^2 + 9x - 162$

Solution

$3x^2 + 9x - 162 = 3(x^2 + 3x - 54)$ First factor out the common factor of 3 from each term of the polynomial.

$\qquad\qquad\qquad\quad = 3(x - 6)(x + 9)$ Then factor the remaining polynomial.

The final answer is $3(x - 6)(x + 9)$. *Be sure you include the 3.*

Check. $3(x - 6)(x + 9) = 3(x^2 + 3x - 54) = 3x^2 + 9x - 162$ ✓

Thus we are sure that the answer is $3(x - 6)(x + 9)$.

Practice Problem 10 Factor. $4x^2 - 8x - 140$

NOTE TO STUDENT: *Fully worked-out solutions to all of the Practice Problems can be found at the back of the text starting at page SP-1*

CAUTION: Don't forget the common factor!

It is quite easy to forget to look for a greatest common factor as the first step of factoring a trinomial. Therefore, it is a good idea to examine your final answer in any factoring problem and ask yourself, "Can I factor out a common factor from any binomial contained inside a set of parentheses?" Often you will be able to see a common factor at that point if you missed it in the first step of the problem.

▲ **EXAMPLE 11** Find a polynomial in factored form for the shaded area in the figure.

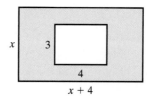

Solution To obtain the shaded area, we find the area of the larger rectangle and subtract from it the area of the smaller rectangle. Thus we have the following:

$$\text{shaded area} = x(x + 4) - (4)(3)$$
$$= x^2 + 4x - 12$$

Now we factor this polynomial to obtain the shaded area $= (x + 6)(x - 2)$.

▲ **Practice Problem 11** Find a polynomial in factored form for the shaded area in the figure.

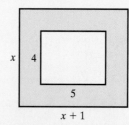

Verbal and Writing Skills

Fill in the blanks.

1. To factor $x^2 + 5x + 6$, find two numbers whose _____ is 6 and whose _____ is 5.

2. To factor $x^2 + 5x - 6$, find two numbers whose _____ is -6 and whose _____ is 5.

Factor.

3. $x^2 + 2x + 1$

4. $x^2 + 11x + 30$

5. $x^2 + 12x + 35$

6. $x^2 + 10x + 21$

7. $x^2 - 4x + 3$

8. $x^2 - 6x + 8$

9. $x^2 - 11x + 28$

10. $x^2 - 13x + 12$

11. $x^2 + 5x - 24$

12. $x^2 + 5x - 36$

13. $x^2 - 13x - 14$

14. $x^2 - 6x - 16$

15. $x^2 + 2x - 35$

16. $x^2 + 4x - 12$

17. $x^2 - 2x - 24$

18. $x^2 - 11x - 26$

19. $x^2 + 15x + 36$

20. $x^2 + 15x + 44$

21. $x^2 - 10x + 24$

22. $x^2 - 13x + 42$

23. $x^2 + 13x + 30$

24. $x^2 + 9x + 20$

25. $x^2 - 6x + 5$

26. $x^2 - 15x + 54$

Mixed Practice   *Look over your answers to exercises 3–26 carefully. Be sure that you are clear on your sign rules. Exercises 27–42 contain a mixture of all the types of problems in this section. Make sure you can do them all. Check your answers by multiplication.*

27. $a^2 + 6a - 16$

28. $a^2 - 13a + 30$

29. $x^2 - 12x + 32$

30. $x^2 - 6x - 27$

31. $x^2 + 4x - 21$

32. $x^2 - 9x + 18$

33. $x^2 + 15x + 56$

34. $x^2 + 20x + 99$

35. $y^2 + 4y - 45$

36. $x^2 + 12x - 45$

37. $x^2 + 9x - 36$

38. $x^2 - 13x + 36$

39. $x^2 - 2xy - 15y^2$

40. $x^2 - 3xy - 28y^2$

41. $x^2 - 16xy + 63y^2$

42. $x^2 + 19xy + 48y^2$

In exercises 43–54, first factor out the greatest common factor from each term. Then factor the remaining polynomial. Refer to Examples 9 and 10.

43. $4x^2 + 24x + 20$

44. $4x^2 + 28x + 40$

45. $6x^2 + 18x + 12$

46. $6x^2 + 24x + 18$

47. $5x^2 - 30x + 25$

48. $2x^2 - 20x + 32$

49. $3x^2 - 6x - 72$

50. $3x^2 - 12x - 63$

51. $7x^2 + 21x - 70$

52. $7x^2 + 7x - 84$

53. $3x^2 - 18x + 15$

54. $5x^2 - 35x + 60$

▲ **55. *Geometry*** Find a polynomial in factored form for the shaded area. Both figures are rectangles with dimensions as labeled.

▲ **56. *Geometry*** How much larger is the perimeter of the large rectangle than the perimeter of the small rectangle?

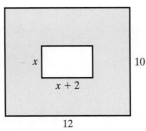

Cumulative Review

57. [3.1.3] Solve for y. $2y - 7x = 6$

58. [2.8.4] Solve $2 - 3x \le 7$ for x. Graph the solution on the number line.

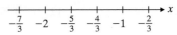

59. [2.6.3] *Travel Speed* A new car that maintains a constant speed travels from Watch Hill, Rhode Island, to Greenwich, Connecticut, in 2 hours. A train, traveling 20 mph faster, makes the trip in $1\frac{1}{2}$ hours. How far is it from Watch Hill to Greenwich? (*Hint:* Let c = the car's speed. First find the speed of the car and the speed of the train. Then be sure to answer the question.)

60. [2.6.2] *Salary* Carla works in an electronics store. She is paid $600 per month plus 4% commission on her sales. If Carla's sales for the month are $80,000, how much will she earn for the month?

Average Temperature *The equation $T = 19 + 2M$ has been used by some meteorologists to predict the monthly average temperature for the small island of Menorca off the coast of Spain during the first 6 months of the year. The variable T represents the average monthly temperature measured in degrees Celsius. The variable M represents the number of months since January.*

61. [1.8.2] What is the average temperature of Menorca during the month of April?

62. [1.8.2] During what month will the average temperature be 29°C?

Quick Quiz 6.3   Factor completely.

1. $x^2 + 17x + 70$

2. $x^2 - 14x + 48$

3. $2x^2 - 4x - 96$

4. Concept Check Explain how you would completely factor $4x^2 - 4x - 120$.

6.4 FACTORING TRINOMIALS OF THE FORM $ax^2 + bx + c$

Using the Trial-and-Error Method

When the coefficient of the x^2-term in a trinomial of the form $ax^2 + bx + c$ is not 1, the trinomial is more difficult to factor. Several possibilities must be considered.

EXAMPLE 1 Factor. $2x^2 + 5x + 3$

Solution In order for the coefficient of the x^2-term of the polynomial to be 2, the coefficients of the x-terms in the factors must be 2 and 1. Thus $2x^2 + 5x + 3 = (2x\ \ )(x\ \ )$.

In order for the last term of the polynomial to be 3, the constants in the factors must be 3 and 1.

Since all signs in the polynomial are positive, we know that each factor in parentheses will contain only positive signs. However, we still have two possibilities. They are as follows:

$$(2x + 1)(x + 3)$$
$$(2x + 3)(x + 1)$$

We check them by multiplying by the FOIL method.

$$(2x + 1)(x + 3) = 2x^2 + 7x + 3 \quad \text{Wrong middle term}$$
$$(2x + 3)(x + 1) = 2x^2 + 5x + 3 \quad \text{Correct middle term}$$

Thus the correct answer is

$$(2x + 3)(x + 1) \quad \text{or} \quad (x + 1)(2x + 3).$$

Practice Problem 1 Factor. $2x^2 + 7x + 5$

Some problems have many more possibilities.

EXAMPLE 2 Factor. $4x^2 - 13x + 3$

Solution

The Different Factorizations of 4 Are:	The Factorization of 3 Is:
(2)(2)	(1)(3)
(1)(4)	

Let us list the possible factoring combinations and compute the middle term by the FOIL method. Note that the signs of the constants in both factors will be negative. Why?

Possible Factors	Middle Term	Correct?
$(2x - 3)(2x - 1)$	$-8x$	No
$(4x - 3)(x - 1)$	$-7x$	No
$(4x - 1)(x - 3)$	$-13x$	Yes

The correct answer is $(4x - 1)(x - 3)$ or $(x - 3)(4x - 1)$.
This method is called the **trial-and-error method.**

Practice Problem 2 Factor. $9x^2 - 64x + 7$

Student Learning Objectives

After studying this section, you will be able to:

 Factor a trinomial of the form $ax^2 + bx + c$ by the trial-and-error method.

 Factor a trinomial of the form $ax^2 + bx + c$ by the grouping method.

 Factor a trinomial of the form $ax^2 + bx + c$ after a common factor has been factored out of each term.

NOTE TO STUDENT: Fully worked-out solutions to all of the Practice Problems can be found at the back of the text starting at page SP-1

EXAMPLE 3 Factor. $3x^2 - 2x - 8$

Solution

Factorization of 3	**Factorizations of 8**
(3)(1)	(8)(1)
	(4)(2)

Let us list only one-half of the possibilities. We'll let the constant in the first factor of each product be positive.

Possible Factors	**Middle Term**	**Correct Factors?**
$(x + 8)(3x - 1)$	$+23x$	No
$(x + 1)(3x - 8)$	$-5x$	No
$(x + 4)(3x - 2)$	$+10x$	No
$(x + 2)(3x - 4)$	$+2x$	No (but only because the sign is wrong)

So we just *reverse* the signs of the constants in the factors.

	Middle Term	**Correct Factors?**
$(x - 2)(3x + 4)$	$-2x$	Yes

The correct answer is

$$(x - 2)(3x + 4) \quad \text{or} \quad (3x + 4)(x - 2).$$

Practice Problem 3 Factor. $3x^2 - x - 14$

NOTE TO STUDENT: Fully worked-out solutions to all of the Practice Problems can be found at the back of the text starting at page SP-1

It takes a good deal of practice to readily factor problems of this type. The more problems you do, the more proficient you will become. The following method will help you factor more quickly.

2 Using the Grouping Method

One way to factor a trinomial of the form $ax^2 + bx + c$ is to write it with four terms and factor by grouping, as we did in Section 6.2. For example, the trinomial $2x^2 + 13x + 20$ can be written as $2x^2 + 5x + 8x + 20$. Using the methods of Section 6.2, we factor it as follows.

$$2x^2 + 5x + 8x + 20 = x(2x + 5) + 4(2x + 5)$$
$$= (2x + 5)(x + 4)$$

We can factor all factorable trinomials of the form $ax^2 + bx + c$ in this way. We will use the following procedure.

GROUPING NUMBER METHOD FOR FACTORING TRINOMIALS OF THE FORM $ax^2 + bx + c$

1. Obtain the grouping number ac.
2. Find the two numbers whose product is the grouping number and whose sum is b.
3. Use those numbers to write bx as the sum of two terms.
4. Factor by grouping.
5. Multiply to check.

Let's try the problem from Example 1.

EXAMPLE 4 Factor by grouping. $2x^2 + 5x + 3$

Solution

1. The grouping number is $(2)(3) = 6$.
2. The factors of 6 are $6 \cdot 1$ and $3 \cdot 2$. We choose the numbers 3 and 2 because their product is 6 and their sum is 5.
3. We write $5x$ as the sum $2x + 3x$.
4. Factor by grouping.

$$
\begin{aligned}
2x^2 + 5x + 3 &= 2x^2 + 2x + 3x + 3 \\
&= 2x(x + 1) + 3(x + 1) \\
&= (x + 1)(2x + 3)
\end{aligned}
$$

5. Multiply to check.

$$
\begin{aligned}
(x + 1)(2x + 3) &= 2x^2 + 3x + 2x + 3 \\
&= 2x^2 + 5x + 3 \; \checkmark
\end{aligned}
$$

Practice Problem 4 Factor by grouping. $2x^2 + 7x + 5$

EXAMPLE 5 Factor by grouping. $4x^2 - 13x + 3$

Solution

1. The grouping number is $(4)(3) = 12$.
2. The factors of 12 are $(12)(1)$ or $(4)(3)$ or $(6)(2)$. Note that the middle term of the polynomial is negative. Thus we choose the numbers -12 and -1 because their product is still 12 and their sum is -13.
3. We write $-13x$ as the sum $-12x + (-1x)$ or $-12x - 1x$.
4. Factor by grouping.

$$
\begin{aligned}
4x^2 - 13x + 3 &= 4x^2 - 12x - 1x + 3 \\
&= 4x(x - 3) - 1(x - 3) \\
&= (x - 3)(4x - 1)
\end{aligned}
$$

Remember to factor out a -1 from the last two terms so that both sets of parentheses contain the same expression.

Practice Problem 5 Factor by grouping. $9x^2 - 64x + 7$

EXAMPLE 6 Factor by grouping. $3x^2 - 2x - 8$

Solution

1. The grouping number is $(3)(-8) = -24$.
2. We want two numbers whose product is -24 and whose sum is -2. They are -6 and 4.
3. We write $-2x$ as the sum $-6x + 4x$.
4. Factor by grouping.

$$
\begin{aligned}
3x^2 - 6x + 4x - 8 &= 3x(x - 2) + 4(x - 2) \\
&= (x - 2)(3x + 4)
\end{aligned}
$$

Practice Problem 6 Factor by grouping. $3x^2 + 4x - 4$

To factor polynomials of the form $ax^2 + bx + c$, use the method, either trial-and-error or grouping, that works best for you.

③ Using the Common Factor Method

Some problems require first factoring out the greatest common factor and then factoring the trinomial by one of the two methods of this section.

EXAMPLE 7 Factor. $9x^2 + 3x - 30$

Solution

$$9x^2 + 3x - 30 = 3(3x^2 + 1x - 10)$$ We first factor out the common factor of 3 from each term of the trinomial.

$$= 3(3x - 5)(x + 2)$$ We then factor the trinomial by the grouping method or by the trial-and-error method.

NOTE TO STUDENT: Fully worked-out solutions to all of the Practice Problems can be found at the back of the text starting at page SP-1

Practice Problem 7 Factor. $8x^2 + 8x - 6$

Be sure to remove the greatest common factor as the very first step.

EXAMPLE 8 Factor. $32x^2 - 40x + 12$

Solution

$$32x^2 - 40x + 12 = 4(8x^2 - 10x + 3)$$ We first factor out the greatest common factor of 4 from each term of the trinomial.

$$= 4(2x - 1)(4x - 3)$$ We then factor the trinomial by the grouping method or by the trial-and-error method.

Practice Problem 8 Factor. $24x^2 - 38x + 10$

Factor by the trial-and-error method. Check your answers using FOIL.

1. $4x^2 + 13x + 3$

2. $3x^2 + 11x + 10$

3. $5x^2 + 7x + 2$

4. $3x^2 + 4x + 1$

5. $4x^2 + 5x - 6$

6. $5x^2 + 13x - 6$

7. $2x^2 - 5x - 3$

8. $2x^2 - x - 6$

Factor by the grouping number method. Check your answers by using FOIL.

9. $9x^2 + 9x + 2$

10. $4x^2 + 11x + 6$

11. $15x^2 - 34x + 15$

12. $10x^2 - 29x + 10$

13. $2x^2 + 3x - 20$

14. $6x^2 + 11x - 10$

15. $8x^2 + 10x - 3$

16. $5x^2 + 2x - 7$

Factor by any method.

17. $6x^2 - 5x - 6$

18. $3x^2 - 13x - 10$

19. $10x^2 + 3x - 1$

20. $6x^2 + x - 5$

21. $7x^2 - 5x - 18$

22. $9x^2 - 22x - 15$

23. $9y^2 - 13y + 4$

24. $5y^2 - 11y + 2$

25. $5a^2 - 13a - 6$

26. $3a^2 - 10a - 8$

27. $14x^2 + 17x - 6$

28. $32x^2 + 36x - 5$

29. $15x^2 + 4x - 4$

30. $8x^2 - 11x + 3$

31. $12x^2 + 28x + 15$

32. $24x^2 + 17x + 3$

33. $12x^2 - 16x - 3$

34. $12x^2 + x - 6$

35. $3x^4 - 14x^2 - 5$

36. $4x^4 + 8x^2 - 5$

37. $2x^2 + 11xy + 15y^2$

38. $15x^2 + 14xy + 3y^2$

39. $5x^2 + 16xy - 16y^2$

40. $12x^2 + 11xy - 5y^2$

Factor by first factoring out the greatest common factor. See Examples 7 and 8.

41. $10x^2 - 25x - 15$

42. $20x^2 - 25x - 30$

43. $6x^3 + 9x^2 - 60x$

44. $6x^3 - 16x^2 - 6x$

Mixed Practice *Factor.*

45. $5x^2 + 3x - 2$

46. $6x^2 + x - 2$

47. $12x^2 - 38x + 20$

48. $20x^2 - 38x + 12$

49. $12x^3 - 20x^2 + 3x$

50. $6x^3 - 23x^2 + 20x$

51. $8x^2 + 16x - 10$

52. $16x^2 + 36x - 10$

53. [2.3.3] Solve. $7x - 3(4 - 2x) = 2(x - 3) - (5 - x)$

54. [2.4.1] Solve. $\dfrac{x}{3} - \dfrac{x}{5} = \dfrac{7}{15}$

Overseas Travelers *The double-bar graph shows the top states visited by foreign travelers (excluding travelers from Canada and Mexico) for the years 2000 and 2005. (Source: www.census.gov.)*

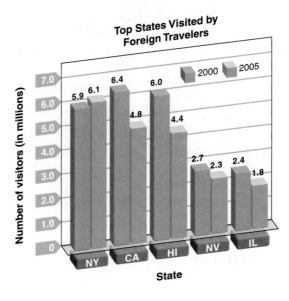

55. [0.5.4] (a) How many overseas travelers visited these five states in 2005?

(b) What percent of these travelers visited Hawaii (HI)?

56. [0.5.4] Of the overseas visitors to these five states in 2000, what percent traveled to California (CA)?

57. [0.5.3] (a) Find the difference in the number of visitors to California (CA) from 2000 to 2005.

(b) This decrease is what percent of the visitors in 2000?

58. [0.5.3] (a) Find the difference in the number of visitors to New York (NY) from 2000 to 2005.

(b) This increase is what percent of the visitors in 2000?

Quick Quiz 6.4 Factor by any method.

1. $12x^2 + 16x - 3$

2. $10x^2 - 21x + 9$

3. $6x^3 - 3x^2 - 30x$

4. Concept Check Explain how you would factor $10x^3 + 18x^2y - 4xy^2$.

How are you doing with your homework assignments in Sections 6.1 to 6.4? Do you feel you have mastered the material so far? Do you understand the concepts you have covered? Before you go further in the textbook, take some time to do each of the following problems.

Factor each of the following problems. Be sure to remove the greatest common factor in each case.

6.1

1. $6xy - 15z + 21$

2. $30x^2 - 45xy - 10x$

3. $7(4x - 5) - b(4x - 5)$

4. $2x(8y + 3z) - 5y(8y + 3z)$

6.2

5. $18 + 3x - 6y - xy$

6. $15x - 9xb + 20w - 12bw$

7. $x^3 - 5x^2 - 3x + 15$

8. $7a + 21b + 2ab + 6b^2$

6.3

9. $x^2 - 15x + 56$

10. $x^2 + 12x - 64$

11. $a^2 + 10ab + 21b^2$

12. $7x^2 - 14x - 245$
(*Hint:* First factor out the greatest common factor.)

6.4

13. $12x^2 + 17x - 5$

14. $3x^2 - 23x + 14$

15. $6x^2 + 17xy + 12y^2$

16. $14x^3 - 20x^2 - 16x$
(*Hint:* First factor out the greatest common factor.)

Now turn to page SA-17 for the answer to each of these problems. Each answer also includes a reference to the objective in which the problem is first taught. If you missed any of these problems, you should stop and review the Examples and Practice Problems in the referenced objective. A little review now will help you master the material in the upcoming sections of the text.

1. _____

2. _____

3. _____

4. _____

5. _____

6. _____

7. _____

8. _____

9. _____

10. _____

11. _____

12. _____

13. _____

14. _____

15. _____

16. _____

Student Learning Objectives

After studying this section, you will be able to:

1. Recognize and factor expressions of the type $a^2 - b^2$ (difference of two squares).

2. Recognize and factor expressions of the type $a^2 + 2ab + b^2$ (perfect-square trinomial).

3. Recognize and factor expressions that require factoring out a common factor and then using a special-case formula.

As we proceed in this section you will be able to reduce the time it takes you to factor polynomials by quickly recognizing and factoring two special types of polynomials: the difference of two squares and perfect-square trinomials.

 Factoring the Difference of Two Squares

Recall the formula from Section 5.5:

$$(a + b)(a - b) = a^2 - b^2.$$

In reverse form we can use it for factoring.

DIFFERENCE OF TWO SQUARES

$$a^2 - b^2 = (a + b)(a - b)$$

We can state it in words in this way: "The difference of two squares can be factored into the sum and difference of those values that were squared."

EXAMPLE 1 Factor. $9x^2 - 1$

Solution We see that the problem is in the form of the difference of two squares. $9x^2$ is a square and 1 is a square. So using the formula we can write the following.

$$9x^2 - 1 = (3x + 1)(3x - 1) \quad \text{Because } 9x^2 = (3x)^2 \text{ and } 1 = (1)^2$$

Practice Problem 1 Factor. $64x^2 - 1$

NOTE TO STUDENT: Fully worked-out solutions to all of the Practice Problems can be found at the back of the text starting at page SP-1

EXAMPLE 2 Factor. $25x^2 - 16$

Solution Again we use the formula for the difference of squares.

$$25x^2 - 16 = (5x + 4)(5x - 4) \quad \text{Because } 25x^2 = (5x)^2 \text{ and } 16 = (4)^2$$

Practice Problem 2 Factor. $36x^2 - 49$

Sometimes the polynomial contains two variables.

EXAMPLE 3 Factor. $4x^2 - 49y^2$

Solution We see that

$$4x^2 - 49y^2 = (2x + 7y)(2x - 7y).$$

Practice Problem 3 Factor. $100x^2 - 81y^2$

CAUTION: Please note that the difference-of-two-squares formula only works if the last term is negative. So if Example 3 had been to factor $4x^2 + 49y^2$, we would **not** have been able to factor the problem. We will examine this in more detail in Section 6.6.

Some problems may involve more than one step.

EXAMPLE 4   Factor. $81x^4 - 1$

Solution   We see that

$$81x^4 - 1 = (9x^2 + 1)(9x^2 - 1). \qquad \text{Because } 81x^4 = (9x^2)^2 \text{ and } 1 = (1)^2$$

Is the factoring complete? We can factor $9x^2 - 1$.

$$81x^4 - 1 = (9x^2 + 1)(3x + 1)(3x - 1) \qquad \text{Because } (9x^2 - 1) = (3x + 1)(3x - 1)$$

Practice Problem 4   Factor. $x^8 - 1$

Factoring Perfect-Square Trinomials

There is a formula that will help us to very quickly factor certain trinomials, called **perfect-square trinomials.** Recall from Section 5.5 the formulas for binomials squared.

$$(a + b)^2 = a^2 + 2ab + b^2$$
$$(a - b)^2 = a^2 - 2ab + b^2$$

We can use these two equations in reverse form for factoring.

PERFECT-SQUARE TRINOMIALS

$$a^2 + 2ab + b^2 = (a + b)^2$$
$$a^2 - 2ab + b^2 = (a - b)^2$$

A perfect-square trinomial is a trinomial that is the result of squaring a binomial. How can we recognize a perfect-square trinomial?

1. The first and last terms are *perfect squares*.
2. The middle term is twice the product of the values whose squares are the first and last terms.

EXAMPLE 5   Factor. $x^2 + 6x + 9$

Solution   This is a perfect-square trinomial.

1. The first and last terms are perfect squares because $x^2 = (x)^2$ and $9 = (3)^2$.
2. The middle term, $6x$, is twice the product of x and 3.

Since $x^2 + 6x + 9$ is a perfect-square trinomial, we can use the formula

$$a^2 + 2ab + b^2 = (a + b)^2$$

with $a = x$ and $b = 3$. So we have

$$x^2 + 6x + 9 = (x + 3)^2.$$

Practice Problem 5 Factor. $x^2 + 10x + 25$

EXAMPLE 6 Factor. $4x^2 - 20x + 25$

Solution This is a perfect-square trinomial. Note that $20x = 2(2x \cdot 5)$. Also note the negative sign. Thus we have the following.

$$4x^2 - 20x + 25 = (2x - 5)^2 \quad \text{Since } a^2 - 2ab + b^2 = (a - b)^2$$

Practice Problem 6 Factor. $25x^2 - 30x + 9$

NOTE TO STUDENT: Fully worked-out solutions to all of the Practice Problems can be found at the back of the text starting at page SP-1

A polynomial may have more than one variable and its exponents may be higher than 2. The same principles apply.

EXAMPLE 7 Factor.

(a) $49x^2 + 42xy + 9y^2$ **(b)** $36x^4 - 12x^2 + 1$

Solution

(a) This is a perfect-square trinomial. Why?

$$49x^2 + 42xy + 9y^2 = (7x + 3y)^2 \quad \text{Because } 49x^2 = (7x)^2, 9y^2 = (3y)^2, \text{ and } 42xy = 2(7x \cdot 3y)$$

(b) This is a perfect-square trinomial. Why?

$$36x^4 - 12x^2 + 1 = (6x^2 - 1)^2 \quad \text{Because } 36x^4 = (6x^2)^2, 1 = (1)^2, \text{ and } 12x^2 = 2(6x^2 \cdot 1)$$

Practice Problem 7 Factor.

(a) $25x^2 + 60xy + 36y^2$ **(b)** $64x^6 - 48x^3 + 9$

Some polynomials appear to be perfect-square trinomials but are not. They were factored in other ways in Section 6.4.

EXAMPLE 8 Factor. $49x^2 + 35x + 4$

Solution   This is *not* a perfect-square trinomial! Although the first and last terms are perfect squares since $(7x)^2 = 49x^2$ and $(2)^2 = 4$, the middle term, $35x$, is not double the product of 2 and $7x$! $35x \neq 28x$! So we must factor by trial and error or by grouping to obtain

$$49x^2 + 35x + 4 = (7x + 4)(7x + 1).$$

Practice Problem 8 Factor. $9x^2 + 15x + 4$

---

3 **Factoring Out a Common Factor Before Using a Special-Case Formula**

For some polynomials, we first need to factor out the greatest common factor. Then we will find an opportunity to use the difference-of-two-squares formula or one of the perfect-square trinomial formulas.

Look carefully at Example 9. Can you identify the greatest common factor?

EXAMPLE 9 Factor. $12x^2 - 48$

Solution

$12x^2 - 48 = 12(x^2 - 4)$ First we factor out the greatest common factor, 12.

$\qquad = 12(x + 2)(x - 2)$ Then we use the difference-of-two-squares formula, $a^2 - b^2 = (a + b)(a - b)$.

Practice Problem 9 Factor. $20x^2 - 45$

Look carefully at Example 10. Can you identify the greatest common factor?

EXAMPLE 10 Factor. $24x^2 - 72x + 54$

Solution

$24x^2 - 72x + 54 = 6(4x^2 - 12x + 9)$ First we factor out the greatest common factor, 6.

$\qquad = 6(2x - 3)^2$ Then we use the perfect-square trinomial formula, $a^2 - 2ab + b^2 = (a - b)^2$.

Practice Problem 10 Factor. $75x^2 - 60x + 12$

Factor by using the difference-of-two-squares formula.

1. $9x^2 - 1$

2. $25x^2 - 1$

3. $81x^2 - 16$

4. $100x^2 - 49$

5. $x^2 - 49$

6. $x^2 - 100$

7. $25x^2 - 81$

8. $49x^2 - 4$

9. $x^2 - 25$

10. $x^2 - 36$

11. $1 - 16x^2$

12. $1 - 64x^2$

13. $16x^2 - 49y^2$

14. $25x^2 - 81y^2$

15. $36x^2 - 169y^2$

16. $64x^2 - 81y^2$

17. $100x^2 - 81$

18. $16a^2 - 25$

19. $25a^2 - 81b^2$

20. $9x^2 - 49y^2$

Factor by using the perfect-square trinomial formula.

21. $9x^2 + 6x + 1$

22. $25x^2 + 10x + 1$

23. $y^2 - 10y + 25$

24. $y^2 - 12y + 36$

25. $36x^2 - 60x + 25$

26. $16x^2 - 72x + 81$

27. $49x^2 + 28x + 4$

28. $25x^2 + 30x + 9$

29. $x^2 + 14x + 49$

30. $x^2 + 8x + 16$

31. $25x^2 - 40x + 16$

32. $64x^2 - 16x + 1$

33. $81x^2 + 36xy + 4y^2$

34. $36x^2 + 60xy + 25y^2$

35. $9x^2 - 30xy + 25y^2$

36. $4x^2 - 28xy + 49y^2$

Mixed Practice *Factor by using either the difference-of-two-squares or the perfect-square trinomial formula.*

37. $16a^2 + 72ab + 81b^2$

38. $169a^2 + 26ab + b^2$

39. $49x^2 - 42xy + 9y^2$

40. $9x^2 - 30xy + 25y^2$

41. $9x^2 + 42x + 49$

42. $25x^2 + 40x + 16$

43. $144x^2 - 1$

44. $16x^2 - 121$

45. $x^4 - 36$

46. $x^4 - 49$

47. $4x^4 - 20x^2 + 25$

48. $36x^4 - 60x^2 + 25$

To Think About Exercises 49–52

49. In Example 4, first we factored $81x^4 - 1$ as $(9x^2 + 1)$ $(9x^2 - 1)$, and then we factored $9x^2 - 1$ as $(3x + 1)$ $(3x - 1)$. Show why you cannot factor $9x^2 + 1$.

50. What two numbers could replace the b in $25x^2 + bx + 16$ so that the resulting trinomial would be a perfect square? (*Hint:* One number is negative.)

51. What value could you give to c so that $16y^2 - 56y + c$ would become a perfect-square trinomial? Is there only one answer or more than one?

52. Jerome says that he can find two values of b so that $100x^2 + bx - 9$ will be a perfect square. Kesha says there is only one that fits, and Larry says there are none. Who is correct and why?

Factor by first looking for a greatest common factor. See Examples 9 and 10.

53. $16x^2 - 36$

54. $27x^2 - 75$

55. $147x^2 - 3y^2$

56. $16y^2 - 100x^2$

57. $16x^2 - 16x + 4$

58. $45x^2 - 60x + 20$

59. $98x^2 + 84x + 18$

60. $128x^2 + 96x + 18$

Mixed Practice *Factor. Be sure to look for common factors first.*

61. $x^2 + 16x + 63$

62. $x^2 - 3x - 40$

63. $2x^2 + 5x - 3$

64. $15x^2 - 11x + 2$

65. $16x^2 - 121$

66. $81x^2 - 36y^2$

67. $9x^2 + 42x + 49$

68. $9x^2 + 30x + 25$

69. $36x^2 - 36x + 9$

70. $6x^2 + 60x + 150$

71. $2x^2 - 32x + 126$

72. $2x^2 - 32x + 110$

Cumulative Review

73. **[5.6.2]** Divide. $(x^3 + x^2 - 2x - 11) \div (x - 2)$

74. **[5.6.2]** Divide. $(6x^3 + 11x^2 - 11x - 20) \div (3x + 4)$

Iguana Diet The green iguana can reach a length of 6 feet and weigh up to 18 pounds. Of the basic diet of the iguana, 40% should consist of greens such as lettuce, spinach, and parsley; 35% should consist of bulk vegetables such as broccoli, zucchini, and carrots; and 25% should consist of fruit.

75. **[2.6.2]** If a certain iguana weighing 150 ounces has a daily diet equal to 2% of its body weight, compose a diet for it in ounces that will meet the iguana's one-day requirement for nutrition.

76. **[2.6.2]** If another iguana weighing 120 ounces has a daily diet equal to 3% of its body weight, compose a diet for it in ounces that will meet the iguana's one-day requirement for nutrition.

Quick Quiz 6.5 Factor the following.

1. $49x^2 - 81y^2$

2. $9x^2 - 48x + 64$

3. $162x^2 - 200$

4. **Concept Check** Explain how to factor the polynomial $24x^2 + 120x + 150$.

Student Learning Objectives

After studying this section, you will be able to:

 1 Identify and factor any polynomial that can be factored.

2 Determine whether a polynomial is prime.

 Identifying and Factoring Polynomials

Often the various types of factoring problems are all mixed together. We need to be able to identify each type of polynomial quickly. The following table summarizes the information we have learned about factoring.

Many polynomials require more than one factoring method. When you are asked to factor a polynomial, it is expected that you will factor it completely. Usually, the first step is factoring out a common factor; then the next step will become apparent.

Carefully go through each example in the following Factoring Organizer. Be sure you understand each step that is involved.

Factoring Organizer

Number of Terms in the Polynomial	Identifying Name and/or Formula	Example
A. Any number of terms	**Common factor** The terms have a common factor consisting of a number, a variable, or both.	$2x^2 - 16x = 2x(x - 8)$ $3x^2 + 9y - 12 = 3(x^2 + 3y - 4)$ $4x^2y + 2xy^2 - wxy + xyz = xy(4x + 2y - w + z)$
B. Two terms	**Difference of two squares** First and last terms are perfect squares. $a^2 - b^2 = (a + b)(a - b)$	$16x^2 - 1 = (4x + 1)(4x - 1)$ $25y^2 - 9x^2 = (5y + 3x)(5y - 3x)$
C. Three terms	**Perfect-square trinomial** First and last terms are perfect squares. $a^2 + 2ab + b^2 = (a + b)^2$ $a^2 - 2ab + b^2 = (a - b)^2$	$25x^2 - 10x + 1 = (5x - 1)^2$ $16x^2 + 24x + 9 = (4x + 3)^2$
D. Three terms	**Trinomial of the form $x^2 + bx + c$** It starts with x^2. The constants of the two factors are numbers whose product is c and whose sum is b.	$x^2 - 7x + 12 = (x - 3)(x - 4)$ $x^2 + 11x - 26 = (x + 13)(x - 2)$ $x^2 - 8x - 20 = (x - 10)(x + 2)$
E. Three terms	**Trinomial of the form $ax^2 + bx + c$** It starts with ax^2, where a is any number but 1.	Use trial-and-error or the grouping number method to factor $12x^2 - 5x - 2$. 1. The grouping number is -24. 2. The two numbers whose product is -24 and whose sum is -5 are -8 and 3. 3. $12x^2 - 5x - 2 = 12x^2 + 3x - 8x - 2$ $\quad = 3x(4x + 1) - 2(4x + 1)$ $\quad = (4x + 1)(3x - 2)$
F. Four terms	**Factor by grouping** Rearrange the order if the first two terms do not have a common factor.	$wx - 6yz + 2wy - 3xz = wx + 2wy - 3xz - 6yz$ $\quad = w(x + 2y) - 3z(x + 2y)$ $\quad = (x + 2y)(w - 3z)$

EXAMPLE 1 Factor.

(a) $25x^3 - 10x^2 + x$

(b) $20x^2y^2 - 45y^2$

(c) $2ax + 4ay + 4x + 8y$

(d) $15x^2 - 3x^3 + 18x$

Solution

(a) $25x^3 - 10x^2 + x = x(25x^2 - 10x + 1)$ Factor out the common factor of x. The other factor is a perfect-square trinomial.

$\qquad\qquad\qquad\quad = x(5x - 1)^2$

(b) $20x^2y^2 - 45y^2 = 5y^2(4x^2 - 9)$

Factor out the common factor of $5y^2$. The other factor is a difference of squares.

$$= 5y^2(2x + 3)(2x - 3)$$

(c) $2ax + 4ay + 4x + 8y = 2[ax + 2ay + 2x + 4y]$

Factor out the common factor of 2.

$$= 2[a(x + 2y) + 2(x + 2y)]$$

Factor the terms inside the bracket by the grouping method.

$$= 2(x + 2y)(a + 2)$$

Factor out the common factor of $(x + 2y)$.

(d) $15x^2 - 3x^3 + 18x = -3x^3 + 15x^2 + 18x$

Rearrange the terms in descending order of powers of x.

$$= -3x(x^2 - 5x - 6)$$

Factor out the common factor of $-3x$.

$$= -3x(x - 6)(x + 1)$$

Factor the trinomial.

Practice Problem 1 Factor. Be careful. These practice problems are mixed.

(a) $3x^2 - 36x + 108$ **(b)** $9x^4y^2 - 9y^2$

(c) $5x^3 - 15x^2y + 10x^2 - 30xy$ **(d)** $12x - 9 - 4x^2$

NOTE TO STUDENT: Fully worked-out solutions to all of the Practice Problems can be found at the back of the text starting at page SP-1

Determining Whether a Polynomial Is Prime

Not all polynomials can be factored using the methods in this chapter. If we cannot factor a polynomial by elementary methods, we will identify it as a **prime** polynomial. If, after you have mastered the factoring techniques in this chapter, you encounter a polynomial that you cannot factor with these methods, you should feel comfortable enough to say, "The polynomial cannot be factored with the methods in this chapter, so it is prime," rather than "I can't do it—I give up!"

EXAMPLE 2 Factor, if possible. $x^2 + 6x + 12$

Solution The factors of 12 are

$$(1)(12) \text{ or } (2)(6) \text{ or } (3)(4).$$

None of these pairs add up to 6, the coefficient of the middle term. Thus the problem cannot be factored by the methods of this chapter. It is prime.

Practice Problem 2 Factor. $x^2 - 9x - 8$

EXAMPLE 3 Factor, if possible. $25x^2 + 4$

Solution We have a formula to factor the difference of two squares. There is no way to factor the sum of two squares. That is, $a^2 + b^2$ cannot be factored. Thus

$$25x^2 + 4 \text{ is prime.}$$

Practice Problem 3 Factor, if possible. $25x^2 + 82x + 4$

Review the six basic types of factoring in the Factoring Organizer on page 396. Each of the six types is included in exercises 1–12. Be sure you can find two of each type.

Factor. Check your answer by multiplying.

1. $5a^2 - 3ab + 8a$

2. $7x^2 + 9xy - 2x$

3. $16x^2 - 25y^2$

4. $64x^2 - 9y^2$

5. $9x^2 - 12xy + 4y^2$

6. $16x^2 + 24xy + 9y^2$

7. $x^2 + 8x + 15$

8. $x^2 + 15x + 54$

9. $15x^2 + 7x - 2$

10. $6x^2 + 13x - 5$

11. $ax - 3cx + 3ay - 9cy$

12. $bx - 2dx + 5by - 10dy$

Mixed Practice *Factor, if possible. Be sure to factor completely. Always factor out the greatest common factor first, if one exists.*

13. $y^2 + 14y + 49$

14. $y^2 + 16y + 64$

15. $4x^2 - 12x + 9$

16. $16x^2 - 40x + 25$

17. $2x^2 - 11x + 12$

18. $3x^2 - 10x + 8$

19. $x^2 - 3xy - 70y^2$

20. $x^2 - 6xy - 16y^2$

21. $ax - 5a + 3x - 15$

22. $by + 7b - 6y - 42$

23. $16x - 4x^3$

24. $8y^2 + 10y - 12$

25. $5x^3y^3 - 10x^2y^3 + 5xy^3$

26. $2x^4y - 12x^3y + 18x^2y$

27. $3xyz^2 - 6xyz - 9xy$

28. $-16x^2 - 2x - 32x^3$

29. $3x^2 + 6x - 105$

30. $4x^2 - 28x - 72$

31. $3x^3 + 3x^2 - 36x$

32. $7x^2 + 3x - 2$

33. $7x^2 - 2x^4 + 4$

34. $14x^2 - x^3 + 32x$

35. $6x^2 - 3x + 2$

36. $4x^3 + 8x^2 - 60x$

Remove the greatest common factor first. Then continue to factor, if possible.

37. $5x^2 + 10xy - 30y$

38. $7a^2 + 21b - 42$

39. $30x^3 + 3x^2y - 6xy^2$

40. $30x^3 - 25x^2y - 30xy^2$

41. $24x^2 - 58x + 30$

42. $12x^2 - 30x + 12$

To Think About

43. A polynomial that cannot be factored by the methods of this chapter is called _____.

44. A binomial of the form $x^2 - d$ can be quickly factored or identified as prime. If it can be factored, what is true of the number d?

Cumulative Review

45. **[0.5.3]** *Salary* When Dave Barry decided to leave the company and work as an independent contractor, he took a pay cut of 14%. He earned $24,080 this year. What did he earn in his previous job?

46. **[2.6.1]** *Antiviral Drug* A major pharmaceutical company is testing a new, powerful antiviral drug. It kills 13 strains of virus every hour. If there are presently 294 live strains of virus in the test container, how many live strains were there 6 hours ago?

47. **[4.3.2]** Solve the system.
$$\frac{1}{2}x - y = 7$$
$$-3x + 2y = -22$$

48. **[2.6.1]** *Personal Library* Gary loves to read. In his living room he has hardcover books, softcover books, and magazines. He has 37 fewer hardcover books than softcover books. He has twice as many softcover books as magazines. If there are 198 total books and magazines in his bookcase, how many of each type does he have?

Quick Quiz 6.6

Completely factor the following, if possible.

1. $6x^2 - 17x + 12$

2. $60x^2 - 9x - 6$

3. $25x^2 + 49$

4. **Concept Check** Explain how to completely factor $2x^2 + 6xw - 5x - 15w$.

Student Learning Objectives

After studying this section, you will be able to:

 Solve quadratic equations by factoring.

 Use quadratic equations to solve applied problems.

 ## Solving Quadratic Equations by Factoring

In Chapter 2, we learned how to solve linear equations such as $3x + 5 = 0$ by finding the root (or value of x) that satisfied the equation. Now we turn to the question of how to solve equations like $3x^2 + 5x + 2 = 0$. Such equations are called **quadratic equations.** A quadratic equation is a polynomial equation in one variable that contains a variable term of degree 2 and no terms of higher degree.

> The *standard form* of a quadratic equation is $ax^2 + bx + c = 0$, where a, b, and c are real numbers and $a \neq 0$.

In this section, we will study quadratic equations in standard form, where a, b, and c are integers.

Many quadratic equations have two real number solutions (also called real **roots**). But how can we find them? The most direct approach is the factoring method. This method depends on a very powerful property.

> **ZERO FACTOR PROPERTY**
> If $a \cdot b = 0$, then $a = 0$ or $b = 0$.

Notice the word *or* in the zero factor property. When we make a statement in mathematics using this word, we intend it to mean *one or the other or both*. Therefore, the zero factor property states that if the product $a \cdot b$ is zero, then a can equal zero or b can equal zero or *both a and b* can equal zero. We can use this principle to solve quadratic equations. Before you start, make sure that the equation is in standard form.

> 1. Make sure the equation is set equal to zero.
> 2. Factor, if possible, the quadratic expression that equals zero.
> 3. Set each factor containing a variable equal to zero.
> 4. Solve the resulting equations to find each root.
> 5. Check each root.

EXAMPLE 1 Solve the equation to find the two roots. $3x^2 + 5x + 2 = 0$

Solution

$$3x^2 + 5x + 2 = 0$$ The equation is in standard form.

$$(3x + 2)(x + 1) = 0$$ Factor the quadratic expression.

$$3x + 2 = 0 \qquad x + 1 = 0$$ Set each factor equal to 0.

$$3x = -2 \qquad x = -1$$ Solve the equations to find the two roots.

$$x = -\frac{2}{3}$$

The two roots (that is, solutions) are $-\frac{2}{3}$ and -1.

Check. We can determine if the two numbers $-\frac{2}{3}$ and -1 are solutions to the equation. Substitute $-\frac{2}{3}$ for x in the *original equation*. If an identity results, $-\frac{2}{3}$ is a solution. Do the same for -1.

$$3x^2 + 5x + 2 = 0 \qquad\qquad 3x^2 + 5x + 2 = 0$$

$$3\left(-\frac{2}{3}\right)^2 + 5\left(-\frac{2}{3}\right) + 2 \stackrel{?}{=} 0 \qquad 3(-1)^2 + 5(-1) + 2 \stackrel{?}{=} 0$$

$$3\left(\frac{4}{9}\right) + 5\left(-\frac{2}{3}\right) + 2 \stackrel{?}{=} 0 \qquad\qquad 3(1) + 5(-1) + 2 \stackrel{?}{=} 0$$

$$\qquad\qquad\qquad\qquad\qquad\qquad\qquad 3 - 5 + 2 \stackrel{?}{=} 0$$

$$\frac{4}{3} - \frac{10}{3} + 2 \stackrel{?}{=} 0 \qquad\qquad\qquad -2 + 2 \stackrel{?}{=} 0$$

$$\qquad\qquad\qquad\qquad\qquad\qquad\qquad\qquad 0 = 0 \ \checkmark$$

$$\frac{4}{3} - \frac{10}{3} + \frac{6}{3} \stackrel{?}{=} 0$$

$$0 = 0 \ \checkmark$$

Thus $-\frac{2}{3}$ and -1 are both roots of the equation $3x^2 + 5x + 2 = 0$.

Practice Problem 1 Solve the equation by factoring to find the two roots and check. $10x^2 - x - 2 = 0$

NOTE TO STUDENT: Fully worked-out solutions to all of the Practice Problems can be found at the back of the text starting at page SP-1

EXAMPLE 2 Solve the equation to find the two roots. $2x^2 + 13x - 7 = 0$

Solution

$$2x^2 + 13x - 7 = 0 \qquad \text{The equation is in standard form.}$$
$$(2x - 1)(x + 7) = 0 \qquad \text{Factor.}$$
$$2x - 1 = 0 \quad x + 7 = 0 \qquad \text{Set each factor equal to 0.}$$
$$2x = 1 \qquad\quad x = -7 \quad \text{Solve the equations to find the two roots.}$$
$$x = \frac{1}{2}$$

The two roots are $\frac{1}{2}$ and -7.

Check. If $x = \frac{1}{2}$, then we have the following.

$$2\left(\frac{1}{2}\right)^2 + 13\left(\frac{1}{2}\right) - 7 = 2\left(\frac{1}{4}\right) + 13\left(\frac{1}{2}\right) - 7$$

$$= \frac{1}{2} + \frac{13}{2} - \frac{14}{2} = 0 \ \checkmark$$

If $x = -7$, then we have the following.

$$2(-7)^2 + 13(-7) - 7 = 2(49) + 13(-7) - 7$$
$$= 98 - 91 - 7 = 0 \ \checkmark$$

Thus $\frac{1}{2}$ and -7 are both roots of the equation $2x^2 + 13x - 7 = 0$.

Practice Problem 2 Solve the equation to find the two roots.
$$3x^2 + 11x - 4 = 0$$

If the quadratic equation $ax^2 + bx + c = 0$ has no visible constant term, then $c = 0$. All such quadratic equations can be solved by factoring out a common factor and then using the zero factor property to obtain two solutions that are real numbers.

EXAMPLE 3   Solve the equation to find the two roots. $7x^2 - 3x = 0$

Solution

$$7x^2 - 3x = 0 \quad \text{The equation is in standard form. Here } c = 0.$$
$$x(7x - 3) = 0 \quad \text{Factor out the common factor.}$$
$$x = 0 \quad 7x - 3 = 0 \quad \text{Set each factor equal to 0 by the zero factor property.}$$
$$7x = 3 \quad \text{Solve the equations to find the two roots.}$$
$$x = \frac{3}{7}$$

The two roots are 0 and $\frac{3}{7}$.

Check.   Verify that 0 and $\frac{3}{7}$ are the roots of $7x^2 - 3x = 0$.

Practice Problem 3   Solve the equation to find the two roots.
$$7x^2 + 11x = 0$$

NOTE TO STUDENT: *Fully worked-out solutions to all of the Practice Problems can be found at the back of the text starting at page SP-1*

If the quadratic equation is not in standard form, we use the same basic algebraic methods we studied in Sections 2.1–2.3 to place the terms on one side and zero on the other so that we can use the zero factor property.

EXAMPLE 4   Solve. $x^2 = 12 - x$

Solution

$$x^2 = 12 - x \quad \text{The equation is not in standard form.}$$
$$x^2 + x - 12 = 0 \quad \text{Add } x \text{ and } -12 \text{ to both sides of the equation so that the left side is equal to zero; we can now factor.}$$
$$(x - 3)(x + 4) = 0 \quad \text{Factor.}$$
$$x - 3 = 0 \quad x + 4 = 0 \quad \text{Set each factor equal to 0 by the zero factor property.}$$
$$x = 3 \quad x = -4 \quad \text{Solve the equations for } x.$$

Check.   If $x = 3$: $(3)^2 \stackrel{?}{=} 12 - 3$ If $x = -4$: $(-4)^2 \stackrel{?}{=} 12 - (-4)$
$$9 \stackrel{?}{=} 12 - 3 \qquad\qquad\qquad 16 \stackrel{?}{=} 12 + 4$$
$$9 = 9 \ \checkmark \qquad\qquad\qquad\qquad 16 = 16 \ \checkmark$$

Both roots check.

Practice Problem 4   Solve. $x^2 - 6x + 4 = -8 + x$

EXAMPLE 5   Solve. $\dfrac{x^2 - x}{2} = 6$

Solution   We must first clear the fractions from the equation.

$$2\left(\frac{x^2 - x}{2}\right) = 2(6) \quad \text{Multiply each side by 2.}$$
$$x^2 - x = 12 \quad \text{Simplify.}$$
$$x^2 - x - 12 = 0 \quad \text{Place in standard form.}$$
$$(x - 4)(x + 3) = 0 \quad \text{Factor.}$$
$$x - 4 = 0 \quad x + 3 = 0 \quad \text{Set each factor equal to zero.}$$
$$x = 4 \quad x = -3 \quad \text{Solve the equations for } x.$$

The check is left to the student.

Practice Problem 5   Solve. $\dfrac{2x^2 - 7x}{3} = 5$

Using Quadratic Equations to Solve Applied Problems

Certain types of word problems—for example, some geometry applications—lead to quadratic equations. We'll show how to solve such word problems in this section.

It is particularly important to check the apparent solutions to the quadratic equation with conditions stated in the word problem. Often a particular solution to the quadratic equation will be eliminated by the conditions of the word problem.

▲ **EXAMPLE 6**   Carlos lives in Mexico City. He has a rectangular brick walkway in front of his house. The length of the walkway is 3 meters longer than twice the width. The area of the walkway is 44 square meters. Find the length and width of the rectangular walkway.

Solution

1. **Understand the problem.**
   Draw a picture.

   Let w = the width in meters.

   Then $2w + 3$ = the length in meters.

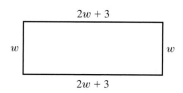

2. **Write an equation.**

   $$\text{area} = (\text{width})(\text{length})$$
   $$44 = w(2w + 3)$$

3. **Solve and state the answer.**

$44 = w(2w + 3)$	
$44 = 2w^2 + 3w$	Remove parentheses.
$0 = 2w^2 + 3w - 44$	Subtract 44 from both sides.
$0 = (2w + 11)(w - 4)$	Factor.
$2w + 11 = 0 \qquad w - 4 = 0$	Set each factor equal to 0.
$2w = -11 \qquad\quad w = 4$	Simplify and solve.
$w = -5\dfrac{1}{2}$	Although $-5\frac{1}{2}$ is a solution to the quadratic equation, it is not a valid solution to the word problem. It would not make sense to have a rectangle with a negative number as a width.

   Since $w = 4$, the width of the walkway is 4 meters. The length is $2w + 3$, so we have $2(4) + 3 = 8 + 3 = 11$. Thus the length of the walkway is 11 meters.

4. **Check.** Is the length 3 meters more than twice the width?

   $$11 \overset{?}{=} 3 + 2(4) \qquad 11 = 3 + 8 \ \checkmark$$

   Is the area of the rectangle 44 square meters?

   $$4 \times 11 \overset{?}{=} 44 \qquad 44 = 44 \ \checkmark$$

▲ **Practice Problem 6**   The length of a rectangle is 2 meters longer than triple the width. The area of the rectangle is 85 square meters. Find the length and width of the rectangle.

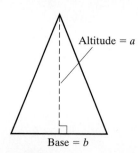

Altitude = a

Base = b

▲ **EXAMPLE 7** The top of a local cable television tower has several small triangular reflectors. The area of each triangle is 49 square centimeters. The altitude of each triangle is 7 centimeters longer than the base. Find the altitude and the base of one of the triangles.

Solution

Let b = the length of the base in centimeters
$b + 7$ = the length of the altitude in centimeters

To find the area of a triangle, we use

$$\text{area} = \frac{1}{2}(\text{altitude})(\text{base}) = \frac{1}{2}ab = \frac{ab}{2}.$$

$\dfrac{ab}{2} = 49$	Write an equation.
$\dfrac{(b + 7)(b)}{2} = 49$	Substitute the expressions for altitude and base.
$\dfrac{b^2 + 7b}{2} = 49$	Simplify.
$b^2 + 7b = 98$	Multiply each side of the equation by 2.
$b^2 + 7b - 98 = 0$	Place the quadratic equation in standard form.
$(b - 7)(b + 14) = 0$	Factor.
$b - 7 = 0 \qquad b + 14 = 0$	Set each factor equal to zero.
$b = 7 \qquad\qquad b = -14$	Solve the equations for b.

We cannot have a base of -14 centimeters, so we reject the negative answer. The only possible solution is 7. So the base is 7 centimeters. The altitude is $b + 7 = 7 + 7 = 14$. The altitude is 14 centimeters. The triangular reflector has a base of 7 centimeters and an altitude of 14 centimeters.

Check. When you do the check, answer the following two questions.

1. Is the altitude 7 centimeters longer than the base?
2. Is the area of a triangle with a base of 7 centimeters and an altitude of 14 centimeters actually 49 square centimeters?

NOTE TO STUDENT: Fully worked-out solutions to all of the Practice Problems can be found at the back of the text starting at page SP-1

▲ **Practice Problem 7** A triangle has an area of 35 square centimeters. The altitude of the triangle is 3 centimeters shorter than the base. Find the altitude and the base of the triangle.

Many problems in the sciences require the use of quadratic equations. You will study these in more detail if you take a course in physics or calculus in college. Often a quadratic equation is given as part of the problem.

When an object is thrown upward, its height (S) in meters is given, approximately, by the quadratic equation

$$S = -5t^2 + vt + h.$$

The letter h represents the initial height in meters. The letter v represents the initial velocity of the object thrown. The letter t represents the time in seconds starting from the time the object is thrown.

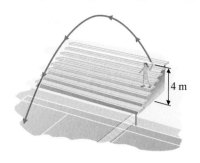

EXAMPLE 8 A tennis ball is thrown upward with an initial velocity of 8 meters/second. Suppose that the initial height above the ground is 4 meters. At what time t will the ball hit the ground?

Solution   In this case $S = 0$ since the ball will hit the ground. The initial upward velocity is $v = 8$ meters/second. The initial height is 4 meters, so $h = 4$.

$S = -5t^2 + vt + h$	Write an equation.
$0 = -5t^2 + 8t + 4$	Substitute all values into the equation.
$5t^2 - 8t - 4 = 0$	Isolate the terms on the left side. (Most students can factor more readily if the squared variable is positive.)
$(5t + 2)(t - 2) = 0$	Factor.
$5t + 2 = 0 \qquad t - 2 = 0$	Set each factor equal to 0.
$5t = -2 \qquad t = 2$	Solve the equations for t.
$t = -\dfrac{2}{5}$	

We want a positive time t in seconds; thus we do not use $t = -\frac{2}{5}$. Therefore, the ball will strike the ground 2 seconds after it is thrown.

Check.   Verify the solution.

Practice Problem 8 A Mexican cliff diver does a dive from a cliff 45 meters above the ocean. This constitutes free fall, so the initial velocity is $v = 0$, and since there is no upward velocity (no springboard), then $h = 45$ meters. How long will it be until he breaks the water's surface?

Using the factoring method, solve for the roots of each quadratic equation. Be sure to place the equation in standard form before factoring. Check your answers.

1. $x^2 - 4x - 21 = 0$

2. $x^2 - x - 20 = 0$

3. $x^2 + 14x + 24 = 0$

4. $x^2 + 10x + 9 = 0$

5. $2x^2 - 7x + 6 = 0$

6. $2x^2 - 11x + 12 = 0$

7. $6x^2 - 13x = -6$

8. $10x^2 + 19x = 15$

9. $x^2 + 13x = 0$

10. $8x^2 - x = 0$

11. $8x^2 = 72$

12. $9x^2 = 81$

13. $5x^2 + 3x = 8x$

14. $6x^2 - 4x = 3x$

15. $6x^2 = 16x - 8$

16. $24x^2 = -10x + 4$

17. $(x - 5)(x + 2) = -4(x + 1)$

18. $(x - 5)(x + 4) = 2(x - 5)$

19. $4x^2 - 3x + 1 = -7x$

20. $9x^2 - 2x + 4 = 10x$

21. $\dfrac{x^2}{3} - 5 + x = -5$

22. $7 + \dfrac{x^2}{2} - x = 7$

23. $\dfrac{x^2 + 10x}{8} = -2$

24. $\dfrac{x^2 + 7x}{5} = 6$

25. $\dfrac{10x^2 - 25x}{12} = 5$

26. $\dfrac{12x^2 - 4x}{5} = 8$

To Think About

27. Why can an equation in standard form with $c = 0$ (that is, an equation of the form $ax^2 + bx = 0$) always be solved?

28. Martha solved $(x + 3)(x - 2) = 14$ as follows:
$$x + 3 = 14 \quad \text{or} \quad x - 2 = 14$$
$$x = 11 \quad \text{or} \quad x = 16$$
Josette said this had to be wrong because these values do not check. Explain what is wrong with Martha's method.

Applications

▲ **29.** *Geometry* The area of a rectangular garden is 140 square meters. The width is 3 meters longer than one-half of the length. Find the length and the width of the garden.

▲ **30.** *Geometry* The area of a triangular sign is 33 square meters. The base of the triangle is 1 meter less than double the altitude. Find the altitude and the base of the sign.

Forming Groups Suppose the number of students in a mathematics class is x. The teacher insists that each student participate in group work each class. The number of possible groups is:

$$G = \frac{x^2 - 3x + 2}{2}$$

31. The class has 13 students. How many possible groups are there?

32. Four students withdraw from the class. How many fewer groups can be formed?

33. A teacher claims each student could be in 36 different groups. How many students are there?

34. The teacher wants each student to participate in a different group each day. There are 45 class days in the semester. How many students must be in the class?

Falling Object Use the following information for exercises 35 and 36. When an object is thrown upward, its height (S), in meters, is given (approximately) by the quadratic equation

$$S = -5t^2 + vt + h,$$

where $v =$ the upward initial velocity in meters/second,
$t =$ the time of flight in seconds, and
$h =$ the height above level ground from which the object is thrown.

35. Johnny is standing on a platform 6 meters high and throws a ball straight up as high as he can at a velocity of 13 meters/second. At what time t will the ball hit the ground? How far from the ground is the ball after 2 seconds have elapsed from the time of the throw? (Assume that the ball is 6 meters from the ground when it leaves Johnny's hand.)

36. You are standing on the edge of a cliff near Acapulco, overlooking the ocean. The place where you stand is 180 meters from the ocean. You drop a pebble into the water. ("Dropping" the pebble implies that there is no initial velocity, so $v = 0$.) How many seconds will it take to hit the water? How far has the pebble dropped after 3 seconds?

Internal Phone Calls The technology and communication office of a local company has set up a new telephone system so that each employee has a separate telephone and extension number. They are studying the possible number of telephone calls that can be made from people in the office to other people in the office. They have discovered that the total number of possible telephone calls T is described by the equation $T = 0.5(x^2 - x)$, where x is the number of people in the office. Use this information to answer exercises 37–38.

37. If 70 people are presently employed at the office, how many possible telephone calls can be made between these 70 people?

38. On the day after Thanksgiving, only a small number of employees were working at the office. It has been determined that on that day, a total of 105 different phone calls could have been made from people working in the office to other people working in the office. How many people worked on the day after Thanksgiving?

The same equation given above is used in the well-known "handshake problem." If there are n people at a party and each person shakes hands with every other person once, the number of handshakes that take place among the n people is $H = 0.5(n^2 - n)$.

39. Barry is hosting a holiday party and has invited 16 friends. If all his friends attend the party and everyone, including Barry, shakes hands with everyone else at the party, how many handshakes will take place?

40. At a mathematics conference, 55 handshakes took place among the organizers during the opening meeting. How many organizers were there?

Cumulative Review *Simplify.*

41. [5.1.1] $(2x^2y^3)(-5x^3y)$

42. [5.1.1] $(3a^4b^5)(4a^6b^8)$

43. [5.1.2] $\dfrac{21a^5b^{10}}{-14ab^{12}}$

44. [5.1.2] $\dfrac{18x^3y^6}{54x^8y^{10}}$

Quick Quiz 6.7 Solve.

1. $15x^2 - 8x + 1 = 0$

2. $4 + x(x - 2) = 7$

3. $4x^2 = 9x + 9$

4. Concept Check Explain how you would solve the following problem: A rectangle has an area of 65 square feet. The length of the rectangle is 3 feet longer than double the width. Find the length and the width of the rectangle.

Putting Your Skills to Work: Use Math to Save Money

CAR REBATES/PAYMENTS

Besides a house, the biggest purchase many people make is a car. When shopping for automobile financing, customers need to take many factors into account, including the amount of the monthly loan payment and whether to choose a lower loan rate or a rebate. Consider the story of Megan.

Megan is interested in buying a $25,000 car and has found a loan that has a term of 5 years and a 9% interest rate. She wants to know what her monthly payments and total cost will be.

To calculate Megan's monthly car payment, we use the amortization formula

$$PMT = \frac{PV}{\left[1 - \frac{1}{(1 + I)^N}\right]\Big/ I}.$$

This formula uses the following variables:

- PMT is the monthly payment.
- I is the interest rate in decimal notation divided by 12, which is the amount of interest paid each month.
- PV is the present value or the amount being financed.
- N is 12 times the number of years of the loan, which is the number of payments needed to pay off the loan.

Here, $PV = \$25,000$, $I = \dfrac{0.09}{12} = 0.0075$, and

$N = 12 \times 5 = 60$.

1. Substitute these values into the amortization formula to find Megan's monthly payment (PMT), rounded to the nearest ten dollars.

2. What is the total amount that Megan will pay in car payments during the 5 years?

3. What is the total amount of interest that Megan will pay for the car?

Suppose the car comes with the following options: (A) a $2000 rebate that Megan deducts from the price of the car before financing the remainder at 9%, or (B) no rebate and a 5-year, 0% interest rate.

4. What would Megan's monthly payment be using option A? Round your answer to the nearest ten dollars.

5. What is the total amount that Megan would pay in car payments using option A?

6. What is the total amount that Megan would pay in car payments using option B?

7. Which option would save Megan the most money?

Extension

8. Look in a newspaper and make the similar comparisons among finance rates and rebates.

Topic	Procedure	Examples
A. Common factor, p. 366.	Factor out the largest common factor from each term.	$2x^2 - 2x = 2x(x - 1)$ $3a^2 + 3ab - 12a = 3a(a + b - 4)$ $8x^4y - 24x^3 = 8x^3(xy - 3)$
Special cases **B. Difference of two squares, p. 390.** **C. Perfect-square trinomials, p. 391.**	If you recognize the special cases, you will be able to factor quickly. $$a^2 - b^2 = (a + b)(a - b)$$ $$a^2 + 2ab + b^2 = (a + b)^2$$ $$a^2 - 2ab + b^2 = (a - b)^2$$	$25x^2 - 36y^2 = (5x + 6y)(5x - 6y)$ $16x^4 - 1 = (4x^2 + 1)(2x + 1)(2x - 1)$ $25x^2 + 10x + 1 = (5x + 1)^2$ $49x^2 - 42xy + 9y^2 = (7x - 3y)^2$
D. Trinomials of the form $x^2 + bx + c$, p. 376.	Factor trinomials of the form $x^2 + bx + c$ by finding two numbers that have a product of c and a sum of b. If each term of the trinomial has a common factor, factor it out as the first step.	$x^2 - 18x + 77 = (x - 7)(x - 11)$ $x^2 + 7x - 18 = (x + 9)(x - 2)$ $5x^2 - 10x - 40 = 5(x^2 - 2x - 8)$ $ = 5(x - 4)(x + 2)$
E. Trinomials of the form $ax^2 + bx + c$, where $a \neq 1$, p. 384.	Factor trinomials of the form $ax^2 + bx + c$ by the grouping number method or by the trial-and-error method.	$6x^2 + 11x - 10$ Grouping number $= -60$ Two numbers whose product is -60 and whose sum is $+11$ are $+15$ and -4. $6x^2 + 15x - 4x - 10 = 3x(2x + 5) - 2(2x + 5)$ $ = (2x + 5)(3x - 2)$
F. Four terms. Factor by grouping, p. 371.	Rearrange the terms if necessary so that the first two terms have a common factor. Then factor out the common factors. $$ax + ay - bx - by = a(x + y) - b(x + y)$$ $$ = (x + y)(a - b)$$	$2ax^2 + 21 + 14x^2 + 3a$ $= 2ax^2 + 14x^2 + 3a + 21$ $= 2x^2(a + 7) + 3(a + 7)$ $= (a + 7)(2x^2 + 3)$
Prime polynomials, p. 397.	A polynomial that is not factorable is called prime.	$x^2 + y^2$ is prime. $x^2 + 5x + 7$ is prime.
Multistep factoring, p. 396.	Many problems require two or three steps of factoring. Always try to factor out the greatest common factor as the first step.	$3x^2 - 21x + 36 = 3(x^2 - 7x + 12)$ $ = 3(x - 4)(x - 3)$ $2x^3 - x^2 - 6x = x(2x^2 - x - 6)$ $ = x(2x + 3)(x - 2)$ $25x^3 - 49x = x(25x^2 - 49)$ $ = x(5x + 7)(5x - 7)$ $8x^2 - 24x + 18 = 2(4x^2 - 12x + 9)$ $ = 2(2x - 3)^2$
Solving quadratic equations by factoring, p. 400.	**1.** Write as $ax^2 + bx + c = 0$. **2.** Factor. **3.** Set each factor equal to 0. **4.** Solve the resulting equations.	Solve. $3x^2 + 5x = 2$ $3x^2 + 5x - 2 = 0$ $(3x - 1)(x + 2) = 0$ $3x - 1 = 0 \qquad x + 2 = 0$ $x = \dfrac{1}{3} \qquad\quad x = -2$

Topic	Procedure	Examples
Using quadratic equations to solve applied problems, p. 403.	Some word problems, like those involving the product of two numbers, area, and formulas with a squared variable, can be solved using the factoring methods we have shown.	The length of a rectangle is 4 less than three times the width. Find the length and width if the area is 55 square inches. Let w = width. Then $3w - 4$ = length. $$55 = w(3w - 4)$$ $$55 = 3w^2 - 4w$$ $$0 = 3w^2 - 4w - 55$$ $$0 = (3w + 11)(w - 5)$$ $$w = -\tfrac{11}{3} \quad \text{or} \quad w = 5$$ $-\tfrac{11}{3}$ is not a valid solution. Thus width = 5 inches and length = 11 inches.

Chapter 6 Review Problems

Section 6.1

Factor out the greatest common factor.

1. $12x^3 - 20x^2 y$

2. $10x^3 - 35x^3 y$

3. $24x^3 y - 8x^2 y^2 - 16x^3 y^3$

4. $45a^3 b^3 + 30a^2 b^3 - 15a^2 b^4$

5. $3a^3 + 6a^2 - 9ab + 12a$

6. $2x - 4y + 6z + 12$

7. $2a(a + 3b) - 5(a + 3b)$

8. $15x^3 y + 6xy^2 + 3xy$

Section 6.2

Factor by grouping.

9. $2ax + 5a - 8x - 20$

10. $a^2 - 4ab + 7a - 28b$

11. $x^2 y + 3y - 2x^2 - 6$

12. $30ax - 15ay + 42x - 21y$

13. $15x^2 - 3x + 10x - 2$

14. $30w^2 - 18w + 5wz - 3z$

Section 6.3

Factor completely. Be sure to factor out any common factors as your first step.

15. $x^2 + 6x - 27$

16. $x^2 + 9x - 10$

17. $x^2 + 14x + 48$

18. $x^2 + 8xy + 15y^2$

19. $x^4 + 13x^2 + 42$

20. $x^4 - 2x^2 - 35$

21. $6x^2 + 30x + 36$

22. $3x^2 + 39x + 36$

23. $2x^2 - 28x + 96$

24. $4x^2 - 44x + 120$

Section 6.4

Factor completely. Be sure to factor out any common factors as your first step.

25. $4x^2 + 7x - 15$

26. $12x^2 + 11x - 5$

27. $15x^2 + 7x - 4$

28. $6x^2 - 13x + 6$

29. $2x^2 - x - 3$

30. $3x^2 + 2x - 8$

31. $20x^2 + 48x - 5$

32. $20x^2 + 21x - 5$

33. $6a^2 + 11a - 10$

34. $6a^2 - 19a + 10$

35. $6x^2 + 4x - 10$

36. $6x^2 - 4x - 10$

37. $4x^2 - 26x + 30$

38. $4x^2 - 20x - 144$

39. $16x^2 + 16x - 60$

40. $24x^2 + 6x - 3$

41. $12x^2 + xy - 6y^2$

42. $6x^2 + 5xy - 25y^2$

Section 6.5

Factor these special cases. Be sure to factor out any common factors.

43. $49x^2 - y^2$

44. $16x^2 - 36y^2$

45. $36x^2 - 60x + 25$

46. $16x^2 + 40x + 25$

47. $25x^2 - 36$

48. $100x^2 - 9$

49. $y^2 - 36x^2$

50. $9y^2 - 25x^2$

51. $36x^2 + 12x + 1$

52. $25x^2 - 20x + 4$

53. $16x^2 - 24xy + 9y^2$

54. $49x^2 - 28xy + 4y^2$

55. $2x^2 - 32$

56. $3x^2 - 27$

57. $28x^2 + 140x + 175$

58. $72x^2 - 192x + 128$

Section 6.6

If possible, factor each polynomial completely. If a polynomial cannot be factored, state that it is prime.

59. $4x^2 - 9y^2$

60. $x^2 + 6x + 9$

61. $x^2 - 9x + 18$

62. $x^2 + 13x - 30$

63. $5x^2 - 13x - 6$

64. $9x^2 - 9x - 4$

65. $24x - 60$

66. $8x^2y^2 - 4xy$

67. $50x^3y^2 + 30x^2y^2 - 10x^2y^2$

68. $26a^3b - 13ab^3 + 52a^2b^4$

69. $x^3 - 16x^2 + 64x$

70. $2x^2 + 40x + 200$

71. $3x^2 - 18x + 27$

72. $25x^3 - 60x^2 + 36x$

73. $7x^2 - 9x - 10$

74. $4x^2 - 13x - 12$

75. $9x^3y - 4xy^3$

76. $3x^3a^3 - 11x^4a^2 - 20x^5a$

77. $12a^2 + 14ab - 10b^2$

78. $121a^2 + 66ab + 9b^2$

79. $7a - 7 - ab + b$

Mixed Practice

If possible, factor each polynomial completely. If a polynomial cannot be factored, state that it is prime.

80. $18b - 42 + 3bc - 7c$

81. $10b + 16 - 24x - 15bx$

82. $5xb - 35x + 4by - 28y$

83. $2a^2x - 15ax + 7x$

84. $x^5 - 17x^3 + 16x$

85. $x^4 - 81y^{12}$

86. $6x^4 - x^2 - 15$

87. $28yz - 16xyz + x^2yz$

88. $12x^3 + 17x^2 + 6x$

89. $16w^2 - 2w - 5$

90. $12w^2 - 12w + 3$

91. $4y^3 + 10y^2 - 6y$

92. $10y^2 + 33y - 7$

93. $8y^{10} - 16y^8$

94. $9x^4 - 144$

95. $x^2 - 6x + 12$

96. $8x^2 - 19x - 6$

97. $8y^5 + 4y^3 - 60y$

98. $9xy^2 + 3xy - 42x$

99. $16x^4y^2 - 56x^2y + 49$

100. $128x^3y - 2xy$

101. $2ax + 5a - 10b - 4bx$

102. $2x^3 - 9 + x^2 - 18x$

Section 6.7

Solve the following equations by factoring.

103. $x^2 + x - 20 = 0$

104. $2x^2 + 11x - 6 = 0$

105. $7x^2 = 15x + x^2$

106. $5x^2 - x = 4x^2 + 12$

107. $2x^2 + 9x - 5 = 0$

108. $x^2 + 11x + 24 = 0$

109. $x^2 + 14x + 45 = 0$

110. $5x^2 = 7x + 6$

111. $3x^2 + 6x = 2x^2 - 9$

112. $4x^2 + 9x - 9 = 0$

113. $5x^2 - 11x + 2 = 0$

Solve.

▲ **114.** *Geometry* The area of a triangle is 25 square inches. The altitude is 5 inches longer than the base. Find the length of the base and the altitude.

▲ **115.** *Geometry* The area of a rectangle is 30 square feet. The length of the rectangle is 4 feet shorter than double the width. Find the length and width of the rectangle.

116. *Rocket Height* The height in feet that a model rocket attains is given by $h = -16t^2 + 80t + 96$, where t is the time measured in seconds. How many seconds will it take until the rocket finally reaches the ground? (*Hint:* At ground level $h = 0$.)

117. *Output Power* An electronic technician is working with a 100-volt electric generator. The output power of the generator is given by the equation $p = -5x^2 + 100x$, where x is the amount of current measured in amperes and p is measured in watts. The technician wants to find the value for x when the power is 480 watts. Can you find the two answers?

How Am I Doing? Chapter 6 Test

Remember to use your Chapter Test Prep Video CD to see the worked-out solutions to the test problems you want to review.

If possible, factor each polynomial completely. If a polynomial cannot be factored, state that it is prime.

1. $x^2 + 12x - 28$

2. $16x^2 - 81$

3. $10x^2 + 27x + 5$

4. $9a^2 - 30a + 25$

5. $7x - 9x^2 + 14xy$

6. $10xy + 15by - 8x - 12b$

7. $6x^3 - 20x^2 + 16x$

8. $5a^2c - 11ac + 2c$

9. $81x^2 - 100$

10. $9x^2 - 15x + 4$

11. $20x^2 - 45$

12. $36x^2 + 1$

13. $3x^3 + 11x^2 + 10x$

14. $60xy^2 - 20x^2y - 45y^3$

15. $81x^2 - 1$

16. $81y^4 - 1$

17. $2ax + 6a - 5x - 15$

18. $aw^2 - 8b + 2bw^2 - 4a$

19. $3x^2 - 3x - 90$

20. $2x^3 - x^2 - 15x$

Solve.

21. $x^2 + 14x + 45 = 0$

22. $14 + 3x(x + 2) = -7x$

23. $2x^2 + x - 10 = 0$

24. $x^2 - 3x - 28 = 0$

Solve using a quadratic equation.

▲ **25.** The park service is studying a rectangular piece of land that has an area of 91 square miles. The length of this piece of land is 1 mile shorter than double the width. Find the length and width of this rectangular piece of land.

1.	
2.	
3.	
4.	
5.	
6.	
7.	
8.	
9.	
10.	
11.	
12.	
13.	
14.	
15.	
16.	
17.	
18.	
19.	
20.	
21.	
22.	
23.	
24.	
25.	

1. _____

2. _____

3. _____

4. _____

5. _____

6. _____

7. _____

8. _____

9. _____

10. _____

11. _____

12. _____

13. _____

14. _____

15. _____

16. _____

17. _____

18. _____

19. _____

20. _____

21. _____

22. _____

23. _____

24. _____

Approximately one-half of this test covers the content of Chapters 0–5. The remainder covers the content of Chapter 6.

1. What percent of 300 is 60?

2. What is 12% of 4.5?

Simplify.

3. $4.3 - 2(3.4 - 7.2) - 6.1$

4. $(-2x^3y^4)(-4xy^6)$

5. $(-3)^4$

6. $(9x - 4)(3x + 2)$

7. $(2x^2 - 6x + 1)(x - 3)$

Solve.

8. $5x - 6 \le 7x + 3$

9. $3x - (7 - 5x) = 3(4x - 5)$

10. $\dfrac{1}{2}x - 3 = \dfrac{1}{4}(3x + 3)$

11. $\dfrac{3}{2}(x - 1) = \dfrac{8}{5}x - 1$

Factor each polynomial completely. If a polynomial cannot be factored, state that it is prime.

12. $6x^2 - 5x + 1$

13. $6x^2 + 5x - 4$

14. $9x^2 + 3x - 2$

15. $121x^2 - 64y^2$

16. $4x + 120 - 80x^2$

17. $x^2 + 5x + 9$

18. $16x^3 + 40x^2 + 25x$

19. $16x^4 - b^4$

20. $2ax - 4bx + 3a - 6b$

21. $15x^2 + 6x - 9$

Solve.

22. $x^2 + 5x - 36 = 0$

23. $3x^2 - 11x + 10 = 0$

Solve using a quadratic equation.

▲ **24.** Mr. Jerome's garden covered 300 square feet. If the width was 5 feet less than the length, find the measurements of the garden.

Calculators are everywhere. The demand for calculators continues to increase throughout the country. Those who open stores specializing in calculators often find that their profits depend greatly on the amount of advertising they do. Could you use your mathematical skills to predict the profits of a calculator store? The mathematics you will learn in this chapter will help you predict the future profitability of a small business.

Rational Expressions and Equations

Student Learning Objective

After studying this section, you will be able to:

 Simplify rational expressions by factoring.

Recall that a rational number is a number that can be written as one integer divided by another integer, such as $3 \div 4$ or $\frac{3}{4}$. We usually use the word *fraction* to mean $\frac{3}{4}$. We can extend this idea to algebraic expressions. A **rational expression** is a polynomial divided by another polynomial, such as

$$(3x + 2) \div (x + 4) \quad \text{or} \quad \frac{3x + 2}{x + 4}.$$

The last fraction is sometimes also called a **fractional algebraic expression.** There is a special restriction for all fractions, including fractional algebraic expressions. The denominator of the fraction cannot be 0. For example, in the expression

$$\frac{3x + 2}{x + 4},$$

the denominator cannot be 0. Therefore, the value of x cannot be -4. The following important restriction will apply throughout this chapter. We state it here to avoid having to mention it repeatedly throughout this chapter.

RESTRICTION

The denominator of a rational expression cannot be zero. Any value of the variable that would make the denominator zero is not allowed.

We have discovered that fractions can be simplified (or reduced) in the following way.

$$\frac{15}{25} = \frac{3 \cdot \cancel{5}}{5 \cdot \cancel{5}} = \frac{3}{5}$$

This is sometimes referred to as the **basic rule of fractions** and can be stated as follows.

BASIC RULE OF FRACTIONS

For any rational expression $\frac{a}{b}$ and any polynomials a, b, and c (where $b \neq 0$ and $c \neq 0$),

$$\frac{ac}{bc} = \frac{a}{b}.$$

We will examine several examples where a, b, and c are real numbers, as well as more involved examples where a, b, and c are polynomials. In either case we shall make extensive use of our factoring skills in this section.

One essential property is revealed by the basic rule of fractions: If the numerator and denominator of a given fraction are multiplied by the same nonzero quantity, an equivalent fraction is obtained. The rule can be used two ways. You can start with $\frac{ac}{bc}$ and end with the equivalent fraction $\frac{a}{b}$. Or, you can start with $\frac{a}{b}$ and end with the equivalent fraction $\frac{ac}{bc}$. In this section we focus on the first process.

EXAMPLE 1 Reduce. $\dfrac{21}{39}$

Solution $\dfrac{21}{39} = \dfrac{7 \cdot \cancel{3}}{13 \cdot \cancel{3}} = \dfrac{7}{13}$ Use the rule $\frac{ac}{bc} = \frac{a}{b}$. Let $c = 3$ because 3 is the greatest common factor of 21 and 39.

Practice Problem 1 Reduce. $\dfrac{28}{63}$

NOTE TO STUDENT: Fully worked-out solutions to all of the Practice Problems can be found at the back of the text starting at page SP-1

Simplifying Rational Expressions by Factoring

The process of reducing fractions shown in Example 1 is sometimes called *dividing out* common factors. When you do this, you are **simplifying the fraction.** Remember, only **factors** of both the numerator and the denominator can be divided out. To apply the basic rule of fractions, it is usually helpful if the numerator and denominator of the fraction are completely factored. You will need to use your factoring skills from Chapter 6 to accomplish this step.

EXAMPLE 2 Simplify. $\dfrac{4x + 12}{5x + 15}$

Solution $\dfrac{4x + 12}{5x + 15} = \dfrac{4(x + 3)}{5(x + 3)}$ Factor 4 from the numerator.

                      Factor 5 from the denominator.

$\qquad\qquad = \dfrac{4\cancel{(x + 3)}}{5\cancel{(x + 3)}}$ Apply the basic rule of fractions.

$\qquad\qquad = \dfrac{4}{5}$

Practice Problem 2 Simplify. $\dfrac{12x - 6}{14x - 7}$

EXAMPLE 3 Simplify. $\dfrac{x^2 + 9x + 14}{x^2 - 4}$

Solution $\qquad = \dfrac{(x + 7)(x + 2)}{(x - 2)(x + 2)}$ Factor the numerator.

                             Factor the denominator.

$\qquad = \dfrac{(x + 7)\cancel{(x + 2)}}{(x - 2)\cancel{(x + 2)}}$ Apply the basic rule of fractions.

$\qquad = \dfrac{x + 7}{x - 2}$

CAUTION: Do not try to remove terms that are added. In Example 3, do not try to remove the x^2 in the top and the x^2 in the bottom of the fraction. The basic rule of fractions applies only to quantities that are factors of both numerator and denominator.

Practice Problem 3 Simplify. $\dfrac{4x^2 - 9}{2x^2 - x - 3}$

Some problems may involve more than one step of factoring. Always remember to factor out any common factors as the first step, if it is possible to do so.

EXAMPLE 4 Simplify. $\dfrac{x^3 - 9x}{x^3 + x^2 - 6x}$

Solution $\quad = \dfrac{x(x^2 - 9)}{x(x^2 + x - 6)}$ Factor out a common factor from the polynomials in the numerator and the denominator.

$\qquad = \dfrac{\cancel{x}\cancel{(x + 3)}(x - 3)}{\cancel{x}\cancel{(x + 3)}(x - 2)}$ Factor each polynomial and apply the basic rule of fractions.

$\qquad = \dfrac{x - 3}{x - 2}$

Practice Problem 4 Simplify. $\dfrac{x^3 - 16x}{x^3 - 2x^2 - 8x}$

When you are simplifying, be on the lookout for the special situation where *a factor in the denominator is the opposite of a factor in the numerator*. In such a case you should factor a negative number from one of the factors so that it becomes equivalent to the other factor and can be divided out. Look carefully at the following two examples.

EXAMPLE 5 Simplify. $\dfrac{5x - 15}{6 - 2x}$

Solution Notice that the variable term in the numerator, $5x$, and the variable term in the denominator, $-2x$, *are opposite in sign*. Likewise, the numerical terms -15 and 6 *are opposite in sign*. Factor out a negative number from the denominator.

$$\frac{5x - 15}{6 - 2x} = \frac{5(x - 3)}{-2(-3 + x)}$$

Factor 5 from the numerator.

Factor -2 from the denominator.

Note that $(x - 3)$ and $(-3 + x)$ are equivalent since $(+x - 3) = (-3 + x)$.

$$= \frac{5\cancel{(x - 3)}}{-2\cancel{(-3 + x)}}$$

Apply the basic rule of fractions.

$$= -\frac{5}{2}$$

Note that $\frac{5}{-2}$ is not considered to be in simple form. We usually avoid leaving a negative number in the denominator. Therefore, to simplify, give the result as $-\frac{5}{2}$ or $\frac{-5}{2}$.

NOTE TO STUDENT: *Fully worked-out solutions to all of the Practice Problems can be found at the back of the text starting at page SP-1*

Practice Problem 5 Simplify. $\dfrac{8x - 20}{15 - 6x}$

EXAMPLE 6 Simplify. $\dfrac{2x^2 - 11x + 12}{16 - x^2}$

Solution

$$= \frac{(x - 4)(2x - 3)}{(4 - x)(4 + x)}$$

Factor the numerator and the denominator. Observe that $(x - 4)$ and $(4 - x)$ are opposites.

$$= \frac{(x - 4)(2x - 3)}{-1(-4 + x)(4 + x)}$$

Factor -1 out of $(+4 - x)$ to obtain $-1(-4 + x)$.

$$= \frac{\cancel{(x - 4)}(2x - 3)}{-1\cancel{(-4 + x)}(4 + x)}$$

Apply the basic rule of fractions since $(x - 4)$ and $(-4 + x)$ are equivalent.

$$= \frac{2x - 3}{-1(4 + x)}$$

$$= -\frac{2x - 3}{4 + x}$$

Practice Problem 6 Simplify. $\dfrac{4x^2 + 3x - 10}{25 - 16x^2}$

After doing Examples 5 and 6, you will notice a pattern. Whenever the factor in the numerator and the factor in the denominator are opposites, the value -1 results. We could actually make this a property.

For all monomials A and B where $A \neq B$, it is true that

$$\frac{A - B}{B - A} = -1.$$

You may use this property when reducing fractions if it is helpful to you. Otherwise, you may use the factoring method shown in Examples 5 and 6.

Some problems will involve two or more variables. In such cases, you will need to factor carefully and make sure that each set of parentheses contains the correct terms.

EXAMPLE 7 Simplify. $\dfrac{x^2 - 7xy + 12y^2}{2x^2 - 7xy - 4y^2}$

Solution

$$= \frac{(x - 4y)(x - 3y)}{(2x + y)(x - 4y)} \quad \text{Factor the numerator.}$$
$$\text{Factor the denominator.}$$

$$= \frac{\cancel{(x - 4y)}(x - 3y)}{(2x + y)\cancel{(x - 4y)}} \quad \text{Apply the basic rule of fractions.}$$

$$= \frac{x - 3y}{2x + y}$$

Practice Problem 7 Simplify. $\dfrac{x^2 - 8xy + 15y^2}{2x^2 - 11xy + 5y^2}$

EXAMPLE 8 Simplify. $\dfrac{6a^2 + ab - 7b^2}{36a^2 - 49b^2}$

Solution

$$= \frac{(6a + 7b)(a - b)}{(6a + 7b)(6a - 7b)} \quad \text{Factor the numerator.}$$
$$\text{Factor the denominator.}$$

$$= \frac{\cancel{(6a + 7b)}(a - b)}{\cancel{(6a + 7b)}(6a - 7b)} \quad \text{Apply the basic rule of fractions.}$$

$$= \frac{a - b}{6a - 7b}$$

Practice Problem 8 Simplify. $\dfrac{25a^2 - 16b^2}{10a^2 + 3ab - 4b^2}$

Simplify.

1. $\dfrac{3a - 9b}{a - 3b}$

2. $\dfrac{5x + 2y}{35x + 14y}$

3. $\dfrac{6x + 18}{x^2 + 3x}$

4. $\dfrac{8x - 12}{2x^3 - 3x^2}$

5. $\dfrac{9x^2 + 6x + 1}{1 - 9x^2}$

6. $\dfrac{4x^2 + 4x + 1}{1 - 4x^2}$

7. $\dfrac{3a^2b(a - 2b)}{6ab^2}$

8. $\dfrac{9ab^2}{6a^2b^2(b + 3a)}$

9. $\dfrac{x^2 + x - 2}{x^2 - x}$

10. $\dfrac{x^2 + x - 12}{x^2 - 3x}$

11. $\dfrac{x^2 - 3x - 10}{3x^2 + 5x - 2}$

12. $\dfrac{4x^2 - 10x + 6}{2x^2 + x - 3}$

13. $\dfrac{x^2 + 4x - 21}{x^3 - 49x}$

14. $\dfrac{x^3 - 3x^2 - 40x}{x^2 + 10x + 25}$

15. $\dfrac{3x^2 - 11x - 4}{x^2 + x - 20}$

16. $\dfrac{x^2 - 9x + 18}{2x^2 - 9x + 9}$

17. $\dfrac{3x^2 - 8x + 5}{4x^2 - 5x + 1}$

18. $\dfrac{3y^2 + 10y + 3}{3y^2 - 14y - 5}$

19. $\dfrac{5x^2 - 27x + 10}{5x^2 + 3x - 2}$

20. $\dfrac{2x^2 + 2x - 12}{x^2 + 3x - 4}$

Mixed Practice   *Take some time to review exercises 1–20 before you proceed with exercises 21–30.*

21. $\dfrac{12 - 3x}{5x^2 - 20x}$

22. $\dfrac{20 - 4ab}{a^2b - 5a}$

23. $\dfrac{2x^2 - 7x - 15}{25 - x^2}$

24. $\dfrac{49 - x^2}{2x^2 - 9x - 35}$

25. $\dfrac{(3x + 4)^2}{9x^2 + 9x - 4}$

26. $\dfrac{12x^2 - 11x - 5}{(4x - 5)^2}$

27. $\dfrac{2x^2 + 9x - 18}{30 - x - x^2}$

28. $\dfrac{4y^2 - y - 3}{8 - 7y - y^2}$

29. $\dfrac{a^2 + 2ab - 3b^2}{2a^2 + 5ab - 3b^2}$

30. $\dfrac{a^2 + 3ab - 10b^2}{3a^2 - 7ab + 2b^2}$

Cumulative Review *Multiply.*

31. **[5.5.2]** $(3x - 7)^2$

32. **[5.5.1]** $(7x + 6y)(7x - 6y)$

33. **[5.4.2]** $(2x + 3)(x - 4)$

34. **[5.5.3]** $(2x + 3)(x - 4)(x - 2)$

35. **[0.3.2]** ***Dividing Acreage*** Walter and Ann Perkins wish to divide $4\frac{7}{8}$ acres of farmland into three equal-sized house lots. What will be the acreage of each lot?

36. **[0.3.1]** ***Roasting a Turkey*** Ron and Mary Larson are planning to cook a $17\frac{1}{2}$-pound turkey. The directions suggest a cooking time of 22 minutes per pound for turkeys that weigh between 16 and 20 pounds. How many hours and minutes should they allow for an approximate cooking time?

Spanish Speakers *The number of people in Mexico who speak Spanish is 12,500,000 more than three times the number of Spanish speakers in the United States. The number of people who speak Spanish in the United States is 17,000,000 less than the number of Spanish speakers in Spain. The number of Spanish speakers in the United States is 22,500,000.*

37. **[2.6.1]** How many people speak Spanish in Spain?

38. **[2.6.1]** How many people speak Spanish in Mexico?

Quick Quiz 7.1 Simplify.

1. $\dfrac{x^3 + 3x^2}{x^3 - 2x^2 - 15x}$

2. $\dfrac{6 - 2ab}{ab^2 - 3b}$

3. $\dfrac{8x^2 + 6x - 5}{16x^2 + 40x + 25}$

4. **Concept Check** Explain why it is important to completely factor both the numerator and the denominator when simplifying

$$\frac{x^2y - y^3}{x^2y + xy^2 - 2y^3}.$$

Student Learning Objectives

After studying this section, you will be able to:

 Multiply rational expressions.

2 **Divide rational expressions.**

1 **Multiplying Rational Expressions**

To multiply two rational expressions, we multiply the numerators and multiply the denominators. As before, the denominators cannot equal zero.

For any two rational expressions $\frac{a}{b}$ and $\frac{c}{d}$ where $b \neq 0$ and $d \neq 0$,

$$\frac{a}{b} \cdot \frac{c}{d} = \frac{ac}{bd}.$$

Simplifying or reducing fractions *prior to multiplying them* usually makes the computations easier to do. Leaving the reducing step until the end makes the simplifying process longer and increases the chance for error. This long approach should be avoided.

As an example, let's do the same problem two ways to see which one is easier. Let's simplify the following problem by multiplying first and then reducing the result.

$$\frac{5}{7} \times \frac{49}{125}$$

$\frac{5}{7} \times \frac{49}{125} = \frac{245}{875}$ Multiply the numerators and multiply the denominators.

$\qquad = \frac{7}{25}$ Reduce the fraction. (*Note:* It can take a bit of trial and error to discover how to reduce it.)

Compare this with the following method, where we reduce the fractions prior to multiplying them.

$$\frac{5}{7} \times \frac{49}{125}$$

$\frac{5}{7} \times \frac{7 \cdot 7}{5 \cdot 5 \cdot 5}$ **Step 1.** It is easier to factor first. We factor the numerator and denominator of the second fraction.

$= \frac{\cancel{5}}{\cancel{7}} \times \frac{\cancel{7} \cdot 7}{\cancel{5} \cdot 5 \cdot 5} = \frac{7}{25}$ **Step 2.** Then we apply the basic rule of fractions to divide out the common factors of 5 and 7 that appear in both a numerator and in a denominator.

A similar approach can be used with the multiplication of rational expressions. We first factor the numerator and denominator of each fraction. Then we divide out any factor that is common to a numerator and a denominator. Finally, we multiply the remaining numerators and the remaining denominators.

EXAMPLE 1   Multiply. $\dfrac{x^2 - x - 12}{x^2 - 16} \cdot \dfrac{2x^2 + 7x - 4}{x^2 - 4x - 21}$

Solution

$\dfrac{(x - 4)(x + 3)}{(x - 4)(x + 4)} \cdot \dfrac{(x + 4)(2x - 1)}{(x + 3)(x - 7)}$ Factoring is always the first step.

$= \dfrac{\cancel{(x - 4)}\cancel{(x + 3)}}{\cancel{(x - 4)}\cancel{(x + 4)}} \cdot \dfrac{\cancel{(x + 4)}(2x - 1)}{\cancel{(x + 3)}(x - 7)}$ Apply the basic rule of fractions. (Three pairs of factors divide out.)

$= \dfrac{2x - 1}{x - 7}$ The final answer.

NOTE TO STUDENT: *Fully worked-out solutions to all of the Practice Problems can be found at the back of the text starting at page SP-1*

Practice Problem 1   Multiply. $\dfrac{6x^2 + 7x + 2}{x^2 - 7x + 10} \cdot \dfrac{x^2 + 3x - 10}{2x^2 + 11x + 5}$

In some cases, it may take several steps to factor a given numerator or denominator. You should always check for a *common factor* as your first step.

EXAMPLE 2 Multiply. $\dfrac{x^4 - 16}{x^3 + 4x} \cdot \dfrac{2x^2 - 8x}{4x^2 + 2x - 12}$

Solution

$= \dfrac{(x^2 + 4)(x^2 - 4)}{x(x^2 + 4)} \cdot \dfrac{2x(x - 4)}{2(2x^2 + x - 6)}$ Factor each numerator and denominator. Factoring out the common factor first is very important.

$= \dfrac{(x^2 + 4)(x + 2)(x - 2)}{x(x^2 + 4)} \cdot \dfrac{2x(x - 4)}{2(x + 2)(2x - 3)}$ Factor again where possible.

$= \dfrac{\cancel{(x^2 + 4)}\cancel{(x + 2)}(x - 2)}{\cancel{x}\cancel{(x^2 + 4)}} \cdot \dfrac{\cancel{2x}(x - 4)}{\cancel{2}\cancel{(x + 2)}(2x - 3)}$ Divide out factors that appear in both a numerator and a denominator. (There are four such pairs of factors.)

$= \dfrac{(x - 2)(x - 4)}{2x - 3}$ or $\dfrac{x^2 - 6x + 8}{2x - 3}$ Write the answer as one fraction. (Usually, if there is more than one factor in a numerator or denominator, the answer is left in factored form.)

Practice Problem 2 Multiply. $\dfrac{2y^2 - 6y - 8}{y^2 - y - 2} \cdot \dfrac{y^2 - 5y + 6}{2y^2 - 32}$

Dividing Rational Expressions

For any two fractions $\frac{a}{b}$ and $\frac{c}{d}$, the operation of division can be performed by inverting the second fraction and multiplying it by the first fraction. When we invert a fraction, we are finding its *reciprocal*. Two numbers are **reciprocals** of each other if their product is 1. The reciprocal of $\frac{3}{5}$ is $\frac{5}{3}$. The reciprocal of 7 is $\frac{1}{7}$. The reciprocal of $\frac{a}{b}$ is $\frac{b}{a}$. Sometimes people state the rule for dividing fractions this way: "To divide two fractions, keep the first fraction unchanged and multiply by the reciprocal of the second fraction."

The definition for division of fractions is

$$\frac{a}{b} \div \frac{c}{d} = \frac{a}{b} \cdot \frac{d}{c}.$$

This property holds whether a, b, c, and d are polynomials or numerical values. (It is assumed, of course, that no denominator is zero.)

In the first step for dividing two rational expressions, invert the second fraction and rewrite the quotient as a product. Then follow the procedure for multiplying rational expressions.

EXAMPLE 3 Divide. $\dfrac{6x + 12y}{2x - 6y} \div \dfrac{9x^2 - 36y^2}{4x^2 - 36y^2}$

Solution

$= \dfrac{6x + 12y}{2x - 6y} \cdot \dfrac{4x^2 - 36y^2}{9x^2 - 36y^2}$ Invert the second fraction and write the problem as the product of two fractions.

$= \dfrac{6(x + 2y)}{2(x - 3y)} \cdot \dfrac{4(x^2 - 9y^2)}{9(x^2 - 4y^2)}$ Factor each numerator and denominator.

$$= \frac{(3)(2)(x + 2y)}{2(x - 3y)} \cdot \frac{(2)(2)(x + 3y)(x - 3y)}{(3)(3)(x + 2y)(x - 2y)}$$ Factor again where possible.

$$= \frac{\cancel{(3)}\cancel{(2)}\cancel{(x + 2y)}}{\cancel{2}\cancel{(x - 3y)}} \cdot \frac{(2)(2)(x + 3y)\cancel{(x - 3y)}}{\cancel{(3)}(3)\cancel{(x + 2y)}(x - 2y)}$$ Divide out factors that appear in both a numerator and a denominator.

$$= \frac{(2)(2)(x + 3y)}{3(x - 2y)}$$ Write the result as one fraction.

$$= \frac{4(x + 3y)}{3(x - 2y)}$$ Simplify. (Usually, answers are left in this form.)

Although it is correct to write this answer as $\dfrac{4x + 12y}{3x - 6y}$, it is customary to leave the answer in factored form to ensure that the final answer is simplified.

Practice Problem 3 Divide. $\dfrac{x^2 + 5x + 6}{x^2 + 8x} \div \dfrac{2x^2 + 5x + 2}{2x^2 + x}$

NOTE TO STUDENT: *Fully worked-out solutions to all of the Practice Problems can be found at the back of the text starting at page SP-1*

A polynomial that is not in fraction form can be written as a fraction if you give it a denominator of 1.

EXAMPLE 4 Divide. $\dfrac{15 - 3x}{x + 6} \div (x^2 - 9x + 20)$

Solution Note that $x^2 - 9x + 20$ can be written as $\dfrac{x^2 - 9x + 20}{1}$.

$$= \frac{15 - 3x}{x + 6} \cdot \frac{1}{x^2 - 9x + 20}$$ Invert and multiply.

$$= \frac{-3(-5 + x)}{x + 6} \cdot \frac{1}{(x - 5)(x - 4)}$$ Factor where possible. Note that we had to factor -3 from the first numerator so that it would have a factor in common with the second denominator.

$$= \frac{-3\cancel{(-5 + x)}}{x + 6} \cdot \frac{1}{\cancel{(x - 5)}(x - 4)}$$ Divide out the common factor. $(-5 + x)$ is equivalent to $(x - 5)$.

$$= \frac{-3}{(x + 6)(x - 4)}$$ The final answer. Note that the answer can be written in several equivalent forms.

or $-\dfrac{3}{(x + 6)(x - 4)}$ or $\dfrac{3}{(x + 6)(4 - x)}$

Practice Problem 4 Divide. $\dfrac{x + 3}{x - 3} \div (9 - x^2)$

CAUTION: It is logical to assume that the problems in Section 7.2 have at least one common factor that can be divided out. Therefore, if after factoring, you do not observe any common factors, you should be somewhat suspicious. In such cases, it would be wise to double-check your factoring steps for errors.

PRACTICE WATCH DOWNLOAD READ REVIEW

Verbal and Writing Skills

1. Before multiplying rational expressions, we should always first try to

_____

_____.

2. Division of two rational expressions is done by keeping the first fraction unchanged and then

_____.

Multiply.

3. $\dfrac{3x + 6}{x - 3} \cdot \dfrac{x^2 + 2x - 15}{x^2 + 4x + 4}$

4. $\dfrac{6x - 6}{x + 3} \cdot \dfrac{x^2 - 3x - 18}{6x^2 - 6}$

5. $\dfrac{24x^3}{4x^2 - 36} \cdot \dfrac{2x^2 + 6x}{16x^2}$

6. $\dfrac{14x^2}{10x + 50} \cdot \dfrac{5x^3 - 125x}{35x^4}$

7. $\dfrac{x^2 + 3x - 10}{x^2 + x - 20} \cdot \dfrac{x^2 - 3x - 4}{x^2 + 4x + 3}$

8. $\dfrac{x^2 - x - 20}{x^2 - 3x - 10} \cdot \dfrac{x^2 + 7x + 10}{x^2 + 4x - 5}$

Divide.

9. $\dfrac{x + 6}{x - 8} \div \dfrac{x + 5}{x^2 - 6x - 16}$

10. $\dfrac{x - 4}{x + 9} \div \dfrac{x - 7}{x^2 + 13x + 36}$

11. $(5x + 4) \div \dfrac{25x^2 - 16}{5x^2 + 11x - 12}$

12. $\dfrac{4x^2 - 25}{4x^2 - 20x + 25} \div (4x + 10)$

13. $\dfrac{3x^2 + 12xy + 12y^2}{x^2 + 4xy + 3y^2} \div \dfrac{4x + 8y}{x + y}$

14. $\dfrac{5x^2 + 10xy + 5y^2}{x^2 + 5xy + 6y^2} \div \dfrac{3x + 3y}{x + 2y}$

Mixed Practice *Perform the operation indicated.*

15. $\dfrac{(x + 5)^2}{3x^2 - 7x + 2} \cdot \dfrac{x^2 - 4x + 4}{x + 5}$

16. $\dfrac{3x^2 - 10x - 8}{(4x + 5)^2} \cdot \dfrac{4x + 5}{(x - 4)^2}$

17. $\dfrac{x^2 + x - 30}{10 - 2x} \div \dfrac{x^2 + 4x - 12}{5x + 15}$

18. $\dfrac{x^2 + 3x - 28}{x^2 + 14x + 49} \div \dfrac{12 - 3x^2}{x^2 + 5x - 14}$

19. $\dfrac{y^2 + 4y - 12}{y^2 + 2y - 24} \cdot \dfrac{y^2 - 16}{y^2 + 2y - 8}$

20. $\dfrac{4y^2 + 11y + 6}{8y^2 + 2y - 3} \cdot \dfrac{4y^2 - 8y + 3}{2y^2 + y - 6}$

21. $\dfrac{x^2 + 7x + 12}{2x^2 + 9x + 4} \div \dfrac{x^2 + 6x + 9}{2x^2 - x - 1}$

22. $\dfrac{x^2 + 7x + 10}{2x^2 - 3x - 2} \cdot \dfrac{2x^2 - 11x - 6}{x^2 - x - 30}$

Cumulative Review

23. **[2.3.2]** Solve. $6x^2 + 3x - 18 = 5x - 2 + 6x^2$

24. **[5.5.3]** Multiply. $(7x^2 - x - 1)(x - 3)$

▲ **25.** **[2.5.2]** *Golden Gate Bridge* The Golden Gate Bridge has a total length (including approaches) of 8981 feet and a road width of 90 feet. The width of the sidewalk is 10.5 feet. (The sidewalk spans the entire length of the bridge.) Assume it would cost $\$x$ per square foot to resurface the road or the sidewalk. Write an expression for how much more it would cost to resurface the road than the sidewalk.

▲ **26.** **[2.7.1]** *Garden Design* Harold Rafton planted a square garden bed. George Avis also planted a garden that was 2 feet less in width, but 3 feet longer in length than Harold's garden. If the area of George's garden was 36 square feet, find the dimensions of each garden.

Quick Quiz 7.2 Multiply.

1. $\dfrac{2x - 10}{x - 4} \cdot \dfrac{x^2 + 5x + 4}{x^2 - 4x - 5}$

2. $\dfrac{3x^2 - 13x - 10}{3x^2 + 2x} \cdot \dfrac{x^2 - 25x}{x^2 - 25}$

Divide.

3. $\dfrac{2x^2 - 18}{3x^2 + 3x} \div \dfrac{x^2 + 6x + 9}{x^2 + 4x + 3}$

4. **Concept Check** Explain how you would divide $\dfrac{21x - 7}{9x^2 - 1} \div \dfrac{1}{3x + 1}$.

Adding and Subtracting Rational Expressions with a Common Denominator

If rational expressions have the same denominator, they can be combined in the same way as arithmetic fractions. The numerators are added or subtracted and the denominator remains the same.

Student Learning Objectives

After studying this section, you will be able to:

1 Add and subtract rational expressions with a common denominator.

2 Determine the LCD for rational expressions with different denominators.

3 Add and subtract rational expressions with different denominators.

ADDING RATIONAL EXPRESSIONS

For any rational expressions $\dfrac{a}{b}$ and $\dfrac{c}{b}$,

$$\frac{a}{b} + \frac{c}{b} = \frac{a + c}{b} \qquad \text{where } b \neq 0.$$

EXAMPLE 1 Add. $\dfrac{5a}{a + 2b} + \dfrac{6a}{a + 2b}$

Solution

$$\frac{5a}{a + 2b} + \frac{6a}{a + 2b} = \frac{5a + 6a}{a + 2b} = \frac{11a}{a + 2b}$$

Note that the denominators are the same. Only add the numerators. Keep the same denominator.
Do not change the denominator.

Practice Problem 1 Add. $\dfrac{2s + t}{2s - t} + \dfrac{s - t}{2s - t}$

SUBTRACTING RATIONAL EXPRESSIONS

For any rational expressions $\dfrac{a}{b}$ and $\dfrac{c}{b}$,

$$\frac{a}{b} - \frac{c}{b} = \frac{a - c}{b} \qquad \text{where } b \neq 0.$$

EXAMPLE 2 Subtract. $\dfrac{3x}{(x + y)(x - 2y)} - \dfrac{8x}{(x + y)(x - 2y)}$

Solution

$$\frac{3x}{(x + y)(x - 2y)} - \frac{8x}{(x + y)(x - 2y)} = \frac{3x - 8x}{(x + y)(x - 2y)} \qquad \text{Write as one fraction.}$$

$$= \frac{-5x}{(x + y)(x - 2y)} \qquad \text{Simplify.}$$

Practice Problem 2 Subtract.

$$\frac{b}{(a - 2b)(a + b)} - \frac{2b}{(a - 2b)(a + b)}$$

② Determining the LCD for Rational Expressions with Different Denominators

How do we add or subtract rational expressions when the denominators are not the same? First we must find the **least common denominator** (LCD). You need to be clear on how to find a least common denominator and how to add and subtract fractions from arithmetic before you attempt this section. Review Sections 0.1 and 0.2 if you have any questions about this topic.

> **HOW TO FIND THE LCD OF TWO OR MORE RATIONAL EXPRESSIONS**
>
> 1. Factor each denominator completely.
>
> 2. The LCD is a product containing each *different factor*.
>
> 3. If a factor occurs more than once in any one denominator, the LCD will contain that factor repeated the greatest number of times that it occurs in any one denominator.

EXAMPLE 3 Find the LCD. $\dfrac{5}{2x - 4}, \dfrac{6}{3x - 6}$

Solution Factor each denominator.

$$2x - 4 = 2(x - 2) \qquad 3x - 6 = 3(x - 2)$$

The different factors are 2, 3, and $(x - 2)$. Since no factor appears more than once in any one denominator, the LCD is the product of these three factors.

$$\text{LCD} = (2)(3)(x - 2) = 6(x - 2)$$

Practice Problem 3 Find the LCD. $\dfrac{7}{6x + 21}, \dfrac{13}{10x + 35}$

EXAMPLE 4 Find the LCD.

(a) $\dfrac{5}{12ab^2c}, \dfrac{13}{18a^3bc^4}$

(b) $\dfrac{8}{x^2 - 5x + 4}, \dfrac{12}{x^2 + 2x - 3}$

Solution If a factor occurs more than once in any one denominator, the LCD will contain that factor repeated the greatest number of times that it occurs in any one denominator.

(a) $12ab^2c = 2 \cdot 2 \cdot 3 \cdot \quad a \cdot \qquad b \cdot b \cdot c$

$18a^3bc^4 = \quad 2 \cdot 3 \cdot 3 \cdot a \cdot a \cdot a \cdot b \cdot \quad c \cdot c \cdot c \cdot c$

$\qquad\qquad 2 \cdot 2 \cdot 3 \cdot 3 \cdot a \cdot a \cdot a \cdot b \cdot b \cdot c \cdot c \cdot c \cdot c$

$\text{LCD} = 2^2 \cdot 3^2 \cdot a^3 \cdot b^2 \cdot c^4 = 36a^3b^2c^4$

(b) $x^2 - 5x + 4 = (x - 4)(x - 1)$

$x^2 + 2x - 3 = \qquad (x - 1)(x + 3)$

$\text{LCD} = (x - 4)(x - 1)(x + 3)$

Practice Problem 4 Find the LCD.

(a) $\dfrac{3}{50xy^2z}, \dfrac{19}{40x^3yz}$

(b) $\dfrac{2}{x^2 + 5x + 6}, \dfrac{6}{3x^2 + 5x - 2}$

Adding and Subtracting Rational Expressions with Different Denominators

If two rational expressions have different denominators, we first change them to equivalent rational expressions with the least common denominator. Then we add or subtract the numerators and keep the common denominator.

EXAMPLE 5 Add. $\dfrac{5}{xy} + \dfrac{2}{y}$

Solution   The denominators are different. We must find the LCD. The two factors are x and y. We observe that the LCD is xy.

$$\frac{5}{xy} + \frac{2}{y} = \frac{5}{xy} + \frac{2}{y} \cdot \frac{x}{x} \qquad \text{Multiply the second fraction by } \frac{x}{x}.$$

$$= \frac{5}{xy} + \frac{2x}{xy} \qquad \text{Now each fraction has a common denominator of } xy.$$

$$= \frac{5 + 2x}{xy} \qquad \text{Write the sum as one fraction.}$$

Practice Problem 5   Add. $\dfrac{7}{a} + \dfrac{3}{abc}$

EXAMPLE 6 Add. $\dfrac{3x}{x^2 - y^2} + \dfrac{5}{x + y}$

Solution   We factor the first denominator so that $x^2 - y^2 = (x + y)(x - y)$. Thus, the factors of the denominators are $(x + y)$ and $(x - y)$. We observe that the LCD $= (x + y)(x - y)$.

$$\frac{3x}{(x + y)(x - y)} + \frac{5}{(x + y)} \cdot \frac{x - y}{x - y} \qquad \text{Multiply the second fraction by } \frac{x - y}{x - y}.$$

$$= \frac{3x}{(x + y)(x - y)} + \frac{5x - 5y}{(x + y)(x - y)} \qquad \begin{array}{l}\text{Now each fraction has a common} \\ \text{denominator of } (x + y)(x - y).\end{array}$$

$$= \frac{3x + 5x - 5y}{(x + y)(x - y)} \qquad \begin{array}{l}\text{Write the sum of the numerators over} \\ \text{the common denominator.}\end{array}$$

$$= \frac{8x - 5y}{(x + y)(x - y)} \qquad \text{Combine like terms.}$$

Practice Problem 6   Add. $\dfrac{2a - b}{a^2 - 4b^2} + \dfrac{2}{a + 2b}$

It is important to remember that the LCD is the smallest algebraic expression into which each denominator can be divided. For rational expressions the LCD must contain *each factor* that appears in any denominator. If the factor is repeated, the LCD must contain that factor the greatest number of times that it appears in any one denominator.

In many cases, the denominators in an addition or subtraction problem are not in factored form. You must factor each denominator to determine the LCD. Combine like terms in the numerator; then determine whether that final numerator can be factored. If so, you may be able to simplify the fraction.

EXAMPLE 7 Add. $\dfrac{5}{x^2 - y^2} + \dfrac{3x}{x^3 + x^2y}$

Solution

$\dfrac{5}{x^2 - y^2} + \dfrac{3x}{x^3 + x^2y}$

$= \dfrac{5}{(x + y)(x - y)} + \dfrac{3x}{x^2(x + y)}$

Factor the two denominators. Observe that the LCD is $x^2(x + y)(x - y)$.

$= \dfrac{5}{(x + y)(x - y)} \cdot \dfrac{x^2}{x^2} + \dfrac{3x}{x^2(x + y)} \cdot \dfrac{x - y}{x - y}$

Multiply each fraction by the appropriate expression to obtain a common denominator of $x^2(x + y)(x - y)$.

$= \dfrac{5x^2}{x^2(x + y)(x - y)} + \dfrac{3x^2 - 3xy}{x^2(x + y)(x - y)}$

$= \dfrac{5x^2 + 3x^2 - 3xy}{x^2(x + y)(x - y)}$

Write the sum of the numerators over the common denominator.

$= \dfrac{8x^2 - 3xy}{x^2(x + y)(x - y)}$

Combine like terms.

$= \dfrac{x(8x - 3y)}{x^2(x + y)(x - y)}$

Divide out the common factor x in the numerator and denominator and simplify.

$= \dfrac{8x - 3y}{x(x + y)(x - y)}$

Practice Problem 7 Add.

$\dfrac{7a}{a^2 + 2ab + b^2} + \dfrac{4}{a^2 + ab}$

NOTE TO STUDENT: Fully worked-out solutions to all of the Practice Problems can be found at the back of the text starting at page SP-1

It is very easy to make a sign mistake when subtracting two fractions. You will find it helpful to place parentheses around the numerator of the second fraction so that you will not forget to subtract the entire numerator.

EXAMPLE 8 Subtract. $\dfrac{3x + 4}{x - 2} - \dfrac{x - 3}{2x - 4}$

Solution Factor the second denominator.

$= \dfrac{3x + 4}{x - 2} - \dfrac{x - 3}{2(x - 2)}$

Observe that the LCD is $2(x - 2)$.

$= \dfrac{2}{2} \cdot \dfrac{3x + 4}{x - 2} - \dfrac{x - 3}{2(x - 2)}$

Multiply the first fraction by $\frac{2}{2}$ so that the resulting fraction will have the common denominator.

$= \dfrac{2(3x + 4) - (x - 3)}{2(x - 2)}$

Write the indicated subtraction as one fraction. Note the parentheses around $x - 3$.

$= \dfrac{6x + 8 - x + 3}{2(x - 2)}$

Remove the parentheses in the numerator.

$= \dfrac{5x + 11}{2(x - 2)}$

Combine like terms.

Practice Problem 8 Subtract.

$\dfrac{x + 7}{3x - 9} - \dfrac{x - 6}{x - 3}$

SIDELIGHT: **Alternate Method** To avoid making errors when subtracting two fractions, some students prefer to change subtraction to addition of the opposite of the second fraction. In other words, we use the property that $\dfrac{a}{b} - \dfrac{c}{b} = \dfrac{a}{b} + \dfrac{-c}{b}$.

Let's revisit Example 8 to see how to use this method.

$$\dfrac{3x + 4}{x - 2} - \dfrac{(x - 3)}{2(x - 2)} \qquad \text{Insert parentheses around } x - 3.$$

$$= \dfrac{2}{2} \cdot \dfrac{3x + 4}{x - 2} + \dfrac{-(x - 3)}{2(x - 2)} \qquad \begin{array}{l}\text{Change subtraction to addition of the opposite, and}\\ \text{multiply the first fraction by } \dfrac{2}{2}.\end{array}$$

$$= \dfrac{2(3x + 4) + -(x - 3)}{2(x - 2)}$$

$$= \dfrac{6x + 8 + (-x) + 3}{2(x - 2)} \qquad \text{Remove parentheses in the numerator.}$$

$$= \dfrac{5x + 11}{2(x - 2)} \qquad \text{Combine like terms.}$$

Try each method and see which one helps you avoid making errors.

EXAMPLE 9 Subtract. $\dfrac{8x}{x^2 - 16} - \dfrac{4}{x - 4}$

Solution

$$\dfrac{8x}{x^2 - 16} - \dfrac{4}{x - 4}$$

$$= \dfrac{8x}{(x + 4)(x - 4)} + \dfrac{-4}{x - 4} \qquad \begin{array}{l}\text{Factor the first denominator. Use the}\\ \text{property that } \dfrac{a}{b} - \dfrac{c}{b} = \dfrac{a}{b} + \dfrac{-c}{b}.\end{array}$$

$$= \dfrac{8x}{(x + 4)(x - 4)} + \dfrac{-4}{x - 4} \cdot \dfrac{x + 4}{x + 4} \qquad \text{Multiply the second fraction by } \dfrac{x + 4}{x + 4}.$$

$$= \dfrac{8x + (-4)(x + 4)}{(x + 4)(x - 4)} \qquad \begin{array}{l}\text{Write the sum of the numerators over the}\\ \text{common denominator.}\end{array}$$

$$= \dfrac{8x - 4x - 16}{(x + 4)(x - 4)} \qquad \text{Remove parentheses.}$$

$$= \dfrac{4x - 16}{(x + 4)(x - 4)} \qquad \begin{array}{l}\text{Combine like terms. Note that the}\\ \text{numerator can be factored.}\end{array}$$

$$= \dfrac{4\cancel{(x - 4)}}{(x + 4)\cancel{(x - 4)}} \qquad \begin{array}{l}\text{Since } (x - 4) \text{ is a } \textit{factor} \text{ of the numerator}\\ \textit{and} \text{ the denominator, we may divide out the}\\ \text{common factor.}\end{array}$$

$$= \dfrac{4}{x + 4}$$

Practice Problem 9 Subtract and simplify.

$$\dfrac{x - 2}{x^2 - 4} - \dfrac{x + 1}{2x^2 + 4x}$$

Verbal and Writing Skills

1. Suppose two rational expressions have denominators of $(x + 3)(x + 5)$ and $(x + 3)^2$. Explain how you would determine the LCD.

2. Suppose two rational expressions have denominators of $(x - 4)^2(x + 7)$ and $(x - 4)^3$. Explain how you would determine the LCD.

Perform the operation indicated. Be sure to simplify.

3. $\dfrac{x}{x + 5} + \dfrac{2x + 1}{5 + x}$

4. $\dfrac{8}{7 + 2x} + \dfrac{x + 3}{2x + 7}$

5. $\dfrac{3x}{x + 3} - \dfrac{x + 5}{x + 3}$

6. $\dfrac{2x + 4}{x - 5} - \dfrac{x + 5}{x - 5}$

7. $\dfrac{8x + 3}{5x + 7} - \dfrac{6x + 10}{5x + 7}$

8. $\dfrac{6x + 5}{4x + 1} - \dfrac{3x + 4}{4x + 1}$

Find the LCD. Do not combine fractions.

9. $\dfrac{10}{3a^2b^3}, \ \dfrac{8}{ab^2}$

10. $\dfrac{12}{5a^2}, \ \dfrac{9}{a^3}$

11. $\dfrac{5}{18x^2y^5}, \ \dfrac{7}{30x^3y^3}$

12. $\dfrac{9}{14xy^3}, \ \dfrac{14}{35x^4y^2}$

13. $\dfrac{9}{2x - 6}, \ \dfrac{5}{9x - 27}$

14. $\dfrac{12}{3x + 12}, \ \dfrac{5}{5x + 20}$

15. $\dfrac{8}{x + 3}, \ \dfrac{15}{x^2 - 9}$

16. $\dfrac{13}{x^2 - 16}, \ \dfrac{7}{x - 4}$

17. $\dfrac{7}{3x^2 + 14x - 5}, \ \dfrac{4}{9x^2 - 6x + 1}$

18. $\dfrac{4}{2x^2 - 9x - 35}, \ \dfrac{3}{4x^2 + 20x + 25}$

Add.

19. $\dfrac{7}{ab} + \dfrac{3}{b}$

20. $\dfrac{8}{cd} + \dfrac{9}{d}$

21. $\dfrac{3}{x + 7} + \dfrac{8}{x^2 - 49}$

22. $\dfrac{5}{x^2 - 2x + 1} + \dfrac{3}{x - 1}$

23. $\dfrac{4y}{y + 1} + \dfrac{y}{y - 1}$

24. $\dfrac{5}{y - 3} + \dfrac{2}{y + 3}$

25. $\dfrac{6}{5a} + \dfrac{5}{3a + 2}$

26. $\dfrac{3}{2a + 5} + \dfrac{4}{7a}$

27. $\dfrac{2}{3xy} + \dfrac{1}{6yz}$

28. $\dfrac{5}{4xy} + \dfrac{5}{12yz}$

Subtract.

29. $\dfrac{5x + 6}{x - 3} - \dfrac{x - 2}{2x - 6}$

30. $\dfrac{7x + 3}{x - 4} - \dfrac{x - 3}{2x - 8}$

31. $\dfrac{3x}{x^2 - 25} - \dfrac{2}{x + 5}$

32. $\dfrac{7x}{x^2 - 9} - \dfrac{6}{x + 3}$

33. $\dfrac{a + 3b}{2} - \dfrac{a - b}{5}$

34. $\dfrac{3b - a}{6} - \dfrac{a + b}{5}$

35. $\dfrac{8}{2x - 3} - \dfrac{6}{x + 2}$

36. $\dfrac{6}{3x - 4} - \dfrac{5}{4x - 3}$

37. $\dfrac{x}{x^2 + 2x - 3} - \dfrac{x}{x^2 - 5x + 4}$

38. $\dfrac{1}{x^2 - 2x} - \dfrac{5}{x^2 - 4x + 4}$

39. $\dfrac{3}{x^2 + 9x + 20} + \dfrac{1}{x^2 + 10x + 24}$

40. $\dfrac{2}{x^2 - 3x - 10} + \dfrac{5}{x^2 - 2x - 15}$

41. $\dfrac{3x - 8}{x^2 - 5x + 6} + \dfrac{x + 2}{x^2 - 6x + 8}$

42. $\dfrac{3x + 5}{x^2 + 4x + 3} + \dfrac{-x + 5}{x^2 + 2x - 3}$

Mixed Practice *Add or subtract.*

43. $\dfrac{6x}{y - 2x} - \dfrac{5x}{2x - y}$

44. $\dfrac{8b}{2b - a} - \dfrac{3b}{a - 2b}$

45. $\dfrac{3y}{8y^2 + 2y - 1} - \dfrac{5y}{2y^2 - 9y - 5}$

46. $\dfrac{2x}{x^2 + 5x + 6} - \dfrac{x + 1}{x^2 + 2x - 3}$

47. $\dfrac{4y}{y^2 + 4y + 3} + \dfrac{2}{y + 1}$

48. $\dfrac{y - 23}{y^2 - y - 20} + \dfrac{2}{y - 5}$

Cumulative Review

49. **[2.4.1]** Solve. $\frac{1}{3}(x - 2) + \frac{1}{2}(x + 3) = \frac{1}{4}(3x + 1)$

50. **[2.3.3]** Solve. $4.8 - 0.6x = 0.8(x - 1)$

51. **[2.8.4]** Solve for x. $x - \frac{1}{5}x > \frac{1}{2} + \frac{1}{10}x$

52. **[5.1.3]** Simplify. $(3x^3 y^4)^4$

53. **[2.6.2]** *Commuting Costs* A subway token costs $1.50. A monthly unlimited-ride subway pass costs $50. How many days per month would you have to use the subway to go to work (assume one subway token to get to work and one subway token to get back home) in order for it to be cheaper to buy a monthly subway pass?

54. **[0.5.3]** *Languages in Finland* In Finland, 91.5% of the population speaks Finnish. The other official language is Swedish, spoken by 5.5% of the population. One of the minority languages is Sámi, spoken by 0.03% of the population. (*Source:* www.stat.fi.) In 2006 there were 5,200,000 people in Finland. How many more people spoke Swedish than Sámi?

Quick Quiz 7.3 Perform the indicated operation. Simplify.

1. $\dfrac{3}{x^2 - 2x - 8} + \dfrac{2}{x - 4}$

2. $\dfrac{2x + y}{xy} - \dfrac{b - y}{by}$

3. $\dfrac{2}{x^2 - 9} + \dfrac{3}{x^2 + 7x + 12}$

4. **Concept Check** Explain how to find the LCD of the fractions $\dfrac{3}{10xy^2 z}$, $\dfrac{6}{25x^2 yz^3}$.

How are you doing with your homework assignments in Sections 7.1 to 7.3? Do you feel you have mastered the material so far? Do you understand the concepts you have covered? Before you go further in the textbook, take some time to do each of the following problems.

Simplify.

7.1

1. $\dfrac{8x - 48}{x^2 - 6x}$

2. $\dfrac{2x^2 - 7x - 15}{x^2 - 12x + 35}$

3. $\dfrac{y^2 + 2y + 1}{2x^2 - 2x^2y^2}$

4. $\dfrac{5x^2 - 23x + 12}{5x^2 + 7x - 6}$

7.2

5. $\dfrac{12a^2}{2x + 10} \cdot \dfrac{8x + 40}{16a^3}$

6. $\dfrac{x - 5}{x^2 + 5x - 14} \cdot \dfrac{x^2 + 12x + 35}{15 - 3x}$

7. $\dfrac{x^2 - 9}{2x + 6} \div \dfrac{2x^2 - 5x - 3}{4x^2 - 1}$

8. $\dfrac{3a^2 + 7a + 2}{4a^2 + 11a + 6} \div \dfrac{6a^2 - 13a - 5}{16a^2 - 9}$

7.3

9. $\dfrac{x - 3y}{xy} - \dfrac{4a - y}{ay}$

10. $\dfrac{7}{2x - 4} + \dfrac{-14}{x^2 - 4}$

11. $\dfrac{2x}{x^2 + 10x + 21} + \dfrac{x - 3}{x + 7}$

12. $\dfrac{4}{x^2 - 2x - 3} - \dfrac{5x}{x^2 + 5x + 4}$

1. _____

2. _____

3. _____

4. _____

5. _____

6. _____

7. _____

8. _____

9. _____

10. _____

11. _____

12. _____

Now turn to page SA-18 for the answer to each of these problems. Each answer also includes a reference to the objective in which the problem is first taught. If you missed any of these problems, you should stop and review the Examples and Practice Problems in the referenced objective. A little review now will help you master the material in the upcoming sections of the text.

Student Learning Objectives

After studying this section, you will be able to:

 Simplify complex rational expressions by adding or subtracting in the numerator and denominator.

 Simplify complex rational expressions using the LCD.

 ### Simplifying Complex Rational Expressions by Adding or Subtracting in the Numerator and Denominator

A **complex rational expression** (also called a **complex fraction**) has a fraction in the numerator or in the denominator, or both.

$$\dfrac{3 + \dfrac{2}{x}}{\dfrac{x}{7} + 2} \qquad \dfrac{\dfrac{x}{y} + 1}{2} \qquad \dfrac{\dfrac{a + b}{3}}{\dfrac{x - 2y}{4}}$$

The bar in a complex rational expression is both a grouping symbol and a symbol for division.

$$\dfrac{\dfrac{a + b}{3}}{\dfrac{x - 2y}{4}} \quad \text{is equivalent to} \quad \left(\dfrac{a + b}{3}\right) \div \left(\dfrac{x - 2y}{4}\right)$$

We need a procedure for simplifying complex rational expressions.

> ### PROCEDURE TO SIMPLIFY A COMPLEX RATIONAL EXPRESSION: ADDING AND SUBTRACTING
>
> 1. Add or subtract so that you have a single fraction in the numerator and in the denominator.
> 2. Divide the fraction in the numerator by the fraction in the denominator. This is done by inverting the fraction in the denominator and multiplying it by the numerator.

EXAMPLE 1 Simplify. $\dfrac{\dfrac{1}{x}}{\dfrac{2}{y^2} + \dfrac{1}{y}}$

Solution

Step 1 Add the two fractions in the denominator.

$$\dfrac{\dfrac{1}{x}}{\dfrac{2}{y^2} + \dfrac{1}{y} \cdot \dfrac{y}{y}} = \dfrac{\dfrac{1}{x}}{\dfrac{2 + y}{y^2}}$$

Step 2 Divide the fraction in the numerator by the fraction in the denominator.

$$\dfrac{1}{x} \div \dfrac{2 + y}{y^2} = \dfrac{1}{x} \cdot \dfrac{y^2}{2 + y} = \dfrac{y^2}{x(2 + y)}$$

Practice Problem 1 Simplify. $\dfrac{\dfrac{1}{a} + \dfrac{1}{a^2}}{\dfrac{2}{b^2}}$

NOTE TO STUDENT: Fully worked-out solutions to all of the Practice Problems can be found at the back of the text starting at page SP-1

A complex rational expression may contain two or more fractions in the numerator and the denominator.

EXAMPLE 2 Simplify. $\dfrac{\dfrac{1}{x} + \dfrac{1}{y}}{\dfrac{3}{a} - \dfrac{2}{b}}$

Solution   We observe that the LCD of the fractions in the numerator is xy. The LCD of the fractions in the denominator is ab.

$= \dfrac{\dfrac{1}{x} \cdot \dfrac{y}{y} + \dfrac{1}{y} \cdot \dfrac{x}{x}}{\dfrac{3}{a} \cdot \dfrac{b}{b} - \dfrac{2}{b} \cdot \dfrac{a}{a}}$ Multiply each fraction by the appropriate value to obtain common denominators.

$= \dfrac{\dfrac{y + x}{xy}}{\dfrac{3b - 2a}{ab}}$ Add the two fractions in the numerator.

Subtract the two fractions in the denominator.

$= \dfrac{y + x}{xy} \cdot \dfrac{ab}{3b - 2a}$ Invert the fraction in the denominator and multiply it by the numerator.

$= \dfrac{ab(y + x)}{xy(3b - 2a)}$ Write the answer as one fraction.

Practice Problem 2

Simplify. $\dfrac{\dfrac{1}{a} + \dfrac{1}{b}}{\dfrac{x}{2} - \dfrac{5}{y}}$

For some complex rational expressions, factoring may be necessary to determine the LCD and to combine fractions.

EXAMPLE 3 Simplify. $\dfrac{\dfrac{1}{x^2 - 1} + \dfrac{2}{x + 1}}{x}$

Solution   We need to factor $x^2 - 1$.

$= \dfrac{\dfrac{1}{(x + 1)(x - 1)} + \dfrac{2}{x + 1} \cdot \dfrac{x - 1}{x - 1}}{x}$ The LCD for the fractions in the numerator is $(x + 1)(x - 1)$.

$= \dfrac{\dfrac{1 + 2x - 2}{(x + 1)(x - 1)}}{x}$ Add the two fractions in the numerator.

$= \dfrac{2x - 1}{(x + 1)(x - 1)} \cdot \dfrac{1}{x}$ Simplify the numerator. Invert the fraction in the denominator and multiply.

$= \dfrac{2x - 1}{x(x + 1)(x - 1)}$ Write the answer as one fraction.

Practice Problem 3

Simplify. $\dfrac{\dfrac{x}{x^2 + 4x + 3} + \dfrac{2}{x + 1}}{x + 1}$

When simplifying complex rational expressions, always check to see if the final fraction can be reduced or simplified.

EXAMPLE 4

Simplify. $\dfrac{\dfrac{3}{a+b} - \dfrac{3}{a-b}}{\dfrac{5}{a^2 - b^2}}$

Solution The LCD of the two fractions in the numerator is $(a+b)(a-b)$.

$$= \dfrac{\dfrac{3}{a+b} \cdot \dfrac{a-b}{a-b} - \dfrac{3}{a-b} \cdot \dfrac{a+b}{a+b}}{\dfrac{5}{a^2 - b^2}}$$

$$= \dfrac{\dfrac{3a - 3b}{(a+b)(a-b)} - \dfrac{3a + 3b}{(a+b)(a-b)}}{\dfrac{5}{a^2 - b^2}}$$

Study carefully how we combine the two fractions in the numerator. Do you see how we obtain $-6b$?

$$= \dfrac{\dfrac{-6b}{(a+b)(a-b)}}{\dfrac{5}{(a+b)(a-b)}} \qquad \text{Factor } a^2 - b^2 \text{ as } (a+b)(a-b).$$

$$= \dfrac{-6b}{(a+b)(a-b)} \cdot \dfrac{(a+b)(a-b)}{5} \qquad \begin{array}{l}\text{Since } (a+b)(a-b) \text{ are factors in both}\\ \text{a numerator and a denominator, they may}\\ \text{be divided out.}\end{array}$$

$$= \dfrac{-6b}{5} \quad \text{or} \quad -\dfrac{6b}{5}$$

NOTE TO STUDENT: *Fully worked-out solutions to all of the Practice Problems can be found at the back of the text starting at page SP-1*

Practice Problem 4

Simplify. $\dfrac{\dfrac{6}{x^2 - y^2}}{\dfrac{1}{x-y} + \dfrac{3}{x+y}}$

② Simplifying Complex Rational Expressions Using the LCD

There is another way to simplify complex rational expressions: Multiply the numerator and denominator of the complex fraction by the least common denominator of all the denominators appearing in the complex fraction.

> **PROCEDURE TO SIMPLIFY A COMPLEX RATIONAL EXPRESSION: MULTIPLYING BY THE LCD**
>
> **1.** Determine the LCD of all individual denominators occurring in the numerator and denominator of the complex rational expression.
>
> **2.** Multiply both the numerator and the denominator of the complex rational expression by the LCD.
>
> **3.** Simplify, if possible.

EXAMPLE 5

Simplify by multiplying by the LCD. $\dfrac{\dfrac{5}{ab^2} - \dfrac{2}{ab}}{3 - \dfrac{5}{2a^2b}}$

Solution   The LCD of all the denominators in the complex rational expression is $2a^2b^2$.

$$= \frac{2a^2b^2\left(\dfrac{5}{ab^2} - \dfrac{2}{ab}\right)}{2a^2b^2\left(3 - \dfrac{5}{2a^2b}\right)}$$

$$= \frac{2a^2b^2\left(\dfrac{5}{ab^2}\right) - 2a^2b^2\left(\dfrac{2}{ab}\right)}{2a^2b^2(3) - 2a^2b^2\left(\dfrac{5}{2a^2b}\right)} \qquad \text{Multiply each term by } 2a^2b^2.$$

$$= \frac{10a - 4ab}{6a^2b^2 - 5b} \qquad \text{Simplify.}$$

Practice Problem 5   Simplify by multiplying by the LCD.

$$\frac{\dfrac{2}{3x^2} - \dfrac{3}{y}}{\dfrac{5}{xy} - 4}$$

So that you can compare the two methods, we will redo Example 4 by multiplying by the LCD.

EXAMPLE 6

Simplify by multiplying by the LCD. $\dfrac{\dfrac{3}{a + b} - \dfrac{3}{a - b}}{\dfrac{5}{a^2 - b^2}}$

Solution   The LCD of all individual fractions contained in the complex fraction is $(a + b)(a - b)$.

$$= \frac{(a + b)(a - b)\left(\dfrac{3}{a + b}\right) - (a + b)(a - b)\left(\dfrac{3}{a - b}\right)}{(a + b)(a - b)\left(\dfrac{5}{(a + b)(a - b)}\right)} \qquad \begin{array}{l}\text{Multiply each term by} \\ \text{the LCD.}\end{array}$$

$$= \frac{3(a - b) - 3(a + b)}{5} \qquad \text{Simplify.}$$

$$= \frac{3a - 3b - 3a - 3b}{5} \qquad \text{Remove parentheses.}$$

$$= -\frac{6b}{5} \qquad \text{Simplify.}$$

Practice Problem 6   Simplify by multiplying by the LCD.

$$\frac{\dfrac{6}{x^2 - y^2}}{\dfrac{7}{x - y} + \dfrac{3}{x + y}}$$

Simplify.

1. $\dfrac{\dfrac{3}{x}}{\dfrac{2}{x^2} + \dfrac{5}{x}}$

2. $\dfrac{\dfrac{4}{x}}{\dfrac{7}{x} + \dfrac{2}{x^2}}$

3. $\dfrac{\dfrac{3}{a} + \dfrac{3}{b}}{\dfrac{3}{ab}}$

4. $\dfrac{\dfrac{1}{a} + \dfrac{2}{b}}{\dfrac{3}{ab}}$

5. $\dfrac{\dfrac{x}{6} - \dfrac{1}{3}}{\dfrac{2}{3x} + \dfrac{5}{6}}$

6. $\dfrac{\dfrac{x}{8} - \dfrac{1}{2}}{\dfrac{3}{8x} + \dfrac{3}{4}}$

7. $\dfrac{\dfrac{7}{5x} - \dfrac{1}{x}}{\dfrac{3}{5} + \dfrac{2}{x}}$

8. $\dfrac{\dfrac{8}{x} - \dfrac{2}{3x}}{\dfrac{2}{3} + \dfrac{5}{x}}$

9. $\dfrac{\dfrac{5}{x} + \dfrac{3}{y}}{3x + 5y}$

10. $\dfrac{\dfrac{1}{x} + \dfrac{1}{y}}{x + y}$

11. $\dfrac{4 - \dfrac{1}{x^2}}{2 + \dfrac{1}{x}}$

12. $\dfrac{1 - \dfrac{36}{x^2}}{1 - \dfrac{6}{x}}$

13. $\dfrac{\dfrac{2}{x + 6}}{\dfrac{2}{x - 6} - \dfrac{2}{x^2 - 36}}$

14. $\dfrac{\dfrac{9}{x^2 - 1}}{\dfrac{4}{x - 1} - \dfrac{2}{x + 1}}$

15. $\dfrac{a + \dfrac{3}{a}}{\dfrac{a^2 + 2}{3a}}$

16. $\dfrac{x + \dfrac{4}{x}}{\dfrac{x^2 + 3}{4x}}$

17. $\dfrac{\dfrac{3}{x - 3}}{\dfrac{1}{x^2 - 9} + \dfrac{2}{x + 3}}$

18. $\dfrac{\dfrac{5}{x + 4}}{\dfrac{1}{x - 4} - \dfrac{2}{x^2 - 16}}$

19. $\dfrac{\dfrac{2}{y - 1} + 2}{\dfrac{2}{y + 1} - 2}$

20. $\dfrac{\dfrac{y}{y + 1} + 1}{\dfrac{2y + 1}{y - 1}}$

To Think About

21. Consider the complex fraction $\dfrac{\dfrac{4}{x + 3}}{\dfrac{5}{x} - 1}$. What values are not allowable replacements for the variable *x*?

22. Consider the complex fraction $\dfrac{\dfrac{5}{x - 2}}{\dfrac{6}{x} + 1}$. What values are not allowable replacements for the variable *x*?

Cumulative Review

23. **[3.1.3]** Solve for y. $4x + 3y = 7$

24. **[2.8.4]** Solve and graph. $7 + x < 11 + 5x$

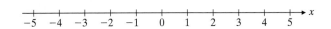

25. **[2.6.1]** When nine is subtracted from double a number, the result is the same as one-half of the same number. What is the number?

26. **[2.6.2]** *Loan Repayment* Manuela is paying back a $4000 loan with no interest from her favorite sister. If Manuela has already paid $125 per month for 17 months, how much of the loan does she have yet to pay?

Quick Quiz 7.4 Simplify.

1. $\dfrac{\dfrac{a}{4b} - \dfrac{1}{3}}{\dfrac{5}{4b} - \dfrac{4}{a}}$

2. $\dfrac{a + b}{\dfrac{1}{a} + \dfrac{1}{b}}$

3. $\dfrac{\dfrac{10}{x^2 - 25}}{\dfrac{3}{x + 5} + \dfrac{2}{x - 5}}$

4. **Concept Check** To simplify the following complex fraction, explain how you would add the two fractions in the numerator.

$$\dfrac{\dfrac{7}{x - 3} + \dfrac{15}{2x - 6}}{\dfrac{2}{x + 5}},$$

Student Learning Objectives

After studying this section, you will be able to:

 Solve equations involving rational expressions that have solutions.

 Determine whether an equation involving rational expressions has no solution.

 ### Solving Equations Involving Rational Expressions That Have Solutions

In Section 2.4 we developed procedures to solve linear equations containing fractions whose denominators were numbers. In this section we use a similar approach to solve equations containing fractions whose denominators are polynomials. It would be wise for you to review Section 2.4 briefly *before you begin this section.* It will be especially helpful to carefully study Examples 1 and 2.

TO SOLVE AN EQUATION CONTAINING RATIONAL EXPRESSIONS

1. Determine the LCD of all the denominators.
2. Multiply each term of the equation by the LCD.
3. Solve the resulting equation.
4. Check your solution. Exclude from your solution any value that would make the LCD equal to zero.

EXAMPLE 1 Solve for x and check your solution. $\dfrac{5}{x} + \dfrac{2}{3} = -\dfrac{3}{x}$

Solution

$$3x\left(\dfrac{5}{x}\right) + 3x\left(\dfrac{2}{3}\right) = 3x\left(-\dfrac{3}{x}\right) \quad \text{Observe that the LCD is } 3x. \text{ Multiply each term by } 3x.$$

$$15 + 2x = -9$$
$$2x = -9 - 15 \quad \text{Subtract 15 from both sides.}$$
$$2x = -24$$
$$x = -12 \quad \text{Divide both sides by 2.}$$

Check.
$$\dfrac{5}{-12} + \dfrac{2}{3} \overset{?}{=} -\dfrac{3}{-12} \quad \text{Replace each } x \text{ by } -12.$$

$$-\dfrac{5}{12} + \dfrac{8}{12} \overset{?}{=} \dfrac{3}{12}$$

$$\dfrac{3}{12} = \dfrac{3}{12} \quad \checkmark \quad \text{It checks.}$$

NOTE TO STUDENT: *Fully worked-out solutions to all of the Practice Problems can be found at the back of the text starting at page SP-1*

Practice Problem 1 Solve for x and check your solution.

$$\dfrac{3}{x} + \dfrac{4}{5} = -\dfrac{2}{x}$$

EXAMPLE 2 Solve and check. $\dfrac{6}{x + 3} = \dfrac{3}{x}$

Solution Observe that the LCD $= x(x + 3)$.

$$x(x + 3)\left(\dfrac{6}{x + 3}\right) = x(x + 3)\left(\dfrac{3}{x}\right) \quad \text{Multiply both sides by } x(x + 3).$$

$$6x = 3(x + 3) \quad \text{Simplify. Do you see how this is done?}$$
$$6x = 3x + 9 \quad \text{Remove parentheses.}$$
$$3x = 9 \quad \text{Subtract } 3x \text{ from both sides.}$$
$$x = 3 \quad \text{Divide both sides by 3.}$$

Check. $\dfrac{6}{3+3} \stackrel{?}{=} \dfrac{3}{3}$ Replace each x by 3.

$\dfrac{6}{6} = \dfrac{3}{3}$ ✓ It checks.

Practice Problem 2 Solve and check.

$$\frac{4}{2x+1} = \frac{6}{2x-1}$$

It is sometimes necessary to factor denominators before the correct LCD can be determined.

EXAMPLE 3 Solve and check. $\dfrac{3}{x+5} - 1 = \dfrac{4-x}{2x+10}$

Solution

$\dfrac{3}{x+5} - 1 = \dfrac{4-x}{2(x+5)}$ Factor $2x+10$. We determine that the LCD is $2(x+5)$.

$2(x+5)\left(\dfrac{3}{x+5}\right) - 2(x+5)(1) = 2(x+5)\left[\dfrac{4-x}{2(x+5)}\right]$ Multiply each term by the LCD.

$2(3) - 2(x+5) = 4 - x$ Simplify.

$6 - 2x - 10 = 4 - x$ Remove parentheses.

$-2x - 4 = 4 - x$ Combine like terms.

$-4 = 4 + x$ Add $2x$ to both sides.

$-8 = x$ Subtract 4 from both sides.

Check. $\dfrac{3}{-8+5} - 1 \stackrel{?}{=} \dfrac{4-(-8)}{2(-8)+10}$ Replace each x in the original equation by -8.

$\dfrac{3}{-3} - 1 \stackrel{?}{=} \dfrac{4+8}{-16+10}$

$-1 - 1 \stackrel{?}{=} \dfrac{12}{-6}$

$-2 = -2$ ✓   It checks. The solution is -8.

Practice Problem 3 Solve and check.

$$\frac{x-1}{x^2-4} = \frac{2}{x+2} + \frac{4}{x-2}$$

NOTE TO STUDENT: Fully worked-out solutions to all of the Practice Problems can be found at the back of the text starting at page SP-1

② Determining Whether an Equation Involving Rational Expressions Has No Solution

Equations containing rational expressions sometimes appear to have solutions when in fact they do not. By this we mean that the "solutions" we get by using completely correct methods are, in actuality, not solutions.

In the case where a value makes a denominator in the equation equal to zero, we say it is not a solution to the equation. Such a value is called an **extraneous solution.** An extraneous solution is an apparent solution that does *not* satisfy the original equation. If all of the apparent solutions of an equation are extraneous solutions, we say that the equation has **no solution.** It is important that you check all apparent solutions in the original equation.

EXAMPLE 4 Solve and check. $\dfrac{y}{y-2} - 4 = \dfrac{2}{y-2}$

Solution Observe that the LCD is $y - 2$.

$$(y-2)\left(\dfrac{y}{y-2}\right) - (y-2)(4) = (y-2)\left(\dfrac{2}{y-2}\right) \qquad \text{Multiply each term by } (y-2).$$

$$y - 4(y-2) = 2 \qquad \text{Simplify. Do you see how this is done?}$$

$$y - 4y + 8 = 2 \qquad \text{Remove parentheses.}$$

$$-3y + 8 = 2 \qquad \text{Combine like terms.}$$

$$-3y = -6 \qquad \text{Subtract 8 from both sides.}$$

$$\dfrac{-3y}{-3} = \dfrac{-6}{-3} \qquad \text{Divide both sides by } -3.$$

$$y = 2 \qquad \text{2 is only an apparent solution.}$$

This equation has no solution.

Why? We can see immediately that $y = 2$ is not a solution of the original equation. When we substitute 2 for y in a denominator, the denominator is equal to zero and the expression is undefined.

Check. $\qquad \dfrac{y}{y-2} - 4 = \dfrac{2}{y-2} \qquad$ Suppose that you try to check the apparent solution by substituting 2 for y.

$$\dfrac{2}{2-2} - 4 \stackrel{?}{=} \dfrac{2}{2-2}$$

$$\dfrac{2}{0} - 4 = \dfrac{2}{0} \qquad \text{This does not check since you do not obtain a real number when you divide by zero.}$$

$$\uparrow \qquad\qquad \uparrow$$

These expressions are not defined.

There is no such number as $2 \div 0$. We see that 2 does *not* check. This equation has **no solution.**

Practice Problem 4 Solve and check.

$$\dfrac{2x}{x+1} = \dfrac{-2}{x+1} + 1$$

Solve and check problems 1–16.

1. $\dfrac{7}{x} + \dfrac{3}{4} = \dfrac{-2}{x}$

2. $\dfrac{8}{x} + \dfrac{2}{5} = \dfrac{-2}{x}$

3. $\dfrac{-1}{4x} + \dfrac{3}{2} = \dfrac{5}{x}$

4. $\dfrac{1}{3x} + \dfrac{5}{6} = \dfrac{2}{x}$

5. $\dfrac{5x + 3}{3x} = \dfrac{7}{3} - \dfrac{9}{x}$

6. $\dfrac{2x + 7}{5x} - \dfrac{1}{x} = \dfrac{8}{5}$

7. $\dfrac{x + 5}{3x} = \dfrac{1}{2}$

8. $\dfrac{x - 6}{5x} = \dfrac{1}{3}$

9. $\dfrac{6}{3x - 5} = \dfrac{3}{2x}$

10. $\dfrac{1}{x + 6} = \dfrac{4}{x}$

11. $\dfrac{2}{2x + 5} = \dfrac{4}{x - 4}$

12. $\dfrac{3}{x + 5} = \dfrac{3}{3x - 2}$

13. $\dfrac{2}{x} + \dfrac{x}{x + 1} = 1$

14. $\dfrac{5}{2} = 3 + \dfrac{2x + 7}{x + 6}$

15. $\dfrac{85 - 4x}{x} = 7 - \dfrac{3}{x}$

16. $\dfrac{63 - 2x}{x} = 2 - \dfrac{5}{x}$

Mixed Practice   *Solve and check. If there is no solution, say so.*

17. $\dfrac{1}{x + 4} - 2 = \dfrac{3x - 2}{x + 4}$

18. $\dfrac{2}{x + 5} - 1 = \dfrac{3x - 4}{x + 5}$

19. $\dfrac{2}{x - 6} - 5 = \dfrac{2(x - 5)}{x - 6}$

20. $7 - \dfrac{x}{x + 5} = \dfrac{5}{5 + x}$

21. $\dfrac{2}{x + 1} - \dfrac{1}{x - 1} = \dfrac{2x}{x^2 - 1}$

22. $\dfrac{8x}{4x^2 - 1} = \dfrac{3}{2x + 1} + \dfrac{3}{2x - 1}$

23. $\dfrac{y + 1}{y^2 + 2y - 3} = \dfrac{1}{y + 3} - \dfrac{1}{y - 1}$

24. $\dfrac{6}{x - 5} + \dfrac{3x + 1}{x^2 - 2x - 15} = \dfrac{5}{x + 3}$

25. $\dfrac{2x}{x + 4} - \dfrac{8}{x - 4} = \dfrac{2x^2 + 32}{x^2 - 16}$

26. $\dfrac{4x}{x + 3} - \dfrac{12}{x - 3} = \dfrac{4x^2 + 36}{x^2 - 9}$

27. $\dfrac{4}{x^2 - 1} + \dfrac{7}{x + 1} = \dfrac{5}{x - 1}$

28. $\dfrac{7}{4x^2 - 1} + \dfrac{2}{2x + 1} = \dfrac{3}{2x - 1}$

29. $\dfrac{x + 11}{x^2 - 5x + 4} + \dfrac{3}{x - 1} = \dfrac{5}{x - 4}$

30. $\dfrac{6}{x - 3} = \dfrac{-5}{x - 2} + \dfrac{-5}{x^2 - 5x + 6}$

To Think About *In each of the following equations, what values are not allowable replacements for the variable x? Do not solve the equation.*

31. $\dfrac{3x}{x - 2} - \dfrac{4x}{x - 4} = \dfrac{3}{x^2 - 6x + 8}$

32. $\dfrac{3x}{x - 1} - \dfrac{5x}{2x + 3} = \dfrac{7}{2x^2 + 5x + 3}$

Cumulative Review

33. [6.4.1] Factor. $8x^2 - 2x - 1$

34. [2.3.3] Solve. $5(x - 2) = 8 - (3 + x)$

▲ **35. [2.6.1]** *Geometry* The perimeter of a rectangular computer monitor is 44 inches. The length is 8 inches less than twice the width. Find the dimensions of the monitor.

36. [3.6.1] Determine the domain and range of the relation. Is the relation a function?

$\{(7, 3), (2, 2), (-2, 0), (2, -2), (7, -3)\}$

Quick Quiz 7.5 Solve. If there is no solution, say so.

1. $\dfrac{3}{4x} - \dfrac{5}{6x} = 2 - \dfrac{1}{2x}$

2. $\dfrac{x}{x - 1} - \dfrac{2}{x} = \dfrac{1}{x - 1}$

3. $\dfrac{6}{x^2 - 2x - 8} + \dfrac{5}{x + 2} = \dfrac{1}{x - 4}$

4. Concept Check Explain how to find the LCD for the following equation. Do not solve the equation.

$$\dfrac{x}{x^2 - 9} + \dfrac{2}{3x - 9} = \dfrac{5}{2x + 6} + \dfrac{3}{2x^2 - 18}$$

Solving Problems Involving Ratio and Proportion

A **ratio** is a comparison of two quantities. You may be familiar with ratios that compare miles to hours or miles to gallons. A ratio is often written as a quotient in the form of a fraction. For example, the ratio of 7 to 9 can be written as $\frac{7}{9}$.

A **proportion** is an equation that states that two ratios are equal. For example,

$$\frac{7}{9} = \frac{21}{27}, \quad \frac{2}{3} = \frac{10}{15}, \quad \text{and} \quad \frac{a}{b} = \frac{c}{d} \quad \text{are proportions.}$$

Let's take a closer look at the last proportion. We can see that the LCD of the fractional equation is bd.

$$(b\ d)\frac{a}{b} = (b\ d)\frac{c}{d} \quad \text{Multiply each side by the LCD.}$$

$$da = bc$$

$$ad = bc \quad \text{Since multiplication is commutative, } da = ad.$$

Thus we have proved the following.

THE PROPORTION EQUATION

If

$$\frac{a}{b} = \frac{c}{d}, \quad \text{then} \quad ad = bc$$

for all real numbers a, b, c, and d, where $b \neq 0$ and $d \neq 0$.

This is sometimes called **cross multiplying.** It can be applied only if you have *one* fraction and nothing else on each side of the equation.

EXAMPLE 1 Michael took 5 hours to drive 245 miles on the turnpike. At the same rate, how many hours will it take him to drive a distance of 392 miles?

Solution

1. *Understand the problem.* Let x = the number of hours it will take to drive 392 miles. If 5 hours are needed to drive 245 miles, then x hours are needed to drive 392 miles.

2. *Write an equation.* We can write this as a proportion. Compare time to distance in each ratio.

$$\begin{array}{c} \text{Time} \longrightarrow \\ \text{Distance} \longrightarrow \end{array} \frac{5 \text{ hours}}{245 \text{ miles}} = \frac{x \text{ hours}}{392 \text{ miles}} \begin{array}{c} \longleftarrow \text{Time} \\ \longleftarrow \text{Distance} \end{array}$$

3. *Solve and state the answer.*

$$5(392) = 245x \quad \text{Cross-multiply.}$$

$$\frac{1960}{245} = x \quad \text{Divide both sides by 245.}$$

$$8 = x$$

It will take Michael 8 hours to drive 392 miles.

4. *Check.* Is $\frac{5}{245} = \frac{8}{392}$? Do the computation and see.

Practice Problem 1 It took Brenda 8 hours to drive 420 miles. At the same rate, how long would it take her to drive 315 miles?

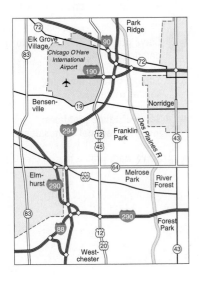

EXAMPLE 2 If $\frac{3}{4}$ inch on a map represents an actual distance of 20 miles, how long is the distance represented by $4\frac{1}{8}$ inches on the same map?

Solution Let $x =$ the distance represented by $4\frac{1}{8}$ inches.

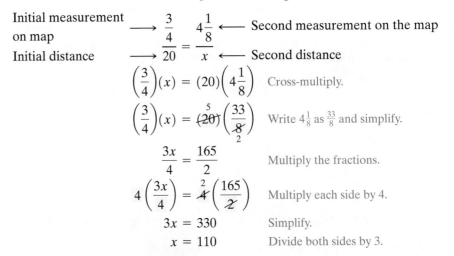

$$\frac{3}{4} = \frac{4\frac{1}{8}}{x}$$

$$\left(\frac{3}{4}\right)(x) = (20)\left(4\frac{1}{8}\right) \quad \text{Cross-multiply.}$$

$$\left(\frac{3}{4}\right)(x) = (20)\left(\frac{33}{8}\right) \quad \text{Write } 4\frac{1}{8} \text{ as } \frac{33}{8} \text{ and simplify.}$$

$$\frac{3x}{4} = \frac{165}{2} \quad \text{Multiply the fractions.}$$

$$4\left(\frac{3x}{4}\right) = 4\left(\frac{165}{2}\right) \quad \text{Multiply each side by 4.}$$

$$3x = 330 \quad \text{Simplify.}$$

$$x = 110 \quad \text{Divide both sides by 3.}$$

$4\frac{1}{8}$ inches on the map represents an actual distance of 110 miles.

Practice Problem 2 If $\frac{5}{8}$ inch on a map represents an actual distance of 30 miles, how long is the distance represented by $2\frac{1}{2}$ inches on the same map?

② Solving Problems Involving Similar Triangles

Similar triangles are triangles that have the same shape, but may be different sizes. For example, if you draw a triangle on a sheet of paper, place the paper in a photocopy machine, and make a copy that is reduced by 25%, you would create a triangle that is similar to the original triangle. The two triangles will have the same shape. The corresponding sides of the triangles will be proportional. The corresponding angles of the triangles will be equal.

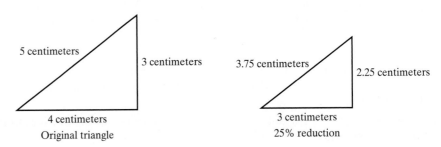

5 centimeters 3 centimeters 3.75 centimeters 2.25 centimeters

4 centimeters 3 centimeters

Original triangle 25% reduction

You can use the proportion equation to show that the corresponding sides of the triangles above are proportional. In fact, you can use the proportion equation to find an unknown length of a side of one of two similar triangles.

▲ **EXAMPLE 3** A ramp is 32 meters long and rises up 15 meters. A ramp at the same angle is 9 meters long. How high is the second ramp?

Solution To answer this question, we find the length of side x in the following two similar triangles.

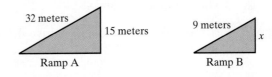

32 meters 15 meters 9 meters x

Ramp A Ramp B

Ramp A, longest side $\longrightarrow$ $\dfrac{32}{9} = \dfrac{15}{x}$ $\longleftarrow$ Shortest side, ramp A
Ramp B, longest side $\longrightarrow$ $\phantom{\dfrac{32}{9}}$ $\phantom{\dfrac{15}{x}}$ $\longleftarrow$ Shortest side, ramp B

$$32x = (9)(15) \quad \text{Cross-multiply.}$$
$$32x = 135 \quad \text{Multiply.}$$
$$x = \frac{135}{32} \quad \text{Divide both sides by 32.}$$
$$\text{or} \quad x = 4\frac{7}{32}$$

The ramp is $4\frac{7}{32}$ meters tall.

▲ **Practice Problem 3**   Triangle C is similar to Triangle D. Find the length of side x. Express your answer as a mixed number.

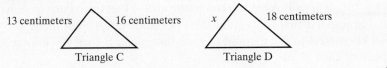

13 centimeters   16 centimeters x 18 centimeters

Triangle C   Triangle D

NOTE TO STUDENT: Fully worked-out solutions to all of the Practice Problems can be found at the back of the text starting at page SP-1

   We can also use similar triangles for indirect measurement—for instance, to find the height of an object that is too tall to measure using standard measuring devices. When the sun shines on two vertical objects at the same time, the shadows and the objects form similar triangles.

▲ **EXAMPLE 4**   A woman who is 5 feet tall casts a shadow that is 8 feet long. At the same time of day, a building casts a shadow that is 72 feet long. How tall is the building?

Solution

1. *Understand the problem.* First we draw a sketch. We do not know the height of the building, so we call it x.

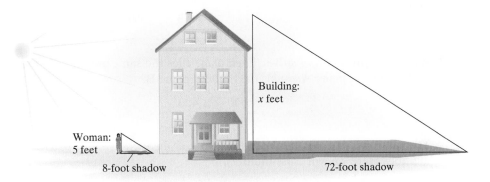

Building:
x feet

Woman:
5 feet

8-foot shadow   72-foot shadow

2. *Write an equation and solve.*

Height of woman $\longrightarrow$ $\dfrac{5}{8} = \dfrac{x}{72}$ $\longleftarrow$ Height of building
Length of woman's shadow $\longrightarrow$ $\phantom{\dfrac{5}{8}}$ $\phantom{\dfrac{x}{72}}$ $\longleftarrow$ Length of building's shadow

$$(5)(72) = 8x \quad \text{Cross-multiply.}$$
$$360 = 8x$$
$$45 = x$$

The height of the building is 45 feet.

▲ **Practice Problem 4**   A man who is 6 feet tall casts a shadow that is 7 feet long. At the same time of day, a large flagpole casts a shadow that is 38.5 feet long. How tall is the flagpole?

In problems such as Example 4, we are assuming that the building and the person are standing exactly perpendicular to the ground. In other words, each triangle is assumed to be a right triangle. In other similar triangle problems, if the triangles are not right triangles you must be careful that the corresponding angles in the two triangles are the same.

3 Solving Distance Problems Involving Rational Expressions

Some distance problems are solved using equations with rational expressions. We will need the formula Distance = Rate × Time, $D = RT$, which we can write in the form $T = \dfrac{D}{R}$. In the United States, distances are usually measured in miles. In European countries, distances are usually measured in kilometers.

Les Ailes
Parce que vous
n'êtes pas un oiseau

EXAMPLE 5   A French commuter airline flies from Paris to Avignon. Plane A flies at a speed that is 50 kilometers per hour faster than plane B. Plane A flies 500 kilometers in the amount of time that plane B flies 400 kilometers. Find the speed of each plane.

Solution

1. **Understand the problem.** Let s = the speed of plane B in kilometers per hour.

   Then $s + 50$ = the speed of plane A in kilometers per hour. Make a simple table for D, R, and T.

   Since $T = \dfrac{D}{R}$, for each plane we divide the expression for D by the expression for R and write it in the table in the column for time.

	D	R	$T = \dfrac{D}{R}$
Plane A	500	$s + 50$	?
Plane B	400	s	?

	D	R	$T = \dfrac{D}{R}$
Plane A	500	$s + 50$	$\dfrac{500}{s + 50}$
Plane B	400	s	$\dfrac{400}{s}$

2. **Write an equation and solve.** Each plane flies the same amount of time. That is, the time for plane A equals the time for plane B.

$$\frac{500}{s + 50} = \frac{400}{s}$$

You can solve this equation using the methods in Section 7.5 or you may cross-multiply. Here we will cross-multiply.

$$500s = (s + 50)(400) \quad \text{Cross-multiply.}$$
$$500s = 400s + 20{,}000 \quad \text{Remove parentheses.}$$
$$100s = 20{,}000 \quad\quad\quad \text{Subtract } 400s \text{ from each side.}$$
$$s = 200 \quad\quad\quad\quad\;\; \text{Divide each side by 100.}$$

Plane B travels 200 kilometers per hour. Since

$$s + 50 = 200 + 50 = 250,$$

plane A travels 250 kilometers per hour.

Practice Problem 5   Two European freight trains traveled toward Paris for the same amount of time. Train A traveled 180 kilometers, while train B traveled 150 kilometers. Train A traveled 10 kilometers per hour faster than train B. What was the speed of each train?

Solving Work Problems

Some applied problems involve the length of time needed to do a job. These problems are often referred to as work problems.

EXAMPLE 6 Reynaldo can sort a huge stack of mail on an old sorting machine in 9 hours. His brother Carlos can sort the same amount of mail using a newer sorting machine in 8 hours. How long would it take them to do the job working together? Express your answer in hours and minutes. Round to the nearest minute.

Solution

1. **Understand the problem.** Let's do a little reasoning.

   If Reynaldo can do the job in 9 hours, then in *1 hour* he could do $\frac{1}{9}$ of the job.

   If Carlos can do the job in 8 hours, then in *1 hour* he could do $\frac{1}{8}$ of the job.

   Let $x =$ the number of hours it takes Reynaldo and Carlos to do the job together. In *1 hour* together they could do $\frac{1}{x}$ of the job.

2. **Write an equation and solve.** The amount of work Reynaldo can do in 1 hour plus the amount of work Carlos can do in 1 hour must be equal to the amount of work they could do together in 1 hour.

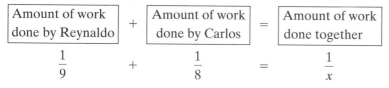

Amount of work done by Reynaldo	$+$	Amount of work done by Carlos	$=$	Amount of work done together
$\frac{1}{9}$	$+$	$\frac{1}{8}$	$=$	$\frac{1}{x}$

Let us solve for x. We observe that the LCD is $72x$.

$$72x\left(\frac{1}{9}\right) + 72x\left(\frac{1}{8}\right) = 72x\left(\frac{1}{x}\right) \quad \text{Multiply each term by the LCD.}$$

$$8x + 9x = 72 \qquad \text{Simplify.}$$
$$17x = 72 \qquad \text{Combine like terms.}$$
$$x = \frac{72}{17} \qquad \text{Divide each side by 17.}$$
$$x = 4\frac{4}{17}$$

To change $\frac{4}{17}$ of an hour to minutes we multiply.

$$\frac{4}{17} \text{ hour} \times \frac{60 \text{ minutes}}{1 \text{ hour}} = \frac{240}{17} \text{ minutes, which is approximately 14.118 minutes}$$

To the nearest minute this is 14 minutes. Thus doing the job together will take 4 hours and 14 minutes.

Practice Problem 6 John Tobey and Dave Wells obtained night custodian jobs at a local factory while going to college part-time. Using the buffer machine, John can buff all the floors in the building in 6 hours. Dave takes a little longer and can do all the floors in the building in 7 hours. Their supervisor bought another buffer machine. How long will it take John and Dave to do all the floors in the building working together, each with his own machine? Express your answer in hours and minutes. Round to the nearest minute.

Calculator

Reciprocals

You can find $\frac{1}{x}$ for any value of x on a scientific calculator by using the key labeled $\boxed{x^{-1}}$ or the key labeled $\boxed{1/x}$. For example, to find $\frac{1}{9}$, we use 9 $\boxed{x^{-1}}$ or 9 $\boxed{1/x}$. The display will read 0.11111111. Therefore we can solve Example 6 as follows:

$$9 \boxed{x^{-1}} \boxed{+} 8 \boxed{x^{-1}} \boxed{=}$$

The display will read 0.2361111. Thus we have obtained the equation $0.2361111 = \frac{1}{x}$. Now this is equivalent to $x = \frac{1}{0.2361111}$. (Do you see why?) Thus we enter 0.2361111 $\boxed{x^{-1}}$, and the display reads 4.2352943. If we round to the nearest hundredth, we have $x \approx 4.24$ hours, which is approximately equal to our answer of $4\frac{4}{17}$ hours.

Solve.

1. $\dfrac{4}{9} = \dfrac{8}{x}$

2. $\dfrac{5}{12} = \dfrac{x}{8}$

3. $\dfrac{x}{17} = \dfrac{12}{5}$

4. $\dfrac{16}{x} = \dfrac{3}{4}$

5. $\dfrac{9.1}{8.4} = \dfrac{x}{6}$

6. $\dfrac{5}{x} = \dfrac{17.5}{4.2}$

7. $\dfrac{7}{x} = \dfrac{40}{130}$

8. $\dfrac{x}{18} = \dfrac{13}{2}$

Applications *Use a proportion to answer exercises 9–16.*

9. *Map Scale* The scale on the AAA map of Colorado is approximately $\frac{3}{4}$ inch to 15 miles. If the distance from Denver to Pueblo measures 5.5 inches on the map, how far apart are the two cities?

10. *Recipe Ratios* Nella Coates's recipe for Shoofly Pie contains $\frac{3}{4}$ cup of unsulfured molasses. This recipe makes a small pie that serves 8 people. If she makes the larger pie that serves 12 people and the ratio of molasses to people remains the same, how much molasses will she need for the larger pie?

11. *Exchange Rates* Robyn spent two months traveling in New Zealand. The day she arrived, the exchange rate was 1.3 New Zealand dollars per U.S. dollar.

   (a) If she exchanged $500 U.S. dollars when she arrived, how many New Zealand dollars did she receive?

   (b) Three days later the value of the New Zealand dollar dropped to 1.15 New Zealand dollars per U.S. dollar. How much less money would she have received had she waited three days to exchange her money?

12. *Exchange Rates* Sean spent a semester studying in Germany. On the day he arrived in Berlin, the exchange rate for the Euro was 0.69 Euro per U.S. dollar. Sean converted $350 to Euros that day.

   (a) How many Euros did Sean receive for his $350?

   (b) On his way home Sean decided to spend a week in London. He had €200 (200 Euros) that he wanted to change into British pounds. If the exchange rate was 0.72 British pounds per Euro, how many pounds did he receive?

13. *Speed Units* Alfonse and Melinda are taking a drive in Mexico. They know that a speed of 100 kilometers per hour is approximately equal to 62 miles per hour. They are now driving on a Mexican road that has a speed limit of 90 kilometers per hour. How many miles per hour is the speed limit? Round to the nearest mile per hour.

14. *Baggage Weight* Dick and Anne took a trip to France. Their suitcases were weighed at the airport and the weight recorded was 39 kilograms. If 50 kilograms is equivalent to 110 pounds, how many pounds did their suitcases weigh? Round to the nearest pound.

15. *Map Scale* On a map the distance between two mountains is $3\frac{1}{2}$ inches. The actual distance between the mountains is 136 miles. Russ is camped at a location that on the map is $\frac{3}{4}$ inch from the base of the mountain. How many miles is he from the base of the mountain? Round to the nearest mile.

16. *Construction Scales* Maria is adding a porch that is 18 feet long to her house. On the drawing done by the carpenter the length is shown as 11 inches. The drawing shows that the width of the porch is 8 inches. How many feet wide will the porch be? Round to the nearest foot.

Geometry *Triangles A and B are similar. Use them to answer exercises 17–20. Leave your answers as fractions.*

Triangle A

Triangle B

▲ **17.** If $x = 20$ in., $y = 29$ in., and $m = 13$ in., find the length of side n.

▲ **18.** If $p = 14$ in., $m = 17$ in., and $z = 23$ in., find the length of side x.

▲ **19.** If $x = 175$ meters, $n = 40$ meters, and $m = 35$ meters, find the length of side y.

▲ **20.** If $z = 18$ cm, $y = 25$ cm, and $n = 9$ cm, find the length of side p.

Geometry *Just as we have discussed similar triangles, other geometric shapes can be similar. Similar geometric shapes will have sides that are proportional. Quadrilaterals abcd and ghjk are similar. Use them to answer exercises 21–24. Leave your answers as fractions.*

▲ **21.** If $a = 5$ ft, $d = 8$ ft, and $g = 7$ ft, find the length of side k.

▲ **22.** If $j = 10$ in., $k = 13$ in., and $c = 8$ in., find the length of side d.

▲ **23.** If $b = 20$ m, $h = 24$ m, and $d = 32$ m, find the length of side k.

▲ **24.** If $a = 16$ cm, $d = 19$ cm, and $k = 23$ cm, find the length of side g.

Use a proportion to solve.

▲ **25.** ***Geometry*** A rectangle whose width-to-length ratio is approximately 5 to 8 is called a **golden rectangle** and is said to be pleasing to the eye. Using this ratio, what should the length of a rectangular picture be if its width is to be 30 inches?

▲ **26.** ***Shadows*** Samantha is 5.5 feet tall and notices that she casts a shadow of 9 feet. At the same time, the new sculpture at the local park casts a shadow of 20 feet. How tall is the sculpture? Round your answer to the nearest foot.

▲ **27.** ***Floral Displays*** Floral designers often create arrangements where the flower height to container ratio is 5 to 3. The FIU Art Museum wishes to create a floral display for the opening of a new show. They know they want to use an antique Chinese vase from their collection that is 13 inches high. How tall will the entire flower arrangement be if they use this standard ratio? (Round your answer to the nearest inch.)

▲ **28. *Securing Wires*** A wire line helps to secure a radio transmission tower. The wire measures 23 meters from the tower to the ground anchor pin. The wire is secured 14 meters up on the tower. If a second wire is secured 130 meters up on the tower and is extended from the tower at the same angle as the first wire, how long would the second wire need to be to reach an anchor pin on the ground? Round to the nearest meter.

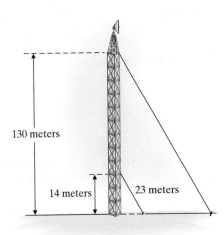

130 meters

14 meters   23 meters

29. *Acceleration* Ben Hale is driving his new Toyota Camry on Interstate 90 at 45 miles per hour. He accelerates at the rate of 3 miles per hour every 2 seconds. How fast will he be traveling after accelerating for 11 seconds?

30. *Braking* Tim Newitt is driving a U-Haul truck to Chicago. He is driving at 55 miles per hour and has to hit the brakes because of heavy traffic. His truck slows at the rate of 2 miles per hour for every 3 seconds. How fast will he be traveling 10 seconds after he hits the brakes?

31. *Flight Speeds* A Montreal commuter airliner travels 40 kilometers per hour faster than the television news helicopter over the city. The commuter airliner travels 1250 kilometers during the same time that the television news helicopter travels only 1050 kilometers. How fast does the commuter airliner fly? How fast does the television news helicopter fly?

32. *Driving Speeds* Melissa drove to Dallas while Marcia drove to Houston in the same amount of time. Melissa drove 360 kilometers, while Marcia drove 280 kilometers. Melissa traveled 20 kilometers per hour faster than Marcia on her trip. What was the average speed in kilometers per hour for each woman?

33. *Fluff Containers* Marshmallow fluff comes in only two sizes, a $7\frac{1}{2}$-oz glass jar and a 16-oz plastic tub. At a local market, the 16-oz tub costs $1.49 and the $7\frac{1}{2}$-oz jar costs $0.79.

  (a) How much does marshmallow fluff in the glass jar cost per oz? (Round your answer to the nearest cent.)

  (b) How much does marshmallow fluff in the plastic tub cost per oz? (Round your answer to the nearest cent.)

  (c) If the marshmallow fluff company decided to add a third size, a 40-oz bucket, how much would the price be if it was at the same unit price as the 16-oz plastic tub? (*Hint:* Set up a proportion. Do *not* use your answer from part (b). Round your answer to the nearest cent.)

34. *Green Tea* Won Ling is a Chinese tea importer in Boston's Chinatown. He charges $12.25 for four sample packs of his famous green tea. The packs are in the following sizes: 25 grams, 40 grams, 50 grams, and 60 grams.

  (a) How much is Won Ling charging per gram for his green tea?

  (b) How much would you pay for a 60-gram pack if he were willing to sell that one by itself?

  (c) How much would you pay for an 800-gram package of green tea if it cost the same amount per gram?

35. *Raking Leaves* When all the leaves have fallen at Fred and Suzie's house in Concord, New Hampshire, Suzie can rake the entire yard in 6 hours. When Fred does it alone, it takes him 8 hours. How long would it take them to rake the yard together? Round to the nearest minute.

36. *Meal Preparation* To celebrate Diwali, a major Indian festival, Deepak and Alpa host a party each year for all their friends. Deepak can decorate and prepare all the food in 6 hours. Alpa takes 5 hours to decorate and prepare the food. How long would it take if they decorated and made the food together? Round your answer to the nearest minute.

Cumulative Review

37. [5.2.2] Write in scientific notation. 0.000892465

38. [5.2.2] Write in decimal notation. 5.82×10^8

39. [5.2.1] Write with positive exponents. $\dfrac{x^{-3}y^{-2}}{z^4 w^{-8}}$

40. [5.2.1] Evaluate. $\left(\dfrac{2}{3}\right)^{-3}$

Quick Quiz 7.6 Solve for x.

1. $\dfrac{16.5}{2.1} = \dfrac{x}{7}$

2. Last week Jeff Slater noted that 164 of the 205 flights to Chicago's O'Hare Airport flying out of Logan Airport departed on time. During the next week 215 flights left Logan Airport for Chicago's O'Hare Airport. If the same ratio holds, how many of those flights would he expect to depart on time?

3. While hiking in the White Mountains, Phil, Melissa, Noah, and Olivia saw a tall tree that cast a shadow 34 feet long. They observed at the same time that a 6-foot-tall person cast a shadow that was 8.5 feet long. How tall is the tree?

4. Concept Check Mike found that his car used 18 gallons of gas to travel 396 miles. He needs to take a trip of 450 miles and wants to know how many gallons of gas it will take. He set up the equation $\dfrac{18}{x} = \dfrac{450}{396}$. Explain what error he made and how he should correctly solve the problem.

Putting Your Skills to Work: Use Math to Save Money

CREDIT CARD FOLLIES

Do you have a credit card? Do you make the minimum payment, or do you try to pay more? Almost all students in college have a major credit card, and most college students pay the monthly minimum payment. What are the implications of making the minimum payment? Consider the story of Adam.

By his junior year in college Adam owed $8000 on his credit card. Each month he made the monthly minimum payment. Some months he needed to charge new items. Adam's math teacher told him that college students often accumulate over $2000 in finance charges on their credit cards a year. Surprised by that amount, Adam decided to add up his monthly finance charges. They are listed below.

January	$138	July	$170
February	$142	August	$183
March	$157	September	$192
April	$158	October	$203
May	$160	November	$210
June	$165	December	$214

1. How much did he accumulate in finance charges during the year?

2. What was his average monthly finance charge?

3. Discuss some ways that Adam can reduce the amount of interest he pays each year.

4. How much do you (or your parents) pay on your credit card each month in finance charges? How much do you (or your parents) pay in a year?

Topic	Procedure	Examples
Simplifying rational expressions, p. 418.	1. Factor the numerator and denominator. 2. Divide out any factor common to both the numerator and denominator.	$$\frac{36x^2 - 16y^2}{18x^2 + 24xy + 8y^2} = \frac{4(3x + 2y)(3x - 2y)}{2(3x + 2y)(3x + 2y)}$$ $$= \frac{2(3x - 2y)}{3x + 2y}$$
Multiplying rational expressions, p. 424.	1. Factor all numerators and denominators. 2. Simplify the resulting rational expression as described above.	$$\frac{x^2 - y^2}{x^2 + 2xy + y^2} \cdot \frac{x^2 + 4xy + 3y^2}{x^2 - 4xy + 3y^2}$$ $$= \frac{(x + y)(x - y)}{(x + y)(x + y)} \cdot \frac{(x + y)(x + 3y)}{(x - y)(x - 3y)}$$ $$= \frac{x + 3y}{x - 3y}$$
Dividing rational expressions, p. 425.	1. Invert the second fraction and rewrite the problem as a product. 2. Multiply the rational expressions.	$$\frac{14x^2 + 17x - 6}{x^2 - 25} \div \frac{4x^2 - 8x - 21}{x^2 + 10x + 25}$$ $$= \frac{(2x + 3)(7x - 2)}{(x + 5)(x - 5)} \cdot \frac{(x + 5)(x + 5)}{(2x - 7)(2x + 3)}$$ $$= \frac{(7x - 2)(x + 5)}{(x - 5)(2x - 7)}$$
Adding rational expressions, p. 429.	1. If the denominators differ, factor them and determine the least common denominator (LCD). 2. Use multiplication to change each fraction into an equivalent one with the LCD as the denominator. 3. Add the numerators; put the answer over the LCD. 4. Simplify as needed.	$$\frac{x - 1}{x^2 - 4} + \frac{x - 1}{3x + 6} = \frac{x - 1}{(x + 2)(x - 2)} + \frac{x - 1}{3(x + 2)}$$ $$\text{LCD} = 3(x + 2)(x - 2)$$ $$\frac{x - 1}{(x + 2)(x - 2)} + \frac{x - 1}{3(x + 2)}$$ $$= \frac{x - 1}{(x + 2)(x - 2)} \cdot \frac{3}{3} + \frac{x - 1}{3(x + 2)} \cdot \frac{x - 2}{x - 2}$$ $$= \frac{3x - 3 + x^2 - 3x + 2}{3(x + 2)(x - 2)}$$ $$= \frac{x^2 - 1}{3(x + 2)(x - 2)}$$ $$= \frac{(x + 1)(x - 1)}{3(x + 2)(x - 2)}$$
Subtracting rational expressions, p. 429.	Move the subtraction sign to the numerator of the second fraction. Add. Simplify if possible. $$\frac{a}{b} - \frac{c}{b} = \frac{a}{b} + \frac{-c}{b}$$	$$\frac{5x}{x - 2} - \frac{3x + 4}{x - 2} = \frac{5x}{x - 2} + \frac{-(3x + 4)}{x - 2}$$ $$= \frac{5x - 3x - 4}{x - 2}$$ $$= \frac{2x - 4}{x - 2}$$ $$= \frac{2(x - 2)}{x - 2} = 2$$

(Continued on next page)

Topic	Procedure	Examples
Simplifying complex rational expressions, p. 438.	1. Add the two fractions in the numerator. 2. Add the two fractions in the denominator. 3. Divide the fraction in the numerator by the fraction in the denominator. This is done by inverting the fraction in the denominator and multiplying by the numerator. 4. Simplify.	$$\frac{\dfrac{x}{x^2-4}+\dfrac{1}{x+2}}{\dfrac{3}{x+2}-\dfrac{4}{x-2}}$$ $$=\frac{\dfrac{x}{(x+2)(x-2)}+\dfrac{1}{x+2}\cdot\dfrac{x-2}{x-2}}{\dfrac{3}{x+2}\cdot\dfrac{x-2}{x-2}-\dfrac{4}{x-2}\cdot\dfrac{x+2}{x+2}}$$ $$=\frac{\dfrac{x+x-2}{(x+2)(x-2)}}{\dfrac{3x-6-4x-8}{(x+2)(x-2)}}$$ $$=\frac{2x-2}{(x+2)(x-2)}\div\frac{-x-14}{(x+2)(x-2)}$$ $$=\frac{2(x-1)}{(x+2)(x-2)}\cdot\frac{(x+2)(x-2)}{-x-14}$$ $$=\frac{2(x-1)}{-x-14}\text{ or }-\frac{2(x-1)}{x+14}\text{ or }\frac{-2(x-1)}{x+14}$$
Solving equations involving rational expressions, p. 444.	1. Determine the LCD of all denominators. 2. Note what values will make the LCD equal to 0. These are excluded from your solutions. 3. Multiply each side by the LCD, distributing as needed. 4. Solve the resulting polynomial equation. 5. Check. Be sure to exclude those values found in step 2.	$$\frac{3}{x-2}=\frac{4}{x+2}$$ $\text{LCD}=(x-2)(x+2).$ $$(x-2)(x+2)\frac{3}{x-2}=\frac{4}{x+2}(x-2)(x+2)$$ $$3(x+2)=4(x-2)$$ $$3x+6=4x-8$$ $$-x=-14$$ $$x=14$$ $$\text{Check: }\frac{3}{14-2}\overset{?}{=}\frac{4}{14+2}$$ $$\frac{3}{12}\overset{?}{=}\frac{4}{16}$$ $$\frac{1}{4}=\frac{1}{4}\checkmark$$
Solving applied problems with proportions, p. 449.	1. Organize the data. 2. Write a proportion equating the respective parts. Let x represent the value that is not known. 3. Solve the proportion.	Renee can make five cherry pies with 3 cups of flour. How many cups of flour does she need to make eight cherry pies? $$\frac{5\text{ cherry pies}}{3\text{ cups flour}}=\frac{8\text{ cherry pies}}{x\text{ cups flour}}$$ $$\frac{5}{3}=\frac{8}{x}$$ $$5x=24$$ $$x=\frac{24}{5}$$ $$x=4\frac{4}{5}$$ $4\frac{4}{5}$ cups of flour are needed for eight cherry pies.

Chapter 7 Review Problems

Section 7.1

Simplify.

1. $\dfrac{bx}{bx - by}$

2. $\dfrac{4x - 4y}{5y - 5x}$

3. $\dfrac{2x^2 + 5x - 3}{2x^2 - 9x + 4}$

4. $\dfrac{3x^2 + 7x + 2}{3x^2 + 13x + 4}$

5. $\dfrac{x^2 - 9}{x^2 - 10x + 21}$

6. $\dfrac{2x^2 + 18x + 40}{3x + 15}$

7. $\dfrac{x^3 - 4x^2}{x^3 - x^2 - 12x}$

8. $\dfrac{2x^2 + 7x - 15}{25 - x^2}$

9. $\dfrac{2x^2 - 2xy - 24y^2}{2x^2 + 5xy - 3y^2}$

10. $\dfrac{4 - y^2}{3y^2 + 5y - 2}$

11. $\dfrac{5x^3 - 10x^2}{25x^4 + 5x^3 - 30x^2}$

12. $\dfrac{16x^2 - 4y^2}{4x - 2y}$

Section 7.2

Multiply or divide.

13. $\dfrac{2x^2 + 6x}{3x^2 - 27} \cdot \dfrac{x^2 + 3x - 18}{4x^2 - 4x}$

14. $\dfrac{y^2 + 8y + 16}{5y^2 + 20y} \div \dfrac{y^2 + 7y + 12}{2y^2 + 5y - 3}$

15. $\dfrac{2y^2 + 3y - 2}{2y^2 + y - 1} \div \dfrac{2y^2 + y - 1}{2y^2 - 3y - 2}$

16. $\dfrac{6y^2 + 13y - 5}{9y^2 + 3y} \div \dfrac{4y^2 + 20y + 25}{12y^2}$

17. $\dfrac{3xy^2 + 12y^2}{2x^2 - 11x + 5} \div \dfrac{2xy + 8y}{8x^2 + 2x - 3}$

18. $\dfrac{11}{x - 2} \cdot \dfrac{2x^2 - 8}{44}$

19. $\dfrac{x^2 - 5xy - 24y^2}{2x^2 - 2xy - 24y^2} \cdot \dfrac{4x^2 + 4xy - 24y^2}{x^2 - 10xy + 16y^2}$

20. $\dfrac{2x^2 + 10x + 2}{8x - 8} \cdot \dfrac{3x - 3}{4x^2 + 20x + 4}$

Section 7.3

Add or subtract.

21. $\dfrac{6}{y + 2} + \dfrac{2}{3y}$

22. $3 + \dfrac{2}{x + 1} + \dfrac{1}{x}$

23. $\dfrac{7}{x + 2} + \dfrac{3}{x - 4}$

24. $\dfrac{2}{x^2 - 9} + \dfrac{x}{x + 3}$

461

25. $\dfrac{x}{y} + \dfrac{3}{2y} + \dfrac{1}{y+2}$

26. $\dfrac{4}{a} + \dfrac{2}{b} + \dfrac{3}{a+b}$

27. $\dfrac{3x+1}{3x} - \dfrac{1}{x}$

28. $\dfrac{x+4}{x+2} - \dfrac{1}{2x}$

29. $\dfrac{27}{x^2-81} + \dfrac{3}{2(x+9)}$

30. $\dfrac{1}{x^2+7x+10} - \dfrac{x}{x+5}$

Section 7.4

Simplify.

31. $\dfrac{\dfrac{4}{3y} - \dfrac{2}{y}}{\dfrac{1}{2y} + \dfrac{1}{y}}$

32. $\dfrac{\dfrac{5}{x} + \dfrac{1}{2x}}{\dfrac{x}{4} + x}$

33. $\dfrac{w - \dfrac{4}{w}}{1 + \dfrac{2}{w}}$

34. $\dfrac{1 - \dfrac{w}{w-1}}{1 + \dfrac{w}{1-w}}$

35. $\dfrac{1 + \dfrac{1}{y^2-1}}{\dfrac{1}{y+1} - \dfrac{1}{y-1}}$

36. $\dfrac{\dfrac{1}{y} + \dfrac{1}{x+y}}{1 + \dfrac{2}{x+y}}$

37. $\dfrac{\dfrac{1}{a+b} - \dfrac{1}{a}}{b}$

38. $\dfrac{\dfrac{2}{a+b} - \dfrac{3}{b}}{\dfrac{1}{a+b}}$

39. $\dfrac{x+5y}{x-6y} \div \left(\dfrac{1}{5y} - \dfrac{1}{x+5y} \right)$

40. $\left(\dfrac{1}{x+2y} - \dfrac{1}{x-y} \right) \div \dfrac{2x-4y}{x^2-3xy+2y^2}$

Section 7.5

Solve for the variable. If there is no solution, say so.

41. $\dfrac{8a-1}{6a+8} = \dfrac{3}{4}$

42. $\dfrac{8}{a-3} = \dfrac{12}{a+3}$

43. $\dfrac{2x-1}{x} - \dfrac{1}{2} = -2$

44. $\dfrac{5-x}{x} - \dfrac{7}{x} = -\dfrac{3}{4}$

45. $\dfrac{5}{2} - \dfrac{2y+7}{y+6} = 3$

46. $\dfrac{5}{4} - \dfrac{1}{2x} = \dfrac{1}{x} + 2$

47. $\dfrac{7}{8x} - \dfrac{3}{4} = \dfrac{1}{4x} + \dfrac{1}{2}$

48. $\dfrac{1}{3x} + 2 = \dfrac{5}{6x} - \dfrac{1}{2}$

49. $\dfrac{3}{y-3} = \dfrac{3}{2} + \dfrac{y}{y-3}$

50. $\dfrac{3x}{x^2-4} - \dfrac{2}{x+2} = -\dfrac{4}{x-2}$

51. $\dfrac{9}{2} - \dfrac{7y-4}{y+2} = -\dfrac{1}{4}$

52. $\dfrac{3y-1}{3y} - \dfrac{6}{5y} = \dfrac{1}{y} - \dfrac{4}{15}$

53. $\dfrac{y+18}{y^2-16} = \dfrac{y}{y+4} - \dfrac{y}{y-4}$

54. $\dfrac{4}{x^2-1} = \dfrac{2}{x-1} + \dfrac{2}{x+1}$

55. $\dfrac{3y+1}{y^2-y} - \dfrac{3}{y-1} = \dfrac{4}{y}$

56. $\dfrac{3}{y-2} + \dfrac{4}{3y+2} = \dfrac{1}{2-y}$

Section 7.6

Solve.

57. $\dfrac{x}{4} = \dfrac{7}{10}$

58. $\dfrac{8}{5} = \dfrac{2}{x}$

59. $\dfrac{33}{10} = \dfrac{x}{8}$

60. $\dfrac{16}{x} = \dfrac{24}{9}$

61. $\dfrac{13.5}{0.6} = \dfrac{360}{x}$

62. $\dfrac{2\frac{1}{2}}{3\frac{1}{4}} = \dfrac{7}{x}$

Use a proportion to answer each question.

63. *Paint Needs* A 5-gallon can of paint will cover 240 square feet. How many gallons of paint will be needed to cover 400 square feet? Round to the nearest tenth of a gallon.

64. *Recipe Ratios* Aunt Lexie uses 3 pounds of sugar to make 100 cookies. How many cookies can she make with 5 pounds of sugar? Round to the nearest whole cookie.

65. *Gas Usage* Leila found that her car used 10 gallons of gas to travel 360 miles. She plans to drive 1400 miles from her home to New York City. How many gallons of gas will she use if her car continues to consume gas at the same rate? Round your answer to the nearest tenth of a gallon.

66. *Map Scale* On a map of Texas, the distance between El Paso and Dallas is 4 inches. The actual distance between these cities is 640 miles. Houston and Dallas are 1.5 inches apart on the same map. How many miles apart are Houston and Dallas?

67. *Travel Speeds* A train travels 180 miles in the same time that a car travels 120 miles. The speed of the train is 20 miles per hour faster than the speed of the car. Find the speed of the train and the speed of the car.

68. *Baking* Sam works at a popular bakery and can make the muffins and bagels for the morning rush in 3 hours. The new baker can do all of the baking in 5 hours. How long would it take Sam and the new baker to prepare for the morning rush if they worked together?

▲ **69.** *Shadows* Mary takes a walk across a canyon in New Mexico. She stands 5.75 feet tall and her shadow is 3 feet long. At the same time, the shadow from the peak of the canyon wall casts a shadow that is 95 feet long. How tall is the peak of the canyon? Round to the nearest foot.

▲ **70.** *Shadows* A flagpole that is 8 feet tall casts a shadow of 3 feet. At the same time of day, a tall office building in the city casts a shadow of 450 feet. How tall is the office building?

71. *Window Cleaning* As part of their spring cleaning routine, Tina and Mathias wash all of their windows, inside and out. Tina can do this job in 4 hours. When Mathias washes the windows, it takes him 6 hours. How long would it take them if they worked together?

72. *Plowing Fields* Sally runs the family farm in Boone, Iowa. She can plow the fields of the farm in 20 hours. Her daughter Brenda can plow the fields of the farm in 30 hours. If they have two identical tractors, how long would it take Brenda and Sally to plow the fields of the farm if they worked together?

Mixed Practice

Perform the indicated operation. Simplify.

73. $\dfrac{a^2 + 2a - 8}{6a^2 - 3a^3}$

74. $\dfrac{4a^3 + 20a^2}{2a^2 + 13a + 15}$

75. $\dfrac{x^2 - y^2}{x^2 + 4xy + 3y^2} \cdot \dfrac{x^2 + xy - 6y^2}{x^2 + xy - 2y^2}$

76. $\dfrac{x^2 - 6xy - 16y^2}{x^2 + 4xy - 21y^2} \cdot \dfrac{x^2 - 8xy + 15y^2}{x^2 - 3xy - 10y^2}$

77. $\dfrac{x}{x + 3} + \dfrac{9x + 18}{x^2 + 3x}$

78. $\dfrac{x - 30}{x^2 - 5x} + \dfrac{x}{x - 5}$

79. $\dfrac{a + b}{ax + ay} - \dfrac{a + b}{bx + by}$

80. $\dfrac{\dfrac{5}{3x} + \dfrac{2}{9x}}{\dfrac{3}{x} + \dfrac{8}{3x}}$

81. $\dfrac{\dfrac{4}{5y} - \dfrac{8}{y}}{y + \dfrac{y}{5}}$

82. $\dfrac{x - 3y}{x + 2y} \div \left(\dfrac{2}{y} - \dfrac{12}{x + 3y} \right)$

83. $\dfrac{7}{x + 2} = \dfrac{4}{x - 4}$

84. $\dfrac{2x - 1}{3x - 8} = \dfrac{5}{8}$

85. $2 + \dfrac{4}{b - 1} = \dfrac{4}{b^2 - b}$

86. *Wire Weight* If 30 meters of wire weighs 8 kilograms, what will 40 meters of the same kind of wire weigh? Round your answer to the nearest tenth.

87. *Driving Rates* Ming drove 270 miles in $4\dfrac{1}{2}$ hours. At that same rate, how far would she drive in 7 hours?

Remember to use your Chapter Test Prep Video CD to see the worked-out solutions to the test problems you want to review.

Perform the operation indicated. Simplify.

1. $\dfrac{2ac + 2ad}{3a^2c + 3a^2d}$

2. $\dfrac{8x^2 - 2x^2y^2}{y^2 + 4y + 4}$

3. $\dfrac{x^2 + 2x}{2x - 1} \cdot \dfrac{10x^2 - 5x}{12x^3 + 24x^2}$

4. $\dfrac{x + 2y}{12y^2} \cdot \dfrac{4y}{x^2 + xy - 2y^2}$

5. $\dfrac{2a^2 - 3a - 2}{a^2 + 5a + 6} \div \dfrac{a^2 - 5a + 6}{a^2 - 9}$

6. $\dfrac{1}{a^2 - a - 2} + \dfrac{3}{a - 2}$

7. $\dfrac{x - y}{xy} - \dfrac{a - y}{ay}$

8. $\dfrac{3x}{x^2 - 3x - 18} - \dfrac{x - 4}{x - 6}$

9. $\dfrac{\dfrac{x}{3y} - \dfrac{1}{2}}{\dfrac{4}{3y} - \dfrac{2}{x}}$

10. $\dfrac{\dfrac{6}{b} - 4}{\dfrac{5}{bx} - \dfrac{10}{3x}}$

11. $\dfrac{2x^2 + 3xy - 9y^2}{4x^2 + 13xy + 3y^2}$

12. $\dfrac{1}{x + 4} - \dfrac{2}{x^2 + 6x + 8}$

In questions 13–18, solve for x. Check your answers. If there is no solution, say so.

13. $\dfrac{4}{3x} - \dfrac{5}{2x} = 5 - \dfrac{1}{6x}$

14. $\dfrac{x - 3}{x - 2} = \dfrac{2x^2 - 15}{x^2 + x - 6} - \dfrac{x + 1}{x + 3}$

15. $3 - \dfrac{7}{x + 3} = \dfrac{x - 4}{x + 3}$

16. $\dfrac{3}{3x - 5} = \dfrac{7}{5x + 4}$

17. $\dfrac{9}{x} = \dfrac{13}{5}$

18. $\dfrac{9.3}{2.5} = \dfrac{x}{10}$

19. A random check of America West air flights last month showed that 113 of the 150 flights checked arrived on time. If the inspectors check 200 flights next month, how many can be expected to be on time? (Round to the nearest whole number.)

20. In northern Michigan the Gunderson family heats their home with firewood. They used $100 worth of wood in 25 days. Mr. Gunderson estimates that he needs to burn wood at that rate for about 92 days during the winter. If that is so, how much will the 92-day supply of wood cost?

▲ 21. A hiking club is trying to construct a rope bridge across a canyon. A 6-foot construction pole held upright casts a 7-foot shadow. At the same time of day, a tree at the edge of the canyon casts a shadow that exactly covers the distance that is needed for the rope bridge. The tree is exactly 87 feet tall. How long should the rope bridge be? Round to the nearest foot.

1. _____

2. _____

3. _____

4. _____

5. _____

6. _____

7. _____

8. _____

9. _____

10. _____

11. _____

12. _____

13. _____

14. _____

15. _____

16. _____

17. _____

18. _____

19. _____

20. _____

21. _____

Cumulative Test for Chapters 0–7

1. _____

2. _____

3. _____

4. _____

5. _____

6. _____

7. _____

8. _____

9. _____

10. (a) _____

(b) _____

11. _____

12. _____

13. _____

14. _____

15. _____

16. _____

17. _____

18. _____

19. _____

20. _____

21. _____

22. _____

Approximately one-half of this test covers the content of Chapters 0–6. The remainder covers the content of Chapter 7.

1. To the nearest thousandth, how much error can be tolerated in the length of a wire that is supposed to be 2.57 centimeters long if specifications allow an error of no more than 0.25%?

2. In 2005, the value of all U.S. agricultural exports was $62,958,000,000. The value of agricultural exports to Canada was $10,570,000,000. What percent of the total value of agricultural exports went to Canada? (*Source: Statistical Abstract of the United States: 2007.*) Round your answer to the nearest tenth of a percent.

3. Henry is thinking about buying a new car for $22,500. His state presently has a sales tax of 6.5%. In a few weeks the state will raise the sales tax to 7%. How much will he save in sales tax if he purchases the car before the sales tax rate is raised?

4. Solve. $5(x - 3) - 2(4 - 2x) = 7(x - 1) - (x - 2)$

5. Solve for h. $A = \pi r^2 h$

6. Solve and graph on a number line. $4(2 - x) < 3$

7. Solve for x. $\dfrac{3}{5}x - 2 \le \dfrac{1}{2}x + \dfrac{3}{5}$

8. Factor. $3ax + 3bx - 2ay - 2by$ 9. Factor. $8a^3 - 38a^2b - 10ab^2$

10. Simplify. Leave the answer in positive exponents.

    (a) $\dfrac{-12x^{-2}y^4}{3x^{-5}y^{-8}}$ (b) $\left(\dfrac{2xy^2}{z^3}\right)^{-2}$

11. Simplify. $\dfrac{4x^2 - 25}{2x^2 + 9x - 35}$

Perform the indicated operations.

12. $\dfrac{x^2 - 4}{x^2 - 25} \cdot \dfrac{3x^2 - 14x - 5}{3x^2 + 6x}$ 13. $\dfrac{2x^2 - 9x + 9}{8x - 12} \div \dfrac{x^2 - 3x}{2x}$

14. $\dfrac{5}{2x + 4} + \dfrac{3}{x - 3}$ 15. $\dfrac{8}{c^2 - 4} - \dfrac{2}{c^2 - 5c + 6}$

Solve for x.

16. $\dfrac{3x - 2}{3x + 2} = 2$ 17. $\dfrac{x - 3}{x} = \dfrac{x + 2}{x + 3}$

In questions 18 and 19, simplify.

18. $\dfrac{\dfrac{1}{x - 3} + \dfrac{5}{x^2 - 9}}{\dfrac{6x}{x + 3}}$ 19. $\dfrac{\dfrac{3}{a} + \dfrac{2}{b}}{\dfrac{5}{a^2} - \dfrac{2}{b^2}}$ 20. Solve for x. $\dfrac{2x + 1}{3} = \dfrac{4}{5}$

21. Jane is looking at a road map. The distance between two cities is 130 miles. This distance is represented by $2\frac{1}{2}$ inches on the map. She sees that the distance she has to drive today is a totally straight interstate highway that is 4 inches long on the map. How many miles does she have to drive?

22. Roberto is working as a telemarketing salesperson for a large corporation. In the last 22 telephone calls he has made, he was able to make a sale five times. His goal is to make 110 sales this month. If his rate for making a sale continues as in the past, how many phone calls must he make?

Sport parachuting has become popular throughout the world. Engaging in this sport requires accuracy and concentration because things happen very quickly. When a person jumps out of a plane and before the parachute is deployed, he or she will fall 2304 feet during the first 12 seconds. How fast could this person travel? How can the speed be calculated? These and other calculations involving sport parachuting require a knowledge of the mathematics covered in this chapter.

Radicals

Student Learning Objectives

After studying this section, you will be able to:

 Evaluate the square root of a perfect square.

 Approximate the square root of a number that is not a perfect square.

Evaluating the Square Root of a Perfect Square

How long is the side of a square whose area is 4? Recall the formula for the area of a square.

area of a square $= s^2$ Our question then becomes, what number times itself is 4?

$$s^2 = 4$$
$$s = 2 \quad \text{because } (2)(2) = 4$$
$$s = -2 \quad \text{because } (-2)(-2) = 4$$

We say that 2 is a **square root** of 4 because $(2)(2) = 4$. We can also say that -2 is a square root of 4 because $(-2)(-2) = 4$. Note that 4 is a **perfect square.** A square root of a perfect square is an integer.

The symbol $\sqrt{}$ is used in mathematics for the **principal square root** of a number, which is the nonnegative square root. The symbol itself $\sqrt{}$ is called the **radical sign.** If we want to find the negative square root of a number, we use the symbol $-\sqrt{}$.

DEFINITION OF PRINCIPAL SQUARE ROOT

For all nonnegative numbers N, the principal square root of N (written $\sqrt{N}$) is defined to be the nonnegative number a if and only if $a^2 = N$.

Notice in the definition that we did not use the words "for all *positive* numbers N" because we also want to include the square root of 0. $\sqrt{0} = 0$ since $0^2 = 0$.

Let us now examine the use of this definition with a few examples.

EXAMPLE 1 Find.   **(a)** $\sqrt{144}$ **(b)** $-\sqrt{9}$

Solution

(a) $\sqrt{144} = 12$ since $(12)^2 = (12)(12) = 144$ and $12 \geq 0$.

(b) The symbol $-\sqrt{9}$ is read "the opposite of the square root of 9."

Because $\sqrt{9} = 3$, we have $-\sqrt{9} = -3$.

Practice Problem 1 Find. **(a)** $\sqrt{64}$ **(b)** $-\sqrt{121}$

The number beneath the radical sign is called the **radicand.** The radicand will not always be an integer.

EXAMPLE 2 Find.

(a) $\sqrt{\dfrac{1}{4}}$ **(b)** $-\sqrt{\dfrac{4}{9}}$

Solution

(a) $\sqrt{\dfrac{1}{4}} = \dfrac{1}{2}$ since $\left(\dfrac{1}{2}\right)^2 = \left(\dfrac{1}{2}\right)\left(\dfrac{1}{2}\right) = \dfrac{1}{4}$.   **(b)** $-\sqrt{\dfrac{4}{9}} = -\dfrac{2}{3}$ since $\sqrt{\dfrac{4}{9}} = \dfrac{2}{3}$.

Practice Problem 2 Find.

(a) $\sqrt{\dfrac{9}{25}}$ **(b)** $-\sqrt{\dfrac{121}{169}}$

EXAMPLE 3 Find. **(a)** $\sqrt{0.09}$ **(b)** $\sqrt{1600}$ **(c)** $\sqrt{225}$

Solution

(a) $\sqrt{0.09} = 0.3$ since $(0.3)(0.3) = 0.09$.

Notice that $\sqrt{0.09}$ is *not* 0.03 because $(0.03)(0.03) = 0.0009$. Remember to count the decimal places when you multiply decimals. When finding the square root of a decimal, you should multiply to check your answer.

(b) $\sqrt{1600} = 40$ since $(40)(40) = 1600$.

(c) $\sqrt{225} = 15$

Practice Problem 3 Find.

(a) $-\sqrt{0.0036}$ **(b)** $\sqrt{2500}$ **(c)** $\sqrt{196}$

2 Approximating the Square Root of a Number That Is Not a Perfect Square

Not all numbers are perfect squares. How can we find the square root of 2? That is, what number times itself equals 2? $\sqrt{2}$ is an irrational number. It cannot be written as $\frac{p}{q}$, where p and q are integers and $q \neq 0$. $\sqrt{2}$ is a nonterminating, nonrepeating decimal, and the best we can do is approximate its value. To do so, we can use the square root key on a calculator.

To use a calculator, enter 2 and push the $\boxed{\sqrt{x}}$ or $\boxed{\sqrt{}}$ key.

$$\boxed{2}\ \boxed{\sqrt{x}}\ \boxed{1.4142136}$$

Remember, this is only an approximation. Most calculators are limited to an eight-digit display.

To use the table of square roots, look under the x column until you find the number 2. Then look across the row to the $\sqrt{x}$ column. The value there is rounded to the nearest thousandth.

$$\sqrt{2} \approx 1.414$$

x	$\sqrt{x}$
1	1.000
2	1.414
3	1.732
4	2.000
5	2.236
6	2.449
7	2.646

EXAMPLE 4 Use a calculator or a table to approximate. Round to the nearest thousandth.

(a) $\sqrt{28}$ **(b)** $\sqrt{191}$

Solution

(a) Using a calculator, we have 28 $\boxed{\sqrt{x}}$ 5.291502622. Thus $\sqrt{28} \approx 5.292$.

(b) Using a calculator, we have 191 $\boxed{\sqrt{x}}$ 13.82027496. Thus $\sqrt{191} \approx 13.820$.

Practice Problem 4 Approximate. Round to the nearest thousandth.

(a) $\sqrt{13}$ **(b)** $\sqrt{35}$ **(c)** $\sqrt{127}$

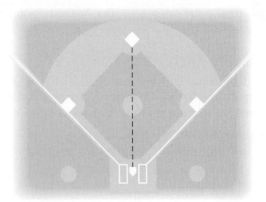

On a professional softball diamond, the distance from home plate to second base is exactly $\sqrt{7200}$ feet. Approximate this value using a calculator.

SIDELIGHT: A Topic for Further Thought

Let's look at the radicand carefully. Does the radicand need to be a nonnegative number? What if we write $\sqrt{-4}$? Is there a number that you can square to get -4?

Obviously, there is *no real number* that you can square to get -4. We know that $(2)^2 = 4$ and $(-2)^2 = 4$. Any real number that is squared will be nonnegative. We therefore conclude that $\sqrt{-4}$ does not represent a real number. Because we want to work with real numbers, our definition of $\sqrt{n}$ requires that n be nonnegative. Thus $\sqrt{-4}$ is *not a real number.*

You may encounter a more sophisticated number system called **complex numbers** in a higher-level mathematics course. In the complex number system, negative numbers have square roots.

Verbal and Writing Skills

1. Define a principal square root in your own words.

2. Is $\sqrt{576} = 24$? Why or why not?

3. Is $\sqrt{0.9} = 0.3$? Why or why not?

4. How would you find $\sqrt{90}$?

Find the two square roots of each number.

5. 9

6. 4

7. 49

8. 36

Find the square root. Do not use a calculator or a table of square roots.

9. $\sqrt{16}$

10. $\sqrt{100}$

11. $\sqrt{81}$

12. $\sqrt{25}$

13. $-\sqrt{36}$

14. $-\sqrt{25}$

15. $\sqrt{0.81}$

16. $\sqrt{0.64}$

17. $\sqrt{\dfrac{36}{121}}$

18. $\sqrt{\dfrac{25}{81}}$

19. $\sqrt{\dfrac{49}{64}}$

20. $\sqrt{\dfrac{64}{169}}$

21. $\sqrt{625}$

22. $\sqrt{1225}$

23. $-\sqrt{10{,}000}$

24. $-\sqrt{40{,}000}$

Mixed Practice   *Find the square root. Do not use a calculator or a table of square roots.*

25. $\sqrt{169}$

26. $\sqrt{225}$

27. $-\sqrt{\dfrac{1}{64}}$

28. $\sqrt{\dfrac{81}{100}}$

29. $\sqrt{\dfrac{9}{16}}$

30. $-\sqrt{\dfrac{16}{49}}$

31. $\sqrt{0.0036}$

32. $\sqrt{0.0009}$

33. $\sqrt{19{,}600}$

34. $\sqrt{28{,}900}$

35. $-\sqrt{289}$

36. $-\sqrt{324}$

Use a calculator or the square root table to approximate to the nearest thousandth.

37. $\sqrt{27}$

38. $\sqrt{33}$

39. $\sqrt{74}$

40. $\sqrt{79}$

41. $-\sqrt{183}$

42. $-\sqrt{123}$

43. $-\sqrt{195}$

44. $-\sqrt{182}$

Applications

▲ 45. *Tree Care* A tree is 12 feet tall. A wire that stretches from a point that is 5 feet from the base of the tree to the top of the tree is $\sqrt{169}$ feet long. Exactly how many feet long is the wire?

▲ 46. *Extension Ladder* A garage is 15 feet tall. A ladder that stretches from a point that is 8 feet from the base of the garage to the top of the garage is $\sqrt{289}$ feet long. Exactly how many feet long is the ladder?

Time for an Object to Fall If we ignore wind resistance, the time in seconds it takes for an object to fall s feet is given by $t = \frac{1}{4}\sqrt{s}$. Thus an object will fall 100 feet in $t = \frac{1}{4}\sqrt{100} = \frac{1}{4}(10) = 2.5$ seconds.

47. How long would it take an object dropped off a building to fall 900 feet?

48. How long would it take an object dropped out of a tower to fall 1600 feet?

To Think About: Finding Higher Roots *In addition to finding the square root of a number, we can also find the cube root of a number. To find the cube root of N (which is written as $\sqrt[3]{N}$), we find a number that, when cubed, will equal N.*

For example, $\sqrt[3]{8} = 2$ since $2^3 = 8$. We can also find the cube root of a negative number—for example, $\sqrt[3]{-8} = -2$ since $(-2)^3 = -8$. The cube root of a positive number is positive. The cube root of a negative number is negative. Therefore, each real number has only one real cube root.

The concept of roots can be extended to fourth and fifth roots. For example, $\sqrt[4]{16} = 2$ since $2^4 = 16$. Also, $\sqrt[5]{32} = 2$ since $2^5 = 32$.

Find each higher root.

49. $\sqrt[3]{27}$

50. $\sqrt[3]{125}$

51. $\sqrt[3]{-64}$

52. $\sqrt[4]{256}$

53. $\sqrt[4]{81}$

54. $\sqrt[4]{625}$

55. $\sqrt[5]{243}$

56. $\sqrt[5]{1024}$

57. Is there a real number that equals $\sqrt[4]{-16}$? Why?

58. Is there a real number that equals $\sqrt[4]{-81}$? Why?

Cumulative Review

59. **[4.4.1]** Solve. $3x + 2y = 8$
$7x - 3y = 11$

60. **[4.4.2]** Solve if possible. $2x - 3y = 1$
$-8x + 12y = 4$

61. **[4.5.1]** *Snowboarding* A snowboard racer bought three new snowboards and two pairs of racing goggles last month for $850. This month, she bought four snowboards and three pairs of goggles for $1150. How much did each snowboard cost? How much did each pair of goggles cost?

62. **[2.6.3]** *Flying with a Wind* With a tailwind, the Qantas flight from Los Angeles to Sydney, Australia, covered 7280 miles in 14 hours, nonstop. The return flight of 7140 miles took 17 hours because of a headwind. If the airspeed of the plane remained constant, and if the return flight encountered a constant headwind equal to the earlier tailwind, what was the plane's airspeed in still air?

Quick Quiz 8.1 Find the square root. Do not use a calculator or a table of square roots.

1. $\sqrt{169}$

2. $\sqrt{\dfrac{25}{64}}$

3. $\sqrt{0.49}$

4. **Concept Check** Explain how you would go about determining $\sqrt{32,400}$ without using a calculator or a table of square roots.

8.2 SIMPLIFYING RADICAL EXPRESSIONS

1 Simplifying a Radical Expression with a Radicand That Is a Perfect Square

We know that $\sqrt{25} = \sqrt{5^2} = 5$, $\sqrt{9} = \sqrt{3^2} = 3$, and $\sqrt{36} = \sqrt{6^2} = 6$. If we write the radicand as a square of a nonnegative number, the square root is this nonnegative base. That is, $\sqrt{x^2} = x$ when x is nonnegative. We can use this idea to simplify radicals.

Student Learning Objectives

After studying this section, you will be able to:

 Simplify a radical expression with a radicand that is a perfect square.

 Simplify a radical expression with a radicand that is not a perfect square.

EXAMPLE 1 Find.

(a) $\sqrt{9^4}$ **(b)** $\sqrt{126^2}$ **(c)** $\sqrt{17^6}$

Solution

(a) Using the law of exponents, we rewrite 9^4 as a square.

$\sqrt{9^4} = \sqrt{(9^2)^2}$ Raise a power to a power: $(9^2)^2 = 9^4$.

$= 9^2$ To check, multiply: $(9^2)(9^2) = 9^4$.

(b) $\sqrt{126^2} = 126$

(c) $\sqrt{17^6} = \sqrt{(17^3)^2}$ Raise a power to a power: $(17^3)^2 = 17^6$.

$= 17^3$ To check, multiply: $(17^3)(17^3) = 17^6$.

Practice Problem 1 Find.

(a) $\sqrt{6^2}$ **(b)** $\sqrt{13^4}$ **(c)** $\sqrt{18^{12}}$

This same concept can be used with variable expressions. We will restrict the values of each variable to be nonnegative if the variable is under the radical sign. Thus *all variable radicands in this chapter* are assumed to *represent positive numbers or zero*.

EXAMPLE 2 Find.

(a) $\sqrt{x^6}$ **(b)** $\sqrt{y^{24}}$

Solution

(a) $\sqrt{x^6} = \sqrt{(x^3)^2}$ By the law of exponents, $(x^3)^2 = x^6$.

$= x^3$

(b) $\sqrt{y^{24}} = y^{12}$

Practice Problem 2 Find.

(a) $\sqrt{y^{18}}$ **(b)** $\sqrt{x^{30}}$

In each of these examples we found the square root of a perfect square, and thus the radical sign disappeared. When you find the square root of a perfect square, do *not* leave a radical sign in your answer.

Often the coefficient of a variable radical will not be written in exponent form. You will want to be able to recognize the squares of numbers from 1 to 15, 20, and 25. If you can, you will be able to find many square roots faster mentally than by using a calculator or a table of square roots.

Before you begin, you need to know the multiplication rule for square roots.

> **MULTIPLICATION RULE FOR SQUARE ROOTS**
>
> For all nonnegative numbers a and b,
>
> $$\sqrt{a} \cdot \sqrt{b} = \sqrt{ab} \quad \text{and} \quad \sqrt{ab} = \sqrt{a} \cdot \sqrt{b}.$$

EXAMPLE 3 Find.

(a) $\sqrt{225x^2}$ **(b)** $\sqrt{x^6 y^{14}}$ **(c)** $\sqrt{169x^8 y^{10}}$

Solution

(a) $\sqrt{225x^2} = 15x$ Using the multiplication rule for square roots,
$\sqrt{225x^2} = \sqrt{225}\sqrt{x^2} = 15x.$

(b) $\sqrt{x^6 y^{14}} = x^3 y^7$

(c) $\sqrt{169x^8 y^{10}} = 13x^4 y^5$

To check each answer, square it. The result should be the expression under the radical sign.

NOTE TO STUDENT: *Fully worked-out solutions to all of the Practice Problems can be found at the back of the text starting at page SP-1*

Practice Problem 3 Find.

(a) $\sqrt{625y^4}$ **(b)** $\sqrt{x^{16} y^{22}}$ **(c)** $\sqrt{121x^{12} y^6}$

② Simplifying a Radical Expression with a Radicand That Is Not a Perfect Square

Most of the time when we encounter a square root, the radicand is not a perfect square. Thus, when we simplify the square root algebraically, we must retain the radical sign. Still, being able to simplify such radical expressions is a useful skill because it allows us to combine them, as we shall see in Section 8.3.

EXAMPLE 4 Simplify.

(a) $\sqrt{20}$ **(b)** $\sqrt{50}$ **(c)** $\sqrt{48}$

Solution To begin, when you look at a radicand, look for perfect squares.

(a) $\sqrt{20} = \sqrt{4 \cdot 5}$ Note that 4 is a perfect square, and we can write 20 as $4 \cdot 5$.

$\quad\quad = \sqrt{4}\sqrt{5}$ Recall that $\sqrt{4} = 2$.

$\quad\quad = 2\sqrt{5}$

(b) $\sqrt{50} = \sqrt{25 \cdot 2}$

$\quad\quad = \sqrt{25}\sqrt{2}$

$\quad\quad = 5\sqrt{2}$

(c) $\sqrt{48} = \sqrt{16 \cdot 3}$

$\quad\quad = \sqrt{16}\sqrt{3}$

$\quad\quad = 4\sqrt{3}$

Practice Problem 4 Simplify.

(a) $\sqrt{98}$ **(b)** $\sqrt{12}$ **(c)** $\sqrt{75}$

The same procedure can be used with square root radical expressions containing variables. The key is to think of squares. That is, think, "How can I rewrite the expression so that it contains a perfect square?"

EXAMPLE 5   Simplify.

(a) $\sqrt{x^3}$

(b) $\sqrt{x^7y^9}$

Solution

(a) Recall that the law of exponents tells us to add the exponents when multiplying two expressions. Because we want one of these expressions to be a perfect square, we think of the exponent 3 as a sum of an even number and 1: $3 = 2 + 1$.

$$\sqrt{x^3} = \sqrt{x^2}\sqrt{x} \quad \text{Because } (x^2)(x) = x^3 \text{ by the law of exponents.}$$
$$= x\sqrt{x}$$

In general, to simplify a square root that has a variable with an exponent in the radicand, first write the square root as a product of the form $\sqrt{x^n}\sqrt{x}$, where n is the largest possible even exponent.

(b) $\sqrt{x^7y^9} = \sqrt{x^6}\sqrt{x}\sqrt{y^8}\sqrt{y}$
$\qquad\quad = x^3\sqrt{x}y^4\sqrt{y}$
$\qquad\quad = x^3y^4\sqrt{xy}$

In the final simplified form, we place the variable factors with exponents first and the radical factor second.

Practice Problem 5   Simplify.

(a) $\sqrt{x^{11}}$

(b) $\sqrt{x^5y^3}$

In summary, to simplify a square root radical, you factor each part of the expression under the radical sign and simplify the perfect squares.

EXAMPLE 6   Simplify.

(a) $\sqrt{12y^5}$

(b) $\sqrt{18x^3y^7w^{10}}$

Solution

(a) $\sqrt{12y^5} = \sqrt{4\cdot3\cdot y^4\cdot y}$
$\qquad\quad = 2y^2\sqrt{3y} \qquad$ Remember, $y^4 = (y^2)^2$ and $\sqrt{(y^2)^2} = y^2$.

(b) $\sqrt{18x^3y^7w^{10}} = \sqrt{9\cdot2\cdot x^2\cdot x\cdot y^6\cdot y\cdot w^{10}}$
$\qquad\qquad\qquad = 3xy^3w^5\sqrt{2xy}$

Practice Problem 6   Simplify.

(a) $\sqrt{48x^{11}}$

(b) $\sqrt{121x^6y^7z^8}$

Throughout the world there has been a significant increase in the production and consumption of energy from alternative sources. These are primarily geothermal, solar, wind, and waste recycling energy production. For example, in 1990 1.7 quadrillion Btu (British thermal units) of energy from alternative sources were produced in the world. By 2000 this had increased to 3 quadrillion Btu (that is, 3,000,000,000,000,000 Btu). Some scientists predict that by 2010 this will increase to $\sqrt{192}$ quadrillion Btu. When simplified this can be written as $8\sqrt{3}$ quadrillion Btu. (*Source:* U.S. Energy Information Administration.)

Verbal and Writing Skills

1. Is $\sqrt{6^2} = \left(\sqrt{6}\right)^2$?

2. Why is it *not correct* to say
$$\sqrt{30} = \sqrt{(-6)(-5)} = \sqrt{-6}\sqrt{-5} ?$$

3. Is $\sqrt{(-9)^2} = -\sqrt{9^2}$?

4. Is $\sqrt{2^4} = \left(\sqrt{2}\right)^4$?

Simplify. Leave the answer in exponent form.

5. $\sqrt{8^2}$

6. $\sqrt{2^2}$

7. $\sqrt{18^4}$

8. $\sqrt{13^4}$

9. $\sqrt{9^6}$

10. $\sqrt{5^{12}}$

11. $\sqrt{33^8}$

12. $\sqrt{51^{16}}$

13. $\sqrt{5^{140}}$

14. $\sqrt{7^{160}}$

Simplify. Assume that all variables represent positive numbers.

15. $\sqrt{x^{12}}$

16. $\sqrt{x^{30}}$

17. $\sqrt{t^{18}}$

18. $\sqrt{w^{22}}$

19. $\sqrt{y^{26}}$

20. $\sqrt{w^{14}}$

21. $\sqrt{36x^8}$

22. $\sqrt{64x^6}$

23. $\sqrt{144x^2}$

24. $\sqrt{196y^{10}}$

25. $\sqrt{x^6y^4}$

26. $\sqrt{x^2y^{18}}$

27. $\sqrt{16x^2y^{20}}$

28. $\sqrt{49a^4b^2}$

29. $\sqrt{100a^{10}b^6}$

30. $\sqrt{81x^6y^{10}}$

31. $\sqrt{24}$

32. $\sqrt{27}$

33. $\sqrt{45}$

34. $\sqrt{20}$

35. $\sqrt{18}$

36. $\sqrt{32}$

37. $\sqrt{72}$

38. $\sqrt{60}$

39. $\sqrt{90}$

40. $\sqrt{125}$

41. $\sqrt{128}$

42. $\sqrt{96}$

43. $\sqrt{8x^3}$

44. $\sqrt{12y^5}$

45. $\sqrt{27w^5}$

46. $\sqrt{25x^5}$

47. $\sqrt{32z^9}$

48. $\sqrt{24y^4}$

49. $\sqrt{28x^5y^7}$

50. $\sqrt{45x^5y^4}$

Mixed Practice *Simplify. Assume all variables represent positive numbers.*

51. $\sqrt{48y^3w}$

52. $\sqrt{12x^2y^3}$

53. $\sqrt{75x^2y^3}$

54. $\sqrt{50x^7y^6}$

55. $\sqrt{64x^{10}}$

56. $\sqrt{49x^6}$

57. $\sqrt{y^7}$

58. $\sqrt{w^9}$

59. $\sqrt{x^5y^{11}}$ **60.** $\sqrt{x^3y^{13}}$ **61.** $\sqrt{135x^5y^7}$ **62.** $\sqrt{180a^5bc^2}$

63. $\sqrt{72a^2b^4c^5}$ **64.** $\sqrt{90ab^3c^6}$ **65.** $\sqrt{169a^3b^8c^5}$ **66.** $\sqrt{121a^8b^5c^{13}}$

To Think About *Simplify. Assume that all variables represent positive numbers.*

67. $\sqrt{(x+5)^2}$ **68.** $\sqrt{x^2 + 10x + 25}$ **69.** $\sqrt{16x^2 + 8x + 1}$

70. $\sqrt{16x^2 + 24x + 9}$ **71.** $\sqrt{x^2y^2 + 14xy^2 + 49y^2}$ **72.** $\sqrt{16x^2 + 16x + 4}$

Cumulative Review

73. **[4.2.1]** Solve. $3x - 2y = 14$
$$y = x - 5$$

74. **[2.4.1]** Solve. $\frac{1}{2}(x - 4) = \frac{1}{3}x - 5$

75. **[6.3.2]** Factor. $2x^2 + 12xy + 18y^2$

76. **[6.3.1]** Factor. $2x^2 - x - 21$

77. **[7.2.2]** Divide. $\dfrac{x^2 + 6x + 9}{2x^2y - 18y} \div \dfrac{6xy + 18y}{3x^2y - 27y}$

78. **[7.3.3]** Add. $\dfrac{6}{2x - 4} + \dfrac{1}{12x + 4}$

Quick Quiz 8.2 Simplify. Assume that all variables represent positive numbers.

1. $\sqrt{120}$

2. $\sqrt{81x^3y^4}$

3. $\sqrt{135x^8y^5}$

4. Concept Check Explain how you would simplify the following expression.
$$\sqrt{y^6z^7}$$

Student Learning Objectives

After studying this section, you will be able to:

 1 Add and subtract radical expressions with like radicals.

2 Add and subtract radical expressions that must first be simplified.

1 Adding and Subtracting Radical Expressions with Like Radicals

Recall that when you simplify an algebraic expression, you combine like terms. $8b + 5b = 13b$ because both terms contain the same variable, b, and you can add the coefficients. This same idea is used when combining radicals. To add or subtract square root radicals, the numbers or the algebraic expressions under the radical signs must be the same. That is, the radicals must be **like radicals.**

EXAMPLE 1 Combine.

(a) $5\sqrt{2} - 8\sqrt{2}$ **(b)** $7\sqrt{a} + 3\sqrt{a} - 5\sqrt{a}$

Solution First check to see if the radicands are the same. If they are, you can combine the radicals.

(a) $5\sqrt{2} - 8\sqrt{2} = (5 - 8)\sqrt{2} = -3\sqrt{2}$

(b) $7\sqrt{a} + 3\sqrt{a} - 5\sqrt{a} = (7 + 3 - 5)\sqrt{a} = 5\sqrt{a}$

Practice Problem 1 Combine.

(a) $7\sqrt{11} + 4\sqrt{11}$ **(b)** $4\sqrt{t} - 7\sqrt{t} + 6\sqrt{t} - 2\sqrt{t}$

Be careful. Not all the terms in an expression will have like radicals.

EXAMPLE 2 Combine. $5\sqrt{2a} + 3\sqrt{a} - 7\sqrt{2} + 3\sqrt{2a}$

Solution The only terms that have the same radicand are $5\sqrt{2a}$ and $3\sqrt{2a}$. These terms may be combined. All other terms stay the same.

$$5\sqrt{2a} + 3\sqrt{a} - 7\sqrt{2} + 3\sqrt{2a} = 8\sqrt{2a} + 3\sqrt{a} - 7\sqrt{2}$$

Practice Problem 2 Combine. $3\sqrt{x} - 2\sqrt{xy} - 5\sqrt{y} + 7\sqrt{xy}$

2 Adding and Subtracting Radical Expressions That Must First Be Simplified

Sometimes it is necessary to simplify one or more radicals before their terms can be combined. Be sure to combine only like radicals.

EXAMPLE 3 Combine.

(a) $2\sqrt{3} + \sqrt{12}$ **(b)** $\sqrt{12} - \sqrt{27} + \sqrt{50}$

Solution

(a) $2\sqrt{3} + \sqrt{12} = 2\sqrt{3} + \sqrt{4 \cdot 3}$ Look for perfect square factors.

$\qquad\qquad\quad = 2\sqrt{3} + 2\sqrt{3}$ Simplify and combine like radicals.

$\qquad\qquad\quad = 4\sqrt{3}$

(b) $\sqrt{12} - \sqrt{27} + \sqrt{50} = \sqrt{4 \cdot 3} - \sqrt{9 \cdot 3} + \sqrt{25 \cdot 2}$

$\qquad\qquad\qquad\qquad = 2\sqrt{3} - 3\sqrt{3} + 5\sqrt{2}$ Only $2\sqrt{3}$ and $-3\sqrt{3}$

$\qquad\qquad\qquad\qquad = -\sqrt{3} + 5\sqrt{2}$ or $5\sqrt{2} - \sqrt{3}$ can be combined.

Practice Problem 3 Combine.

(a) $\sqrt{50} - \sqrt{18} + \sqrt{98}$ **(b)** $\sqrt{12} + \sqrt{18} - \sqrt{50} + \sqrt{27}$

EXAMPLE 4 Combine. $\sqrt{2a} + \sqrt{8a} + \sqrt{27a}$

Solution

$$\sqrt{2a} + \sqrt{8a} + \sqrt{27a} = \sqrt{2a} + \sqrt{4 \cdot 2a} + \sqrt{9 \cdot 3a}$$ Look for perfect square factors.
$$= \sqrt{2a} + 2\sqrt{2a} + 3\sqrt{3a}$$ Be careful. $\sqrt{2a}$ and $\sqrt{3a}$ are **not** like radicals. The expressions under the radical signs must be the same.
$$= 3\sqrt{2a} + 3\sqrt{3a}$$

Practice Problem 4 Combine.
$$\sqrt{9x} + \sqrt{8x} - \sqrt{4x} + \sqrt{50x}$$

Special care should be taken if the original radical has a numerical coefficient. If the radical can be simplified, the two resulting numerical coefficients should be multiplied.

EXAMPLE 5 Combine. $3\sqrt{20} + 4\sqrt{45} - 2\sqrt{80}$

Solution

$$3\sqrt{20} + 4\sqrt{45} - 2\sqrt{80} = 3 \cdot \sqrt{4} \cdot \sqrt{5} + 4 \cdot \sqrt{9} \cdot \sqrt{5} - 2 \cdot \sqrt{16} \cdot \sqrt{5}$$
$$= 3 \cdot 2 \cdot \sqrt{5} + 4 \cdot 3 \cdot \sqrt{5} - 2 \cdot 4 \cdot \sqrt{5}$$
$$= 6\sqrt{5} + 12\sqrt{5} - 8\sqrt{5}$$
$$= 10\sqrt{5}$$

Practice Problem 5 Combine.
$$\sqrt{27} - 4\sqrt{3} + 2\sqrt{75}$$

EXAMPLE 6 Combine. $3a\sqrt{8a} + 2\sqrt{50a^3}$

Solution

$$3a\sqrt{8a} + 2\sqrt{50a^3} = 3a\sqrt{4}\sqrt{2a} + 2\sqrt{25a^2}\sqrt{2a}$$
$$= 3a \cdot 2 \cdot \sqrt{2a} + 2 \cdot 5a \cdot \sqrt{2a}$$
$$= 6a\sqrt{2a} + 10a\sqrt{2a}$$
$$= 16a\sqrt{2a}$$

(*Note:* If you are unsure of the last step, show the use of the distributive property in performing the addition.)

$$6a\sqrt{2a} + 10a\sqrt{2a} = (6 + 10)a\sqrt{2a}$$
$$= 16a\sqrt{2a}$$

Practice Problem 6 Combine.
$$2a\sqrt{12a} + 3\sqrt{27a^3}$$

Verbal and Writing Skills

1. List the steps involved in simplifying an expression such as $\sqrt{75x} + \sqrt{48x} + \sqrt{16x^2}$.

2. Mario said that, since $\sqrt{12}$ and $\sqrt{3}$ are not like radicals, $\sqrt{12} + \sqrt{3}$ cannot be simplified. What do you think?

Combine, if possible. Do not use a calculator or a table of square roots.

3. $3\sqrt{5} - \sqrt{5} + 4\sqrt{5}$

4. $\sqrt{7} + 2\sqrt{7} - 6\sqrt{7}$

5. $\sqrt{2} + 8\sqrt{3} - 5\sqrt{3} + 4\sqrt{2}$

6. $\sqrt{5} - \sqrt{6} + 3\sqrt{5} - 2\sqrt{6}$

7. $\sqrt{5} + \sqrt{45}$

8. $\sqrt{7} - \sqrt{63}$

9. $\sqrt{80} + 4\sqrt{20}$

10. $\sqrt{80} + 2\sqrt{20}$

11. $3\sqrt{8} - 5\sqrt{2}$

12. $2\sqrt{27} - 4\sqrt{3}$

13. $-\sqrt{2} + \sqrt{18} + \sqrt{98}$

14. $-\sqrt{54} + \sqrt{24} + 5\sqrt{6}$

15. $\sqrt{25} + \sqrt{72} + 3\sqrt{12}$

16. $4\sqrt{12} + 2\sqrt{16} - \sqrt{44}$

17. $3\sqrt{20} - \sqrt{50} + \sqrt{18}$

18. $\sqrt{50} + \sqrt{28} - \sqrt{9}$

19. $8\sqrt{2x} + \sqrt{50x}$

20. $6\sqrt{5y} - \sqrt{20y}$

21. $1.2\sqrt{3x} - 0.5\sqrt{12x}$

22. $-1.5\sqrt{2a} + 0.2\sqrt{8a}$

23. $\sqrt{20y} + 2\sqrt{45y} - \sqrt{5y}$

24. $\sqrt{72w} - 3\sqrt{2w} - \sqrt{50w}$

Mixed Practice *Combine if possible. Do not use a calculator or a table of square roots.*

25. $\sqrt{50} - 3\sqrt{8}$

26. $\sqrt{75} - 5\sqrt{27}$

27. $3\sqrt{28x} - 5x\sqrt{63x}$

28. $-2b\sqrt{45b} + 5\sqrt{80b}$

29. $5\sqrt{8x^3} - 3x\sqrt{50x}$

30. $-2\sqrt{27y^3} + y\sqrt{12y}$

31. $5x\sqrt{48} - 2x\sqrt{75}$

32. $2x\sqrt{45} - 7x\sqrt{80}$

33. $2\sqrt{6y^3} - 2y\sqrt{54}$

34. $2x\sqrt{12x^3} - 6\sqrt{5x^2}$

35. $5x\sqrt{8x} - 24\sqrt{50x^3}$

36. $6\sqrt{18x^3} - 5x\sqrt{72x}$

Applications

▲ **37.** *Mars Mission* A planned unmanned mission to Mars may possibly require a small vehicle to make scientific measurements as it travels along the perimeter of a region like the one shown below. If the region is rectangular and the dimensions are as labeled, find the exact distance to be traveled by the vehicle in one trip around the perimeter.

$(\sqrt{5} + 2\sqrt{3})$ miles

$(3\sqrt{5} + \sqrt{3})$ miles

▲ **38.** *Mars Mission* Suppose that on the second day of the mission to Mars, the vehicle must travel along the perimeter of a right triangle with dimensions as labeled below. Find the exact distance to be traveled by the vehicle in one trip around the perimeter.

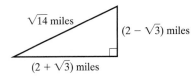

$\sqrt{14}$ miles

$(2 - \sqrt{3})$ miles

$(2 + \sqrt{3})$ miles

Cumulative Review

39. [2.8.4] Solve and graph on a number line. $7 - 3x \le 11$

$$\begin{array}{ccccccc} \hline -3 & -2 & -1 & 0 & 1 & 2 & 3 \end{array}$$

Simplify.

40. [5.2.1] $\left(\dfrac{2x^2y}{3xy^3}\right)^{-2}$

41. [5.1.3] $(-3a^2b^0x^4)^3$

Quick Quiz 8.3 Combine if possible. Do not use a calculator or a table of square roots.

1. $\sqrt{98} + 3\sqrt{8}$

2. $5\sqrt{3} + 2\sqrt{75} - 3\sqrt{48}$

3. $\sqrt{9a} + \sqrt{8a} + \sqrt{49a} + \sqrt{32a}$

4. Concept Check Explain how you would combine $3\sqrt{8x^3} + 5x\sqrt{98x}$.

Student Learning Objectives

After studying this section, you will be able to:

 Multiply monomial radical expressions.

2 Multiply a monomial radical expression by a polynomial.

3 Multiply two polynomial radical expressions.

1 Multiplying Monomial Radical Expressions

Recall the basic rule for multiplication of square root radicals,

$$\sqrt{a}\sqrt{b} = \sqrt{ab}.$$

We will use this concept to multiply square root radical expressions. Note that the direction "multiply" means to express the product in simplest form.

EXAMPLE 1 Multiply. $\sqrt{7}\sqrt{14x}$

Solution $\sqrt{7}\sqrt{14x} = \sqrt{98x}$

We do *not* stop here, because the radical $\sqrt{98x}$ can be simplified.

$$\sqrt{98x} = \sqrt{49 \cdot 2x} = \sqrt{49}\sqrt{2x} = 7\sqrt{2x}$$

Practice Problem 1 Multiply. $\sqrt{3a}\sqrt{6a}$

EXAMPLE 2 Multiply.

(a) $\left(2\sqrt{3}\right)\left(5\sqrt{7}\right)$ **(b)** $\left(2a\sqrt{3a}\right)\left(3\sqrt{6a}\right)$

Solution

(a) $\left(2\sqrt{3}\right)\left(5\sqrt{7}\right) = 10\sqrt{21}$ Multiply the coefficients: $(2)(5) = 10$.
Multiply the radicals: $\sqrt{3}\sqrt{7} = \sqrt{21}$.

(b) $\left(2a\sqrt{3a}\right)\left(3\sqrt{6a}\right) = 6a\sqrt{18a^2}$
$$= 6a\sqrt{9}\sqrt{2}\sqrt{a^2} \quad \text{Simplify } \sqrt{18a^2}.$$
$$= 18a^2\sqrt{2} \quad\quad\quad \text{Multiply } (6a)(3)(a).$$

Practice Problem 2 Multiply.

(a) $\left(2\sqrt{3}\right)\left(5\sqrt{5}\right)$ **(b)** $\left(4\sqrt{3x}\right)\left(2x\sqrt{6x}\right)$

EXAMPLE 3 Find the area of a farm field that measures $\sqrt{9400}$ feet long and $\sqrt{4800}$ feet wide. Express your answer as a simplified radical.

Solution We multiply.

$$\left(\sqrt{9400}\right)\left(\sqrt{4800}\right) = \left(10\sqrt{94}\right)\left(40\sqrt{3}\right)$$
$$= 400\sqrt{282} \text{ square feet (ft}^2\text{)}$$

Practice Problem 3 Find the area of a computer chip that measures $\sqrt{180}$ millimeters long and $\sqrt{150}$ millimeters wide.

Multiplying a Monomial Radical Expression by a Polynomial

Recall that a binomial consists of two terms. $\sqrt{2} + \sqrt{5}$ is a binomial. We use the distributive property when multiplying a binomial by another factor.

EXAMPLE 4 Multiply and simplify. $\sqrt{5}(\sqrt{2} + 3\sqrt{7})$

Solution
$$= \sqrt{5}\sqrt{2} + 3\sqrt{5}\sqrt{7}$$
$$= \sqrt{10} + 3\sqrt{35}$$

In similar fashion, we can multiply a trinomial by another factor.

Practice Problem 4 Multiply and simplify.
$$\sqrt{6}(\sqrt{3} + 2\sqrt{8})$$

We discover that $\sqrt{5} \cdot \sqrt{5} = 5$ and $\sqrt{3} \cdot \sqrt{3} = 3$. These are specific cases of the multiplication rule for square root radicals.

> For any nonnegative real number a,
> $$\sqrt{a} \cdot \sqrt{a} = a.$$

EXAMPLE 5 Multiply. $\sqrt{a}(3\sqrt{a} - 2\sqrt{5})$

Solution
$$= 3\sqrt{a}\sqrt{a} - 2\sqrt{5}\sqrt{a} \quad \text{Use the distributive property.}$$
$$= 3a - 2\sqrt{5a}$$

Practice Problem 5 Multiply.
$$2\sqrt{x}(4\sqrt{x} - x\sqrt{2})$$

Be sure to simplify all radicals after multiplying.

 Multiplying Two Polynomial Radical Expressions

Recall the FOIL method used to multiply two binomials. The same method can be used for radical expressions.

Algebraic Expressions
$$(2x + y)(x - 2y) = 2x^2 - 4xy + xy - 2y^2$$
$$= 2x^2 - 3xy - 2y^2$$

Radical Expressions
$$(2\sqrt{3} + \sqrt{5})(\sqrt{3} - 2\sqrt{5}) = 2\sqrt{9} - 4\sqrt{15} + \sqrt{15} - 2\sqrt{25}$$
$$= 2(3) - 3\sqrt{15} - 2(5)$$
$$= 6 - 3\sqrt{15} - 10$$
$$= -4 - 3\sqrt{15}$$

Let's look at this procedure more closely.

EXAMPLE 6 Multiply. $\left(\sqrt{2} + 5\right)\left(\sqrt{2} - 3\right)$

Solution

$$\left(\sqrt{2} + 5\right)\left(\sqrt{2} - 3\right)$$

$$= \sqrt{4} - 3\sqrt{2} + 5\sqrt{2} - 15 \qquad \text{Multiply to obtain the four products.}$$
$$= 2 + 2\sqrt{2} - 15 \qquad \text{Simplify } \sqrt{4}; \text{ combine the middle terms.}$$
$$= -13 + 2\sqrt{2}$$

Practice Problem 6 Multiply.
$$\left(\sqrt{2} + \sqrt{6}\right)\left(2\sqrt{2} - \sqrt{6}\right)$$

Be especially watchful if there is a number outside the radical sign. You must multiply the numerical coefficients very carefully.

EXAMPLE 7 Multiply. $\left(\sqrt{2} - 3\sqrt{6}\right)\left(\sqrt{2} + \sqrt{6}\right)$

Solution

$$\left(\sqrt{2} - 3\sqrt{6}\right)\left(\sqrt{2} + \sqrt{6}\right)$$

$$= \sqrt{4} + \sqrt{12} - 3\sqrt{12} - 3\sqrt{36} \qquad \text{Multiply to obtain the four products.}$$
$$= 2 - 2\sqrt{12} - 18 \qquad \text{Simplify; combine the middle terms.}$$
$$= -16 - 4\sqrt{3} \qquad \text{Simplify } -2\sqrt{12}.$$

Practice Problem 7 Multiply.
$$\left(\sqrt{6} + \sqrt{5}\right)\left(\sqrt{2} + 2\sqrt{5}\right)$$

If an expression with radicals is squared, write the expression as the product of two binomials and multiply.

EXAMPLE 8 Multiply. $\left(2\sqrt{3} - \sqrt{6}\right)^2$

Solution
$$= \left(2\sqrt{3} - \sqrt{6}\right)\left(2\sqrt{3} - \sqrt{6}\right)$$
$$= 4\sqrt{9} - 2\sqrt{18} - 2\sqrt{18} + \sqrt{36}$$
$$= 12 - 4\sqrt{18} + 6$$
$$= 18 - 12\sqrt{2}$$

Practice Problem 8 Multiply.
$$\left(3\sqrt{5} - \sqrt{10}\right)^2$$

Multiply. Be sure to simplify any radicals in your answer. Do not use a calculator or a table of square roots.

1. $\sqrt{3}\sqrt{10}$

2. $\sqrt{5}\sqrt{7}$

3. $\sqrt{2}\sqrt{22}$

4. $\sqrt{3}\sqrt{15}$

5. $\sqrt{18}\sqrt{3}$

6. $\sqrt{28}\sqrt{3}$

7. $(4\sqrt{5})(3\sqrt{2})$

8. $(2\sqrt{7})(3\sqrt{3})$

9. $\sqrt{5}\sqrt{10a}$

10. $\sqrt{5b}\sqrt{20}$

11. $(3\sqrt{10x})(2\sqrt{5x})$

12. $(3\sqrt{10x})(4\sqrt{5x})$

13. $(-4\sqrt{8a})(-2\sqrt{3a})$

14. $(4\sqrt{12a})(2\sqrt{6a})$

15. $(-3\sqrt{ab})(2\sqrt{b})$

16. $(-2\sqrt{x})(5\sqrt{xy})$

17. $\sqrt{5}(\sqrt{2}+\sqrt{3})$

18. $\sqrt{10}(\sqrt{2}+\sqrt{3})$

19. $\sqrt{3}(2\sqrt{6}+5\sqrt{15})$

20. $\sqrt{6}(5\sqrt{12}-4\sqrt{3})$

21. $2\sqrt{x}(\sqrt{x}-8\sqrt{5})$

22. $-3\sqrt{b}(2\sqrt{a}+3\sqrt{b})$

23. $\sqrt{6}(\sqrt{2}-3\sqrt{6}+2\sqrt{10})$

24. $\sqrt{10}(\sqrt{5}-3\sqrt{10}+5\sqrt{2})$

25. $(2\sqrt{3}+\sqrt{6})(\sqrt{3}-2\sqrt{6})$

26. $(3\sqrt{5}+\sqrt{3})(\sqrt{5}+\sqrt{3})$

27. $(5+3\sqrt{2})(3+\sqrt{2})$

28. $(7-2\sqrt{3})(4+\sqrt{3})$

29. $(2\sqrt{7}-3\sqrt{3})(\sqrt{7}+\sqrt{3})$

30. $(5\sqrt{6}-\sqrt{2})(\sqrt{6}+3\sqrt{2})$

31. $(\sqrt{3}+2\sqrt{6})(2\sqrt{3}-\sqrt{6})$

32. $(\sqrt{2}+3\sqrt{10})(2\sqrt{2}-\sqrt{10})$

33. $(3\sqrt{7}-\sqrt{8})(\sqrt{8}+2\sqrt{7})$

34. $(\sqrt{12}-\sqrt{5})(\sqrt{5}+2\sqrt{12})$

35. $(2\sqrt{6}-1)^2$

36. $(3\sqrt{2}+2)^2$

37. $(\sqrt{3}+5\sqrt{2})^2$

38. $(\sqrt{3}-2\sqrt{7})^2$

39. $(3\sqrt{5}-\sqrt{3})^2$

40. $(2\sqrt{3}+5\sqrt{5})^2$

Mixed Practice *Multiply. Be sure to simplify any radicals in your answer.*

41. $(3\sqrt{5})(2\sqrt{8})$

42. $(5\sqrt{3})(\sqrt{24})$

43. $\sqrt{7}(\sqrt{2}+3\sqrt{14}-\sqrt{6})$

44. $\sqrt{6}(\sqrt{6}+5\sqrt{10}-\sqrt{3})$

45. $(5\sqrt{2}-3\sqrt{7})(3\sqrt{2}+\sqrt{7})$

46. $(6\sqrt{5}+7\sqrt{6})(2\sqrt{5}-\sqrt{6})$

47. $2\sqrt{a}\left(3\sqrt{b} + \sqrt{ab} - 2\sqrt{a}\right)$

48. $3\sqrt{y}\left(2\sqrt{6} - \sqrt{x} + 3\sqrt{y}\right)$

Multiply the following binomials. Be sure to simplify your answer. Do not use a calculator or a table of square roots.

49. $\left(3x\sqrt{y} + \sqrt{5}\right)\left(3x\sqrt{y} - \sqrt{5}\right)$

50. $\left(4a\sqrt{5} - 2\sqrt{2}\right)\left(4a\sqrt{5} + 2\sqrt{2}\right)$

Applications *Ramp Construction* *A wheelchair access ramp is constructed with the dimensions as listed on the diagram. The gray shaded triangle is a right triangle.*

▲ **51.** Find the area of the shaded portion of the wheelchair ramp.

▲ **52.** Find the perimeter of the shaded portion of the wheelchair ramp. Leave your answer as a radical in simplified form.

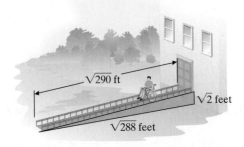

√290 ft

√2 feet

√288 feet

Cumulative Review

53. [6.5.1] Factor. $64a^2 - 25b^2$

54. [6.5.2] Factor. $4x^2 - 20x + 25$

55. [2.6.3] *Speed in Knots* Ships often measure their speed in knots (or nautical miles per hour). The regular mile (statute mile) that we are more familiar with is 5280 feet. A nautical mile is about 6076 feet. Write a formula of the form $m = ak$, where m is the number of statute miles, k is the number of nautical miles, and a is a constant number (in this case a fraction). Use the formula to find out how fast a Coast Guard cutter is going in miles per hour if it is traveling at 35 knots.

56. [0.5.3] *Car Insurance* Phil and Melissa LaBelle found that the collision and theft portion of their car insurance could be reduced by 15% if they purchased an auto security device. They purchased and had such a device installed for $350. The yearly collision and theft portion of their car insurance is $280. How many years will it take for the security device to pay for itself?

Quick Quiz 8.4 Multiply. Be sure to simplify any radicals in your answer.

1. $\left(2\sqrt{x}\right)\left(4\sqrt{y}\right)\left(3\sqrt{xy}\right)$

2. $\sqrt{3}\left(\sqrt{8} + 2\sqrt{3} - 4\sqrt{5}\right)$

3. $\left(4\sqrt{2} - 5\sqrt{3}\right)\left(2\sqrt{2} + 3\sqrt{3}\right)$

4. Concept Check Explain how to simplify $\left(3\sqrt{3} - 2\right)^2$.

How are you doing with your homework assignments in Sections 8.1 to 8.4? Do you feel you have mastered the material so far? Do you understand the concepts you have covered? Before you go further in the textbook, take some time to do each of the following problems.

Assume all variables are positive.

8.1

Find the square roots, if possible. (If it is impossible to obtain a real number for an answer, say so.) Approximate to the nearest thousandth.

1. $\sqrt{49}$

2. $\sqrt{\dfrac{25}{64}}$

3. $\sqrt{81}$

4. $-\sqrt{100}$

5. $\sqrt{5}$

6. $-\sqrt{121}$

8.2

Simplify.

7. $\sqrt{0.04}$

8. $\sqrt{0.36}$

9. $\sqrt{x^4}$

10. $\sqrt{25x^4y^6}$

11. $\sqrt{50}$

12. $\sqrt{x^5}$

13. $\sqrt{144a^2b^3}$

14. $\sqrt{32x^5}$

8.3

Combine, if possible.

15. $\sqrt{98} + \sqrt{128}$

16. $3\sqrt{2} - \sqrt{8} + \sqrt{18}$

17. $2\sqrt{28} + 3\sqrt{63} - \sqrt{49}$

18. $3\sqrt{8} + \sqrt{12} + \sqrt{50} - 4\sqrt{75}$

19. $5\sqrt{2x} - 7\sqrt{8x}$

20. $-4x\sqrt{75} - 2x\sqrt{27}$

8.4

Multiply. Simplify your answer.

21. $\left(5\sqrt{2}\right)\left(3\sqrt{15}\right)$

22. $\left(4x\sqrt{6}\right)\left(x\sqrt{10}\right)$

23. $\sqrt{2}\left(\sqrt{6} + 3\sqrt{2}\right)$

24. $\left(\sqrt{7} - 3\right)^2$

25. $\left(\sqrt{11} - \sqrt{10}\right)\left(\sqrt{11} + \sqrt{10}\right)$

26. $\left(2\sqrt{3} - \sqrt{5}\right)^2$

27. $\left(\sqrt{x} + 5\right)\left(\sqrt{x} - 2\right)$

28. $\left(3\sqrt{7} + 4\sqrt{3}\right)\left(5\sqrt{7} - 2\sqrt{3}\right)$

Now turn to page SA-21 for the answer to each of these problems. Each answer also includes a reference to the objective in which the problem is first taught. If you missed any of these problems, you should stop and review the Examples and Practice Problems in the referenced objective. A little review now will help you master the material in the upcoming sections of the text.

1. _____
2. _____
3. _____
4. _____
5. _____
6. _____
7. _____
8. _____
9. _____
10. _____
11. _____
12. _____
13. _____
14. _____
15. _____
16. _____
17. _____
18. _____
19. _____
20. _____
21. _____
22. _____
23. _____
24. _____
25. _____
26. _____
27. _____
28. _____

1 Using the Quotient Rule for Square Roots to Simplify a Fraction Involving Radicals

Just as there is a multiplication rule for square roots, there is a quotient rule for square roots. The rules are similar.

QUOTIENT RULE FOR SQUARE ROOTS

For all positive numbers a and b,
$$\frac{\sqrt{a}}{\sqrt{b}} = \sqrt{\frac{a}{b}} \quad \text{and} \quad \sqrt{\frac{a}{b}} = \frac{\sqrt{a}}{\sqrt{b}}.$$

This can be used to divide square root radicals or to simplify square root radical expressions involving division. We will be using both parts of the quotient rule.

EXAMPLE 1 Simplify.

(a) $\dfrac{\sqrt{75}}{\sqrt{3}}$

(b) $\sqrt{\dfrac{25}{36}}$

Solution

(a) We notice that 3 is a factor of 75. We use the quotient rule to rewrite the expression.

$$\frac{\sqrt{75}}{\sqrt{3}} = \sqrt{\frac{75}{3}}$$
$$= \sqrt{25} \quad \text{Divide.}$$
$$= 5 \quad \text{Simplify.}$$

(b) Since both 25 and 36 are perfect squares, we will rewrite this as the quotient of square roots.

$$\sqrt{\frac{25}{36}} = \frac{\sqrt{25}}{\sqrt{36}}$$
$$= \frac{5}{6}$$

Practice Problem 1 Simplify.

(a) $\dfrac{\sqrt{98}}{\sqrt{2}}$

(b) $\sqrt{\dfrac{81}{100}}$

EXAMPLE 2 Simplify. $\sqrt{\dfrac{20}{x^6}}$

Solution

$$\sqrt{\frac{20}{x^6}} = \frac{\sqrt{20}}{\sqrt{x^6}} = \frac{\sqrt{4}\sqrt{5}}{x^3} = \frac{2\sqrt{5}}{x^3} \quad \text{Don't forget to simplify } \sqrt{20} \text{ as } 2\sqrt{5}.$$

Practice Problem 2 Simplify.

$$\sqrt{\frac{50}{a^4}}$$

Rationalizing the Denominator of a Fraction with a Square Root in the Denominator

Sometimes, when calculating with fractions that contain radicals, it is advantageous to have an integer in the denominator. If a fraction has a radical in the denominator, we **rationalize the denominator.** That is, we multiply to change the fraction to an equivalent one that has an integer in the denominator. Remember, we do not want to change the value of the fraction. Thus we will multiply the numerator and the denominator by the same number.

EXAMPLE 3 Simplify. $\dfrac{3}{\sqrt{2}}$

Solution Think, "What times 2 will make a perfect square?"

$$\frac{3}{\sqrt{2}} = \frac{3}{\sqrt{2}} \times 1 = \frac{3}{\sqrt{2}} \times \frac{\sqrt{2}}{\sqrt{2}} = \frac{3\sqrt{2}}{\sqrt{4}} = \frac{3\sqrt{2}}{2}$$

Practice Problem 3 Simplify.

$$\frac{9}{\sqrt{7}}$$

(Note that a fraction is not simplified unless the denominator is rationalized.)

When rationalizing a denominator containing a square root radical, we will want to use the smallest possible radical that will yield the square root of a perfect square. Very often we will not use the radical that is in the denominator.

EXAMPLE 4 Simplify.

(a) $\dfrac{\sqrt{7}}{\sqrt{8}}$

(b) $\dfrac{3}{\sqrt{x^3}}$

Solution

(a) Think, "What times 8 will make a perfect square?"

$$\frac{\sqrt{7}}{\sqrt{8}} = \frac{\sqrt{7}}{\sqrt{8}} \times \frac{\sqrt{2}}{\sqrt{2}} = \frac{\sqrt{14}}{\sqrt{16}} = \frac{\sqrt{14}}{4}$$

(b) Think, "What times x^3 will give an even exponent?"

$$\frac{3}{\sqrt{x^3}} \times \frac{\sqrt{x}}{\sqrt{x}} = \frac{3\sqrt{x}}{\sqrt{x^4}} = \frac{3\sqrt{x}}{x^2}$$

We could have simplified $\sqrt{x^3}$ before we rationalized the denominator.

$$\frac{3}{\sqrt{x^3}} = \frac{3}{x\sqrt{x}} \times \frac{\sqrt{x}}{\sqrt{x}} = \frac{3\sqrt{x}}{x\sqrt{x^2}} = \frac{3\sqrt{x}}{(x)(x)} = \frac{3\sqrt{x}}{x^2}$$

Practice Problem 4 Simplify.

(a) $\dfrac{\sqrt{2}}{\sqrt{12}}$

(b) $\dfrac{6a}{\sqrt{a^7}}$

EXAMPLE 5 Simplify. $\dfrac{\sqrt{2}}{\sqrt{27x}}$

Solution Since it is not apparent what we should multiply 27 by to obtain a perfect square, we will begin by simplifying the denominator.

$$\frac{\sqrt{2}}{\sqrt{27x}} = \frac{\sqrt{2}}{3\sqrt{3x}}$$

Now it is easy to see that we multiply by $\dfrac{\sqrt{3x}}{\sqrt{3x}}$ to rationalize the denominator.

$$= \frac{\sqrt{2}}{3\sqrt{3x}} \times \frac{\sqrt{3x}}{\sqrt{3x}}$$

$$= \frac{\sqrt{6x}}{3\sqrt{9x^2}}$$

$$= \frac{\sqrt{6x}}{9x}$$

NOTE TO STUDENT: *Fully worked-out solutions to all of the Practice Problems can be found at the back of the text starting at page SP-1*

Practice Problem 5 Simplify.

$$\frac{\sqrt{5x}}{\sqrt{8x}}$$

③ Rationalizing the Denominator of a Fraction with a Binomial Denominator Containing at Least One Square Root

Sometimes the denominator of a fraction is a binomial with a square root radical term. $\dfrac{1}{5 - 3\sqrt{2}}$ is such a fraction. How can we eliminate the radical term in the denominator? Recall that when we multiply $(a + b)(a - b)$, we obtain the square of a minus the square of b: $a^2 - b^2$. For example, $(x + 3)(x - 3) = x^2 - 9$. The resulting terms are perfect squares. We can use this idea to eliminate the radical in $5 - 3\sqrt{2}$.

$$\left(5 - 3\sqrt{2}\right)\left(5 + 3\sqrt{2}\right) = 5^2 + 15\sqrt{2} - 15\sqrt{2} - \left(3\sqrt{2}\right)^2 = 5^2 - \left(3\sqrt{2}\right)^2 = 25 - 18 = 7$$

Expressions like $\left(5 - 3\sqrt{2}\right)$ and $\left(5 + 3\sqrt{2}\right)$ are called **conjugates.**

EXAMPLE 6 Simplify.

(a) $\dfrac{2}{\sqrt{3} - 4}$

(b) $\dfrac{\sqrt{x}}{\sqrt{5} + \sqrt{3}}$

Solution

(a) The conjugate of $\sqrt{3} - 4$ is $\sqrt{3} + 4$.

$$\frac{2}{\sqrt{3} - 4} \cdot \frac{\sqrt{3} + 4}{\sqrt{3} + 4} = \frac{2\sqrt{3} + 8}{\left(\sqrt{3}\right)^2 + 4\sqrt{3} - 4\sqrt{3} - 4^2}$$

$$= \frac{2\sqrt{3} + 8}{3 - 16}$$

$$= \frac{2\sqrt{3} + 8}{-13}$$

$$= -\frac{2\sqrt{3} + 8}{13}$$

(b) The conjugate of $\sqrt{5} + \sqrt{3}$ is $\sqrt{5} - \sqrt{3}$.

$$\frac{\sqrt{x}}{\sqrt{5} + \sqrt{3}} \cdot \frac{\sqrt{5} - \sqrt{3}}{\sqrt{5} - \sqrt{3}} = \frac{\sqrt{5x} - \sqrt{3x}}{(\sqrt{5})^2 - \sqrt{15} + \sqrt{15} - (\sqrt{3})^2}$$

$$= \frac{\sqrt{5x} - \sqrt{3x}}{5 - 3}$$

$$= \frac{\sqrt{5x} - \sqrt{3x}}{2}$$

Be careful not to combine $\sqrt{5x}$ and $\sqrt{3x}$ in the numerator. They are not like radicals.

Practice Problem 6 Simplify.

(a) $\dfrac{4}{\sqrt{3} + \sqrt{5}}$

(b) $\dfrac{\sqrt{a}}{\sqrt{10} - 3}$

Note that in Example 7 the numerator is a binomial.

EXAMPLE 7 Rationalize the denominator. $\dfrac{\sqrt{3} + \sqrt{2}}{\sqrt{3} - \sqrt{2}}$

Solution   The conjugate of $\sqrt{3} - \sqrt{2}$ is $\sqrt{3} + \sqrt{2}$.

$$\frac{\sqrt{3} + \sqrt{2}}{\sqrt{3} - \sqrt{2}} \cdot \frac{\sqrt{3} + \sqrt{2}}{\sqrt{3} + \sqrt{2}}$$

$$= \frac{\sqrt{9} + \sqrt{6} + \sqrt{6} + \sqrt{4}}{(\sqrt{3})^2 - (\sqrt{2})^2} \qquad \text{Multiply.}$$

$$= \frac{3 + 2\sqrt{6} + 2}{3 - 2} \qquad \text{Simplify and combine like terms.}$$

$$= \frac{5 + 2\sqrt{6}}{1}$$

$$= 5 + 2\sqrt{6}$$

Practice Problem 7 Rationalize the denominator.

$$\frac{\sqrt{3} + \sqrt{5}}{\sqrt{3} - \sqrt{5}}$$

Note that in Examples 6 and 7 we used a raised dot $\cdot$ to indicate multiplication instead of the $\times$. You need to be familiar with both forms of multiplication notation.

8.5 EXERCISES

MyMathLab

Math XP

 PRACTICE

 WATCH

DOWNLOAD

READ

 REVIEW

Simplify. Be sure to rationalize all denominators. Do not use a calculator or a table of square roots.

1. $\dfrac{\sqrt{12}}{\sqrt{3}}$

2. $\dfrac{\sqrt{3}}{\sqrt{27}}$

3. $\dfrac{\sqrt{7}}{\sqrt{63}}$

4. $\dfrac{\sqrt{52}}{\sqrt{13}}$

5. $\dfrac{\sqrt{98}}{\sqrt{2}}$

6. $\dfrac{\sqrt{5}}{\sqrt{80}}$

7. $\dfrac{\sqrt{6}}{\sqrt{x^4}}$

8. $\dfrac{\sqrt{12}}{\sqrt{x^2}}$

9. $\dfrac{\sqrt{18}}{\sqrt{a^4}}$

10. $\dfrac{\sqrt{24}}{\sqrt{b^6}}$

11. $\dfrac{3}{\sqrt{7}}$

12. $\dfrac{2}{\sqrt{10}}$

13. $\dfrac{8}{\sqrt{15}}$

14. $\dfrac{9}{\sqrt{14}}$

15. $\dfrac{x\sqrt{x}}{\sqrt{2}}$

16. $\dfrac{\sqrt{3y}}{\sqrt{6}}$

17. $\dfrac{\sqrt{8}}{\sqrt{x}}$

18. $\dfrac{\sqrt{18}}{\sqrt{y}}$

19. $\dfrac{6}{\sqrt{28}}$

20. $\dfrac{3}{\sqrt{27}}$

21. $\dfrac{6}{\sqrt{a}}$

22. $\dfrac{8}{\sqrt{b}}$

23. $\dfrac{x}{\sqrt{2x^5}}$

24. $\dfrac{y}{\sqrt{5x^3}}$

25. $\dfrac{\sqrt{18}}{\sqrt{2x^3}}$

26. $\dfrac{\sqrt{24}}{\sqrt{6x^7}}$

27. $\sqrt{\dfrac{3}{5}}$

28. $\sqrt{\dfrac{1}{10}}$

29. $\sqrt{\dfrac{10}{21}}$

30. $\sqrt{\dfrac{15}{17}}$

31. $\dfrac{9}{\sqrt{32x}}$

32. $\dfrac{3}{\sqrt{50x}}$

Rationalize the denominator. Simplify your answer. Do not use a calculator or a table of square roots.

33. $\dfrac{4}{\sqrt{3}-1}$

34. $\dfrac{2}{\sqrt{6}+1}$

35. $\dfrac{8}{\sqrt{10}+\sqrt{2}}$

36. $\dfrac{5}{\sqrt{11}-\sqrt{6}}$

37. $\dfrac{\sqrt{6}}{\sqrt{6}-\sqrt{3}}$

38. $\dfrac{\sqrt{3}}{\sqrt{5}+\sqrt{3}}$

39. $\dfrac{\sqrt{7}}{\sqrt{8}+\sqrt{7}}$

40. $\dfrac{\sqrt{5}}{\sqrt{5}+\sqrt{6}}$

41. $\dfrac{3x}{2\sqrt{2}-\sqrt{5}}$

42. $\dfrac{4x}{2\sqrt{7}+2\sqrt{6}}$

43. $\dfrac{\sqrt{x}}{\sqrt{6}+\sqrt{2}}$

44. $\dfrac{\sqrt{x}}{\sqrt{11}+\sqrt{5}}$

45. $\dfrac{\sqrt{10}-\sqrt{3}}{\sqrt{10}+\sqrt{3}}$

46. $\dfrac{\sqrt{7}+\sqrt{3}}{\sqrt{7}-\sqrt{3}}$

47. $\dfrac{4\sqrt{7}+3}{\sqrt{5}-\sqrt{2}}$

48. $\dfrac{4\sqrt{3}+2\sqrt{5}}{\sqrt{5}-\sqrt{3}}$

49. $\dfrac{4\sqrt{3}+2}{\sqrt{8}-\sqrt{6}}$

50. $\dfrac{3\sqrt{5}+4}{\sqrt{15}-\sqrt{3}}$

51. $\dfrac{x-25}{\sqrt{x}+5}$

52. $\dfrac{x-36}{\sqrt{x}-6}$

Applications

▲ **53.** *Geometry* A pyramid has a volume V and a height h. The length of one side of the square base is $s = \sqrt{\dfrac{3V}{h}}$.

(a) Simplify this expression by rationalizing the denominator.

(b) Find the length of the side of a pyramid with a volume of 36 cubic inches and a height of 12 inches.

▲ **54.** *Geometry* A cylinder has a volume V and a height h. The radius of the cylinder is $r = \sqrt{\dfrac{V}{\pi h}}$.

(a) Simplify this expression by rationalizing the denominator.

(b) Find the radius of a cylindrical storage silo of volume 6250 cubic feet and height 10π feet.

▲ **55.** *Relay Tower* A reflector sheet for a microwave relay tower is shaped like a rectangle. The area of the rectangle is 3 square meters. Ideally the length should be exactly $\left(\sqrt{5} + 2\right)$ meters long. What should be the width of the rectangle? Express the answer exactly using radical expressions.

▲ **56.** *Relay Tower* An engineer needs the approximate dimensions of the rectangle in exercise 55. Approximate the length and the width of that rectangle to the nearest thousandth.

Cumulative Review

57. [2.3.3] Solve for x. Round your answer to the nearest hundredth. $-2(x - 5) = 8x - 3(x + 1) + 18$

58. [1.8.1] Evaluate $a^2 - b \div a + a^3 - b$ for $a = -1$ and $b = 4$.

Quick Quiz 8.5 Simplify. Be sure to rationalize all denominators.

1. $\dfrac{3}{\sqrt{10}}$

2. $\dfrac{5x}{3\sqrt{2} + \sqrt{5}}$

3. $\dfrac{\sqrt{6} + \sqrt{5}}{\sqrt{6} - \sqrt{5}}$

4. **Concept Check** Explain how you would simplify $\dfrac{x - 4}{\sqrt{x} + 2}$.

Student Learning Objectives

After studying this section, you will be able to:

 Use the Pythagorean Theorem to solve applied problems.

 Solve radical equations.

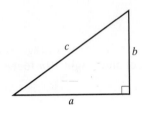

1 Using the Pythagorean Theorem to Solve Applied Problems

In ancient Greece, mathematicians studied a number of properties of right triangles. They proved that the square of the length of the longest side of a right triangle is equal to the sum of the squares of the lengths of the other two sides. This property is called the Pythagorean Theorem, in honor of the Greek mathematician Pythagoras (ca. 590 B.C.). The shorter sides, a and b, are referred to as the **legs** of the right triangle. The longest side, c, is called the **hypotenuse** of the right triangle. We state the theorem as follows.

> **PYTHAGOREAN THEOREM**
>
> In any right triangle, if c is the length of the hypotenuse and a and b are the lengths of the two legs, then
> $$c^2 = a^2 + b^2.$$

If we know any two sides of a right triangle, we can find the third side using this theorem.

Notice that the hypotenuse is always opposite the right angle. The smallest side is opposite the smallest angle.

Recall that in Section 8.1 we showed that if $s^2 = 4$, then $s = 2$ or -2. In general, if $x^2 = a$, then $x = \pm\sqrt{a}$. This is sometimes called "taking the square root of each side of an equation." The abbreviation $\pm\sqrt{a}$ means $+\sqrt{a}$ or $-\sqrt{a}$. It is read "plus or minus $\sqrt{a}$."

▲ **EXAMPLE 1** The ramp to Tony Pitkin's barn rises 5 feet over a horizontal distance of 12 feet. How long is the ramp? That is, find the length of the hypotenuse of a right triangle whose legs are 5 feet and 12 feet.

Solution

1. *Understand the problem.*
   Draw and label a diagram.

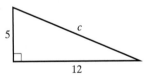

2. *Write an equation.* Write the Pythagorean Theorem. Substitute the known values into the equation.
$$c^2 = a^2 + b^2$$
$$c^2 = 5^2 + 12^2$$

3. *Solve and state the answer.*

$c^2 = 25 + 144$

$c^2 = 169$ Take the square root of each side of the equation.

$c = \pm\sqrt{169}$

$\quad = \pm 13$

The hypotenuse is 13 feet. We do not use -13 because length is not negative. Thus the length of the ramp is 13 feet.

4. *Check.* Substitute the values for a, b, and c into the Pythagorean Theorem and evaluate. Does $13^2 = 5^2 + 12^2$? Yes. The solution checks. ✓

NOTE TO STUDENT: Fully worked-out solutions to all of the Practice Problems can be found at the back of the text starting at page SP-1

▲ **Practice Problem 1** Find the length of the hypotenuse of a right triangle with legs of 9 centimeters and 12 centimeters.

Sometimes you will need to find one of the legs of a right triangle given the hypotenuse and the other leg.

▲ **EXAMPLE 2** Find the unknown leg of a right triangle that has a hypotenuse of 10 yards and one leg of $\sqrt{19}$ yards.

Solution Draw a diagram. The diagram is in the margin on the right.

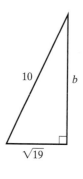

$c^2 = a^2 + b^2$ Write the Pythagorean Theorem.

$10^2 = \left(\sqrt{19}\right)^2 + b^2$ Substitute the known values into the equation.

$100 = 19 + b^2$

$81 = b^2$

$\pm 9 = b$ Take the square root of each side of the equation.

The leg is 9 yards long.

▲ **Practice Problem 2** The hypotenuse of a right triangle is $\sqrt{17}$ meters. One leg is 1 meter long. Find the length of the other leg.

Sometimes the answer to a problem will be an irrational number such as $\sqrt{33}$. It is best to leave the answer in radical form unless you are asked for an approximate answer. To find the approximate value, use a calculator or a square root table. $\sqrt{33}$ is an exact answer.

▲ **EXAMPLE 3** A 25-foot ladder is placed against a building. The foot of the ladder is 8 feet from the wall. At approximately what height does the top of the ladder touch the building? Round to the nearest tenth.

Solution
$$c^2 = a^2 + b^2$$
$$25^2 = 8^2 + b^2$$
$$625 = 64 + b^2$$
$$561 = b^2$$
$$\pm\sqrt{561} = b$$

We want only the positive value for the distance, so $b = \sqrt{561}$. Using a calculator, we have 561 $\boxed{\sqrt{}}$ 23.68543856. Rounding, we obtain $b \approx 23.7$.

If you do not have a calculator, you will need to use the square root table in Appendix A. However, this square root table goes only as far as $\sqrt{200}$! To approximate larger radicals with the square root table, you will have to write the radical in a different fashion.

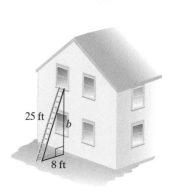

$b = \sqrt{3}\sqrt{187}$ Since $\sqrt{ab} = \sqrt{a}\sqrt{b}$ and $3 \cdot 187 = 561$.

$b \approx (1.732)(13.675)$ Replace each radical by the decimal approximation from the table.

$b \approx 23.7$ Multiply and round to the nearest tenth.

The ladder touches the building at a height of approximately 23.7 feet.

▲ **Practice Problem 3** A support line is placed 3 meters away from the base of an 8-meter pole. If the support line is attached to the top of the pole and pulled tight (assume that it is a straight line), how long is the support line from the ground to the pole? Round to the nearest tenth.

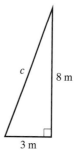

② Solving Radical Equations

A **radical equation** is an equation with a variable in one or more of the radicands. A square root radical equation can be simplified by squaring each side of the equation. The apparent solution to a radical equation *must* be checked by substitution into the *original* equation.

For some square root radical equations, you will need to *isolate the radical before* you square each side of the equation.

EXAMPLE 4 Solve and check. $1 + \sqrt{5x - 4} = 5$

Solution

$1 + \sqrt{5x - 4} = 5$	We want to isolate the radical first.
$\sqrt{5x - 4} = 4$	Subtract 1 from each side to isolate the radical.
$\left(\sqrt{5x - 4}\right)^2 = (4)^2$	Square each side.
$5x - 4 = 16$	Simplify and solve for x.
$5x = 20$	
$x = 4$	

Check.
$$1 + \sqrt{5(4) - 4} \stackrel{?}{=} 5$$
$$1 + \sqrt{20 - 4} \stackrel{?}{=} 5$$
$$1 + \sqrt{16} \stackrel{?}{=} 5$$
$$1 + 4 = 5 \checkmark$$

Thus 4 is the solution.

Practice Problem 4 Solve and check. $\sqrt{3x - 2} - 7 = 0$

NOTE TO STUDENT: Fully worked-out solutions to all of the Practice Problems can be found at the back of the text starting at page SP-1

Some problems will have an apparent solution, but that solution does not always check. An apparent solution that does not satisfy the original equation is called an **extraneous root.**

EXAMPLE 5 Solve and check. $\sqrt{x + 3} = -7$

Solution

$\sqrt{x + 3} = -7$	
$\left(\sqrt{x + 3}\right)^2 = (-7)^2$	Square each side.
$x + 3 = 49$	Simplify and solve for x.
$x = 46$	

Check.
$$\sqrt{46 + 3} \stackrel{?}{=} -7$$
$$\sqrt{49} \stackrel{?}{=} -7$$
$$7 \neq -7! \quad \text{Does not check! The apparent solution is an extraneous root.}$$

There is no solution to Example 5.

Practice Problem 5 Solve and check. $\sqrt{5x + 4} + 2 = 0$

Wait a minute! Is the original equation in Example 5 possible? No. $\sqrt{x + 3} = -7$ is an impossible statement. We defined the $\sqrt{}$ symbol to mean the positive square root of the radicand. Thus, it cannot be equal to a negative number. If we had noticed this in the first step, we could have written down "no solution" immediately.

When you square both sides of an equation you may obtain a quadratic equation. In such cases, you may obtain two apparent solutions, and you must check both of them.

EXAMPLE 6 Solve and check. $\sqrt{3x + 1} = x + 1$.

Solution

$$\sqrt{3x + 1} = x + 1$$
$$\left(\sqrt{3x + 1}\right)^2 = (x + 1)^2 \qquad \text{Square each side.}$$
$$3x + 1 = x^2 + 2x + 1 \quad \text{Simplify and set the equation equal to 0.}$$
$$0 = x^2 - x$$
$$0 = x(x - 1) \qquad \text{Factor the quadratic equation.}$$
$$x = 0 \quad \text{or} \quad x - 1 = 0 \qquad \text{Set each factor equal to 0 and solve.}$$
$$x = 1$$

Check. If $x = 0$: $\sqrt{3(0) + 1} \stackrel{?}{=} 0 + 1$ If $x = 1$: $\sqrt{3(1) + 1} \stackrel{?}{=} 1 + 1$

$\qquad\qquad\qquad \sqrt{0 + 1} \stackrel{?}{=} 0 + 1 \qquad\qquad\qquad\qquad \sqrt{3 + 1} \stackrel{?}{=} 2$

$\qquad\qquad\qquad\qquad \sqrt{1} \stackrel{?}{=} 1 \qquad\qquad\qquad\qquad\qquad\quad \sqrt{4} \stackrel{?}{=} 2$

$\qquad\qquad\qquad\qquad 1 = 1 \;\checkmark \qquad\qquad\qquad\qquad\qquad\quad 2 = 2 \;\checkmark$

Thus 0 and 1 are both solutions.

It is possible in these cases for both apparent solutions to check, for both to be extraneous, or for one to check and the other to be extraneous.

Practice Problem 6 Solve and check.

$$-2 + \sqrt{6x - 1} = 3x - 2$$

EXAMPLE 7 Solve and check. $\sqrt{2x - 1} = x - 2$

Solution

$$\sqrt{2x - 1} = x - 2$$
$$\left(\sqrt{2x - 1}\right)^2 = (x - 2)^2 \qquad \text{Square each side.}$$
$$2x - 1 = x^2 - 4x + 4$$
$$0 = x^2 - 6x + 5 \qquad \text{Simplify and set the equation to 0.}$$
$$0 = (x - 5)(x - 1) \qquad \text{Solve for } x.$$
$$x - 5 = 0 \quad \text{or} \quad x - 1 = 0$$
$$x = 5 \qquad\qquad\quad x = 1$$

Check. If $x = 5$: $\sqrt{2(5) - 1} \stackrel{?}{=} 5 - 2$ If $x = 1$: $\sqrt{2(1) - 1} \stackrel{?}{=} 1 - 2$

$$\sqrt{10 - 1} \stackrel{?}{=} 3$$
$$\sqrt{9} \stackrel{?}{=} 3$$
$$3 = 3 \checkmark$$

$$\sqrt{2 - 1} \stackrel{?}{=} -1$$
$$\sqrt{1} \stackrel{?}{=} -1$$
$$1 \neq -1$$

It does not check. In this case 1 is an extraneous root.

Thus only 5 is a solution to this equation.

NOTE TO STUDENT: *Fully worked-out solutions to all of the Practice Problems can be found at the back of the text starting at page SP-1*

Practice Problem 7 Solve and check. $2 - x + \sqrt{x + 4} = 0$

Example 8 is somewhat different in format. Take a minute to see how it compares to Examples 6 and 7.

EXAMPLE 8 Solve and check. $\sqrt{3x + 3} = \sqrt{5x - 1}$

Solution Here there are two radicals. Each radical is already isolated.

$$\sqrt{3x + 3} = \sqrt{5x - 1}$$
$$\left(\sqrt{3x + 3}\right)^2 = \left(\sqrt{5x - 1}\right)^2 \quad \text{Square each side.}$$
$$3x + 3 = 5x - 1 \quad \text{Simplify and solve for } x.$$
$$3 = 2x - 1$$
$$4 = 2x$$
$$2 = x$$

Check.
$$\sqrt{3(2) + 3} \stackrel{?}{=} \sqrt{5(2) - 1}$$
$$\sqrt{6 + 3} \stackrel{?}{=} \sqrt{10 - 1}$$
$$\sqrt{9} = \sqrt{9} \checkmark$$

Thus 2 is the solution.

Practice Problem 8 Solve and check. $\sqrt{2x + 1} = \sqrt{x - 10}$

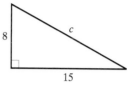
Use the Pythagorean Theorem to find the length of the third side of each right triangle. Leave any irrational answers in radical form.

▲ **1.**

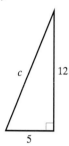

▲ **2.**

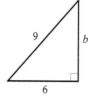

▲ **3.**

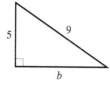

▲ **4.**

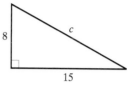

Exercises 5–14 refer to a right triangle with legs a and b and hypotenuse c. Find the exact length of the missing side.

▲ **5.** $a = 7, b = 7$ Find c.

▲ **6.** $a = 8, b = 4$ Find c.

▲ **7.** $a = \sqrt{14}, b = 7$ Find c.

▲ **8.** $a = \sqrt{21}, b = 5$ Find c.

▲ **9.** $c = 15, b = 13$ Find a.

▲ **10.** $c = 19, b = 16$ Find a.

▲ **11.** $c = \sqrt{82}, a = 5$ Find b.

▲ **12.** $a = \sqrt{5}, c = 7$ Find b.

▲ **13.** $c = 13.8, b = 9.42$
 Find a to the nearest hundredth.

▲ **14.** $a = 4.5, b = 6.34$
 Find c to the nearest hundredth.

Applications *Draw a diagram and use the Pythagorean Theorem to solve. Round to the nearest tenth.*

▲ **15.** *Tenting* A guy rope is attached to the top of a tent pole. The tent pole is 8 feet tall. If the guy rope is pegged into the ground 5 feet from the tent, how long is the guy rope?

▲ **16.** *Baseball* A baseball diamond is a square. Each side of the square is 90 feet long. How far is it from home plate to second base? *Hint:* Draw the diagonal.

▲ **17.** *Flying a Kite* A kite is flying on a string that is 100 feet long and is fastened to the ground at the other end. Assume that the string is a straight line. How high is the kite if it is flying above a point that is 20 feet away from where it is fastened to the ground?

▲ **18.** *Boat Mooring* A boat must be moored 20 feet away from a dock. The boat will be 10 feet below the level of the dock at low tide. What is the minimum length of rope needed to go from the boat to the dock at low tide?

Radio Tower Construction *To estimate the cost of erecting a radio tower, we need to know how many feet of support cable are needed.*

▲ **19.** Ground anchor #2 is exactly 54 feet from the base of the tower (dashed line in diagram). The bottom ground support cable from level 6 to ground anchor #2 is attached to the tower at a height of 130 feet. The top ground support cable from level 6 to ground anchor #2 is attached to the tower at a height of 135 feet. How long is each support cable to level 6? (Round to the nearest hundredth of a foot.)

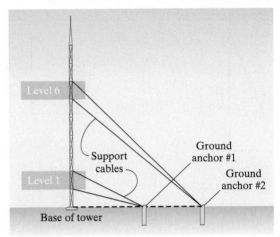

▲ **20.** Ground anchor #1 is exactly 28 feet from the base of the tower (dashed line in diagram). The bottom ground support cable from level 1 to ground anchor #1 is attached to the tower at a height of 42 feet. The top ground support cable from level 1 to ground anchor #1 is attached to the tower at a height of 47 feet. How long is each support cable to level 1? (Round to the nearest hundredth of a foot.)

Solve for the variable. Check your solutions.

21. $\sqrt{x + 2} = 3$

22. $\sqrt{x - 8} = 4$

23. $\sqrt{2x + 7} = 5$

24. $\sqrt{2x - 5} = 7$

25. $\sqrt{2x + 2} = \sqrt{3x - 5}$

26. $\sqrt{5x - 5} = \sqrt{4x + 1}$

27. $\sqrt{2x} - 5 = 4$

28. $\sqrt{13x} - 2 = 5$

29. $\sqrt{3x + 10} = x$

30. $\sqrt{6x - 5} = x$

31. $\sqrt{5y + 1} = y + 1$

32. $\sqrt{2y + 9} = y + 3$

33. $\sqrt{x + 3} = 3x - 1$

34. $\sqrt{2x + 3} = 2x - 9$

35. $\sqrt{3y + 1} - y = 1$

36. $\sqrt{3y - 8} + 2 = y$

37. $\sqrt{6y + 1} - 3y = y$

38. $\sqrt{2x + 3} - 2 = x$

To Think About *Solve for x.*

39. $\sqrt{12x + 1} = 2\sqrt{6x} - 1$

40. $\sqrt{x - 2} + 3 = \sqrt{4x + 1}$

Cumulative Review

41. **[7.5.1]** Solve. $\dfrac{5x}{x - 4} = 5 + \dfrac{4x}{x - 4}$

42. **[5.5.2]** Multiply mentally. $(7x - 2)^2$

43. **[3.6.4]** $f(x) = 2x^2 - 3x + 6$
Find $f(-2)$.

44. **[3.6.3]** Is $x^2 + y^2 = 9$ a relation or a function?

Quick Quiz 8.6

1. In a right triangle with legs a and b and a hypotenuse of c, find the exact length of hypotenuse c if $a = \sqrt{3}$ and $b = 5$.

Solve. Verify your solutions.

2. $8 - \sqrt{5x - 6} = 0$

3. $\sqrt{6x + 1} = x - 1$

4. **Concept Check** A student attempted to solve the equation $2 + \sqrt{1 - 8x} = x$ and found values of $x = -3$ and $x = -1$. Explain how you would check these values to see if they are solutions to the original equation.

1 Solving Problems Involving Direct Variation

Often in daily life there is a relationship between two measurable quantities. For example, we turn up the thermostat and the heating bill increases. We say that the heating bill varies directly as the temperature on the thermostat. If y **varies directly as** x, then $y = kx$, where k is a constant. The constant is often called the **constant of variation.** Consider the following example.

EXAMPLE 1 Cliff works part-time in a local supermarket while going to college. His salary varies directly as the number of hours worked. Last week he earned $33.60 for working 7 hours. This week he earned $52.80. How many hours did he work?

Solution Let S = his salary,

h = the number of hours he worked, and

k = the constant of variation.

Since his salary varies directly as the number of hours he worked, we write

$$S = k \cdot h.$$

We can find the constant k by substituting the known values of $S = 33.60$ and $h = 7$.

$$33.60 = k \cdot 7 = 7k$$

$$\frac{33.60}{7} = \frac{7k}{7} \qquad \text{Solve for } k.$$

$$4.80 = k \qquad \text{The constant of variation is 4.80.}$$

$$S = 4.80h \qquad \text{Replace } k \text{ in the variation equation by 4.80.}$$

How many hours did he work to earn $52.80?

$$S = 4.80h \qquad \text{The direct variation equation with } k = 4.80.$$

$$52.80 = 4.80h \qquad \text{Substitute 52.80 for } S \text{ and solve for } h.$$

$$\frac{52.80}{4.80} = \frac{4.80h}{4.80}$$

$$11 = h$$

Cliff worked 11 hours this week.

Practice Problem 1 The *change* in temperature measured on the Celsius scale varies directly as the change measured on the Fahrenheit scale. The change in temperature from freezing water to boiling water is 100° on the Celsius scale and 180° on the Fahrenheit scale. If the Fahrenheit temperature drops 20°, what will be the change in temperature on the Celsius scale?

NOTE TO STUDENT: Fully worked-out solutions to all of the Practice Problems can be found at the back of the text starting at page SP-1

SOLVING A DIRECT VARIATION PROBLEM

1. Write the direct variation equation.
2. Solve for the constant k by substituting in known values.
3. Replace k in the direct variation equation by the value obtained in step 2.
4. Solve for the desired value.

EXAMPLE 2   If y varies directly as x, and $y = 20$ when $x = 3$, find the value of y when $x = 21$.

Solution

Step 1 $y = kx$

Step 2 To find k, we substitute $y = 20$ and $x = 3$.

$$20 = k(3)$$
$$20 = 3k$$
$$\frac{20}{3} = k$$

Step 3 We now write the variation equation with k replaced by $\frac{20}{3}$.

$$y = \frac{20}{3}x$$

Step 4 We replace x by 21 and find y.

$$y = \left(\frac{20}{\cancel{3}}\right)(\cancel{21}^{\,7}) = (20)(7) = 140$$

Thus $y = 140$ when $x = 21$.

Practice Problem 2   If y varies directly as x, and $y = 18$ when $x = 5$, find the value of y when $x = \frac{20}{23}$.

A number of real-life situations can be described by direct variation equations. The following table shows some of the more common forms of the direct variation equation. In each case, $k =$ the constant of variation.

SAMPLE DIRECT VARIATION SITUATIONS

Verbal Description	*Variation Equation*
y varies directly as x	$y = kx$
b varies directly as the square of c	$b = kc^2$
l varies directly as the cube of m	$l = km^3$
V varies directly as the square root of h	$V = k\sqrt{h}$

EXAMPLE 3   In a certain class of racing cars, the maximum speed varies directly as the square root of the horsepower of the engine. If a car with 225 horsepower can achieve a maximum speed of 120 mph, what speed could it achieve with 256 horsepower?

Solution   Let $V =$ the maximum speed,
$h =$ the horsepower of the engine, and
$k =$ the constant of variation.

Step 1 Since the maximum speed (V) varies directly as the square root of the horsepower of the engine,

$$V = k\sqrt{h}.$$

Step 2 $120 = k\sqrt{225}$ Substitute known values of V and h.
$120 = k \cdot 15$ Solve for k.
$\quad\ 8 = k$

Step 3 Now we can write the direct variation equation with the known value for k.

$$V = 8\sqrt{h}$$

NOTE TO STUDENT: Fully worked-out solutions to all of the Practice Problems can be found at the back of the text starting at page SP-1

Step 4 $V = 8\sqrt{256}$ Substitute the value of $h = 256$.
$\quad\quad\quad = (8)(16)$ Solve for V.
$\quad\quad\quad = 128$

Thus a car with 256 horsepower could achieve a maximum speed of 128 mph.

Practice Problem 3   A certain type of car has a stopping distance that varies directly as the square of its speed. If this car is traveling at a speed of 20 mph on an ice-covered road, it can stop in 60 feet. If the car is traveling at 40 mph on an icy road, what will its stopping distance be?

 Solving Problems Involving Inverse Variation

If one variable is a constant multiple of the reciprocal of another, the two variables are said to **vary inversely.** If y varies inversely as x, we express this by the equation $y = \dfrac{k}{x}$, where k is the constant of variation. Inverse variation problems can be solved by a four-step procedure similar to that used for direct variation problems.

SOLVING AN INVERSE VARIATION PROBLEM

1. Write the inverse variation equation.
2. Solve for the constant k by substituting in known values.
3. Replace k in the inverse variation equation by the value obtained in step 2.
4. Solve for the desired value.

EXAMPLE 4   If y varies inversely as x, and $y = 12$ when $x = 7$, find the value of y when $x = \dfrac{2}{3}$.

Solution

Step 1 $y = \dfrac{k}{x}$

Step 2 $12 = \dfrac{k}{7}$ Substitute known values of x and y to find k.
$\quad\quad\quad 84 = k$

Step 3 $y = \dfrac{84}{x}$

Step 4   To find y when $x = \dfrac{2}{3}$, we substitute.

$$y = \frac{84}{\dfrac{2}{3}}$$
$$= \frac{84}{1} \div \frac{2}{3}$$
$$= \frac{\overset{42}{\cancel{84}}}{1} \cdot \frac{3}{\cancel{2}}$$
$$= 42 \cdot 3 = 126$$

Thus $y = 126$ when $x = \dfrac{2}{3}$.

Practice Problem 4   If y varies inversely as x, and $y = 8$ when $x = 15$, find the value of y when $x = \dfrac{3}{5}$.

EXAMPLE 5 A car manufacturer is thinking of reducing the size of the wheel used in a subcompact car. The number of times a car wheel must turn to cover a given distance varies inversely as the radius of the wheel. (Notice that this says that the smaller the wheel, the more times it must turn to cover a given distance.) A wheel with a radius of 0.35 meter must turn 400 times to cover a specified distance on a test track. How many times would it have to turn if the radius were reduced to 0.30 meter (see the sketch)?

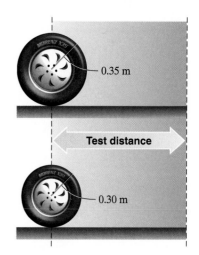

Solution   Let n = the number of times the car wheel turns,

r = the radius of the wheel, and

k = the constant of variation.

Step 1   Since the number of turns varies inversely as the radius, we can write the following.

$$n = \frac{k}{r} \quad \text{Write the variation equation.}$$

Step 2 $400 = \dfrac{k}{0.35}$ Substitute known values of n and r to find k.

$140 = k$

Step 3 $n = \dfrac{140}{r}$ Use the variation equation where k is known.

How many times must the wheel turn if the radius is 0.30 meter?

Step 4 $n = \dfrac{140}{0.30}$ Substitute 0.30 for r.

$n = 466\dfrac{2}{3}$

The wheel would have to turn $466\dfrac{2}{3}$ times to cover the same distance if the radius were only 0.30 meter.

Practice Problem 5   Over the last three years, the market research division of a calculator company found that the volume of sales of scientific calculators varies inversely as the price of the calculator. One year 120,000 calculators were sold at $30 each. How many calculators were sold the next year when the price was $24 for each calculator?

The following table contains various forms of the inverse variation equation. In each case, k = the constant of variation.

SAMPLE INVERSE VARIATION SITUATIONS

Verbal Description	*Variation Equation*
y varies inversely as x	$y = \dfrac{k}{x}$
b varies inversely as the square of c	$b = \dfrac{k}{c^2}$
l varies inversely as the cube of m	$l = \dfrac{k}{m^3}$
d varies inversely as the square root of t	$d = \dfrac{k}{\sqrt{t}}$

EXAMPLE 6 The illumination of a light source varies inversely as the square of the distance from the source. The illumination measures 25 candlepower when a certain light is 4 meters away. Find the illumination when the light is 8 meters away (see figure).

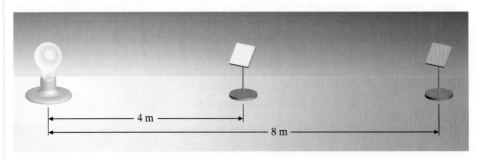

Solution Let I = the measurement of illumination,

d = the distance from the light source, and

k = the constant of variation.

Step 1 Since the illumination varies inversely as the square of the distance,

$$I = \frac{k}{d^2}.$$

Step 2 We evaluate the constant by substituting the given values.

$$25 = \frac{k}{4^2} \quad \text{Substitute } I = 25 \text{ and } d = 4.$$

$$25 = \frac{k}{16} \quad \text{Simplify and solve for } k.$$

$$400 = k$$

Step 3 We may now write the variation equation with the constant evaluated.

$$I = \frac{400}{d^2}$$

Step 4 $I = \dfrac{400}{8^2}$ Substitute a distance of 8 meters.

$= \dfrac{400}{64}$ Square 8.

$= \dfrac{25}{4} = 6.25$

The illumination is 6.25 candlepower when the light source is 8 meters away.

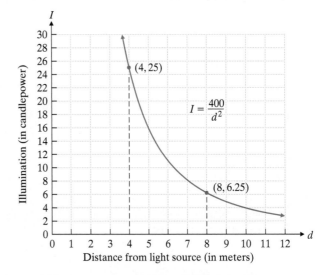

Distance from light source (in meters)

Practice Problem 6 If the amount of power in an electrical circuit is held constant, the resistance in the circuit varies inversely as the square of the amount of current. If the amount of current is 0.01 ampere, the resistance is 800 ohms. What is the resistance if the amount of current is 0.02 ampere?

NOTE TO STUDENT: Fully worked-out solutions to all of the Practice Problems can be found at the back of the text starting at page SP-1

 Identifying the Graphs of Variation Equations

The graphs of direct variation equations are shown in the following chart. Notice that as x increases, y also increases.

Variation Statement	Equation	Graph
1. y varies directly as x	$y = kx$	
2. y varies directly as x^2	$y = kx^2$	
3. y varies directly as x^3	$y = kx^3$	
4. y varies directly as the square root of x	$y = k\sqrt{x}$	

The graphs of inverse variation equations are shown in the following chart. Notice that as x increases, y decreases.

Variation Statement	Equation	Graph
5. y varies inversely as x	$y = \dfrac{k}{x}$	
6. y varies inversely as x^2	$y = \dfrac{k}{x^2}$	

PRACTICE   WATCH   DOWNLOAD   READ   REVIEW

1. If y varies directly as x, and $y = 9$ when $x = 2$, find y when $x = 16$.

2. If y varies directly as x, and $y = 5$ when $x = 3$, find y when $x = 18$.

3. If y varies directly as the cube of x, and $y = 12$ when $x = 2$, find y when $x = 7$.

4. If y varies directly as the square root of x, and $y = 7$ when $x = 4$, find y when $x = 25$.

5. If y varies directly as the square of x, and $y = 900$ when $x = 25$, find y when $x = 30$.

6. If y varies directly as the square of x, and $y = 8000$ when $x = 16$, find y when $x = 5$.

Applications

7. *Food Nutrition Facts* There are 100 calories in 20 grams of potato chips. The number of calories varies directly with the weight of the potato chips. How many calories will 45 grams of potato chips have?

8. *Stretching a Spring* Hooke's Law states that the force needed to stretch a spring varies directly with the amount the spring is stretched. If a force of 50 pounds is needed to stretch a spring 5 inches, how far will a force of 120 pounds stretch the same spring?

▲ **9.** *Storage Cube* The time it takes to fill a storage cube with sand varies directly as the cube of the side of the box. A storage cube that is 2.0 meters on each side (inside dimensions) can be filled in 7 minutes by a sand loader. How long will it take to fill a storage cube that is 4.0 meters on each side (inside dimensions)?

10. *Weight on the Moon* The weight of an object on the surface of the moon varies directly as the weight of the object on the surface of Earth. An astronaut with his protective suit weighs 175 lb on Earth's surface; on the moon and wearing the same suit, he will weigh 29 lb. If his moon rover vehicle weighs 2000 pounds on Earth, how much will it weigh on the moon?

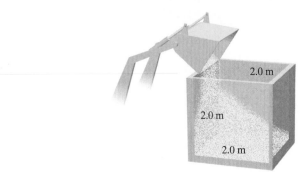

2.0 m
2.0 m
2.0 m

11. If y varies inversely as x, and $y = 18$ when $x = 5$, find y when $x = 8$.

12. If y varies inversely as x, and $y = 21$ when $x = 4$, find y when $x = 18$.

13. If y varies inversely as x, and $y = \frac{1}{3}$ when $x = 12$, find y when $x = 4$.

14. If y varies inversely as x, and $y = \frac{1}{10}$ when $x = 30$, find y when $x = 1$.

15. If y varies inversely as the square of x, and $y = 30$ when $x = 2$, find y when $x = 9$.

16. If y varies inversely as the cube of x, and $y = \frac{1}{9}$ when $x = 3$, find y when $x = \frac{1}{2}$.

Applications

17. *Ice Cube Melting* The amount of time in minutes that it takes for an ice cube to melt varies inversely as the temperature of the water that the ice cube is placed in. When an ice cube is placed in 60°F water, it takes 2.3 minutes to melt. How long would it take for an ice cube of this size to melt if it were placed in 40°F water?

18. *Textile Mills* The textile mills in Lowell, Massachusetts, were operated with water power. The Merrimack River and its canals powered huge turbines that generated the electricity for the mills. In each mill room, a single drive-shaft supplied power to all machines by means of belts that ran to pulleys on the machines. The driveshaft turned at a constant speed, but the pulleys were different diameters, which provided for the different speeds of each machine. The speed of each machine varied inversely as the diameter of its pulley. If a machine with a 60-centimeter pulley turned at 1000 rpm, what was the speed of a machine with an 80-centimeter pulley?

19. *Variation in Weight on Earth* The weight of an object near Earth's surface varies inversely as the square of its distance from the center of Earth. An object weighs 1000 pounds on Earth's surface. This is approximately 4000 miles from the center of Earth. How much will the object weigh 6000 miles from the center of Earth?

20. *Radio Waves* The wavelength, w, of a radio wave varies inversely as the frequency, f. A radio wave with a frequency of 1600 kilohertz has a length of 225 meters. What is the length of a radio wave with a frequency of 900 kilohertz?

21. *Distance Vision in a Boat* A person on the mast of a sailing ship 81 meters above the water can see approximately 12.3 kilometers out to sea. In general, the distance, d, you can see varies directly as the square root of your height, h, above the water. In this case, the equation is $d = \frac{15}{11}\sqrt{h}$, where d is measured in kilometers and h is measured in meters.

(a) Find the distances you can see when $h = 121, 36$, and 4. Round to the nearest tenth.

(b) Graph the equation using values of h of 121, 81, 36, and 4.

22. *Computer Files* Suppose a computer has a hard drive that can hold 2000 files if each file is 100,000 bytes long. The number of files, f, the hard drive can hold varies inversely as the size, s, of a file. In this case, the equation is $f = \dfrac{200,000,000}{s}$, where f is the number of files and s is the size of a file in bytes.

(a) Find the number of files that can be held if $s = 400,000$, $s = 200,000$, and $s = 50,000$.

(b) Graph the equation using the values $s = 400,000$, $s = 200,000$, $s = 100,000$, and $s = 50,000$.

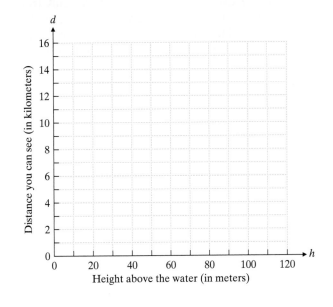

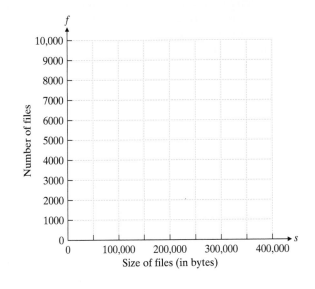

To Think About *Write an equation to describe the variation and solve.*

23. **Planetary Motion** Kepler's Third Law of Planetary Motion states that the square of the time it takes a planet to complete one revolution of the sun varies directly with the cube of the distance between the planet and the sun. Astronomers use 1 astronomical unit (au) as a measure of the distance from Earth to the sun.

    (a) Using this astronomical unit, find the constant k for the direct variation equation that represents Kepler's Third Law.

    (b) Pluto is 39.5 times as far from the sun as Earth is. Find the approximate length of the Plutonian year in terms of Earth years.

24. **Sailboat Design** An important element of sailboat design is the calculation of the force of wind on the sail. The force varies directly with the square of the wind speed and directly with the area of the sail. For a certain class of racing boat, the force on 100 square feet of sail is 280 pounds when the wind speed is 6 mph. What would be the force on the sail if it were rigged for a storm (the amount of sail exposed to the wind is reduced by lashing part of the sail to the boom) so that only 60 square feet of sail were exposed and the wind speed were 40 mph?

Cumulative Review

25. **[7.5.1]** Solve for a. $\dfrac{80{,}000}{320} = \dfrac{120{,}000}{320 + a}$

26. **[3.2.1]** Graph the equation $S = kh$, where $k = 4.80$. Use h for the horizontal axis and S for the vertical. Assume that h and S are nonnegative.

Quick Quiz 8.7

1. The intensity of illumination from a light source varies inversely with the square of the distance from the light source. When a photoelectric cell is placed 2 inches from a light source the intensity is 45 lumens. Then it is moved so that it receives only 5 lumens. How far is it from the light source?

2. The number of calories in a breakfast bar varies directly with the weight of the bar. There are 150 calories in a bar weighing 40 grams. How many calories are in a bar weighing 75 grams?

3. The areas of similar triangle-shaped supports vary directly as the squares of their perimeters. A support that has a perimeter of 8 centimeters has an area of 6.4 square centimeters. A similar triangle-shaped support has a perimeter of 20 centimeters. What is its area?

4. **Concept Check** The distance a body falls from rest varies directly as the square of the time it falls. If an object falls 64 feet in two seconds, explain how you would write an equation to describe this relationship. How would you find the constant k in this equation?

Putting Your Skills to Work: Use Math to Save Money

FOOD AND RICE PRICES

If you've noticed an increase in food prices during recent months, you are not alone. Food prices are rising throughout the world. Reasons for this include bad weather that hampered crops, an increased demand for food, particularly in countries throughout Asia, and the global increase in fuel prices.

During the spring of 2008, food prices in the United States for staples such as bread, milk, eggs, and flour rose sharply. Milk prices jumped 26% from the previous year, while eggs jumped 50% from $1.45 per dozen to $2.18 per dozen (U.S. Labor Department). Overall food prices in the Dallas-Fort Worth area rose 7.1% between June of 2007 and June of 2008.

Rice is another food item that has seen a sharp increase in price. During the spring of 2008, two U.S. warehouse retail chains, Sam's Club and Costco, moved to limit how much rice customers could buy because of concerns about global supply and demand. The cost of rice in some locations increased by 50% from approximately $10 to $15 for a 25-pound bag.

Food Prices and You

Has the increase in food prices—or the increase in the cost of rice—had an impact on you and your family? How much do you spend on food each week? How about each month? How does that compare with what you were spending on food each month a year ago? Consider the story of Lucy.

Lucy and her family enjoy rice as part of their dinner four times a week. Lucy has a family of four that consumes approximately $\frac{2}{3}$ cup of rice with each meal.

1. How many cups of rice does Lucy prepare a week?

2. If a cup of raw rice weighs approximately 21 ounces, how many ounces of raw rice does Lucy's family consume each week?

3. There are 16 ounces in a pound. Compute the number of pounds of raw rice Lucy's family eats each week.

4. Last year she bought a 25-pound bag of rice for $9.25. She recently purchased a 25-pound bag of rice for $15.73.

   (a) What is the price increase per pound?

   (b) What is the percent of increase?

5. Approximately how much in today's prices is the cost per week of the rice consumed by Lucy's family?

6. How much rice do you eat per week? Has the rise in rice prices influenced your purchase of rice? Why or why not?

7. In what other food prices have you noticed an increase? What are some things you can do as a consumer to keep your costs low?

Topic	Procedure	Examples
Square roots, p. 468.	1. The square roots of perfect squares up to $\sqrt{225} = 15$ (or so) should be memorized. Approximate values of others can be found in tables or by calculator. 2. Negative numbers do not have real number square roots.	$\sqrt{196} = 14$ $\sqrt{169} = 13$ $\sqrt{200} \approx 14.142$ $\sqrt{13} \approx 3.606$ $\sqrt{-5}$ This is not a real number.
Simplifying radicals, p. 473.	Assuming that a and b are nonnegative, $$\sqrt{ab} = \sqrt{a}\sqrt{b}$$ Square root radicals are simplified by taking the square roots of all factors that are perfect squares and leaving all other factors under the radical sign.	$\sqrt{48} = \sqrt{16}\sqrt{3} = 4\sqrt{3}$ $\sqrt{12x^2y^3z} = \sqrt{2^2x^2y^2}\sqrt{3yz}$ $= 2xy\sqrt{3yz}$ $\sqrt{a^{11}} = \sqrt{a^{10} \cdot a} = \sqrt{a^{10}}\sqrt{a} = a^5\sqrt{a}$
Adding and subtracting radicals, p. 478.	Simplify all radicals and then add or subtract like radicals.	$\sqrt{12} + \sqrt{18} + \sqrt{27} + \sqrt{50}$ $= \sqrt{4}\sqrt{3} + \sqrt{9}\sqrt{2} + \sqrt{9}\sqrt{3} + \sqrt{25}\sqrt{2}$ $= 2\sqrt{3} + 3\sqrt{2} + 3\sqrt{3} + 5\sqrt{2}$ $= 5\sqrt{3} + 8\sqrt{2}$
Multiplying radicals, p. 482.	Square root radicals are multiplied like polynomials. Assuming that a and b are nonnegative, apply the rule $\sqrt{a}\sqrt{b} = \sqrt{ab}$ and simplify all results.	$(2\sqrt{6} + \sqrt{3})(\sqrt{6} - 3\sqrt{3})$ $= 2\sqrt{36} - 6\sqrt{18} + \sqrt{18} - 3\sqrt{9}$ $= 2(6) - 5\sqrt{18} - 3(3)$ $= 12 - 5\sqrt{9}\sqrt{2} - 9$ $= 12 - 5(3)\sqrt{2} - 9$ $= 3 - 15\sqrt{2}$
Dividing radicals, p. 488.	For positive numbers a and b, $$\frac{\sqrt{a}}{\sqrt{b}} = \sqrt{\frac{a}{b}} \quad \text{and} \quad \sqrt{\frac{a}{b}} = \frac{\sqrt{a}}{\sqrt{b}}.$$	$\frac{\sqrt{80}}{\sqrt{5}} = \sqrt{\frac{80}{5}} = \sqrt{16} = 4$ $\sqrt{\frac{144}{25}} = \frac{\sqrt{144}}{\sqrt{25}} = \frac{12}{5}$
Conjugates, p. 490.	Note that the product of the sum and difference of two square root radicals is equal to the difference of their squares and thus contains no radicals. This sum and difference are called conjugates.	$(3\sqrt{6} + \sqrt{10})$ and $(3\sqrt{6} - \sqrt{10})$ are conjugates. $(3\sqrt{6} + \sqrt{10})(3\sqrt{6} - \sqrt{10})$ $= 9\sqrt{6}^2 - \sqrt{10}^2$ $= 9 \cdot 6 - 10 = 54 - 10 = 44$
Rationalizing denominators, pp. 489–490.	To simplify a fraction with a square root radical in the denominator, the radical (in the denominator) must be removed. 1. Multiply the denominator by itself if it is a monomial. Multiply the numerator by the same quantity. 2. Multiply the denominator by its conjugate if it is a binomial. Multiply the numerator by the same quantity.	$\frac{\sqrt{10}}{\sqrt{6}} = \frac{\sqrt{10} \cdot \sqrt{6}}{\sqrt{6} \cdot \sqrt{6}} = \frac{2\sqrt{15}}{6} = \frac{\sqrt{15}}{3}$ $\frac{\sqrt{3}}{\sqrt{3} - \sqrt{2}} = \frac{\sqrt{3}(\sqrt{3} + \sqrt{2})}{(\sqrt{3} - \sqrt{2})(\sqrt{3} + \sqrt{2})}$ $= \frac{\sqrt{9} + \sqrt{6}}{3 - 2} = 3 + \sqrt{6}$
Radicals with fractions, p. 489.	Note that $\sqrt{\frac{5}{3}}$ is the same as $\frac{\sqrt{5}}{\sqrt{3}}$ and so must be rationalized.	$\sqrt{\frac{5}{3}} = \frac{\sqrt{5}}{\sqrt{3}} \cdot \frac{\sqrt{3}}{\sqrt{3}} = \frac{\sqrt{15}}{3}$
Pythagorean Theorem, p. 494.	In any right triangle with hypotenuse of length c and legs of lengths a and b, $$c^2 = a^2 + b^2.$$	Find c to the nearest tenth if $a = 7$ and $b = 9$. $c^2 = 7^2 + 9^2 = 49 + 81 = 130$ $c = \sqrt{130} \approx 11.4$

Topic	Procedure	Examples
Radical equations, *p. 496.*	To solve an equation containing a square root radical: **1.** Isolate the radical by itself on one side of the equation. **2.** Square both sides. **3.** Solve the resulting equation. **4.** Check all apparent solutions. Extraneous roots may have been introduced in step 2.	Solve. $\sqrt{3x + 2} - 4 = 8$ $$\sqrt{3x + 2} = 12$$ $$(\sqrt{3x + 2})^2 = 12^2$$ $$3x + 2 = 144$$ $$3x = 142$$ $$x = \frac{142}{3}$$ *Check:* $\sqrt{3\left(\dfrac{142}{3}\right) + 2} \stackrel{?}{=} 12$ $$\sqrt{144} = 12 \ \checkmark$$ So $x = \dfrac{142}{3}$.
Variation: direct, *p. 502.*	If y varies directly with x, there is a constant of variation k such that $y = kx$. Once k is determined, the value of y or x can easily be computed.	y varies directly with x. When $x = 2$, $y = 7$. $$y = kx$$ $$7 = k(2) \quad \text{Substituting.}$$ $$k = \frac{7}{2} \quad \text{Solving.}$$ $$y = \frac{7}{2}x$$ What is y when $x = 8$? $$y = \frac{7}{2}x = \frac{7}{2} \cdot 8 = 28$$
Variation: inverse, *p. 504.*	If y varies inversely with x, there is a constant of variation k such that $$y = \frac{k}{x}.$$	y varies inversely with x. When x is 5, y is 12. What is y when x is 15? $$y = \frac{k}{x}$$ $$12 = \frac{k}{5} \quad \text{Substituting.}$$ $$k = 60 \quad \text{Solving.}$$ $$y = \frac{60}{x} \quad \text{Substituting.}$$ $$x = 15 \quad y = \frac{60}{15} = 4$$

Chapter 8 Review Problems

Section 8.1

Simplify or evaluate, if possible.

1. $\sqrt{36}$

2. $\sqrt{25}$

3. $\sqrt{169}$

4. $\sqrt{196}$

5. $-\sqrt{81}$

6. $\sqrt{64}$

7. $\sqrt{49}$

8. $-\sqrt{121}$

9. $-\sqrt{144}$

10. $\sqrt{289}$

11. $\sqrt{0.04}$

12. $\sqrt{0.49}$

13. $\sqrt{\dfrac{1}{100}}$

14. $\sqrt{\dfrac{64}{81}}$

Approximate, using the square root table in Appendix A or a calculator. Round to the nearest thousandth, if necessary.

15. $\sqrt{105}$ **16.** $\sqrt{198}$ **17.** $\sqrt{77}$ **18.** $\sqrt{88}$

Section 8.2

Simplify.

19. $\sqrt{27}$ **20.** $\sqrt{52}$ **21.** $\sqrt{28}$

22. $\sqrt{125}$ **23.** $\sqrt{40}$ **24.** $\sqrt{80}$

25. $\sqrt{x^8}$ **26.** $\sqrt{y^{10}}$ **27.** $\sqrt{x^5 y^6}$

28. $\sqrt{a^3 b^4}$ **29.** $\sqrt{16x^3 y^5}$ **30.** $\sqrt{98x^4 y^6}$

31. $\sqrt{12x^5}$ **32.** $\sqrt{72x^9}$ **33.** $\sqrt{48x^8}$

34. $\sqrt{125x^{12}}$ **35.** $\sqrt{120a^3 b^4 c^5}$ **36.** $\sqrt{121a^6 b^4 c}$

37. $\sqrt{56x^7 y^9}$ **38.** $\sqrt{99x^{13} y^7}$

Section 8.3

Simplify.

39. $\sqrt{5} - \sqrt{20} + \sqrt{80}$ **40.** $5\sqrt{6} - \sqrt{24} + 2\sqrt{54}$

41. $x\sqrt{3} + 3x\sqrt{3} + \sqrt{27x^2}$ **42.** $a\sqrt{2} + \sqrt{12a^2} + a\sqrt{98}$

43. $5\sqrt{5} - 6\sqrt{20} + 2\sqrt{10}$ **44.** $2\sqrt{40} - 2\sqrt{90} + 3\sqrt{28}$

Section 8.4

Simplify.

45. $\left(2\sqrt{x}\right)\left(3\sqrt{x^3}\right)$ **46.** $\left(-5\sqrt{a}\right)\left(2\sqrt{ab}\right)$

47. $\left(\sqrt{6a^2}\right)\left(\sqrt{6b^5}\right)$ **48.** $\left(-4x\sqrt{x}\right)\left(2x\sqrt{x}\right)$

49. $\sqrt{6}\left(\sqrt{24} - 5\sqrt{6}\right)$ **50.** $\sqrt{3}\left(\sqrt{48} - 8\sqrt{3}\right)$

51. $\sqrt{2}\left(\sqrt{5} - \sqrt{3} - 2\sqrt{2}\right)$ **52.** $\sqrt{5}\left(\sqrt{6} - 2\sqrt{5} + \sqrt{10}\right)$

53. $\left(\sqrt{11} + 2\right)\left(2\sqrt{11} - 1\right)$

54. $\left(\sqrt{10} + 3\right)\left(3\sqrt{10} - 1\right)$

55. $\left(2 + 3\sqrt{6}\right)\left(4 - 2\sqrt{3}\right)$

56. $\left(5 - \sqrt{2}\right)\left(3 - \sqrt{12}\right)$

57. $\left(2\sqrt{3} + 3\sqrt{6}\right)^2$

58. $\left(5\sqrt{2} - 2\sqrt{6}\right)^2$

Section 8.5

Rationalize the denominator.

59. $\dfrac{1}{\sqrt{3x}}$

60. $\dfrac{2y}{\sqrt{5}}$

61. $\dfrac{x^2 y}{\sqrt{8}}$

62. $\dfrac{3ab}{\sqrt{2b}}$

63. $\sqrt{\dfrac{5}{6}}$

64. $\sqrt{\dfrac{6}{11}}$

65. $\dfrac{\sqrt{x^3}}{\sqrt{5x}}$

66. $\dfrac{\sqrt{a^5}}{\sqrt{6a}}$

67. $\dfrac{3}{\sqrt{5} + \sqrt{2}}$

68. $\dfrac{2}{\sqrt{6} - \sqrt{3}}$

69. $\dfrac{1 - \sqrt{5}}{2 + \sqrt{5}}$

70. $\dfrac{1 - \sqrt{3}}{3 + \sqrt{3}}$

Section 8.6

In exercises 71–77, use the Pythagorean Theorem to find the length of the missing side of a right triangle with legs a and b and hypotenuse c.

▲ **71.** $a = 5, b = 8$

▲ **72.** $c = \sqrt{11}, b = 3$

▲ **73.** $c = 5, a = 3.5$

▲ **74.** If $a = 2.400$ and $b = 2.000$, find c and round to the nearest thousandth.

▲ **75.** *Flagpole* A flagpole is 24 meters tall. A man stands 18 meters from the base of the pole. How far is it from the feet of the man to the top of the pole?

▲ **76.** *Distances Between Cities* Kingman is a city in Arizona that is 143 miles west of Flagstaff. Fountain Hills is 120 miles south of Flagstaff. How far is it from Kingman to Fountain Hills? Round your answer to the nearest tenth of a mile.

▲ **77.** *Softball Diamond* The four bases in a fast-pitch softball field form a square. The length of the diagonal (the distance from home plate to second base) is $\sqrt{7200}$ feet. How long is each side of the square?

Solve. Be sure to verify your answers.

78. $\sqrt{x - 4} = 5$

79. $\sqrt{5x - 1} = 8$

80. $\sqrt{1 - 5x} = \sqrt{9 - x}$

81. $\sqrt{-5 + 2x} = \sqrt{1 + x}$ **82.** $\sqrt{10x + 9} = -1 + 2x$ **83.** $\sqrt{2x - 5} = 10 - x$

84. $6 - \sqrt{5x - 1} = x + 1$ **85.** $4x + \sqrt{x + 2} = 5x - 4$

Section 8.7

86. If y varies directly with the square of x, and $y = 27$ when $x = 3$, find the value of y when $x = 4$.

87. If y varies inversely with the square root of x, and $y = 2$ when $x = 25$, find the value of y when $x = 100$.

88. If y varies inversely with the cube of x, and $y = 4$ when $x = 2$, find the value of y when $x = 4$.

89. *Illumination from a Light Source* The intensity of illumination from a light source varies inversely as the square of the distance from the light source. When an object is 4 meters from a light source, the intensity is 45 lumens. What is the intensity at a distance of 12 meters?

90. *Skid Marks of a Car* When an automobile driver slams on the brakes, the length of skid marks on the road varies directly with the square of the speed of the car. At 30 mph a certain car had skid marks 40 feet long. How long will the skid marks be if the car travels at 55 mph?

91. *Horsepower of a Boat* The horsepower that is needed to drive a racing boat through water varies directly with the cube of the speed of the boat. What will happen to the horsepower requirement if someone wants to double the maximum speed of a given boat?

Mixed Practice *Find the square root.*

92. $\sqrt{\dfrac{36}{121}}$

93. $\sqrt{0.0004}$

Simplify.

94. $\sqrt{98x^6y^3}$

Combine.

95. $3\sqrt{27} - 2\sqrt{75} + \sqrt{48}$

Multiply and simplify your answer.

96. $\left(\sqrt{3} - 2\sqrt{5}\right)\left(2\sqrt{5} + 3\sqrt{2}\right)$

97. $\left(4\sqrt{3} + 3\right)^2$

Rationalize the denominator.

98. $\dfrac{5}{\sqrt{12}}$

99. $\dfrac{\sqrt{3} + \sqrt{6}}{2\sqrt{3} + \sqrt{2}}$

Solve.

100. $\sqrt{2x - 3} = 9$

101. $\sqrt{10x + 5} = 2x + 1$

How Am I Doing? Chapter 8 Test

Remember to use your Chapter Test Prep Video CD to see the worked-out solutions to the test problems you want to review.

Evaluate.

1. $\sqrt{121}$

2. $\sqrt{\dfrac{9}{100}}$

Simplify.

3. $\sqrt{48x^2y^7}$

4. $\sqrt{100x^3yz^4}$

Combine and simplify.

5. $8\sqrt{3} + 5\sqrt{27} - 5\sqrt{48}$

6. $\sqrt{4a} + \sqrt{8a} + \sqrt{36a} + \sqrt{18a}$

Multiply.

7. $(2\sqrt{a})(3\sqrt{b})(2\sqrt{ab})$

8. $\sqrt{5}(\sqrt{10} + 2\sqrt{3} - 3\sqrt{5})$

9. $(2\sqrt{3} + 5)^2$

10. $(4\sqrt{2} - \sqrt{5})(3\sqrt{2} + \sqrt{5})$

In questions 11–14, simplify.

11. $\sqrt{\dfrac{x}{5}}$

12. $\dfrac{3}{\sqrt{12}}$

13. $\dfrac{\sqrt{3} + 4}{5 + \sqrt{3}}$

14. $\dfrac{3a}{\sqrt{5} + \sqrt{2}}$

15. Use the table of square roots or a calculator to approximate $\sqrt{156}$. Round your answer to the nearest hundredth.

Find the missing side by using the Pythagorean Theorem.

▲ **16.**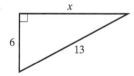

▲ **17.**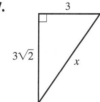

Solve. Verify your solutions.

18. $6 - \sqrt{2x + 1} = 0$

19. $x = 5 + \sqrt{x + 7}$

Solve.

20. The intensity of illumination from a light source varies inversely with the square of the distance from the light source. When a photoelectric cell is placed 8 inches from a light source, the intensity is 12 lumens. Then it is moved so that it receives only 3 lumens. How far is it from the light source?

21. Commission on the sale of office supplies varies directly with the amount of the sale. If a salesman earns $23.40 on sales of $780, what does he earn on sales of $2859?

22. The area of an equilateral triangle varies directly as the square of its perimeter. If the perimeter is 12 centimeters, the area is 6.93 square centimeters. What is the area of an equilateral triangle whose perimeter is 21 centimeters? Round to the nearest tenth.

1. _____

2. _____

3. _____

4. _____

5. _____

6. _____

7. _____

8. _____

9. _____

10. _____

11. _____

12. _____

13. _____

14. _____

15. _____

16. _____

17. _____

18. _____

19. _____

20. _____

21. _____

22. _____

Cumulative Test for Chapters 0–8

1. _____

2. _____

3. _____

4. _____

5. _____

6. _____

7. _____

8. _____

9. _____

10. _____

11. _____

12. _____

13. _____

14. _____

15. _____

16. _____

17. _____

18. _____

19. _____

20. _____

21. _____

22. _____

23. _____

24. _____

25. _____

26. _____

Approximately one-half of this test covers the content of Chapters 0–7. The remainder covers the content of Chapter 8.

In questions 1–3, simplify.

1. $\dfrac{28}{42}$

2. $5\dfrac{1}{3} + 1\dfrac{3}{4}$

3. $6\dfrac{1}{4} \div 6\dfrac{2}{3}$

4. Write 0.07% as a decimal.

Simplify.

5. $-11 - 16 + 8 + 4 - 13 + 31$

6. $(-3)^2 \cdot 4 - 8 \div 2 - (3 - 2)^3$

7. $4a^2bc^3 - 2abc^2 - 7a^2bc^3$

8. $(x - 1)(x - 4)(x + 4)$

9. $(4x - 5)^2$

10. $(3x - 11)(3x + 11)$

11. $\dfrac{4x}{x + 2} + \dfrac{4}{x + 2}$

12. $\dfrac{x^2 - 5x + 6}{x^2 - 5x - 6} \div \dfrac{x^2 - 4}{x^2 + 2x + 1}$

Simplify, if possible.

13. $\sqrt{98x^5 y^6}$

14. $\sqrt{50} - \sqrt{98} + \sqrt{162}$

15. $\left(\sqrt{6} - \sqrt{3}\right)^2$

16. $\dfrac{\sqrt{3} + \sqrt{2}}{\sqrt{3} - \sqrt{2}}$

17. $\left(3\sqrt{6} - \sqrt{2}\right)\left(\sqrt{6} + 4\sqrt{2}\right)$

18. $\dfrac{2}{2 - \sqrt{5}}$

19. $-\sqrt{25}$

20. $\sqrt{-16}$

Use the Pythagorean Theorem to find the missing side of each triangle. Be sure to simplify your answer.

21. $a = 4, b = \sqrt{7}$

▲ **22.** $a = 19, c = 21$

In questions 23 and 24, solve for the variable. Be sure to verify your answer(s).

23. $\sqrt{4x - 3} = x$

24. $\sqrt{6y + 13} - 1 = y$

25. If y varies inversely as x, and $y = 4$ when $x = 5$, find the value of y when $x = 2$.

26. The surface area of a balloon varies directly with the square of the cross-sectional radius. When the radius is 2 meters, the surface area is 1256 square meters. Find the surface area if the radius is 0.5 meter.

518

CHAPTER

9

Firefighters need to use mathematics to determine water flow in fighting large fires. Calculations are made about the angle of the hose and the pressure needed to pump the water to the necessary height. Firefighters need to pass a mathematics exam in order to be appointed to city fire departments. Much of the mathematics that is needed to solve these problems is covered in this chapter.

Quadratic Equations

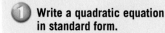 **Writing a Quadratic Equation in Standard Form**

In Section 6.7 we introduced quadratic equations. A **quadratic equation** is a polynomial equation of degree two. For example,

$$3x^2 + 5x - 7 = 0, \quad \frac{1}{2}x^2 - 5x = 0, \quad 2x^2 = 5x + 9, \quad \text{and} \quad 3x^2 = 8$$

are all quadratic equations.

> The **standard form of a quadratic equation** is $ax^2 + bx + c = 0$, where a, b, and c are real numbers, and $a \neq 0$.

You will need to be able to recognize the real numbers that represent a, b, and c in specific equations. It will be easier to do so if these equations are placed in standard form. Note that it is usually easier to work with quadratic equations if a is positive, so in cases where a is negative, we recommend that you multiply the entire equation by -1. This is a suggestion and not a requirement.

EXAMPLE 1 Place each quadratic equation in standard form and identify the real numbers a, b, and c.

(a) $5x^2 - 6x + 3 = 0$ — This equation is in standard form.

Solution $ax^2 + bx + c = 0$ — Match each term to the standard form.
$$a = 5, \quad b = -6, \quad c = 3$$

(b) $2x^2 + 5x = 4$ — The right-hand side is not zero. It is not in standard form.

Solution $2x^2 + 5x - 4 = 0$ — Add -4 to each side of the equation.
$ax^2 + bx + c = 0$ — Match each term to the standard form.
$$a = 2, \quad b = 5, \quad c = -4$$

(c) $-2x^2 + 15x + 4 = 0$ — It is easier to work with quadratic equations if the first term is not negative.

Solution $2x^2 - 15x - 4 = 0$ — Multiply each term on both sides of the equation by -1.
$ax^2 + bx + c = 0$
$$a = 2, \quad b = -15, \quad c = -4$$

(d) $7x^2 - 9x = 0$ — This equation is in standard form.

Solution $ax^2 + bx + c = 0$ — Note that the constant term is missing, so we know $c = 0$.
$$a = 7, \quad b = -9, \quad c = 0$$

Practice Problem 1 Place each quadratic equation in standard form and identify the real numbers a, b, and c. If $a < 0$, you can multiply each term of the equation by -1 to obtain an equivalent equation. This is not required, but most students find it helpful to have the first term be a positive number.

(a) $2x^2 + 12x - 9 = 0$ **(b)** $7x^2 = 6x - 8$
(c) $-x^2 - 6x + 3 = 0$ **(d)** $10x^2 - 12x = 0$

Solving Quadratic Equations of the Form $ax^2 + bx = 0$ by Factoring

Notice that the terms in a quadratic equation of the form $ax^2 + bx = 0$ both have x as a factor. To solve such equations, begin by factoring out the x and any common numerical factor. Thus you may be able to factor out an x or an expression such as $5x$. Remember that you want to remove the *greatest common factor*. So if $5x$ is a common factor of each term, be sure to factor it out. Then use the zero factor property discussed in Section 6.7. This property states that if $a \cdot b = 0$, then either $a = 0$ or $b = 0$. Once each factor is set equal to 0, solve for x.

EXAMPLE 2 Solve. $7x^2 + 9x - 2 = -8x - 2$

Solution

$$7x^2 + 9x - 2 = -8x - 2 \quad \text{The equation is not in standard form.}$$
$$7x^2 + 9x - 2 + 8x + 2 = 0 \quad \text{Add } 8x + 2 \text{ to each side.}$$
$$7x^2 + 17x = 0 \quad \text{Combine like terms.}$$
$$x(7x + 17) = 0 \quad \text{Factor.}$$
$$x = 0 \qquad 7x + 17 = 0 \quad \text{Set each factor equal to zero.}$$
$$7x = -17$$
$$x = -\frac{17}{7}$$

One of the solutions to the preceding equation is 0. This will always be true of equations of this form. One root will be zero. The other root will be a nonzero real number.

Practice Problem 2 Solve. $2x^2 - 7x - 6 = 4x - 6$

 ## Solving Quadratic Equations of the Form $ax^2 + bx + c = 0$ by Factoring

If an equation you are trying to solve contains fractions, clear the fractions by multiplying each term by the least common denominator of all the fractions in the equation. Sometimes an equation does not look like a quadratic equation. However, it takes the quadratic form once the fractions have been cleared. A note of caution: Always check a possible solution in the original equation. Any apparent solution that would make the denominator of any fraction in the original equation 0 is not a valid solution.

EXAMPLE 3 Solve and check. $8x - 6 + \dfrac{1}{x} = 0$

Solution

$$8x - 6 + \frac{1}{x} = 0 \quad \text{The equation has a fractional term.}$$

$$x(8x) - (x)(6) + x\left(\frac{1}{x}\right) = x(0) \quad \text{Multiply each term by the LCD, which is } x.$$

$$8x^2 - 6x + 1 = 0 \quad \text{Simplify. The equation is now in standard form.}$$
$$(4x - 1)(2x - 1) = 0 \quad \text{Factor.}$$
$$4x - 1 = 0 \qquad 2x - 1 = 0 \quad \text{Set each factor equal to 0.}$$
$$4x = 1 \qquad 2x = 1 \quad \text{Solve each equation for } x.$$
$$x = \frac{1}{4} \qquad x = \frac{1}{2}$$

Check. Checking fractional roots is more difficult, but you should be able to do it if you work carefully.

$$\text{If } x = \frac{1}{4}: \quad 8\left(\frac{1}{4}\right) - 6 + \frac{1}{\frac{1}{4}} = 2 - 6 + 4 = 0 \quad \left(\text{since } 1 \div \frac{1}{4} = 1 \cdot \frac{4}{1} = 4\right)$$

$$0 = 0 \ \checkmark$$

$$\text{If } x = \frac{1}{2}: \quad 8\left(\frac{1}{2}\right) - 6 + \frac{1}{\frac{1}{2}} = 4 - 6 + 2 = 0 \quad \left(\text{since } 1 \div \frac{1}{2} = 1 \cdot \frac{2}{1} = 2\right)$$

$$0 = 0 \ \checkmark$$

Both roots check, so $\frac{1}{2}$ and $\frac{1}{4}$ are the two roots that satisfy the original equation

$$8x - 6 + \frac{1}{x} = 0.$$

It is also correct to write the answers in decimal form as 0.5 and 0.25. This will speed up the checking process, especially if you use a scientific calculator when performing the check.

NOTE TO STUDENT: *Fully worked-out solutions to all of the Practice Problems can be found at the back of the text starting at page SP-1*

Practice Problem 3 Solve and check.

$$3x + 10 - \frac{8}{x} = 0$$

If a quadratic equation is given to us in standard form, we examine it to see if it can be factored. If it is possible to factor the quadratic equation, then we use the zero factor property to find the solutions.

EXAMPLE 4 Solve and check. $6x^2 + 7x - 10 = 0$

Solution

$$(6x - 5)(x + 2) = 0 \qquad \text{Factor the quadratic equation.}$$
$$6x - 5 = 0 \qquad x + 2 = 0 \qquad \text{Set each factor equal to zero and solve for } x.$$
$$6x = 5 \qquad\qquad x = -2$$
$$x = \frac{5}{6}$$

Thus the solutions to the equation are $\frac{5}{6}$ and -2.

Check each solution in the original equation.

$$6\left(\frac{5}{6}\right)^2 + 7\left(\frac{5}{6}\right) - 10 \overset{?}{=} 0 \qquad 6(-2)^2 + 7(-2) - 10 \overset{?}{=} 0$$

$$6\left(\frac{25}{36}\right) + 7\left(\frac{5}{6}\right) - 10 \overset{?}{=} 0 \qquad 6(4) + 7(-2) - 10 \overset{?}{=} 0$$

$$\frac{25}{6} + \frac{35}{6} - 10 \overset{?}{=} 0 \qquad\qquad 24 + (-14) - 10 \overset{?}{=} 0$$

$$\frac{60}{6} - 10 \overset{?}{=} 0 \qquad\qquad\qquad 10 - 10 = 0 \ \checkmark$$

$$10 - 10 = 0 \ \checkmark$$

Practice Problem 4 Solve and check. $2x^2 + 9x - 18 = 0$

It is important to place a quadratic equation in standard form before factoring. This will sometimes involve removing parentheses and combining like terms.

EXAMPLE 5 Solve. $x + (x - 6)(x - 2) = 2(x - 1)$

Solution

$$
\begin{array}{ll}
x + x^2 - 2x - 6x + 12 = 2x - 2 & \text{Remove parentheses.} \\
x^2 - 7x + 12 = 2x - 2 & \text{Combine like terms.} \\
x^2 - 9x + 14 = 0 & \text{Add } -2x + 2 \text{ to each side.} \\
(x - 7)(x - 2) = 0 & \text{Factor.} \\
x = 7, \quad x = 2 & \text{Use the zero factor property.}
\end{array}
$$

The solutions to the equation are 7 and 2. The check is left to the student.

Practice Problem 5 Solve. $-4x + (x - 5)(x + 1) = 5(1 - x)$

EXAMPLE 6 Solve. $\dfrac{8}{x + 1} - \dfrac{x}{x - 1} = \dfrac{2}{x^2 - 1}$

Solution   The LCD $= (x + 1)(x - 1)$. We multiply each term by the LCD.

$$(x + 1)(x - 1)\left(\frac{8}{x + 1}\right) - (x + 1)(x - 1)\left(\frac{x}{x - 1}\right) = (x + 1)(x - 1)\left[\frac{2}{(x + 1)(x - 1)}\right]$$

$$
\begin{array}{ll}
8(x - 1) - x(x + 1) = 2 & \text{Simplify.} \\
8x - 8 - x^2 - x = 2 & \text{Remove parentheses.} \\
-x^2 + 7x - 8 = 2 & \text{Combine like terms.} \\
-x^2 + 7x - 8 - 2 = 0 & \text{Subtract 2 from each side.} \\
-x^2 + 7x - 10 = 0 & \text{Simplify.} \\
x^2 - 7x + 10 = 0 & \text{Multiply each term by } -1. \\
(x - 5)(x - 2) = 0 & \text{Factor and solve for } x. \\
x = 5, \quad x = 2 & \text{Use the zero factor property.}
\end{array}
$$

Check.   The check is left to the student.

Practice Problem 6 Solve.

$$\frac{10x + 18}{x^2 + x - 2} = \frac{3x}{x + 2} + \frac{2}{x - 1}$$

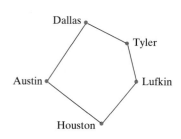

 Solving Applied Problems Involving Quadratic Equations

A manager of a truck delivery service is determining the possible routes he may have to send trucks on during a given day. His delivery service must have the capability to send trucks between any two of the following Texas cities: Austin, Dallas, Tyler, Lufkin, and Houston. He made a rough map of the cities.

What is the maximum number of routes he may have to send trucks on in a given day, if there is a route between every two cities? One possible way to figure this out is to draw a straight line between every two cities. Let us assume that no three cities lie on a straight line. Therefore, the route between any two cities is distinct from all other routes. The figure to the right shows that we can draw exactly 10 lines, so the answer is that there are 10 truck routes. What if the manager has to service 18 cities? How many truck routes will there be then?

As the number of cities increases, it becomes quite difficult (and time consuming) to use a drawing to determine the number of routes. In a more advanced math

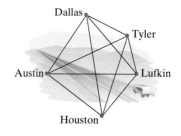

course, it can be proved that the maximum number of truck routes (t) can be found by using the number of cities (n) in the following equation.

$$t = \frac{n^2 - n}{2}$$

Thus, to find the number of truck routes for 18 cities, we merely substitute 18 for n.

$$t = \frac{(18)^2 - (18)}{2}$$
$$= \frac{324 - 18}{2}$$
$$= \frac{306}{2}$$
$$= 153$$

Thus 153 truck routes are needed to service 18 cities if no three of those cities lie on a straight line.

EXAMPLE 7 A truck delivery company can handle a maximum of 36 truck routes in one day, and every two cities have a distinct truck route between them. How many separate cities can the truck company service?

Solution

$$t = \frac{n^2 - n}{2}$$

$$36 = \frac{n^2 - n}{2}$$ Substitute 36 for t, the number of truck routes.

$$72 = n^2 - n$$ Multiply both sides of the equation by 2.

$$0 = n^2 - n - 72$$ Subtract 72 from each side to obtain the standard form of a quadratic equation.

$$0 = (n - 9)(n + 8)$$ Factor.

$$n - 9 = 0 \qquad n + 8 = 0$$ Set each factor equal to zero.

$$n = 9 \qquad\qquad n = -8$$ Solve for n.

We reject -8 as a meaningless solution for this problem. We cannot have a negative number of cities. Thus our answer is that the company can service 9 cities.

Check. If the truck company services $n = 9$ cities, does that really result in a maximum number of $t = 36$ truck routes?

$$t = \frac{n^2 - n}{2}$$ The equation in which $t =$ the number of truck routes and $n =$ the number of cities to be connected.

$$36 \stackrel{?}{=} \frac{(9)^2 - (9)}{2}$$ Substitute $t = 36$ and $n = 9$.

$$36 \stackrel{?}{=} \frac{81 - 9}{2}$$

$$36 \stackrel{?}{=} \frac{72}{2}$$

$$36 = 36 \ ✓$$

NOTE TO STUDENT: Fully worked-out solutions to all of the Practice Problems can be found at the back of the text starting at page SP-1

Practice Problem 7 How many cities can the company service with 28 truck routes?

PRACTICE   WATCH   DOWNLOAD   READ   REVIEW

Write in standard form. Determine the values of a, b, and c.

1. $x^2 + 8x + 7 = 0$

2. $x^2 + 6x - 5 = 0$

3. $8x^2 - 11x = 0$

4. $5x^2 + 15x = 0$

5. $x^2 + 15x - 7 = 12x + 8$

6. $4x^2 + 13x + 8 = x - 12$

Using the factoring method, solve for the roots of each quadratic equation. Be sure to place your equation in standard form before factoring.

7. $27x^2 - 9x = 0$

8. $6x + 18x^2 = 0$

9. $4x^2 - x = -9x$

10. $3x^2 - x = 8x$

11. $11x^2 - 13x = 8x - 3x^2$

12. $60x^2 + 14x = 9x^2 - 20x$

13. $x^2 - 3x - 28 = 0$

14. $x^2 + 3x - 40 = 0$

15. $2x^2 + 11x - 6 = 0$

16. $2x^2 - 9x - 5 = 0$

17. $3x^2 - 16x + 5 = 0$

18. $5x^2 - 21x + 4 = 0$

19. $x^2 = 5x + 14$

20. $x^2 = 5x + 24$

21. $12x^2 + 17x + 6 = 0$

22. $14x^2 + 29x + 12 = 0$

23. $a^2 + 9a + 39 = 7 - 3a$

24. $m^2 + 15m + 33 = 5 - m$

25. $6y^2 = -7y + 3$

26. $8y^2 = -2y + 3$

27. $25x^2 - 60x + 36 = 0$

Mixed Practice

28. $9x^2 - 24x + 16 = 0$

29. $(x - 5)(x + 2) = -10$

30. $(x - 6)(x + 1) = 8$

31. $3x^2 + 8x - 10 = -6x - 10$

32. $5x^2 - 7x + 12 = -8x + 12$

33. $y(y - 7) = 2(5 - 2y)$

34. $x(x - 4) = 3(x - 2)$

35. $(x - 6)(x - 4) = 3$

36. $(y + 6)(y - 1) = 8$

37. $2x(x + 3) = (3x + 1)(x + 1)$

38. $2(2x + 1)(x + 1) = x(7x + 1)$

Find the LCD and multiply each term by the LCD. Then solve. Check your solutions.

39. $x + \dfrac{8}{x} = 6$

40. $x - 1 = \dfrac{20}{x}$

41. $\dfrac{x}{7} - \dfrac{2}{x} = -\dfrac{5}{7}$

42. $\dfrac{x}{5} - \dfrac{7}{x} = \dfrac{2}{5}$

43. $\dfrac{3y + 5}{2} = \dfrac{-3}{y - 2}$

44. $\dfrac{4x - 1}{3} = \dfrac{-1}{x - 2}$

45. $\dfrac{4}{x} + \dfrac{3}{x + 5} = 2$

46. $\dfrac{7}{x} + \dfrac{6}{x - 1} = 2$

47. $\dfrac{24}{x^2 - 4} = 1 + \dfrac{2x - 6}{x - 2}$

48. $\dfrac{5x + 7}{x^2 - 9} = 1 + \dfrac{2x - 8}{x - 3}$

To Think About

49. Why can a quadratic equation in standard form with $c = 0$ always be solved?

50. Martha solved $(x + 3)(x - 2) = 14$ as follows.

$$x + 3 = 14 \qquad x - 2 = 14$$
$$x = 11 \qquad\quad x = 16$$

Josette said this had to be wrong because these values do not check. Explain what is wrong with Martha's method.

51. The equation $ax^2 - 7x + c = 0$ has the roots $-\dfrac{3}{2}$ and 6. Find the values of a and c.

52. The equation $2x^2 + bx - 15 = 0$ has the roots -3 and $\dfrac{5}{2}$. Find the value of b.

Applications *Business Revenue The revenue for producing the label for the new Starblazer video game is given by the formula R = −2.5n² + 80n − 280, where R is the revenue (in dollars) and n is the number of labels produced (in hundreds).*

53. If the revenue for producing labels is $320, what are the two values of the two different numbers of labels that may be produced?

54. If the revenue for producing labels is $157.50, what are the two values of the different numbers of labels that may be produced?

55. What is the average of the two answers you obtained in problem 53? Use the revenue equation to find the revenue for producing that number of labels. What is significant about that answer?

56. What is the average of the two answers you obtained in problem 54? Use the revenue equation to find the revenue for producing that number of labels. What is significant about that answer?

57. Find the revenue for producing 1500 and 1700 labels. Compare the two revenues with the results from 53 and 55. What can you conclude?

58. Find the revenue for producing 1400 and 1800 labels. Compare the two revenues with the results from 54 and 56. What can you conclude?

Cumulative Review

59. **[7.4.1]** Simplify. $\dfrac{\dfrac{1}{x} + \dfrac{3}{x-2}}{\dfrac{4}{x^2-4}}$

60. **[7.3.3]** Add the fractions. $\dfrac{2x}{3x-5} + \dfrac{2}{3x^2-11x+10}$

Quick Quiz 9.1 Using the factoring method, solve for the roots of each quadratic equation.

1. $6x^2 - x = 15$

2. $x^2 = 2(7x - 24)$

3. $x^2 = 2 - \dfrac{7}{3}x$

4. **Concept Check** In the following problem, explain how you would place the quadratic equation in standard form.

$$2 = \frac{5}{x+1} + \frac{3}{x-1}$$

Student Learning Objectives

After studying this section, you will be able to:

 Solve quadratic equations using the square root property.

 Solve quadratic equations by completing the square.

 Solving Quadratic Equations Using the Square Root Property

Recall that the quadratic equation $x^2 = a$ has two possible solutions, $x = \sqrt{a}$ or $x = -\sqrt{a}$, where a is a nonnegative real number. That is, $x = \pm\sqrt{a}$. This basic idea is called the **square root property.**

> **SQUARE ROOT PROPERTY**
>
> If $x^2 = a$, then $x = \sqrt{a}$ or $x = -\sqrt{a}$, for all nonnegative real numbers a.

EXAMPLE 1 Solve.

(a) $x^2 = 49$ **(b)** $x^2 = 20$ **(c)** $5x^2 = 125$

Solution

(a) $x^2 = 49$
$x = \pm\sqrt{49}$
$x = \pm 7$

(b) $x^2 = 20$
$x = \pm\sqrt{20}$
$x = \pm 2\sqrt{5}$

(c) $5x^2 = 125$ Divide both sides by 5 before taking the square root.
$x^2 = 25$
$x = \pm\sqrt{25}$
$x = \pm 5$

Practice Problem 1 Solve.

(a) $x^2 = 1$ **(b)** $x^2 - 50 = 0$ **(c)** $3x^2 = 81$

NOTE TO STUDENT: Fully worked-out solutions to all of the Practice Problems can be found at the back of the text starting at page SP-1

EXAMPLE 2 Solve. $3x^2 + 5x = 18 + 5x + x^2$

Solution Simplify the equation by placing all the variable terms on the left and the constants on the right.

$$3x^2 + 5x = 18 + 5x + x^2$$
$$2x^2 = 18$$
$$x^2 = 9$$
$$x = \pm 3$$

Practice Problem 2 Solve. $4x^2 - 5 = 319$

The square root property can also be used if the term that is squared is a binomial.

Sometimes when we use the square root property we obtain an irrational number. This will occur if the number on the right side of the equation is not a perfect square.

EXAMPLE 3   Solve. $(3x + 1)^2 = 8$

Solution $(3x + 1)^2 = 8$

$3x + 1 = \pm\sqrt{8}$ Take the square root of both sides.

$3x + 1 = \pm 2\sqrt{2}$ Simplify as much as possible.

Now we must solve the two equations expressed by the plus or minus statement.

$$3x + 1 = +2\sqrt{2} \qquad\qquad 3x + 1 = -2\sqrt{2}$$

$$3x = -1 + 2\sqrt{2} \qquad\qquad 3x = -1 - 2\sqrt{2}$$

$$x = \frac{-1 + 2\sqrt{2}}{3} \qquad\qquad x = \frac{-1 - 2\sqrt{2}}{3}$$

The roots of this quadratic equation are irrational numbers. They are

$$\frac{-1 + 2\sqrt{2}}{3} \quad \text{and} \quad \frac{-1 - 2\sqrt{2}}{3} \quad \text{or} \quad \frac{-1 \pm 2\sqrt{2}}{3}$$

We cannot simplify these roots further, so we leave them in this form.

Practice Problem 3   Solve.

$$(2x - 3)^2 = 12$$

2   Solving Quadratic Equations by Completing the Square

If a quadratic equation is not in a form where we can use the square root property, we can rewrite the equation by a method called **completing the square** to make it so. That is, we rewrite the equation with a perfect square on the left side so that we have an equation of the form $(ax + b)^2 = c$.

EXAMPLE 4   Solve. $x^2 + 12x = 4$

Solution   Write out the equation $x^2 + 12x = 4$. Think of a way to determine what number to add to each side.

$x^2 + 12x \qquad\qquad = 4$

$(x + 6)(x + 6) = 4 + 36$ We need a +36 to complete the square.

$x^2 + 12x + 36 = 4 + 36$ Add +36 to both sides of the equation.

$(x + 6)^2 = 40$ Write the left side as a perfect square.

$x + 6 = \pm\sqrt{40}$ Solve for x.

$$x + 6 = +2\sqrt{10} \qquad\qquad x + 6 = -2\sqrt{10}$$

$$x = -6 + 2\sqrt{10} \qquad\qquad x = -6 - 2\sqrt{10}$$

The two roots are $-6 + 2\sqrt{10}$ and $-6 - 2\sqrt{10}$ or $-6 \pm 2\sqrt{10}$.

Practice Problem 4   Solve. $x^2 + 10x = 3$

Completing the square is a little more difficult when the coefficient of x is an odd number. Let's see what happens.

EXAMPLE 5 Solve. $x^2 - 3x - 1 = 0$

Solution Write the equation so that all the x-terms are on the left and the constants are on the right.

$$x^2 - 3x \qquad = 1$$

$(x \quad)(x \quad)$

The missing number when added to itself must be -3.

Since $-\dfrac{3}{2} + \left(-\dfrac{3}{2}\right) = -\dfrac{6}{2} = -3$, the missing number is $-\dfrac{3}{2}$.

$$\left(x - \frac{3}{2}\right)\left(x - \frac{3}{2}\right)$$

Complete the pattern.

$$x^2 - 3x + \frac{9}{4} = 1 + \frac{9}{4}$$

Complete the square by adding $\left(-\dfrac{3}{2}\right)\left(-\dfrac{3}{2}\right) = \dfrac{9}{4}$ to each side.

$$\left(x - \frac{3}{2}\right)^2 = \frac{13}{4}$$

Solve for x.

$$x - \frac{3}{2} = \pm\sqrt{\frac{13}{4}}$$

$$x - \frac{3}{2} = \pm\frac{\sqrt{13}}{2}$$

$$x - \frac{3}{2} = +\frac{\sqrt{13}}{2} \qquad x - \frac{3}{2} = -\frac{\sqrt{13}}{2}$$

$$x = \frac{3}{2} + \frac{\sqrt{13}}{2} \qquad x = \frac{3}{2} - \frac{\sqrt{13}}{2}$$

$$x = \frac{3 + \sqrt{13}}{2} \qquad x = \frac{3 - \sqrt{13}}{2}$$

The two roots are $\dfrac{3 \pm \sqrt{13}}{2}$.

Practice Problem 5 Solve.

$$x^2 - 5x = 7$$

NOTE TO STUDENT: Fully worked-out solutions to all of the Practice Problems can be found at the back of the text starting at page SP-1

If a, the coefficient of the squared variable, is not 1, we divide all terms of the equation by a so that the coefficient of the squared variable will be 1.

Let us summarize for future reference the steps we perform in order to solve a quadratic equation by completing the square.

COMPLETING THE SQUARE

1. Put the equation in the form $ax^2 + bx = c$. If $a \neq 1$, divide each term by a.

2. Square $\dfrac{b}{2}$ and add the result to both sides of the equation.

3. Factor the left side (a perfect square trinomial).

4. Use the square root property.

5. Solve the equations.

6. Check the solutions in the original equation.

EXAMPLE 6   Solve. $4x^2 + 4x - 3 = 0$

Solution

Step 1 $4x^2 + 4x = 3$ Place the constant on the right. This puts the equation in the form $ax^2 + bx = c$.

$x^2 + x = \dfrac{3}{4}$ Divide all terms by 4.

Step 2 $x^2 + 1x = \dfrac{3}{4}$ Take one-half the coefficient of x and square it. $\left(\dfrac{1}{2}\right)^2 = \dfrac{1}{4}$.

$x^2 + x + \dfrac{1}{4} = \dfrac{3}{4} + \dfrac{1}{4}$ Add $\dfrac{1}{4}$ to each side.

Step 3 $\left(x + \dfrac{1}{2}\right)^2 = 1$ Factor the left side.

Step 4 $x + \dfrac{1}{2} = \pm\sqrt{1}$ Use the square root property.

Step 5 $x + \dfrac{1}{2} = \pm 1$

$x + \dfrac{1}{2} = 1$ \qquad\qquad $x + \dfrac{1}{2} = -1$

$x = \dfrac{2}{2} - \dfrac{1}{2}$ \qquad\qquad $x = -\dfrac{2}{2} - \dfrac{1}{2}$

$x = \dfrac{1}{2}$ \qquad\qquad $x = -\dfrac{3}{2}$

Thus the two roots are $\dfrac{1}{2}$ and $-\dfrac{3}{2}$.

Step 6 $4\left(\dfrac{1}{2}\right)^2 + 4\left(\dfrac{1}{2}\right) - 3 \overset{?}{=} 0$ Check.

$4\left(\dfrac{1}{4}\right) + 4\left(\dfrac{1}{2}\right) - 3 \overset{?}{=} 0$

$1 + 2 - 3 = 0$ ✓

$4\left(-\dfrac{3}{2}\right)^2 + 4\left(-\dfrac{3}{2}\right) - 3 \overset{?}{=} 0$

$4\left(\dfrac{9}{4}\right) + 4\left(-\dfrac{3}{2}\right) - 3 \overset{?}{=} 0$

$9 - 6 - 3 = 0$ ✓

Both values check.

Practice Problem 6   Solve.

$$8x^2 + 2x - 1 = 0$$

This method of completing the square will enable us to solve any quadratic equation that has real number roots. It is usually faster, however, to factor the quadratic equation if the polynomial is factorable.

9.2 EXERCISES

MyMathLab

PRACTICE

WATCH

DOWNLOAD

READ

REVIEW

Solve using the square root property.

1. $x^2 = 64$

2. $x^2 = 36$

3. $x^2 = 98$

4. $x^2 = 72$

5. $x^2 - 28 = 0$

6. $x^2 - 45 = 0$

7. $5x^2 = 45$

8. $3x^2 = 48$

9. $6x^2 = 120$

10. $3x^2 = 150$

11. $3x^2 - 375 = 0$

12. $4x^2 - 252 = 0$

13. $5x^2 + 13 = 73$

14. $6x^2 + 5 = 53$

15. $13x^2 + 17 = 82$

16. $6x^2 - 13 = 23$

17. $(x - 7)^2 = 16$

18. $(x - 6)^2 = 100$

19. $(x + 4)^2 = 6$

20. $(x - 5)^2 = 36$

21. $(2x + 5)^2 = 2$

22. $(3x - 4)^2 = 6$

23. $(3x - 1)^2 = 7$

24. $(2x + 3)^2 = 10$

25. $(5x + 1)^2 = 18$

26. $(8x + 5)^2 = 20$

27. $(4x - 5)^2 = 54$

28. $(5x - 3)^2 = 50$

Solve by completing the square.

29. $x^2 - 6x = 11$

30. $x^2 - 2x = 4$

31. $x^2 + 8x - 20 = 0$

32. $x^2 - 10x + 16 = 0$

33. $x^2 - 12x - 5 = 0$

34. $x^2 + 16x + 30 = 0$

35. $x^2 - 7x = 0$

36. $x^2 - 9x = 0$

37. $5x^2 - 25x = 0$

38. $2x^2 - 3x = 20$

39. $2x^2 - 7x = 9$

40. $2x^2 + 7x = 4$

Cumulative Review

41. [4.4.1] Solve. $3a - 5b = 8$
$$5a - 7b = 8$$

42. [7.5.1] Solve. $\dfrac{x}{x+3} - \dfrac{2}{x} = \dfrac{x-2}{x^2+3x}$

43. [1.8.2] *Hunting Knife* While hiking in the Grand Canyon, Will had to use his hunting knife to clear a fallen branch from the path. Find the pressure on the blade of his knife if the knife is 8 in. long and sharpened to an edge 0.01 in. wide and he applies a force of 16 pounds. Use the formula $P = \frac{F}{A}$, where P is the pressure on the blade of the knife, F is the force in pounds placed on the knife, and A is the area exposed to an object by the sharpened edge of the knife.

44. [2.6.2] *Tire Defects* Darlene inspected a sample of 100 new steel-belted radial tires. 13 of the tires had defects in workmanship, while 9 of the tires had defects in materials. Of those defective tires, 6 had defects in both workmanship and materials. How many of the 100 tires had any kind of defect? (Assume that the only defects found were those in materials or workmanship.)

Quick Quiz 9.2 Solve using the square root property.

1. $4x^2 + 13 = 37$

2. $(3x - 2)^2 = 40$

Solve by completing the square.

3. $x^2 + 8x - 10 = 0$

4. Concept Check Explain the first three steps of how you would solve the following problem by completing the square.
$$5x^2 - 10x + 2 = 0$$

Student Learning Objectives

After studying this section, you will be able to:

1 Solve a quadratic equation using the quadratic formula.

2 Find a decimal approximation for the real roots of a quadratic equation.

3 Determine whether a quadratic equation has no real solutions.

1 Solving a Quadratic Equation Using the Quadratic Formula

To find solutions of a quadratic equation, you can factor or you can complete the square when the trinomial is not factorable. If we use the method of completing the square on the general equation $ax^2 + bx + c = 0$, we can derive a general formula for the solution of any quadratic equation. This is called the **quadratic formula.**

> **QUADRATIC FORMULA**
>
> The roots of any quadratic equation of the form $ax^2 + bx + c = 0$, where a, b, and c are real numbers and $a \neq 0$, are
>
> $$x = \frac{-b \pm \sqrt{b^2 - 4ac}}{2a}.$$

All you will need to know are the values of a, b, and c to solve a quadratic equation using the quadratic formula.

EXAMPLE 1 Solve using the quadratic formula. $3x^2 + 10x + 7 = 0$

Solution In our given equation, $3x^2 + 10x + 7 = 0$, we have

$a = 3$ (the coefficient of x^2), $\quad b = 10$ (the coefficient of x), and
$c = 7$ (the constant term).

$$x = \frac{-b \pm \sqrt{b^2 - 4ac}}{2a} = \frac{-10 \pm \sqrt{(10)^2 - 4(3)(7)}}{2(3)}$$ Write the quadratic formula and substitute values for a, b, and c.

$$= \frac{-10 \pm \sqrt{100 - 84}}{6}$$ Simplify.

$$= \frac{-10 \pm \sqrt{16}}{6} = \frac{-10 \pm 4}{6}$$

$$x = \frac{-10 + 4}{6} = \frac{-6}{6} = -1$$ Using the positive sign.

$$x = \frac{-10 - 4}{6} = \frac{-14}{6} = -\frac{7}{3}$$ Using the negative sign.

Thus the two solutions are -1 and $-\frac{7}{3}$.

[*Note:* Here we obtain rational roots. We would obtain the same answer by factoring $3x^2 + 10x + 7 = 0$ as $(3x + 7)(x + 1) = 0$ and setting each factor $= 0$.]

Practice Problem 1 Solve using the quadratic formula. $x^2 - 7x + 6 = 0$

Often the roots will be irrational numbers. Sometimes these roots can be simplified. You should always leave your answer in simplest form.

EXAMPLE 2 Solve. $2x^2 - 4x - 9 = 0$

Solution $\quad a = 2, \quad b = -4, \quad c = -9$

$$x = \frac{-b \pm \sqrt{b^2 - 4ac}}{2a} = \frac{-(-4) \pm \sqrt{(-4)^2 - 4(2)(-9)}}{2(2)}$$

$$= \frac{4 \pm \sqrt{16 + 72}}{4} = \frac{4 \pm \sqrt{88}}{4}$$ Do not stop here!

Notice that we can simplify $\sqrt{88}$.

$$x = \frac{4 \pm 2\sqrt{22}}{4} = \frac{2(2 \pm \sqrt{22})}{4} = \frac{\overset{1}{\cancel{2}}(2 \pm \sqrt{22})}{\underset{2}{\cancel{4}}} = \frac{2 \pm \sqrt{22}}{2}$$

Be careful. Here we were able to divide the numerator and denominator by 2 because 2 was a factor of every term.

Practice Problem 2 Solve. $3x^2 - 8x + 3 = 0$

A quadratic equation *must* be written in the standard form $ax^2 + bx + c = 0$ *before* the quadratic formula can be used. Several algebraic steps may be needed to accomplish this objective. Also, since it is much easier to use the formula if a, b, and c are integers, we will avoid using fractional values.

EXAMPLE 3 Solve. $x^2 = 5 - \frac{3}{4}x$

Solution   First we obtain an equivalent equation that does not have fractions.

$$x^2 = 5 - \frac{3}{4}x$$

$$4(x^2) = 4(5) - 4\left(\frac{3}{4}x\right) \qquad \text{Multiply each term by the LCD of 4.}$$

$$4x^2 = 20 - 3x \qquad\qquad \text{Simplify.}$$

$$4x^2 + 3x - 20 = 0 \qquad\qquad \text{Add } 3x - 20 \text{ to each side.}$$

$$a = 4, \quad b = 3, \quad c = -20 \qquad \text{Substitute } a = 4, b = 3, \text{ and } c = -20 \text{ into}$$
$$\text{the quadratic formula.}$$

$$x = \frac{-3 \pm \sqrt{(3)^2 - 4(4)(-20)}}{2(4)}$$

$$x = \frac{-3 \pm \sqrt{9 + 320}}{8}$$

$$x = \frac{-3 \pm \sqrt{329}}{8}$$

Practice Problem 3 Solve.

$$x^2 = 7 - \frac{3}{5}x$$

Finding a Decimal Approximation for the Real Roots of a Quadratic Equation

EXAMPLE 4 Find the roots of $3x^2 - 5x = 7$. Approximate to the nearest thousandth.

Solution

$$3x^2 - 5x = 7$$

$$3x^2 - 5x - 7 = 0 \qquad \text{Place the equation in standard form.}$$

$$a = 3, \quad b = -5, \quad c = -7$$

$$x = \frac{-(-5) \pm \sqrt{(-5)^2 - 4(3)(-7)}}{2(3)} = \frac{5 \pm \sqrt{25 + 84}}{6}$$

$$= \frac{5 \pm \sqrt{109}}{6}$$

Using most scientific calculators, we can find $\dfrac{5 + \sqrt{109}}{6}$ with these keystrokes.

$$5\ \boxed{+}\ 109\ \boxed{\sqrt{\ }}\ \boxed{=}\ \boxed{\div}\ 6\ \boxed{=}\ \ 2.5733844$$

Rounding to the nearest thousandth, we have $x \approx 2.573$.

To find $\dfrac{5 - \sqrt{109}}{6}$, we use the following keystrokes.

$$5\ \boxed{-}\ 109\ \boxed{\sqrt{\ }}\ \boxed{=}\ \boxed{\div}\ 6\ \boxed{=}\ \ -0.9067178$$

Rounding to the nearest thousandth, we have $x \approx -0.907$.
If you do not have a calculator with a square root key, look up $\sqrt{109}$ in the square root table. $\sqrt{109} \approx 10.440$. Using this result, we have

$$x \approx \frac{5 + 10.440}{6} = \frac{15.440}{6} \approx 2.573 \quad \text{(rounded to the nearest thousandth)}$$

and

$$x \approx \frac{5 - 10.440}{6} = \frac{-5.440}{6} \approx -0.907 \quad \text{(rounded to the nearest thousandth)}.$$

If you have a graphing calculator, there is an alternate method for finding a decimal approximation of the roots of a quadratic equation. In Section 9.4, you will cover graphing quadratic equations. You will discover that finding the roots of a quadratic equation of the form $ax^2 + bx + c = 0$ is equivalent to finding the values of x on the graph of $y = ax^2 + bx + c$, where $y = 0$. These may be approximated quite quickly with a graphing calculator.

NOTE TO STUDENT: *Fully worked-out solutions to all of the Practice Problems can be found at the back of the text starting at page SP-1*

Practice Problem 4 Find the roots of $2x^2 = 13x + 5$. Approximate to the nearest thousandth.

③ Determining Whether a Quadratic Equation Has No Real Solutions

EXAMPLE 5 Solve. $2x^2 + 5 = -3x$

Solution The equation is not in standard form.

$$2x^2 + 5 = -3x \qquad \text{Add } 3x \text{ to both sides.}$$
$$2x^2 + 3x + 5 = 0 \qquad \text{We can now find } a, b, \text{ and } c.$$
$$a = 2, \quad b = 3, \quad c = 5 \qquad \text{Substitute these values into the quadratic formula.}$$

$$x = \frac{-3 \pm \sqrt{(3)^2 - 4(2)(5)}}{2(2)} = \frac{-3 \pm \sqrt{9 - 40}}{4} = \frac{-3 \pm \sqrt{-31}}{4}$$

There is no real number that is $\sqrt{-31}$. Since we are using only real numbers in this book, we say there is no solution to the problem. (*Note:* In more advanced math courses, these types of numbers, called complex numbers, will be studied.)

Practice Problem 5 Solve. $5x^2 + 2x = -3$

We can tell whether the roots of any given quadratic equation are real numbers. Look at the quadratic formula.

$$\frac{-b \pm \sqrt{b^2 - 4ac}}{2a}$$

The expression under the radical sign is called the **discriminant.**

$b^2 - 4ac$ is the discriminant.

If the discriminant is a negative number, the roots are not real numbers, and there is no real number solution (no real roots) to the equation.

    If the discriminant is a positive number, the roots are real numbers, and there are two real number solutions (two real roots) to the equation. In addition, if $b^2 - 4ac$ is a perfect square, the two real number solutions are rational numbers. If $b^2 - 4ac$ is not a perfect square, the two real number solutions are irrational. If the discriminant is equal to 0, there is one real number solution (one root) to the equation.

EXAMPLE 6 Determine whether $3x^2 = 5x - 4$ has real number solution(s).

Solution   First, we place the equation in standard form. Then we need only check the discriminant.

$$3x^2 = 5x - 4$$
$$3x^2 - 5x + 4 = 0$$
$a = 3, \quad b = -5, \quad c = 4$ Substitute the values for a, b, and c into the discriminant and evaluate.

$$b^2 - 4ac = (-5)^2 - 4(3)(4) = 25 - 48 = -23$$

The discriminant is negative. Thus $3x^2 = 5x - 4$ has no real number solution(s).

Practice Problem 6   Determine whether each equation has real number solution(s).

(a) $5x^2 = 3x + 2$ **(b)** $2x^2 + 5 = 4x$

Verbal and Writing Skills

1. In the equation $3x^2 + 4x - 7 = 0$, what can we learn from the discriminant?

2. In the equation $7x^2 - 6x + 2 = 0$, what can we learn from the discriminant?

3. Is the equation $4x^2 = -5x + 6$ in standard form? If not, place it in standard form and find a, b, and c.

4. Is the equation $8x = -9x^2 + 2$ in standard form? If not, place it in standard form and find a, b, and c.

Solve using the quadratic formula. If there are no real roots, say so.

5. $x^2 + 3x - 10 = 0$

6. $x^2 - 5x + 4 = 0$

7. $x^2 - 3x - 8 = 0$

8. $x^2 - x - 8 = 0$

9. $4x^2 + 7x - 2 = 0$

10. $3x^2 + 7x - 6 = 0$

11. $2x^2 = 3x + 20$

12. $9x - 2 = 3x^2$

13. $6x^2 - 3x = 1$

14. $3 = 7x - 4x^2$

15. $x + \dfrac{3}{2} = 3x^2$

16. $\dfrac{2}{3}x^2 + x = \dfrac{1}{2}$

17. $\dfrac{x}{2} + \dfrac{5}{x} = \dfrac{7}{2}$

18. $\dfrac{1}{4} + \dfrac{5}{x} = \dfrac{x}{4}$

19. $5x^2 + 6x + 2 = 0$

20. $3x^2 - 4x + 2 = 0$

21. $4y^2 = 10y - 3$

22. $8y^2 = 8y - 1$

23. $\dfrac{d^2}{2} + \dfrac{5d}{6} - 2 = 0$

24. $\dfrac{k^2}{4} - \dfrac{5k}{6} = \dfrac{2}{3}$

25. $4x^2 + 8x + 11 = 0$

26. $5x^2 + 7x + 9 = 0$

27. $x^2 + 12x + 36 = 0$

28. $16x^2 - 8x + 1 = 0$

Use the quadratic formula to find the roots. Find a decimal approximation to the nearest thousandth.

29. $x^2 + 5x - 2 = 0$

30. $x^2 + 8x - 8 = 0$

31. $2x^2 - 7x - 5 = 0$

32. $3x^2 + 4x - 6 = 0$

33. $5x^2 + 10x + 1 = 0$

34. $2x^2 + 9x + 2 = 0$

Mixed Practice *Solve for the variable. Choose the method you feel will work best.*

35. $6x^2 - 13x + 6 = 0$

36. $6x^2 - 7x - 3 = 0$

37. $3(x^2 + 1) = 10x$

38. $3 + x(x + 2) = 18$

39. $(t + 5)(t - 3) = 7$

40. $(b + 4)(b - 7) = 4$

41. $y^2 - \dfrac{2}{5}y = 2$

42. $y^2 - 3 = \dfrac{8}{3}y$

43. $3x^2 - 13 = 0$

44. $5x^2 + 9x = 0$

45. $x(x - 2) = 7$

46. $(4x - 1)^2 = 24$

To Think About

▲ **47.** *Swimming Pool Design* John and Chris Maney have designed a new in-ground swimming pool for their backyard. The pool is rectangular and measures 30 feet by 20 feet. They have also designed a tile border to go around the pool. In the diagram below, this border is x feet wide. The tile border covers 216 square feet in area. What is the width x of this tile border? (*Hint:* Find an expression for the area of the large rectangle and subtract an expression for the area of the small rectangle.)

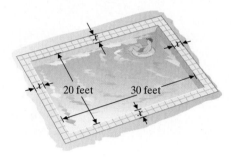

▲ **48.** *Lynn Campus Entrance* The Alumni Association of North Shore Community College has designed a new garden for the front entrance of the Lynn campus. It is rectangular and measures 40 feet by 30 feet. The association has designed a cement walkway to go around this rectangular garden. In the diagram below, this cement walkway is x feet wide. The entire cement walkway covers 456 square feet in area. What is the width x of this cement walkway? (*Hint:* Find an expression for the area of the large rectangle and subtract an expression for the area of the small rectangle.)

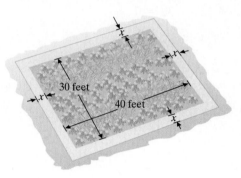

Cumulative Review

49. [5.6.2] Divide. $(x^3 + 8x^2 + 17x + 6) \div (x + 3)$

50. [3.4.2] Find the equation of the line passing through $(-8, 2)$ and $(5, -3)$.

Quick Quiz 9.3 Solve by using the quadratic formula.

1. $9x^2 + 6x - 1 = 0$

2. $4x^2 - 4x = 19$

3. $2 + x = 3x(x + 1)$

4. **Concept Check** Explain how you would determine if the quadratic equation $5x^2 - 8x + 9 = 0$ has real solutions or has no real solutions.

How are you doing with your homework assignments in Sections 9.1 to 9.3? Do you feel you have mastered the material so far? Do you understand the concepts you have covered? Before you go further in the textbook, take some time to do each of the following problems.

9.1

Solve by factoring.

1. $x^2 - 13x - 48 = 0$

2. $5x^2 + 7x = 14x$

3. $5x^2 = 22x - 8$

4. $-x + 1 = 6x^2$

5. $x(x + 9) = 4(x + 6)$

6. $\dfrac{5}{x + 2} = \dfrac{2x - 1}{5}$

Simplify all answers in problems 7–14.

9.2

Solve by taking the square root of each side of the equation.

7. $x^2 - 18 = 0$

8. $3x^2 + 1 = 76$

Solve by completing the square.

9. $x^2 + 6x + 1 = 0$

10. $2x^2 + 3x - 7 = 0$

9.3

Solve if possible using the quadratic formula.

11. $2x^2 + 4x - 5 = 0$

12. $4x^2 = x + 5$

13. $3x^2 + 8x + 1 = 0$

14. $5x^2 + 3 = 4x$

Now turn to page SA-23 for the answer to each of these problems. Each answer also includes a reference to the objective in which the problem is first taught. If you missed any of these problems, you should stop and review the Examples and Practice Problems in the referenced objective. A little review now will help you master the material in the upcoming sections of the text.

1. _____

2. _____

3. _____

4. _____

5. _____

6. _____

7. _____

8. _____

9. _____

10. _____

11. _____

12. _____

13. _____

14. _____

1 Graphing Equations of the Form $y = ax^2 + bx + c$ Using Ordered Pairs

Recall that the graph of a linear equation is always a straight line. What is the graph of a quadratic equation? What does the graph look like if one of the variable terms is squared? We will use point plotting to look for a pattern. We begin with a table of values.

EXAMPLE 1 Graph. **(a)** $y = x^2$ **(b)** $y = -x^2$

Solution

(a) $y = x^2$

x	y
−3	9
−2	4
−1	1
0	0
1	1
2	4
3	9

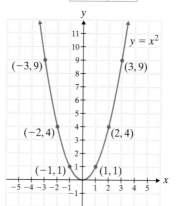

(b) $y = -x^2$

x	y
−3	−9
−2	−4
−1	−1
0	0
1	−1
2	−4
3	−9

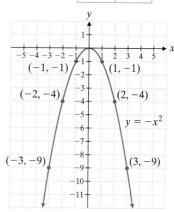

Practice Problem 1 Graph. **(a)** $y = 2x^2$ **(b)** $y = -2x^2$

(a) $y = 2x^2$

x	y
−2	
−1	
0	
1	
2	

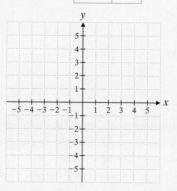

(b) $y = -2x^2$

x	y
−2	
−1	
0	
1	
2	

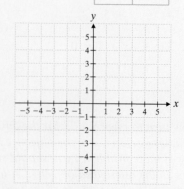

TO THINK ABOUT: When the Coefficient of x^2 Is Negative   What happens to the graph when the coefficient of x^2 is negative? What happens when the coefficient of x^2 is an integer greater than 1 or less than -1? Graph $y = 3x^2$ and $y = -3x^2$. What happens when $-1 < a < 1$? Graph $y = \frac{1}{3}x^2$ and $y = -\frac{1}{3}x^2$.

The curves that you have been graphing are called **parabolas.** What happens to the graph of a parabola as the quadratic equation that describes it changes?

In the preceding example, the equations we worked with were relatively simple. They provided us with an understanding of the general shape of the graph of a quadratic equation. We have seen how changing the equation affects the graph of that equation. Let's look at an equation that is slightly more difficult to graph.

The **vertex** is the lowest point on a parabola that opens upward, or the highest point on a parabola that opens downward.

> **Graphing Calculator**
>
> **Quadratic Equations**
>
> Graph $y = x^2 - 4x$.
>
> Use the Trace and Zoom features. Use the feature that calculates the minimum point to find the vertex.
>
>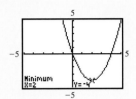
>
> Thus the vertex is at $(2, -4)$.

EXAMPLE 2   Graph $y = x^2 - 2x$. Identify the coordinates of the vertex.

Solution

x	y	
-2	8	
-1	3	
0	0	
1	-1	vertex
2	0	
3	3	
4	8	

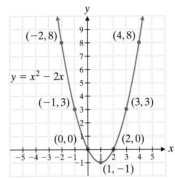

The vertex is at $(1, -1)$.

Notice that the x-intercepts are at $(0, 0)$ and $(2, 0)$. The x-coordinates are the solutions to the equation $x^2 - 2x = 0$.

$$x^2 - 2x = 0$$
$$x(x - 2) = 0$$
$$x = 0, \quad x = 2$$

Practice Problem 2   Graph. For each table of values, use x-values from -4 to 2. Identify the coordinates of the vertex.

(a) $y = x^2 + 2x$

(b) $y = -x^2 - 2x$

(a)

x	y
-4	
-3	
-2	
-1	
0	
1	
2	

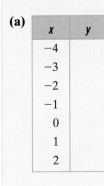

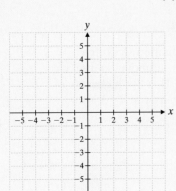

(b)

x	y
-4	
-3	
-2	
-1	
0	
1	
2	

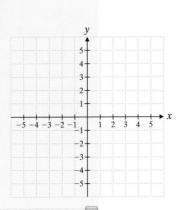

② Graphing Equations of the Form $y = ax^2 + bx + c$ Using the Vertex Formula

Finding the vertex is the key to graphing a parabola. Let's take a look at some equations and identify the vertex of each parabola.

Equation: $\quad y = x^2 \quad\quad y = x^2 - 3 \quad\quad y = (x - 1)^2 \quad\quad\quad y = (x + 2)^2 \quad\quad\quad\quad y = x^2 - 2x$
$$y = x^2 - 2x + 1 \quad\quad y = x^2 + 4x + 4$$

Vertex: $\quad\quad (0, 0) \quad\quad (0, -3) \quad\quad\quad (1, 0) \quad\quad\quad\quad\quad (-2, 0) \quad\quad\quad\quad\quad (1, -1)$

Notice that the x-coordinate of the vertex of the parabola for each of the first two equations is 0. You may also notice that these two equations do not have middle terms. That is, for $y = ax^2 + bx + c$, $b = 0$. This is not true of the last three equations. As you may have already guessed, there is a relationship between the quadratic equation and the x-coordinate of the vertex of the parabola it describes.

VERTEX OF A PARABOLA

The x-coordinate of the vertex of the parabola described by $y = ax^2 + bx + c$ is

$$x = \frac{-b}{2a}.$$

Once we have the x-coordinate, we can find the y-coordinate using the equation. We can plot the point and draw a general sketch of the curve. The sign of the squared term will tell us whether the graph opens upward or downward.

EXAMPLE 3 $\quad y = -x^2 - 2x + 3$. Determine the vertex and the x-intercepts. Then sketch the graph.

Solution We will first determine the coordinates of the vertex. We begin by finding the x-coordinate.

$$x = \frac{-b}{2a} = \frac{-(-2)}{2(-1)} = \frac{2}{-2} = -1$$

$$y = -(-1)^2 - 2(-1) + 3 \quad\quad \text{Determine } y \text{ when } x \text{ is } -1.$$
$$= 4 \quad\quad\quad\quad\quad\quad\quad\quad\quad \text{The point } (-1, 4) \text{ is the vertex of the parabola.}$$

Now we will determine the x-intercepts, those points where the graph crosses the x-axis. This occurs when $y = 0$. These are the solutions to the equation $-x^2 - 2x + 3 = 0$. Note that this equation is factorable.

$$-x^2 - 2x + 3 = 0 \quad\quad \text{Find the solutions to the equation.}$$
$$-(x^2 + 2x - 3) = 0$$
$$-(x - 1)(x + 3) = 0 \quad\quad \text{Factor.}$$
$$x - 1 = 0 \quad\quad x + 3 = 0 \quad\quad \text{Solve for } x.$$
$$x = 1 \quad\quad\quad x = -3 \quad\quad \text{The } x\text{-intercepts are at } (1, 0) \text{ and at } (-3, 0).$$

We now have enough information to sketch the graph. Since $a = -1$, the parabola opens downward and the vertex $(-1, 4)$ is the highest point. The graph crosses the x-axis at $(-3, 0)$ and at $(1, 0)$.

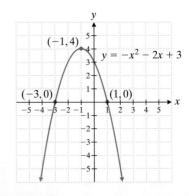

Practice Problem 3 $y = x^2 - 4x - 5.$ Determine the vertex and the x-intercepts. Then sketch the graph.

NOTE TO STUDENT: Fully worked-out solutions to all of the Practice Problems can be found at the back of the text starting at page SP-1

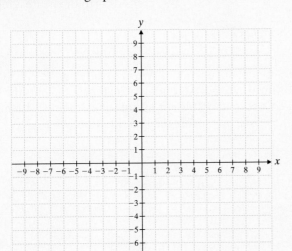

EXAMPLE 4 A person throws a ball into the air and its distance from the ground in feet is given by the equation $h = -16t^2 + 64t + 5$, where t is measured in seconds.

(a) Graph the equation $h = -16t^2 + 64t + 5$.

(b) What physical significance does the vertex of the graph have?

(c) After how many seconds will the ball hit the ground?

Solution

(a) Since the equation is a quadratic equation, the graph is a parabola. a is negative. Thus the parabola opens downward and the vertex is the highest point. We begin by finding the coordinates of the vertex. Our equation has ordered pairs (t, h) instead of (x, y). The t-coordinate of the vertex is

$$t = \frac{-b}{2a} = \frac{-64}{2(-16)} = \frac{-64}{-32} = 2.$$

Substitute $t = 2$ into the equation to find the h-coordinate of the vertex.

$$h = -16(2)^2 + 64(2) + 5 = -64 + 128 + 5 = 69$$

The vertex (t, h) is $(2, 69)$.

To sketch this graph, it would be helpful to know the h-intercept, that is, the point where $t = 0$.

$$h = -16(0)^2 + 64(0) + 5$$
$$= 5$$

The h-intercept is at $(0, 5)$.

The t-intercept is where $h = 0$. We will use the quadratic formula to find t when h is 0, that is, to find the roots of the quadratic equation $-16t^2 + 64t + 5 = 0$.

$$-16t^2 + 64t + 5 = 0$$

$$t = \frac{-64 \pm \sqrt{(64)^2 - 4(-16)(5)}}{2(-16)}$$

$$t = -\frac{64 \pm \sqrt{4096 + 320}}{-32} = -\frac{64 \pm \sqrt{4416}}{-32} = \frac{8 \pm \sqrt{69}}{4}$$

Using a calculator or a square root table, we have the following.

$$t = \frac{8 + 8.307}{4} \approx 4.077, \qquad t \approx 4.1 \quad \text{(to the nearest tenth)}$$

$$t = \frac{8 - 8.307}{4} \approx -0.077$$

We will not use negative t-values. We are studying the height of the ball from the time it is thrown ($t = 0$ seconds) to the time that it hits the ground (approximately $t = 4.1$ seconds). It is not useful to find points with negative values of t because they have no meaning in this situation.

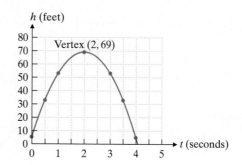

(b) The vertex represents the greatest height of the ball. The ball is 69 feet above the ground 2 seconds after it is thrown.

(c) The ball will hit the ground approximately 4.1 seconds after it is thrown. This is the point at which the graph intersects the t-axis, the positive root of the equation.

NOTE TO STUDENT: Fully worked-out solutions to all of the Practice Problems can be found at the back of the text starting at page SP-1

Practice Problem 4   A ball is thrown upward with a speed of 32 feet per second from a distance 10 feet above the ground. The height of the ball in feet is given by the equation $h = -16t^2 + 32t + 10$, where t is measured in seconds.

(a) Graph the equation $h = -16t^2 + 32t + 10$.

(b) What is the greatest height of the ball?

(c) After how many seconds will the ball hit the ground?

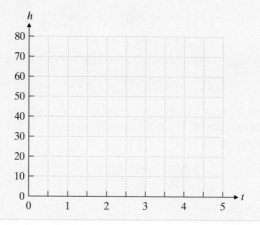

Verbal and Writing Skills

1. How many x-intercepts can a parabola have?

2. How many vertices can a parabola have?

3. For a parabola in the form $y = ax^2 + bx + c$, what happens to the equation $x = \dfrac{-b}{2a}$ if $b = 0$?

4. For a parabola in the form $y = ax^2 + bx + c$, what can be said about the parabola if $a < 0$?

Graph each equation.

5. $y = x^2 + 2$

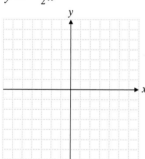

6. $y = x^2 - 4$

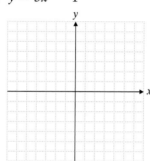

7. $y = -\frac{1}{3}x^2$

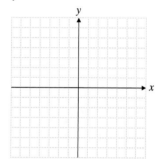

8. $y = -\frac{1}{2}x^2$

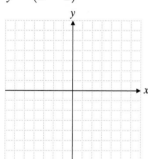

9. $y = 3x^2 - 1$

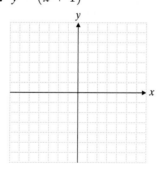

10. $y = 2x^2 + 3$

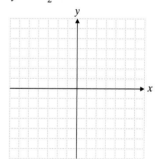

11. $y = (x - 2)^2$

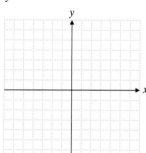

12. $y = (x + 1)^2$

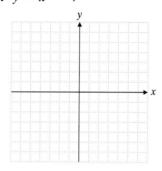

13. $y = -\frac{1}{2}x^2 + 4$

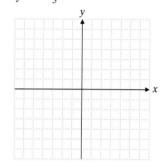

14. $y = -\frac{1}{3}x^2 + 6$

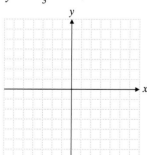

15. $y = \frac{1}{2}(x - 3)^2$

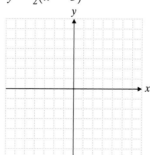

16. $y = \frac{1}{2}(x + 2)^2$

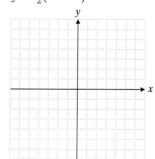

Graph each equation. Identify the coordinates of the vertex.

17. $y = x^2 + 4x$

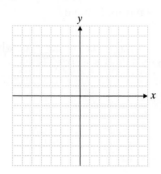

18. $y = -x^2 + 4x$

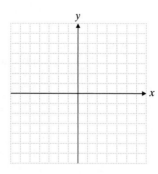

19. $y = -x^2 - 4x$

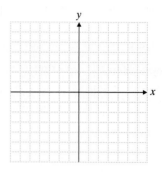

20. $y = x^2 - 4x$

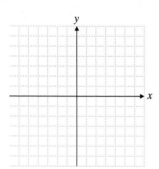

21. $y = -2x^2 + 8x$

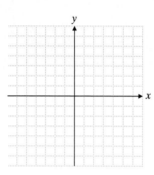

22. $y = x^2 + 6x$

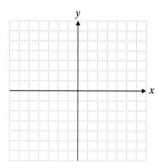

Determine the vertex and the x-intercepts. Then sketch the graph.

23. $y = x^2 + 2x - 3$

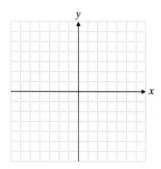

24. $y = x^2 - 4x + 3$

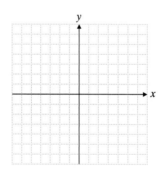

25. $y = -x^2 - 6x - 5$

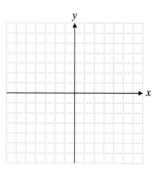

26. $y = -x^2 + 8x - 15$

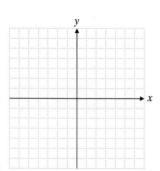

27. $y = x^2 + 6x + 9$

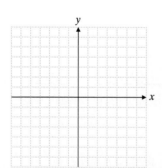

28. $y = -x^2 + 2x - 1$

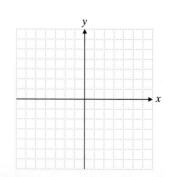

Applications

29. *Miniature Rockets* A miniature rocket is launched so that its height h in meters after t seconds is given by the equation $h = -4.9t^2 + 39.2t + 4$.

(a) Draw a graph of the equation.

(b) How high is the rocket after 2 seconds?

(c) What is the maximum height the rocket will attain?

(d) How long after the launch will the rocket hit the ground?

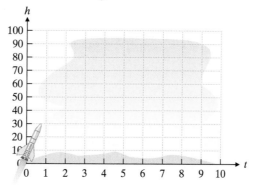

30. *Miniature Rockets* A miniature rocket is launched so that its height h in meters after t seconds is given by the equation $h = -4.9t^2 + 19.6t + 10$.

(a) Draw a graph of the equation.

(b) How high is the rocket after 3 seconds?

(c) What is the maximum height the rocket will attain?

(d) How long after the launch will the rocket hit the ground?

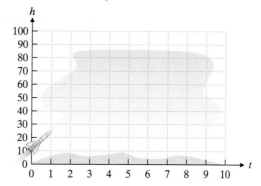

31. *Mosquito Population* The number of mosquitoes in some areas varies with the amount of rain. The number, N, of mosquitoes measured in millions in Hamilton, Massachusetts, in May can be predicted by the equation $N = 9x - x^2$, where x is the number of inches of rain during that month. (*Source*: Massachusetts Environmental Protection Agency.)

(a) Graph the equation.

(b) How many inches of rain produce the maximum number of mosquitoes in May?

(c) What is the maximum number of mosquitoes?

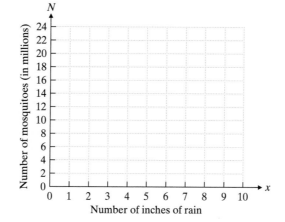

Use your graphing calculator to find the vertex, the y-intercept, and the x-intercepts of the parabolas described by the following equations. Round to the nearest thousandth when necessary.

32. $y = -156x^2 - 289x + 1133$

33. $y = -301x^2 - 167x + 1528$

Cumulative Review

34. [8.7.2] If y varies inversely as x, and $y = \dfrac{1}{3}$ when $x = 21$, find y when $x = 7$.

35. [8.7.1] If y varies directly as x^2, and $y = 12$ when $x = 2$, find y when $x = 5$.

36. [7.6.3] *Football* A football player picks up a ball from the field and runs 60 yards. He runs the first half of this distance at 20 feet per second. He runs the second half at 24 feet per second. How long does it take him to run the 60 yards?

Quick Quiz 9.4 Consider the equation $y = x^2 - 6x + 5$.

1. What is the vertex of this curve?

2. What are the x-intercepts of this curve?

3. Graph the equation.

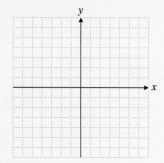

4. **Concept Check** Explain how you would find the vertex of the equation $y = 4x^2 + 16x - 2$.

① Solving Applied Problems Using Quadratic Equations

We can use what we have learned about quadratic equations to solve word problems. Some word problems will involve geometry.

Recall that the Pythagorean Theorem states that in a right triangle, $a^2 + b^2 = c^2$, where a and b are the lengths of the legs and c is the length of the hypotenuse. We will use this theorem to solve word problems involving right triangles.

Student Learning Objective

After studying this section, you will be able to:

① Solve applied problems using quadratic equations.

▲ **EXAMPLE 1** The hypotenuse of a right triangle is 25 meters in length. One leg is 17 meters longer than the other. Find the length of each leg.

Solution

1. *Understand the problem.* Draw a picture.
2. *Write an equation.* The problem involves a right triangle. Use the Pythagorean Theorem.

$$a^2 + b^2 = c^2$$
$$x^2 + (x + 17)^2 = 25^2$$

3. *Solve and state the answer.*

$$x^2 + (x + 17)^2 = 25^2$$
$$x^2 + x^2 + 34x + 289 = 625$$
$$2x^2 + 34x - 336 = 0$$
$$x^2 + 17x - 168 = 0$$
$$(x + 24)(x - 7) = 0$$
$$x + 24 = 0 \qquad x - 7 = 0$$
$$x = -24 \qquad x = 7$$

(Note that $x = -24$ is not a valid solution for this particular word problem.)

One leg is 7 meters in length. The other leg is $x + 17 = 7 + 17 = 24$ meters in length.

4. *Check.* Check the conditions in the original problem.
Do the two legs differ by 17 meters?

$$24 - 7 \overset{?}{=} 17 \qquad 17 = 17 \ \checkmark$$

Do the sides of the triangle satisfy the Pythagorean Theorem?

$$7^2 + 24^2 \overset{?}{=} 25^2$$
$$49 + 576 \overset{?}{=} 625$$
$$625 = 625 \ \checkmark$$

▲ **Practice Problem 1** The hypotenuse of a right triangle is 30 meters in length. One leg is 6 meters shorter than the other leg. Find the length of each leg.

NOTE TO STUDENT: Fully worked-out solutions to all of the Practice Problems can be found at the back of the text starting at page SP-1

Sometimes it may help to use two variables in the initial part of the problem. Then one variable can be eliminated by substitution.

EXAMPLE 2 The ski club is renting a bus to travel to Mount Snow. The members agreed to share the cost of $180 equally. On the day of the ski trip, three members were sick with the flu and could not go. This increased the share of each person going on the trip by $10. How many people originally planned to attend?

Solution

1. *Understand the problem.*

   Let s = the number of students in the ski club

   c = the cost for each student in the original group

2. *Write an equation(s).*

   number of students × cost per student = total cost

   $$s \cdot c = 180 \qquad \textbf{(1)}$$

   If three people are sick, the number of students drops by three, but the cost for each increases by $10. The total is still $180. Therefore, we have the following.

   $$(s - 3)(c + 10) = 180 \qquad \textbf{(2)}$$

3. *Solve and state the answer.*

   $$c = \frac{180}{s} \qquad \text{Solve equation \textbf{(1)} for } c.$$

   $$(s - 3)\left(\frac{180}{s} + 10\right) = 180 \qquad \text{Substitute } \frac{180}{s} \text{ for } c \text{ in equation \textbf{(2)}.}$$

   $$(s - 3)\left(\frac{180 + 10s}{s}\right) = 180 \qquad \text{Add } \frac{180}{s} + 10. \text{ Use the LCD, which is } s.$$

   $$(s - 3)(180 + 10s) = 180s \qquad \text{Multiply both sides of the equation by } s.$$

   $$10s^2 + 150s - 540 = 180s \qquad \text{Multiply the binomials.}$$

   $$10s^2 - 30s - 540 = 0$$

   $$s^2 - 3s - 54 = 0$$

   $$(s - 9)(s + 6) = 0$$

   $$s = 9 \qquad s = -6$$

   The number of students originally going on the ski trip was 9.

4. *Check.* Would the cost increase by $10 if the number of students dropped from nine to six? Nine people in the club would mean that each would pay $20 ($180 \div 9 = 20$).

   If the number of students was reduced by three, there were six people who took the trip. Their cost was $30 each ($180 \div 6 = 30$). The increase is $10.

   $$20 + 10 \overset{?}{=} 30$$

   $$30 = 30 \quad \checkmark$$

NOTE TO STUDENT: Fully worked-out solutions to all of the Practice Problems can be found at the back of the text starting at page SP-1

Practice Problem 2 Several friends from the University of New Mexico decided to charter a sport fishing boat for the afternoon while on a vacation at the coast. They agreed to share the rental cost of $240 equally. At the last minute, two of them could not go on the trip. This raised the cost for each person in the group by $4. How many people originally planned to charter the boat?

▲ EXAMPLE 3   Minette is fencing a garden that borders the back of a large barn. She has 120 feet of fencing. She would like a rectangular garden that measures 1350 square feet in area. She wants to use the back of the barn, so she needs to use fencing on three sides only. What dimensions should she use for her garden?

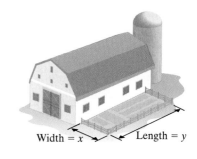

Width = x Length = y

Solution

1. *Understand the problem.* Draw a picture.

   Let x = the width of the garden in feet and
      y = the length of the garden in feet.

2. *Write an equation(s).* Look at the drawing. The 120 feet of fencing will be needed for the width twice (two sides) and the length once (one side).

$$120 = 2x + y \quad \textbf{(1)}$$

   The area formula is $A = (\text{width})(\text{length})$.

$$1350 = xy \quad \textbf{(2)}$$

3. *Solve and state the answer.*

$$y = 120 - 2x \qquad \text{Solve for } y \text{ in equation } \textbf{(1)}.$$
$$1350 = x(120 - 2x) \qquad \text{Substitute for } y \text{ in equation } \textbf{(2)}.$$
$$1350 = 120x - 2x^2 \qquad \text{Solve for } x.$$

$$2x^2 - 120x + 1350 = 0$$
$$x^2 - 60x + 675 = 0$$
$$(x - 15)(x - 45) = 0$$
$$x = 15 \qquad\qquad x = 45$$

First Solution: If the width is 15 feet, the length is 90 feet.

Check: $15 + 15 + 90 = 120$ ✓
             $(15)(90) = 1350$ ✓

First solution

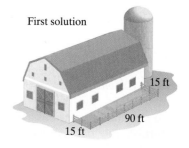

15 ft

90 ft

15 ft

Second Solution: If the width is 45 feet, the length is 30 feet.

Check: $45 + 45 + 30 = 120$ ✓
             $(45)(30) = 1350$ ✓

Thus we see that Minette has two choices that satisfy the given requirements (see the figures). She would probably choose the shape that is the most practical for her. (If her barn is shorter than 90 feet in length, she could not use the first solution!)

   Remark: Is there another way to do this problem? Yes. Use only the variable x to represent the length and width. If you let x = the width and $120 - 2x$ = the length, you can immediately write equation **(3)**, which is

Second solution

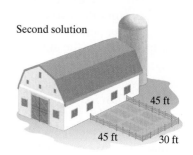

45 ft

45 ft   30 ft

$$1350 = x(120 - 2x). \quad \textbf{(3)}$$

However, many students find that this is difficult to do, and they would prefer to use two variables. Work these problems in the way that is easiest for you.

▲ Practice Problem 3   An Arizona rancher has a small fenced-in area against a canyon wall. Therefore, he only needs three sides of fencing to make a rectangular holding area for cattle. He has 160 feet of fencing available. He wants the rectangular area to measure 3150 square feet. What dimensions should he use for the rectangular holding area?

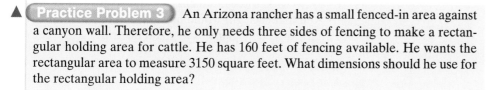

Applications *Solve.*

▲ **1.** *Geometry* The base of a triangle is 5 centimeters shorter than the altitude. The area of the triangle is 88 square centimeters. Find the lengths of the base and altitude.

▲ **2.** *Geometry* The area of a rectangle is 96 square meters. If the length is 4 more than the width, what are the dimensions of the rectangle?

▲ **3.** *Garden Expansion* Emily wishes to expand the size of her front garden, which is square. Emily measures her front yard and decides she can increase the width by 4 feet and the length by 9 feet. If the area of the newly expanded garden is 266 square feet, what were the dimensions of the old garden?

▲ **4.** *Addition to a House* Mike and Norma have a home with a very small rectangular room on the first floor that they use for a living room. This room is 2 feet longer than it is wide. They have decided to put on an addition to the house that will give them a larger rectangular room to meet the need for an adequate living room for their family. The length of the living room will increase by 10 feet. The width of the living room will increase by 8 feet. The area of the new, expanded living room will be 252 square feet. What are the dimensions of the present living room before the addition is built?

5. *Gift Expense* To help out Nina and Tom with their new twins, the employees at Tom's office decided to give the new parents a gift certificate worth $420 to furnish the babies' room. The employees planned to share the cost equally. At the last minute, seven people from the executive staff chipped in, lowering the cost per person by $5. How many people originally planned to give the gift?

6. *Septic Systems* Everyone in the School Street Townhouse Association agreed to equally share the cost of a new $6000 septic system. However, before the system was installed, five additional units were built and the five new families were asked to contribute their share. This lowered the contribution of each family by $200. How many families originally agreed to share the cost of the new septic system?

7. *Music Cost* A fraternity was sponsoring a spring fling, for which the $400 DJ cost was to be split evenly among those members attending. If 20 more members than were expected came, reducing the cost per person by $1, find the number of fraternity members that attended.

8. *Beach House Rental* The Cornell University *Wild Roses* Ultimate team took a spring break trip to Savannah, Georgia, where they all agreed to chip in and rent a beach house for the week. The cost of the rental for the week was $1500, and each girl was assessed an equal amount. Just before they left New York, three more girls decided to go. The captain said that each girl would now pay $25 less. How many girls originally planned to go on the trip?

▲ **9. Fenway Park** A security fence encloses a rectangular area on one side of Fenway Park in Boston. Three sides of fencing are used, since the fourth side of the area is formed by a building. The enclosed area measures 1250 square feet. Exactly 100 feet of fencing is used to fence in three sides of this rectangle. What are the possible dimensions that could have been used to construct this area?

▲ **10. Garden Fencing** Henry Redbone needs to fence in his garden, to protect it from the rabbits who love to feast on his lettuce. He plans on planting the garden next to the barn, so that he only needs to fence in three sides of the garden. If Henry has 80 feet of fencing available, and he wants the garden to have an area of 750 square feet, what are the possible dimensions for the garden?

11. Experimental Jet A pilot is testing a new experimental craft. The cruising speed is classified information. In a test the jet traveled 2400 miles. The pilot revealed that if he had increased his speed 200 mph, the total trip would have taken 1 hour less. Determine the cruising speed of the jet.

▲ **12. Gymnasium Floor Replacement** A gymnasium floor is being covered by square shock-absorbing tiles. The old gym floor required 864 square tiles. The new tiles are 2 inches larger in both length and width than the old tiles. The new flooring will require only 600 tiles. What is the length of a side of one of the new shock-absorbing tiles? (*Hint:* Since the area is the same in each case, write an expression for area with old tiles and one with new tiles. Set the two expressions equal to each other.)

Use the following information to solve exercises 13–18.

Business Expansion *Two young college graduates opened a chain of Quick Print stores in southern California. The chain expanded during the late 1990s and early 2000s and many new stores were opened. The number of Quick Print Centers in operation during any year from 1998 to 2008 is given by the equation $y = -2.5x^2 + 22.5x + 50$, where x is the number of years since 1998.*

13. How many Quick Print Centers were in operation in 1998?

14. How many Quick Print Centers were in operation in 2001?

15. How many more Quick Print Centers were in operation in 2002 than in 2000?

16. How many fewer Quick Print Centers were in operation in 2007 than in 2005?

17. During what two years were there 100 Quick Print Centers?

18. During what two years were there 70 Quick Print Centers?

Cannon Shell Path Big Bertha, *an infamous World War I cannon, fired shells over an amazingly long distance. The path of a shell from this monstrous cannon can be modeled by the formula $y = -0.018x^2 + 1.336x$, where x is the horizontal distance the cannon shell traveled in miles and y is the height in miles the shell would reach on its flight.*

19. How far could *Big Bertha* send a shell? Round to the nearest whole number.

20. (a) How far into its journey would the shell achieve maximum height? Round to the nearest whole number.
   (b) What was the maximum height in miles? Round to the nearest whole number.

Cumulative Review

21. [2.8.4] Graph on a number line the solution to $3x + 11 \geq 9x - 4$.

22. [8.5.3] Rationalize the denominator. $\dfrac{\sqrt{3}}{\sqrt{5} + 2}$

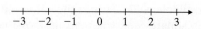

▲ **23. [4.5.1]** *Recycling Bin*   A recycling bin that has a length of 3 feet, a width of 2 feet, and a height of 2 feet has three separate sections inside for sorting. Section 1 is for paper, section 2 is for glass and plastic, and section 3 is for metal. Sections 1 and 2 have identical measurements, but section 3 has $\frac{2}{3}$ the volume of section 1 or 2. If all three sections have the same width and height, how long is section 3?

Quick Quiz 9.5

1. The hypotenuse of a right triangle is 13 yards long. One leg is 7 yards longer than the other leg. Find the length of each leg.

When an object is thrown upward, its height (S) in meters is given approximately by the quadratic equation
$$S = -5t^2 + vt + h$$
where $v =$ the initial upward velocity in meters per second,

      $t =$ the time of flight in seconds, and

      $h =$ the height above level ground from which the object is thrown.

Suppose a ball is thrown upward with a velocity of 25 meters per second at a height of 30 meters above the ground.

2. How long after it is thrown will the ball hit the ground?

3. What is the maximum height of the ball?

4. Concept Check   The hiking club needs to raise $720 through dues paid by each member. However, 4 people dropped out of the club so the dues went up by $6 for each member. Explain how you could set up an equation to find out how many members were in the club prior to the dropout.

Putting Your Skills to Work: Use Math to Save Money

ADJUST THE THERMOSTAT

Some Helpful Information

According to the U.S. Department of Energy, 45% of a typical home utility bill is for heating and/or cooling. We may not have any control over how fuel efficient our home was constructed, but we can control the setting on the thermostat. Consumer Reports suggests that every 1° change in our thermostat setting has a 2% impact on our utility bill. Consider the story of Maria and Josef.

A Specific Example

Maria and Josef use electricity to heat and cool their home, and their electricity bill averages $205 per month. They generally set their thermostat at 72° in the colder months and 78° in the warmer months.

Quick Calculations and Facts

1. If Maria and Josef set their thermostat at 68° in the colder months, how much per month could they expect to save?

2. If they adjust their thermostat to 84° in the warmer months, how much could they expect to save per month?

3. On the average, if Maria and Josef adjusted their thermostat setting 5° (either upward in the warmer months or downward in the colder

months), how much could they expect to save on their utilities per year?

Making Personal Applications to Your Own Life

4. Where do you have your thermostat set?

5. How much could you save on your utility bill by adjusting the settings either up or down?

HOME HEATING OIL PRICES

In parts of the United States where the climate turns cold during the winter, many homes are heated by oil. Mark lives just outside of Boston in a house heated by oil. During the winter of 2007, Mark paid $2.95 per gallon for home heating oil. The price of home heating oil has since increased to $4.45 per gallon. Looking ahead to the winter of 2008, Mark wants to budget enough money for his oil bills.

6. The oil company will only deliver 100 gallons of heating oil or more. Determine the cost of 100 gallons of home heating oil in December 2007 and November 2008. What is the difference in cost?

7. Mark knows he will need a 100-gallon oil delivery every month during the winter (November and December 2008, and January, February, and March 2009). What can Mark expect to pay in home heating oil costs for the five months of winter from November 2008 through March 2009?

8. Mark typically keeps his thermostat set at 72° during the winter months. How much money would Mark save on his heating oil costs if he lowers his thermostat to 68° for the entire winter?

9. How much money would Mark save if he sets his thermostat to 64° during the winter?

10. How low would Mark need to set his thermostat for the winter in order to save one month's worth of heating costs ($445)?

Topic	Procedure	Examples
Solving quadratic equations by factoring, p. 521.	**1.** Clear the equation of fractions, if necessary. **2.** Write as $ax^2 + bx + c = 0$. **3.** Factor. **4.** Set each factor equal to 0. **5.** Solve the resulting equations.	Solve. $x + \dfrac{1}{2} = \dfrac{3}{2x}$ Multiply each term by LCD $= 2x$. $$2x(x) + 2x\left(\dfrac{1}{2}\right) = 2x\left(\dfrac{3}{2x}\right)$$ $$2x^2 + x = 3$$ $$2x^2 + x - 3 = 0$$ $$(2x + 3)(x - 1) = 0$$ $$2x + 3 = 0 \qquad x - 1 = 0$$ $$2x = -3 \qquad x = 1$$ $$x = -\dfrac{3}{2}$$
Solving quadratic equations: taking the square root of each side of the equation, p. 528.	Begin solving quadratic equations of the form $ax^2 - c = 0$ by solving for x^2. Then use the property that if $x^2 = a$, $x = \pm\sqrt{a}$. This amounts to taking the square root of each side of the equation.	Solve. $2x^2 + 1 = 99$ $$2x^2 = 99 - 1$$ $$2x^2 = 98$$ $$\dfrac{2x^2}{2} = \dfrac{98}{2}$$ $$x^2 = 49$$ $$x = \pm\sqrt{49} = \pm 7$$
Solving quadratic equations: completing the square, p. 529.	**1.** Put the equation in the form $ax^2 + bx = c$. If $a \neq 1$, divide each term by a. **2.** Square $\dfrac{b}{2}$ and add the result to both sides of the equation. **3.** Factor the left side (a perfect square trinomial). **4.** Use the square root property. **5.** Solve the equations. **6.** Check the solutions in the original equation.	Solve. $5x^2 - 3x - 7 = 0$ $$5x^2 - 3x = 7$$ $$x^2 - \dfrac{3}{5}x = \dfrac{7}{5}$$ $$x^2 - \dfrac{3}{5}x + \dfrac{9}{100} = \dfrac{7}{5} + \dfrac{9}{100}$$ $$\left(x - \dfrac{3}{10}\right)^2 = \dfrac{149}{100}$$ $$x - \dfrac{3}{10} = \pm\dfrac{\sqrt{149}}{10}$$ $$x = \dfrac{3}{10} \pm \dfrac{\sqrt{149}}{10}$$ $$x = \dfrac{3 \pm \sqrt{149}}{10}$$
Solving quadratic equations: quadratic formula, p. 534.	**1.** Put the equation in standard form: $ax^2 + bx + c = 0$. **2.** Carefully determine the values of a, b, and c. **3.** Substitute these into the formula $$x = \dfrac{-b \pm \sqrt{b^2 - 4ac}}{2a}.$$ **4.** Simplify.	Solve. $3x^2 - 4x - 8 = 0$ $$a = 3, \ b = -4, \ c = -8$$ $$x = \dfrac{-(-4) \pm \sqrt{(-4)^2 - 4(3)(-8)}}{2(3)}$$ $$= \dfrac{4 \pm \sqrt{16 + 96}}{6}$$ $$= \dfrac{4 \pm \sqrt{112}}{6} = \dfrac{4 \pm \sqrt{16 \cdot 7}}{6}$$ $$= \dfrac{4 \pm 4\sqrt{7}}{6} = \dfrac{\cancel{2}\left(2 \pm 2\sqrt{7}\right)}{\cancel{2} \cdot 3} = \dfrac{2 \pm 2\sqrt{7}}{3}$$

Topic	Procedure	Examples
The discriminant, **p. 537.**	The *discriminant* in the quadratic formula is $b^2 - 4ac$. If it is negative, the quadratic equation has no real solutions (since it is not possible to take the square root of a negative number and get a real number).	What kinds of solutions does this equation have? $$3x^2 + 2x + 1 = 0$$ $$b^2 - 4ac = 2^2 - 4 \cdot 3 \cdot 1$$ $$= 4 - 12 = -8$$ This equation has no real solutions.
Properties of **parabolas, p. 544.**	**1.** The graph of $y = ax^2 + bx + c$ is a parabola. It opens up if $a > 0$ and down if $a < 0$. **2.** The vertex of a parabola is its lowest (if $a > 0$) or highest (if $a < 0$) point. Its x-coordinate is $x = \dfrac{-b}{2a}$.	The vertex of $y = 3x^2 + 4x - 11$ has x-coordinate $$x = \frac{-b}{2a} = \frac{-4}{6} = -\frac{2}{3}.$$ The parabola opens upward.
Graphing parabolas, **p. 544.**	Graph the vertex of a parabola and several other points to get a good idea of its graph. Usually, we find the x-intercepts if any exist. The following procedure is helpful. **1.** Determine if $a < 0$ or $a > 0$. **2.** Find the vertex. **3.** Find the x-intercepts. **4.** Plot one or two extra points if necessary. **5.** Draw the graph.	Graph. $y = x^2 - 2x - 8$ $a = 1, b = -2, c = -8; a > 0$, so the parabola opens upward. The x-coordinate of the vertex is $$x = \frac{-b}{2a} = \frac{-(-2)}{2} = 1.$$ If $x = 1, y = (1)^2 - 2(1) - 8 = -9$. The vertex is at $(1, -9)$. To find the x-intercepts, let $y = 0$. $$0 = x^2 - 2x - 8$$ $$0 = (x - 4)(x + 2)$$ $$x = 4, \quad x = -2$$ The x-intercepts are $(4, 0)$ and $(-2, 0)$.
Formulas and applied **problems, p. 551.**	Some word problems require the use of a quadratic equation. You may want to follow this four-step procedure for problem solving. **1.** Understand the problem. **2.** Write an equation. **3.** Solve and state the answer. **4.** Check.	The area of a rectangle is 48 square inches. The length is three times the width. Find the dimensions of the rectangle. **1.** Draw a picture. **2.** $(3w)w = 48$ **3.** $3w^2 = 48$ $w^2 = 16$ $w = 4$ inches, $l = 12$ inches **4.** $(4)(12) = 48$

Chapter 9 Review Problems

Section 9.1 *Write in standard form. Determine the values of a, b, and c. Do not solve.*

1. $6x^2 = 5x - 8$

2. $9x^2 + 3x = -5x^2 + 16$

3. $(x + 3)(2x - 3) = x^2 + 5$

4. $3x(4x + 1) = -x(5 - x)$

5. $\dfrac{3}{x^2} - \dfrac{6}{x} - 2 = 0$

6. $\dfrac{x}{x + 2} - \dfrac{3}{x - 2} = 5$

Place in standard form. Solve by factoring.

7. $x^2 + 26x + 25 = 0$

8. $x^2 + 16x + 64 = 0$

9. $x^2 + 6x - 55 = 0$

10. $4x^2 - 16x + 15 = 0$

11. $9x^2 - 24x + 16 = 0$

12. $15x^2 = 26x - 8$

13. $6x^2 = 13x - 5$

14. $4x^2 = -5x + 9$

15. $x^2 + \dfrac{1}{6}x - 2 = 0$

16. $\dfrac{1}{2}x^2 = \dfrac{3}{4}x - \dfrac{1}{4}$

17. $x^2 + \dfrac{2}{15}x = \dfrac{1}{15}$

18. $1 + \dfrac{13}{12x} - \dfrac{1}{3x^2} = 0$

19. $x^2 + 9x + 5 = x + 5$

20. $x^2 - 2x + 7 = 7 + x$

21. $1 + \dfrac{2}{3x + 4} = \dfrac{3}{3x + 2}$

22. $2 - \dfrac{5}{x + 1} = \dfrac{3}{x - 1}$

23. $5 + \dfrac{24}{2 - x} = \dfrac{24}{2 + x}$

24. $\dfrac{4}{9} = \dfrac{5x}{3} - x^2$

Section 9.2 *Solve by taking the square root of each side.*

25. $x^2 - 8 = 41$

26. $x^2 + 11 = 92$

27. $x^2 - 5 = 17$

28. $x^2 + 11 = 50$

29. $2x^2 - 1 = 15$

30. $3x^2 + 4 = 154$

31. $3x^2 + 6 = 60$

32. $2x^2 - 5 = 43$

33. $(x - 4)^2 = 7$

34. $(x - 2)^2 = 3$

35. $(3x + 2)^2 = 28$

36. $(2x - 1)^2 = 32$

Solve by completing the square.

37. $x^2 + 8x + 7 = 0$

38. $x^2 + 14x + 33 = 0$

39. $2x^2 - 8x - 90 = 0$

40. $-5x^2 + 30x - 35 = 0$

41. $3x^2 + 6x - 6 = 0$

42. $2x^2 + 10x - 3 = 0$

Section 9.3 *Solve using the quadratic formula.*

43. $x^2 + 4x - 6 = 0$

44. $x^2 + 4x - 8 = 0$

45. $2x^2 - 7x + 4 = 0$

46. $2x^2 + 5x - 6 = 0$

47. $3x^2 - 8x - 4 = 0$

48. $4x^2 - 2x - 11 = 0$

49. $5x^2 - x = 2$

50. $3x^2 + 3x = 1$

Mixed Practice *Solve by any method. If there is no real number solution, say so.*

51. $2x^2 - 9x + 10 = 0$

52. $4x^2 - 4x - 3 = 0$

53. $25x^2 + 10x + 1 = 0$

54. $2x^2 - 11x + 12 = 0$

55. $3x^2 - 6x + 2 = 0$

56. $5x^2 - 7x = 8$

57. $4x^2 + 4x = x^2 + 5$

58. $5x^2 + 7x + 1 = 0$

59. $x^2 = 9x + 3$

60. $2x^2 = 9x + 5$

61. $-20x^2 - 15x = -8x^2 - 18$

62. $2x^2 - 1 = 35$

63. $\dfrac{(y-2)^2}{20} + 3 + y = 0$

64. $\dfrac{(y+2)^2}{5} + 2y = -9$

65. $3x^2 + 1 = 6 - 8x$

66. $4x^2 + 11x = 5x + 1$

67. $8y(y+1) = 7y + 9$

68. $\dfrac{3y-2}{4} = \dfrac{y^2-2}{y}$

69. $\dfrac{y^2+5}{2y} = \dfrac{2y-1}{3}$

70. $\dfrac{5x^2}{2} = x - \dfrac{7x^2}{2}$

Section 9.4 *Graph each quadratic equation. Label the vertex.*

71. $y = 2x^2$

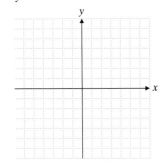

72. $y = x^2 + 4$

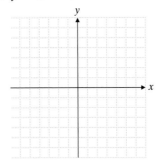

73. $y = x^2 - 3$

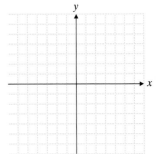

74. $y = -\dfrac{1}{2}x^2$

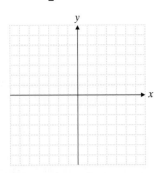

75. $y = x^2 - 3x - 4$

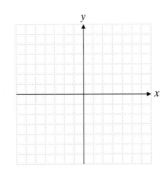

76. $y = \dfrac{1}{2}x^2 - 2$

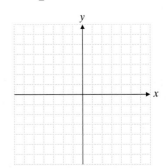

77. $y = -2x^2 + 12x - 17$

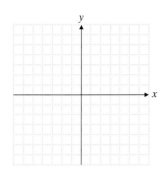

78. $y = -3x^2 - 2x + 4$

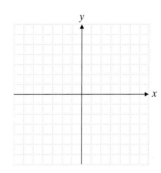

Section 9.5 *Solve.*

▲ **79.** ***Dog Pen Fencing*** Ken is building a dog pen against his house. He has 15 feet of fencing and wants to pen in an area of 28 square feet. What are the possible dimensions of the pen?

▲ **80.** ***Geometry*** The hypotenuse of a right triangle is 20 centimeters. One leg of the right triangle is 4 centimeters shorter than the other. What is the length of each leg of the right triangle?

▲ **81.** ***TV Screen Dimensions*** A television designer believes that the dimensions (length and width) of a TV screen are as important to sales as the size (measured by the diagonal) of the TV. If the diagonal of a newly designed TV is 30 inches and the length is 6 inches more than the width, what are the dimensions of the TV?

82. ***Space Probe*** Suppose an advanced alien civilization sends a space ship to orbit around our planet. They send a probe to Earth to do research. The equation for the height of the probe is $h = -225t^2 + 202,500$, where h is the distance in feet from the surface of Earth t seconds after the probe leaves the mother ship. How long will it take the probe to reach Earth from the mother ship?

83. ***Golf Club*** Last year the golf club on campus raised $720 through dues assigned equally to each member. This year there were four fewer members. As a result, the dues for each member went up by $6. How many members were in the golf club last year?

84. ***Automobile Trip*** Alice drove 90 miles to visit her cousin. Her average speed on the trip home was 15 mph faster than her average speed on the trip there. Her total travel time for both trips was 3.5 hours. What was her average rate of speed on each trip?

Use the following information to solve exercises 85–88.

Lacorazza Pizza Chain *The Lacorazza family opened a chain of pizza stores in New England in 1980. The number of pizza stores in operation during any year is given by the equation* $y = -0.05x^2 + 2.5x + 60$*, where x is the number of years since 1980.*

85. According to the equation, how many pizza stores will still be in operation in 2030?

86. How many fewer pizza stores will still be in operation in 2040 than in 2030?

87. During what two years will there be exactly 80 pizza stores in operation?

88. During what two years will there be exactly 90 pizza stores in operation?

How Am I Doing? Chapter 9 Test

Remember to use your Chapter Test Prep Video CD to see the worked-out solutions to the test problems you want to review.

Solve by any desired method. If there is no real number solution, say so.

1. $5x^2 + 7x = 4$

2. $3x^2 + 13x = 10$

3. $2x^2 = 2x - 5$

4. $4x^2 - 19x = 4x - 15$

5. $12x^2 + 11x = 5$

6. $18x^2 + 32 = 48x$

7. $2x^2 - 11x + 3 = 5x + 3$

8. $5x^2 + 7 = 52$

9. $2x(x - 6) = 6 - x$

10. $x^2 - x = \dfrac{3}{4}$

Graph each quadratic equation. Locate the vertex.

11. $y = 3x^2 - 6x$

12. $y = -x^2 + 8x - 12$

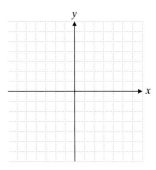

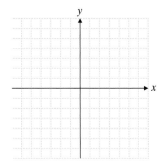

Solve.

▲ **13.** The hypotenuse of a right triangle is 15 meters in length. One leg is 3 meters longer than the other leg. Find the length of each leg.

14. When an object is thrown upward, its height (S) in meters is given (approximately) by the quadratic equation

$$S = -5t^2 + vt + h,$$

where v = the initial upward velocity in meters per second,

t = the time of flight in seconds, and

h = the height above level ground from which the object is thrown.

Suppose that a ball is thrown upward with a velocity of 33 meters/second at a height of 14 meters above the ground. How long after it is thrown will it strike the ground?

1. _____

2. _____

3. _____

4. _____

5. _____

6. _____

7. _____

8. _____

9. _____

10. _____

11. _____

12. _____

13. _____

14. _____

Practice Final Examination

The following questions cover the content of Chapters 0–9. Follow the directions for each question and simplify your answers.

1. Simplify.
$$-2x + 3y\{7 - 2[x - (4x + y)]\}$$

2. Evaluate if $x = -2$ and $y = 3$.
$$2x^2 - 3xy - 4y$$

3. Simplify. $(-3x^2y)(-6x^3y^4)$

4. Combine like terms.
$$5x^2y - 6xy + 8xy - 3x^2y - 10xy$$

5. $2(x + 4) - 6 = 4x - 3$

6. Solve for x.
$$\frac{1}{2}(x + 4) - \frac{2}{3}(x - 7) = 4x$$

7. Solve for x and graph the resulting inequality on a number line.
$$5x + 3 - (4x - 2) \le 6x - 8$$

8. Multiply. $(2x + 3)^2$

9. Multiply. $(2x + y)(x^2 - 3xy + 2y^2)$

10. Factor completely. $4x^2 - 18x - 10$

11. Factor completely. $3x^3 - 9x^2 - 30x$

12. Combine. $\dfrac{2}{x - 3} - \dfrac{3}{x^2 - x - 6} + \dfrac{4}{x + 2}$

13. Simplify. $\dfrac{\dfrac{3}{x} + \dfrac{5}{2x}}{1 + \dfrac{2}{x + 2}}$

14. Solve for x. $\dfrac{2}{x + 2} = \dfrac{4}{x - 2} + \dfrac{3x}{x^2 - 4}$

15. Find the slope of the line and then graph the line. $5x - 2y - 3 = 0$

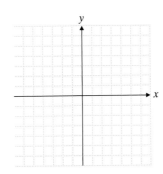

16. Find the equation of the line that has a slope of $-\dfrac{3}{4}$ and passes through the point $(-2, 5)$.

1. _____

2. _____

3. _____

4. _____

5. _____

6. _____

7. _____

8. _____

9. _____

10. _____

11. _____

12. _____

13. _____

14. _____

15. _____

16. _____

17. _____

18. _____

19. _____

20. _____

21. _____

22. _____

23. _____

24. _____

25. _____

26. _____

27. _____

28. _____

29. _____

30. _____

▲ **17.** Two quarter-circles are attached to a rectangle. The dimensions are indicated in the figure. Find the area of the region. Use 3.14 as an approximation for π.

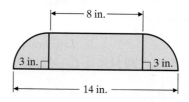

18. Solve for x and y.

$$2x + 7y = 4$$
$$-3x - 5y = 5$$

19. Solve for a and b.

$$a - \frac{3}{4}b = \frac{1}{4}$$
$$\frac{3}{2}a + \frac{1}{2}b = -\frac{9}{2}$$

20. Simplify and combine.
$$\sqrt{45x^3} + 2x\sqrt{20x} - 6\sqrt{5x^3}$$

21. Multiply and simplify.
$$\sqrt{6}\left(3\sqrt{2} - 2\sqrt{6} + 4\sqrt{3}\right)$$

22. Simplify by rationalizing the denominator. $\dfrac{\sqrt{3} + \sqrt{7}}{\sqrt{5} - \sqrt{7}}$

23. Solve for x. $12x^2 - 5x - 2 = 0$

24. Solve for y. $2y^2 = 6y - 1$

25. Solve for x. $4x^2 + 3 = 19$

▲ **26.** In the right triangle shown, sides b and c are given. Find the length of side a.

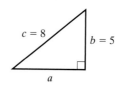

27. A number is tripled and then increased by 6. The result is 21. What is the original number?

▲ **28.** A rectangular region has a perimeter of 38 meters. The length of the rectangle is 2 meters shorter than double the width. What are the dimensions of the rectangle?

29. A benefit concert was held on campus to raise scholarship money. A total of 360 tickets were sold. Admission prices were $5 for reserved seats and $3 for general admission. Total receipts were $1480. How many reserved-seat tickets were sold? How many general admission tickets were sold?

▲ **30.** The area of a triangle is 68 square meters. The altitude of the triangle is one meter longer than double the base of the triangle. Find the altitude and base of the triangle.

Appendix A Table of Square Roots

x	$\sqrt{x}$	x	$\sqrt{x}$	x	$\sqrt{x}$	x	$\sqrt{x}$	x	$\sqrt{x}$
1	1.000	41	6.403	81	9.000	121	11.000	161	12.689
2	1.414	42	6.481	82	9.055	122	11.045	162	12.728
3	1.732	43	6.557	83	9.110	123	11.091	163	12.767
4	2.000	44	6.633	84	9.165	124	11.136	164	12.806
5	2.236	45	6.708	85	9.220	125	11.180	165	12.845
6	2.449	46	6.782	86	9.274	126	11.225	166	12.884
7	2.646	47	6.856	87	9.327	127	11.269	167	12.923
8	2.828	48	6.928	88	9.381	128	11.314	168	12.961
9	3.000	49	7.000	89	9.434	129	11.358	169	13.000
10	3.162	50	7.071	90	9.487	130	11.402	170	13.038
11	3.317	51	7.141	91	9.539	131	11.446	171	13.077
12	3.464	52	7.211	92	9.592	132	11.489	172	13.115
13	3.606	53	7.280	93	9.644	133	11.533	173	13.153
14	3.742	54	7.348	94	9.695	134	11.576	174	13.191
15	3.873	55	7.416	95	9.747	135	11.619	175	13.229
16	4.000	56	7.483	96	9.798	136	11.662	176	13.266
17	4.123	57	7.550	97	9.849	137	11.705	177	13.304
18	4.243	58	7.616	98	9.899	138	11.747	178	13.342
19	4.359	59	7.681	99	9.950	139	11.790	179	13.379
20	4.472	60	7.746	100	10.000	140	11.832	180	13.416
21	4.583	61	7.810	101	10.050	141	11.874	181	13.454
22	4.690	62	7.874	102	10.100	142	11.916	182	13.491
23	4.796	63	7.937	103	10.149	143	11.958	183	13.528
24	4.899	64	8.000	104	10.198	144	12.000	184	13.565
25	5.000	65	8.062	105	10.247	145	12.042	185	13.601
26	5.099	66	8.124	106	10.296	146	12.083	186	13.638
27	5.196	67	8.185	107	10.344	147	12.124	187	13.675
28	5.292	68	8.246	108	10.392	148	12.166	188	13.711
29	5.385	69	8.307	109	10.440	149	12.207	189	13.748
30	5.477	70	8.367	110	10.488	150	12.247	190	13.784
31	5.568	71	8.426	111	10.536	151	12.288	191	13.820
32	5.657	72	8.485	112	10.583	152	12.329	192	13.856
33	5.745	73	8.544	113	10.630	153	12.369	193	13.892
34	5.831	74	8.602	114	10.677	154	12.410	194	13.928
35	5.916	75	8.660	115	10.724	155	12.450	195	13.964
36	6.000	76	8.718	116	10.770	156	12.490	196	14.000
37	6.083	77	8.775	117	10.817	157	12.530	197	14.036
38	6.164	78	8.832	118	10.863	158	12.570	198	14.071
39	6.245	79	8.888	119	10.909	159	12.610	199	14.107
40	6.325	80	8.944	120	10.954	160	12.649	200	14.142

Unless the value of $\sqrt{x}$ ends in 000, all values are rounded to the nearest thousandth.

Appendix B Using the Mathematics Blueprint for Problem Solving

After studying this section, you will be able to:

 Use the Mathematics Blueprint to solve real-life problems.

Using the Mathematics Blueprint to Solve Real-Life Problems

When a builder constructs a new home or office building, he or she often has a blueprint. This accurate drawing shows the basic form of the building. It also shows the dimensions of the structure to be built. This blueprint serves as a useful reference throughout the construction process.

Similarly, when solving real-life problems, it is helpful to have a "mathematics blueprint." This is a simple way to organize the information provided in the word problem in a chart or in a graph. You can record the facts you need to use. You can determine what it is you are trying to find and how you can go about actually finding it. You can record other information that you think will be helpful as you work through the problem.

As we solve real-life problems, we will use three steps.

Step 1 *Understand the problem.* Here we will read through the problem. We draw a picture if it will help, and use the Mathematics Blueprint as a guide to assist us in thinking through the steps needed to solve the problem.

Step 2 *Solve and state the answer.* We will use arithmetic or algebraic procedures along with problem-solving strategies to find a solution.

Step 3 *Check.* We will use a variety of techniques to see if the answer in step 2 is the solution to the word problem. This will include estimating to see if the answer is reasonable, repeating our calculation, and working backward from the answer to see if we arrive at the original conditions of the problem.

▲ **EXAMPLE 1** Nancy and John want to install wall-to-wall carpeting in their living room. The floor of the rectangular living room is $11\frac{2}{3}$ feet wide and $19\frac{1}{2}$ feet long. How much will it cost if the carpet is $18.00 per square yard?

Solution

1. *Understand the problem.* First, read the problem carefully. Drawing a sketch of the living room may help you see what is required. The carpet will cover the floor of the living room, so we need to find the area. Now we fill in the Mathematics Blueprint.

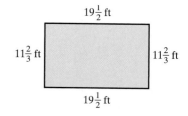

Mathematics Blueprint for Problem Solving

Gather the Facts	What Am I Solving For?	What Must I Calculate?	Key Points to Remember
The living room measures $11\frac{2}{3}$ ft by $19\frac{1}{2}$ ft. The carpet costs $18.00 per square yard.	(a) the area of the room in square feet (b) the area of the room in square yards (c) the cost of the carpet	(a) Multiply $11\frac{2}{3}$ ft by $19\frac{1}{2}$ ft to get area in square feet. (b) Divide the number of square feet by 9 to get the number of square yards. (c) Multiply the number of square yards by $18.00.	There are 9 square feet, 3 feet × 3 feet, in 1 square yard; therefore, we must divide the number of square feet by 9 to obtain square yards.

2. *Solve and state the answer.*

   (a) To find the area of a rectangle, we multiply the length times the width.

   $$11\frac{2}{3} \times 19\frac{1}{2} = \frac{35}{3} \times \frac{39}{2}$$

   $$= \frac{455}{2} = 227\frac{1}{2}$$

   A minimum of $227\frac{1}{2}$ square feet of carpet will be needed. We say a minimum because some carpet may be wasted in cutting. Carpet is sold by the square yard. We will want to know the amount of carpet needed in square yards.

   (b) To determine the area in square yards, we divide $227\frac{1}{2}$ by 9. (9 sq ft = 1 sq yd)

   $$227\frac{1}{2} \div 9 = \frac{455}{2} \div \frac{9}{1}$$

   $$= \frac{455}{2} \times \frac{1}{9} = \frac{455}{18} = 25\frac{5}{18}$$

   A minimum of $25\frac{5}{18}$ square yards of carpet will be needed.

   (c) Since the carpet costs $18.00 per square yard, we will multiply the number of square yards needed by $18.00.

   $$25\frac{5}{18} \times 18 = \frac{455}{18} \times \frac{18}{1} = 455$$

   The carpet will cost a minimum of $455.00 for this room.

3. *Check.* We will estimate to see if our answers are reasonable.

   (a) We will estimate by rounding each number to the nearest 10.

   $$11\frac{2}{3} \times 19\frac{1}{2} \longrightarrow 10 \times 20 = 200$$

   This is close to our answer of $227\frac{1}{2}$ sq ft. Our answer is reasonable. ✓

   (b) We will estimate by rounding to the nearest hundred and ten, respectively.

   $$227\frac{1}{2} \div 9 \longrightarrow 200 \div 10 = 20$$

   This is close to our answer of $25\frac{5}{18}$ sq yd. Our answer is reasonable. ✓

   (c) We will estimate by rounding each number to the nearest 10.

   $$25\frac{5}{18} \times 18 \longrightarrow 30 \times 20 = 600$$

   This is close to our answer of $455. Our answer seems reasonable. ✓

"Remember to estimate. It will save you time and money!"

NOTE TO STUDENT: *Fully worked-out solutions to all of the Practice Problems can be found at the back of the text starting at page SP-1*

▲ **Practice Problem 1** Jeff went to help Abby pick out wall-to-wall carpet for her new house. Her rectangular living room measures $16\frac{1}{2}$ feet by $10\frac{1}{2}$ feet. How much will it cost to carpet the room if the carpet costs $20 per square yard?

Mathematics Blueprint for Problem Solving

Gather the Facts	What Am I Solving For?	What Must I Calculate?	Key Points to Remember

TO THINK ABOUT: Example 1 Follow-Up   Assume that the carpet in Example 1 comes in a standard width of 12 feet. How much carpet will be wasted if it is laid out on the living room floor in one strip that is $19\frac{1}{2}$ feet long? How much carpet will be wasted if it is laid in two sections side by side that are each $11\frac{2}{3}$ feet long? Assuming you have to pay for wasted carpet, what is the minimum cost to carpet the room?

EXAMPLE 2   The following chart shows the 2006 sales of Micropower Computer Software for each of the four regions of the United States. Use the chart to answer the following questions (round all answers to the nearest whole percent):

(a) What percent of the sales personnel are assigned to the Northeast?
(b) What percent of the volume of sales is attributed to the Northeast?
(c) What percent of the sales personnel are assigned to the Southeast?
(d) What percent of the volume of sales is attributed to the Southeast?
(e) Which of these two regions of the country has sales personnel that appear to be more effective in terms of the volume of sales?

Region of the U.S.	Number of Sales Personnel	Dollar Volume of Sales
Northeast	12	1,560,000
Southeast	18	4,300,000
Northwest	10	3,660,000
Southwest	15	3,720,000
Total	55	13,240,000

Solution

1. **Understand the problem.** We will only need to deal with figures from the Northeast region and the Southeast region.

Mathematics Blueprint for Problem Solving

Gather the Facts	What Am I Solving For?	What Must I Calculate?	Key Points to Remember

Personnel:
  12 Northeast
  18 Southeast
  55 total
Sales Volume:
  $1,560,000 NE
  $4,300,000 SE
  $13,240,000 Total
	(a) the percent of the total personnel in the Northeast
	(b) the percent of the total sales in the Northeast
	(c) the percent of the total personnel in the Southeast
	(d) the percent of the total sales in the Southeast
	(e) compare the percentages from the two regions
		(a) 12 is what percent of 55? Divide. $12 \div 55$
		(b) 1,560,000 is what percent of 13,240,000? $1,560,000 \div 13,240,000$
		(c) $18 \div 55$
		(d) $4,300,000 \div 13,240,000$
			We do not need to use the numbers for the Northwest or the Southwest.

2. *Solve and state the answer.*

(a) $\dfrac{12}{55} = 0.21818\ldots$

$\approx 22\%$

(b) $\dfrac{1{,}560{,}000}{13{,}240{,}000} = \dfrac{156}{1324} \approx 0.1178$

$\approx 12\%$

(c) $\dfrac{18}{55} = 0.32727\ldots$

$\approx 33\%$

(d) $\dfrac{4{,}300{,}000}{13{,}240{,}000} = \dfrac{430}{1324} \approx 0.3248$

$\approx 32\%$

(e) Let us suppose for a minute that we assume that effectiveness is uniform in the sales force. This would indicate we might expect 1% of the sales force to be responsible for 1% of the sales. We would expect 5% of the sales force to be responsible for 5% of the sales. This is about what we see in the Southeast. In that region 33% of the sales force made 32% of the sales. However, in sharp contrast, in the Northeast 23% of the sales force only achieved 12% of the sales. This is much less than we would expect. We must be cautious here. *If there are no other significant factors,* it would appear that the Southeast sales force is more effective. (There may be other significant factors affecting sales, such as a recession in the Northeast, new and inexperienced sales personnel, or fewer competing companies in the Southeast.)

3. *Check.* You may want to use a calculator to check the division in step 2, or you may use estimation.

(a) $\dfrac{12}{55} \rightarrow \dfrac{10}{60} \approx 0.17$

$= 17\%$ ✓

(b) $\dfrac{1{,}560{,}000}{13{,}240{,}000} \rightarrow \dfrac{1{,}600{,}000}{13{,}000{,}000} \approx 0.12$

$= 12\%$ ✓

(c) $\dfrac{18}{55} \rightarrow \dfrac{20}{60} \approx 0.33$

$= 33\%$ ✓

(d) $\dfrac{4{,}300{,}000}{13{,}240{,}000} \rightarrow \dfrac{4{,}300{,}000}{13{,}000{,}000} \approx 0.33$

$= 33\%$ ✓

Practice Problem 2 Using the chart for Example 2, answer the following questions. (Round all answers to the nearest whole percent.)

(a) What percent of the sales personnel are assigned to the Northwest?

(b) What percent of the sales volume is attributed to the Northwest?

(c) What percent of the sales personnel are assigned to the Southwest?

(d) What percent of the sales volume is attributed to the Southwest?

(e) Which of these two regions of the country has sales personnel that appear to be more effective in terms of volume of sales?

NOTE TO STUDENT: Fully worked-out solutions to all of the Practice Problems can be found at the back of the text starting at page SP-1

Mathematics Blueprint for Problem Solving

Gather the Facts	What Am I Solving For?	What Must I Calculate?	Key Points to Remember

TO THINK ABOUT: Example 2 Follow-Up Suppose in 2007 the number of sales personnel (55) increased by 60%. What would the new number of sales personnel be? Suppose in 2008 that the number of sales personnel decreased by 60% from the number of sales personnel in 2007. What would the new number be? Why is this number not 55, since we have increased the number by 60% and then decreased the result by 60%? Explain.

Applications

Use the Mathematics Blueprint for Problem Solving to help you solve each of the following exercises.

▲ **1.** *Patio Pavers* Shaun wants to place pavers to make a new backyard patio. The patio will measure $3\frac{1}{3}$ yards by $4\frac{1}{3}$ yards. If the pavers are each 1 square foot and cost $2.35 each, how much will the pavers cost?

▲ **2.** *Decking Costs* The Carters need to replace their deck floor. The deck is $18\frac{1}{2}$ feet by $10\frac{1}{2}$ feet. If the new decking costs $5.25 per square foot, how much will it cost them to replace the deck? (Round your answer to the nearest cent.)

▲ **3.** *Garden Design* Jenna White has room for a rectangular garden that can measure as much as $15\frac{1}{2}$ feet wide and $25\frac{2}{3}$ feet long.

   (a) She decides the garden should be as large as possible, but wants to fence it off to protect the vegetables from the neighborhood raccoons. How much fencing does she need?

   (b) She finds that she can buy a prepackaged 90 feet of fencing for $155.00 or she can get a cut-to-order length for $2.10 per foot. Which should she buy? How much money does she save?

▲ **4.** *Pond Design* The Lee family is installing a backyard pond for their prize koi. (A koi is a kind of fish bred in Japan. They are noted for their large size and variety of color.)

   (a) If the pond measures $11\frac{1}{2}$ feet by 7 feet, and is $3\frac{1}{2}$ feet deep, how much water will it take to fill the pond?

   (b) If each cubic foot is 7.5 gallons, how many gallons of water will the pond hold?

   (c) If at least 200 gallons of water is recommended for each koi, how many koi can the Lee family place in their pond?

Exercise Training   *The following directions are posted on the wall at the gym.*

Beginning exercise training schedule

On day 1, each athlete will begin the morning as follows:

Jog.............. $1\frac{1}{2}$ miles

Walk........... $1\frac{3}{4}$ miles

Rest............. $2\frac{1}{2}$ minutes

Walk........... 1 mile

▲ **5.** Betty's athletic trainer told her to follow the beginning exercise training schedule on day 1. On day 2, she is to increase all distances and times by $\frac{1}{3}$ that of day 1. On day 3, she is to increase all distances and times by $\frac{1}{3}$ that of day 2. What will be her training schedule on day 3?

▲ **6.** Melinda's athletic trainer told her to follow the beginning exercise training schedule on day 1. On day 2, she is to increase all distances and times by $\frac{1}{3}$ that of day 1. On day 3, she is to once again increase all distances and times by $\frac{1}{3}$ that of day 1. What will be her training schedule on day 3?

To Think About

Refer to exercises 5 and 6 in working exercises 7–10.

7. Who will have a more demanding schedule on day 3, Betty or Melinda? Why?

8. If Betty kept up the same type of increase day after day, how many miles would she be jogging on day 5?

9. If Melinda kept up the same type of increase day after day, how many miles would she be jogging on day 7?

10. Which athletic trainer would appear to have the best plan for training athletes if they used this plan for 14 days? Why?

11. *House Prices* In 1985, the median selling price of an existing single-family home in Atlanta, Georgia, was $66,200. Between 1985 and 1990, the average price increased by 30%. Between 1990 and 2005 the average price increased again, this time by 15%. What was the median house price in Atlanta in 2005?

12. *Egg Weight* Chicken eggs are classified by weight per dozen eggs. Large eggs weigh 24 ounces per dozen and medium eggs weigh 21 ounces per dozen.

(a) If you do not include the shell, which is 12% of the total weight of an egg, how many ounces of eggs do you get from a dozen large eggs? From a dozen medium eggs?

(b) At a local market, large eggs sell for $1.79 a dozen, and medium eggs for $1.39 a dozen. If you do not include the shell, which is a better buy, large or medium eggs?

Family Budgets   *For the following problems, use the chart below.*

The Whittier family has created the following budget:

Rent	20%	Clothing	10%
Food	31%	Medical	12%
Utilities	5%	Savings	10%
Entertainment	6%	Miscellaneous	6%

13. The income for this family comes from Mrs. Whittier's annual salary of $62,500.

(a) If 28% of her salary is withheld for various taxes, how much money does the family have available for their budget?

(b) How much of Mrs. Whittier's take-home pay is budgeted for food?

14. With one child ready to head off to college, Mrs. Whittier is looking for ways to save money.

(a) She figures out that if they plant a garden this year, the family can save 12% of their food costs. How much will she save in food costs? (This does not account for the cost of planting a garden.) What is the new food budget amount?

(b) What percentage of Mrs. Whittier's budget is the new food amount?

Paycheck Stub *Use the following information from a paycheck stub to solve exercises 15–18.*

TOBEY & SLATER INC.				Check Number	Payroll Period		Pay Date
5000 Stillwell Avenue					From Date	To Date	
Queens, NY 10001				495885	10-30-99	11-30-99	12-01-99

Name	Social Security No.	I.D. Number	File Number	Rate/Salary	Department	MS	DEP	Res
Fred J. Gilliani	012-34-5678	01	1379	1150.00	0100	M	5	NY

	Current	Year to Date		Current	Year to Date
GROSS	1,150.00	6,670.00	STATE	67.76	388.45
FEDERAL	138.97	781.07	LOCAL	5.18	30.04
FICA	87.98	510.28	DIS-SUI	.00	.00
W-2 GROSS		6,670.00	NET	790.47	4,960.16

Earnings						Deductions/Specials		
No.	Type	Hours	Rate	Amount	Dept/Job No.	No.	Description	Amount
96	REGULAR			1,150.00	0100	82	Retirement	12.56
						75	Medical	36.28
						56	Union Dues	10.80

Gross pay is the pay an employee receives for his or her services before deductions. Net pay is the pay the employee actually gets to take home. You may round each amount to the nearest whole percent for exercises 15–18.

15. What percent of Fred's gross pay is deducted for federal, state, and local taxes?

16. What percent of Fred's gross pay is deducted for retirement and medical?

17. What percent of Fred's gross pay does he actually get to take home?

18. What percent of Fred's deductions are special deductions?

Appendix C Practice with Operations of Whole Numbers

Addition Practice

1. 23
    +14

2. 42
    +33

3. 50
    +44

4. 83
    +16

5. 51
    +27

6. 16
    +13

7. 32
    +29

8. 64
    +17

9. 327
    + 42

10. 223
    + 54

11. 463
    + 28

12. 504
    + 96

13. 739
    +682

14. 567
    +485

15. 840 + 60

16. 364 + 37

17. 915 + 796

18. 420 + 899

19. 213 + 46 + 30

20. 326 + 21 + 52

21. 132 + 441 + 16

22. 671 + 204 + 12

23. 139 + 61 + 222

24. 524 + 73 + 195

25. 701 + 166 + 24 + 11

26. 439 + 365 + 45 + 81

Subtraction Practice

1. 32
    −11

2. 87
    −25

3. 56
    −34

4. 73
    −30

5. 93
    −25

6. 21
    −16

7. 40
    −11

8. 60
    −15

9. 576
    − 45

10. 294
    − 71

11. 780
    − 54

12. 208
    − 17

13. 406
    − 28

14. 100
    − 34

15. 635
    −126

16. 375
    −147

17. 500
    −244

18. 200
    −137

19. 922
    −739

20. 646
    −377

21. 1729
    − 856

22. 2382
    − 490

23. 7806
    − 327

24. 3024
    − 156

25. 8200
    −6134

26. 2004
    −1326

Multiplication Practice

1. 23 $\times\ 3$	**2.** 13 $\times\ 2$	**3.** 54 $\times\ 7$	**4.** 67 $\times\ 9$	**5.** 74 $\times 21$	**6.** 53 $\times 31$
7. 92 $\times 40$	**8.** 70 $\times 52$	**9.** 82 $\times 95$	**10.** 69 $\times 39$	**11.** 212 $\times\ 43$	**12.** 341 $\times\ 22$
13. 295 $\times\ 41$	**14.** 419 $\times\ 72$	**15.** 304 $\times\ 68$	**16.** 620 $\times\ 39$	**17.** 261 $\times 144$	**18.** 124 $\times 433$
19. 545 $\times 522$	**20.** 634 $\times 799$	**21.** 391 $\times 609$	**22.** 817 $\times 460$	**23.** 3844 $\times\ 209$	**24.** 7409 $\times\ 106$

25. 72,499(683)   **26.** 86,243(725)

Division Practice

1. $8\overline{)128}$ **2.** $3\overline{)168}$ **3.** $7\overline{)415}$ **4.** $6\overline{)287}$

5. $9\overline{)1116}$ **6.** $4\overline{)1184}$ **7.** $6\overline{)1404}$ **8.** $3\overline{)1701}$

9. $8\overline{)4174}$ **10.** $5\overline{)3697}$ **11.** $17\overline{)5468}$ **12.** $13\overline{)9795}$

13. $146\overline{)12,994}$ **14.** $163\overline{)14,833}$ **15.** $1728 \div 54$ **16.** $3813 \div 93$

17. $3701 \div 34$ **18.** $6052 \div 49$ **19.** $15,836 \div 74$ **20.** $23,256 \div 68$

21. $30,632 \div 27$ **22.** $85,069 \div 79$ **23.** $30,752 \div 248$ **24.** $49,878 \div 326$

25. $271,125 \div 241$ **26.** $546,924 \div 357$

Appendix D The Point–Slope Form of a Line

 Using the Point–Slope Form of the Equation of a Line

Recall from Section 3.1 that we defined a line as an equation that can be written in the form $Ax + By = C$. This form is called the **standard form of the equation of a line** and although this form tells us that the graph is a straight line, it reveals little about the line. A more useful form of the equation was introduced in Section 3.3 called the **slope–intercept form**, $y = mx + b$. This form immediately reveals the slope and y-intercept of a line, and also allows us to easily write the equation of a line when we know its slope and y-intercept. But, what happens if we know the slope of a line and a point on the line that is not the y-intercept? Can we write the equation of the line? By the definition of slope, we have the following:

$$m = \frac{y - y_1}{x - x_1}$$

$$m(x - x_1) = y - y_1$$

That is, $y - y_1 = m(x - x_1)$.

This is the point–slope form of the equation of a line.

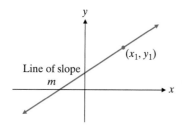

POINT–SLOPE FORM

The **point–slope form** of the equation of a line is $y - y_1 = m(x - x_1)$, where m is the slope and (x_1, y_1) are the coordinates of a known point on the line.

Write the Equation of a Line Given Its Slope and One Point on the Line

EXAMPLE 1 Find an equation of the line that has slope $-\frac{3}{4}$ and passes through the point $(-6, 1)$. Express your answer in standard form.

Solution Since we don't know the y-intercept, we can't use the slope–intercept form easily. Therefore, we use the point–slope form.

$$y - y_1 = m(x - x_1)$$

$$y - 1 = -\frac{3}{4}[x - (-6)] \qquad \text{Substitute the given values.}$$

$$y - 1 = -\frac{3}{4}x - \frac{9}{2} \qquad \text{Simplify. (Do you see how we did this?)}$$

$$4y - 4(1) = 4\left(-\frac{3}{4}x\right) - 4\left(\frac{9}{2}\right) \qquad \text{Multiply each term by the LCD 4.}$$

$$4y - 4 = -3x - 18 \qquad \text{Simplify.}$$

$$3x + 4y = -18 + 4 \qquad \text{Add } 3x + 4 \text{ to each side.}$$

$$3x + 4y = -14 \qquad \text{Add like terms.}$$

The equation in standard form is $3x + 4y = -14$.

Student Learning Objectives

After studying this section, you will be able to:

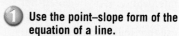 Use the point–slope form of the equation of a line.

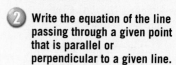 Write the equation of the line passing through a given point that is parallel or perpendicular to a given line.

Graphing Calculator

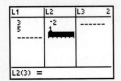

Using Linear Regression to Find an Equation

Many graphing calculators, such as the TI-84 Plus, will find the equation of a line in slope–intercept form if you enter the points as a collection of data and use the Regression feature. We would enter the data from Example 2 as follows:

L1	L2	L3	2
3	‾²		
5		------	

L2(3) =

The output of the calculator uses the notation $y = ax + b$ instead of $y = mx + b$.

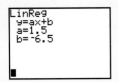

LinReg
y=ax+b
a=1.5
b=‾6.5

Thus, our answer to Example 2 using the graphing calculator would be $y = 1.5x - 6.5$.

NOTE TO STUDENT: Fully worked-out solutions to all of the Practice Problems can be found at the back of the text starting at page SP-1

Practice Problem 1 Find an equation of the line that passes through $(5, -2)$ and has a slope of $\frac{3}{4}$. Express your answer in standard form.

Write the Equation of a Line Given Two Points on the Line We can use the point–slope form to find the equation of a line if we are given two points. Carefully study the following example. Be sure you understand each step. You will encounter this type of problem frequently.

EXAMPLE 2 Find an equation of the line that passes through $(3, -2)$ and $(5, 1)$. Express your answer in slope–intercept form.

Solution First we find the slope.

$$m = \frac{y_2 - y_1}{x_2 - x_1} = \frac{1 - (-2)}{5 - 3} = \frac{1 + 2}{2} = \frac{3}{2}$$

Now we substitute the value of the slope and the coordinates of either point into the point–slope equation. Let's use $(5, 1)$.

$$y - y_1 = m(x - x_1)$$

$$y - 1 = \frac{3}{2}(x - 5) \qquad \text{Substitute } m = \frac{3}{2} \text{ and } (x_1, y_1) = (5, 1).$$

$$y - 1 = \frac{3}{2}x - \frac{15}{2} \qquad \text{Remove parentheses.}$$

$$y = \frac{3}{2}x - \frac{15}{2} + 1 \qquad \text{Add 1 to each side of the equation.}$$

$$y = \frac{3}{2}x - \frac{15}{2} + \frac{2}{2} \qquad \text{Add the two fractions.}$$

$$y = \frac{3}{2}x - \frac{13}{2} \qquad \text{Simplify.}$$

Practice Problem 2 Find an equation of the line that passes through $(-4, 1)$ and $(-2, -3)$. Express your answer in slope–intercept form.

Before we go further, we want to point out that these various forms of the equation of a straight line are just that—*forms* for convenience. We are *not* using different equations each time, nor should you simply try to memorize the different variations without understanding when to use them. They can easily be derived from the definition of slope, as we have seen. And remember, you can *always* use the definition of slope to find the equation of a line. You may find it helpful to review Examples 1 and 2 for a few minutes before going ahead to Example 3. It is important to see how each example is different.

 ## Writing the Equation of a Parallel or Perpendicular Line

Let us now look at parallel and perpendicular lines. If we are given the equation of a line and a point not on the line, we can find the equation of a second line that passes through the given point and is parallel or perpendicular to the first line. We can do this because we know that the slopes of parallel lines are equal and that the slopes of perpendicular lines are negative reciprocals of each other.

We begin by finding the slope of the given line. Then we use the point–slope form to find the equation of the second line. Study each step of the following example carefully.

EXAMPLE 3 Find an equation of the line passing through the point $(-2, -4)$ and parallel to the line $2x + 5y = 8$. Express the answer in standard form.

Solution First we need to find the slope of the line $2x + 5y = 8$. We do this by writing the equation in slope–intercept form.

$$5y = -2x + 8$$
$$y = -\frac{2}{5}x + \frac{8}{5}$$

The slope of the given line is $-\frac{2}{5}$. Since parallel lines have the same slope, the slope of the unknown line is also $-\frac{2}{5}$. Now we substitute $m = -\frac{2}{5}$ and the coordinates of the point $(-2, -4)$ into the point–slope form of the equation of a line.

$$y - y_1 = m(x - x_1)$$

$$y - (-4) = -\frac{2}{5}[x - (-2)] \qquad \text{Substitute.}$$

$$y + 4 = -\frac{2}{5}(x + 2) \qquad \text{Simplify.}$$

$$y + 4 = -\frac{2}{5}x - \frac{4}{5} \qquad \text{Remove parentheses.}$$

$$5y + 5(4) = 5\left(-\frac{2}{5}x\right) - 5\left(\frac{4}{5}\right) \qquad \text{Multiply each term by the LCD 5.}$$

$$5y + 20 = -2x - 4 \qquad \text{Simplify.}$$

$$2x + 5y = -4 - 20 \qquad \text{Add } 2x - 20 \text{ to each side.}$$

$$2x + 5y = -24 \qquad \text{Simplify.}$$

$2x + 5y = -24$ is an equation of the line passing through the point $(-2, -4)$ and parallel to the line $2x + 5y = 8$.

Practice Problem 3 Find an equation of the line passing through $(4, -5)$ and parallel to the line $5x - 3y = 10$. Express the answer in standard form.

NOTE TO STUDENT: Fully worked-out solutions to all of the Practice Problems can be found at the back of the text starting at page SP-1

Some extra steps are needed if the desired line is to be perpendicular to the given line. Carefully note the approach in Example 4.

EXAMPLE 4 Find an equation of the line that passes through the point $(2, -3)$ and is perpendicular to the line $3x - y = -12$. Express the answer in standard form.

Solution   To find the slope of the line $3x - y = -12$, we rewrite it in slope–intercept form.

$$-y = -3x - 12$$
$$y = 3x + 12$$

This line has a slope of 3. Therefore, the slope of a line perpendicular to this line is the negative reciprocal $-\frac{1}{3}$.

Now substitute the slope $m = -\frac{1}{3}$ and the coordinates of the point $(2, -3)$ into the point–slope form of the equation.

$$y - y_1 = m(x - x_1)$$

$$y - (-3) = -\frac{1}{3}(x - 2) \qquad \text{Substitute.}$$

$$y + 3 = -\frac{1}{3}(x - 2) \qquad \text{Simplify.}$$

$$y + 3 = -\frac{1}{3}x + \frac{2}{3} \qquad \text{Remove parentheses.}$$

$$3y + 3(3) = 3\left(-\frac{1}{3}x\right) + 3\left(\frac{2}{3}\right) \qquad \text{Multiply each term by the LCD 3.}$$

$$3y + 9 = -x + 2 \qquad \text{Simplify.}$$
$$x + 3y = 2 - 9 \qquad \text{Add } x - 9 \text{ to each side.}$$
$$x + 3y = -7 \qquad \text{Simplify.}$$

$x + 3y = -7$ is an equation of the line that passes through the point $(2, -3)$ and is perpendicular to the line $3x - y = -12$.

Practice Problem 4 Find an equation of the line that passes through $(-4, 3)$ and is perpendicular to the line $6x + 3y = 7$. Express the answer in standard form.

Find an equation of the line that passes through the given point and has the given slope. Express your answer in slope–intercept form.

1. $(6, 4)$, $m = -\dfrac{2}{3}$

2. $(4, 6)$, $m = -\dfrac{1}{2}$

3. $(-7, -2)$, $m = 5$

4. $(8, 0)$, $m = -3$

5. $(6, 0)$, $m = -\dfrac{1}{5}$

6. $(0, -1)$, $m = -\dfrac{5}{3}$

Find an equation of the line passing through the pair of points. Write the equation in slope–intercept form.

7. $(-4, -1)$ and $(3, 4)$

8. $(7, -2)$ and $(-1, -3)$

9. $\left(\dfrac{1}{2}, -3\right)$ and $\left(\dfrac{7}{2}, -5\right)$

10. $\left(\dfrac{7}{6}, 1\right)$ and $\left(-\dfrac{1}{3}, 0\right)$

11. $(12, -3)$ and $(7, -3)$

12. $(4, 8)$ and $(-3, 8)$

Find an equation of the line satisfying the conditions given. Express your answer in standard form.

13. Parallel to $5x - y = 4$ and passing through $(-2, 0)$

14. Parallel to $3x - y = -5$ and passing through $(-1, 0)$

15. Parallel to $x = 3y - 8$ and passing through $(5, -1)$

16. Parallel to $2y + x = 7$ and passing through $(-5, -4)$

17. Perpendicular to $2y = -3x$ and passing through $(6, -1)$

18. Perpendicular to $y = 5x$ and passing through $(4, -2)$

19. Perpendicular to $x + 7y = -12$ and passing through $(-4, -1)$

20. Perpendicular to $x - 4y = 2$ and passing through $(3, -1)$

To Think About

Without graphing determine whether the following pairs of lines are (a) parallel, (b) perpendicular, or (c) neither parallel nor perpendicular.

21. $-3x + 5y = 40$
$5y + 3x = 17$

22. $5x - 6y = 19$
$6x + 5y = -30$

23. $y = -\dfrac{3}{4}x - 2$
$6x + 8y = -5$

24. $y = \dfrac{2}{3}x + 6$
$-2x - 3y = -12$

25. $y = \dfrac{5}{6}x - \dfrac{1}{3}$
$6x + 5y = -12$

26. $y = \dfrac{3}{7}x - \dfrac{1}{14}$
$14y + 6x = 3$

Solutions to Practice Problems

Chapter 0

0.1 Practice Problems

1. (a) $\dfrac{10}{16} = \dfrac{2 \times 5}{2 \times 2 \times 2 \times 2} = \dfrac{5}{2 \times 2 \times 2} = \dfrac{5}{8}$

(b) $\dfrac{24}{36} = \dfrac{2 \times 2 \times 2 \times 3}{2 \times 2 \times 3 \times 3} = \dfrac{2}{3}$

(c) $\dfrac{36}{42} = \dfrac{2 \times 2 \times 3 \times 3}{2 \times 3 \times 7} = \dfrac{2 \times 3}{7} = \dfrac{6}{7}$

2. (a) $\dfrac{4}{12} = \dfrac{2 \times 2 \times 1}{2 \times 2 \times 3} = \dfrac{1}{3}$

(b) $\dfrac{25}{125} = \dfrac{5 \times 5 \times 1}{5 \times 5 \times 5} = \dfrac{1}{5}$

(c) $\dfrac{73}{146} = \dfrac{73 \times 1}{73 \times 2} = \dfrac{1}{2}$

3. (a) $\dfrac{18}{6} = \dfrac{2 \times 3 \times 3}{2 \times 3 \times 1} = 3$

(b) $\dfrac{146}{73} = \dfrac{73 \times 2}{73 \times 1} = 2$

(c) $\dfrac{28}{7} = \dfrac{2 \times 2 \times 7}{7} = 2 \times 2 = 4$

4. 56 out of $154 = \dfrac{56}{154} = \dfrac{2 \times 7 \times 4}{2 \times 7 \times 11} = \dfrac{4}{11}$

5. (a) $\dfrac{12}{7} = 12 \div 7 \quad 7\overline{)12}$

$\underline{7}$

$5 \quad$ Remainder

$\dfrac{12}{7} = 1\dfrac{5}{7}$

(b) $\dfrac{20}{5} = 20 \div 5 \quad 5\overline{)20}$

$\underline{20}$

$0 \quad$ Remainder

$\dfrac{20}{5} = 4$

6. (a) $3\dfrac{2}{5} = \dfrac{(3 \times 5) + 2}{5} = \dfrac{15 + 2}{5} = \dfrac{17}{5}$

(b) $1\dfrac{3}{7} = \dfrac{(1 \times 7) + 3}{7} = \dfrac{7 + 3}{7} = \dfrac{10}{7}$

(c) $2\dfrac{6}{11} = \dfrac{(2 \times 11) + 6}{11} = \dfrac{22 + 6}{11} = \dfrac{28}{11}$

(d) $4\dfrac{2}{3} = \dfrac{(4 \times 3) + 2}{3} = \dfrac{12 + 2}{3} = \dfrac{14}{3}$

7. (a) $\dfrac{3}{8} = \dfrac{?}{24}; \quad 8 \times 3 = 24$

$\dfrac{3 \times 3}{8 \times 3} = \dfrac{9}{24}$

(b) $\dfrac{5}{6} = \dfrac{?}{30}; \quad 6 \times 5 = 30$

$\dfrac{5 \times 5}{6 \times 5} = \dfrac{25}{30}$

(c) $\dfrac{2}{7} = \dfrac{?}{56}; \quad 7 \times 8 = 56$

$\dfrac{2 \times 8}{7 \times 8} = \dfrac{16}{56}$

0.2 Practice Problems

1. (a) $\dfrac{3}{6} + \dfrac{2}{6} = \dfrac{3 + 2}{6} = \dfrac{5}{6}$

(b) $\dfrac{3}{11} + \dfrac{8}{11} = \dfrac{3 + 8}{11} = \dfrac{11}{11} = 1$

(c) $\dfrac{1}{8} + \dfrac{2}{8} + \dfrac{1}{8} = \dfrac{1 + 2 + 1}{8} = \dfrac{4}{8} = \dfrac{1}{2}$

(d) $\dfrac{5}{9} + \dfrac{8}{9} = \dfrac{5 + 8}{9} = \dfrac{13}{9}$ or $1\dfrac{4}{9}$

2. (a) $\dfrac{11}{13} - \dfrac{6}{13} = \dfrac{11 - 6}{13} = \dfrac{5}{13}$

(b) $\dfrac{8}{9} - \dfrac{2}{9} = \dfrac{8 - 2}{9} = \dfrac{6}{9} = \dfrac{2}{3}$

3. The LCD is 24, since 24 is exactly divisible by 8 and 12. There is no smaller number that is exactly divisible by 8 and 12.

4. Find the LCD using prime factors.

$\dfrac{8}{35}$ and $\dfrac{6}{15}$

$35 = 7 \cdot 5$

$15 = 5 \cdot 3$

$\text{LCD} = 7 \cdot 5 \cdot 3 = 105$

5. Find the LCD of $\dfrac{5}{12}$ and $\dfrac{7}{30}$.

$12 = 3 \cdot 2 \cdot 2$

$30 = 3 \cdot 2 \cdot 5$

$\text{LCD} = 3 \cdot 2 \cdot 2 \cdot 5 = 60$

6. Find the LCD of $\dfrac{2}{27}, \dfrac{1}{18},$ and $\dfrac{5}{12}$.

$27 = 3 \cdot 3 \cdot 3$

$18 = 3 \cdot 3 \cdot 2$

$12 = 3 \cdot 2 \cdot 2$

$\text{LCD} = 3 \cdot 3 \cdot 3 \cdot 2 \cdot 2 = 108$

7. From Practice Problem 3, the LCD is 24.

$\dfrac{5}{12} = \dfrac{5 \times 2}{12 \times 2} = \dfrac{10}{24} \qquad \dfrac{1}{8} = \dfrac{1 \times 3}{8 \times 3} = \dfrac{3}{24}$

$\dfrac{5}{12} + \dfrac{1}{8} = \dfrac{10}{24} + \dfrac{3}{24} = \dfrac{13}{24}$

$\dfrac{13}{24}$ of the farm fields were planted in corn or soybeans.

8. We can see by inspection that both 5 and 25 divide exactly into 50. Thus 50 is the LCD.

$\dfrac{3}{5} = \dfrac{3 \times 10}{5 \times 10} = \dfrac{30}{50} \qquad \dfrac{4}{25} = \dfrac{4 \times 2}{25 \times 2} = \dfrac{8}{50}$

$\dfrac{3}{5} + \dfrac{4}{25} + \dfrac{1}{50} = \dfrac{30}{50} + \dfrac{8}{50} + \dfrac{1}{50} = \dfrac{30 + 8 + 1}{50} = \dfrac{39}{50}$

9. Add $\dfrac{1}{49} + \dfrac{3}{14}$

First find the LCD.

$49 = 7 \cdot 7$

$14 = 7 \cdot 2$

$\text{LCD} = 7 \cdot 7 \cdot 2 = 98$

Then change to equivalent fractions and add.

$\dfrac{1}{49} = \dfrac{1 \times 2}{49 \times 2} = \dfrac{2}{98} \qquad \dfrac{3}{14} = \dfrac{3 \times 7}{14 \times 7} = \dfrac{21}{98}$

$\dfrac{1}{49} + \dfrac{3}{14} = \dfrac{2}{98} + \dfrac{21}{98} = \dfrac{23}{98}$

10. $\dfrac{1}{12} - \dfrac{1}{30}$

First find the LCD.

$$12 = 2 \cdot 2 \cdot 3$$
$$30 = \;| \; 2 \cdot 3 \cdot 5$$

$$LCD = 2 \cdot 2 \cdot 3 \cdot 5 = 60$$

Then change to equivalent fractions and subtract.

$$\dfrac{1}{12} = \dfrac{1 \times 5}{12 \times 5} = \dfrac{5}{60} \qquad\qquad \dfrac{1}{30} = \dfrac{1 \times 2}{30 \times 2} = \dfrac{2}{60}$$

$$\dfrac{1}{12} - \dfrac{1}{30} = \dfrac{5}{60} - \dfrac{2}{60} = \dfrac{3}{60} = \dfrac{1}{20}$$

11. $\dfrac{2}{3} + \dfrac{3}{4} - \dfrac{3}{8}$

First find the LCD.

$$3 = 3$$
$$4 = \;| \; 2 \cdot 2$$
$$8 = \;| \; 2 \cdot 2 \cdot 2$$

$$LCD = 3 \cdot 2 \cdot 2 \cdot 2 = 3 \cdot 8 = 24$$

$$\dfrac{2}{3} = \dfrac{2 \times 8}{3 \times 8} = \dfrac{16}{24} \qquad \dfrac{3}{4} = \dfrac{3 \times 6}{4 \times 6} = \dfrac{18}{24} \qquad \dfrac{3}{8} = \dfrac{3 \times 3}{8 \times 3} = \dfrac{9}{24}$$

Now combine the fractions.

$$\dfrac{2}{3} + \dfrac{3}{4} - \dfrac{3}{8} = \dfrac{16}{24} + \dfrac{18}{24} - \dfrac{9}{24} = \dfrac{25}{24} \text{ or } 1\dfrac{1}{24}$$

12. (a) The LCD of 3 and 5 is 15.

$$1\dfrac{2}{3} = \dfrac{5}{3} = \dfrac{5 \times 5}{3 \times 5} = \dfrac{25}{15} \qquad\qquad 2\dfrac{4}{5} = \dfrac{14}{5} = \dfrac{14 \times 3}{5 \times 3} = \dfrac{42}{15}$$

$$1\dfrac{2}{3} + 2\dfrac{4}{5} = \dfrac{25}{15} + \dfrac{42}{15} = \dfrac{67}{15} = 4\dfrac{7}{15}$$

(b) The LCD of 4 and 3 is 12.

$$5\dfrac{1}{4} = \dfrac{21}{4} = \dfrac{21 \times 3}{4 \times 3} = \dfrac{63}{12} \qquad\qquad 2\dfrac{2}{3} = \dfrac{8}{3} = \dfrac{8 \times 4}{3 \times 4} = \dfrac{32}{12}$$

$$5\dfrac{1}{4} - 2\dfrac{2}{3} = \dfrac{63}{12} - \dfrac{32}{12} = \dfrac{31}{12} = 2\dfrac{7}{12}$$

13.

Rectangle with top $6\dfrac{1}{2}$ cm, bottom $6\dfrac{1}{2}$ cm, left side $4\dfrac{1}{5}$ cm, right side $4\dfrac{1}{5}$ cm.

$$4\dfrac{1}{5} + 6\dfrac{1}{2} + 4\dfrac{1}{5} + 6\dfrac{1}{2}$$

$$= \dfrac{21}{5} + \dfrac{13}{2} + \dfrac{21}{5} + \dfrac{13}{2}$$

$$= \dfrac{42}{10} + \dfrac{65}{10} + \dfrac{42}{10} + \dfrac{65}{10}$$

$$= \dfrac{214}{10} = \dfrac{107}{5} = 21\dfrac{2}{5}$$

The perimeter is $21\dfrac{2}{5}$ cm.

0.3 Practice Problems

1. (a) $\dfrac{2}{7} \times \dfrac{5}{11} = \dfrac{2 \cdot 5}{7 \cdot 11} = \dfrac{10}{77}$

(b) $\dfrac{1}{5} \times \dfrac{7}{10} = \dfrac{1 \cdot 7}{5 \cdot 10} = \dfrac{7}{50}$

(c) $\dfrac{9}{5} \times \dfrac{1}{4} = \dfrac{9 \cdot 1}{5 \cdot 4} = \dfrac{9}{20}$

(d) $\dfrac{8}{9} \times \dfrac{3}{10} = \dfrac{8 \cdot 3}{9 \cdot 10} = \dfrac{24}{90} = \dfrac{4}{15}$

2. (a) $\dfrac{3}{5} \times \dfrac{4}{3} = \dfrac{3 \cdot 4}{5 \cdot 3} = \dfrac{4 \cdot 3}{5 \cdot 3} = \dfrac{4}{5}$

(b) $\dfrac{9}{10} \times \dfrac{5}{12} = \dfrac{3 \cdot 3}{2 \cdot 5} \times \dfrac{5}{2 \cdot 2 \cdot 3} = \dfrac{3}{8}$

3. (a) $4 \times \dfrac{2}{7} = \dfrac{4}{1} \times \dfrac{2}{7} = \dfrac{4 \cdot 2}{1 \cdot 7} = \dfrac{8}{7} \text{ or } 1\dfrac{1}{7}$

(b) $12 \times \dfrac{3}{4} = \dfrac{12}{1} \times \dfrac{3}{4} = \dfrac{4 \cdot 3}{1} \times \dfrac{3}{4} = \dfrac{9}{1} = 9$

4. Multiply. $5\dfrac{3}{5}$ times $3\dfrac{3}{4}$

$$5\dfrac{3}{5} \times 3\dfrac{3}{4} = \dfrac{28}{5} \times \dfrac{15}{4} = \dfrac{4 \cdot 7}{5} \times \dfrac{3 \cdot 5}{4} = \dfrac{21}{1} = 21$$

The area of the field is 21 square miles.

5. $3\dfrac{1}{2} \times \dfrac{1}{14} \times 4 = \dfrac{7}{2} \times \dfrac{1}{14} \times \dfrac{4}{1} = \dfrac{7}{2} \times \dfrac{1}{2 \cdot 7} \times \dfrac{2 \cdot 2}{1} = 1$

6. (a) $\dfrac{2}{5} \div \dfrac{1}{3} = \dfrac{2}{5} \times \dfrac{3}{1} = \dfrac{6}{5} \text{ or } 1\dfrac{1}{5}$

(b) $\dfrac{12}{13} \div \dfrac{4}{3} = \dfrac{12}{13} \times \dfrac{3}{4} = \dfrac{4 \cdot 3}{13} \times \dfrac{3}{4} = \dfrac{9}{13}$

7. (a) $\dfrac{3}{7} \div 6 = \dfrac{3}{7} \div \dfrac{6}{1} = \dfrac{3}{7} \times \dfrac{1}{6} = \dfrac{3}{7} \times \dfrac{1}{2 \cdot 3} = \dfrac{1}{14}$

(b) $8 \div \dfrac{2}{3} = \dfrac{8}{1} \div \dfrac{2}{3} = \dfrac{8}{1} \times \dfrac{3}{2} = \dfrac{2 \cdot 4}{1} \times \dfrac{3}{2} = \dfrac{12}{1} = 12$

8. (a) $\dfrac{\frac{3}{11}}{\frac{5}{7}} = \dfrac{3}{11} \div \dfrac{5}{7} = \dfrac{3}{11} \times \dfrac{7}{5} = \dfrac{21}{55}$

(b) $\dfrac{\frac{12}{5}}{\frac{8}{15}} = \dfrac{12}{5} \div \dfrac{8}{15} = \dfrac{12}{5} \times \dfrac{15}{8} = \dfrac{3 \cdot 4}{5} \times \dfrac{3 \cdot 5}{2 \cdot 4} = \dfrac{9}{2} \text{ or } 4\dfrac{1}{2}$

9. (a) $1\dfrac{2}{5} \div 2\dfrac{1}{3} = \dfrac{7}{5} \div \dfrac{7}{3} = \dfrac{7}{5} \times \dfrac{3}{7} = \dfrac{3}{5}$

(b) $4\dfrac{2}{3} \div 7 = \dfrac{14}{3} \div \dfrac{7}{1} = \dfrac{14}{3} \times \dfrac{1}{7} = \dfrac{2 \cdot 7}{3} \times \dfrac{1}{7} = \dfrac{2}{3}$

(c) $\dfrac{1\frac{1}{5}}{1\frac{2}{7}} = 1\dfrac{1}{5} \div 1\dfrac{2}{7} = \dfrac{6}{5} \div \dfrac{9}{7} = \dfrac{6}{5} \times \dfrac{7}{9} = \dfrac{2 \cdot 3}{5} \times \dfrac{7}{3 \cdot 3} = \dfrac{14}{15}$

10. $64 \div 5\dfrac{1}{3} = \dfrac{64}{1} \div \dfrac{16}{3} = \dfrac{64}{1} \times \dfrac{3}{16} = \dfrac{4 \cdot 16}{1} \times \dfrac{3}{16} = \dfrac{12}{1} = 12$

He can fill 12 jars.

11. $126 \div 5\dfrac{1}{4} = \dfrac{126}{1} \div \dfrac{21}{4} = \dfrac{126}{1} \times \dfrac{4}{21} = \dfrac{6 \cdot 21}{1} \times \dfrac{4}{21} = 24$

The car got 24 miles per gallon.

0.4 Practice Problems

1. (a) $0.9 = \dfrac{9}{10} =$ nine tenths

(b) $0.09 = \dfrac{9}{100} =$ nine hundredths.

(c) $0.731 = \dfrac{731}{1000} =$ seven hundred thirty-one thousandths

(d) $1.371 = 1\dfrac{371}{1000} =$ one and three hundred seventy-one thousandths

(e) $0.0005 = \dfrac{5}{10,000} =$ five ten-thousandths

2. (a) $\dfrac{3}{8} = 0.375$

$$\begin{array}{r} 0.375 \\ 8\overline{)3.000} \\ \underline{2\,4} \\ 60 \\ \underline{56} \\ 40 \\ \underline{40} \\ 0 \end{array}$$

(b) $\dfrac{7}{200} = 0.035$

$$\begin{array}{r} 0.035 \\ 200\overline{)7.000} \\ \underline{6\,00} \\ 1\,000 \\ \underline{1\,000} \\ 0 \end{array}$$

(c) $\dfrac{33}{20} = 1.65$

$$
\begin{array}{r}
1.65 \\
20\overline{\smash{)}33.00} \\
\underline{20} \\
13\,0 \\
\underline{12\,0} \\
1\,00 \\
\underline{1\,00} \\
0
\end{array}
$$

3. (a) $\dfrac{1}{6} = 0.1666\ldots$ or $0.1\overline{6}$

$$
\begin{array}{r}
0.166 \\
6\overline{\smash{)}1.000} \\
\underline{6} \\
40 \\
\underline{36} \\
40 \\
\underline{36} \\
4
\end{array}
$$

(b) $\dfrac{5}{11} = 0.454545\ldots$ or $0.\overline{45}$

$$
\begin{array}{r}
0.4545 \\
11\overline{\smash{)}5.0000} \\
\underline{4\,4} \\
60 \\
\underline{55} \\
50 \\
\underline{44} \\
60 \\
\underline{55} \\
5
\end{array}
$$

4. (a) $0.8 = \dfrac{8}{10} = \dfrac{4}{5}$

(b) $0.88 = \dfrac{88}{100} = \dfrac{22}{25}$

(c) $0.45 = \dfrac{45}{100} = \dfrac{9}{20}$

(d) $0.148 = \dfrac{148}{1000} = \dfrac{37}{250}$

(e) $0.612 = \dfrac{612}{1000} = \dfrac{153}{250}$

(f) $0.016 = \dfrac{16}{1000} = \dfrac{2}{125}$

5. (a)
$$
\begin{array}{r}
3.12 \\
5.08 \\
+1.42 \\
\hline
9.62
\end{array}
$$
(b)
$$
\begin{array}{r}
152.003 \\
-136.118 \\
\hline
15.885
\end{array}
$$
(c)
$$
\begin{array}{r}
1.1 \\
3.16 \\
+5.123 \\
\hline
9.383
\end{array}
$$
(d)
$$
\begin{array}{r}
1.0052 \\
-0.1234 \\
\hline
0.8818
\end{array}
$$

6. (a)
$$
\begin{array}{r}
0.0610 \\
5.0008 \\
+1.3000 \\
\hline
6.3618
\end{array}
$$
(b)
$$
\begin{array}{r}
18.000 \\
-0.126 \\
\hline
17.874
\end{array}
$$

7.
$$
\begin{array}{r}
0.5 \\
\times\,0.3 \\
\hline
0.15
\end{array}
$$
(one decimal place)
(one decimal place)
(two decimal places)

8.
$$
\begin{array}{r}
0.12 \\
\times\,0.4 \\
\hline
0.048
\end{array}
$$
(two decimal places)
(one decimal place)
(three decimal places)

9. (a)
$$
\begin{array}{r}
1.23 \\
\times\,0.005 \\
\hline
0.00615
\end{array}
$$
(two decimal places)
(three decimal places)
(five decimal places)

(b)
$$
\begin{array}{r}
0.003 \\
\times\,0.00002 \\
\hline
0.00000006
\end{array}
$$
(three decimal places)
(five decimal places)
(eight decimal places)

10.
$$
\begin{array}{r}
5.26 \\
6\overline{\smash{)}31.56} \\
\underline{30} \\
1\,5 \\
\underline{1\,2} \\
36 \\
\underline{36} \\
0
\end{array}
$$
Each box of paper costs $5.26.

11
$$
\begin{array}{r}
300\,00. \\
.06\,\overline{\smash{)}1800.00} \\
\underline{18} \\
00000
\end{array}
$$
Thus, $1800 \div 0.06 = 30{,}000$.

12.
$$
\begin{array}{r}
0.0036 \\
4.9\,\overline{\smash{)}0.0\,1764} \\
\underline{147} \\
294 \\
\underline{294} \\
0
\end{array}
$$
Thus, $0.01764 \div 4.9 = 0.0036$.

13. (a) $0.0016 \times 100 = 0.16$
Move decimal point 2 places to the right.
(b) $2.34 \times 1000 = 2340$
Move decimal point 3 places to the right.
(c) $56.75 \times 10{,}000 = 567{,}500$
Move decimal point 4 places to the right.

14. (a) $\dfrac{5.82}{10} = 0.582$ (Move decimal point 1 place to the left.)

(b) $\dfrac{123.4}{1000} = 0.1234$ (Move decimal point 3 places to the left.)

(c) $\dfrac{0.00614}{10{,}000} = 0.000000614$ (Move decimal point 4 places to the left.)

0.5 Practice Problems

1. (a) $0.92 = 92\%$
(b) $0.0736 = 7.36\%$
(c) $0.7 = 0.70 = 70\%$
(d) $0.0003 = 0.03\%$

2. (a) $3.04 = 304\%$
(b) $5.186 = 518.6\%$
(c) $2.1 = 2.10 = 210\%$

3. (a) $7\% = 0.07$
(b) $9.3\% = 0.093$
(c) $131\% = 1.31$
(d) $0.04\% = 0.0004$

4. Change the percents to decimals and multiply.
(a) 18% of $50 = 0.18 \times 50 = 9$
(b) 4% of $64 = 0.04 \times 64 = 2.56$
(c) 156% of $35 = 1.56 \times 35 = 54.6$

5. (a) 4.2% of $38{,}000 = 0.042 \times 38{,}000 = 1596$
His raise is $1596.
(b) $38{,}000 + 1596 = 39{,}596$
His new salary is $39,596.

6. $\dfrac{37}{148}$ reduces to $\dfrac{37 \cdot 1}{37 \cdot 4} = \dfrac{1}{4} = 0.25 = 25\%$

7. (a) $\dfrac{24}{48} = \dfrac{1}{2} = 0.5 = 50\%$ **(b)** $\dfrac{4}{25} = 0.16 = 16\%$

8. Round 128,621 to 100,000. Round 378 to 400.
$100{,}000 \times 400 = 40{,}000{,}000$

9. (a) Round 422.8 miles to 400 miles. Round 19.3 gallons to 20 gallons.
$$
\begin{array}{r}
20 \\
20\overline{\smash{)}400}
\end{array}
$$
Roberta's truck gets about 20 miles per gallon.
(b) Round 3862 miles to 4000 miles.
$$
\begin{array}{r}
200 \\
20\overline{\smash{)}4000}
\end{array}
$$
She will use about 200 gallons of gas for her trip.
Round $3.69\dfrac{9}{10}$ to $4.00.
$200 \times \$4 = \$800$
The estimated cost is $800.

10. $56.93\% = 0.5693$
Round 0.5693 to 0.6. Round $293,567.12 to $300,000.
$0.6 \times \$300{,}000 = \$180{,}000$
An estimate for the answer is $180,000.

Chapter 1 1.1 Practice Problems

1.

	Number	Integer	Rational Number	Irrational Number	Real Number
(a)	$-\frac{2}{5}$		X		X
(b)	$1.515151\ldots$		X		X
(c)	-8	X	X		X
(d)	π			X	X

2. (a) Population growth of 1259 is $+1259$.
   (b) Depreciation of \$763 is -763.
   (c) Wind-chill factor of minus 10°F is -10.

3. (a) The additive inverse of $\frac{2}{5}$ is $-\frac{2}{5}$.
   (b) The additive inverse of -1.92 is $+1.92$.
   (c) The opposite of a loss of 12 yards on a football play is a gain of 12 yards on the play.

4. (a) $|-7.34| = 7.34$
   (b) $\left|\frac{5}{8}\right| = \frac{5}{8}$
   (c) $\left|\frac{0}{2}\right| = \frac{0}{2} = 0$

5. (a) $37 + 19$
     $37 + 19 = 56$
     $37 + 19 = +56$
   (b) $-23 + (-35)$
     $23 + 35 = 58$
     $-23 + (-35) = -58$

6. $-\frac{3}{5} + \left(-\frac{4}{7}\right)$

$-\frac{21}{35} + \left(-\frac{20}{35}\right)$

$\frac{21}{35} + \frac{20}{35} = \frac{41}{35} \text{ or } 1\frac{6}{35}$

$-\frac{21}{35} + \left(-\frac{20}{35}\right) = -\frac{41}{35} \text{ or } -1\frac{6}{35}$

7. $-12.7 + (-9.38)$
    $12.7 + 9.38 = 22.08$
    $-12.7 + (-9.38) = -22.08$

8. $-7 + (-11) + (-33)$
    $= -18 + (-33)$
    $= -51$

9. $-9 + 15$
    $15 - 9 = 6$
    $-9 + 15 = 6$

10. $-\frac{5}{12} + \frac{7}{12} + \left(-\frac{11}{12}\right)$

$= \frac{2}{12} + \left(-\frac{11}{12}\right) = -\frac{9}{12} = -\frac{3}{4}$

11. $-6.3 + (-8.0) + 3.5$
    $= -14.3 + 3.5$
    $= -10.8$

12.
-6	$+5$
-7	$+5$
-2	$+3$
-15	13

$-6 + 5 + (-7) + (-2) + 5 + 3 = -15 + 13 = -2$

13. (a) $-2.9 + (-5.7) = -8.6$
   (b) $\frac{2}{3} + \left(-\frac{1}{4}\right)$

$= \frac{8}{12} + \left(-\frac{3}{12}\right) = \frac{5}{12}$

1.2 Practice Problems

1. $9 - (-3) = 9 + (+3) = 12$

2. $-12 - (-5) = -12 + (+5) = -7$

3. (a) $\frac{5}{9} - \frac{7}{9} = \frac{5}{9} + \left(-\frac{7}{9}\right) = -\frac{2}{9}$

   (b) $-\frac{5}{21} - \left(-\frac{3}{7}\right) = -\frac{5}{21} + \frac{3}{7} = -\frac{5}{21} + \frac{9}{21} = \frac{4}{21}$

4. $-17.3 - (-17.3) = -17.3 + 17.3 = 0$

5. (a) $-21 - 9$
     $= -21 + (-9)$
     $= -30$
   (b) $17 - 36$
     $= 17 + (-36)$
     $= -19$
   (c) $12 - (-15)$
     $= 12 + 15$
     $= 27$
   (d) $\frac{3}{5} - 2$
     $= \frac{3}{5} + (-2)$
     $= \frac{3}{5} + \left(-\frac{10}{5}\right) = -\frac{7}{5} \text{ or } -1\frac{2}{5}$

6. $350 - (-186) = 350 + 186 = 536$
The helicopter is 536 feet from the sunken vessel.

1.3 Practice Problems

1. (a) $(-6)(-2) = 12$
   (b) $(7)(9) = 63$
   (c) $\left(-\frac{3}{5}\right)\left(\frac{2}{7}\right) = -\frac{6}{35}$
   (d) $\left(\frac{5}{6}\right)(-7) = \left(\frac{5}{6}\right)\left(-\frac{7}{1}\right) = -\frac{35}{6} \text{ or } -5\frac{5}{6}$

2. $(-5)(-2)(-6) = (+10)(-6) = -60$

3. (a) Positive; $-2(-3) = 6$
   (b) Negative; $(-1)(-3)(-2) = 3(-2) = -6$
   (c) Positive;
     $-4\left(-\frac{1}{4}\right)(-2)(-6) = 1(-2)(-6) = -2(-6) = +12 \text{ or } 12$

4. (a) $-36 \div (-2) = 18$
   (b) $-49 \div 7 = -7$
   (c) $\frac{50}{-10} = -5$
   (d) $\frac{-39}{13} = -3$

5. (a) The numbers have the same sign, so the result will be positive. Divide the absolute values.

$$1.8\overline{\smash{)}12.6}\quad\begin{array}{c}7.\\\underline{12\,6}\end{array}$$

    $-12.6 \div (-1.8) = 7$

   (b) The numbers have different signs, so the result will be negative. Divide the absolute values.

$$0.9\overline{\smash{)}0.4\,5}\quad\begin{array}{c}0.5\\\underline{4\,5}\end{array}$$

    $0.45 \div (-0.9) = -0.5$

6. $-\frac{5}{16} \div \left(-\frac{10}{13}\right) = \left(-\frac{5}{16}\right)\left(-\frac{13}{10}\right) = \left(-\frac{\overset{1}{5}}{16}\right)\left(-\frac{13}{\underset{2}{10}}\right) = \frac{13}{32}$

7. (a) $\dfrac{-12}{-\frac{4}{5}} = -\frac{12}{1} \div \left(-\frac{4}{5}\right) = -\frac{\overset{3}{12}}{1}\left(-\frac{5}{\underset{1}{4}}\right) = 15$

   (b) $\dfrac{-\frac{2}{9}}{\frac{8}{13}} = -\frac{2}{9} \div \frac{8}{13} = -\frac{\overset{1}{2}}{9}\left(\frac{13}{\underset{4}{8}}\right) = -\frac{13}{36}$

8. (a) $6(-10) = -60$

The team lost approximately 60 yards with plays that were considered medium losses.

(b) $7(15) = 105$

The team gained approximately 105 yards with plays that were considered medium gains.

(c) $-60 + 105 = 45$

A total of 45 yards were gained during plays that were medium losses and medium gains.

1.4 Practice Problems

1. (a) $6(6)(6)(6) = 6^4$

(b) $-2(-2)(-2)(-2)(-2) = (-2)^5$

(c) $108(108)(108) = 108^3$

(d) $-11(-11)(-11)(-11)(-11)(-11) = (-11)^6$

(e) $(w)(w)(w) = w^3$

(f) $(z)(z)(z)(z) = z^4$

2. (a) $3^5 = (3)(3)(3)(3)(3) = 243$

(b) $2^2 = (2)(2) = 4$

$3^3 = (3)(3)(3) = 27$

$2^2 + 3^3 = 4 + 27 = 31$

3. (a) $(-3)^3 = -27$

(b) $(-2)^6 = 64$

(c) $-2^4 = -(2^4) = -16$

(d) $-(3^6) = -729$

4. (a) $\left(\frac{1}{3}\right)^3 = \left(\frac{1}{3}\right)\left(\frac{1}{3}\right)\left(\frac{1}{3}\right) = \frac{1}{27}$

(b) $(0.3)^4 = (0.3)(0.3)(0.3)(0.3) = 0.0081$

(c) $\left(\frac{3}{2}\right)^4 = \left(\frac{3}{2}\right)\left(\frac{3}{2}\right)\left(\frac{3}{2}\right)\left(\frac{3}{2}\right) = \frac{81}{16}$

(d) $3^4 = (3)(3)(3)(3) = 81$

$4^2 = (4)(4) = 16$

$(3)^4(4)^2 = (81)(16) = 1296$

(e) $4^2 - 2^4 = 16 - 16 = 0$

1.5 Practice Problems

1. $25 \div 5 \cdot 6 + 2^3$

$= 25 \div 5 \cdot 6 + 8$

$= 5 \cdot 6 + 8$

$= 30 + 8$

$= 38$

2. $(-4)^3 - 2^6 = -64 - 64 = -128$

3. $6 - (8 - 12)^2 + 8 \div 2$

$= 6 - (-4)^2 + 8 \div 2$

$= 6 - (16) + 8 \div 2$

$= 6 - 16 + 4$

$= -10 + 4$

$= -6$

4. $\left(-\frac{1}{7}\right)\left(-\frac{14}{5}\right) + \left(-\frac{1}{2}\right) \div \frac{3}{4}$

$= \left(-\frac{1}{7}\right)\left(-\frac{14}{5}\right) + \left(-\frac{1}{2}\right)\frac{4}{3}$

$= \frac{2}{5} + \left(-\frac{2}{3}\right)$

$= \frac{2 \cdot 3}{5 \cdot 3} + \left(-\frac{2 \cdot 5}{3 \cdot 5}\right)$

$= \frac{6}{15} + \left(-\frac{10}{15}\right)$

$= -\frac{4}{15}$

1.6 Practice Problems

1. (a) $3(x + 2y) = 3x + 3(2y) = 3x + 6y$

(b) $-2(a - 3b) = -2(a) + (-2)(-3b) = -2a + 6b$

2. (a) $-(-3x + y) = (-1)(-3x + y)$

$= (-1)(-3x) + (-1)(y)$

$= 3x - y$

3. (a) $\frac{3}{5}(a^2 - 5a + 25) = \left(\frac{3}{5}\right)(a^2) + \left(\frac{3}{5}\right)(-5a) + \left(\frac{3}{5}\right)(25)$

$= \frac{3}{5}a^2 - 3a + 15$

(b) $2.5(x^2 - 3.5x + 1.2)$

$= (2.5)(x^2) + (2.5)(-3.5x) + (2.5)(1.2)$

$= 2.5x^2 - 8.75x + 3$

4. $-4x(x - 2y + 3) = (-4)(x)(x) + (-4)(x)(-2)(y) + (-4)(x)(3)$

$= -4x^2 + 8xy - 12x$

5. $(3x^2 - 2x)(-4) = (3x^2)(-4) + (-2x)(-4) = -12x^2 + 8x$

6. $400(6x + 9y) = 400(6x) + 400(9y)$

$= 2400x + 3600y$

The area of the field in square feet is $2400x + 3600y$.

1.7 Practice Problems

1. (a) $5a$ and $8a$ are like terms.

$2b$ and $-4b$ are like terms.

(b) y^2 and $-7y^2$ are like terms. These are the only like terms.

2. (a) $16y^3 + 9y^3 = (16 + 9)y^3 = 25y^3$

(b) $5a + 7a + 4a = (5 + 7 + 4)a = 16a$

3. $-8y^2 - 9y^2 + 4y^2 = (-8 - 9 + 4)y^2 = -13y^2$

4. (a) $1.3x + 3a - 9.6x + 2a = -8.3x + 5a$

(b) $5ab - 2ab^2 - 3a^2b + 6ab = 11ab - 2ab^2 - 3a^2b$

(c) There are no like terms in the expression $7x^2y - 2xy^2 - 3x^2y^2 - 4xy$, so no terms can be combined.

5. $5xy - 2x^2y + 6xy^2 - xy - 3xy^2 - 7x^2y$

$= 5xy - xy - 2x^2y - 7x^2y + 6xy^2 - 3xy^2$

$= 4xy - 9x^2y + 3xy^2$

6. $\frac{1}{7}a^2 + 2a^2 = \frac{1}{7}a^2 + \frac{2}{1}a^2 = \frac{1}{7}a^2 + \frac{2 \cdot 7}{1 \cdot 7}a^2$

$= \frac{1}{7}a^2 + \frac{14}{7}a^2 = \frac{15}{7}a^2$

$-\frac{5}{12}b - \frac{1}{3}b = -\frac{5}{12}b - \frac{1 \cdot 4}{3 \cdot 4}b = -\frac{5}{12}b - \frac{4}{12}b$

$= -\frac{9}{12}b = -\frac{3}{4}b$

Thus, our solution is $\frac{15}{7}a^2 - \frac{3}{4}b$.

7. $5a(2 - 3b) - 4(6a + 2ab) = 10a - 15ab - 24a - 8ab$

$= -14a - 23ab$

1.8 Practice Problems

1. $4 - \frac{1}{2}x = 4 - \frac{1}{2}(-8)$

$= 4 + 4$

$= 8$

2. (a) $4x^2 = 4(-3)^2 = 4(9) = 36$

(b) $(4x)^2 = [4(-3)]^2 = [-12]^2 = 144$

3. $2x^2 - 3x = 2(-2)^2 - 3(-2)$

$= 2(4) - 3(-2)$

$= 8 + 6$

$= 14$

4. Area of a triangle is $A = \frac{1}{2}ab$

altitude = 3 meters (m)

base = 7 meters (m)

$$A = \frac{1}{2}(3\,\text{m})(7\,\text{m})$$

$$= \frac{1}{2}(3)(7)(\text{m})(\text{m})$$

$$= \left(\frac{3}{2}\right)(7)(\text{m})^2$$

$$= \frac{21}{2}(\text{m})^2$$

$$= 10.5 \text{ square meters}$$

5. Area of a circle is

$A = \pi r^2$

$r = 3$ meters

$A \approx 3.14(3\,\text{m})^2$

$\quad = 3.14(9)(\text{m})^2$

$\quad = 28.26$ square meters

6. Formula $C = \frac{5}{9}(F - 32)$

$$= \frac{5}{9}(68 - 32)$$

$$= \frac{5}{9}(36)$$

$$= 5(4)$$

$$= 20° \text{ Celsius}$$

7. Use the formula.

$k \approx 1.61(r)$

$\quad = 1.61(35)$ Replace r by 35.

$\quad = 56.35$

The truck is traveling at approximately 56.35 kilometers per hour. It is violating the minimum speed limit.

1.9 Practice Problems

1. $5[4x - 3(y - 2)]$

$= 5[4x - 3y + 6]$

$= 20x - 15y + 30$

2. $-3[2a - (3b - c) + 4d]$

$= -3[2a - 3b + c + 4d]$

$= -6a + 9b - 3c - 12d$

3. $3[4x - 2(1 - x)] - [3x + (x - 2)]$

$= 3[4x - 2 + 2x] - [3x + x - 2]$

$= 3[6x - 2] - [4x - 2]$

$= 18x - 6 - 4x + 2$

$= 14x - 4$

4. $-2\{5x - 3[2x - (3 - 4x)]\}$

$= -2\{5x - 3[2x - 3 + 4x]\}$

$= -2\{5x - 3[6x - 3]\}$

$= -2\{5x - 18x + 9\}$

$= -2\{-13x + 9\}$

$= 26x - 18$

Chapter 2 2.1 Practice Problems

1. $x + 14 = 23$

$x + 14 + (-14) = 23 + (-14)$

$\qquad x + 0 = 9$

$\qquad x = 9$

Check. $x + 14 = 23$

$\qquad 9 + 14 \stackrel{?}{=} 23$

$\qquad 23 = 23$ ✓

2. $17 = x - 5$

$17 + 5 = x - 5 + 5$

$\quad 22 = x + 0$

$\quad 22 = x$

Check. $17 = x - 5$

$\qquad 17 \stackrel{?}{=} 22 - 5$

$\qquad 17 = 17$ ✓

3. $0.5 - 1.2 = x - 0.3$

$\qquad -0.7 = x - 0.3$

$-0.7 + 0.3 = x - 0.3 + 0.3$

$\qquad -0.4 = x$

Check. $0.5 - 1.2 = x - 0.3$

$0.5 - 1.2 \stackrel{?}{=} -0.4 - 0.3$

$\qquad -0.7 = -0.7$ ✓

4. $x + 8 = -22 + 6$

$-2 + 8 \stackrel{?}{=} -22 + 6$

$\qquad 6 \neq -16$ This is not true.

Thus $x = -2$ is not a solution. Solve to find the solution.

$\qquad x + 8 = -22 + 6$

$\qquad x + 8 = -16$

$x + 8 - 8 = -16 - 8$

$\qquad x = -24$

5. $\dfrac{1}{20} - \dfrac{1}{2} = x + \dfrac{3}{5}$

$\dfrac{1}{20} - \dfrac{1 \cdot 10}{2 \cdot 10} = x + \dfrac{3 \cdot 4}{5 \cdot 4}$

$\dfrac{1}{20} - \dfrac{10}{20} = x + \dfrac{12}{20}$

$-\dfrac{9}{20} = x + \dfrac{12}{20}$

$-\dfrac{9}{20} + \left(-\dfrac{12}{20}\right) = x + \dfrac{12}{20} + \left(-\dfrac{12}{20}\right)$

$-\dfrac{21}{20} = x$

$x = -\dfrac{21}{20}$ or $-1\dfrac{1}{20}$

Check.

$\dfrac{1}{20} - \dfrac{1}{2} = x + \dfrac{3}{5}$

$\dfrac{1}{20} - \dfrac{1}{2} \stackrel{?}{=} -\dfrac{21}{20} + \dfrac{3}{5}$

$\dfrac{1}{20} - \dfrac{10}{20} \stackrel{?}{=} -\dfrac{21}{20} + \dfrac{12}{20}$

$-\dfrac{9}{20} = -\dfrac{9}{20}$ ✓

2.2 Practice Problems

1. $\dfrac{1}{8}x = -2$

$8\left(\dfrac{1}{8}x\right) = 8(-2)$

$\left(\dfrac{8}{1}\right)\left(\dfrac{1}{8}\right)x = -16$

$\qquad x = -16$

2. $9x = 72$

$\dfrac{9x}{9} = \dfrac{72}{9}$

$x = 8$

3. $6x = 50$

$\dfrac{6x}{6} = \dfrac{50}{6}$

$x = \dfrac{25}{3}$

4. $-27x = 54$

$\dfrac{-27x}{-27} = \dfrac{54}{-27}$

$x = -2$

5. $-x = 36$

$-1x = 36$

$\dfrac{-1x}{-1} = \dfrac{36}{-1}$

$x = -36$

6. $-51 = -6x$

$\dfrac{-51}{-6} = \dfrac{-6x}{-6}$

$\dfrac{17}{2} = x$

7. $16.2 = 5.2x - 3.4x$

$16.2 = 1.8x$

$\dfrac{16.2}{1.8} = \dfrac{1.8x}{1.8}$

$9 = x$

2.3 Practice Problems

1. $9x + 2 = 38$

$9x + 2 + (-2) = 38 + (-2)$

$\qquad 9x = 36$

$\qquad \dfrac{9x}{9} = \dfrac{36}{9}$

$\qquad x = 4$

Check. $9(4) + 2 \stackrel{?}{=} 38$

$\qquad 36 + 2 \stackrel{?}{=} 38$

$\qquad 38 = 38$ ✓

2. $13x = 2x - 66$

$13x + (-2x) = 2x + (-2x) - 66$

$\qquad 11x = -66$

$\qquad \dfrac{11x}{11} = \dfrac{-66}{11}$

$\qquad x = -6$

3.
$$3x + 2 = 5x + 2$$
$$3x + (-3x) + 2 = 5x + (-3x) + 2$$
$$2 = 2x + 2$$
$$2 + (-2) = 2x + 2 + (-2)$$
$$0 = 2x$$
$$\frac{0}{2} = \frac{2x}{2}$$
$$0 = x$$

Check. $3(0) + 2 \overset{?}{=} 5(0) + 2$
$$2 = 2 \checkmark$$

4.
$$-z + 8 - z = 3z + 10 - 3$$
$$-2z + 8 = 3z + 7$$
$$-2z + 2z + 8 = 3z + 2z + 7$$
$$8 = 5z + 7$$
$$8 + (-7) = 5z + 7 + (-7)$$
$$1 = 5z$$
$$\frac{1}{5} = \frac{5z}{5}$$
$$\frac{1}{5} = z$$

5. $4x - (x + 3) = 12 - 3(x - 2)$
$$4x - x - 3 = 12 - 3x + 6$$
$$3x - 3 = -3x + 18$$
$$3x + 3x - 3 = -3x + 3x + 18$$
$$6x - 3 = 18$$
$$6x - 3 + 3 = 18 + 3$$
$$6x = 21$$
$$\frac{6x}{6} = \frac{21}{6}$$
$$x = \frac{7}{2}$$

Check. $4\left(\frac{7}{2}\right) - \left(\frac{7}{2} + 3\right) \overset{?}{=} 12 - 3\left(\frac{7}{2} - 2\right)$
$$14 - \frac{13}{2} \overset{?}{=} 12 - 3\left(\frac{3}{2}\right)$$
$$\frac{28}{2} - \frac{13}{2} \overset{?}{=} \frac{24}{2} - \frac{9}{2}$$
$$\frac{15}{2} = \frac{15}{2} \checkmark$$

6. $4(-2x - 3) = -5(x - 2) + 2$
$$-8x - 12 = -5x + 10 + 2$$
$$-8x - 12 = -5x + 12$$
$$-8x + 8x - 12 = -5x + 8x + 12$$
$$-12 = 3x + 12$$
$$-12 - 12 = 3x + 12 - 12$$
$$-24 = 3x$$
$$\frac{-24}{3} = \frac{3x}{3}$$
$$-8 = x$$

7. $0.3x - 2(x + 0.1) = 0.4(x - 3) - 1.1$
$$0.3x - 2x - 0.2 = 0.4x - 1.2 - 1.1$$
$$-1.7x - 0.2 = 0.4x - 2.3$$
$$-1.7x + 1.7x - 0.2 = 0.4x + 1.7x - 2.3$$
$$-0.2 = 2.1x - 2.3$$
$$-0.2 + 2.3 = 2.1x - 2.3 + 2.3$$
$$2.1 = 2.1x$$
$$\frac{2.1}{2.1} = \frac{2.1x}{2.1}$$
$$1 = x$$

8. $5(2z - 1) + 7 = 7z - 4(z + 3)$
$$10z - 5 + 7 = 7z - 4z - 12$$
$$10z + 2 = 3z - 12$$
$$10z - 3z + 2 = 3z - 3z - 12$$
$$7z + 2 = -12$$
$$7z + 2 - 2 = -12 - 2$$
$$7z = -14$$
$$\frac{7z}{7} = \frac{-14}{7}$$
$$z = -2$$

Check. $5[2(-2) - 1] + 7 \overset{?}{=} 7(-2) - 4[-2 + 3]$
$$5[-4 - 1] + 7 \overset{?}{=} 7(-2) - 4[1]$$
$$5(-5) + 7 \overset{?}{=} -14 - 4$$
$$-25 + 7 \overset{?}{=} -18$$
$$-18 = -18 \checkmark$$

2.4 Practice Problems

1.
$$\frac{3}{8}x - \frac{3}{2} = \frac{1}{4}x$$
$$8\left(\frac{3}{8}x - \frac{3}{2}\right) = 8\left(\frac{1}{4}x\right)$$
$$\left(\frac{8}{1}\right)\left(\frac{3}{8}\right)(x) - \left(\frac{8}{1}\right)\left(\frac{3}{2}\right) = \left(\frac{8}{1}\right)\left(\frac{1}{4}\right)(x)$$
$$3x - 12 = 2x$$
$$3x + (-3x) - 12 = 2x + (-3x)$$
$$-12 = -x$$
$$12 = x$$

2.
$$\frac{5x}{4} - 1 = \frac{3x}{4} + \frac{1}{2}$$
$$4\left(\frac{5x}{4}\right) - 4(1) = 4\left(\frac{3x}{4}\right) + 4\left(\frac{1}{2}\right)$$
$$5x - 4 = 3x + 2$$
$$5x - 3x - 4 = 3x - 3x + 2$$
$$2x - 4 = 2$$
$$2x - 4 + 4 = 2 + 4$$
$$2x = 6$$
$$\frac{2x}{2} = \frac{6}{2}$$
$$x = 3$$

Check.
$$\frac{5(3)}{4} - 1 \overset{?}{=} \frac{3(3)}{4} + \frac{1}{2}$$
$$\frac{15}{4} - 1 \overset{?}{=} \frac{9}{4} + \frac{1}{2}$$
$$\frac{15}{4} - \frac{4}{4} \overset{?}{=} \frac{9}{4} + \frac{2}{4}$$
$$\frac{11}{4} = \frac{11}{4} \checkmark$$

3.
$$\frac{x + 6}{9} = \frac{x}{6} + \frac{1}{2}$$
$$\frac{x}{9} + \frac{6}{9} = \frac{x}{6} + \frac{1}{2}$$
$$18\left(\frac{x}{9}\right) + 18\left(\frac{6}{9}\right) = 18\left(\frac{x}{6}\right) + 18\left(\frac{1}{2}\right)$$
$$2x + 12 = 3x + 9$$
$$2x - 2x + 12 = 3x - 2x + 9$$
$$12 = x + 9$$
$$12 - 9 = x + 9 - 9$$
$$3 = x$$

4.
$$\frac{1}{2}(x + 5) = \frac{1}{5}(x - 2) + \frac{1}{2}$$
$$\frac{x}{2} + \frac{5}{2} = \frac{x}{5} - \frac{2}{5} + \frac{1}{2}$$
$$10\left(\frac{x}{2}\right) + 10\left(\frac{5}{2}\right) = 10\left(\frac{x}{5}\right) - 10\left(\frac{2}{5}\right) + 10\left(\frac{1}{2}\right)$$
$$5x + 25 = 2x - 4 + 5$$
$$5x + 25 = 2x + 1$$
$$5x - 2x + 25 = 2x - 2x + 1$$
$$3x + 25 = 1$$
$$3x + 25 - 25 = 1 - 25$$
$$3x = -24$$
$$\frac{3x}{3} = \frac{-24}{3}$$
$$x = -8$$

Check. $\frac{1}{2}(-8 + 5) = \frac{1}{5}(-8 - 2) + \frac{1}{2}$
$$\frac{1}{2}(-3) = \frac{1}{5}(-10) + \frac{1}{2}$$
$$-\frac{3}{2} = -2 + \frac{1}{2}$$
$$-\frac{3}{2} = -\frac{4}{2} + \frac{1}{2}$$
$$-\frac{3}{2} = -\frac{3}{2} \checkmark$$

5.
$$2.8 = 0.3(x - 2) + 2(0.1x - 0.3)$$
$$2.8 = 0.3x - 0.6 + 0.2x - 0.6$$
$$10(2.8) = 10(0.3x) - 10(0.6) + 10(0.2x) - 10(0.6)$$
$$28 = 3x - 6 + 2x - 6$$
$$28 = 5x - 12$$
$$28 + 12 = 5x - 12 + 12$$
$$40 = 5x$$
$$\frac{40}{5} = \frac{5x}{5}$$
$$8 = x$$

2.5 Practice Problems

1. (a) $x + 4$ **(b)** $3x$ **(c)** $x - 8$ **(d)** $\frac{1}{4}x$

2. (a) $3x + 8$ **(b)** $3(x + 8)$ **(c)** $\frac{1}{3}(x + 4)$

3. Let a = Ann's hours per week.
Then $a - 17$ = Marie's hours per week.

4.

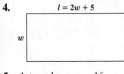

width = w
length = $2w + 5$

5. 1st angle = $s - 16$
2nd angle = s
3rd angle = $2s$

Third angle
$2s$
Second angle s $s - 16$ First angle

6. Let x = the number of students in the fall.
$\frac{2}{3}x$ = the number of students in the spring.
$\frac{1}{5}x$ = the number of students in the summer.

2.6 Practice Problems

1. Let x = the unknown number.
$$\frac{3}{4}x = -81$$
$$4\left(\frac{3}{4}x\right) = 4(-81)$$
$$3x = -324$$
$$\frac{3x}{3} = \frac{-324}{3}$$
$$x = -108$$
The number is -108.

2. Let x = the unknown number.
$$3x - 2 = 49$$
$$3x - 2 + 2 = 49 + 2$$
$$3x = 51$$
$$\frac{3x}{3} = \frac{51}{3}$$
$$x = 17$$
The number is 17.

3. Let x = the first number.
$$x + (3x - 12) = 24$$
$$4x - 12 = 24$$
$$4x = 36$$
$$x = 9$$
First number is 9. Second number = $3(9) - 12 = 15$.

4. Let rainfall in Canada = x
Then rainfall in Texas = $3x - 14$
$$3x - 14 = 43$$
$$3x = 57$$
$$x = 19$$
19 inches of rain was recorded in Canada.

5. (a) $d = rt$
$$220 = 4r$$
$$55 = r$$
Leaving the city, her average speed was 55 mph.
(b) $d = rt$
$$225 = 4.5r$$
$$50 = r$$
On the return trip, her average speed was 50 mph.
(c) She traveled 5 mph faster on the trip leaving the city.

6. Let x = her final exam score. Since the final counts as two tests, divide her total score by 6.
$$\frac{78 + 80 + 100 + 96 + x + x}{6} = 90$$
$$\frac{354 + 2x}{6} = 90$$
$$6\left(\frac{354 + 2x}{6}\right) = 6(90)$$
$$354 + 2x = 540$$
$$2x = 186$$
$$x = 93$$
She needs a 93 on the final exam.

2.7 Practice Problems

1.
$$A = lw$$
$$\frac{A}{w} = \frac{lw}{w}$$
$$\frac{A}{w} = l$$
$$\frac{120\ (\text{yd})(\text{yd})}{8\ \text{yd}} = l$$
$$15\ \text{yd} = l$$

2.
$$A = \frac{1}{2}a(b_1 + b_2)$$
$$256\ (\text{ft})^2 = \frac{1}{2}a(12\ \text{ft} + 20\ \text{ft})$$
$$256\ (\text{ft})(\text{ft}) = \frac{1}{2}(a)(32\ \text{ft})$$
$$256\ (\text{ft})(\text{ft}) = (16\ \text{ft})a$$
$$16\ \text{ft} = a$$
The altitude is 16 feet.

3. $C = 2\pi r$
$$C = 2(3.14)(15)$$
$$C = 94.2$$
Rounded to the nearest meter, the circumference is approximately 94 meters.

4. P = the sum of the three sides.
$P = 15\ \text{cm} + 15\ \text{cm} + 15\ \text{cm}$ An equilateral triangle has three equal sides.
$= 45\ \text{cm}$
The perimeter of the triangle is 45 cm.

5.
$132°$
x x
$$132° + x + x = 180°$$
$$132° + 2x = 180°$$
$$2x = 48°$$
$$x = 24°$$
Both angles measure 24°.

6. Let x = the measure of the first angle. Then $2x + 5$ = the measure of the second angle, and $\frac{1}{2}x$ = the measure of the third angle.

$$x + 2x + 5° + \frac{1}{2}x = 180°$$

$$\frac{7}{2}x + 5° = 180°$$

$$\frac{7}{2}x = 175°$$

$$x = \frac{2}{7}(175°)$$

$$x = 50°$$

The measure of the first angle is 50°, the measure of the second angle is $2(50°) + 5° = 105°$, and the measure of the third angle is $\frac{1}{2}(50°) = 25°$.

7. Surface Area $= 4\pi r^2$

$\qquad\qquad = 4(3.14)(5\,\text{m})^2$

$\qquad\qquad = 314\,\text{m}^2$

Rounded to the nearest square meter, the surface area is 314 square meters.

8. $V = \pi r^2 h$

$\quad = (3.14)(3\,\text{ft})^2(4\,\text{ft})$

$\quad = 113.04\,\text{ft}^3$

Rounded to the nearest cubic foot, 113 cubic feet of sand can be stored in the drum.

9. Calculate the area of the pool.

$A = lw$

$A = (12\,\text{ft})(8\,\text{ft}) = 96\,\text{ft}^2.$

Now add 6 feet to the length and 6 feet to the width of the pool and calculate the area.

$A = lw$

$A = (18\,\text{ft})(14\,\text{ft}) = 252\,\text{ft}^2.$

Now subtract the areas.

$252\,\text{ft}^2 - 96\,\text{ft}^2 = 156\,\text{ft}^2$ at \$12 per square foot.

$\qquad 156 \times 12 = 1872$

The cost would be \$1872.

10. $V = lwh$

$\quad V = (6\,\text{ft})(5\,\text{ft})(8\,\text{ft}) = 240\,\text{ft}^3$

$\quad \text{Weight} = 240\,\text{ft}^3 \times \dfrac{62.4\,\text{lb}}{1\,\text{ft}^3} = 14{,}976\,\text{lb} \approx 15{,}000\,\text{lb}$

2.8 Practice Problems

1. (a) $7 > 2$ **(b)** $-2 > -4$ **(c)** $-1 < 2$
    (d) $-8 < -5$ **(e)** $0 > -2$ **(f)** $5 > -3$

2. (a) x is greater than 5

    (b) x is less than or equal to -2

    (c) x is less than 3 (or 3 is greater than x)

    (d) x is greater than or equal to $-\dfrac{3}{2}$

3. (a) Since the temperature can never exceed 180 degrees, then the temperature must always be less than or equal to 180 degrees. Thus, $t \le 180$.

    (b) Since the debt must be less than 15,000, we have $d < 15{,}000$.

4. (a) $\quad 7 > 2$ **(b)** $-3 < -1$
$\qquad -14 < -4$ $3 > 1$

    (c) $-10 \ge -20$ **(d)** $-15 \le -5$
$\qquad 1 \le 2$ $3 \ge 1$

5. $\qquad 8x - 2 < 3$

$\quad 8x - 2 + 2 < 3 + 2$

$\qquad\qquad 8x < 5$

$\qquad\qquad \dfrac{8x}{8} < \dfrac{5}{8}$

$\qquad\qquad x < \dfrac{5}{8}$

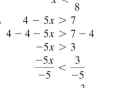

6. $\qquad 4 - 5x > 7$

$\quad 4 - 4 - 5x > 7 - 4$

$\qquad\qquad -5x > 3$

$\qquad\qquad \dfrac{-5x}{-5} < \dfrac{3}{-5}$

$\qquad\qquad x < -\dfrac{3}{5}$

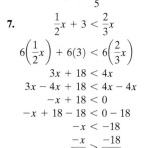

7. $\qquad \dfrac{1}{2}x + 3 < \dfrac{2}{3}x$

$\quad 6\left(\dfrac{1}{2}x\right) + 6(3) < 6\left(\dfrac{2}{3}x\right)$

$\qquad\qquad 3x + 18 < 4x$

$\quad 3x - 4x + 18 < 4x - 4x$

$\qquad\qquad -x + 18 < 0$

$\quad -x + 18 - 18 < 0 - 18$

$\qquad\qquad -x < -18$

$\qquad\qquad \dfrac{-x}{-1} > \dfrac{-18}{-1}$

$\qquad\qquad x > 18$

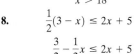

8. $\qquad \dfrac{1}{2}(3 - x) \le 2x + 5$

$\qquad \dfrac{3}{2} - \dfrac{1}{2}x \le 2x + 5$

$\quad 2\left(\dfrac{3}{2}\right) - 2\left(\dfrac{1}{2}x\right) \le 2(2x) + 2(5)$

$\qquad\qquad 3 - x \le 4x + 10$

$\quad 3 - x - 4x \le 4x - 4x + 10$

$\qquad\qquad 3 - 5x \le 10$

$\quad 3 - 3 - 5x \le 10 - 3$

$\qquad\qquad -5x \le 7$

$\qquad\qquad \dfrac{-5x}{-5} \ge \dfrac{7}{-5}$

$\qquad\qquad x \ge -\dfrac{7}{5}$

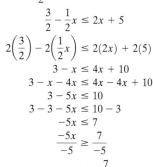

9. $\qquad 2000n - 700{,}000 \ge 2{,}500{,}000$

$\quad 2000n - 700{,}000 + 700{,}000 \ge 2{,}500{,}000 + 700{,}000$

$\qquad\qquad 2000n \ge 3{,}200{,}000$

$\qquad\qquad \dfrac{2000n}{2000} \ge \dfrac{3{,}200{,}000}{2000}$

$\qquad\qquad n \ge 1600$

Chapter 3   3.1 Practice Problems

1. Point B is 3 units to the right on the x-axis and 4 units up from the point where we stopped on the x-axis.

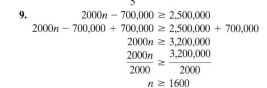

2. (a) Begin by counting 2 squares to the left, starting at the origin. Since the y-coordinate is negative, count 4 units down from the point where we stopped on the x-axis. Label the point I.

    (b) Begin by counting 4 squares to the left of the origin. Then count 5 units up because the y-coordinate is positive. Label the point J.

(c) Begin by counting 4 units to the right of the origin. Then count 2 units down because the y-coordinate is negative. Label the point K.

3. The points are plotted in the figure.

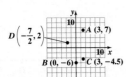

4. Move along the x-axis to get as close as possible to B. We end up at 2. Thus the first number of the ordered pair is 2. Then count 7 units upward on a line parallel to the y-axis to reach B. So the second number of the ordered pair is 7. Thus, point B is represented by $(2, 7)$.

5. $A = (-2, -1)$; $B = (-1, 3)$; $C = (0, 0)$; $D = (2, -1)$; $E = (3, 1)$

6. **(a)**

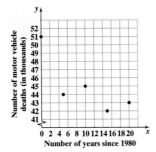

(b) Motor vehicle deaths were significantly high in 1980. During 1985–2000, the number of motor vehicle deaths was relatively stable.

7. **(a)** Replace x with 3 and y with -1.

$$3x + 2y = 5$$
$$3(3) + 2(-1) \stackrel{?}{=} 5$$
$$9 - 2 \stackrel{?}{=} 5$$
$$7 = 5 \quad \text{False}$$

The ordered pair $(3, -1)$ is not a solution to $3x + 2y = 5$.

(b) Replace x with 2 and y with $-\dfrac{1}{2}$.

$$3x + 2y = 5$$
$$3(2) + 2\left(-\frac{1}{2}\right) \stackrel{?}{=} 5$$
$$6 + (-1) \stackrel{?}{=} 5$$
$$5 = 5 \quad \checkmark \text{ True}$$

The ordered pair $\left(2, -\dfrac{1}{2}\right)$ is a solution to $3x + 2y = 5$.

8. **(a)** Replace x by 0 in the equation.
$$3x - 4y = 12$$
$$3(0) - 4y = 12$$
$$0 - 4y = 12$$
$$y = -3$$

The ordered pair is $(0, -3)$.

(b) Replace the variable y by 3.
$$3x - 4y = 12$$
$$3x - 4(3) = 12$$
$$3x - 12 = 12$$
$$3x = 24$$
$$x = 8$$

The ordered pair is $(8, 3)$.

(c) Replace the variable y by -6.
$$3x - 4y = 12$$
$$3x - 4(-6) = 12$$
$$3x + 24 = 12$$
$$3x = -12$$
$$x = -4$$

The ordered pair is $(-4, -6)$.

9. First solve the equation for y.

$$8 - 2y + 3x = 0$$
$$-2y = -3x - 8$$
$$y = \frac{-3x - 8}{-2}$$
$$y = \frac{-3x}{-2} + \frac{-8}{-2}$$
$$y = \frac{3}{2}x + 4$$

(a) Replace x with 0.

$$y = \frac{3}{2}(0) + 4$$
$$y = 4$$

(b) Replace x with 2.

$$y = \frac{3}{2}(2) + 4$$
$$y = 3 + 4$$
$$y = 7$$

Thus, we have the ordered pairs $(0, 4)$ and $(2, 7)$.

3.2 Practice Problems

1. Graph $x + y = 10$.
   Let $x = 0$.
   $$x + y = 10$$
   $$0 + y = 10$$
   $$y = 10$$

   Let $x = 3$.
   $$x + y = 10$$
   $$3 + y = 10$$
   $$y = 7$$

   Let $x = 10$.
   $$x + y = 10$$
   $$10 + y = 10$$
   $$y = 0$$

   Plot the ordered pairs $(0, 10)$, $(3, 7)$, and $(10, 0)$.

2. $$7x + 3 = -2y + 3$$
    $$7x + 3 - 3 = -2y + 3 - 3$$
    $$7x = -2y$$
    $$7x + 2y = -2y + 2y$$
    $$7x + 2y = 0$$

   Let $x = 0$.
   $$7(0) + 2y = 0$$
   $$2y = 0$$
   $$y = 0$$

   Let $x = -2$.
   $$7(-2) + 2y = 0$$
   $$-14 + 2y = 0$$
   $$2y = 14$$
   $$y = 7$$

   Let $x = 2$.
   $$7(2) + 2y = 0$$
   $$14 + 2y = 0$$
   $$2y = -14$$
   $$y = -7$$

   Graph the ordered pairs $(0, 0)$, $(-2, 7)$, and $(2, -7)$.

3. $2y - x = 6$
   Find the two intercepts.

   Let $y = 0$.   Let $x = 0$.
   $$2(0) - x = 6 \qquad 2y - 0 = 6$$
   $$-x = 6 \qquad\qquad 2y = 6$$
   $$x = -6 \qquad\qquad y = 3$$

   The x-intercept is $(-6, 0)$, the y-intercept is $(0, 3)$.

Find a third point.
Let $y = 1$.
$2(1) - x = 6$
$2 - x = 6$
$-x = 4$
$x = -4$
Graph the ordered pairs $(-6, 0)$, $(0, 3)$,
and $(-4, 1)$.

4. $2y - 3 = 0$
Solve for y.
$2y = 3$
$y = \dfrac{3}{2}$
This line is parallel to the x-axis. It is a
horizontal line $1\frac{1}{2}$ units above the x-axis.

5. $x + 3 = 0$
Solve for x.
$x = -3$
This line is parallel to the y-axis. It is a vertical
line 3 units to the left of the y-axis.

3.3 Practice Problems

1. $m = \dfrac{y_2 - y_1}{x_2 - x_1} = \dfrac{-1 - 1}{-4 - 6} = \dfrac{-2}{-10} = \dfrac{1}{5}$

2. $m = \dfrac{y_2 - y_1}{x_2 - x_1} = \dfrac{1 - 0}{-1 - 2} = \dfrac{1}{-3} = -\dfrac{1}{3}$

3. (a) $m = \dfrac{3 - 6}{-5 - (-5)} = \dfrac{-3}{0}$

$\dfrac{-3}{0}$ is undefined. Therefore there is no slope and the line is a
vertical line through $x = -5$.

(b) $m = \dfrac{-11 - (-11)}{3 - (-7)} = \dfrac{0}{10} = 0$

$m = 0$. The line is a horizontal line through $y = -11$.

4. Solve for y.
$4x - 2y = -5$
$-2y = -4x - 5$
$y = \dfrac{-4x - 5}{-2}$
$y = 2x + \dfrac{5}{2}$ Slope $= 2$ y-intercept $= \left(0, \dfrac{5}{2}\right)$

5. (a) $y = mx + b$

$m = -\dfrac{3}{7}$ y-intercept $= \left(0, \dfrac{2}{7}\right)$, $b = \dfrac{2}{7}$

$y = -\dfrac{3}{7}x + \dfrac{2}{7}$

(b) $y = -\dfrac{3}{7}x + \dfrac{2}{7}$

$7(y) = 7\left(-\dfrac{3}{7}x\right) + 7\left(\dfrac{2}{7}\right)$

$7y = -3x + 2$

$3x + 7y = 2$

6. y-intercept $= (0, -1)$. Thus the coordinates of the y-intercept for this
line are $(0, -1)$. Plot the point. Slope is $\dfrac{\text{rise}}{\text{run}}$. Since the slope for this
line is $\dfrac{3}{4}$, we will go up (rise) 3 units and go over (run) 4 units to the
right from the point $(0, -1)$. This is the point $(4, 2)$.

7. $y = -\dfrac{2}{3}x + 5$

The y-intercept is $(0, 5)$ since $b = 5$. Plot the point $(0, 5)$. The slope is
$-\dfrac{2}{3} = \dfrac{-2}{3}$. Begin at $(0, 5)$, go down 2 units and to the right 3 units.
This is the point $(3, 3)$. Draw the line that connects the points $(0, 5)$
and $(3, 3)$.

8. (a) Parallel lines have the same slope. Line j has a slope of $\dfrac{1}{4}$.

(b) Perpendicular lines have slopes whose product is -1.
$m_1 m_2 = -1$
$\dfrac{1}{4}m_2 = -1$
$4\left(\dfrac{1}{4}\right)m_2 = -1(4)$
$m_2 = -4$

Thus line k has a slope of -4.

9. (a) The slope of line n is $\dfrac{2}{3}$. The slope of a line that is parallel to
line n is $\dfrac{2}{3}$.

(b) $m_1 m_2 = -1$
$\dfrac{2}{3}m_2 = -1$
$m_2 = -\dfrac{3}{2}$

The slope of a line that is perpendicular to line n is $-\dfrac{3}{2}$.

3.4 Practice Problems

1. $y = mx + b$
$12 = -\dfrac{3}{4}(-8) + b$
$12 = 6 + b$
$6 = b$

An equation of the line is $y = -\dfrac{3}{4}x + 6$.

2. Find the slope.
$m = \dfrac{y_2 - y_1}{x_2 - x_1} = \dfrac{1 - 5}{-1 - 3} = \dfrac{-4}{-4} = 1$

Using either of the two points given, substitute x and y values into the
equation $y = mx + b$.
$m = 1$ $x = 3$ and $y = 5$.
$y = mx + b$
$5 = 1(3) + b$
$5 = 3 + b$
$2 = b$

An equation of the line is $y = x + 2$.

3. The y-intercept is $(0, 1)$. Thus $b = 1$. Look for another point on the
line. We choose $(6, 2)$. Count the number of vertical units from 1 to 2
(rise). Count the number of horizontal units from 0 to 6 (run). $m = \dfrac{1}{6}$

Now we can write an equation of the line.
$y = mx + b$
$y = \dfrac{1}{6}x + 1$

3.5 Practice Problems

1. Graph $x - y \geq -10$.

Begin by graphing the line $x - y = -10$. Use any method discussed previously. Since there is an equals sign in the inequality, draw a solid line to indicate that the line is part of the solution set. The easiest test point is $(0, 0)$. Substitute $x = 0$, $y = 0$ in the inequality.

$$x - y \geq -10$$
$$0 - 0 \geq -10$$
$$0 \geq -10 \quad \text{true}$$

Therefore, shade the side of the line that includes the point $(0, 0)$.

$x - y \geq -10$

2. Step 1 Graph $2y = x$. Since $>$ is used, the line should be a dashed line.

Step 2 The line passes through $(0, 0)$.

Step 3 Choose another test point, say $(-1, 1)$.

$$2y > x$$
$$2(1) > -1$$
$$2 > -1 \quad \text{true}$$

Shade the region that includes $(-1, 1)$, that is, the region above the line.

$2y > x$

3. Step 1 Graph $y = -3$. Since $\geq$ is used, the line should be solid.

Step 2 Test $(0, 0)$ in the inequality.

$$y \geq -3$$
$$0 \geq -3 \quad \text{true}$$

Shade the region that includes $(0, 0)$, that is, the region above the line $y = -3$.

$y \geq -3$

3.6 Practice Problems

1. The domain is $\{-3, 0, 3, 20\}$. The range is $\{-5, 5\}$.

2. (a) Look at the ordered pairs. No two ordered pairs have the same first coordinate. Thus this set of ordered pairs defines a function.

   (b) Look at the ordered pairs. Two different ordered pairs, $(60, 30)$ and $(60, 120)$, have the same first coordinate. Thus this relation is not a function.

3. (a) Looking at the table, we see that no two different ordered pairs have the same first coordinate. The cost of gasoline is a function of the distance traveled.

   Note that cost depends on distance. Thus distance is the independent variable. Since a negative distance does not make sense, the domain is {all nonnegative real numbers}.

   The range is {all nonnegative real numbers}.

   (b) Looking at the table, we see two ordered pairs, $(5, 20)$ and $(5, 30)$, have the same first coordinate. Thus this relation is not a function.

4. Construct a table, plot the ordered pairs, and connect the points.

x	$y = x^2 - 2$	y
-2	$y = (-2)^2 - 2 = 2$	2
-1	$y = (-1)^2 - 2 = -1$	-1
0	$y = (0)^2 - 2 = -2$	-2
1	$y = (1)^2 - 2 = -1$	-1
2	$y = (2)^2 - 2 = 2$	2

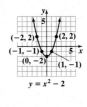

$y = x^2 - 2$

5. Select values of y and then substitute them into the equation to obtain values of x.

y	$x = y^2 - 1$	x	y
-2	$x = (-2)^2 - 1 = 3$	3	-2
-1	$x = (-1)^2 - 1 = 0$	0	-1
0	$x = (0)^2 - 1 = -1$	-1	0
1	$x = (1)^2 - 1 = 0$	0	1
2	$x = (2)^2 - 1 = 3$	3	2

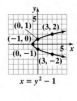

$x = y^2 - 1$

6. $y = \dfrac{6}{x}$

x	$y = \dfrac{6}{x}$	y
-3	$y = \dfrac{6}{-3} = -2$	-2
-2	$y = \dfrac{6}{-2} = -3$	-3
-1	$y = \dfrac{6}{-1} = -6$	-6
0	We cannot divide by 0.	
1	$y = \dfrac{6}{1} = 6$	6
2	$y = \dfrac{6}{2} = 3$	3
3	$y = \dfrac{6}{3} = 2$	2

$y = \dfrac{6}{x}$

7. (a) The graph of a vertical line is not a function.

   (b) This curve is a function. Any vertical line will cross the curve in only one location.

   (c) This curve is not the graph of a function. There exist vertical lines that will cross the curve in more than one place.

8. $f(x) = -2x^2 + 3x - 8$

   (a) $f(2) = -2(2)^2 + 3(2) - 8$
   $$= -2(4) + 3(2) - 8$$
   $$= -8 + 6 - 8$$
   $$= -10$$

   (b) $f(-3) = -2(-3)^2 + 3(-3) - 8$
   $$= -2(9) + 3(-3) - 8$$
   $$= -18 - 9 - 8$$
   $$= -35$$

   (c) $f(0) = -2(0)^2 + 3(0) - 8$
   $$= -2(0) + 3(0) - 8$$
   $$= 0 + 0 - 8$$
   $$= -8$$

Chapter 4   4.1 Practice Problems

1. $x + y = 12$
      $-x + y = 4$
Graph the equations on the same coordinate system.
The lines intersect at the point $x = 4$, $y = 8$. Thus the solution to the system of equations is $(4, 8)$.

2. $4x + 2y = 8$
      $-6x - 3y = 6$
Graph both equations on the same coordinate system.
These lines are parallel. They do not intersect. Hence there is no solution to this system of equations.

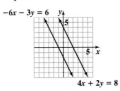

3. $3x - 9y = 18$
      $-4x + 12y = -24$
Graph both equations on the same coordinate system.
Notice both equations represent the same line. Thus there is an infinite number of solutions to this system.

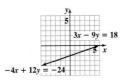

4. (a) Bill Tupper's Electrical Service charges $100 for a house call and $30 per hour. Thus we obtain the first equation, $y = 100 + 30x$. Wire for Hire charges $50 for a house call and $40 per hour. Thus we obtain the second equation, $y = 50 + 40x$.

(b) **Bill Tupper's Electrical Service** $y = 100 + 30x$
   Let $x = 0$. $y = 100 + 30(0) = 100$
   Let $x = 4$. $y = 100 + 30(4) = 220$
   Let $x = 8$. $y = 100 + 30(8) = 340$

x	y
0	100
4	220
8	340

   Wire for Hire $y = 50 + 40x$
   Let $x = 0$. $y = 50 + 40(0) = 50$
   Let $x = 4$. $y = 50 + 40(4) = 210$
   Let $x = 8$. $y = 50 + 40(8) = 370$

x	y
0	50
4	210
8	370

(c) We see that the graphs of the two lines intersect at $(5, 250)$. Thus the two companies will charge the same if 5 hours of electrical repairs are required.

(d) We draw a dashed line at $x = 6$. We see the line representing Wire for Hire is higher than the line representing Bill Tupper's Electrical Service after 5 hours. Thus the cost would be less if we use Bill Tupper's Electrical Service for 6 hours of work.

4.2 Practice Problems

1. $5x + 3y = 19$ **(1)**
    $2x - y = 12$ **(2)**

Step 1   Solve equation (2) for y.
       $2x - y = 12$ **(2)**
         $-y = -2x + 12$
          $y = 2x - 12$

Step 2   Now substitute $2x - 12$ for y in equation (1).
          $5x + 3y = 19$ **(1)**
        $5x + 3(2x - 12) = 19$

Step 3   Solve this equation.
        $5x + 6x - 36 = 19$
          $11x - 36 = 19$
            $11x = 55$
             $x = 5$

Step 4   Now obtain the value for the second variable.
         $2x - y = 12$ **(2)**
        $2(5) - y = 12$
         $10 - y = 12$
          $-y = 2$
           $y = -2$
The solution is $(5, -2)$.

Step 5   Check.

$5x + 3y = 19$ **(1)**	$2x - y = 12$ **(2)**
$5(5) + 3(-2) \stackrel{?}{=} 19$	$2(5) - (-2) \stackrel{?}{=} 12$
$25 - 6 \stackrel{?}{=} 19$	$10 + 2 \stackrel{?}{=} 12$
$19 = 19$ ✓	$12 = 12$ ✓

2. $\frac{1}{3}x - \frac{1}{2}y = 1$ **(1)**
    $x + 4y = -8$ **(2)**
Clear equation **(1)** of fractions. We observe that the LCD is 6.
$$6\left(\frac{1}{3}x\right) - 6\left(\frac{1}{2}y\right) = 6(1)$$
        $2x - 3y = 6$ **(3)**
The new system is:
       $2x - 3y = 6$ **(3)**
        $x + 4y = -8$ **(2)**

Step 1   Solve equation (2) for x.
       $x + 4y = -8$ **(2)**
         $x = -4y - 8$

Step 2   Substitute $-4y - 8$ for x in equation **(3)**
          $2x - 3y = 6$ **(3)**
        $2(-4y - 8) - 3y = 6$

Step 3   Solve this equation.
        $-8y - 16 - 3y = 6$
          $-11y - 16 = 6$
            $-11y = 22$
             $y = -2$

Step 4   Obtain the value of the second variable.

$$x + 4y = -8 \quad \textbf{(2)}$$
$$x + 4(-2) = -8$$
$$x - 8 = -8$$
$$x = 0$$

The solution is $(0, -2)$.

Step 5   Check.

$$\frac{1}{3}x - \frac{1}{2}y = 1 \quad \textbf{(1)} \qquad\qquad x + 4y = -8 \quad \textbf{(2)}$$

$$\frac{1}{3}(0) - \frac{1}{2}(-2) \overset{?}{=} 1 \qquad\qquad 0 + 4(-2) \overset{?}{=} -8$$

$$0 + 1 \overset{?}{=} 1 \qquad\qquad\qquad\qquad -8 = -8 \;\checkmark$$

$$1 = 1 \;\checkmark$$

4.3 Practice Problems

1. $3x + y = 7$ **(1)**
$5x - 2y = 8$ **(2)**
Multiply equation **(1)** by 2.
$2(3x) + 2(y) = 2(7)$
$6x + 2y = 14$ **(3)**
$\underline{5x - 2y = 8} \quad \textbf{(2)}$
$11x = 22$
$x = 2$
$5x - 2y = 8 \quad \textbf{(2)}$
$5(2) - 2y = 8$
$10 - 2y = 8$
$-2y = -2$
$y = 1$
The solution is $(2, 1)$.

2. $4x + 5y = 17$ **(1)**
$3x + 7y = 12$ **(2)**
Multiply equation **(1)** by -3 and equation **(2)** by 4.
$-3(4x) + (-3)(5y) = (-3)(17)$
$4(3x) + 4(7y) = 4(12)$
$-12x - 15y = -51$ **(3)**
$\underline{12x + 28y = 48} \quad \textbf{(4)}$
$13y = -3$

$$y = -\frac{3}{13}$$

Substitute $y = -\dfrac{3}{13}$ into one of the original equations.

$$4x + 5\left(-\frac{3}{13}\right) = 17 \quad \textbf{(1)}$$

$$4x - \frac{15}{13} = 17$$

$$13(4x) - 13\left(\frac{15}{13}\right) = 13(17)$$

$$52x - 15 = 221$$
$$52x = 236$$

$$x = \frac{59}{13}$$

The solution is $\left(\dfrac{59}{13}, -\dfrac{3}{13}\right)$.

3. $\dfrac{2}{3}x - \dfrac{3}{4}y = 3$ **(1)**

$-2x + y = 6$ **(2)**

Multiply equation **(1)** by 12.

$$12\left(\frac{2}{3}x\right) - 12\left(\frac{3}{4}y\right) = 12(3)$$

$$8x - 9y = 36$$

We now have an equivalent system without fractions.

$8x - 9y = 36$ **(3)**
$-2x + y = 6$ **(2)**

Multiply equation **(2)** by 4 to eliminate x.

$4(-2x) + 4(y) = 4(6)$
$8x - 9y = 36$ **(3)**
$\underline{-8x + 4y = 24} \quad \textbf{(4)}$
$-5y = 60$
$y = -12$

Substitute $y = -12$ into one of the original equations.

$-2x + (-12) = 6$
$-2x - 12 = 6$
$-2x = 18$
$x = -9$

The solution is $(-9, -12)$.

4. $0.2x + 0.3y = -0.1$ **(1)**
$0.5x - 0.1y = -1.1$ **(2)**

Multiply each term of each equation by 10.

$10(0.2x) + 10(0.3y) = 10(-0.1)$
$10(0.5x) - 10(0.1y) = 10(-1.1)$
$2x + 3y = -1$ **(3)**
$5x - y = -11$ **(4)**

Multiply equation **(4)** by 3 to eliminate y.

$2x + 3y = -1$ **(3)**
$\underline{15x - 3y = -33} \quad \textbf{(5)}$
$17x = -34$
$x = -2$

Substitute $x = -2$ into equation **(3)**.

$2(-2) + 3y = -1$
$-4 + 3y = -1$
$3y = 3$
$y = 1$

The solution is $(-2, 1)$.

4.4 Practice Problems

1. (a) $3x + 5y = 1485$
$x + 2y = 564$
Solve for x in the second equation and solve using the substitution method.
$x = -2y + 564$
$3(-2y + 564) + 5y = 1485$
$-6y + 1692 + 5y = 1485$
$-y = -207$
$y = 207$
Substitute $y = 207$ into the second equation and solve for x.
$x + 2(207) = 564$
$x + 414 = 564$
$x = 150$
The solution is $(150, 207)$.

(b) $7x + 6y = 45$
$6x - 5y = -2$
Using the addition method, multiply the first equation by -6 and the second equation by 7.
$-6(7x) + (-6)(6y) = (-6)(45)$
$7(6x) - 7(5y) = 7(-2)$
$-42x - 36y = -270$
$\underline{42x - 35y = -14}$
$-71y = -284$
$y = 4$
Substitute $y = 4$ into one of the original equations and solve for x.
$7x + 6(4) = 45$
$7x + 24 = 45$
$7x = 21$
$x = 3$
The solution is $(3, 4)$.

2. $4x + 2y = 2$
$-6x - 3y = 6$

Use the addition method. Multiply the first equation by 3 and the second equation by 2.

$12x + 6y = 6$
$\underline{-12x - 6y = 12}$
$0 = 18$

The statement $0 = 18$ is not true. There is no solution to this system of equations.

3. $3x - 9y = 18$
$-4x + 12y = -24$

Use the addition method. Multiply the first equation by 4 and the second equation by 3.

$12x - 36y = 72$
$\underline{-12x + 36y = -72}$
$0 = 0$

The statement $0 = 0$ is true and identifies the equations as dependent. There is an infinite number of solutions to the system.

4.5 Practice Problems

1. Let $x =$ number of gallons/hour pumped by the small pump and $y =$ number of gallons/hour pumped by the large pump.

1st day $8x + 5y = 49,000$
2nd day $5x + 3y = 30,000$

Use the addition method.
Multiply the first equation by 5 and the second equation by -8.

$40x + 25y = 245,000$
$\underline{-40x - 24y = -240,000}$
$y = 5000$

Substitute $y = 5000$ into one of the original equations and solve for x.

$8x + 5(5000) = 49,000$
$8x + 25,000 = 49,000$
$8x = 24,000$
$x = 3000$

The small pump removes 3000 gallons per hour.
The large pump removes 5000 gallons per hour.

2. Let $x =$ cost/gallon of unleaded premium gasoline and $y =$ cost/gallon of unleaded regular gasoline.

Last week's purchase $7x + 8y = 47.15$
This week's purchase $8x + 4y = 38.20$

Use the addition method. Multiply the first equation by 8 and the second equation by -7.

$56x + 64y = 377.20$
$\underline{-56x - 28y = -267.40}$
$36y = 109.8$
$y = 3.05$

Substitute $y = 3.05$ into one of the original equations and solve for x.

$7x + 8(3.05) = 47.15$
$7x + 24.40 = 47.15$
$7x = 22.75$
$x = 3.25$

Unleaded premium gasoline costs \$3.25/gallon.
Unleaded regular gasoline costs \$3.05/gallon.

3. Let $x =$ the amount of 20% solution and $y =$ the amount of 80% solution.

Total amount of 65% solution
$x + y = 4000$

Amount of acid in 65% solution
$0.2x + 0.8y = 0.65(4000)$

Use the substitution method. Solve for x in the first equation.
$x = 4000 - y$

Substitute $x = 4000 - y$ into the second equation.
$0.2(4000 - y) + 0.8y = 0.65(4000)$
$800 - 0.2y + 0.8y = 2600$
$800 + 0.6y = 2600$
$0.6y = 1800$
$y = 3000$

Now substitute $y = 3000$ into the first equation.

$x + y = 4000$
$x + 3000 = 4000$
$x = 1000$

Therefore he will need 1000 liters of 20% solution and 3000 liters of 80% solution.

4. To go downstream with the current, we have as the rate of travel
$$r = \frac{d}{t} = \frac{72}{3} = 24 \text{ miles/hour}$$

To go upstream against the current, we have as the rate of travel
$$r = \frac{d}{t} = \frac{72}{4} = 18 \text{ miles/hour}$$

Let $x =$ the speed of the boat in miles/hour and $y =$ the speed of the current in miles/hour.
Thus we have the equations $x + y = 24$ and $x - y = 18$. Use the addition method to solve.

$x + y = 24$
$\underline{x - y = 18}$
$2x \quad\ = 42$
$x = 21$

Substitute $x = 21$ into one of the original equations and solve for y.
$21 + y = 24$
$y = 3$

Thus the speed of the boat in still water is 21 miles/hour and the speed of the current is 3 miles/hour.

Chapter 5 5.1 Practice Problems

1. (a) $a^7 \cdot a^5 = a^{7+5} = a^{12}$ **(b)** $w^{10} \cdot w = w^{10+1} = w^{11}$

2. (a) $x^3 \cdot x^9 = x^{3+9} = x^{12}$ **(b)** $3^7 \cdot 3^4 = 3^{7+4} = 3^{11}$
(c) $a^3 \cdot b^2 = a^3 \cdot b^2$ (cannot be simplified)

3. (a) $(-a^8)(a^4) = (-1 \cdot 1)(a^8 \cdot a^4)$
$= -1(a^8 \cdot a^4)$
$= -1a^{12}$
$= -a^{12}$

(b) $(3y^2)(-2y^3) = (3)(-2)(y^2 \cdot y^3) = -6y^5$
(c) $(-4x^3)(-5x^2) = (-4)(-5)(x^3 \cdot x^2) = 20x^5$

4. $(2xy)\left(-\dfrac{1}{4}x^2y\right)(6xy^3) = (2)\left(-\dfrac{1}{4}\right)(6)(x \cdot x^2 \cdot x)(y \cdot y \cdot y^3)$
$= -3x^4y^5$

5. (a) $\dfrac{10^{13}}{10^7} = 10^{13-7} = 10^6$ **(b)** $\dfrac{x^{11}}{x} = x^{11-1} = x^{10}$

(c) $\dfrac{y^{18}}{y^8} = y^{18-8} = y^{10}$

6. (a) $\dfrac{c^3}{c^4} = \dfrac{1}{c^{4-3}} = \dfrac{1}{c^1} = \dfrac{1}{c}$ **(b)** $\dfrac{10^{31}}{10^{56}} = \dfrac{1}{10^{56-31}} = \dfrac{1}{10^{25}}$

(c) $\dfrac{z^{15}}{z^{21}} = \dfrac{1}{z^{21-15}} = \dfrac{1}{z^6}$

7. (a) $\dfrac{-7x^7}{-21x^9} = \dfrac{1}{3x^{9-7}} = \dfrac{1}{3x^2}$ **(b)** $\dfrac{15x^{11}}{-3x^4} = -5x^{11-4} = -5x^7$

(c) $\dfrac{23x^8}{46x^9} = \dfrac{1}{2x^{9-8}} = \dfrac{1}{2x}$

8. (a) $\dfrac{x^7y^9}{y^{10}} = \dfrac{x^7}{y}$ **(b)** $\dfrac{12x^5y^6}{-24x^3y^8} = -\dfrac{x^2}{2y^2}$

9. (a) $\dfrac{10^7}{10^7} = 1$ **(b)** $\dfrac{12a^4}{15a^4} = \dfrac{4}{5}\left(\dfrac{a^4}{a^4}\right) = \dfrac{4}{5}(1) = \dfrac{4}{5}$

10. (a) $\dfrac{-20a^3b^8c^4}{28a^3b^7c^5} = -\dfrac{5a^0b}{7c} = -\dfrac{5(1)b}{7c} = -\dfrac{5b}{7c}$

(b) $\dfrac{5x^0y^6}{10x^4y^8} = \dfrac{5(1)y^6}{10x^4y^8} = \dfrac{1}{2x^4y^2}$

11. $\dfrac{(-6ab^5)(3a^2b^4)}{16a^5b^7} = \dfrac{-18a^3b^9}{16a^5b^7} = -\dfrac{9b^2}{8a^2}$

12. (a) $(a^4)^3 = a^{4 \cdot 3} = a^{12}$ **(b)** $(10^5)^2 = 10^{5 \cdot 2} = 10^{10}$

    (c) $(-1)^{15} = -1$

13. (a) $(3xy)^3 = (3)^3 x^3 y^3 = 27x^3 y^3$

    (b) $(yz)^{37} = y^{37} z^{37}$

    (c) $(-3x^3)^2 = (-3)^2 (x^3)^2 = 9x^6$

14. (a) $\left(\dfrac{x}{5}\right)^3 = \dfrac{x^3}{5^3} = \dfrac{x^3}{125}$ **(b)** $\left(\dfrac{4a}{b}\right)^6 = \dfrac{(4a)^6}{b^6} = \dfrac{4^6 a^6}{b^6} = \dfrac{4096 a^6}{b^6}$

15. $\left(\dfrac{-2x^3 y^0 z}{4xz^2}\right)^5 = \left(\dfrac{-x^2}{2z}\right)^5 = \dfrac{(-1)^5 (x^2)^5}{2^5 z^5} = -\dfrac{x^{10}}{32 z^5}$

5.2 Practice Problems

1. (a) $x^{-12} = \dfrac{1}{x^{12}}$ **(b)** $w^{-5} = \dfrac{1}{w^5}$ **(c)** $z^{-2} = \dfrac{1}{z^2}$

2. (a) $4^{-3} = \dfrac{1}{4^3} = \dfrac{1}{64}$ **(b)** $2^{-4} = \dfrac{1}{2^4} = \dfrac{1}{16}$

3. (a) $\dfrac{3}{w^{-4}} = 3w^4$ **(b)** $\dfrac{x^{-6} y^4}{z^{-2}} = \dfrac{y^4 z^2}{x^6}$ **(c)** $x^{-6} y^{-5} = \dfrac{1}{x^6 y^5}$

4. (a) $(2x^4 y^{-5})^{-2} = 2^{-2} x^{-8} y^{10} = \dfrac{y^{10}}{2^2 x^8} = \dfrac{y^{10}}{4x^8}$

    (b) $\dfrac{y^{-3} z^{-4}}{y^2 z^{-6}} = \dfrac{z^6}{y^2 y^3 z^4} = \dfrac{z^6}{y^5 z^4} = \dfrac{z^2}{y^5}$

5. (a) $78{,}200 = 7.82 \times 10{,}000 = 7.82 \times 10^4$

    Notice we moved the decimal point 4 places to the left.

    (b) $4{,}786{,}000 = 4.786 \times 1{,}000{,}000 = 4.786 \times 10^6$

6. (a) $0.98 = 9.8 \times 10^{-1}$

    (b) $0.000092 = 9.2 \times 10^{-5}$

7. (a) $1.93 \times 10^6 = 1.93 \times 1{,}000{,}000 = 1{,}930{,}000$

    (b) $8.562 \times 10^{-5} = 8.562 \times \dfrac{1}{100{,}000} = 0.00008562$

8. $30{,}900{,}000{,}000{,}000{,}000 \text{ meters} = 3.09 \times 10^{16} \text{ meters}$

9. (a) $(56{,}000)(1{,}400{,}000{,}000) = (5.6 \times 10^4)(1.4 \times 10^9)$

$$= (5.6)(1.4)(10^4)(10^9)$$

$$= 7.84 \times 10^{13}$$

    (b) $\dfrac{0.000111}{0.00000037} = \dfrac{1.11 \times 10^{-4}}{3.7 \times 10^{-7}}$

$$= \dfrac{1.11}{3.7} \times \dfrac{10^{-4}}{10^{-7}}$$

$$= \dfrac{1.11}{3.7} \times \dfrac{10^7}{10^4}$$

$$= 0.3 \times 10^3$$

$$= 3.0 \times 10^2$$

10. $159 \text{ parsecs} = (159 \text{ parsecs}) \dfrac{(3.09 \times 10^{13} \text{ kilometers})}{1 \text{ parsec}}$

$$= 491.31 \times 10^{13} \text{ kilometers}$$

$$d = r \times t$$

$$491.31 \times 10^{13} \text{ km} = \dfrac{50{,}000 \text{ km}}{1 \text{ hr}} \times t$$

$$4.9131 \times 10^{15} \text{ km} = \dfrac{5 \times 10^4 \text{ km}}{1 \text{ hr}} \times t$$

$$\dfrac{4.9131 \times 10^{15} \text{ km}}{\dfrac{5 \times 10^4 \text{ km}}{1 \text{ hr}}} = t$$

$$\dfrac{4.9131 \times 10^{15} \text{ km} (1 \text{ hr})}{5.0 \times 10^4 \text{ km}} = t$$

$$0.98262 \times 10^{11} \text{ hr} = t$$

    It would take the probe about 9.83×10^{10} hours.

5.3 Practice Problems

1. (a) This polynomial is of degree 5. It has two terms, so it is a binomial.

    (b) This polynomial is of degree 7, since the sum of the exponents is $3 + 4 = 7$. It has one term, so it is a monomial.

    (c) This polynomial is of degree 3. It has three terms, so it is a trinomial.

2. $(-8x^3 + 3x^2 + 6) + (2x^3 - 7x^2 - 3)$

$$= [-8x^3 + 2x^3] + [3x^2 + (-7x^2)] + [6 + (-3)]$$

$$= [(-8 + 2)x^3] + [(3 - 7)x^2] + [6 - 3]$$

$$= -6x^3 + (-4x^2) + 3$$

$$= -6x^3 - 4x^2 + 3$$

3. $\left(-\dfrac{1}{3}x^2 - 6x - \dfrac{1}{12}\right) + \left(\dfrac{1}{4}x^2 + 5x - \dfrac{1}{3}\right)$

$$= \left[-\dfrac{1}{3}x^2 + \dfrac{1}{4}x^2\right] + [-6x + 5x] + \left[-\dfrac{1}{12} + \left(-\dfrac{1}{3}\right)\right]$$

$$= \left[\left(-\dfrac{1}{3} + \dfrac{1}{4}\right)x^2\right] + [(-6 + 5)x] + \left[-\dfrac{1}{12} + \left(-\dfrac{1}{3}\right)\right]$$

$$= \left[\left(-\dfrac{4}{12} + \dfrac{3}{12}\right)x^2\right] + [-x] + \left[-\dfrac{1}{12} - \dfrac{4}{12}\right]$$

$$= -\dfrac{1}{12}x^2 - x - \dfrac{5}{12}$$

4. $(3.5x^3 - 0.02x^2 + 1.56x - 3.5) + (-0.08x^2 - 1.98x + 4)$

$$= 3.5x^3 + (-0.02 - 0.08)x^2 + (1.56 - 1.98)x + (-3.5 + 4)$$

$$= 3.5x^3 - 0.1x^2 - 0.42x + 0.5$$

5. $(5x^3 - 15x^2 + 6x - 3) - (-4x^3 - 10x^2 + 5x + 13)$

$$= (5x^3 - 15x^2 + 6x - 3) + (4x^3 + 10x^2 - 5x - 13)$$

$$= (5 + 4)x^3 + (-15 + 10)x^2 + (6 - 5)x + (-3 - 13)$$

$$= 9x^3 - 5x^2 + x - 16$$

6. $(x^3 - 7x^2 y + 3xy^2 - 2y^3) - (2x^3 + 4xy - 6y^3)$

$$= (x^3 - 7x^2 y + 3xy^2 - 2y^3) + (-2x^3 - 4xy + 6y^3)$$

$$= (1 - 2)x^3 - 7x^2 y + 3xy^2 - 4xy + (-2 + 6)y^3$$

$$= -x^3 - 7x^2 y + 3xy^2 - 4xy + 4y^3$$

7. (a) 1974 is 4 years later than 1970, so $x = 4$.

    $0.03(4) + 5.4 = 0.12 + 5.4 = 5.52$

    We estimate that the average truck in 1974 obtained 5.52 miles per gallon.

    (b) 2011 is 41 years later than 1970, so $x = 41$.

    $0.03(41) + 5.4 = 1.23 + 5.4 = 6.63$

    We predict that the average truck in 2011 will obtain 6.63 miles per gallon.

5.4 Practice Problems

1. $4x^3(-2x^2 + 3x) = 4x^3(-2x^2) + 4x^3(3x)$

$$= 4(-2)(x^3 \cdot x^2) + (4 \cdot 3)(x^3 \cdot x)$$

$$= -8x^5 + 12x^4$$

2. (a) $-3x(x^2 + 2x - 4) = -3x^3 - 6x^2 + 12x$

    (b) $6xy(x^3 + 2x^2 y - y^2) = 6x^4 y + 12x^3 y^2 - 6xy^3$

3. $(-6x^3 + 4x^2 - 2x)(-3xy) = 18x^4 y - 12x^3 y + 6x^2 y$

4. $(5x - 1)(x - 2) = 5x^2 - 10x - x + 2 = 5x^2 - 11x + 2$

5. $(8a - 5b)(3a - b) = 24a^2 - 8ab - 15ab + 5b^2$

$$= 24a^2 - 23ab + 5b^2$$

6. $(3a + 2b)(2a - 3c) = 6a^2 - 9ac + 4ab - 6bc$

7. $(3x - 2y)(3x - 2y) = 9x^2 - 6xy - 6xy + 4y^2$

$$= 9x^2 - 12xy + 4y^2$$

8. $(2x^2 + 3y^2)(5x^2 + 6y^2) = 10x^4 + 12x^2 y^2 + 15x^2 y^2 + 18y^4$

$$= 10x^4 + 27x^2 y^2 + 18y^4$$

9. $A = (\text{length})(\text{width}) = (7x + 3)(2x - 1)$

$$= 14x^2 - 7x + 6x - 3$$

$$= 14x^2 - x - 3$$

    There are $(14x^2 - x - 3)$ square feet in the room.

5.5 Practice Problems

1. $(6x + 7)(6x - 7) = (6x)^2 - (7)^2 = 36x^2 - 49$

2. $(3x - 5y)(3x + 5y) = (3x)^2 - (5y)^2 = 9x^2 - 25y^2$

3. (a) $(4a - 9b)^2 = (4a)^2 - 2(4a)(9b) + (9b)^2$

$$= 16a^2 - 72ab + 81b^2$$

    (b) $(5x + 4)^2 = (5x)^2 + 2(5x)(4) + (4)^2 = 25x^2 + 40x + 16$

$y = -4x + 9$

x	y
0	9
1	5
2	1

$y = -4(0) + 9 \quad y = 9$

$y = -4(1) + 9 \quad y = 5$

$y = -4(2) + 9 \quad y = 1$

$x - 2y = 0$

x	y
0	0
2	1
4	-2

$0 - 2y = 0 \quad y = 0$

$-1 - 2y = 0 \quad y = .5$

$\dfrac{-2}{-2}y = \dfrac{-1}{-2}$

$2 - 2y = 0$

$\dfrac{8}{+8} - 2y = \dfrac{0}{-3} \quad y = 1$

$\dfrac{4}{-4} - 2y = \dfrac{0}{-4}$

$\dfrac{-2}{-2}y = \dfrac{4}{-2}$

$y = -2$

$x + 4y = 8$

x	y
0	4
2	
4	

$0 + 4y = 8$

$\dfrac{4}{4}y = \dfrac{8}{2}$

$y = 4$

$\dfrac{-8}{-8} + 4y = \dfrac{8}{2}$

$4y = 6$

$$6x + 1 = y$$

$$6(0) + 1 = y$$

$$1 = y$$

$$\underset{+9x}{9x} - 18y = 90 - 9x$$

$$\frac{-18y}{-18} = \frac{90 - 9x}{-18}$$

$$\underset{-6x}{6x} - 9y = 81 - 6x$$

$$\frac{-9y}{-9} = \frac{81 - 6x}{-9}$$

$$y = \frac{81 - 6x}{-9}$$

$$y = -9 + \frac{2}{3}x$$

$$y = 7x + 6$$

$$y = 7(0) + 6$$

$$y = 6$$

$$2x - 3y = 6$$

$$2x - 3(-4) = 6$$

$$2x + 12 = 6$$
$$\underset{-12}{} \quad \underset{-12}{}$$

$$\frac{2x}{2} = \frac{-6}{2}$$

$$x = -3$$

$$y = 7(2) + 6$$

$$y = 20$$

$$2x - 3y = 6$$

$$2(3) - 3y = 6$$

$$\frac{6}{6} - 3y = \frac{6}{6}$$

$$\frac{-3y}{-3} = 0$$

$$6y + 7(6) = -42$$

$$6y - 42 = -42$$
$$\underset{+42}{} \quad \underset{+42}{}$$

$$6y = 0$$

$$6(7) + 7x = -42$$

$$42 + 7x = -42$$
$$\underset{+42}{} \quad \underset{-42}{}$$

$$\frac{7x}{7} = \frac{-84}{7}$$

$$x = -12$$

x	y
0	9
-2	12
-3	18

$$y = 3(0) + 9 \quad y = 0$$

$$y = 3(-2) + 9 \quad y = 3$$

$$y = 3(-3) + 9 \quad y = 0$$

$$y = -3(0) + 9$$

4.
$$
\begin{array}{r}
4x^3 - 2x^2 + x \\
\underline{x^2 + 3x - 2} \\
-8x^3 + 4x^2 - 2x \\
12x^4 - 6x^3 + 3x^2 \\
\underline{4x^5 - 2x^4 + x^3} \\
4x^5 + 10x^4 - 13x^3 + 7x^2 - 2x
\end{array}
$$

5. $(2x^2 + 5x + 3)(x^2 - 3x - 4)$
$$
\begin{aligned}
&= 2x^2(x^2 - 3x - 4) + 5x(x^2 - 3x - 4) + 3(x^2 - 3x - 4) \\
&= 2x^4 - 6x^3 - 8x^2 + 5x^3 - 15x^2 - 20x + 3x^2 - 9x - 12 \\
&= 2x^4 - x^3 - 20x^2 - 29x - 12
\end{aligned}
$$

6. $(3x - 2)(2x + 3)(3x + 2) = (3x - 2)(3x + 2)(2x + 3)$
$$
\begin{aligned}
&= [(3x)^2 - 2^2](2x + 3) \\
&= (9x^2 - 4)(2x + 3)
\end{aligned}
$$

$$
\begin{array}{r}
9x^2 - 4 \\
\underline{2x + 3} \\
27x^2 + 0x - 12 \\
\underline{18x^3 + 0x^2 - 8x} \\
18x^3 + 27x^2 - 8x - 12
\end{array}
$$

Thus we have
$(3x - 2)(2x + 3)(3x + 2) = 18x^3 + 27x^2 - 8x - 12.$

5.6 Practice Problems

1. $\dfrac{15y^4 - 27y^3 - 21y^2}{3y^2} = \dfrac{15y^4}{3y^2} - \dfrac{27y^3}{3y^2} - \dfrac{21y^2}{3y^2} = 5y^2 - 9y - 7$

2.
$$
\begin{array}{r}
x^2 + 6x + 7 \\
x + 4 \overline{) x^3 + 10x^2 + 31x + 35} \\
\underline{x^3 + 4x^2} \\
6x^2 + 31x \\
\underline{6x^2 + 24x} \\
7x + 35 \\
\underline{7x + 28} \\
7
\end{array}
$$
Ans: $x^2 + 6x + 7 + \dfrac{7}{x + 4}$

3.
$$
\begin{array}{r}
2x^2 + x + 1 \\
x - 1 \overline{) 2x^3 - x^2 + 0x + 1} \\
\underline{2x^3 - 2x^2} \\
x^2 + 0x \\
\underline{x^2 - x} \\
x + 1 \\
\underline{x - 1} \\
2
\end{array}
$$
Ans: $2x^2 + x + 1 + \dfrac{2}{x - 1}$

4.
$$
\begin{array}{r}
5x^2 + x - 2 \\
4x - 3 \overline{) 20x^3 - 11x^2 - 11x + 6} \\
\underline{20x^3 - 15x^2} \\
4x^2 - 11x \\
\underline{4x^2 - 3x} \\
-8x + 6 \\
\underline{-8x + 6} \\
0
\end{array}
$$
Ans: $5x^2 + x - 2$

Check. $(4x - 3)(5x^2 + x - 2)$
$$
\begin{aligned}
&= 20x^3 + 4x^2 - 8x - 15x^2 - 3x + 6 \\
&= 20x^3 - 11x^2 - 11x + 6
\end{aligned}
$$

Chapter 6 6.1 Practice Problems

1. (a) $21a - 7b = 7(3a - b)$ because $7(3a - b) = 21a - 7b$.
  (b) $5xy + 8x = x(5y + 8)$ because $x(5y + 8) = 5xy + 8x$.

2. 4 is the greatest numerical common factor and a is a factor of each term. Thus, $4a$ is the greatest common factor.
$12a^2 + 16ab^2 - 12a^2b = 4a(3a + 4b^2 - 3ab)$

3. (a) The largest integer common to both terms is 8.
  $16a^3 - 24b^3 = 8(2a^3 - 3b^3)$

  (b) r^3 is common to all the terms.
  $r^3s^2 - 4r^4s + 7r^5 = r^3(s^2 - 4rs + 7r^2)$

4. We can factor $9, a, b^2,$ and c out of each term.
  $18a^3b^2c - 27ab^3c^2 - 45a^2b^2c^2 = 9ab^2c(2a^2 - 3bc - 5ac)$

5. We can factor $6, x,$ and y^2 out of each term.
  $30x^3y^2 - 24x^2y^2 + 6xy^2 = 6xy^2(5x^2 - 4x + 1)$
  Check: $6xy^2(5x^2 - 4x + 1) = 30x^3y^2 - 24x^2y^2 + 6xy^2$

6. $3(a + 5b) + x(a + 5b) = (a + 5b)(3 + x)$

7. $8y(9y^2 - 2) - (9y^2 - 2) = 8y(9y^2 - 2) - 1(9y^2 - 2)$
$$
= (9y^2 - 2)(8y - 1)
$$

8. $\pi b^2 - \pi a^2 = \pi(b^2 - a^2)$

6.2 Practice Problems

1. $3y(2x - 7) - 8(2x - 7) = (2x - 7)(3y - 8)$

2. $6x^2 - 15x + 4x - 10 = 3x(2x - 5) + 2(2x - 5)$
$$
= (2x - 5)(3x + 2)
$$

3. $ax + 2a + 4bx + 8b = a(x + 2) + 4b(x + 2)$
$$
= (x + 2)(a + 4b)
$$

4. $6a^2 + 5bc + 10ab + 3ac = 6a^2 + 10ab + 3ac + 5bc$
$$
\begin{aligned}
&= 2a(3a + 5b) + c(3a + 5b) \\
&= (3a + 5b)(2a + c)
\end{aligned}
$$

5. $6xy + 14x - 15y - 35 = 2x(3y + 7) - 15y - 35$
$$
\begin{aligned}
&= 2x(3y + 7) - 5(3y + 7) \\
&= (3y + 7)(2x - 5)
\end{aligned}
$$

6. $3x + 6y - 5ax - 10ay = 3(x + 2y) - 5a(x + 2y)$
$$
= (x + 2y)(3 - 5a)
$$

7. $10ad + 27bc - 6bd - 45ac = 10ad - 6bd - 45ac + 27bc$
$$
\begin{aligned}
&= 2d(5a - 3b) - 9c(5a - 3b) \\
&= (5a - 3b)(2d - 9c)
\end{aligned}
$$
  Check: $(5a - 3b)(2d - 9c) = 10ad - 45ac - 6bd + 27bc$
$$
= 10ad + 27bc - 6bd - 45ac
$$

6.3 Practice Problems

1. The two numbers that you can multiply to get 12 and add to get 8 are 6 and 2. $x^2 + 8x + 12 = (x + 6)(x + 2)$

2. The two numbers that have a product of 30 and a sum of 17 are 2 and 15. $x^2 + 17x + 30 = (x + 2)(x + 15)$

3. The two numbers that have a product of $+18$ and a sum of -11 must both be negative. The numbers are -9 and -2.
  $x^2 - 11x + 18 = (x - 9)(x - 2)$

4. The two numbers whose product is 24 and whose sum is -11 are -8 and -3. $x^2 - 11x + 24 = (x - 8)(x - 3)$ or $(x - 3)(x - 8)$

5. The two numbers whose product is -24 and whose sum is -5 are -8 and $+3$. $x^2 - 5x - 24 = (x - 8)(x + 3)$ or $(x + 3)(x - 8)$

6. The two numbers whose product is -60 and whose sum is $+17$ are $+20$ and -3. $y^2 + 17y - 60 = (y + 20)(y - 3)$
  Check: $(y + 20)(y - 3) = y^2 - 3y + 20y - 60 = y^2 + 17y - 60$

7. List the possible factors of 60.

Factors of 60	Difference
60 and 1	59
30 and 2	28
20 and 3	17
15 and 4	11
12 and 5	7 ← Desired Value
10 and 6	4

For the coefficient of the middle term to be -7, we must add -12 and $+5$. $x^2 - 7x - 60 = (x - 12)(x + 5)$

8. $a^4 = (a^2)(a^2)$
  The two numbers whose product is -42 and whose sum is 1 are 7 and -6. $a^4 + a^2 - 42 = (a^2 + 7)(a^2 - 6)$

9. $3x^2 + 45x + 150 = 3(x^2 + 15x + 50)$
$$
= 3(x + 5)(x + 10)
$$

10. $4x^2 - 8x - 140 = 4(x^2 - 2x - 35)$
$$
= 4(x + 5)(x - 7)
$$

11. $x(x + 1) - 4(5) = x^2 + x - 20$
$$
= (x + 5)(x - 4)
$$

6.4 Practice Problems

1. To get a first term of $2x^2$, the coefficients of x in the factors must be 2 and 1. To get a last term of 5, the constants in the factors must be 1 and 5. Possibilities:

$(2x + 1)(x + 5) = 2x^2 + 11x + 5$

$(2x + 5)(x + 1) = 2x^2 + 7x + 5$

Thus $2x^2 + 7x + 5 = (2x + 5)(x + 1)$ or $(x + 1)(2x + 5)$.

2. The different factorizations of 9 are $(3)(3)$ and $(1)(9)$. The only factorization of 7 is $(1)(7)$.

Possible Factors	Middle Term	Correct?
$(3x - 7)(3x - 1)$	$-24x$	No
$(9x - 7)(x - 1)$	$-16x$	No
$(9x - 1)(x - 7)$	$-64x$	Yes

The correct answer is $(9x - 1)(x - 7)$ or $(x - 7)(9x - 1)$.

3. The only factorization of 3 is $(3)(1)$. The different factorizations of 14 are $(14)(1)$ and $(7)(2)$.

Possible Factors	Middle Term	Correct Factors?
$(3x + 14)(x - 1)$	$+11x$	No
$(3x + 1)(x - 14)$	$-41x$	No
$(3x + 7)(x - 2)$	$+x$	No (wrong sign)
$(3x + 2)(x - 7)$	$-19x$	No

To get the correct sign on the middle term, reverse the signs of the constants in the factors.

The correct answer is $(3x - 7)(x + 2)$ or $(x + 2)(3x - 7)$.

4. The grouping number is $2 \cdot 5 = 10$. The two numbers whose product is 10 and whose sum is 7 are 5 and 2. Write $7x$ as the sum $5x + 2x$.

$2x^2 + 7x + 5 = 2x^2 + 5x + 2x + 5$

$\qquad = x(2x + 5) + 1(2x + 5)$

$\qquad = (2x + 5)(x + 1)$

5. The grouping number is $9 \cdot 7 = 63$. The two numbers with a product of 63 and a sum of -64 are -63 and -1. Write $-64x$ as the sum $-63x - 1x$.

$9x^2 - 64x + 7 = 9x^2 - 63x - x + 7$

$\qquad = 9x(x - 7) - 1(x - 7)$

$\qquad = (x - 7)(9x - 1)$

6. The grouping number is $3(-4) = -12$. The two numbers with a product of -12 and a sum of 4 are 6 and -2. Write $4x$ as the sum $6x - 2x$.

$3x^2 + 4x - 4 = 3x^2 + 6x - 2x - 4$

$\qquad = 3x(x + 2) - 2(x + 2)$

$\qquad = (x + 2)(3x - 2)$

7. $8x^2 + 8x - 6 = 2(4x^2 + 4x - 3)$

$\qquad = 2(2x - 1)(2x + 3)$

8. $24x^2 - 38x + 10 = 2(12x^2 - 19x + 5)$

$\qquad = 2(4x - 5)(3x - 1)$

6.5 Practice Problems

1. $64x^2 - 1 = (8x + 1)(8x - 1)$ because $64x^2 = (8x)^2$ and $1 = (1)^2$.

2. $36x^2 - 49 = (6x + 7)(6x - 7)$ because $36x^2 = (6x)^2$ and $49 = (7)^2$.

3. $100x^2 - 81y^2 = (10x + 9y)(10x - 9y)$

4. $x^8 - 1 = (x^4 + 1)(x^4 - 1)$

$\qquad = (x^4 + 1)(x^2 + 1)(x^2 - 1)$

$\qquad = (x^4 + 1)(x^2 + 1)(x + 1)(x - 1)$

5. The first and last terms are perfect squares: $x^2 = (x)^2$ and $25 = (5)^2$. The middle term, $10x$, is twice the product of x and 5.

$x^2 + 10x + 25 = (x + 5)^2$

6. $30x = 2(5x \cdot 3)$

Also note the negative sign. $25x^2 - 30x + 9 = (5x - 3)^2$

7. (a) $25x^2 = (5x)^2$, $36y^2 = (6y)^2$, and $60xy = 2(5x \cdot 6y)$.

$25x^2 + 60xy + 36y^2 = (5x + 6y)^2$

(b) $64x^6 = (8x^3)^2$, $9 = (3)^2$, and $48x^3 = 2(8x^3 \cdot 3)$.

$64x^6 - 48x^3 + 9 = (8x^3 - 3)^2$

8. $9x^2 = (3x)^2$ and $4 = (2)^2$, but $15x \neq 2(3x \cdot 2) = 12x$.

$9x^2 + 15x + 4 = (3x + 1)(3x + 4)$

9. $20x^2 - 45 = 5(4x^2 - 9)$

$\qquad = 5(2x + 3)(2x - 3)$

10. $75x^2 - 60x + 12 = 3(25x^2 - 20x + 4)$

$\qquad = 3(5x - 2)^2$

6.6 Practice Problems

1. (a) $3x^2 - 36x + 108$

$\qquad = 3(x^2 - 12x + 36)$

$\qquad = 3(x - 6)^2$

(b) $9x^4y^2 - 9y^2$

$\qquad = 9y^2(x^4 - 1)$

$\qquad = 9y^2(x^2 + 1)(x^2 - 1)$

$\qquad = 9y^2(x^2 + 1)(x + 1)(x - 1)$

(c) $5x^3 - 15x^2y + 10x^2 - 30xy = 5x(x^2 - 3xy + 2x - 6y)$

$\qquad = 5x[x(x - 3y) + 2(x - 3y)]$

$\qquad = 5x(x - 3y)(x + 2)$

(d) $12x - 9 - 4x^2 = -4x^2 + 12x - 9$

$\qquad = -(4x^2 - 12x + 9)$

$\qquad = -(2x - 3)^2$

2. $x^2 - 9x - 8$

The factorizations of -8 are $(-2)(4)$, $(2)(-4)$, $(-8)(1)$, and $(-1)(8)$.

None of these pairs will add up to be the coefficient of the middle term. Thus the polynomial cannot be factored. It is prime.

3. $25x^2 + 82x + 4$

Check to see if this is a perfect square trinomial.

$2[(5)(2)] = 2(10) = 20$

This is not the coefficient of the middle term. The grouping number is 100. No factors add to 82. It is prime.

6.7 Practice Problems

1. $10x^2 - x - 2 = 0$

$(5x + 2)(2x - 1) = 0$

$5x + 2 = 0 \qquad 2x - 1 = 0$

$5x = -2 \qquad\quad 2x = 1$

$x = -\dfrac{2}{5} \qquad\quad x = \dfrac{1}{2}$

Check: $10\left(-\dfrac{2}{5}\right)^2 - \left(-\dfrac{2}{5}\right) - 2 \overset{?}{=} 0 \qquad 10\left(\dfrac{1}{2}\right)^2 - \dfrac{1}{2} - 2 \overset{?}{=} 0$

$10\left(\dfrac{4}{25}\right) + \dfrac{2}{5} - 2 \overset{?}{=} 0 \qquad 10\left(\dfrac{1}{4}\right) - \dfrac{1}{2} - 2 \overset{?}{=} 0$

$\dfrac{8}{5} + \dfrac{2}{5} - 2 \overset{?}{=} 0 \qquad\qquad \dfrac{5}{2} - \dfrac{1}{2} - 2 \overset{?}{=} 0$

$\dfrac{10}{5} - \dfrac{10}{5} \overset{?}{=} 0 \qquad\qquad\quad \dfrac{4}{2} - 2 \overset{?}{=} 0$

$0 = 0 \qquad\qquad\qquad\qquad 2 - 2 \overset{?}{=} 0$

$\qquad\qquad\qquad\qquad\qquad\qquad 0 = 0$

Thus $-\dfrac{2}{5}$ and $\dfrac{1}{2}$ are both roots of the equation.

2. $3x^2 + 11x - 4 = 0$

$(3x - 1)(x + 4) = 0$

$3x - 1 = 0 \qquad x + 4 = 0$

$3x = 1 \qquad\qquad x = -4$

$x = \dfrac{1}{3}$

Thus $\dfrac{1}{3}$ and -4 are both roots of the equation.

3. $7x^2 + 11x = 0$

$x(7x + 11) = 0$

$x = 0 \qquad 7x + 11 = 0$

$\qquad\qquad\quad 7x = -11$

$\qquad\qquad\quad x = \dfrac{-11}{7}$

Thus 0 and $-\frac{11}{7}$ are both roots of the equation.

4. $\quad x^2 - 6x + 4 = -8 + x$

$\quad x^2 - 7x + 12 = 0$

$(x - 3)(x - 4) = 0$

$x - 3 = 0 \qquad x - 4 = 0$

$\quad x = 3 \qquad\quad x = 4$

The roots are 3 and 4.

5. $\qquad \dfrac{2x^2 - 7x}{3} = 5$

$3\left(\dfrac{2x^2 - 7x}{3}\right) = 3(5)$

$2x^2 - 7x = 15$

$2x^2 - 7x - 15 = 0$

$(2x + 3)(x - 5) = 0$

$2x + 3 = 0 \qquad x - 5 = 0$

$\quad x = -\dfrac{3}{2} \qquad\quad x = 5$

The roots are $-\frac{3}{2}$ and 5.

6. Let w = width, then

$3w + 2$ = length.

$(3w + 2)w = 85$

$3w^2 + 2w = 85$

$3w^2 + 2w - 85 = 0$

$(3w + 17)(w - 5) = 0$

$3w + 17 = 0 \qquad w - 5 = 0$

$\quad w = -\dfrac{17}{3} \qquad\quad w = 5$

The only valid answer is width = 5 meters

$\qquad\qquad$ length = $3(5) + 2 = 17$ meters

7. Let b = base.

$b - 3$ = altitude.

$\dfrac{b(b - 3)}{2} = 35$

$\dfrac{b^2 - 3b}{2} = 35$

$b^2 - 3b = 70$

$b^2 - 3b - 70 = 0$

$(b + 7)(b - 10) = 0$

$b + 7 = 0$

$\quad b = -7$

This is not a valid answer.

$b - 10 = 0$

$\quad b = 10$

Thus the base = 10 centimeters

altitude = $10 - 3 = 7$ centimeters

8. $\quad -5t^2 + 45 = 0$

$-5(t^2 - 9) = 0$

$\quad t^2 - 9 = 0$

$(t + 3)(t - 3) = 0$

$t + 3 = 0 \qquad t - 3 = 0$

$\quad t = -3 \qquad\quad t = 3$

$t = -3$ is not a valid answer.

Thus it will be 3 seconds before he breaks the water's surface.

Chapter 7 7.1 Practice Problems

1. $\dfrac{28}{63} = \dfrac{7 \cdot 4}{7 \cdot 9} = \dfrac{4}{9}$

2. $\dfrac{12x - 6}{14x - 7} = \dfrac{6(2x - 1)}{7(2x - 1)} = \dfrac{6}{7}$

3. $\dfrac{4x^2 - 9}{2x^2 - x - 3} = \dfrac{(2x - 3)(2x + 3)}{(2x - 3)(x + 1)} = \dfrac{2x + 3}{x + 1}$

4. $\dfrac{x^3 - 16x}{x^3 - 2x^2 - 8x} = \dfrac{x(x^2 - 16)}{x(x^2 - 2x - 8)} = \dfrac{x(x + 4)(x - 4)}{x(x + 2)(x - 4)} = \dfrac{x + 4}{x + 2}$

5. $\dfrac{8x - 20}{15 - 6x} = \dfrac{4(2x - 5)}{-3(-5 + 2x)} = \dfrac{4}{-3} = -\dfrac{4}{3}$

6. $\dfrac{4x^2 + 3x - 10}{25 - 16x^2} = \dfrac{(4x - 5)(x + 2)}{(5 + 4x)(5 - 4x)} = \dfrac{(4x - 5)(x + 2)}{-1(-5 + 4x)(5 + 4x)}$

$\qquad = \dfrac{x + 2}{-1(5 + 4x)} = -\dfrac{x + 2}{5 + 4x}$

7. $\dfrac{x^2 - 8xy + 15y^2}{2x^2 - 11xy + 5y^2} = \dfrac{(x - 3y)(x - 5y)}{(x - 5y)(2x - y)} = \dfrac{x - 3y}{2x - y}$

8. $\dfrac{25a^2 - 16b^2}{10a^2 + 3ab - 4b^2} = \dfrac{(5a + 4b)(5a - 4b)}{(5a + 4b)(2a - b)} = \dfrac{5a - 4b}{2a - b}$

7.2 Practice Problems

1. $\dfrac{6x^2 + 7x + 2}{x^2 - 7x + 10} \cdot \dfrac{x^2 + 3x - 10}{2x^2 + 11x + 5}$

$\quad = \dfrac{(2x + 1)(3x + 2)}{(x - 5)(x - 2)} \cdot \dfrac{(x + 5)(x - 2)}{(2x + 1)(x + 5)} = \dfrac{3x + 2}{x - 5}$

2. $\dfrac{2y^2 - 6y - 8}{y^2 - y - 2} \cdot \dfrac{y^2 - 5y + 6}{2y^2 - 32}$

$\quad = \dfrac{2(y^2 - 3y - 4)}{(y - 2)(y + 1)} \cdot \dfrac{(y - 3)(y - 2)}{2(y^2 - 16)}$

$\quad = \dfrac{2(y - 4)(y + 1)}{(y - 2)(y + 1)} \cdot \dfrac{(y - 3)(y - 2)}{2(y + 4)(y - 4)} = \dfrac{y - 3}{y + 4}$

3. $\dfrac{x^2 + 5x + 6}{x^2 + 8x} \div \dfrac{2x^2 + 5x + 2}{2x^2 + x}$

$\quad = \dfrac{x^2 + 5x + 6}{x^2 + 8x} \cdot \dfrac{2x^2 + x}{2x^2 + 5x + 2}$

$\quad = \dfrac{(x + 2)(x + 3)}{x(x + 8)} \cdot \dfrac{x(2x + 1)}{(2x + 1)(x + 2)} = \dfrac{x + 3}{x + 8}$

4. $\dfrac{x + 3}{x - 3} \div (9 - x^2) = \dfrac{x + 3}{x - 3} \div \dfrac{9 - x^2}{1} = \dfrac{x + 3}{x - 3} \cdot \dfrac{1}{9 - x^2}$

$\quad = \dfrac{x + 3}{x - 3} \cdot \dfrac{1}{(3 + x)(3 - x)}$

$\quad = \dfrac{1}{(x - 3)(3 - x)}$

7.3 Practice Problems

1. $\dfrac{2s + t}{2s - t} + \dfrac{s - t}{2s - t} = \dfrac{2s + t + s - t}{2s - t} = \dfrac{3s}{2s - t}$

2. $\dfrac{b}{(a - 2b)(a + b)} - \dfrac{2b}{(a - 2b)(a + b)}$

$\quad = \dfrac{b - 2b}{(a - 2b)(a + b)}$

$\quad = \dfrac{-b}{(a - 2b)(a + b)}$

3. $\dfrac{7}{6x + 21}, \quad \dfrac{13}{10x + 35}$

$6x + 21 = 3(2x + 7)$

$10x + 35 = 5(2x + 7)$

$\text{LCD} = 3 \cdot 5 \cdot (2x + 7) = 15(2x + 7)$

4. (a) $\dfrac{3}{50xy^2z}, \dfrac{19}{40x^3yz}$

$50xy^2z = 2 \cdot \quad 5 \cdot 5 \cdot x \cdot \quad\quad y \cdot y \cdot z$

$40x^3yz = 2 \cdot 2 \cdot 2 \cdot 5 \cdot \Big| \; x \cdot x \cdot x \cdot y \cdot \Big| \; z$

$$\downarrow\downarrow\downarrow\downarrow\quad\downarrow\downarrow\downarrow\downarrow$$

$$2 \cdot 2 \cdot 2 \cdot 5 \cdot 5 \cdot x \cdot x \cdot x \cdot y \cdot y \cdot z$$

$\text{LCD} = 2^3 \cdot 5^2 \cdot x^3 \cdot y^2 \cdot z = 200x^3y^2z$

(b) $\dfrac{2}{x^2 + 5x + 6}, \dfrac{6}{3x^2 + 5x - 2}$

$x^2 + 5x + 6 = (x + 3)(x + 2)$

$3x^2 + 5x - 2 = \quad\Big|\quad (x + 2)(3x - 1)$

$$\downarrow\quad\quad\downarrow\quad\quad\downarrow$$

$\text{LCD} = (x + 3)(x + 2)(3x - 1)$

5. $\text{LCD} = abc$

$\dfrac{7}{a} + \dfrac{3}{abc} = \dfrac{7}{a} \cdot \dfrac{bc}{bc} + \dfrac{3}{abc} = \dfrac{7bc}{abc} + \dfrac{3}{abc} = \dfrac{7bc + 3}{abc}$

6. $a^2 - 4b^2 = (a + 2b)(a - 2b)$

$\text{LCD} = (a + 2b)(a - 2b)$

$\dfrac{2a - b}{a^2 - 4b^2} + \dfrac{2}{a + 2b}$

$= \dfrac{2a - b}{(a + 2b)(a - 2b)} + \dfrac{2}{(a + 2b)} \cdot \dfrac{a - 2b}{a - 2b}$

$= \dfrac{2a - b}{(a + 2b)(a - 2b)} + \dfrac{2a - 4b}{(a + 2b)(a - 2b)}$

$= \dfrac{2a - b + 2a - 4b}{(a + 2b)(a - 2b)}$

$= \dfrac{4a - 5b}{(a + 2b)(a - 2b)}$

7. $\dfrac{7a}{a^2 + 2ab + b^2} + \dfrac{4}{a^2 + ab}$

$= \dfrac{7a}{(a + b)^2} + \dfrac{4}{a(a + b)}$ $\text{LCD} = a(a + b)^2$

$= \dfrac{7a^2}{a(a + b)^2} + \dfrac{4(a + b)}{a(a + b)^2} = \dfrac{7a^2 + 4a + 4b}{a(a + b)^2}$

8. $\dfrac{x + 7}{3x - 9} - \dfrac{x - 6}{x - 3} = \dfrac{x + 7}{3(x - 3)} - \dfrac{x - 6}{x - 3}$

$= \dfrac{x + 7}{3(x - 3)} - \dfrac{x - 6}{x - 3} \cdot \dfrac{3}{3} = \dfrac{x + 7 - 3(x - 6)}{3(x - 3)}$

$= \dfrac{x + 7 - 3x + 18}{3(x - 3)} = \dfrac{-2x + 25}{3(x - 3)}$

9. $\dfrac{x - 2}{x^2 - 4} - \dfrac{x + 1}{2x^2 + 4x} = \dfrac{x - 2}{(x + 2)(x - 2)} - \dfrac{x + 1}{2x(x + 2)}$

$= \dfrac{x - 2}{(x + 2)(x - 2)} \cdot \dfrac{2x}{2x} - \dfrac{x + 1}{2x(x + 2)} \cdot \dfrac{x - 2}{x - 2}$

$= \dfrac{2x(x - 2) - (x + 1)(x - 2)}{2x(x + 2)(x - 2)} = \dfrac{2x^2 - 4x - x^2 + x + 2}{2x(x + 2)(x - 2)}$

$= \dfrac{x^2 - 3x + 2}{2x(x + 2)(x - 2)} = \dfrac{(x - 1)(x - 2)}{2x(x + 2)(x - 2)} = \dfrac{x - 1}{2x(x + 2)}$

7.4 Practice Problems

1. $\dfrac{\dfrac{1}{a} + \dfrac{1}{a^2}}{\dfrac{2}{b^2}} = \dfrac{\dfrac{1}{a} \cdot \dfrac{a}{a} + \dfrac{1}{a^2}}{\dfrac{2}{b^2}} = \dfrac{\dfrac{a + 1}{a^2}}{\dfrac{2}{b^2}} = \dfrac{a + 1}{a^2} \div \dfrac{2}{b^2}$

$= \dfrac{a + 1}{a^2} \cdot \dfrac{b^2}{2} = \dfrac{b^2(a + 1)}{2a^2}$

2. $\dfrac{\dfrac{1}{a} + \dfrac{1}{b}}{\dfrac{x}{2} - \dfrac{5}{y}} = \dfrac{\dfrac{1}{a} \cdot \dfrac{b}{b} + \dfrac{1}{b} \cdot \dfrac{a}{a}}{\dfrac{x}{2} \cdot \dfrac{y}{y} - \dfrac{5}{y} \cdot \dfrac{2}{2}} = \dfrac{\dfrac{b + a}{ab}}{\dfrac{xy - 10}{2y}} = \dfrac{b + a}{ab} \cdot \dfrac{2y}{xy - 10} = \dfrac{2y(b + a)}{ab(xy - 10)}$

3. $\dfrac{\dfrac{x}{x^2 + 4x + 3} + \dfrac{2}{x + 1}}{\dfrac{x + 1}{x + 2x + 6}} = \dfrac{\dfrac{x}{(x + 1)(x + 3)} + \dfrac{2}{x + 1} \cdot \dfrac{x + 3}{x + 3}}{x + 1}$

$= \dfrac{\dfrac{x + 2x + 6}{(x + 1)(x + 3)}}{(x + 1)} = \dfrac{3x + 6}{(x + 1)(x + 3)} \cdot \dfrac{1}{(x + 1)}$

$= \dfrac{3(x + 2)}{(x + 1)^2(x + 3)}$

4. $\dfrac{\dfrac{6}{x^2 - y^2}}{\dfrac{1}{x - y} + \dfrac{3}{x + y}} = \dfrac{\dfrac{6}{x^2 - y^2}}{\dfrac{1}{x - y} \cdot \dfrac{x + y}{x + y} + \dfrac{3}{x + y} \cdot \dfrac{x - y}{x - y}}$

$= \dfrac{\dfrac{6}{x^2 - y^2}}{\dfrac{x + y}{(x + y)(x - y)} + \dfrac{3x - 3y}{(x + y)(x - y)}} = \dfrac{\dfrac{6}{(x + y)(x - y)}}{\dfrac{4x - 2y}{(x + y)(x - y)}}$

$= \dfrac{6}{(x + y)(x - y)} \cdot \dfrac{(x + y)(x - y)}{2(2x - y)} = \dfrac{2 \cdot 3}{2(2x - y)} = \dfrac{3}{2x - y}$

5. The LCD of all the denominators is $3x^2y$.

$\dfrac{\dfrac{2}{3x^2} - \dfrac{3}{y}}{\dfrac{5}{xy} - 4} = \dfrac{3x^2y\left(\dfrac{2}{3x^2} - \dfrac{3}{y}\right)}{3x^2y\left(\dfrac{5}{xy} - 4\right)} = \dfrac{3x^2y\left(\dfrac{2}{3x^2}\right) - 3x^2y\left(\dfrac{3}{y}\right)}{3x^2y\left(\dfrac{5}{xy}\right) - 3x^2y(4)} = \dfrac{2y - 9x^2}{15x - 12x^2y}$

6. The LCD of all the denominators is $(x + y)(x - y)$.

$\dfrac{\dfrac{6}{x^2 - y^2}}{\dfrac{7}{x - y} + \dfrac{3}{x + y}}$

$= \dfrac{(x + y)(x - y)\left(\dfrac{6}{(x + y)(x - y)}\right)}{(x + y)(x - y)\left(\dfrac{7}{x - y}\right) + (x + y)(x - y)\left(\dfrac{3}{x + y}\right)}$

$= \dfrac{6}{7(x + y) + 3(x - y)}$

$= \dfrac{6}{7x + 7y + 3x - 3y}$

$= \dfrac{6}{10x + 4y}$

$= \dfrac{2 \cdot 3}{2(5x + 2y)}$

$= \dfrac{3}{5x + 2y}$

7.5 Practice Problems

1. $\dfrac{3}{x} + \dfrac{4}{5} = -\dfrac{2}{x}$ $\text{LCD} = 5x$

$5x\left(\dfrac{3}{x}\right) + 5x\left(\dfrac{4}{5}\right) = 5x\left(-\dfrac{2}{x}\right)$

$15 + 4x = -10$

$4x = -10 - 15$

$4x = -25$

$x = -\dfrac{25}{4}$

Check.

$\dfrac{3}{-\dfrac{25}{4}} + \dfrac{4}{5} \overset{?}{=} -\dfrac{2}{-\dfrac{25}{4}}$

$-\dfrac{12}{25} + \dfrac{4}{5} \overset{?}{=} \dfrac{8}{25}$

$-\dfrac{12}{25} + \dfrac{20}{25} \overset{?}{=} \dfrac{8}{25}$

$\dfrac{8}{25} = \dfrac{8}{25}$ ✓

2. LCD $= (2x + 1)(2x - 1)$

$$\frac{4}{2x + 1} = \frac{6}{2x - 1}$$

$$(2x + 1)(2x - 1)\left(\frac{4}{2x + 1}\right) = (2x + 1)(2x - 1)\left(\frac{6}{2x - 1}\right)$$

$$4(2x - 1) = 6(2x + 1)$$
$$8x - 4 = 12x + 6$$
$$-4x - 4 = 6$$
$$-4x = 10$$
$$x = -\frac{5}{2}$$

Check.

$$\frac{4}{2\left(-\dfrac{5}{2}\right) + 1} \stackrel{?}{=} \frac{6}{2\left(-\dfrac{5}{2}\right) - 1}$$

$$\frac{4}{-5 + 1} \stackrel{?}{=} \frac{6}{-5 - 1}$$

$$\frac{4}{-4} \stackrel{?}{=} \frac{6}{-6}$$

$$-1 = -1 \checkmark$$

3.
$$\frac{x - 1}{x^2 - 4} = \frac{2}{x + 2} + \frac{4}{x - 2}$$

$$\frac{x - 1}{(x + 2)(x - 2)} = \frac{2}{x + 2} + \frac{4}{x - 2}$$

$$(x + 2)(x - 2)\left[\frac{x - 1}{(x + 2)(x - 2)}\right]$$

$$= (x + 2)(x - 2)\left(\frac{2}{x + 2}\right) + (x + 2)(x - 2)\left(\frac{4}{x - 2}\right)$$

$$x - 1 = 2(x - 2) + 4(x + 2)$$
$$x - 1 = 2x - 4 + 4x + 8$$
$$x - 1 = 6x + 4$$
$$-5x - 1 = 4$$
$$-5x = 5$$
$$x = -1$$

Check.
$$\frac{-1 - 1}{(-1)^2 - 4} \stackrel{?}{=} \frac{2}{-1 + 2} + \frac{4}{-1 - 2}$$

$$\frac{-2}{-3} \stackrel{?}{=} \frac{2}{1} + \frac{4}{-3}$$

$$\frac{2}{3} \stackrel{?}{=} \frac{6}{3} - \frac{4}{3}$$

$$\frac{2}{3} = \frac{2}{3} \checkmark$$

4. The LCD is $x + 1$.

$$\frac{2x}{x + 1} = \frac{-2}{x + 1} + 1$$

$$(x + 1)\left(\frac{2x}{x + 1}\right) = (x + 1)\left(\frac{-2}{x + 1}\right) + (x + 1)(1)$$

$$2x = -2 + x + 1$$
$$2x = x - 1$$
$$x = -1 \text{ (but see the check)}$$

Check.
$$\frac{2(-1)}{-1 + 1} \stackrel{?}{=} \frac{-2}{-1 + 1} + 1$$

$$\frac{-2}{0} \stackrel{?}{=} \frac{-2}{0} + 1$$

These expressions are not defined; therefore, there is no solution to this equation.

7.6 Practice Problems

1. Let $x =$ the number of hours it will take to drive 315 miles.

$$\frac{8}{420} = \frac{x}{315}$$
$$8(315) = 420x$$
$$\frac{2520}{420} = x$$
$$6 = x$$

It would take Brenda 6 hours to drive 315 miles.

2. Let $x =$ the distance represented by $2\frac{1}{2}$ inches.

$$\frac{\dfrac{5}{8}}{30} = \frac{2\dfrac{1}{2}}{x}$$

$$\frac{5}{8}x = 30\left(2\frac{1}{2}\right)$$

$$\frac{5}{8}x = 30\left(\frac{5}{2}\right)$$

$$\frac{5}{8}x = 75$$

$$8\left(\frac{5}{8}x\right) = 8(75)$$

$$5x = 600$$
$$x = 120$$

Therefore $2\frac{1}{2}$ inches represents 120 miles.

3.
$$\frac{13}{x} = \frac{16}{18}$$
$$13(18) = 16x$$
$$234 = 16x$$
$$\frac{234}{16} = x$$

Side x has length
$$\frac{117}{8} = 14\frac{5}{8} \text{ cm.}$$

4.
$$\frac{6}{7} = \frac{x}{38.5}$$
$$6(38.5) = 7x$$
$$231 = 7x$$
$$x = 33 \text{ feet}$$

5. Let $x =$ the speed of train B. Then

train A time $= \dfrac{180}{x + 10}$ and train B time $= \dfrac{150}{x}$.

$$\frac{180}{x + 10} = \frac{150}{x}$$
$$180x = 150(x + 10)$$
$$180x = 150x + 1500$$
$$30x = 1500$$
$$x = 50$$

Train B traveled 50 kilometers per hour. Train A traveled $50 + 10 = 60$ kilometers per hour.

6.

	Number of Hours	Part of the Job Done in One Hour
John	6 hours	$\dfrac{1}{6}$
Dave	7 hours	$\dfrac{1}{7}$
John & Dave Together	x	$\dfrac{1}{x}$

LCD $= 42x$

$$\frac{1}{6} + \frac{1}{7} = \frac{1}{x}$$

$$42x\left(\frac{1}{6}\right) + 42x\left(\frac{1}{7}\right) = 42x\left(\frac{1}{x}\right)$$

$$7x + 6x = 42$$
$$13x = 42$$

$$x = 3\frac{3}{13}$$

$$\frac{3}{13} \text{ hour} \times \frac{60 \text{ min}}{1 \text{ hour}} = \frac{180}{13} \text{ min} \approx 13.846 \text{ min}$$

Thus, doing the job together will take 3 hours and 14 minutes.

Chapter 8 8.1 Practice Problems

1. (a) $\sqrt{64} = 8$ because $8^2 = 64$.

(b) $-\sqrt{121} = -11$ since $\sqrt{121} = 11$.

2. (a) $\sqrt{\dfrac{9}{25}} = \dfrac{3}{5}$ since $\left(\dfrac{3}{5}\right)^2 = \left(\dfrac{3}{5}\right)\left(\dfrac{3}{5}\right) = \dfrac{9}{25}$.

(b) $-\sqrt{\dfrac{121}{169}} = -\dfrac{11}{13}$ since $\sqrt{\dfrac{121}{169}} = \dfrac{11}{13}$.

3. (a) $-\sqrt{0.0036} = -0.06$ since $(0.06)(0.06) = 0.0036$.

(b) $\sqrt{2500} = 50$ since $(50)(50) = 2500$.

(c) $\sqrt{196} = 14$

4. (a) $\sqrt{13} \approx 3.606$

(b) $\sqrt{35} \approx 5.916$

(c) $\sqrt{127} \approx 11.269$

8.2 Practice Problems

1. (a) $\sqrt{6^2} = 6$

(b) $\sqrt{13^4} = \sqrt{(13^2)^2} = 13^2$

(c) $\sqrt{18^{12}} = \sqrt{(18^6)^2} = 18^6$

2. (a) $\sqrt{y^{18}} = \sqrt{(y^9)^2} = y^9$

(b) $\sqrt{x^{30}} = \sqrt{(x^{15})^2} = x^{15}$

3. (a) $\sqrt{625y^4} = \sqrt{625}\sqrt{y^4} = 25y^2$

(b) $\sqrt{x^{16}y^{22}} = x^8y^{11}$

(c) $\sqrt{121x^{12}y^6} = 11x^6y^3$

4. (a) $\sqrt{98} = \sqrt{49\cdot2} = \sqrt{49}\sqrt{2} = 7\sqrt{2}$

(b) $\sqrt{12} = \sqrt{4\cdot3} = \sqrt{4}\sqrt{3} = 2\sqrt{3}$

(c) $\sqrt{75} = \sqrt{25\cdot3} = \sqrt{25}\sqrt{3} = 5\sqrt{3}$

5. (a) $\sqrt{x^{11}} = \sqrt{x^{10}}\sqrt{x} = x^5\sqrt{x}$

(b) $\sqrt{x^5y^3} = \sqrt{x^4}\sqrt{x}\sqrt{y^2}\sqrt{y} = x^2\sqrt{x}\,y\sqrt{y} = x^2y\sqrt{xy}$

6. (a) $\sqrt{48x^{11}} = \sqrt{16\cdot3\cdot x^{10}\cdot x} = 4x^5\sqrt{3x}$

(b) $\sqrt{121x^6y^7z^8} = \sqrt{121\cdot x^6\cdot y^6\cdot y\cdot z^8} = 11x^3y^3z^4\sqrt{y}$

8.3 Practice Problems

1. (a) $7\sqrt{11} + 4\sqrt{11} = (7+4)\sqrt{11} = 11\sqrt{11}$

(b) $4\sqrt{t} - 7\sqrt{t} + 6\sqrt{t} - 2\sqrt{t}$
$= (4 - 7 + 6 - 2)\sqrt{t}$
$= 1\sqrt{t}$
$= \sqrt{t}$

2. $3\sqrt{x} - 2\sqrt{xy} - 5\sqrt{y} + 7\sqrt{xy}$
$= 3\sqrt{x} + 5\sqrt{xy} - 5\sqrt{y}$

3. (a) $\sqrt{50} - \sqrt{18} + \sqrt{98}$
$= \sqrt{25\cdot2} - \sqrt{9\cdot2} + \sqrt{49\cdot2}$
$= 5\sqrt{2} - 3\sqrt{2} + 7\sqrt{2} = 9\sqrt{2}$

(b) $\sqrt{12} + \sqrt{18} - \sqrt{50} + \sqrt{27}$
$= \sqrt{4\cdot3} + \sqrt{9\cdot2} - \sqrt{25\cdot2} + \sqrt{9\cdot3}$
$= 2\sqrt{3} + 3\sqrt{2} - 5\sqrt{2} + 3\sqrt{3} = 5\sqrt{3} - 2\sqrt{2}$

4. $\sqrt{9x} + \sqrt{8x} - \sqrt{4x} + \sqrt{50x}$
$= \sqrt{9\cdot x} + \sqrt{4\cdot2x} - \sqrt{4\cdot x} + \sqrt{25\cdot2x}$
$= 3\sqrt{x} + 2\sqrt{2x} - 2\sqrt{x} + 5\sqrt{2x}$
$= \sqrt{x} + 7\sqrt{2x}$

5. $\sqrt{27} - 4\sqrt{3} + 2\sqrt{75}$
$= \sqrt{9}\cdot\sqrt{3} - 4\cdot\sqrt{3} + 2\sqrt{25}\sqrt{3}$
$= 3\cdot\sqrt{3} - 4\sqrt{3} + 2\cdot5\cdot\sqrt{3}$
$= 3\sqrt{3} - 4\sqrt{3} + 10\sqrt{3} = 9\sqrt{3}$

6. $2a\sqrt{12a} + 3\sqrt{27a^3}$
$= 2a\sqrt{4}\sqrt{3a} + 3\sqrt{9a^2}\sqrt{3a}$
$= 2a\cdot2\cdot\sqrt{3a} + 3\cdot3a\cdot\sqrt{3a}$
$= 4a\sqrt{3a} + 9a\sqrt{3a}$
$= 13a\sqrt{3a}$

8.4 Practice Problems

1. $\sqrt{3a}\sqrt{6a} = \sqrt{18a^2} = \sqrt{9a^2\cdot2} = 3a\sqrt{2}$

2. (a) $(2\sqrt{3})(5\sqrt{5}) = 10\sqrt{15}$

(b) $(4\sqrt{3x})(2x\sqrt{6x}) = 8x\sqrt{18x^2}$
$= 8x\sqrt{9}\sqrt{2}\sqrt{x^2} = 24x^2\sqrt{2}$

3. $(\sqrt{180})(\sqrt{150}) = (6\sqrt{5})(5\sqrt{6})$
$= 30\sqrt{30} \text{ mm}^2$

4. $\sqrt{6}(\sqrt{3} + 2\sqrt{8})$
$= \sqrt{6}\sqrt{3} + \sqrt{6}2\sqrt{8} = \sqrt{18} + 2\sqrt{48}$
$= \sqrt{9}\sqrt{2} + 2\sqrt{16}\sqrt{3}$
$= 3\sqrt{2} + 8\sqrt{3}$

5. $2\sqrt{x}(4\sqrt{x} - x\sqrt{2})$
$= 8\sqrt{x}\sqrt{x} - 2x\sqrt{x}\sqrt{2}$
$= 8x - 2x\sqrt{2x}$

6. $(\sqrt{2} + \sqrt{6})(2\sqrt{2} - \sqrt{6})$
$= 2\sqrt{4} - \sqrt{12} + 2\sqrt{12} - \sqrt{36}$
$= 2(2) + \sqrt{12} - 6$
$= 4 + 2\sqrt{3} - 6$
$= -2 + 2\sqrt{3}$

7. $(\sqrt{6} + \sqrt{5})(\sqrt{2} + 2\sqrt{5})$
$= \sqrt{12} + 2\sqrt{30} + \sqrt{10} + 2\sqrt{25}$
$= 2\sqrt{3} + 2\sqrt{30} + \sqrt{10} + 10$

8. $(3\sqrt{5} - \sqrt{10})^2$
$= (3\sqrt{5} - \sqrt{10})(3\sqrt{5} - \sqrt{10})$
$= 9\sqrt{25} - 3\sqrt{50} - 3\sqrt{50} + \sqrt{100}$
$= 45 - 6\sqrt{50} + 10$
$= 55 - 30\sqrt{2}$

8.5 Practice Problems

1. (a) $\dfrac{\sqrt{98}}{\sqrt{2}} = \sqrt{\dfrac{98}{2}} = \sqrt{49} = 7$

(b) $\sqrt{\dfrac{81}{100}} = \dfrac{\sqrt{81}}{\sqrt{100}} = \dfrac{9}{10}$

2. $\sqrt{\dfrac{50}{a^4}} = \dfrac{\sqrt{50}}{\sqrt{a^4}} = \dfrac{\sqrt{25}\sqrt{2}}{a^2} = \dfrac{5\sqrt{2}}{a^2}$

3. $\dfrac{9}{\sqrt{7}} = \dfrac{9}{\sqrt{7}}\cdot\dfrac{\sqrt{7}}{\sqrt{7}} = \dfrac{9\sqrt{7}}{\sqrt{49}} = \dfrac{9\sqrt{7}}{7}$

4. (a) $\dfrac{\sqrt{2}}{\sqrt{12}} = \dfrac{\sqrt{2}}{\sqrt{12}}\cdot\dfrac{\sqrt{3}}{\sqrt{3}} = \dfrac{\sqrt{6}}{\sqrt{36}} = \dfrac{\sqrt{6}}{6}$

(b) $\dfrac{6a}{\sqrt{a^7}} = \dfrac{6a}{\sqrt{a^7}}\cdot\dfrac{\sqrt{a}}{\sqrt{a}} = \dfrac{6a\sqrt{a}}{\sqrt{a^8}} = \dfrac{6a\sqrt{a}}{a^4} = \dfrac{6\sqrt{a}}{a^3}$

5. $\dfrac{\sqrt{5x}}{\sqrt{8x}} = \dfrac{\sqrt{5x}}{2\sqrt{2x}} = \dfrac{\sqrt{5x}}{2\sqrt{2x}} \cdot \dfrac{\sqrt{2x}}{\sqrt{2x}} = \dfrac{\sqrt{10x^2}}{2\sqrt{4x^2}} = \dfrac{x\sqrt{10}}{4x} = \dfrac{\sqrt{10}}{4}$

6. (a) $\dfrac{4}{\sqrt{3} + \sqrt{5}} = \dfrac{4}{\sqrt{3} + \sqrt{5}} \cdot \dfrac{\sqrt{3} - \sqrt{5}}{\sqrt{3} - \sqrt{5}}$

$= \dfrac{4\sqrt{3} - 4\sqrt{5}}{(\sqrt{3})^2 - \sqrt{15} + \sqrt{15} - (\sqrt{5})^2}$

$= \dfrac{4\sqrt{3} - 4\sqrt{5}}{3 - 5} = \dfrac{4\sqrt{3} - 4\sqrt{5}}{-2}$

$= \dfrac{4(\sqrt{3} - \sqrt{5})}{-2} = -2(\sqrt{3} - \sqrt{5})$

(b) $\dfrac{\sqrt{a}}{\sqrt{10} - 3} = \dfrac{\sqrt{a}}{\sqrt{10} - 3} \cdot \dfrac{\sqrt{10} + 3}{\sqrt{10} + 3}$

$= \dfrac{\sqrt{10a} + 3\sqrt{a}}{(\sqrt{10})^2 + 3\sqrt{10} - 3\sqrt{10} - 3^2}$

$= \dfrac{\sqrt{10a} + 3\sqrt{a}}{10 - 9} = \dfrac{\sqrt{10a} + 3\sqrt{a}}{1} = \sqrt{10a} + 3\sqrt{a}$

7. $\dfrac{\sqrt{3} + \sqrt{5}}{\sqrt{3} - \sqrt{5}} = \dfrac{\sqrt{3} + \sqrt{5}}{\sqrt{3} - \sqrt{5}} \cdot \dfrac{\sqrt{3} + \sqrt{5}}{\sqrt{3} + \sqrt{5}}$

$= \dfrac{\sqrt{9} + \sqrt{15} + \sqrt{15} + \sqrt{25}}{(\sqrt{3})^2 - (\sqrt{5})^2}$

$= \dfrac{3 + 2\sqrt{15} + 5}{3 - 5} = \dfrac{8 + 2\sqrt{15}}{-2}$

$= \dfrac{-2(-4 - \sqrt{15})}{-2} = -4 - \sqrt{15}$

8.6 Practice Problems

1. $c^2 = a^2 + b^2$
$c^2 = 9^2 + 12^2$
$c^2 = 81 + 144$
$c^2 = 225$
$c = \pm\sqrt{225}$
$c = \pm 15$
The hypotenuse is 15 centimeters because length is not negative.

2. $c^2 = a^2 + b^2$
$(\sqrt{17})^2 = (1)^2 + b^2$
$17 = 1 + b^2$
$16 = b^2$
$\pm 4 = b$
The leg is 4 meters long.

3. $c^2 = 3^2 + 8^2$
$c^2 = 9 + 64$
$c^2 = 73$
$c = \pm\sqrt{73}$
The support line is $\sqrt{73} \approx 8.5$ meters long.

4. $\sqrt{3x - 2} - 7 = 0$
$\sqrt{3x - 2} = 7$
$(\sqrt{3x - 2})^2 = (7)^2$
$3x - 2 = 49$
$3x = 51$
$x = 17$
The solution is 17.
Check.
$\sqrt{3(17) - 2} - 7 \overset{?}{=} 0$
$\sqrt{51 - 2} - 7 \overset{?}{=} 0$
$\sqrt{49} - 7 \overset{?}{=} 0$
$7 - 7 \overset{?}{=} 0$
$0 = 0$ ✓

5. $\sqrt{5x + 4} + 2 = 0$
$\sqrt{5x + 4} = -2$
$(\sqrt{5x + 4})^2 = (-2)^2$
$5x + 4 = 4$
$5x = 0$
$x = 0$
Check.
$\sqrt{5(0) + 4} + 2 \overset{?}{=} 0$
$\sqrt{4} + 2 \overset{?}{=} 0$
$2 + 2 \overset{?}{=} 0$
$4 \neq 0$
No, this does not check. There is no solution.

6. $-2 + \sqrt{6x - 1} = 3x - 2$
$\sqrt{6x - 1} = 3x$
$(\sqrt{6x - 1})^2 = (3x)^2$
$6x - 1 = 9x^2$
$0 = 9x^2 - 6x + 1$
$0 = (3x - 1)^2$
$0 = 3x - 1$
$x = \dfrac{1}{3}$
Check.
$-2 + \sqrt{6\left(\dfrac{1}{3}\right) - 1} \overset{?}{=} 3\left(\dfrac{1}{3}\right) - 2$
$-2 + \sqrt{2 - 1} \overset{?}{=} 1 - 2$
$-2 + \sqrt{1} \overset{?}{=} -1$
$-2 + 1 \overset{?}{=} -1$
$-1 = -1$ ✓
The solution is $\dfrac{1}{3}$.

7. $2 - x + \sqrt{x + 4} = 0$
$\sqrt{x + 4} = x - 2$
$(\sqrt{x + 4})^2 = (x - 2)^2$
$x + 4 = x^2 - 4x + 4$
$0 = x^2 - 5x$
$0 = x(x - 5)$
$x = 0$ or $x - 5 = 0$
$x = 5$
Check.
$2 - 0 + \sqrt{0 + 4} \overset{?}{=} 0$
$2 + \sqrt{4} \overset{?}{=} 0$
$2 + 2 \overset{?}{=} 0$
$4 \neq 0$
$x = 0$ does not check.
$2 - 5 + \sqrt{5 + 4} \overset{?}{=} 0$
$-3 + \sqrt{9} \overset{?}{=} 0$
$-3 + 3 \overset{?}{=} 0$
$0 = 0$ ✓
The only solution is 5.

8. $\sqrt{2x + 1} = \sqrt{x - 10}$
$(\sqrt{2x + 1})^2 = (\sqrt{x - 10})^2$
$2x + 1 = x - 10$
$x = -11$
Check.
$\sqrt{2(-11) + 1} \overset{?}{=} \sqrt{-11 - 10}$
$\sqrt{-22 + 1} \overset{?}{=} \sqrt{-21}$
$\sqrt{-21}$ is not a real number.
Thus, $x = -11$ is extraneous. There is no solution.

8.7 Practice Problems

1. Change of 100°C = change of 180°F.

Let C = the change on the Celsius scale

f = the change on the Fahrenheit scale, and

k = the constant of variation.

$$C = kf$$
$$100 = k \cdot 180$$
$$\frac{5}{9} = k$$

Thus, $C = \frac{5}{9}f$.

Let $f = -20$.

$$C = \frac{5}{9}(-20)$$
$$C = -11\frac{1}{9}$$

The temperature drops $11\frac{1}{9}$ degrees Celsius.

2. $y = kx$

To find k we substitute $y = 18$ and $x = 5$.

$$18 = k(5)$$
$$\frac{18}{5} = k$$

We now write the variation equation with k replaced by $\frac{18}{5}$.

$$y = \frac{18}{5}x$$

Replace x by $\frac{20}{23}$ and solve for y.

$$y = \frac{18}{5} \cdot \frac{20}{23}$$
$$y = \frac{72}{23}$$

Thus $y = \frac{72}{23}$ when $x = \frac{20}{23}$.

3. Let d = the distance to stop the car,

s = the speed of the car, and

k = the constant of variation.

Since the distance varies directly as the square of the speed, we have

$$d = ks^2.$$

To evaluate k we substitute the known distance and speed.

$$60 = k(20)^2$$
$$60 = k(400)$$
$$\frac{3}{20} = k$$

Now write the variation equation with the known value for k.

$$d = \frac{3}{20}s^2$$
$$d = \frac{3}{20}(40)^2$$
$$d = \frac{3}{20}(1600)$$
$$d = 240$$

The distance to stop the car going 40 mph on an ice-covered road is 240 feet.

4. $y = \frac{k}{x}$

$8 = \frac{k}{15}$ Substitute known values of x and y to find k.

$120 = k$

$y = \frac{120}{x}$ Write the variation equation with k replaced by 120.

$y = \frac{120}{\frac{3}{5}} = \frac{120}{1} \cdot \frac{5}{3} = 200$ Find y when $x = \frac{3}{5}$.

Thus $y = 200$ when $x = \frac{3}{5}$.

5. Let V = the volume of sales,

p = the price of the calculator, and

k = the constant of variation.

Since volume varies inversely as the price, then

$$V = \frac{k}{p}.$$

$120{,}000 = \frac{k}{30}$ Evaluate k by substituting known values for V and p.

$3{,}600{,}000 = k$

$V = \frac{3{,}600{,}000}{24}$ Write the variation equation with the evaluated constant and substitute $p = 24$.

$V = 150{,}000$

Thus the volume increased to 150,000 calculators sold per year when the price was reduced to $24 per calculator.

6. Let R = the resistance in the circuit,

i = the amount of current, and

k = the constant of variation.

Since the resistance in the circuit varies inversely as the square of the amount of current,

$$R = \frac{k}{i^2}.$$

To evaluate k, substitute the known values for R and i.

$$800 = \frac{k}{(0.01)^2}$$
$$0.08 = k$$

$R = \frac{0.08}{(0.02)^2}$ Write the variation equation with the evaluated constant of $k = 0.08$ and substitute $i = 0.02$.

$$R = \frac{0.08}{0.0004}$$
$$R = 200$$

Thus the resistance is 200 ohms if the amount of current is increased to 0.02 ampere.

Chapter 9   9.1 Practice Problems

1. (a) $2x^2 + 12x - 9 = 0$

$a = 2, b = 12, c = -9$

(b) $7x^2 = 6x - 8$

$7x^2 - 6x + 8 = 0$

$a = 7, b = -6, c = 8$

(c) $-x^2 - 6x + 3 = 0$

$x^2 + 6x - 3 = 0$

$a = 1, b = 6, c = -3$

(d) $10x^2 - 12x = 0$

$a = 10, b = -12, c = 0$

2. $2x^2 - 7x - 6 = 4x - 6$

$2x^2 - 7x - 6 - 4x + 6 = 0$

$2x^2 - 11x = 0$

$x(2x - 11) = 0$

$x = 0$ $2x - 11 = 0$

$2x = 11$

$x = \frac{11}{2}$

The two roots are 0 and $\frac{11}{2}$.

3.
$$3x + 10 - \frac{8}{x} = 0$$

$$x(3x) + x(10) - x\left(\frac{8}{x}\right) = x(0)$$

$$3x^2 + 10x - 8 = 0$$

$$(3x - 2)(x + 4) = 0$$

$$3x - 2 = 0 \qquad\qquad x + 4 = 0$$

$$3x = 2$$

$$x = \frac{2}{3} \qquad\qquad\qquad x = -4$$

Check.

$$3\left(\frac{2}{3}\right) + 10 - \frac{8}{\frac{2}{3}} \overset{?}{=} 0 \qquad 3(-4) + 10 - \frac{8}{-4} \overset{?}{=} 0$$

$$2 + 10 - 12 \overset{?}{=} 0 \qquad\qquad -12 + 10 + 2 \overset{?}{=} 0$$

$$0 = 0 \ \checkmark \qquad\qquad\qquad 0 = 0 \ \checkmark$$

Both roots check, so $x = \frac{2}{3}$ and $x = -4$ are the two roots that satisfy the equation $3x + 10 - \frac{8}{x} = 0$.

4.
$$2x^2 + 9x - 18 = 0$$

$$(2x - 3)(x + 6) = 0$$

$$2x - 3 = 0 \qquad x + 6 = 0$$

$$2x = 3$$

$$x = \frac{3}{2} \qquad\qquad x = -6$$

Check.

$$2\left(\frac{3}{2}\right)^2 + 9\left(\frac{3}{2}\right) - 18 \overset{?}{=} 0 \qquad 2(-6)^2 + 9(-6) - 18 \overset{?}{=} 0$$

$$2\left(\frac{9}{4}\right) + \frac{27}{2} - 18 \overset{?}{=} 0 \qquad\qquad 2(36) - 54 - 18 \overset{?}{=} 0$$

$$\frac{9}{2} + \frac{27}{2} - 18 \overset{?}{=} 0 \qquad\qquad 72 - 54 - 18 \overset{?}{=} 0$$

$$\frac{36}{2} - 18 \overset{?}{=} 0 \qquad\qquad\qquad 18 - 18 \overset{?}{=} 0$$

$$18 - 18 \overset{?}{=} 0 \qquad\qquad\qquad\qquad 0 = 0 \ \checkmark$$

$$0 = 0 \ \checkmark$$

Both roots check, so $x = \frac{3}{2}$ and $x = -6$ are the two roots that satisfy the equation $2x^2 + 9x - 18 = 0$.

5. $-4x + (x - 5)(x + 1) = 5(1 - x)$

$$-4x + x^2 - 4x - 5 = 5 - 5x$$

$$x^2 - 8x - 5 = 5 - 5x$$

$$x^2 - 3x - 10 = 0$$

$$(x + 2)(x - 5) = 0$$

$$x + 2 = 0 \qquad x - 5 = 0$$

$$x = -2 \qquad x = 5$$

6.
$$\frac{10x + 18}{x^2 + x - 2} = \frac{3x}{x + 2} + \frac{2}{x - 1}$$

$$\frac{10x + 18}{(x - 1)(x + 2)} = \frac{3x}{x + 2} + \frac{2}{x - 1} \qquad \text{Multiply by the LCD.}$$

$$10x + 18 = 3x(x - 1) + 2(x + 2)$$

$$10x + 18 = 3x^2 - 3x + 2x + 4$$

$$10x + 18 = 3x^2 - x + 4$$

$$0 = 3x^2 - 11x - 14$$

$$0 = (3x - 14)(x + 1)$$

$$3x - 14 = 0 \qquad x + 1 = 0$$

$$x = \frac{14}{3} \qquad\qquad x = -1$$

7. $t =$ the number of truck routes

$n =$ the number of cities they can service

$$t = \frac{n^2 - n}{2}$$

$$28 = \frac{n^2 - n}{2}$$

$$56 = n^2 - n$$

$$0 = n^2 - n - 56$$

$$0 = (n + 7)(n - 8)$$

$$n + 7 = 0 \qquad\qquad n - 8 = 0$$

$$n = -7 \qquad\qquad n = 8$$

We reject $n = -7$. There cannot be a negative number of cities. Thus our answer is 8 cities.

9.2 Practice Problems

1. (a) $x^2 = 1$

$$x = \pm\sqrt{1}$$

$$x = \pm 1$$

(b) $x^2 - 50 = 0$

$$x^2 = 50$$

$$x = \pm\sqrt{50}$$

$$x = \pm 5\sqrt{2}$$

(c) $3x^2 = 81$

$$x^2 = 27$$

$$x = \pm\sqrt{27}$$

$$x = \pm 3\sqrt{3}$$

2. $4x^2 - 5 = 319$

$$4x^2 = 324$$

$$x^2 = 81$$

$$x = \pm 9$$

3. $(2x - 3)^2 = 12$

$$2x - 3 = \pm\sqrt{12}$$

$$2x - 3 = \pm 2\sqrt{3}$$

$$2x - 3 = +2\sqrt{3} \qquad\qquad 2x - 3 = -2\sqrt{3}$$

$$2x = 3 + 2\sqrt{3} \qquad\qquad 2x = 3 - 2\sqrt{3}$$

$$x = \frac{3 + 2\sqrt{3}}{2} \qquad\qquad x = \frac{3 - 2\sqrt{3}}{2}$$

The roots are $\frac{3 \pm 2\sqrt{3}}{2}$.

4.
$$x^2 + 10x = 3$$

$$x^2 + 10x + (5)^2 = 3 + 25$$

$$(x + 5)^2 = 28$$

$$x + 5 = \pm\sqrt{28}$$

$$x + 5 = +2\sqrt{7} \qquad\qquad x + 5 = -2\sqrt{7}$$

$$x = -5 + 2\sqrt{7} \qquad\qquad x = -5 - 2\sqrt{7}$$

The roots are $-5 \pm 2\sqrt{7}$.

5.
$$x^2 - 5x = 7$$

$$x^2 - 5x + \left(\frac{5}{2}\right)^2 = 7 + \left(\frac{5}{2}\right)^2$$

$$\left(x - \frac{5}{2}\right)^2 = 7 + \frac{25}{4}$$

$$\left(x - \frac{5}{2}\right)^2 = \frac{53}{4}$$

$$x - \frac{5}{2} = \pm\sqrt{\frac{53}{4}}$$

$$x - \frac{5}{2} = \pm\frac{\sqrt{53}}{2}$$

$$x - \frac{5}{2} = +\frac{\sqrt{53}}{2} \qquad\qquad x - \frac{5}{2} = -\frac{\sqrt{53}}{2}$$

$$x = \frac{5}{2} + \frac{\sqrt{53}}{2} \qquad\qquad x = \frac{5}{2} - \frac{\sqrt{53}}{2}$$

$$x = \frac{5 + \sqrt{53}}{2} \qquad\qquad x = \frac{5 - \sqrt{53}}{2}$$

The roots are $\frac{5 \pm \sqrt{53}}{2}$.

6. $8x^2 + 2x - 1 = 0$

$8x^2 + 2x = 1$

$x^2 + \dfrac{1}{4}x = \dfrac{1}{8}$

$x^2 + \dfrac{1}{4}x + \dfrac{1}{64} = \dfrac{1}{8} + \dfrac{1}{64}$

$\left(x + \dfrac{1}{8}\right)^2 = \dfrac{9}{64}$

$x + \dfrac{1}{8} = \pm\sqrt{\dfrac{9}{64}}$

$x + \dfrac{1}{8} = \pm\dfrac{3}{8}$

$x + \dfrac{1}{8} = \dfrac{3}{8}$ $\qquad$ $x + \dfrac{1}{8} = -\dfrac{3}{8}$

$x = \dfrac{3}{8} - \dfrac{1}{8}$ $\qquad$ $x = -\dfrac{3}{8} - \dfrac{1}{8}$

$x = \dfrac{2}{8}$ $\qquad\qquad$ $x = -\dfrac{4}{8}$

$x = \dfrac{1}{4}$ $\qquad\qquad$ $x = -\dfrac{1}{2}$

The roots are $\dfrac{1}{4}$ and $-\dfrac{1}{2}$.

9.3 Practice Problems

1. $x^2 - 7x + 6 = 0$

$a = 1, b = -7, c = 6$

$x = \dfrac{-b \pm \sqrt{b^2 - 4ac}}{2a} = \dfrac{-(-7) \pm \sqrt{(-7)^2 - 4(1)(6)}}{2(1)}$

$= \dfrac{7 \pm \sqrt{49 - 24}}{2}$

$= \dfrac{7 \pm \sqrt{25}}{2}$

$= \dfrac{7 \pm 5}{2}$

$x = \dfrac{7 + 5}{2} = \dfrac{12}{2} = 6$ $\qquad$ $x = \dfrac{7 - 5}{2} = \dfrac{2}{2} = 1$

The solutions are 6 and 1.

2. $3x^2 - 8x + 3 = 0$

$a = 3, b = -8, c = 3$

$x = \dfrac{-b \pm \sqrt{b^2 - 4ac}}{2a} = \dfrac{-(-8) \pm \sqrt{(-8)^2 - 4(3)(3)}}{2(3)}$

$= \dfrac{8 \pm \sqrt{64 - 36}}{6}$

$= \dfrac{8 \pm \sqrt{28}}{6}$

$= \dfrac{8 \pm 2\sqrt{7}}{6}$

$x = \dfrac{2\left(4 \pm \sqrt{7}\right)}{6} = \dfrac{4 \pm \sqrt{7}}{3}$

3. $\qquad\qquad x^2 = 7 - \dfrac{3}{5}x$

$5x^2 = 35 - 3x$

$5x^2 + 3x - 35 = 0$

$a = 5, b = 3, c = -35$

$x = \dfrac{-3 \pm \sqrt{(3)^2 - 4(5)(-35)}}{2(5)}$

$x = \dfrac{-3 \pm \sqrt{9 + 700}}{10}$

$x = \dfrac{-3 \pm \sqrt{709}}{10}$

4. $\qquad\qquad 2x^2 = 13x + 5$

$2x^2 - 13x - 5 = 0$

$a = 2, b = -13, c = -5$

$x = \dfrac{-(-13) \pm \sqrt{(-13)^2 - 4(2)(-5)}}{2(2)}$

$= \dfrac{13 \pm \sqrt{169 + 40}}{4} = \dfrac{13 \pm \sqrt{209}}{4}$

$x = \dfrac{13 + \sqrt{209}}{4}$ $\qquad$ $x = \dfrac{13 - \sqrt{209}}{4}$

$x \approx 6.864$ $\quad$ and $\quad x \approx -0.364$

5. $\qquad\qquad 5x^2 + 2x = -3$

$5x^2 + 2x + 3 = 0$

$a = 5, b = 2, c = 3$

$x = \dfrac{-2 \pm \sqrt{(2)^2 - 4(5)(3)}}{2(5)}$

$= \dfrac{-2 \pm \sqrt{4 - 60}}{10} = \dfrac{-2 \pm \sqrt{-56}}{10}$

There is no real number that is $\sqrt{-56}$.
There is no solution to this problem.

6. (a) $\qquad\qquad 5x^2 = 3x + 2$

$5x^2 - 3x - 2 = 0$

$b^2 - 4ac = (-3)^2 - 4(5)(-2)$

$= 9 + 40 = 49$

Because $49 > 0$ there will be two real roots that are rational numbers.

(b) $\qquad\qquad 2x^2 + 5 = 4x$

$2x^2 - 4x + 5 = 0$

$a = 2, b = -4, c = 5$

$b^2 - 4ac = (-4)^2 - 4(2)(5)$

$= 16 - 40$

$= -24$

Because $-24 < 0$, the equation $2x^2 + 5 = 4x$ has no real number solutions(s).

9.4 Practice Problems

1. (a) $y = 2x^2$

x	y
-2	8
-1	2
0	0
1	2
2	8

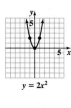

$y = 2x^2$

(b) $y = -2x^2$

x	y
-2	-8
-1	-2
0	0
1	-2
2	-8

$y = -2x^2$

2. (a) $y = x^2 + 2x$

x	y
−4	8
−3	3
−2	0
−1	−1
0	0
1	3
2	8

vertex (at −1, −1)

(b) $y = -x^2 - 2x$

x	y
−4	−8
−3	−3
−2	0
−1	1
0	0
1	−3
2	−8

vertex (at −1, 1)

3. $y = x^2 - 4x - 5$

$$x = \frac{-b}{2a} = \frac{-(-4)}{2(1)} = \frac{4}{2} = 2$$

If $x = 2$, then
$y = (2)^2 - 4(2) - 5 = -9$
The point $(2, -9)$ is the vertex.
x-intercepts:
$$x^2 - 4x - 5 = 0$$
$$(x + 1)(x - 5) = 0$$
$$x + 1 = 0 \qquad x - 5 = 0$$
$$x = -1 \qquad x = 5$$
The x-intercepts are the points $(-1, 0)$ and $(5, 0)$.

4. $h = -16t^2 + 32t + 10$

(a) Since the equation is a quadratic equation, the graph is a parabola. a is negative, therefore the parabola opens downward and the vertex is the highest point.

$$t = \frac{-b}{2a} = \frac{-32}{2(-16)} = \frac{-32}{-32} = 1$$

If $t = 1$, then $h = -16(1)^2 + 32(1) + 10$
$$h = -16 + 32 + 10$$
$$h = 26$$

The vertex (t, h) is $(1, 26)$.
Now find the h-intercept, the point where $t = 0$.
$$h = -16t^2 + 32t + 10$$
$$h = 10$$

The h-intercept is at $(0, 10)$.

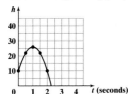

Table Values

t	h
0	10
0.5	22
1	26
1.5	22
2	10

(b) The ball is 26 feet above the ground one second after it is thrown. This is the greatest height of the ball.

(c) The ball hits the ground when $h = 0$.

$$0 = -16t^2 + 32t + 10$$
$$t = \frac{-32 \pm \sqrt{(32)^2 - 4(-16)(10)}}{2(-16)}$$
$$t = \frac{-32 \pm \sqrt{1024 + 640}}{-32}$$
$$t = \frac{-32 \pm \sqrt{1664}}{-32}$$
$$t = \frac{4 \pm \sqrt{26}}{4}$$
$$t \approx \frac{4 - 5.099}{4} \quad \text{and} \quad t \approx \frac{4 + 5.099}{4}$$
$$t \approx -0.275 \qquad t \approx 2.275$$

In this situation it is not useful to find points for negative values of t. Therefore $t \approx 2.3$ seconds. The ball will hit the ground approximately 2.3 seconds after it is thrown.

9.5 Practice Problems

1. $a^2 + b^2 = c^2$ Let $a = x$, then $b = x - 6$.
$$x^2 + (x - 6)^2 = 30^2$$
$$x^2 + x^2 - 12x + 36 = 900$$
$$2x^2 - 12x - 864 = 0$$
$$x^2 - 6x - 432 = 0$$
$$(x + 18)(x - 24) = 0$$
$$x + 18 = 0 \qquad x - 24 = 0$$
$$x = -18 \qquad x = 24$$

One leg is 24 meters, the other leg is $x - 6$ or $24 - 6 = 18$ meters.

2. Let c = cost for each person in the original group
s = number of people

Number of people × cost per person = total cost
$$s \cdot c = 240$$

If 2 people cannot go, then the number of people drops by 2 but the cost for each increases by \$4.
$$(s - 2)(c + 4) = 240$$

If $s \cdot c = 240$, then
$$c = \frac{240}{s}$$
$$(s - 2)\left(\frac{240}{s} + 4\right) = 240$$
$$(s - 2)\left(\frac{240 + 4s}{s}\right) = 240$$
$$(s - 2)(240 + 4s) = 240s$$
$$4s^2 + 232s - 480 = 240s$$
$$4s^2 - 8s - 480 = 0$$
$$s^2 - 2s - 120 = 0$$
$$(s - 12)(s + 10) = 0$$
$$s = 12 \qquad s = -10$$

The number of people originally going on the trip was 12.

Check. If 12 people originally planned the trip then
$$12 \times c = 240.$$
$$c = \$20 \text{ per person}$$

If 2 people dropped out, does this mean they must increase the cost of the trip by \$4?
$$10 \times c = 240$$
$$c = \$24.00 \text{ Yes.}$$

3. Let x = width

y = length

$160 = 2x + y$

The area of a rectangle is (width)(length).

$A = xy$ Substitute $y = 160 - 2x$.

$3150 = x(160 - 2x)$

$3150 = 160x - 2x^2$

$2x^2 - 160x + 3150 = 0$

$x^2 - 80x + 1575 = 0$

$(x - 45)(x - 35) = 0$

$x = 45$ and $x = 35$

First solution:

If the width is 45 feet then the length is

$y = 160 - 2(45)$

$= 160 - 90$

$= 70$ feet

Check. $45 + 45 + 70 = 160$ ✓

$(45)(70) = 3150$ ✓

Second solution:

If the width is 35 feet then the length is

$y = 160 - 2(35) = 160 - 70 = 90$ feet

Check. $35 + 35 + 90 = 160$ ✓

$(35)(90) = 3150$ ✓

Appendix B
Practice Problems

1.

Mathematics Blueprint for Problem Solving			
Gather the Facts	**What Am I Solving for?**	**What Must I Calculate?**	**Key Points to Remember**
Living room measures $16\frac{1}{2}$ ft × $10\frac{1}{2}$ ft. The carpet costs $20.00 per square yard.	Area of room in square feet. Area of room in square yards. Cost of the carpet.	Multiply $16\frac{1}{2}$ ft by $10\frac{1}{2}$ ft to get the area in square feet. Divide the number of square feet by 9 to get the number of square yards. Multiply the number of square yards by $20.00.	9 sq feet = 1 sq yard

$16\frac{1}{2} \times 10\frac{1}{2} = 173\frac{1}{4}$ square feet

$173\frac{1}{4} \div 9 = 19\frac{1}{4}$ square yards

$19\frac{1}{4} \times 20 = \$385.00$ total cost of carpet

Check:

Estimate area of room $16 \times 10 = 160$ square feet

Estimate area in square yards $160 \div 10 = 16$

Estimate the cost $16 \times 20 = \$320.00$

This is close to our answer of $385.00. Our answer seems reasonable.

2. (a) $\dfrac{10}{55} \approx 0.181818 \approx 18\%$

(b) $\dfrac{3,660,000}{13,240,000} = \dfrac{366}{1324} \approx 0.276 \approx 28\%$

(c) $\dfrac{15}{55} \approx 0.272727 \approx 27\%$

(d) $\dfrac{3,720,000}{13,240,000} = \dfrac{372}{1324} \approx 0.281 \approx 28\%$

(e) We notice that 18% of the company's sales force is located in the Northwest, and they were responsible for 28% of the sales volume. The percent of sales compared to the percent of sales force is about 150%. 27% of the company's sales force is located in the Southwest, and they were responsible for 28% of the sales volume. The percent of sales compared to the percent of sales force is approximately 100%. It would appear that the Northwest sales force is more effective.

Appendix D
Practice Problems

1. $y - y_1 = m(x - x_1)$

$y - (-2) = \dfrac{3}{4}(x - 5)$

$y + 2 = \dfrac{3}{4}x - \dfrac{15}{4}$

$4y + 4(2) = 4\left(\dfrac{3}{4}x\right) - 4\left(\dfrac{15}{4}\right)$

$4y + 8 = 3x - 15$

$-3x + 4y = -15 - 8$

$3x - 4y = 23$

2. $(-4, 1)$ and $(-2, -3)$

$m = \dfrac{y_2 - y_1}{x_2 - x_1} = \dfrac{-3 - 1}{-2 - (-4)} = \dfrac{-4}{-2 + 4} = \dfrac{-4}{2} = -2$

Substitute $m = -2$ and $(x_1, y_1) = (-4, 1)$ into the point–slope equation.

$y - y_1 = m(x - x_1)$

$y - 1 = -2[x - (-4)]$

$y - 1 = -2(x + 4)$

$y - 1 = -2x - 8$

$y = -2x - 7$

3. First, we need to find the slope of the line $5x - 3y = 10$. We do this by writing the equation in slope–intercept form

$$5x - 3y = 10$$
$$-3y = -5x + 10$$
$$y = \frac{5}{3}x - \frac{10}{3}$$

The slope is $\frac{5}{3}$. A line parallel to this passing through $(4, -5)$ would have an equation

$$y - (-5) = \frac{5}{3}(x - 4)$$
$$y + 5 = \frac{5}{3}x - \frac{20}{3}$$
$$3y + 3(5) = 3\left(\frac{5}{3}x\right) - 3\left(\frac{20}{3}\right)$$
$$3y + 15 = 5x - 20$$
$$-5x + 3y = -35$$
$$5x - 3y = 35$$

4. Find the slope of the line $6x + 3y = 7$ by rewriting it in slope–intercept form.

$$6x + 3y = 7$$
$$3y = -6x + 7$$
$$y = -2x + \frac{7}{3}$$

The slope is -2. A line perpendicular to this passing through $(-4, 3)$ would have a slope of $\frac{1}{2}$, and would have the equation

$$y - 3 = \frac{1}{2}[x - (-4)]$$
$$y - 3 = \frac{1}{2}(x + 4)$$
$$y - 3 = \frac{1}{2}x + 2$$
$$2y - 2(3) = 2\left(\frac{1}{2}x\right) + 2(2)$$
$$2y - 6 = x + 4$$
$$-x + 2y = 10$$
$$x - 2y = -10$$

Answers to Selected Exercises

Chapter 0 0.1 Exercises 1. 12   3. When two or more numbers are multiplied, each number that is multiplied is called a factor.

In 2×3, 2 and 3 are factors.   5. 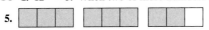   7. $\frac{3}{4}$ 9. $\frac{1}{3}$ 11. 5   13. $\frac{2}{3}$ 15. $\frac{6}{17}$ 17. $\frac{7}{9}$

19. $2\frac{5}{6}$ 21. $9\frac{2}{5}$ 23. $5\frac{3}{7}$ 25. $20\frac{1}{2}$ 27. $6\frac{2}{5}$ 29. $12\frac{1}{3}$ 31. $\frac{16}{5}$ 33. $\frac{33}{5}$ 35. $\frac{11}{9}$ 37. $\frac{59}{7}$ 39. $\frac{97}{4}$ 41. $\frac{47}{3}$ 43. 24

45. 21   47. 12   49. 21   51. 15   53. 70   55. $15\frac{3}{4}$ 57. $\frac{33}{160}$ 59. $\frac{1}{4}$ 61. $\frac{1}{2}$ 63. $\frac{4}{9}$ 65. $\frac{13}{16}$

Quick Quiz 0.1 1. $\frac{21}{23}$ 2. $\frac{75}{11}$ 3. $4\frac{19}{21}$ 4. See Student Solutions Manual

0.2 Exercises 1. Answers may vary. A sample answer is: 8 is exactly divisible by 4.   3. 80   5. 20   7. 54   9. 105   11. 120

13. 120   15. 546   17. 90   19. $\frac{5}{8}$ 21. $\frac{2}{7}$ 23. $\frac{29}{24}$ or $1\frac{5}{24}$ 25. $\frac{31}{63}$ 27. $\frac{11}{15}$ 29. $\frac{35}{36}$ 31. $\frac{2}{45}$ 33. $\frac{1}{2}$ 35. $\frac{53}{56}$ 37. $\frac{3}{2}$ or $1\frac{1}{2}$

39. $\frac{11}{30}$ 41. $\frac{1}{4}$ 43. $\frac{1}{2}$ 45. $7\frac{11}{15}$ 47. $1\frac{35}{72}$ 49. $4\frac{11}{12}$ 51. $6\frac{13}{28}$ 53. $5\frac{19}{24}$ 55. $4\frac{3}{7}$ 57. 9   59. $\frac{23}{24}$ 61. $7\frac{9}{16}$ 63. $\frac{10}{21}$

65. $2\frac{7}{10}$ 67. $19\frac{11}{21}$ 69. $1\frac{13}{24}$ 71. $12\frac{1}{12}$ 73. $30\frac{7}{9}$ 75. $24\frac{2}{3}$ mi   77. $4\frac{1}{12}$ hr   79. $A = 12$ in.; $B = 15\frac{7}{8}$ in.   81. $1\frac{5}{8}$ in.

83. $\frac{9}{11}$ 84. $\frac{133}{5}$

Quick Quiz 0.2 1. $\frac{5}{3}$ or $1\frac{2}{3}$ 2. $7\frac{8}{15}$ 3. $2\frac{5}{18}$ 4. See Student Solutions Manual

0.3 Exercises 1. First, change each number to an improper fraction. Look for a common factor in the numerator and denominator to divide by, and, if one is found, perform the division. Multiply the numerators. Multiply the denominators.   3. $\frac{20}{7}$ or $2\frac{6}{7}$ 5. $\frac{17}{30}$ 7. $\frac{6}{25}$ 9. $\frac{12}{5}$ or $2\frac{2}{5}$

11. $\frac{1}{6}$ 13. $\frac{6}{7}$ 15. $\frac{18}{5}$ or $3\frac{3}{5}$ 17. $\frac{3}{5}$ 19. $\frac{1}{7}$ 21. 14   23. $\frac{8}{7}$ or $1\frac{1}{7}$ 25. $\frac{7}{27}$ 27. $\frac{7}{6}$ or $1\frac{1}{6}$ 29. $\frac{15}{14}$ or $1\frac{1}{14}$ 31. $\frac{8}{35}$

33. $1\frac{1}{3}$ 35. $8\frac{2}{3}$ 37. $6\frac{1}{4}$ 39. $\frac{8}{15}$ 41. 6   43. $\frac{5}{4}$ or $1\frac{1}{4}$ 45. $54\frac{3}{4}$ 47. 40   49 28   51. $\frac{3}{16}$ 53. (a) $\frac{5}{63}$ (b) $\frac{7}{125}$

55. (a) $\frac{7}{6}$ (b) $\frac{8}{21}$ 57. 26 shirts   59. 136 square miles   61. (a) $3\frac{1}{7}$ (b) $3\frac{3}{5}$ 62. (a) $\frac{37}{3}$ (b) $\frac{79}{8}$

Quick Quiz 0.3 1. $\frac{5}{6}$ 2. $14\frac{5}{8}$ 3. $1\frac{8}{25}$ 4. See Student Solutions Manual

How Am I Doing? Sections 0.1–0.3 1. $\frac{3}{11}$ (obj. 0.1.2)   2. $\frac{2}{5}$ (obj. 0.1.2)   3. $3\frac{3}{4}$ (obj. 0.1.3)   4. $\frac{33}{7}$ (obj. 0.1.3)   5. 24 (obj. 0.1.4)

6. 45 (obj. 0.1.4)   7. LCD $= 120$ (obj. 0.2.2)   8. $\frac{5}{7}$ (obj. 0.2.1)   9. $\frac{19}{42}$ (obj. 0.2.3)   10. $8\frac{5}{12}$ (obj. 0.2.4)   11. $\frac{7}{18}$ (obj. 0.2.3)

12. $\frac{4}{21}$ (obj. 0.2.3)   13. $1\frac{33}{40}$ (obj. 0.2.4)   14. $\frac{10}{9}$ or $1\frac{1}{9}$ (obj. 0.3.1)   15. 42 (obj. 0.3.1)   16. $\frac{7}{2}$ or $3\frac{1}{2}$ (obj. 0.3.2)   17. $\frac{28}{39}$ (obj. 0.3.2)

18. $\frac{4}{15}$ (obj. 0.3.2)   19. 20 mi^2 (obj. 0.3.1)

0.4 Exercises 1. 10, 100, 1000, 10,000, and so on   3. 3, left   5. 0.625   7. 0.2   9. $0.\overline{63}$ 11. $\frac{4}{5}$ 13. $\frac{1}{4}$ 15. $\frac{5}{8}$ 17. $\frac{3}{50}$

19. $\frac{17}{5}$ or $3\frac{2}{5}$ 21. $\frac{11}{2}$ or $5\frac{1}{2}$ 23. 2.09   25. 10.82   27. 261.208   29. 131.79   31. 51.443   33. 122.63   35. 30.282   37. 0.0032

39. 0.10575   41. 87.3   43. 0.0565   45. 2.64   47. 261.5   49. 0.508   51. 3450   53. 0.0076   55. 73,600   57. 0.73892

59. 14.98   61. 855,400   63. 8.22   65. 16.378   67. 2.12   69. 768.3   71. 23.65   73. 1.537   75. 2.6026 L   77. 21 shifts; $4

79. $\frac{2}{3}$ 80. $\frac{1}{6}$ 81. $\frac{93}{100}$ 82. $\frac{11}{10}$ or $1\frac{1}{10}$

Quick Quiz 0.4 1. 5.7078   2. 3.522   3. 28.8   4. See Student Solutions Manual

0.5 Exercises 1. Answers may vary. Sample answers follow. 19% means 19 out of 100 parts. Percent means per 100. 19% is really a fraction with a denominator of 100. In this case it would be $\frac{19}{100}$.   3. 28%   5. 56.8%   7. 7.6%   9. 239%   11. 0.03   13. 0.004   15. 2.5

17. 0.074   19. 5.2   21. 13   23. 72.8   25. 150%   27. 5%   29. 250%   31. 85%   33. $4.92 tip; $37.72 total bill   35. 21%

37. 540 gifts   39. (a) $29,640   (b) $35,040   41. 210,000   43. 18,000,000   45. 240   47. 20,000   49. 0.1   51. $4000.00

53. $6400   55. 25 mi/gal   57. $250,000   59. $15,000   61. 10%   63. $538.13   64. $6220.50   65. 22 mi/gal   66. 4.0 in.

Quick Quiz 0.5 1. 15%   2. 96.9   3. 20   4. See Student Solutions Manual

Chapter 0 Putting Your Skills to Work 1. $100, $300, $2000, $8000, $8000, $8000, $12,000   **2.** $365   **3.** 7 months

Chapter 0 Review Problems 1. $\frac{3}{4}$ **2.** $\frac{3}{10}$ **3.** $\frac{18}{41}$ **4.** $\frac{4}{7}$ **5.** $\frac{23}{5}$ **6.** $6\frac{4}{5}$ **7.** $6\frac{1}{2}$ **8.** 15   **9.** 5   **10.** 40   **11.** 22

12. $\frac{17}{20}$ **13.** $\frac{29}{24}$ or $1\frac{5}{24}$ **14.** $\frac{4}{15}$ **15.** $\frac{13}{30}$ **16.** $5\frac{23}{30}$ **17.** $3\frac{19}{20}$ **18.** $2\frac{29}{36}$ **19.** $1\frac{11}{12}$ **20.** $\frac{30}{11}$ or $2\frac{8}{11}$ **21.** $10\frac{1}{2}$ **22.** $2\frac{3}{8}$

23. $\frac{20}{7}$ or $2\frac{6}{7}$ **24.** $\frac{1}{16}$ **25.** $\frac{24}{5}$ or $4\frac{4}{5}$ **26.** $\frac{3}{20}$ **27.** 6   **28.** 7.201   **29.** 7.737   **30.** 29.561   **31.** 4.436   **32.** 0.03745

33. 362,341   **34.** 0.07956   **35.** 10.368   **36.** 0.00186   **37.** 0.07132   **38.** 1.3075   **39.** 90   **40.** 1.82   **41.** 0.5   **42.** 0.375

43. $\frac{9}{25}$ **44.** 0.014   **45.** 0.361   **46.** 0.0002   **47.** 1.253   **48.** 0.25%   **49.** 32.5%   **50.** 90%   **51.** 10%   **52.** 510

53. 3.96   **54.** 64%   **55.** 12.5%   **56.** 60%   **57.** 17,839,233 people   **58.** 75%   **59.** 400,000,000,000   **60.** 2500   **61.** 300,000

62. 19   **63.** $12,000   **64.** 30   **65.** $320   **66.** $300

How Am I Doing? Chapter 0 Test 1. $\frac{8}{9}$ (obj. 0.1.2)   **2.** $\frac{4}{3}$ (obj. 0.1.2)   **3.** $\frac{45}{7}$ (obj. 0.1.3)   **4.** $11\frac{2}{3}$ (obj. 0.1.3)

5. $\frac{15}{8}$ or $1\frac{7}{8}$ (obj. 0.2.3)   **6.** $4\frac{7}{8}$ (obj. 0.2.4)   **7.** $\frac{5}{6}$ (obj. 0.2.4)   **8.** $\frac{4}{3}$ or $1\frac{1}{3}$ (obj. 0.3.1)   **9.** $\frac{5}{24}$ (obj. 0.3.1)   **10.** $1\frac{21}{22}$ (obj. 0.3.2)

11. $8\frac{1}{8}$ (obj. 0.3.1)   **12.** $\frac{7}{2}$ or $3\frac{1}{2}$ (obj. 0.3.2)   **13.** 14.64 (obj. 0.4.4)   **14.** 3.9897 (obj. 0.4.4)   **15.** 1.312 (obj. 0.4.5)   **16.** 73.85 (obj. 0.4.7)

17. 230 (obj. 0.4.6)   **18.** 263.259 (obj. 0.4.7)   **19.** 7.3% (obj. 0.5.2)   **20.** 1.965 (obj. 0.5.3)   **21.** 6.3 (obj. 0.5.4)   **22.** 0.336 (obj. 0.5.4)

23. 6% (obj. 0.5.5)   **24.** 30% (obj. 0.5.5)   **25.** 18 computer chips (obj. 0.3.2)   **26.** 100 (obj. 0.6.1)   **27.** 700 (obj. 0.6.1)

Chapter 1

1.1 Exercises 1. Integer, rational number, real number   **3.** Irrational number, real number   **5.** Rational number, real number   **7.** Rational number, real number   **9.** Irrational number, real number   **11.** −20,000   **13.** $-37\frac{1}{2}$ **15.** +7   **17.** −8

19. 2.73   **21.** 1.3   **23.** $\frac{5}{6}$ **25.** −11   **27.** −50   **29.** $\frac{3}{10}$ **31.** $-\frac{7}{13}$ **33.** $\frac{1}{35}$ **35.** −19.2   **37.** 0.4   **39.** −14.16   **41.** −6

43. 0   **45.** $\frac{9}{20}$ **47.** −8   **49.** −3   **51.** 59   **53.** $\frac{7}{18}$ **55.** $\frac{2}{5}$ **57.** −2.21   **59.** 12   **61.** 0   **63.** 15.94   **65.** $167 profit

67. −$3800   **69.** 9-yd gain   **71.** 3500   **73.** $32,000,000   **75.** 18   **77.** $\frac{19}{16}$ or $1\frac{3}{16}$ **78.** $\frac{2}{3}$ **79.** $\frac{1}{12}$ **80.** $\frac{25}{34}$ **81.** 1.52

82. 0.65   **83.** 1.141   **84.** 0.26

Quick Quiz 1.1 1. −34   **2.** 0.5   **3.** $-\frac{13}{24}$ **4.** See Student Solutions Manual

1.2 Exercises 1. First change subtracting −3 to adding a positive three. Then use the rules for addition of two real numbers with different signs. Thus, $-8 - (-3) = -8 + 3 = -5$.   **3.** −17   **5.** −4   **7.** −11   **9.** 8   **11.** 5   **13.** 0   **15.** 3   **17.** $-\frac{2}{5}$ **19.** $\frac{27}{20}$ or $1\frac{7}{20}$

21. $-\frac{19}{12}$ or $-1\frac{7}{12}$ **23.** −0.9   **25.** 4.47   **27.** $-\frac{17}{5}$ or $-3\frac{2}{5}$ **29.** $-\frac{14}{3}$ or $-4\frac{2}{3}$ **31.** −53   **33.** −73   **35.** 7.1   **37.** $\frac{35}{4}$ or $8\frac{3}{4}$

39. $-\frac{37}{6}$ or $-6\frac{1}{6}$ **41.** $-\frac{38}{35}$ or $-1\frac{3}{35}$ **43.** −8.5   **45.** $-\frac{29}{5}$ or $-5\frac{4}{5}$ **47.** 6.06   **49.** −5.047   **51.** 7   **53.** −48   **55.** −2   **57.** 0

59. −62   **61.** 1.6   **63.** $149   **65.** −21   **66.** −51   **67.** −19   **68.** 15°F   **69.** $6\frac{2}{3}$ mi

Quick Quiz 1.2 1. 7   **2.** −1.9   **3.** $\frac{51}{56}$ **4.** See Student Solutions Manual

1.3 Exercises 1. To multiply two real numbers, multiply the absolute values. The sign of the result is positive if both numbers have the same sign, but negative if the two numbers have opposite signs.   **3.** −20   **5.** 0   **7.** 49   **9.** 0.264   **11.** −4.5   **13.** $-\frac{3}{2}$ or $-1\frac{1}{2}$ **15.** $\frac{9}{11}$

17. $-\frac{5}{26}$ **19.** 4   **21.** 6   **23.** 15   **25.** −12   **27.** −130   **29.** −0.6   **31.** −0.9   **33.** $-\frac{3}{10}$ **35.** $\frac{20}{3}$ or $6\frac{2}{3}$ **37.** $\frac{7}{10}$ **39.** 14

41. $-\frac{5}{4}$ or $-1\frac{1}{4}$ **43.** $-\frac{2}{3}$ **45.** −24   **47.** −72   **49.** −16   **51.** −0.3   **53.** $-\frac{8}{35}$ **55.** $\frac{2}{27}$ **57.** 9   **59.** 4   **61.** 17   **63.** −72

65. −1   **67.** He gave $4.40 to each person and to himself.   **69.** $235.60   **71.** Approximately 20 yd   **73.** Approximately 70 yd

75. A gain of 20 yd   **76.** −6.69   **77.** $-\frac{11}{6}$ or $-1\frac{5}{6}$ **78.** −15   **79.** −88

Quick Quiz 1.3 1. $-\frac{15}{8}$ or $-1\frac{7}{8}$ **2.** −120   **3.** 4   **4.** See Student Solutions Manual

1.4 Exercises **1.** The base is 4 and the exponent is 4. Thus you multiply $(4)(4)(4)(4) = 256$.   **3.** The answer is negative. When you raise a negative number to an odd power the result is always negative.   **5.** If you have parentheses surrounding the -2, then the base is -2 and the exponent is 4. The result is 16. If you do not have parentheses, then the base is 2. You evaluate to obtain 16 and then take the opposite of 16, which is -16. Thus $(-2)^4 = 16$ but $-2^4 = -16$.   **7.** 5^7 **9.** w^2 **11.** p^4 **13.** $(3q)^3$ or 3^3q^3 **15.** 27   **17.** 81   **19.** 216   **21.** -27 **23.** 16
25. -25 **27.** $\dfrac{1}{16}$ **29.** $\dfrac{8}{125}$ **31.** 4.41   **33.** 0.0016   **35.** 256   **37.** -256 **39.** 161   **41.** -21 **43.** -128 **45.** 23   **47.** -576
48. -19 **49.** $-\dfrac{5}{3}$ or $-1\dfrac{2}{3}$ **50.** -8 **51.** 2.52   **52.** \$1696

Quick Quiz 1.4 **1.** 256   **2.** 3.24   **3.** $\dfrac{27}{64}$ **4.** See Student Solutions Manual

1.5 Exercises **1.** $3(4) + 6(5)$ **3.** **(a)** 90 **(b)** 42   **5.** 12   **7.** 5   **9.** -29 **11.** 24   **13.** 21   **15.** 13   **17.** -6 **19.** 42
21. $\dfrac{9}{4}$ or $2\dfrac{1}{4}$ **23.** 0.848   **25.** $-\dfrac{7}{16}$ **27.** 5   **29.** $-\dfrac{23}{2}$ or $-11\dfrac{1}{2}$ **31.** 7.56   **33.** $\dfrac{1}{4}$ **35.** $2(-2) + 4(-1) + 10(0) + 2(+1)$
37. 2 above par   **39.** 0.125   **40.** $-\dfrac{19}{12}$ or $-1\dfrac{7}{12}$ **41.** -1 **42.** $\dfrac{72}{125}$

Quick Quiz 1.5 **1.** -77 **2.** 16.96   **3.** 27   **4.** See Student Solutions Manual

How Am I Doing? Sections 1.1–1.5 **1.** -9 (obj. 1.1.4)   **2.** $-\dfrac{41}{24}$ or $-1\dfrac{17}{24}$ (obj. 1.1.3)   **3.** 1.24 (obj. 1.1.3)   **4.** 4.8 (obj. 1.1.4)
5. 11 (obj. 1.2.1)   **6.** $-\dfrac{29}{30}$ (obj. 1.2.1)   **7.** 12.3 (obj. 1.2.1)   **8.** 10 (obj. 1.2.1)   **9.** -96 (obj. 1.3.1)   **10.** $\dfrac{10}{11}$ (obj. 1.3.1)
11. -0.9 (obj. 1.3.3)   **12.** $-\dfrac{10}{17}$ (obj. 1.3.3)   **13.** 0.343 (obj. 1.4.2)   **14.** 256 (obj. 1.4.2)   **15.** -256 (obj. 1.4.2)   **16.** $\dfrac{8}{27}$ (obj. 1.4.2)
17. 54 (obj. 1.4.2)   **18.** 24 (obj. 1.5.1)   **19.** 10 (obj. 1.5.1)   **20.** 11 (obj. 1.5.1)   **21.** -3.3 (obj. 1.5.1)   **22.** $-\dfrac{9}{20}$ (obj. 1.5.1)

1.6 Exercises **1.** variable   **3.** Here we are multiplying 4 by x by x. Since we know from the definition of exponents that x multiplied by x is x^2, this gives us an answer of $4x^2$.
5. Yes, $a(b - c)$ can be written as $a[b + (-c)]$.
$$3(10 - 2) = (3 \times 10) - (3 \times 2)$$
$$3 \times 8 = 30 - 6$$
$$24 = 24$$
7. $3x - 6y$ **9.** $-8a + 6b$ **11.** $9x + 3y$ **13.** $-8m - 24n$ **15.** $-x + 3y$ **17.** $-81x + 45y - 72$ **19.** $-10x + 2y - 12$
21. $10x^2 - 20x + 15$ **23.** $\dfrac{x^2}{5} + 2xy - \dfrac{4x}{5}$ **25.** $5x^2 + 10xy + 5xz$ **27.** $13.5x - 15$ **29.** $18x^2 + 3xy - 3x$ **31.** $-3x^2y - 2xy^2 + xy$
33. $-5a^2b - 10ab^2 + 20ab$ **35.** $2a^2 - 4a + \dfrac{8}{3}$ **37.** $0.36x^3 + 0.09x^2 - 0.15x$ **39.** $-1.32q^3 - 0.28qr - 4q$
41. $800(5x + 14y) = 4000x + 11{,}200y$ ft^2 **43.** $4x(3000 - 2y) = 12{,}000x - 8xy$ ft^2 **44.** -16 **45.** 64   **46.** 14   **47.** 4   **48.** 10

Quick Quiz 1.6 **1.** $-15a - 35b$ **2.** $-2x^2 + 8xy - 16x$ **3.** $-12a^2b + 15ab^2 + 27ab$ **4.** See Student Solutions Manual

1.7 Exercises **1.** A term is a number, a variable, or a product of numbers and variables.   **3.** The two terms $5x$ and $-8x$ are like terms because they both have the variable x with the exponent of one.   **5.** The only like terms are $7xy$ and $-14xy$ because the other two have different exponents even though they have the same variables.   **7.** $-25b^2$ **9.** $6a^3 - 7a^2$ **11.** $-5x - 5y$ **13.** $7.1x - 3.5y$ **15.** $-2x - 8.7y$
17. $5p + q - 18$ **19.** $5bc - 6ac$ **21.** $x^2 - 10x + 3$ **23.** $-10y^2 - 16y + 12$ **25.** $-\dfrac{1}{15}x - \dfrac{2}{21}y$ **27.** $\dfrac{11}{20}a^2 - \dfrac{5}{6}b$ **29.** $-2rs + 2r$
31. $28a - 20b$ **33.** $-27ab - 11b^2$ **35.** $-2c - 10d^2$ **37.** $32x + 23$ **39.** $7a + 9b$ **41.** $14x - 8$ ft   **43.** $-\dfrac{13}{12}$ or $-1\dfrac{1}{12}$ **44.** $-\dfrac{3}{8}$
45. $\dfrac{23}{50}$ **46.** $-\dfrac{15}{98}$

Quick Quiz 1.7 **1.** $\dfrac{13}{6}xy + \dfrac{5}{3}x^2y$ **2.** $0.6a^2b - 4.4ab^2$ **3.** $20x - 2y$ **4.** See Student Solutions Manual

1.8 Exercises **1.** -5 **3.** -12 **5.** $\dfrac{25}{2}$ or $12\dfrac{1}{2}$ **7.** -26 **9.** -1.3 **11.** $\dfrac{25}{4}$ or $6\dfrac{1}{4}$ **13.** 10   **15.** 5   **17.** -24 **19.** -20
21. 9   **23.** 39   **25.** -2 **27.** 44   **29.** -2 **31.** -9 **33.** 29   **35.** 32   **37.** 32   **39.** $-\dfrac{1}{2}$ **41.** 352 ft^2 **43.** 1.24 cm^2
45. 32 in.^{2} **47.** 56,000 ft^2 **49.** 28.26 ft^2 **51.** $-78.5°$C   **53.** \$2340.00   **55.** 122°F; $-67°$F   **57.** 16   **58.** $-x^2 + 2x - 4y$

Quick Quiz 1.8 **1.** 2   **2.** $\dfrac{9}{2}$ or $4\dfrac{1}{2}$ **3.** 33   **4.** See Student Solutions Manual

1.9 Exercises **1.** $-(3x + 2y)$ **3.** distributive   **5.** $3x + 6y$ **7.** $2c - 16d$ **9.** $x - 7y$ **11.** $8x^2 - 4x + 12$ **13.** $-2x + 26y$
15. $2x + 11y + 10$ **17.** $15a - 60ab$ **19.** $12a^2 - 21a - 18$ **21.** $3a^2 + 16b + 12b^2$ **23.** $-13a + 9b - 1$ **25.** $9b^2 + 36b - 12$
27. $12a^2 - 8b$ **29.** $1947.52°F$ **30.** $453,416 \text{ ft}^2$ **31.** 54 to 67.5 kg   **32.** 4.05 to 6.3 kg

Quick Quiz 1.9 **1.** $-14x - 4y$ **2.** $-6x + 15y - 36$ **3.** $-8a - 24ab + 8b$ **4.** See Student Solutions Manual

Putting Your Skills to Work **1.** $1500   **2.** $3450   **3.** $6900   **4.** About 55 months, or 4 years and 7 months
5. About $288   **6.** $250; about 28 months, or 2 years and 4 months   **7.** 2 months   **8.** $25,500   **9.** Answers may vary
10. Answers may vary   **11.** Answers may vary

Chapter 1 Review Problems **1.** -8 **2.** -4.2 **3.** -9 **4.** 1.9 **5.** $-\dfrac{1}{3}$ **6.** $-\dfrac{7}{22}$ **7.** $\dfrac{1}{6}$ **8.** $-\dfrac{1}{2}$ **9.** 8 **10.** 13

11. -33 **12.** 9.2 **13.** $-\dfrac{13}{8}$ or $-1\dfrac{5}{8}$ **14.** $\dfrac{11}{24}$ **15.** -22.7 **16.** -88 **17.** -3 **18.** 13 **19.** 32 **20.** $\dfrac{5}{6}$ **21.** $-\dfrac{25}{7}$ or $-3\dfrac{4}{7}$

22. -72 **23.** 60 **24.** -30 **25.** -4 **26.** 16 **27.** -29 **28.** 1 **29.** $-\dfrac{1}{2}$ **30.** $-\dfrac{4}{7}$ **31.** -30 **32.** -5 **33.** -9.1

34. 0.9 **35.** 10.1 **36.** -1.2 **37.** 1.9 **38.** -1.3 **39.** 24 yd   **40.** $-22°F$ **41.** $14,776$ ft   **42.** Stock B   **43.** -243 **44.** 64

45. 625 **46.** $-\dfrac{8}{27}$ **47.** -81 **48.** 0.36 **49.** $\dfrac{25}{36}$ **50.** $\dfrac{27}{64}$ **51.** -44 **52.** 30 **53.** 3 **54.** $-21x + 7y$ **55.** $18x - 3x^2 + 9xy$

56. $-7x^2 + 3x - 11$ **57.** $-6xy^3 - 3xy^2 + 3y^3$ **58.** $-5a^2b + 3bc$ **59.** $-3x - 4y$ **60.** $-5x^2 - 35x - 9$ **61.** $10x^2 - 8x - \dfrac{1}{2}$

62. -55 **63.** 1 **64.** -4 **65.** -3 **66.** 0 **67.** 17 **68.** 15 **69.** $810   **70.** $68°F$ to $77°F$ **71.** $75.36   **72.** $8580.00
73. $100,000 \text{ ft}^2$; $200,000   **74.** 10.45 ft^2; $689.70   **75.** $-2x + 42$ **76.** $-17x - 18$ **77.** $-2 + 10x$ **78.** $-12x^2 + 63x$
79. $5xy^3 - 6x^3y - 13x^2y^2 - 6x^2y$ **80.** $x - 10y + 35 - 15xy$ **81.** $-31a + 2b$ **82.** $15a - 15b - 10ab$ **83.** $-3x - 9xy + 18y^2$

84. $10x + 8xy - 32y$ **85.** -2.3 **86.** 8 **87.** $-\dfrac{22}{15}$ or $-1\dfrac{7}{15}$ **88.** $-\dfrac{1}{8}$ **89.** -1 **90.** -0.5 **91.** $\dfrac{81}{40}$ or $2\dfrac{1}{40}$ **92.** -8

93. 240 **94.** -25.42 **95.** $600   **96.** 0.0081 **97.** -0.0625 **98.** 10 **99.** $-4.9x + 4.1y$ **100.** $-\dfrac{1}{9}$ **101.** $-\dfrac{2}{3}$
102. No; the dog's temperature is below the normal temperature of $101.48°F$.   **103.** $3y^2 + 12y - 7x - 28$ **104.** $-12x + 6y + 12xy$

How Am I Doing? Chapter 1 Test **1.** -0.3 (obj. 1.1.4)   **2.** 2 (obj. 1.2.1)   **3.** $-\dfrac{14}{3}$ or $-4\dfrac{2}{3}$ (obj. 1.3.1)   **4.** -70 (obj. 1.3.1)

5. 4 (obj. 1.3.3)   **6.** -3 (obj. 1.3.3)   **7.** -64 (obj. 1.4.2)   **8.** 2.56 (obj. 1.4.2)   **9.** $\dfrac{16}{81}$ (obj. 1.4.2)   **10.** 6.8 (obj. 1.5.1)   **11.** -25 (obj. 1.5.1)

12. $-5x^2 - 10xy + 35x$ (obj. 1.6.1)   **13.** $6a^2b^2 + 4ab^3 - 14a^2b^3$ (obj. 1.6.1)   **14.** $2a^2b + \dfrac{15}{2}ab$ (obj. 1.7.2)   **15.** $-1.8x^2y - 4.7xy^2$ (obj. 1.7.2)

16. $5a + 30$ (obj. 1.7.2)   **17.** $14x - 16y$ (obj. 1.7.2)   **18.** 122 (obj. 1.8.1)   **19.** 37 (obj. 1.8.1)   **20.** $\dfrac{13}{6}$ or $2\dfrac{1}{6}$ (obj. 1.8.1)

21. 96.6 km/hr (obj. 1.8.2)   **22.** $22,800 \text{ ft}^2$ (obj. 1.8.2)   **23.** $23.12 (obj. 1.8.2)   **24.** 3 cans (obj. 1.8.2)   **25.** $3x - 6xy - 21y^2$ (obj. 1.9.1)
26. $-3a - 9ab + 3b^2 - 3ab^2$ (obj. 1.9.1)

Chapter 2 **2.1 Exercises** **1.** equals, equal   **3.** solution   **5.** Answers vary   **7.** $x = 4$ **9.** $x = 11$ **11.** $x = 17$ **13.** $x = -5$
15. $x = -13$ **17.** $x = 62$ **19.** $x = 15$ **21.** $x = 21$ **23.** $x = 0$ **25.** $x = 0$ **27.** $x = 21$ **29.** No; $x = 9$ **31.** No; $x = -14$
33. Yes   **35.** Yes   **37.** $x = -1.8$ **39.** $x = 0.6$ **41.** $x = 1$ **43.** $x = -\dfrac{1}{4}$ **45.** $x = -7$ **47.** $x = \dfrac{17}{6}$ or $2\dfrac{5}{6}$ **49.** $x = \dfrac{13}{12}$ or $1\dfrac{1}{12}$
51. $x = 5.2$ **53.** $x = 20.2$ **55.** $-2x - 4y$ **56.** $-2y^2 - 4y + 4$

Quick Quiz 2.1 **1.** $x = 14.3$ **2.** $x = -3.5$ **3.** $x = -8$ **4.** See Student Solutions Manual

2.2 Exercises **1.** 6 **3.** 7 **5.** $x = 36$ **7.** $x = -30$ **9.** $x = 80$ **11.** $x = -15$ **13.** $x = 4$ **15.** $x = 8$ **17.** $x = -\dfrac{8}{3}$

19. $x = 50$ **21.** $x = 15$ **23.** $x = -7$ **25.** $x = 0.2$ or $\dfrac{1}{5}$ **27.** $x = 4$ **29.** No; $x = -7$ **31.** Yes   **33.** $y = -0.03$ **35.** $t = \dfrac{8}{3}$

37. $y = -0.7$ **39.** $x = 3$ **41.** $x = -4$ **43.** $x = \dfrac{5}{12}$ **45.** $x = 1$ **47.** $m = 2$ **49.** $x = 27$ **51.** $x = -10.5$ **53.** $48 - 6 = 42$
54. $-27 - 10 = -37$ **55.** $5 + 16 = 21$ **56.** 22.5 tons   **57.** 27 earthquakes

Quick Quiz 2.2 **1.** $x = -38$ **2.** $x = 14$ **3.** $x = -12$ **4.** See Student Solutions Manual

2.3 Exercises **1.** $x = 2$ **3.** $x = 6$ **5.** $x = -4$ **7.** $x = 13$ **9.** $x = 3.1$ **11.** $x = 28$ **13.** $x = -27$ **15.** $x = 8$

17. $x = 3$ **19.** $x = \dfrac{11}{2}$ **21.** $x = -9$ **23.** Yes   **25.** No; $x = -11$ **27.** $x = -1$ **29.** $x = 7$ **31.** $y = -1$ **33.** $x = 16$

35. $y = 3$ **37.** $x = 4$ **39.** $x = \dfrac{1}{4}$ or 0.25 **41.** $x = 2.5$ or $\dfrac{5}{2}$ **43.** $x = 6.5$ **45.** $a = 0$ **47.** $x = -\dfrac{16}{5}$ or -3.2 **49.** $y = 2$

51. $x = -4$ **53.** $z = 5$ **55.** $a = -6.5$ **57.** $x = -\dfrac{2}{3}$ **59.** $x = -0.25$ **61.** $x = 8$ **63.** $x = -4.5$ **65.** $9xy - 8y^2$

66. $x - 9$ **67.** $1023.94   **68. (a)** $629.30   **(b)** $647.28

Quick Quiz 2.3 **1.** $x = -\dfrac{4}{11}$ **2.** $x = 4$ **3.** $x = -\dfrac{8}{7}$ **4.** See Student Solutions Manual

2.4 Exercises **1.** $x = -1$ **3.** $x = 1$ **5.** $x = 1$ **7.** $x = 12$ **9.** $y = 20$ **11.** $x = 3$ **13.** $x = \dfrac{7}{3}$ **15.** $x = -3.5$

17. Yes **19.** No **21.** $x = 1$ **23.** $x = 8$ **25.** $x = 2$ **27.** $x = -5$ **29.** $y = 4$ **31.** $x = -22$ **33.** $x = 2$ **35.** $x = -12$

37. $x = -\dfrac{5}{3}$ **39.** $x = -6$ **41.** No solution **43.** Infinite number of solutions **45.** $x = 0$ **47.** No solution **49.** $-\dfrac{52}{3}$ or $-17\dfrac{1}{3}$

50. $\dfrac{22}{5}$ or $4\dfrac{2}{5}$ **51.** $572 - 975$ g **52.** 3173 seats

Quick Quiz 2.4 **1.** $x = -\dfrac{7}{5}$ **2.** $x = \dfrac{19}{31}$ **3.** $x = \dfrac{11}{26}$ **4.** See Student Solutions Manual

How Am I Doing? Sections 2.1–2.4 **1.** $x = -9$ (obj. 2.1.1) **2.** $x = 7.5$ (obj. 2.1.1) **3.** $x = 9$ (obj. 2.2.2)

4. $x = -8$ (obj. 2.2.2) **5.** $x = \dfrac{2}{3}$ (obj. 2.3.1) **6.** $x = \dfrac{1}{8}$ or 0.125 (obj. 2.3.2) **7.** $x = \dfrac{3}{5}$ or 0.6 (obj. 2.3.3) **8.** $x = 3.75$ (obj. 2.3.3)

9. $x = -1.6$ or $-\dfrac{8}{5}$ (obj. 2.3.3) **10.** $x = \dfrac{10}{7}$ (obj. 2.4.1) **11.** $x = -3$ (obj. 2.4.1) **12.** $x = \dfrac{9}{11}$ (obj. 2.4.1) **13.** $x = 5$ (obj. 2.4.1)

14. $x = -\dfrac{1}{3}$ (obj. 2.4.1)

2.5 Exercises **1.** $x + 5$ **3.** $x - 12$ **5.** $\dfrac{1}{8}x$ or $\dfrac{x}{8}$ **7.** $2x$ **9.** $3 + \dfrac{1}{2}x$ **11.** $2x + 9$ **13.** $\dfrac{1}{3}(x + 7)$ **15.** $\dfrac{1}{3}x - 2x$

17. $5x - 11$ **19.** $x = $ value of a share of AT&T stock **21.** $w = $ width
$x + 74.50 = $ value of a share of IBM stock $2w + 7 = $ length

23. $x = $ number of boxes sold by Keiko **25.** 1st angle $= s - 16$ **27.** $v = $ value of exports of Canada **29.** 1st angle $= 3x$
$x - 43 = $ number of boxes sold by Sarah 2nd angle $= s$ $2v = $ value of exports of Japan 2nd angle $= x$
$x + 53 = $ number of boxes sold by Imelda 3rd angle $= 2s$ 3rd angle $= x - 14$

31. $x = $ men aged 16–24 **33.** $x = 7$ **34.** $x = -\dfrac{5}{2}$ or $-2\dfrac{1}{2}$
$x + 82 = $ men aged 25–34
$x - 25 = $ men aged 35–44
$\dfrac{1}{2}x$ or $x - 110 = $ men aged 45 and above

Quick Quiz 2.5 **1.** $x + 10$ or $10 + x$ **2.** $2x - 5$ **3.** First angle: $x + 15$ **4.** See Student Solutions Manual
Second angle: x
Third angle: $2x$

2.6 Exercises **1.** 1261 **3.** 2368 **5.** 182 **7.** 43 **9.** 9 **11.** -4 **13.** 12 **15.** 15 used bikes **17.** 48,840 wildfires
19. 6 CDs **21.** 4 items **23.** 320.2 mi/hr **25.** 550 lb **27.** 5 mi **29.** She traveled 52 mph on the mountain road. It was 12 mph
faster on the highway route. **31.** 90 **33. (a)** $F - 40 = \dfrac{x}{4}$ **(b)** 200 chirps **(c)** 77°F **34.** $10x^3 - 30x^2 - 15x$

35. $-2a^2b + 6ab - 10a^2$ **36.** $-5x - 6y$ **37.** $-4x^2y - 7xy^2 - 8xy$

Quick Quiz 2.6 **1.** 17 **2.** 48 **3.** 74 **4.** See Student Solutions Manual

2.7 Exercises **1.** distance around **3.** surface **5.** 180° **7.** 156 in.^{2} **9.** 98 in.^{2} **11.** 6 ft **13.** 9 in. **15.** 153.86 ft^2
17. 9520.48 mi **19.** 24 in. **21.** 134,400 m^2 **23.** 5 cm **25.** 200.96 cm **27.** 74° **29.** Equal angles $= 72°$ **31.** 30° each
3rd angle $= 36°$

33. 1st angle $= 46°$ **35.** 19 ft **37.** 6 ft **39.** 2512 in.^{3} **41.** Yes
2nd angle $= 23°$
3rd angle $= 111°$
43. (a) $V = 452.16$ cm^3 **45.** 163.28 ft^2 **47. (a)** 22.77 yd^2 **49.** 73.96 in. **51.** 9.3 feet wide and 34.2 feet long **53.** 6.28 ft
(b) $S = 376.8$ cm^2 **(b)** $56.93

54. $3x^2 + 8x - 24$ **55.** $-11x - 17$ **56.** 213 billion **57. (a)** $56,282$ **(b)** 155%

Quick Quiz 2.7 **1.** 116 ft^2 **2.** 113 in.^{3} **3.** $2280 **4.** See Student Solutions Manual

2.8 Exercises **1.** Yes, both statements imply 5 is to the right of -6 on a number line. **3.** $>$ **5.** $>$ **7.** $<$ **9. (a)** $<$ **(b)** $>$

11. (a) $>$ **(b)** $<$ **13.** $>$ **15.** $>$ **17.** $>$ **19.** $<$ **21.** $<$ **23.** $<$ **25.** $\begin{array}{ccccc} & & \circ & & \\ \hline 5 & 6 & 7 & 8 & 9 \end{array}$

27. 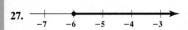 **29.** 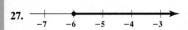 **31.**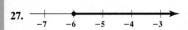

33. 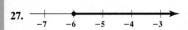   **35.** $x \geq -\dfrac{2}{3}$ **37.** $x < -20$ **39.** $x \leq 3.7$ **41.** $c \geq 12$

43. $h \geq 48$ **45.**   **47.** $x \leq -3$

49. $x \leq 5$ **51.** $x > -9$

53. $x \geq 8$ **55.** $x < -12$

57. $x < -1$ **59.** $x < \dfrac{3}{2}$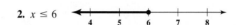

61. $x > -6$ 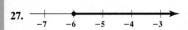   **63.** $x > \dfrac{1}{3}$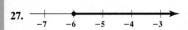

65. $3 > 1$ Adding any number to both sides of an inequality doesn't reverse the direction. **67.** $x > 3$ **69.** $x \geq 4$ **71.** $x < -1$

73. $x \leq 14$ **75.** $x < -3$ **77.** 76 or greater **79.** 8 days or more **81.** 6.08 **82.** 15% **83.** 2% **84.** 37.5%

Quick Quiz 2.8 1. 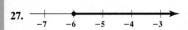   **2.** $x \leq 6$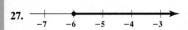

3. $-5 < x$ 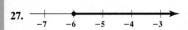   **4.** See Student Solutions Manual

Putting Your Skills to Work 1. SHELL **2.** SHELL **3.** ARCO **4.** ARCO **5.** 3.75 gallons **6.** Answers may vary **7.** Answers may vary **8.** Answers may vary

Chapter 2 Review Problems 1. $x = -7$ **2.** $x = -3$ **3.** $x = 2$ **4.** $x = -3$ **5.** $x = 40.4$ **6.** $x = -7$ **7.** $x = -2$

8. $x = -27$ **9.** $x = 20$ **10.** $x = -9$ **11.** $x = -3$ **12.** $x = 3$ **13.** $x = 4$ **14.** $x = -7$ **15.** $x = 3$ **16.** $x = -\dfrac{7}{2}$ or -3.5

17. $x = 5$ **18.** $x = 1$ **19.** $x = 20$ **20.** $x = \dfrac{2}{3}$ **21.** $x = 5$ **22.** $x = \dfrac{35}{11}$ **23.** $x = 4$ **24.** $x = -17$ **25.** $x = \dfrac{2}{5}$ or 0.4

26. $x = 32$ **27.** $x = \dfrac{26}{7}$ **28.** $x = 0$ **29.** $x = 0$ **30.** $x = \dfrac{3}{4}$ **31.** $x = -17$ **32.** $x = -32$ **33.** $x = -\dfrac{17}{5}$ or -3.4

34. $x = 9$ **35.** $x + 19$ **36.** $\dfrac{2}{3}x$ **37.** $3(x + 4)$ **38.** $2x - 3$ **39.** $r =$ the number of retired people; $4r =$ the number of working people; $0.5r =$ the number of unemployed people **40.** $w =$ the width; $3w + 5 =$ the length **41.** $b =$ the number of degrees in angle B; $2b =$ the number of degrees in angle A; $b - 17 =$ the number of degrees in angle C **42.** $a =$ the number of students in algebra; $a + 29 =$ the number of students in biology; $0.5a =$ the number of students in geology **43.** $x = 3$ **44.** $x = -7$ **45.** \$40 **46.** 16 years old **47.** 13.3 hr; 12.3 hr **48.** 88 **49.** 21.23 in.^{2} **50.** $P = 24$ ft **51.** $71°$ **52.** 42 mi^2 **53.** 462 ft^3 **54.** 381.51 in.^{3} **55.** 415.27 ft^2 **56.** 12 in. **57.** \$3440 **58.** \$23,864 **59.** $x \leq -1$ 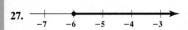 **60.** $x \geq 3$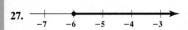

61. $x < -4$ **62.** $x > -3$

63. $x \geq 6$ **64.** $x < 5$

65. $x < 10$ **66.** $x > -3$ **67.** $h \leq 32$ hr **68.** $n \leq 22$

69. $x = \dfrac{13}{6}$ **70.** $x = \dfrac{11}{8}$ **71.** $x = 0$ **72.** $x = 4$ **73.** $x = 4$ **74.** $x = -5$

75. $x < 2$ **76.** $x \leq -8$

77. $x \geq \dfrac{19}{7}$ **78.** $x \geq -15$

How Am I Doing? Chapter 2 Test **1.** $x = 2$ (obj. 2.3.1) **2.** $x = \frac{1}{3}$ (obj. 2.3.2) **3.** $y = -\frac{7}{2}$ or -3.5 (obj. 2.3.3)

4. $y = 8.4$ or $\frac{42}{5}$ (obj. 2.4.1) **5.** $x = 1$ (obj. 2.3.3) **6.** $x = -\frac{6}{5}$ or -1.2 (obj. 2.3.3) **7.** $y = 7$ (obj. 2.4.1) **8.** $y = \frac{7}{3}$ or $2\frac{1}{3}$ (obj. 2.3.3)

9. $x = 13$ (obj. 2.3.3) **10.** $x = 20$ (obj. 2.3.3) **11.** $x = 10$ (obj. 2.3.3) **12.** $x = -4$ (obj. 2.3.3) **13.** $x = 12$ (obj. 2.3.3)

14. $x = -\frac{1}{5}$ or -0.2 (obj. 2.4.1) **15.** $x = 3$ (obj. 2.4.1) **16.** $x = 2$ (obj. 2.4.1) **17.** $x = -2$ (obj. 2.4.1)

18. $x \le -3$ (obj. 2.8.4)

19. $x > -\frac{5}{4}$ (obj. 2.8.4) **20.** $x < 2$ (obj. 2.8.4)

21. $x \ge \frac{1}{2}$ (obj. 2.8.4) **22.** 35 (obj. 2.5.1) **23.** 36 (obj. 2.5.1) **24.** -4 (obj. 2.5.1)

25. First side $= 20$ m; second side $= 30$ m; third side $= 16$ m (obj. 2.6.1) **26.** Width $= 20$ m; length $= 47$ m (obj. 3.2.6) **27.** 213.52 in. (obj. 2.7.1)
28. 192 in.^{2} (obj. 2.7.1) **29.** 4187 in.^{3} (obj. 2.7.2) **30.** 96 cm^2 (obj. 2.7.1) **31.** $450 (obj. 2.7.3)

Cumulative Test for Chapters 0–2 **1.** 3.69 **2.** $\frac{31}{24}$ or $1\frac{7}{24}$ **3.** $-34x + 10y - 20$ **4.** 21 **5.** 30

6. $-2st - 7s^2t + 14st^2$ **7.** $25x^2$ **8.** $6x - 40 + 48y - 24xy$ **9.** $x = \frac{5}{6}$ **10.** $x = 4$ **11.** $y = 1$ **12.** $x \le 2$

13. $x \le 3$ **14.** $x > -15$

15. $x \le -1$ **16.** $x \le 1$ **17.** 92 **18.** -7

19. Literature students $= 42$; sociology students $= 54$ **20.** Width $= 7$ cm; length $= 32$ cm **21.** $A = 162.5$ m^2; $731.25

22. $V = 113.04$ in.^{3}; 169.56 lb

Chapter 3 3.1 Exercises **1.** 0 **3.** The order in which you write the numbers matters. The graph of $(5, 1)$ is not the same as

the graph of $(1, 5)$. **5.**

7. $R: (-3, -5)$ **9.** $(-4, -1)$ **11.** No **13.** Yes
$S: \left(-4\frac{1}{2}, 0\right)$ $(-3, -2)$ **15.** $y = -\frac{2}{3}x + 4$ **17.** $y = -4x + 11$
$X: (3, -5)$ $(-2, -3)$
$Y: \left(2\frac{1}{2}, 6\right)$ $(-1, -5)$ **19.** $y = \frac{2}{3}x - 2$ **21.** (a) $(3, 8)$
$(0, -3)$ (b) $(6, 13)$
$(2, -1)$ (c) Answers may vary

23. (a) $(0, 7)$ (b) $(2, 15)$ **25.** (a) $(-1, 11)$ (b) $(3, -13)$ **27.** (a) $(-3, -5)$ (b) $(5, 1)$ **29.** (a) $(-2, 0)$ (b) $(-4, 3)$

31. (a) $(7, 3)$ (b) $\left(-1, \frac{5}{7}\right)$ **33.** (a) $\left(-\frac{1}{2}, 18\right)$ (b) $\left(\frac{8}{3}, -1\right)$ **35.** B5 **37.** E1 **39.** D3

41. (a)

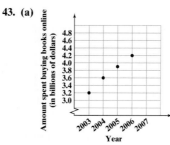

(b) The number of CDs shipped decreased overall
between 2001 and 2006, with a slight increase
from 2003 to 2004.

43. (a)

(b) An estimated $4.5 billion was spent buying books online in 2007.

45. 1133.54 yd^2 **46.** 12

Quick Quiz 3.1 **1.**

2. (a) $(-2, 3)$ (b) $(3, -22)$ (c) $(0, -7)$
3. (a) $(3, 8)$ (b) $(-9, -8)$ (c) $(4.5, 10)$
4. See Student Solutions Manual

3.2 Exercises **1.** No, replacing x by -2 and y by 5 in the equation does not result in a true statement. **3.** x-axis

5. $y = -2x + 1$
$(0, 1)$
$(-2, 5)$
$(1, -1)$

7. $y = x - 4$
$(0, -4)$
$(2, -2)$
$(4, 0)$

9. $y = -2x + 3$
$(0, 3)$
$(2, -1)$
$(4, -5)$

11. $y = 3x + 2$
$(-1, -1)$
$(0, 2)$
$(1, 5)$

13. $3x - 2y = 0$

15. $y = -\frac{3}{4}x + 3$

17. $4x + 3y = 12$

19. $y = 6 - 2x$

21. $x + 3 = 6y$

23. $y - 2 = 3y$

25. $2x + 9 = 5x$

27. $2x + 5y - 2 = -12$

29.

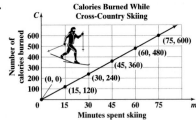

31.

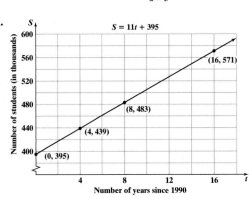

33. $x = -2$ **34.** $x \geq -\frac{14}{3}$

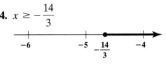

Quick Quiz 3.2 **1.**

$y = -2x + 4$

2.

$y = \frac{2}{5}x - 3$

3.

$y = \frac{1}{4}x + 2$

4. See Student Solutions Manual

3.3 Exercises **1.** No, division by zero is impossible, so the slope is undefined. **3.** 3 **5.** -5 **7.** $\frac{3}{5}$ **9.** $-\frac{1}{4}$ **11.** $-\frac{4}{3}$ **13.** $-\frac{16}{5}$

15. $m = 8$; $(0, 9)$ **17.** $m = -3$; $(0, 4)$ **19.** $m = -\frac{8}{7}$; $\left(0, \frac{3}{4}\right)$ **21.** $m = -6$; $(0, 0)$ **23.** $m = -6$; $\left(0, \frac{4}{5}\right)$ **25.** $m = -\frac{3}{4}$; $\left(0, \frac{3}{2}\right)$

27. $m = \frac{7}{3}$; $\left(0, -\frac{4}{3}\right)$ **29.** $y = \frac{3}{5}x + 3$ **31.** $y = 4x - 5$ **33.** $y = -\frac{5}{4}x - \frac{3}{4}$

35.

$y = \frac{3}{4}x - 4$

37.

$y = -\frac{5}{3}x + 2$

39.

$y = \frac{2}{3}x + 2$

41.

$y + 2x = 3$

43.

$y = 2x$

45. (a) $\dfrac{5}{6}$ **(b)** $-\dfrac{6}{5}$ **47. (a)** 6 **(b)** $-\dfrac{1}{6}$ **49. (a)** $\dfrac{2}{3}$ **(b)** $-\dfrac{3}{2}$ **51.** Yes; $2x - 3y = 18$

53. (a) $y = 35x + 625$ **(b)** $m = 35;\ (0, 625)$ **(c)** The amount of increase (in thousands) in the number of home health aides in the U.S. per year.

55. $\dfrac{27}{20}$ or $1\dfrac{7}{20}$ **56.** $\dfrac{5}{6}$ **57.** $x > 4$ **58.** $x < \dfrac{12}{5}$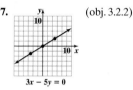

59. $x \le 24$ **60.** $x \le -22$

Quick Quiz 3.3 **1.** $\dfrac{1}{2}$ **2.** $m = 3;\ \left(0, -\dfrac{3}{2}\right)$ **3.** $y = -\dfrac{5}{7}x - 5$ **4.** See Student Solutions Manual

How Am I Doing? Sections 3.1–3.3 **1.** (obj. 3.1.1) **2.** $A\ (3, 9)$
$B\ (7, -7)$
$C\ (-7, -8)$ (obj. 3.1.2)
$D\ (-4, 3)$

3. $(4, -25);\ (0, 3);\ (-2, 17)$ (obj. 3.1.3)

4. (obj. 3.2.1) $y = -4x - 3$
5. (obj. 3.2.1) $y = \dfrac{3}{4}x - 1$
6. (obj. 3.2.2)
7. (obj. 3.2.2) $3x - 5y = 0$

8. $x = 3$ (obj. 3.2.3) **9.** -2 (obj. 3.3.1) **10.** Slope $= 2$
$y\text{-intercept} = \left(0, -\dfrac{4}{3}\right)$
(obj. 3.3.2) **11.** $-\dfrac{4}{3}$ (obj. 3.3.5) **12.** $\dfrac{5}{3}$ (obj. 3.3.5)

3.4 Exercises **1.** $y = 4x + 12$ **3.** $y = -2x + 11$ **5.** $y = -3x + \dfrac{7}{2}$ **7.** $y = \dfrac{1}{4}x + 4$ **9.** $y = -2x - 6$ **11.** $y = -4x + 2$

13. $y = 5x - 10$ **15.** $y = \dfrac{1}{3}x + \dfrac{1}{2}$ **17.** $y = -3x$ **19.** $y = -3x + 3$ **21.** $y = -\dfrac{2}{3}x + 1$ **23.** $y = \dfrac{2}{3}x - 4$ **25.** $y = -\dfrac{2}{3}x$

27. $y = -2$ **29.** $y = -2$ **31.** $x = 4$ **33.** $y = \dfrac{1}{3}x + 5$ **35.** $y = -\dfrac{1}{2}x + 4$ **37.** $y = 2.4x + 227$ **39.** $x < -4$ **40.** $x \le 3$

41. $x \ge \dfrac{14}{3}$ **42.** $x > -4$ **43. (a)** 2920 gal **(b)** $24.82 **44.** Width $= 550$ ft; length $= 1170$ ft

Quick Quiz 3.4 **1.** $y = \dfrac{2}{3}x - 7$ **2.** $y = 6x + 19$ **3.** $x = 4$; the slope is undefined **4.** See Student Solutions Manual

3.5 Exercises **1.** No, all points in one region will be solutions to the inequality while all points in the other region will not be solutions. Thus testing any point will give the same result, as long as the point is not on the boundary line.

3. $y < 2x - 4$
5. $2x - 3y < 6$
7. $2x - y \ge 3$
9. $y \ge 4x$
11. $y < -\dfrac{1}{2}x$

13. $x \ge 2$
15. $2x - 3y + 6 \ge 0$
17. $2x > -3y$
19. $2x > 3 - y$
21. $x > -2y$

23. 12 **24.** -17 **25.** -28 **26.** -25 **27.** $95.20 **28.** $19,040

Quick Quiz 3.5   **1.** Use a dashed line. If the inequality has a $<$ or a $>$ symbol, the points on the line itself are not included. This is indicated by a dashed line.   **2.**

$3y \leq -7x$

3.

$-5x + 2y > -3$

  4. See Student Solutions Manual

3.6 Exercises   **1.** Using a table of values, an algebraic equation, or a graph   **3.** possible values of the independent variable
5. If a vertical line can intersect the graph more than once, the relation is not a function. If no such line exists, then the relation is a function.

7. (a) Domain $= \left\{ -3, \dfrac{2}{5}, 3 \right\}$ **(b)** Not a function   **9. (a)** Domain $= \{0, 3, 6\}$ **(b)** Function

        Range $= \left\{ -1, \dfrac{2}{5}, 4 \right\}$ Range $= \{0.5, 1.5, 2.5\}$

11. (a) Domain $= \{1, 9, 12, 14\}$ **(b)** Function   **13. (a)** Domain $= \{3, 5, 7\}$ **(b)** Not a function
      Range $= \{1, 3, 12\}$ Range $= \{75, 85, 95, 100\}$

15.

$y = x^2 + 3$

17.

$y = 2x^2$

19.

$x = -2y^2$

21.

$x = y^2 - 4$

23.

$y = \dfrac{2}{x}$

25.

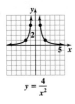

$y = \dfrac{4}{x^2}$

27.

$x = (y + 1)^2$

29.

$y = \dfrac{4}{x - 2}$

31. Function   **33.** Not a function   **35.** Function
37. Not a function   **39. (a)** 26   **(b)** 2   **(c)** -4
41. (a) 3   **(b)** 24   **(c)** 9
43. $f(0) = 31.6, f(4) = 32.24, f(10) = 34.4$
The curve slopes more steeply for larger values of x. The growth rate is increasing as x gets larger.

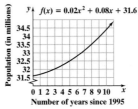

45. $-8x^3 + 12x^2 - 32x$ **46.** $5a^2b + 30ab - 10a^2$ **47.** $-19x + 2y - 2$ **48.** $9x^2y - 6xy^2 + 7xy$

Quick Quiz 3.6   **1.** No. Two different ordered pairs have the same first coordinate.   **2. (a)** 41   **(b)** 34   **3. (a)** -7 **(b)** $-\dfrac{7}{8}$
4. See Student Solutions Manual

Putting Your Skills to Work   **1.** Approximately \$620,000   **2.** \$84,000   **3.** Approximately \$536,000   **4.** Answers may vary
5. Answers may vary   **6.** Answers may vary

Chapter 3 Review Problems   **1.**

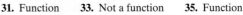

2.

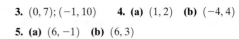

3. $(0, 7); (-1, 10)$ **4. (a)** $(1, 2)$ **(b)** $(-4, 4)$
5. (a) $(6, -1)$ **(b)** $(6, 3)$

6.

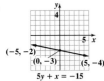

$5y + x = -15$

7.

8.

$3y = 2x + 6$

9. $m = -\dfrac{5}{6}$ **10.** $m = \dfrac{9}{11}; \left(0, \dfrac{15}{11} \right)$ **11.** $y = -\dfrac{1}{2}x + 3$

12. $m = -\dfrac{5}{3}$ **13.**

$y = -\dfrac{1}{2}x + 3$

14.

$2x - 3y = -12$

15.

$6x - 2y = 6 + 6x$

16. $y = x + 3$ **17.** $y = -6x + 14$
18. $y = -\dfrac{1}{3}x + \dfrac{11}{3}$ **19.** $y = 7$ **20.** $-\dfrac{2}{3}$ **21.** -3
22. $y = \dfrac{2}{3}x - 3$ **23.** $y = -3x + 1$ **24.** $x = 5$

25.

$y < \frac{1}{3}x + 2$

26.

$3y + 2x \geq 12$

27.

$x \leq 2$

28. Domain: $\{-6, -5, 5\}$
Range: $\{-6, 5\}$
not a function

29. Domain: $\{-2, 2, 5, 6\}$
Range: $\{-3, 4\}$
function

30. Function   **31.** Not a function
32. Function

33.

$y = x^2 - 5$

34.

$x = y^2 + 3$

35.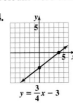

$y = (x - 3)^2$

36. (a) 7 (b) 31   **37.** (a) -1 (b) -5 **38.** (a) -11 (b) 1

39. (a) 1 (b) $\frac{1}{5}$ **40.** (a) 0 (b) 4   **41.** $210   **42.** $174

43. $y = 0.09x + 30$; $(0, 30)$ tells us that if Russ and Norma use no electricity, the minimum cost will be $30.   **44.** $m = 0.09$; the electric bill increases $0.09 for each kilowatt-hour of use.   **45.** 1300 kilowatt-hours

46. 2400 kilowatt-hours   **47.** 77.1 million barrels   **48.** 94.35 million barrels   **49.** $m = 1.15$. This means that for each additional year, the world uses 1.15 million more barrels of petroleum.   **50.** y-intercept $= (0, 65.6)$. This means that when $x = 0$, which is the year 1990, a total of 65.6 million barrels of petroleum were consumed worldwide.   **51.** 2017   **52.** 2020

53.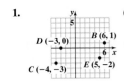

$5x + 3y = -15$

54.

$y = \frac{3}{4}x - 3$

55. $-\frac{2}{5}$ **56.** $m = -\frac{7}{6}$; y-intercept $= \left(0, \frac{5}{3}\right)$ **57.** $y = \frac{2}{3}x - 7$

58.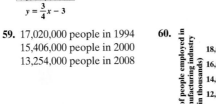

$y < -2x + 1$

59. 17,020,000 people in 1994
15,406,000 people in 2000
13,254,000 people in 2008

60.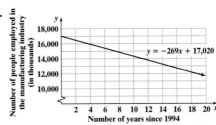

$y = -269x + 17,020$

Number of people employed in the manufacturing industry (in thousands)

Number of years since 1994

61. The slope is -269. The slope tells us that the number of people employed in manufacturing decreases each year by 269 thousand. In other words, employment in manufacturing goes down 269,000 people each year.   **62.** The y-intercept is $(0, 17,020)$. This tells us that in the year 1994, the number of manufacturing jobs was 17,020 thousand, which is 17,020,000.   **63.** 2014   **64.** 2019

How Am I Doing? Chapter 3 Test

1. 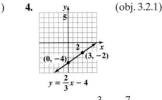 (obj. 3.1.1)

$D(-3, 0)$ $B(6, 1)$ $E(5, -2)$ $C(-4, -3)$

2.   (obj. 3.2.1)

$(1, 1)$ $(-1, 2)$ $(3, 0)$ $6x - 3 = 5x - 2y$

3.   (obj. 3.2.1)

$8x + 2y = -4$

4.   (obj. 3.2.1)

$(0, -4)$ $(3, -2)$ $y = \frac{2}{3}x - 4$

5. $m = -\frac{3}{2}$; $\left(0, \frac{5}{2}\right)$ (obj. 3.3.2)   **6.** $m = 1$ (obj. 3.3.1)   **7.** $y = \frac{1}{2}x - 4$ (obj. 3.3.3)   **8.** $m = 0$ (obj. 3.3.1)   **9.** $y = -\frac{3}{2}x + \frac{7}{2}$ (obj. 3.4.2)

10. $x = 2$; undefined (obj. 3.4.2)

11.   (obj. 3.5.1)

$4y \leq 3x$

12. (obj. 3.5.1)

$-3x - 2y > 10$

13. No, two different ordered pairs have the same first coordinate. (obj. 3.6.1)   **14.** Yes, any vertical line passes through no more than one point on the graph. (obj. 3.6.3)

15. (obj. 3.6.2)

$y = 2x^2 - 3$

16. (a) -3 (b) -3 (obj. 3.6.4)   **17.** (a) -3 (b) $-\frac{1}{5}$ (obj. 3.6.4)

Cumulative Test for Chapters 0–3 **1.** $-\frac{22}{75}$ **2.** $-25x^2 + 60x$ **3.** $x = 7$ **4.** $x = -4$ **5.** Height $= 12$ ft; length $= 33$ ft

6. -1.514 **7.** 8.52   **8.** -8.6688 **9.** $y = \frac{-2x + 8}{3}$ **10.** $y = 2x - 7$

11. $x = 7$ **12.** $y = -\dfrac{2}{3}x - 3$ **13.** $m = 0$ **14.** $m = \dfrac{4}{3}$

15. $y = \dfrac{2}{3}x - 4$

16. $3x + 8 = 5x$

17. $2x + 5y \le -10$

18. No $x = y^2 + 3$

19. Yes **20. (a)** 1 **(b)** 10

Chapter 4 4.1 Exercises

1. The lines are parallel. One line intersects the y-axis at -5. The other line intersects the y-axis at 6. There is no solution. The system is inconsistent. **3.** intersect, one **5.** The lines intersect at the y-intercept. The y-intercept is the solution of the system.

7. $x + y = 5$; $x - y = 3$; $(4, 1)$

9. $y = 2x + 5$ $y = -3x$; $(-1, 3)$

11. $3x - 2y = 12$; $4x + y = 5$; $(2, -3)$

13. $y = 2x + 3$; $y = -4x - 3$; $(-1, 1)$

15. $2x + 3y = 14$; $3x - 2y = -18$; $(-2, 6)$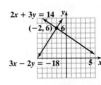

17. 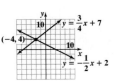 $y = \dfrac{3}{4}x + 7$; $y = -\dfrac{1}{2}x + 2$; $(-4, 4)$

19. 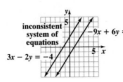 inconsistent system of equations; $-9x + 6y = -9$; $3x - 2y = -4$

21. 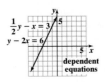 $\dfrac{1}{2}y - x = 3$; $y - 2x = 6$; dependent equations

23. 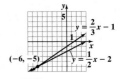 $y = \dfrac{2}{3}x - 1$; $y = \dfrac{1}{2}x - 2$; $(-6, -5)$

25. (a) Newtown: $y = 7x$
Tryus: $y = 36 + 3x$

(b)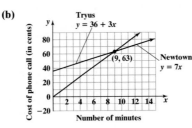

x	1	4	8	12
$y = 7x$	7	28	56	84

x	1	4	8	12
$y = 36 + 3x$	39	48	60	72

(c) 9 min **(d)** Tryus **27.** $(47.52, -61.38)$ **29.** $(-33, -40)$ **31.** $x \ge \dfrac{1}{8}$ **32.** $9 - x$ **33.** $x = \dfrac{2}{3}y + 2$ **34.** $y = -2x + 8$

Quick Quiz 4.1

1. 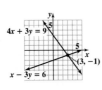 $4x + 3y = 9$; $x - 3y = 6$; $(3, -1)$

2. The slope of each line is $\dfrac{3}{5}$. The y-intercept of the first line is $(0, 2)$. The y-intercept of the second line is $(0, 7)$. Since the lines have the same slopes but different y-intercepts, they are parallel lines. Therefore there is no solution. It is an inconsistent system of equations.

3. When graphed, the two equations yield the same line. The equations are dependent. There is an infinite number of solutions.
4. See Student Solutions Manual

4.2 Exercises
1. $(3, -1)$ **3.** $(1, 2)$ **5.** $(3, -5)$ **7.** $(0, 3)$ **9.** $(6, 4)$ **11.** $(2, 3)$ **13.** $\left(\dfrac{2}{3}, \dfrac{5}{3}\right)$ **15.** $(2, -3)$

17. $(-3, 6)$ **19.** $(3, -2)$ **21.** $(6, -5)$ **23.** $(-2, 5)$ **25.** $(-4, -12)$ **27.** $(8, 1)$ **29.** $3, 7$ **31.** both

33. No solution, parallel lines have no points in common **35. (a)** $y = 90x + 50$ **(b)** 5 times plowing; $500 **(c)** 13 times; Adirondack Plowing
$y = 80x + 100$

36. $m = -\dfrac{7}{11}; \left(0, \dfrac{19}{11}\right)$ **37.** $6x + 3y \ge 9$ **38.** $10.50 **39.** About 64%; about $71.30

Quick Quiz 4.2 **1.** $\left(\dfrac{1}{2}, 1\right)$ **2.** $(4, -3)$ **3.** $(0, 5)$ **4.** See Student Solutions Manual

4.3 Exercises **1.** Eliminate y by multiplying the second equation by 2, then adding the equations.
3. Eliminate y by multiplying the first equation by 4, multiplying the second equation by 3, then adding the equations. **5.** $(-1, -4)$ **7.** $(2, -1)$

9. $\left(-\dfrac{1}{2}, 4\right)$ **11.** $\left(3, \dfrac{2}{5}\right)$ **13.** $(1, 0)$ **15.** $\left(-\dfrac{1}{2}, \dfrac{1}{3}\right)$ **17.** $(-10, 4)$ **19.** $(-2, 3)$ **21.** $\left(\dfrac{1}{3}, 1\right)$ **23.** Multiply equation (1) by 4.

25. Multiply equation (1) by 6.   **27.** $\left(2, \dfrac{1}{5}\right)$ **29.** $(3, -2)$ **31.** $\left(-2, \dfrac{4}{3}\right)$ **33.** $(-3, 3)$
Multiply equation (2) by 20.

35. Multiply equation (1) by 10.   **37.** Multiply equation (1) by 10.   **39.** $(-1, 2)$ **41.** $(-4, -7)$ **43.** $(8, -10)$ **45.** $\left(-\dfrac{2}{3}, 6\right)$
$5x - 3y = 1$ Multiply equation (2) by 100.
$5x + 3y = 6$ $40x + 5y = 90$
                        $20x - 5y = 100$

47. $(12, 16)$ **49.** $(0, 2)$ **51.** $\left(-\dfrac{4}{3}, \dfrac{2}{3}\right)$ **52.** between 550 and 770 airplanes   **53.** \$720   **54.** $x = \dfrac{8}{7}$ **55.** $y = \dfrac{5}{3}$

Quick Quiz 4.3   1. $(3, -2)$ **2.** $(4, -1)$ **3.** $(5, 1)$ **4.** See Student Solutions Manual

How Am I Doing? Sections 4.1–4.3   1. (obj. 4.1.1)   **2.** (obj. 4.1.1)   **3.** $(6, -4)$ (obj. 4.2.1)

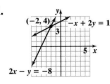

4. $(0, -2)$ (obj. 4.2.1)   **5.** $(-2, 3)$ (obj. 4.2.2)   **6.** $(1, -2)$ (obj. 4.2.1)   **7.** $(7, 1)$ (obj. 4.3.1)   **8.** $\left(-3, \dfrac{1}{2}\right)$ (obj. 4.3.1)

9. $(-9, -12)$ (obj. 4.3.2)   **10.** $(1, -3)$ (obj. 4.3.3)

4.4 Exercises   1. parallel lines, inconsistent   **3.** intersect, independent, consistent   **5.** $(3, -7)$ **7.** $(18, 13)$ **9.** No solution

11. Infinite number of solutions   **13.** $(5, 4)$ **15.** $(50, 40)$ **17.** $(8, 5)$ **19.** $\left(\dfrac{1}{2}, \dfrac{2}{3}\right)$ **21.** $\left(\dfrac{7}{3}, \dfrac{8}{3}\right)$ **23.** $(1, -2)$ **25.** $\left(\dfrac{3}{5}, \dfrac{3}{5}\right)$

27. No solution. The system is inconsistent.   **29.** $(-4, -3)$ **31.** $x = -\dfrac{2}{9}$ **32.** $x = 0.1$ **33.** \$3748.75   **34.** More; \$384.25 refund

Quick Quiz 4.4   1. $(9, -8)$ **2.** No solution   **3.** $(3, 2)$ **4.** See Student Solutions Manual

4.5 Exercises   1. 8 1st class   **3.** 5 @ \$35   **5.** 9 hrs day shift   **7.** Original length = 10 ft   **9.** 8 qts of 50% antifreeze originally.
6 coach class   7 @ \$40   14 hrs night shift   Original width = 9 ft   She added 8 quarts of 80% solution.
                                              New length = 14 ft
                                              New width = 14 ft

11. 30 lb nuts   **13.** 50 km/hr = wind speed   **15.** Jewelry wants 10 \$10 bills and 7 \$20 bills.   **17.** \$0.69 per lb = oranges
20 lb raisins   550 km/hr = plane's speed   Cosmetics wants 18 \$10 bills and 3 \$20 bills.   \$0.79 per lb = apples

19. 132 general admission   **21.** $x = 19$, which represents 2024; 6,227,000 people   **23.** $-\dfrac{1}{2}$ **24.** slope $= -\dfrac{3}{4}$; y-intercept $= (0, -2)$
210 student admission

25. $y = \dfrac{5}{4}x + \dfrac{7}{2}$ **26.** $y = -2x + 10$

Quick Quiz 4.5   1. 10,000 people purchased student tickets; 8500 people purchased general admission tickets   **2.** each Dell PC cost \$800; each
Apple iMac cost \$1200   **3.** 400 mi/hr = speed of plane; 100 mi/hr = speed of jet stream   **4.** See Student Solutions Manual

Putting Your Skills to Work   1. Store A   **2. (a)** Approximately \$3   **(b)** Approximately \$6   **(c)** Approximately \$9
3. (a) Approximately \$12   **(b)** Approximately \$16   **(c)** Approximately \$19   **4.** Approximately \$24   **5.** Approximately \$27   **6.** Store B
7. Answers may vary

Chapter 4 Review Problems   1.   **2.**   **3.**   **4.**

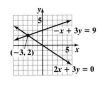   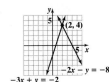

5. $(3, 3)$ **6.** $(3, 5)$ **7.** $(15, 21)$ **8.** $(8, 12)$ **9.** $(2, 1)$ **10.** $(-1, 0)$ **11.** $\left(0, -\dfrac{1}{2}\right)$ **12.** $\left(\dfrac{1}{3}, -4\right)$ **13.** $(5, -3)$

14. $(12, 1)$ **15.** $\left(\dfrac{4}{5}, -\dfrac{6}{5}\right)$ **16.** $\left(\dfrac{33}{13}, -\dfrac{28}{13}\right)$ **17.** $(6, -4)$ **18.** $(-1, -4)$ **19.** No solution, inconsistent system   **20.** $(9, 4)$

21. $\left(-3, -\dfrac{2}{3}\right)$ **22.** Infinite number of solutions, dependent equations   **23.** $(-2, -2)$ **24.** $(-1, -2)$ **25.** $(4, -3)$ **26.** $\left(-\dfrac{24}{41}, \dfrac{2}{41}\right)$

27. Infinite number of solutions, dependent equations   **28.** $(1960, 1040)$ **29.** $(8, -5)$ **30.** $(-4, -7)$ **31.** $\left(\dfrac{29}{28}, \dfrac{1}{7}\right)$ **32.** $(4, 10)$

33. $(10, 8)$ **34.** $(-12, -15)$ **35.** No solution, inconsistent system   **36.** Infinite number of solutions, dependent equations   **37.** $\left(\dfrac{33}{14}, \dfrac{16}{7}\right)$

38. $\left(-\dfrac{7}{32}, \dfrac{33}{16}\right)$ **39.** $(-5, 4)$ **40.** $(9, -8)$ **41.** $(2, 4)$ **42.** $(3, 1)$ **43.** $(3, 2)$ **44.** $(12, -12)$ **45.** 126 adults; 60 students

46. 21 2-lb bags; 42 5-lb bags   **47.** 275 mph in still air; 25 mph wind speed   **48.** 20 L of 20% solution; 20 L of 30% solution
49. 54 balsam firs; 25 Norwegian pines   **50.** 8 tons of 15% salt; 16 tons of 30% salt   **51.** Speed of the boat, 19 km/hr; speed of the current, 4 km/hr
52. $3 for a gram of copper; $4 hourly labor rate   **53.** 11 Level Outfield seats; 14 Upper Skydeck seats
54. 12 Field Level Bases seats; 10 Field Level Infield seats

How Am I Doing? Chapter 4 Test
1. $(-3, -4)$ (obj. 4.2.1)   **2.** $(-3, 4)$ (obj. 4.3.1)   **3.** 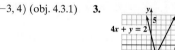 (obj. 4.1.1)   **4.** $(6, 2)$ (obj. 4.4.1)

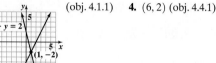

5. $(4, 3)$ (obj. 4.4.1)   **6.** $(5, 1)$ (obj. 4.4.1)   **7.** $(3, 0)$ (obj. 4.4.1)   **8.** No solution, inconsistent system (obj. 4.4.1, 4.4.2)
9. Infinite number of solutions, dependent equations (obj. 4.4.1, 4.4.2)   **10.** $(-2, 3)$ (obj. 4.4.1)   **11.** $(-5, 3)$ (obj. 4.4.1)
12. $(-3.5, -4.5)$ (obj. 4.4.1)   **13.** $5, -3$ (obj. 4.5.1)   **14.** 16,000 of the $12 tickets; 14,500 of the $8 tickets (obj. 4.5.1)
15. A shirt costs $20 and a pair of slacks costs $24. (obj. 4.5.1)   **16.** 1000 booklets; $450 total cost (obj. 4.5.1)   **17.** 50 km/hr; 450 km/hr (obj. 4.5.1)

Cumulative Test for Chapters 0–4
1. $1\dfrac{5}{8}$ **2.** $-\dfrac{7}{5}$ or $-1\dfrac{2}{5}$ **3.** -23 **4.** $22x + 12$ **5.** $y = -\dfrac{1}{2}x + \dfrac{5}{2}$ **6.** $x = 68$

7.   **8.** $m = \dfrac{1}{10}$ **9.** $x > -10$ **10.** $x \le 11$ **11.** $(8, -8)$

12. $(24, 8)$ **13.** Infinite number of solutions; dependent equations   **14.** $(5, 1)$ **15.** Inconsistent system; no solution   **16.** 36 kg of water
17. 1 mi/hr   **18.** The printer that broke prints 1200 labels/hour; the other printer prints 1800 labels/hour.

Chapter 5
5.1 Exercises
1. When multiplying exponential expressions with the same base, keep the base the same and add the exponents.

3. $\dfrac{2^2}{2^3} = \dfrac{2 \cdot 2}{2 \cdot 2 \cdot 2} = \dfrac{1}{2} = \dfrac{1}{2^{3-2}}$ **5.** 6; x and y; 11 and 1   **7.** $2^2 a^3 b$ **9.** $-5x^3 y^2 z^2$ **11.** 7^{10} **13.** 5^{26} **15.** x^{12} **17.** t^{16} **19.** $-20x^6$

21. $50x^3$ **23.** $18x^3 y^8$ **25.** $\dfrac{2}{15}x^3 y^5$ **27.** $-2.75x^3 yz$ **29.** 0 **31.** $80x^3 y^7$ **33.** $-24x^4 y^7$ **35.** 0 **37.** $-24a^4 b^3 x^2 y^5$ **39.** $-30x^3 y^4 z^6$

41. y^7 **43.** $\dfrac{1}{y^3}$ **45.** $\dfrac{1}{11^{12}}$ **47.** 2^7 **49.** $\dfrac{a^8}{4}$ **51.** $\dfrac{x^7}{y^9}$ **53.** $2x^4$ **55.** $-\dfrac{x^3}{2y^2}$ **57.** $\dfrac{f^2}{30g^5}$ **59.** $17x^4 y^3$ **61.** $\dfrac{y^2}{16x^3}$ **63.** $\dfrac{3a}{4}$

65. $\dfrac{2c^2}{7ab}$ **67.** x^{12} **69.** $x^{15} y^5$ **71.** $r^6 s^{12}$ **73.** $27a^9 b^6 c^3$ **75.** $9a^8$ **77.** $\dfrac{x^7}{128m^{28}}$ **79.** $\dfrac{25x^2}{49y^4}$ **81.** $81a^8 b^{12}$ **83.** $-8x^9 z^3$

85. $\dfrac{9}{x}$ **87.** $25a^5 b^7$ **89.** $\dfrac{64}{y^{10}}$ **91.** $\dfrac{16x^4}{y^{12}}$ **92.** -11 **93.** -46 **94.** $\dfrac{2}{25}$ **95.** -4 **96.** $\approx 39.9\%$ of Brazil's land area

97. $\approx 78{,}206$ km^2 of rain forest lost

Quick Quiz 5.1
1. $-10x^3 y^7$ **2.** $-\dfrac{4x^3}{5y^2}$ **3.** $81x^{12} y^{20}$ **4.** See Student Solutions Manual

5.2 Exercises
1. $\dfrac{1}{x^4}$ **3.** $\dfrac{1}{81}$ **5.** y^8 **7.** $\dfrac{z^6}{x^4 y^5}$ **9.** $\dfrac{a^3}{b^2}$ **11.** $\dfrac{x^9}{8}$ **13.** $\dfrac{3}{x^2}$ **15.** $\dfrac{1}{9x^2 y^4}$ **17.** $\dfrac{3xz^3}{y^2}$ **19.** $3x$ **21.** $\dfrac{b^3 d}{ac^4}$ **23.** $\dfrac{1}{8}$

25. $\dfrac{z^8}{9y^4}$ **27.** $\dfrac{1}{x^6 y}$ **29.** 1.2378×10^5 **31.** 6.3×10^{-2} **33.** 8.8961×10^{11} **35.** 3.42×10^{-6} **37.** $302{,}000$ **39.** 0.00047

41. $983{,}000$ **43.** 2.37×10^{-5} mph   **45.** $149{,}600{,}000$ km   **47.** 6.3×10^{15} **49.** 1.0×10^1 **51.** 8.1×10^{-11} **53.** 4.5×10^5

55. 2.82×10^4 dollars   **57.** 6.6×10^{-5} mi   **59.** About 3.68×10^8 mi/yr   **61.** 159.3% **63.** -0.8 **64.** -1 **65.** $-\dfrac{1}{28}$

Quick Quiz 5.2
1. $\dfrac{3y^2}{x^3 z^4}$ **2.** $\dfrac{a^8}{2b}$ **3.** 8.76×10^{-3} **4.** See Student Solutions Manual

5.3 Exercises
1. A polynomial in x is the sum of a finite number of terms of the form ax^n, where a is any real number and n is a whole number. An example is $3x^2 - 5x - 9$.   **3.** The degree of a polynomial in x is the largest exponent of x in any of the terms of the polynomial.

5. Degree 4; monomial   **7.** Degree 5; trinomial   **9.** Degree 6; binomial   **11.** $5x - 28$ **13.** $-2x^2 + 2x - 1$ **15.** $\frac{5}{6}x^2 + \frac{1}{2}x - 9$

17. $3.4x^3 + 2.2x^2 - 13.2x - 5.4$ **19.** $5x - 24$ **21.** $\frac{1}{15}x^2 - \frac{1}{14}x + 11$ **23.** $3x^3 - x^2 + 8x$ **25.** $-4.7x^4 - 0.7x^2 - 1.6x + 0.4$

27. $6x - 6$ **29.** $-3x^2y + 6xy^2 - 4$ **31.** $x^4 - 3x^3 - 4x^2 - 24$ **33.** 743,000   **35.** 90,400   **37.** $3x^2 + 12x$ **39.** $y = \frac{8}{3}x + \frac{2}{3}$

40. $B = \frac{3A}{2CD}$ **41.** $x = 7$ **42.** $x = 2$

Quick Quiz 5.3   **1.** $-4x^2 - 11x + 5$ **2.** $6x^2 - 9x - 16$ **3.** $-3x + 8$ **4.** See Student Solutions Manual

How Am I Doing? Sections 5.1–5.3   **1.** $-24x^3y^5$ (obj. 5.1.1)   **2.** $-\frac{7y^3}{5x^7}$ (obj. 5.1.2)   **3.** $\frac{4x^5}{y^9}$ (obj. 5.1.2)   **4.** $81x^{20}y^4$ (obj. 5.1.3)

5. $\frac{x^6}{16y^8}$ (obj. 5.2.1)   **6.** $\frac{x^5}{3y^5}$ (obj. 5.2.1)   **7.** 5.874×10^4 (obj. 5.2.2)   **8.** 9.362×10^{-5} (obj. 5.2.2)   **9.** $2.3x^2 - 0.4x - 5.4$ (obj. 5.3.2)

10. $4x^2 + 2x - 8$ (obj. 5.3.3)   **11.** $\frac{1}{6}x^3 + \frac{3}{8}x^2 + 3x$ (obj. 5.3.3)   **12.** $\frac{5}{16}x^2 - \frac{3}{10}x - \frac{3}{8}$ (obj. 5.3.2)

5.4 Exercises   **1.** $-12x^4 + 2x^2$ **3.** $21x^3 - 9x^2$ **5.** $-4x^6 + 10x^4 - 2x^3$ **7.** $x + \frac{3}{2}x^2 + \frac{5}{2}x^3$ **9.** $-2x^5y + 4x^4y - 5x^3y$

11. $9x^4y + 3x^3y - 24x^2y$ **13.** $3x^4 - 9x^3 + 15x^2 - 6x$ **15.** $-2x^3y^3 + 12x^2y^3 - 16xy$ **17.** $-28x^5y + 12x^4y + 8x^3y - 4x^2y$
19. $-6c^2d^5 + 8c^2d^3 - 12c^2d$ **21.** $12x^7 - 6x^5 + 18x^4 + 54x^3$ **23.** $-16x^6 + 10x^5 - 12x^4$ **25.** $x^2 + 12x + 35$ **27.** $x^2 + 8x + 12$
29. $x^2 - 6x - 16$ **31.** $x^2 - 9x + 20$ **33.** $-20x^2 - 7x + 6$ **35.** $2x^2 + 6xy - 5x - 15y$ **37.** $15x^2 - 5xy + 6x - 2y$ **39.** $20y^2 - 7y - 3$
41. The signs are incorrect. The result is $-3x + 6$.   **43.** $20x$ **45.** $20x^2 - 23xy + 6y^2$ **47.** $49x^2 - 28x + 4$ **49.** $16a^2 + 16ab + 4b^2$

51. $0.8x^2 + 11.94x - 0.9$ **53.** $\frac{1}{4}x^2 + \frac{1}{24}x - \frac{1}{12}$ **55.** $2ax - 10ab - 3bx + 15b^2$ **57.** $10x^2 - 11x - 6$ **59.** $x = -10$

60. $w = -\frac{25}{7}$ or $-3\frac{4}{7}$ **61.** 9 twenties; 8 tens; 23 fives   **62.** $240 billion   **63.** $505.5 billion   **64.** $594 billion   **65.** $664.8 billion

Quick Quiz 5.4   **1.** $8x^3y^4 - 12x^2y^3 + 16xy^2$ **2.** $6x^2 - x - 15$ **3.** $12a^2 - 26ab + 12b^2$ **4.** See Student Solutions Manual

5.5 Exercises   **1.** binomial   **3.** The middle term is missing. The answer should be $16x^2 - 56x + 49$.   **5.** $y^2 - 49$ **7.** $x^2 - 64$

9. $36x^2 - 25$ **11.** $4x^2 - 49$ **13.** $25x^2 - 9y^2$ **15.** $0.36x^2 - 9$ **17.** $4y^2 + 20y + 25$ **19.** $25x^2 - 40x + 16$ **21.** $49x^2 + 42x + 9$

23. $9x^2 - 42x + 49$ **25.** $\frac{4}{9}x^2 + \frac{1}{3}x + \frac{1}{16}$ **27.** $81x^2y^2 + 72xyz + 16z^2$ **29.** $49x^2 - 9y^2$ **31.** $9c^2 - 30cd + 25d^2$ **33.** $81a^2 - 100b^2$

35. $25x^2 + 90xy + 81y^2$ **37.** $x^3 - 4x^2 + 8x - 15$ **39.** $4x^4 - 7x^3 + 2x^2 - 3x - 1$ **41.** $a^4 + a^3 - 13a^2 + 17a - 6$
43. $3x^3 - 2x^2 - 25x + 24$ **45.** $2x^3 - x^2 - 16x + 15$ **47.** $2a^3 + 3a^2 - 50a - 75$ **49.** $24x^3 + 14x^2 - 11x - 6$ **51.** 19; 41
52. Width = 7 m; length = 10 m

Quick Quiz 5.5   **1.** $49x^2 - 144y^2$ **2.** $6x^3 - x^2 - 19x - 6$ **3.** $15x^4 - 16x^3 - 8x^2 + 17x - 6$ **4.** See Student Solutions Manual

5.6 Exercises   **1.** $5x^3 - 3x + 4$ **3.** $2y^2 - 3y - 1$ **5.** $9x^4 - 4x^2 - 7$ **7.** $8x^4 - 9x + 6$ **9.** $3x + 5$ **11.** $x - 3 + \frac{-32}{x - 5}$

13. $3x^2 - 4x + 8 + \frac{-10}{x + 1}$ **15.** $2x^2 - 3x - 2 + \frac{-5}{2x + 5}$ **17.** $2x^2 + 3x - 1$ **19.** $4y^2 - 2y + 5$ **21.** $y^2 - 4y - 1 + \frac{-9}{y + 3}$

23. $y^3 + 2y^2 - 5y - 10 + \frac{-25}{y - 2}$ **25.** 2000: $2.83; 2005: $3.20   **26.** 170 and 171   **27. (a)** 6.5   **(b)** 7.5   **(c)** 15.4%   **(d)** 8.7

Quick Quiz 5.6   **1.** $5x^3 - 16x^2 - 2x$ **2.** $4x^2 - 5x - 2$ **3.** $x^2 + 2x + 8 + \frac{13}{x - 2}$ **4.** See Student Solutions Manual

Putting Your Skills to Work   **1.** $445   **2.** $388.30   **3.** $310.64   **4.** 30.2%   **5.** Answers may vary

Chapter 5 Review Problems   **1.** $-18a^7$ **2.** 5^{23} **3.** $6x^4y^6$ **4.** $-14x^4y^9$ **5.** $\frac{1}{7^{12}}$ **6.** $\frac{1}{x^5}$ **7.** y^{14} **8.** $\frac{1}{9^{11}}$ **9.** $-\frac{3}{5x^5y^4}$

10. $-\frac{2a}{3b^6}$ **11.** x^{24} **12.** b^{30} **13.** $9a^6b^4$ **14.** $81x^{12}y^4$ **15.** $\frac{25a^2b^4}{c^6}$ **16.** $\frac{y^9}{64w^{15}z^6}$ **17.** $\frac{b^5}{a^3}$ **18.** $\frac{m^8}{p^5}$ **19.** $\frac{2y^3}{x^6}$ **20.** $\frac{x^{15}}{8y^3}$

21. $\frac{1}{36a^8b^{10}}$ **22.** $\frac{3y^2}{x^3}$ **23.** $\frac{4w^2}{x^5y^6z^8}$ **24.** $\frac{b^5c^3d^4}{27a^2}$ **25.** 1.563402×10^{11} **26.** 1.79632×10^5 **27.** 9.2×10^{-4} **28.** 1.74×10^{-6}

29. 120,000   **30.** 6,034,000   **31.** 3,000,000   **32.** 0.25   **33.** 0.0000432   **34.** 0.000000006   **35.** 2.0×10^{13} **36.** 9.36×10^{19}
37. 9.6×10^{-10} **38.** 7.8×10^{-11} **39.** 7×10^8 dollars   **40.** 7.94×10^{14} cycles   **41.** 6×10^9 operations   **42.** $5.5x^2 - x - 2.3$

43. $5.6x^2 - 1.1x - 4$ **44.** $-x^3 + 2x^2 - x + 8$ **45.** $7x^3 - 3x^2 - 6x + 4$ **46.** $\frac{1}{10}x^2y - \frac{13}{21}x + \frac{5}{12}$ **47.** $\frac{1}{4}x^2 - \frac{1}{4}x + \frac{1}{10}$

48. $2x^2 + 3x - 5$ **49.** $-3x^2 - 15$ **50.** $15x^2 + 2x - 1$ **51.** $28x^2 - 29x + 6$ **52.** $20x^2 + 48x + 27$ **53.** $10x^3 - 30x^2 + 15x$
54. $-15x^4 - 20x^3 + 25x^2 - 40x$ **55.** $-4x^2y^4 - 20x^2y^3 + 24xy^2$ **56.** $5a^2 - 8ab - 21b^2$ **57.** $8x^4 - 10x^2y - 12x^2 + 15y$
58. $-15x^6y^2 - 9x^4y + 6x^2y$ **59.** $16x^2 + 24x + 9$ **60.** $a^2 - 25b^2$ **61.** $49x^2 - 36y^2$ **62.** $25a^2 - 20ab + 4b^2$ **63.** $64x^2 + 144xy + 81y^2$
64. $4x^3 + 27x^2 + 5x - 3$ **65.** $2x^3 - 7x^2 - 42x + 72$ **66.** $2y^2 + 3y + 4$ **67.** $6x^3 + 7x^2 - 18x$ **68.** $4x^2 - 6x + 8$
69. $53x^3 - 12x^2 + 19x + 13$ **70.** $3x - 2$ **71.** $3x - 7$ **72.** $3x^2 + 2x + 4 + \dfrac{9}{2x - 1}$ **73.** $2x^2 - 5x + 13 - \dfrac{27}{x + 2}$ **74.** $5x + 2$
75. $3x - 4 + \dfrac{3}{2x + 3}$ **76.** $x^2 + 3x + 8$ **77.** $2x^2 + 4x + 5 + \dfrac{11}{x - 2}$ **78.** 4^{29} **79.** 6^{15} **80.** $\dfrac{y^4}{4x^6}$ **81.** $\dfrac{-3ac^2}{b^2}$ **82.** $\dfrac{y^4z^6}{9}$
83. 9.22×10^{-4} **84.** $-4x^3 - 12x + 17$ **85.** $-12x^5 - 30x^4 + 12x^2$ **86.** $14x^3 + 15x^2 - 23x + 6$ **87.** $2x^2 + 5x + 3 + \dfrac{4}{3x - 2}$
88. About \$9.33   **89.** 1.34×10^9 people   **90.** 2.733×10^{-23} g   **91.** 4.76×10^7 lb   **92.** $3xy + 2x$ **93.** $2x^2 - 4y^2$

How Am I Doing? Chapter 5 Test
1. 3^{34} (obj. 5.1.1)   **2.** $\dfrac{1}{25^{16}}$ (obj. 5.1.2)   **3.** 8^{24} (obj. 5.1.3)   **4.** $12x^4y^{10}$ (obj. 5.1.1)
5. $-\dfrac{7x^3}{5}$ (obj. 5.1.2)   **6.** $-125x^3y^{18}$ (obj. 5.1.3)   **7.** $\dfrac{49a^{14}b^4}{9}$ (obj. 5.1.3)   **8.** $\dfrac{3x^4}{4}$ (obj. 5.1.3)   **9.** $\dfrac{1}{64}$ (obj. 5.2.1)   **10.** $\dfrac{6c^5}{a^4b^3}$ (obj. 5.2.1)
11. $3xy^7$ (obj. 5.2.1)   **12.** 5.482×10^{-4} (obj. 5.2.2)   **13.** $582,000,000$ (obj. 5.2.2)   **14.** 2.4×10^{-6} (obj. 5.2.2)   **15.** $-2x^2 + 5x$ (obj. 5.3.2)
16. $-11x^3 - 4x^2 + 7x - 8$ (obj. 5.3.3)   **17.** $-21x^5 + 28x^4 - 42x^3 + 14x^2$ (obj. 5.4.1)   **18.** $15x^4y^3 - 18x^3y^2 + 6x^2y$ (obj. 5.4.1)
19. $10a^2 + 7ab - 12b^2$ (obj. 5.4.2)   **20.** $6x^3 - 11x^2 - 19x - 6$ (obj. 5.5.3)   **21.** $49x^4 + 28x^2y^2 + 4y^4$ (obj. 5.5.2)   **22.** $25s^2 - 121t^2$ (obj. 5.5.1)
23. $12x^4 - 14x^3 + 25x^2 - 29x + 10$ (obj. 5.5.3)   **24.** $3x^4 + 4x^3y - 15x^2y^2$ (obj. 5.4.2)   **25.** $3x^3 - x + 5$ (obj. 5.6.1)
26. $2x^2 - 7x + 4$ (obj. 5.6.2)   **27.** $2x^2 + 6x + 12$ (obj. 5.6.2)   **28.** 3.04×10^9 barrels per year (obj. 5.2.2)   **29.** 4.18×10^6 mi (obj. 5.2.2)

Cumulative Test for Chapters 0–5
1. $\dfrac{1}{30}$ **2.** -0.74 **3.** $-\dfrac{6}{7}$ **4.** $2x^2 - 13x$ **5.** 110.55 **6.** 66 **7.** $x = -\dfrac{9}{2}$
8. $x = 15$ **9.** $x < -3$ **10.** $x \geq \dfrac{3}{2}$ **11.** $12,400$ employees   **12.** $20x^2 - 21x - 5$ **13.** $9x^2 - 30x + 25$ **14.** $6x^3 - 17x^2 - 26x - 8$
15. $-20x^5y^8$ **16.** $-\dfrac{2x^3}{3y^9}$ **17.** $16x^{12}y^8$ **18.** $\dfrac{9z^8}{w^2x^3y^4}$ **19.** 4.25×10^{10} **20.** 5.6×10^{-4} **21.** 4.0×10^{-35} **22.** $5x^3 + 7x^2 - 6x + 50$
23. $-36x^3y^2 + 18x^2y^3 - 48xy^4$ **24.** $6x^3 - 19x^2 + 18x - 5$ **25.** $x + 5 + \dfrac{3}{x - 3}$

Chapter 6
6.1 Exercises
1. factors   **3.** No; $6a^3 + 3a^2 - 9a$ has a common factor of $3a$.   **5.** $3a(a + 1)$ **7.** $7ab(3 - 2b)$
9. $2\pi r(h + r)$ **11.** $5x(x^2 + 5x - 3)$ **13.** $4(3ab - 7bc + 5ac)$ **15.** $8x^2(2x^3 + 3x - 4)$ **17.** $7x(2xy - 5y - 9)$
19. $9x(6x - 5y + 2)$ **21.** $y(3xy - 2a + 5x - 2)$ **23.** $8xy(3x - 5y)$ **25.** $7x^2y^2(x + 3)$ **27.** $8x^2y(x^2 + 3y - 4)$
29. $(x + 2y)(7a - b)$ **31.** $(x - 4)(3x - 2)$ **33.** $(2a - 3c)(6b - 5d)$ **35.** $(b - a^2)(7c - 5d + 2f)$ **37.** $(xy - 3)(2a - 4 - z)$
39. $(a - 3b)(4a^3 + 1)$ **41.** $(a + 2)(1 - x)$ **43.** $17, 19, 21$ **44.** $856,000$ metric tons   **45.** $2,621,500$ metric tons
46. About 22 lb/person   **47.** About 30 lb/person

Quick Quiz 6.1
1. $x(3 - 4x + 2y)$ **2.** $5x(4x^2 - 5x - 1)$ **3.** $(a + 3b)(8a - 7b)$ **4.** See Student Solutions Manual

6.2 Exercises
1. We must remove a common factor of 5 from the last two terms. This will give us $3x(x - 2y) + 5(x - 2y)$. Then our final answer is $(x - 2y)(3x + 5)$.   **3.** $(b - 3)(a + 4)$ **5.** $(x - 4)(x^2 + 3)$ **7.** $(a + 3b)(2x - y)$ **9.** $(3a + b)(x - 2)$ **11.** $(a + 2b)(5 + 6c)$
13. $(c - 2d)(6 + x)$ **15.** $(y - 2)(y - 3)$ **17.** $(9 - y)(6 + y)$ **19.** $(3x + y)(2a - 1)$ **21.** $(x + 4)(2x - 3)$ **23.** $(t - 1)(t^2 + 1)$
25. $(7x + 2y^2)(4x + 3w)$ **27.** We must rearrange the terms in a different order so that the expression in parentheses is the same in each case.
We use the order $6a^2 - 8ad + 9ab - 12bd$ to factor $2a(3a - 4d) + 3b(3a - 4d) = (3a - 4d)(2a + 3b)$.   **29.** 36.2%   **30.** Instructor: \$40,100;
professor: \$91,400   **31.** About 622,000,000 metric tons

Quick Quiz 6.2
1. $(7x + 12)(a - 2)$ **2.** $(y^2 + 3)(2x - 5)$ **3.** $(10y - 3)(x + 4b)$ **4.** See Student Solutions Manual

6.3 Exercises
1. product, sum   **3.** $(x + 1)(x + 1)$ **5.** $(x + 5)(x + 7)$ **7.** $(x - 3)(x - 1)$ **9.** $(x - 7)(x - 4)$
11. $(x + 8)(x - 3)$ **13.** $(x - 14)(x + 1)$ **15.** $(x + 7)(x - 5)$ **17.** $(x - 6)(x + 4)$ **19.** $(x + 12)(x + 3)$ **21.** $(x - 6)(x - 4)$
23. $(x + 3)(x + 10)$ **25.** $(x - 5)(x - 1)$ **27.** $(a + 8)(a - 2)$ **29.** $(x - 4)(x - 8)$ **31.** $(x + 7)(x - 3)$ **33.** $(x + 7)(x + 8)$
35. $(y + 9)(y - 5)$ **37.** $(x + 12)(x - 3)$ **39.** $(x + 3y)(x - 5y)$ **41.** $(x - 7y)(x - 9y)$ **43.** $4(x + 5)(x + 1)$ **45.** $6(x + 1)(x + 2)$
47. $5(x - 1)(x - 5)$ **49.** $3(x + 4)(x - 6)$ **51.** $7(x + 5)(x - 2)$ **53.** $3(x - 1)(x - 5)$ **55.** $(12 + x)(10 - x)$ **57.** $y = \dfrac{7}{2}x + 3$
58. $x \geq -\dfrac{5}{3}$ **59.** 120 mi   **60.** \$3800   **61.** 25°C   **62.** June

Quick Quiz 6.3
1. $(x + 7)(x + 10)$ **2.** $(x - 6)(x - 8)$ **3.** $2(x + 6)(x - 8)$ **4.** See Student Solutions Manual

6.4 Exercises
1. $(4x + 1)(x + 3)$ **3.** $(5x + 2)(x + 1)$ **5.** $(4x - 3)(x + 2)$ **7.** $(2x + 1)(x - 3)$ **9.** $(3x + 1)(3x + 2)$
11. $(3x - 5)(5x - 3)$ **13.** $(2x - 5)(x + 4)$ **15.** $(4x - 1)(2x + 3)$ **17.** $(3x + 2)(2x - 3)$ **19.** $(5x - 1)(2x + 1)$
21. $(x - 2)(7x + 9)$ **23.** $(9y - 4)(y - 1)$ **25.** $(5a + 2)(a - 3)$ **27.** $(7x - 2)(2x + 3)$ **29.** $(5x - 2)(3x + 2)$
31. $(6x + 5)(2x + 3)$ **33.** $(6x + 1)(2x - 3)$ **35.** $(3x^2 + 1)(x^2 - 5)$ **37.** $(2x + 5y)(x + 3y)$ **39.** $(5x - 4y)(x + 4y)$

41. $5(2x + 1)(x - 3)$ **43.** $3x(2x - 5)(x + 4)$ **45.** $(5x - 2)(x + 1)$ **47.** $2(3x - 2)(2x - 5)$ **49.** $x(6x - 1)(2x - 3)$

51. $2(2x - 1)(2x + 5)$ **53.** $x = \dfrac{1}{10}$ **54.** $x = \dfrac{7}{2}$ **55. (a)** 19,400,000 travelers   **(b)** $\approx 22.7\%$ **56.** $\approx 27.4\%$

57. (a) 1.6 million   **(b)** 25%   **58. (a)** 0.2 million   **(b)** $\approx 3.4\%$

Quick Quiz 6.4 **1.** $(2x + 3)(6x - 1)$ **2.** $(5x - 3)(2x - 3)$ **3.** $3x(2x - 5)(x + 2)$ **4.** See Student Solutions Manual

How Am I Doing? Sections 6.1–6.4 **1.** $3(2xy - 5z + 7)$ (obj. 6.1.1)   **2.** $5x(6x - 9y - 2)$ (obj. 6.1.1)   **3.** $(4x - 5)(7 - b)$ (obj. 6.1.1)
4. $(8y + 3z)(2x - 5y)$ (obj. 6.1.1)   **5.** $(6 + x)(3 - y)$ (obj. 6.2.1)   **6.** $(3x + 4w)(5 - 3b)$ (obj. 6.2.1)   **7.** $(x - 5)(x^2 - 3)$ (obj. 6.2.1)
8. $(a + 3b)(7 + 2b)$ (obj. 6.2.1)   **9.** $(x - 7)(x - 8)$ (obj. 6.3.1)   **10.** $(x + 16)(x - 4)$ (obj. 6.3.1)   **11.** $(a + 3b)(a + 7b)$ (obj. 6.3.1)
12. $7(x + 5)(x - 7)$ (obj. 6.3.2)   **13.** $(4x - 1)(3x + 5)$ (obj. 6.4.1)   **14.** $(x - 7)(3x - 2)$ (obj. 6.4.1)   **15.** $(2x + 3y)(3x + 4y)$ (obj. 6.4.2)
16. $2x(7x + 4)(x - 2)$ (obj. 6.4.3)

6.5 Exercises **1.** $(3x + 1)(3x - 1)$ **3.** $(9x + 4)(9x - 4)$ **5.** $(x + 7)(x - 7)$ **7.** $(5x + 9)(5x - 9)$ **9.** $(x + 5)(x - 5)$
11. $(1 + 4x)(1 - 4x)$ **13.** $(4x + 7y)(4x - 7y)$ **15.** $(6x + 13y)(6x - 13y)$ **17.** $(10x + 9)(10x - 9)$ **19.** $(5a + 9b)(5a - 9b)$
21. $(3x + 1)^2$ **23.** $(y - 5)^2$ **25.** $(6x - 5)^2$ **27.** $(7x + 2)^2$ **29.** $(x + 7)^2$ **31.** $(5x - 4)^2$ **33.** $(9x + 2y)^2$ **35.** $(3x - 5y)^2$
37. $(4a + 9b)^2$ **39.** $(7x - 3y)^2$ **41.** $(3x + 7)^2$ **43.** $(12x + 1)(12x - 1)$ **45.** $(x^2 + 6)(x^2 - 6)$ **47.** $(2x^2 - 5)^2$
49. No two binomials can be multiplied to obtain $9x^2 + 1$.   **51.** 49; one answer   **53.** $4(2x + 3)(2x - 3)$ **55.** $3(7x + y)(7x - y)$
57. $4(2x - 1)^2$ **59.** $2(7x + 3)^2$ **61.** $(x + 9)(x + 7)$ **63.** $(2x - 1)(x + 3)$ **65.** $(4x + 11)(4x - 11)$ **67.** $(3x + 7)^2$

69. $9(2x - 1)^2$ **71.** $2(x - 9)(x - 7)$ **73.** $x^2 + 3x + 4 + \dfrac{-3}{x - 2}$ **74.** $2x^2 + x - 5$ **75.** 1.2 oz of greens; 1.05 oz of bulk vegetables;

0.75 oz of fruit   **76.** 1.44 oz of greens; 1.26 oz of bulk vegetables; 0.9 oz of fruit

Quick Quiz 6.5 **1.** $(7x + 9y)(7x - 9y)$ **2.** $(3x - 8)^2$ **3.** $2(9x + 10)(9x - 10)$ **4.** See Student Solutions Manual

6.6 Exercises **1.** $a(5a - 3b + 8)$ **3.** $(4x + 5y)(4x - 5y)$ **5.** $(3x - 2y)^2$ **7.** $(x + 5)(x + 3)$ **9.** $(3x + 2)(5x - 1)$
11. $(a - 3c)(x + 3y)$ **13.** $(y + 7)^2$ **15.** $(2x - 3)^2$ **17.** $(2x - 3)(x - 4)$ **19.** $(x - 10y)(x + 7y)$ **21.** $(a + 3)(x - 5)$
23. $4x(2 + x)(2 - x)$ **25.** $5xy^3(x - 1)^2$ **27.** $3xy(z + 1)(z - 3)$ **29.** $3(x + 7)(x - 5)$ **31.** $3x(x + 4)(x - 3)$
33. $-1(2x^2 + 1)(x + 2)(x - 2)$ **35.** Prime   **37.** $5(x^2 + 2xy - 6y)$ **39.** $3x(2x + y)(5x - 2y)$ **41.** $2(3x - 5)(4x - 3)$
43. Prime   **44.** The number d must be a perfect square such as 1, 4, 9, 16, 25, 36, 49, 64, 81, 100. . . .   **45.** \$28,000   **46.** 372 live strains
47. $(4, -5)$ **48.** 57 hardcover books, 94 softcover books, 47 magazines

Quick Quiz 6.6 **1.** $(2x - 3)(3x - 4)$ **2.** $3(4x + 1)(5x - 2)$ **3.** Prime   **4.** See Student Solutions Manual

6.7 Exercises **1.** $x = -3, x = 7$ **3.** $x = -12, x = -2$ **5.** $x = \dfrac{3}{2}, x = 2$ **7.** $x = \dfrac{2}{3}, x = \dfrac{3}{2}$ **9.** $x = 0, x = -13$ **11.** $x = 3, x = -3$

13. $x = 0, x = 1$ **15.** $x = \dfrac{2}{3}, x = 2$ **17.** $x = -3, x = 2$ **19.** $x = -\dfrac{1}{2}$ **21.** $x = 0, x = -3$ **23.** $x = -8, x = -2$

25. $x = -\dfrac{3}{2}, x = 4$ **27.** You can always factor out x.   **29.** $L = 14$ m; $W = 10$ m   **31.** 66 groups   **33.** 10 students

35. 12 m above ground after 2 sec   **37.** 2415 telephone calls   **39.** 136 handshakes   **41.** $-10x^5y^4$ **42.** $12a^{10}b^{13}$ **43.** $-\dfrac{3a^4}{2b^2}$ **44.** $\dfrac{1}{3x^5y^4}$

Quick Quiz 6.7 **1.** $x = \dfrac{1}{5}, x = \dfrac{1}{3}$ **2.** $x = 3, x = -1$ **3.** $x = 3, x = -\dfrac{3}{4}$ **4.** See Student Solutions Manual

Putting Your Skills to Work **1.** \$520   **2.** About \$31,200   **3.** About \$6200   **4.** \$480   **5.** About \$28,800   **6.** \$25,000
7. Option B   **8.** Answers may vary

Chapter 6 Review Problems **1.** $4x^2(3x - 5y)$ **2.** $5x^3(2 - 7y)$ **3.** $8x^2y(3x - y - 2xy^2)$ **4.** $15a^2b^3(3a + 2 - b)$
5. $3a(a^2 + 2a - 3b + 4)$ **6.** $2(x - 2y + 3z + 6)$ **7.** $(a + 3b)(2a - 5)$ **8.** $3xy(5x^2 + 2y + 1)$ **9.** $(2x + 5)(a - 4)$
10. $(a - 4b)(a + 7)$ **11.** $(x^2 + 3)(y - 2)$ **12.** $3(2x - y)(5a + 7)$ **13.** $(5x - 1)(3x + 2)$ **14.** $(5w - 3)(6w + z)$
15. $(x + 9)(x - 3)$ **16.** $(x + 10)(x - 1)$ **17.** $(x + 6)(x + 8)$ **18.** $(x + 3y)(x + 5y)$ **19.** $(x^2 + 7)(x^2 + 6)$ **20.** $(x^2 - 7)(x^2 + 5)$
21. $6(x + 2)(x + 3)$ **22.** $3(x + 1)(x + 12)$ **23.** $2(x - 6)(x - 8)$ **24.** $4(x - 5)(x - 6)$ **25.** $(4x - 5)(x + 3)$ **26.** $(3x - 1)(4x + 5)$
27. $(5x + 4)(3x - 1)$ **28.** $(3x - 2)(2x - 3)$ **29.** $(2x - 3)(x + 1)$ **30.** $(3x - 4)(x + 2)$ **31.** $(10x - 1)(2x + 5)$
32. $(5x - 1)(4x + 5)$ **33.** $(3a - 2)(2a + 5)$ **34.** $(3a - 2)(2a - 5)$ **35.** $2(x - 1)(3x + 5)$ **36.** $2(x + 1)(3x - 5)$
37. $2(2x - 3)(x - 5)$ **38.** $4(x - 9)(x + 4)$ **39.** $4(2x - 3)(2x + 5)$ **40.** $3(4x - 1)(2x + 1)$ **41.** $(4x + 3y)(3x - 2y)$
42. $(3x - 5y)(2x + 5y)$ **43.** $(7x + y)(7x - y)$ **44.** $4(2x + 3y)(2x - 3y)$ **45.** $(6x - 5)^2$ **46.** $(4x + 5)^2$ **47.** $(5x + 6)(5x - 6)$
48. $(10x + 3)(10x - 3)$ **49.** $(y + 6x)(y - 6x)$ **50.** $(3y + 5x)(3y - 5x)$ **51.** $(6x + 1)^2$ **52.** $(5x - 2)^2$ **53.** $(4x - 3y)^2$
54. $(7x - 2y)^2$ **55.** $2(x + 4)(x - 4)$ **56.** $3(x + 3)(x - 3)$ **57.** $7(2x + 5)^2$ **58.** $8(3x - 4)^2$ **59.** $(2x + 3y)(2x - 3y)$
60. $(x + 3)^2$ **61.** $(x - 3)(x - 6)$ **62.** $(x + 15)(x - 2)$ **63.** $(x - 3)(5x + 2)$ **64.** $(3x - 4)(3x + 1)$ **65.** $12(2x - 5)$
66. $4xy(2xy - 1)$ **67.** $10x^2y^2(5x + 2)$ **68.** $13ab(2a^2 - b^2 + 4ab^3)$ **69.** $x(x - 8)^2$ **70.** $2(x + 10)^2$ **71.** $3(x - 3)^2$
72. $x(5x - 6)^2$ **73.** $(7x + 5)(x - 2)$ **74.** $(4x + 3)(x - 4)$ **75.** $xy(3x + 2y)(3x - 2y)$ **76.** $x^3a(3a + 4x)(a - 5x)$
77. $2(3a + 5b)(2a - b)$ **78.** $(11a + 3b)^2$ **79.** $(a - 1)(7 - b)$ **80.** $(3b - 7)(6 + c)$ **81.** $(5b + 8)(2 - 3x)$ **82.** $(b - 7)(5x + 4y)$
83. $x(2a - 1)(a - 7)$ **84.** $x(x + 4)(x - 4)(x + 1)(x - 1)$ **85.** $(x^2 + 9y^6)(x + 3y^3)(x - 3y^3)$ **86.** $(3x^2 - 5)(2x^2 + 3)$
87. $yz(14 - x)(2 - x)$ **88.** $x(3x + 2)(4x + 3)$ **89.** $(2w + 1)(8w - 5)$ **90.** $3(2w - 1)^2$ **91.** $2y(2y - 1)(y + 3)$
92. $(5y - 1)(2y + 7)$ **93.** $8y^8(y^2 - 2)$ **94.** $9(x^2 + 4)(x - 2)(x + 2)$ **95.** Prime   **96.** Prime   **97.** $4y(2y^2 - 5)(y^2 + 3)$

98. $3x(3y + 7)(y - 2)$ **99.** $(4x^2y - 7)^2$ **100.** $2xy(8x + 1)(8x - 1)$ **101.** $(2x + 5)(a - 2b)$ **102.** $(2x + 1)(x + 3)(x - 3)$

103. $x = -5, x = 4$ **104.** $x = -6, x = \dfrac{1}{2}$ **105.** $x = 0, x = \dfrac{5}{2}$ **106.** $x = 4, x = -3$ **107.** $x = -5, x = \dfrac{1}{2}$ **108.** $x = -8, x = -3$

109. $x = -5, x = -9$ **110.** $x = -\dfrac{3}{5}, x = 2$ **111.** $x = -3$ **112.** $x = -3, x = \dfrac{3}{4}$ **113.** $x = \dfrac{1}{5}, x = 2$ **114.** Base = 5 in., altitude = 10 in.

115. Width = 5 ft, length = 6 ft **116.** 6 sec **117.** 8 amperes, 12 amperes

How Am I Doing? Chapter 6 Test
1. $(x + 14)(x - 2)$ (obj. 6.3.1) **2.** $(4x + 9)(4x - 9)$ (obj. 6.5.1) **3.** $(5x + 1)(2x + 5)$ (obj. 6.4.2)
4. $(3a - 5)^2$ (obj. 6.5.2) **5.** $x(7 - 9x + 14y)$ (obj. 6.1.1) **6.** $(2x + 3b)(5y - 4)$ (obj. 6.4.2) **7.** $2x(3x - 4)(x - 2)$ (obj. 6.6.1)
8. $c(5a - 1)(a - 2)$ (obj. 6.6.1) **9.** $(9x + 10)(9x - 10)$ (obj. 6.5.1) **10.** $(3x - 1)(3x - 4)$ (obj. 6.6.1) **11.** $5(2x + 3)(2x - 3)$ (obj. 6.6.1)
12. Prime (obj. 6.6.2) **13.** $x(3x + 5)(x + 2)$ (obj. 6.6.1) **14.** $-5y(2x - 3y)^2$ (obj. 6.6.1) **15.** $(9x + 1)(9x - 1)$ (obj. 6.5.1)
16. $(9y^2 + 1)(3y + 1)(3y - 1)$ (obj. 6.5.1) **17.** $(x + 3)(2a - 5)$ (obj. 6.6.1) **18.** $(a + 2b)(w + 2)(w - 2)$ (obj. 6.6.1)

19. $3(x - 6)(x + 5)$ (obj. 6.6.1) **20.** $x(2x + 5)(x - 3)$ (obj. 6.6.1) **21.** $x = -5, x = -9$ (obj. 6.7.1) **22.** $x = -\dfrac{7}{3}, x = -2$ (obj. 6.7.1)

23. $x = -\dfrac{5}{2}, x = 2$ (obj. 6.7.1) **24.** $x = 7, x = -4$ (obj. 6.7.1) **25.** Width = 7 mi, length = 13 mi (obj. 6.7.2)

Cumulative Test for Chapters 0–6
1. 20% **2.** 0.54 **3.** 5.8 **4.** $8x^4y^{10}$ **5.** 81 **6.** $27x^2 + 6x - 8$ **7.** $2x^3 - 12x^2 + 19x - 3$

8. $x \geq -\dfrac{9}{2}$ **9.** $x = 2$ **10.** $x = -15$ **11.** $x = -5$ **12.** $(3x - 1)(2x - 1)$ **13.** $(3x + 4)(2x - 1)$ **14.** $(3x + 2)(3x - 1)$

15. $(11x + 8y)(11x - 8y)$ **16.** $-4(5x + 6)(4x - 5)$ **17.** Prime **18.** $x(4x + 5)^2$ **19.** $(4x^2 + b^2)(2x + b)(2x - b)$

20. $(2x + 3)(a - 2b)$ **21.** $3(5x - 3)(x + 1)$ **22.** $x = -9, x = 4$ **23.** $x = \dfrac{5}{3}, x = 2$ **24.** Length = 20 ft, width = 15 ft

Chapter 7
7.1 Exercises
1. 3 **3.** $\dfrac{6}{x}$ **5.** $\dfrac{3x + 1}{1 - 3x}$ **7.** $\dfrac{a(a - 2b)}{2b}$ **9.** $\dfrac{x + 2}{x}$ **11.** $\dfrac{x - 5}{3x - 1}$ **13.** $\dfrac{x - 3}{x(x - 7)}$ **15.** $\dfrac{3x + 1}{x + 5}$

17. $\dfrac{3x - 5}{4x - 1}$ **19.** $\dfrac{x - 5}{x + 1}$ **21.** $-\dfrac{3}{5x}$ **23.** $\dfrac{-2x - 3}{5 + x}$ **25.** $\dfrac{3x + 4}{3x - 1}$ **27.** $\dfrac{2x - 3}{5 - x}$ **29.** $\dfrac{a - b}{2a - b}$ **31.** $9x^2 - 42x + 49$

32. $49x^2 - 36y^2$ **33.** $2x^2 - 5x - 12$ **34.** $2x^3 - 9x^2 - 2x + 24$ **35.** $1\dfrac{5}{8}$ acres **36.** 6 hr, 25 min **37.** 39,500,000 people
38. 80,000,000 people

Quick Quiz 7.1
1. $\dfrac{x}{x - 5}$ **2.** $-\dfrac{2}{b}$ **3.** $\dfrac{2x - 1}{4x + 5}$ **4.** See Student Solutions Manual

7.2 Exercises
1. factor the numerator and denominator completely and divide out common factors **3.** $\dfrac{3(x + 5)}{x + 2}$ **5.** $\dfrac{3x^2}{4(x - 3)}$ **7.** $\dfrac{x - 2}{x + 3}$
9. $\dfrac{(x + 6)(x + 2)}{x + 5}$ **11.** $x + 3$ **13.** $\dfrac{3(x + 2y)}{4(x + 3y)}$ **15.** $\dfrac{(x + 5)(x - 2)}{3x - 1}$ **17.** $\dfrac{-5(x + 3)}{2(x - 2)}$ or $\dfrac{5(x + 3)}{-2(x - 2)}$ or $-\dfrac{5(x + 3)}{2(x - 2)}$ **19.** 1

21. $\dfrac{x - 1}{x + 3}$ **23.** $x = -8$ **24.** $7x^3 - 22x^2 + 2x + 3$ **25.** $79.5(8981)x = \$713,989.5x$ **26.** Harold's is 6 ft by 6 ft. George's is 9 ft by 4 ft.

Quick Quiz 7.2
1. $\dfrac{2(x + 4)}{x - 4}$ **2.** $\dfrac{x - 25}{x + 5}$ **3.** $\dfrac{2(x - 3)}{3x}$ **4.** See Student Solutions Manual

7.3 Exercises
1. The LCD would be a product that contains each factor. However, any repeated factor in any one denominator must be repeated the greatest number of times it occurs in any one denominator. So the LCD would be $(x + 5)(x + 3)^2$. **3.** $\dfrac{3x + 1}{x + 5}$ **5.** $\dfrac{2x - 5}{x + 3}$

7. $\dfrac{2x - 7}{5x + 7}$ **9.** $3a^2b^3$ **11.** $90x^3y^5$ **13.** $18(x - 3)$ **15.** $x^2 - 9$ **17.** $(x + 5)(3x - 1)^2$ **19.** $\dfrac{7 + 3a}{ab}$ **21.** $\dfrac{3x - 13}{(x + 7)(x - 7)}$

23. $\dfrac{5y^2 - 3y}{(y + 1)(y - 1)}$ **25.** $\dfrac{43a + 12}{5a(3a + 2)}$ **27.** $\dfrac{4z + x}{6xyz}$ **29.** $\dfrac{9x + 14}{2(x - 3)}$ **31.** $\dfrac{x + 10}{(x + 5)(x - 5)}$ **33.** $\dfrac{3a + 17b}{10}$ **35.** $\dfrac{-4x + 34}{(2x - 3)(x + 2)}$

37. $\dfrac{-7x}{(x + 3)(x - 1)(x - 4)}$ **39.** $\dfrac{4x + 23}{(x + 4)(x + 5)(x + 6)}$ **41.** $\dfrac{4x - 13}{(x - 4)(x - 3)}$ **43.** $\dfrac{11x}{y - 2x}$ **45.** $\dfrac{-17y^2 - 10y}{(4y - 1)(2y + 1)(y - 5)}$

47. $\dfrac{6}{y + 3}$ **49.** $x = -7$ **50.** $x = 4$ **51.** $x > \dfrac{5}{7}$ **52.** $81x^{12}y^{16}$ **53.** At least 17 days **54.** 284,440 people

Quick Quiz 7.3
1. $\dfrac{2x + 7}{(x + 2)(x - 4)}$ **2.** $\dfrac{bx + by + xy}{bxy}$ **3.** $\dfrac{5x - 1}{(x + 3)(x - 3)(x + 4)}$ **4.** See Student Solutions Manual

How Am I Doing? Sections 7.1–7.3
1. $\dfrac{8}{x}$ (obj. 7.1.1) **2.** $\dfrac{2x + 3}{x - 7}$ (obj. 7.1.1) **3.** $\dfrac{y + 1}{2x^2(1 - y)}$ (obj. 7.1.1)

4. $\dfrac{x - 4}{x + 2}$ (obj. 7.1.1) **5.** $\dfrac{3}{a}$ (obj. 7.2.1) **6.** $\dfrac{x + 5}{-3(x - 2)}$ or $-\dfrac{x + 5}{3(x - 2)}$ (obj. 7.2.1) **7.** $\dfrac{2x - 1}{2}$ (obj. 7.2.2) **8.** $\dfrac{4a - 3}{2a - 5}$ (obj. 7.2.2)

9. $\dfrac{xy - 3ax - 3ay}{axy}$ (obj. 7.3.3)   **10.** $\dfrac{7}{2(x + 2)}$ (obj. 7.3.3)   **11.** $\dfrac{x^2 + 2x - 9}{(x + 3)(x + 7)}$ (obj. 7.3.3)   **12.** $\dfrac{-5x^2 + 19x + 16}{(x - 3)(x + 1)(x + 4)}$ (obj. 7.3.3)

7.4 Exercises 1. $\dfrac{3x}{2 + 5x}$ **3.** $a + b$ **5.** $\dfrac{x^2 - 2x}{4 + 5x}$ **7.** $\dfrac{2}{3x + 10}$ **9.** $\dfrac{1}{xy}$ **11.** $\dfrac{2x - 1}{x}$ **13.** $\dfrac{x - 6}{x + 5}$ **15.** $\dfrac{3a^2 + 9}{a^2 + 2}$ **17.** $\dfrac{3x + 9}{2x - 5}$

19. $\dfrac{y + 1}{-y + 1}$ **21.** No expression in any denominator can be zero because division by zero is undefined. So $-3, 5,$ and 0 are not allowed.

23. $y = \dfrac{-4x + 7}{3}$ **24.** $x > -1$ ⟶   **25.** 6   **26.** \$1875

Quick Quiz 7.4 1. $\dfrac{a(3a - 4b)}{3(5a - 16b)}$ **2.** ab **3.** $\dfrac{2}{x - 1}$ **4.** See Student Solutions Manual

7.5 Exercises 1. $x = -12$ **3.** $x = \dfrac{7}{2}$ or $3\dfrac{1}{2}$ or 3.5   **5.** $x = 15$ **7.** $x = 10$ **9.** $x = -5$ **11.** $x = -\dfrac{14}{3}$ or $-4\dfrac{2}{3}$ **13.** $x = -2$

15. $x = 8$ **17.** $x = -1$ **19.** No solution   **21.** $x = -3$ **23.** $y = -5$ **25.** No solution   **27.** $x = 4$ **29.** No solution

31. $x = 2, x = 4$ **33.** $(4x + 1)(2x - 1)$ **34.** $x = \dfrac{5}{2}$ or $2\dfrac{1}{2}$ or 2.5   **35.** Width = 10 in.; length = 12 in.

36. Domain = $\{7, 2, -2\}$; Range = $\{3, 2, 0, -2, -3\}$; not a function

Quick Quiz 7.5 1. $x = \dfrac{5}{24}$ **2.** $x = 2$ **3.** No solution   **4.** See Student Solutions Manual

7.6 Exercises 1. $x = 18$ **3.** $x = \dfrac{204}{5}$ or $40\dfrac{4}{5}$ or 40.8   **5.** $x = 6.5$ **7.** $x = \dfrac{91}{4}$ or $22\dfrac{3}{4}$ or 22.75   **9.** 110 mi

11. (a) 650 New Zealand dollars   **(b)** 75 New Zealand dollars less   **13.** 56 mph   **15.** 29 mi   **17.** $\dfrac{377}{20}$ or $18\dfrac{17}{20}$ in.   **19.** 200 m

21. $k = \dfrac{56}{5}$ or $11\dfrac{1}{5}$ ft   **23.** $\dfrac{768}{20}$ or $38\dfrac{2}{5}$ m   **25.** 48 in.   **27.** 35 in.   **29.** 61.5 mph   **31.** Commuter = 250 km/hr   **33. (a)** \$0.11
helicopter = 210 km/hr

(b) \$0.09   **(c)** \$3.73   **35.** $3\dfrac{3}{7}$ hr or 3 hr, 26 min   **37.** 8.92465×10^{-4} **38.** 582,000,000   **39.** $\dfrac{w^8}{x^3y^2z^4}$ **40.** $\dfrac{27}{8}$ or $3\dfrac{3}{8}$

Quick Quiz 7.6 1. $x = 55$ **2.** 172 flights   **3.** 24 ft   **4.** See Student Solutions Manual

Putting Your Skills to Work 1. \$2092   **2.** \$174.33   **3.** Answers may vary   **4.** Answers may vary

Chapter 7 Review Problems 1. $\dfrac{x}{x - y}$ **2.** $-\dfrac{4}{5}$ **3.** $\dfrac{x + 3}{x - 4}$ **4.** $\dfrac{x + 2}{x + 4}$ **5.** $\dfrac{x + 3}{x - 7}$ **6.** $\dfrac{2(x + 4)}{3}$ **7.** $\dfrac{x}{x + 3}$ **8.** $\dfrac{2x - 3}{5 - x}$

9. $\dfrac{2(x - 4y)}{2x - y}$ **10.** $\dfrac{2 - y}{3y - 1}$ **11.** $\dfrac{x - 2}{(5x + 6)(x - 1)}$ **12.** $4x + 2y$ **13.** $\dfrac{x + 6}{6(x - 1)}$ **14.** $\dfrac{2y - 1}{5y}$ **15.** $\dfrac{(y + 2)(2y + 1)(y - 2)}{(2y - 1)(y + 1)^2}$

16. $\dfrac{4y(3y - 1)}{(3y + 1)(2y + 5)}$ **17.** $\dfrac{3y(4x + 3)}{2(x - 5)}$ **18.** $\dfrac{x + 2}{2}$ **19.** $\dfrac{2(x + 3y)}{x - 4y}$ **20.** $\dfrac{3}{16}$ **21.** $\dfrac{4(5y + 1)}{3y(y + 2)}$ **22.** $\dfrac{3x^2 + 6x + 1}{x(x + 1)}$

23. $\dfrac{10x - 22}{(x + 2)(x - 4)}$ **24.** $\dfrac{(x - 1)(x - 2)}{(x + 3)(x - 3)}$ **25.** $\dfrac{2xy + 4x + 5y + 6}{2y(y + 2)}$ **26.** $\dfrac{(2a + b)(a + 4b)}{ab(a + b)}$ **27.** $\dfrac{3x - 2}{3x}$ **28.** $\dfrac{2x^2 + 7x - 2}{2x(x + 2)}$

29. $\dfrac{3}{2(x - 9)}$ **30.** $\dfrac{1 - 2x - x^2}{(x + 5)(x + 2)}$ **31.** $-\dfrac{4}{9}$ **32.** $\dfrac{22}{5x^2}$ **33.** $w - 2$ **34.** 1   **35.** $-\dfrac{y^2}{2}$ **36.** $\dfrac{x + 2y}{y(x + y + 2)}$

37. $\dfrac{-1}{a(a + b)}$ or $-\dfrac{1}{a(a + b)}$ **38.** $\dfrac{-3a - b}{b}$ or $-\dfrac{3a + b}{b}$ **39.** $\dfrac{5y(x + 5y)^2}{x(x - 6y)}$ **40.** $\dfrac{-3y}{2(x + 2y)}$ or $-\dfrac{3y}{2(x + 2y)}$ **41.** $a = 2$ **42.** $a = 15$

43. $x = \dfrac{2}{7}$ **44.** $x = -8$ **45.** $y = -4$ **46.** $x = -2$ **47.** $x = \dfrac{1}{2}$ **48.** $x = \dfrac{1}{5}$ **49.** No solution   **50.** $x = -\dfrac{12}{5}$ or $-2\dfrac{2}{5}$ or -2.4

51. $y = 6$ **52.** $y = 2$ **53.** $y = -2$ **54.** No solution   **55.** $y = \dfrac{5}{4}$ or $1\dfrac{1}{4}$ or 1.25   **56.** $y = 0$ **57.** $x = \dfrac{14}{5}$ or $2\dfrac{4}{5}$ or 2.8

58. $x = \dfrac{5}{4}$ or $1\dfrac{1}{4}$ or 1.25   **59.** $x = \dfrac{132}{5}$ or $26\dfrac{2}{5}$ or 26.4   **60.** $x = 6$ **61.** $x = 16$ **62.** $x = \dfrac{91}{10}$ or $9\dfrac{1}{10}$ or 9.1   **63.** 8.3 gal

64. 167 cookies   **65.** 38.9 gal   **66.** 240 mi   **67.** Train = 60 mph   **68.** $1\dfrac{7}{8}$ hr or about 1 hr, 53 min   **69.** 182 ft   **70.** 1200 ft
car = 40 mph

71. $2\dfrac{2}{5}$ hr or 2 hr, 24 min   **72.** 12 hr   **73.** $-\dfrac{a + 4}{3a^2}$ **74.** $\dfrac{4a^2}{2a + 3}$ **75.** $\dfrac{x - 2y}{x + 2y}$ **76.** $\dfrac{x - 8y}{x + 7y}$ **77.** $\dfrac{x + 6}{x}$ **78.** $\dfrac{x + 6}{x}$

79. $\dfrac{b^2 - a^2}{ab(x + y)}$ **80.** $\dfrac{1}{3}$ **81.** $-\dfrac{6}{y^2}$ **82.** $\dfrac{y(x + 3y)}{2(x + 2y)}$ **83.** $x = 12$ **84.** $x = -32$ **85.** $b = -2$ **86.** 10.7 kg   **87.** 420 mi

How Am I Doing? Chapter 7 Test   **1.** $\dfrac{2}{3a}$ (obj. 7.1.1)   **2.** $\dfrac{2x^2(2 - y)}{(y + 2)}$ (obj. 7.1.1)   **3.** $\dfrac{5}{12}$ (obj. 7.2.1)   **4.** $\dfrac{1}{3y(x - y)}$ (obj. 7.2.1)

5. $\dfrac{2a + 1}{a + 2}$ (obj. 7.2.2)   **6.** $\dfrac{3a + 4}{(a + 1)(a - 2)}$ (obj. 7.3.3)   **7.** $\dfrac{x - a}{ax}$ (obj. 7.3.3)   **8.** $-\dfrac{x + 2}{x + 3}$ (obj. 7.3.3)   **9.** $\dfrac{x}{4}$ (obj. 7.4.2)

10. $\dfrac{6x}{5}$ (obj. 7.4.1)   **11.** $\dfrac{2x - 3y}{4x + y}$ (obj. 7.1.1)   **12.** $\dfrac{x}{(x + 2)(x + 4)}$ (obj. 7.3.3)   **13.** $x = -\dfrac{1}{5}$ (obj. 7.5.1)   **14.** $x = 4$ (obj. 7.5.1)

15. No solution (obj. 7.5.2)   **16.** $x = \dfrac{47}{6}$ (obj. 7.5.1)   **17.** $x = \dfrac{45}{13}$ (obj. 7.6.1)   **18.** $x = 37.2$ (obj. 7.6.1)   **19.** 151 flights (obj. 7.6.1)

20. \$368 (obj. 7.6.1)   **21.** 102 ft (obj. 7.6.2)

Cumulative Test for Chapters 0–7   **1.** 0.006 cm   **2.** 16.8%   **3.** \$112.50   **4.** $x = 6$ **5.** $h = \dfrac{A}{\pi r^2}$

6. $x > 1.25$ **7.** $x \le 26$ **8.** $(a + b)(3x - 2y)$ **9.** $2a(4a + b)(a - 5b)$ **10. (a)** $-4x^3y^{12}$ **(b)** $\dfrac{z^6}{4x^2y^4}$

11. $\dfrac{2x + 5}{x + 7}$ **12.** $\dfrac{(x - 2)(3x + 1)}{3x(x + 5)}$ **13.** $\dfrac{1}{2}$ **14.** $\dfrac{11x - 3}{2(x + 2)(x - 3)}$ **15.** $\dfrac{6c - 28}{(c + 2)(c - 2)(c - 3)}$ **16.** $x = -2$

17. $x = -\dfrac{9}{2}$ or $-4\dfrac{1}{2}$ or -4.5 **18.** $\dfrac{x + 8}{6x(x - 3)}$ **19.** $\dfrac{3ab^2 + 2a^2b}{5b^2 - 2a^2}$ **20.** $x = \dfrac{7}{10}$ or 0.7 **21.** 208 mi   **22.** 484 phone calls

Chapter 8   **8.1 Exercises**   **1.** The principal square root of N, where $N \ge 0$, is a nonnegative number a that has the property $a^2 = N$.

3. No; $(0.3)(0.3) = 0.09$ **5.** ± 3 **7.** ± 7 **9.** 4   **11.** 9   **13.** -6 **15.** 0.9   **17.** $\dfrac{6}{11}$ **19.** $\dfrac{7}{8}$ **21.** 25   **23.** -100 **25.** 13

27. $-\dfrac{1}{8}$ **29.** $\dfrac{3}{4}$ **31.** 0.06   **33.** 140   **35.** -17 **37.** 5.196   **39.** 8.602   **41.** -13.528 **43.** -13.964 **45.** 13 ft   **47.** 7.5 sec

49. 3   **51.** -4 **53.** 3   **55.** 3   **57.** No, a 4th power must be nonnegative.   **59.** $(2, 1)$ **60.** No solution, inconsistent system

61. Snowboard: \$250; goggles: \$50   **62.** 470 mph

Quick Quiz 8.1   **1.** 13   **2.** $\dfrac{5}{8}$ **3.** 0.7   **4.** See Student Solutions Manual

8.2 Exercises   **1.** Yes, $\sqrt{6^2} = \sqrt{6 \cdot 6} = \sqrt{36} = 6$ **3.** No, $\sqrt{(-9)^2} = \sqrt{(-9)(-9)} = \sqrt{81} = 9$ **5.** 8   **7.** 18^2 **9.** 9^3 **11.** 33^4
$\left(\sqrt{6}\right)^2 = \left(\sqrt{6}\right)\left(\sqrt{6}\right) = \sqrt{36} = 6$ but $-\sqrt{9^2} = -\sqrt{9 \cdot 9} = -\sqrt{81} = -9$

13. 5^{70} **15.** x^6 **17.** t^9 **19.** y^{13} **21.** $6x^4$ **23.** $12x$ **25.** x^3y^2 **27.** $4xy^{10}$ **29.** $10a^5b^3$ **31.** $2\sqrt{6}$ **33.** $3\sqrt{5}$ **35.** $3\sqrt{2}$

37. $6\sqrt{2}$ **39.** $3\sqrt{10}$ **41.** $8\sqrt{2}$ **43.** $2x\sqrt{2x}$ **45.** $3w^2\sqrt{3w}$ **47.** $4z^4\sqrt{2z}$ **49.** $2x^2y^3\sqrt{7xy}$ **51.** $4y\sqrt{3yw}$ **53.** $5xy\sqrt{3y}$

55. $8x^5$ **57.** $y^3\sqrt{y}$ **59.** $x^2y^5\sqrt{xy}$ **61.** $3x^2y^3\sqrt{15xy}$ **63.** $6ab^2c^2\sqrt{2c}$ **65.** $13ab^4c^2\sqrt{ac}$ **67.** $x + 5$ **69.** $4x + 1$

71. $y(x + 7)$ **73.** $(4, -1)$ **74.** $x = -18$ **75.** $2(x + 3y)^2$ **76.** $(2x - 7)(x + 3)$ **77.** $\dfrac{x + 3}{4y}$ **78.** $\dfrac{37x + 10}{4(x - 2)(3x + 1)}$

Quick Quiz 8.2   **1.** $2\sqrt{30}$ **2.** $9xy^2\sqrt{x}$ **3.** $3x^4y^2\sqrt{15y}$ **4.** See Student Solutions Manual

8.3 Exercises   **1.** 1. Simplify each radical term.   **3.** $6\sqrt{5}$ **5.** $5\sqrt{2} + 3\sqrt{3}$ **7.** $4\sqrt{5}$ **9.** $12\sqrt{5}$ **11.** $\sqrt{2}$ **13.** $9\sqrt{2}$
                        2. Combine like radicals.

15. $5 + 6\sqrt{2} + 6\sqrt{3}$ **17.** $6\sqrt{5} - 2\sqrt{2}$ **19.** $13\sqrt{2x}$ **21.** $0.2\sqrt{3x}$ **23.** $7\sqrt{5y}$ **25.** $-\sqrt{2}$ **27.** $6\sqrt{7x} - 15x\sqrt{7x}$

29. $-5x\sqrt{2x}$ **31.** $10x\sqrt{3}$ **33.** $2y\sqrt{6y} - 6y\sqrt{6}$ **35.** $-110x\sqrt{2x}$ **37.** $\left(8\sqrt{5} + 6\sqrt{3}\right)$ mi

39. $x \ge -\dfrac{4}{3}$ **40.** $\dfrac{9y^4}{4x^2}$ **41.** $-27a^6x^{12}$

Quick Quiz 8.3   **1.** $13\sqrt{2}$ **2.** $3\sqrt{3}$ **3.** $10\sqrt{a} + 6\sqrt{2a}$ **4.** See Student Solutions Manual

8.4 Exercises   **1.** $\sqrt{30}$ **3.** $2\sqrt{11}$ **5.** $3\sqrt{6}$ **7.** $12\sqrt{10}$ **9.** $5\sqrt{2a}$ **11.** $30x\sqrt{2}$ **13.** $16a\sqrt{6}$ **15.** $-6b\sqrt{a}$

17. $\sqrt{10} + \sqrt{15}$ **19.** $6\sqrt{2} + 15\sqrt{5}$ **21.** $2x - 16\sqrt{5x}$ **23.** $2\sqrt{3} - 18 + 4\sqrt{15}$ **25.** $-6 - 9\sqrt{2}$ **27.** $21 + 14\sqrt{2}$

29. $5 - \sqrt{21}$ **31.** $-6 + 9\sqrt{2}$ **33.** $2\sqrt{14} + 34$ **35.** $25 - 4\sqrt{6}$ **37.** $53 + 10\sqrt{6}$ **39.** $48 - 6\sqrt{15}$ **41.** $12\sqrt{10}$

43. $\sqrt{14} + 21\sqrt{2} - \sqrt{42}$ **45.** $9 - 4\sqrt{14}$ **47.** $6\sqrt{ab} + 2a\sqrt{b} - 4a$ **49.** $9x^2y - 5$ **51.** 12 ft^2 **53.** $(8a + 5b)(8a - 5b)$

54. $(2x - 5)^2$ **55.** $m = \left(\dfrac{1519}{1320}\right)k$; approx. 40.3 mph   **56.** $8\dfrac{1}{3}$ yr

Quick Quiz 8.4   **1.** $24xy$ **2.** $2\sqrt{6} + 6 - 4\sqrt{15}$ **3.** $-29 + 2\sqrt{6}$ **4.** See Student Solutions Manual

How Am I Doing? Sections 8.1–8.4 **1.** 7 (obj. 8.1.1) **2.** $\frac{5}{8}$ (obj. 8.1.1) **3.** 9 (obj. 8.1.1) **4.** −10 (obj. 8.1.1)

5. 2.236 (obj. 8.1.2) **6.** −11 (obj. 8.1.2) **7.** 0.2 (obj. 8.2.1) **8.** 0.6 (obj. 8.2.1) **9.** x^2 (obj. 8.2.1) **10.** $5x^2y^3$ (obj. 8.2.1)

11. $5\sqrt{2}$ (obj. 8.2.2) **12.** $x^2\sqrt{x}$ (obj. 8.2.2) **13.** $12ab\sqrt{b}$ (obj. 8.2.2) **14.** $4x^2\sqrt{2x}$ (obj. 8.2.2) **15.** $15\sqrt{2}$ (obj. 8.3.2)

16. $4\sqrt{2}$ (obj. 8.3.2) **17.** $13\sqrt{7} - 7$ (obj. 8.3.2) **18.** $11\sqrt{2} - 18\sqrt{3}$ (obj. 8.3.2) **19.** $-9\sqrt{2x}$ (obj. 8.3.2) **20.** $-26x\sqrt{3}$ (obj. 8.3.2)

21. $15\sqrt{30}$ (obj. 8.4.1) **22.** $8x^2\sqrt{15}$ (obj. 8.4.1) **23.** $2\sqrt{3} + 6$ (obj. 8.4.2) **24.** $16 - 6\sqrt{7}$ (obj. 8.4.3) **25.** 1 (obj. 8.4.3)

26. $17 - 4\sqrt{15}$ (obj. 8.4.3) **27.** $x + 3\sqrt{x} - 10$ (obj. 8.4.3) **28.** $81 + 14\sqrt{21}$ (obj. 8.4.3)

8.5 Exercises **1.** 2 **3.** $\frac{1}{3}$ **5.** 7 **7.** $\frac{\sqrt{6}}{x^2}$ **9.** $\frac{3\sqrt{2}}{a^2}$ **11.** $\frac{3\sqrt{7}}{7}$ **13.** $\frac{8\sqrt{15}}{15}$ **15.** $\frac{x\sqrt{2x}}{2}$ **17.** $\frac{2\sqrt{2x}}{x}$ **19.** $\frac{3\sqrt{7}}{7}$ **21.** $\frac{6\sqrt{a}}{a}$

23. $\frac{\sqrt{2x}}{2x^2}$ **25.** $\frac{3\sqrt{x}}{x^2}$ **27.** $\frac{\sqrt{15}}{5}$ **29.** $\frac{\sqrt{210}}{21}$ **31.** $\frac{9\sqrt{2x}}{8x}$ **33.** $2\sqrt{3} + 2$ **35.** $\sqrt{10} - \sqrt{2}$ **37.** $2 + \sqrt{2}$ **39.** $2\sqrt{14} - 7$

41. $x(2\sqrt{2} + \sqrt{5})$ **43.** $\frac{\sqrt{6x} - \sqrt{2x}}{4}$ **45.** $\frac{13 - 2\sqrt{30}}{7}$ **47.** $\frac{4\sqrt{35} + 3\sqrt{5} + 3\sqrt{2} + 4\sqrt{14}}{3}$ **49.** $5\sqrt{6} + 8\sqrt{2}$ **51.** $\sqrt{x} - 5$

53. (a) $s = \frac{1}{h}\sqrt{3Vh}$ **(b)** $s = 3$ in. **55.** $(3\sqrt{5} - 6)$ m **57.** $x \approx -0.71$ **58.** 0

Quick Quiz 8.5 **1.** $\frac{3\sqrt{10}}{10}$ **2.** $\frac{15x\sqrt{2} - 5x\sqrt{5}}{13}$ **3.** $11 + 2\sqrt{30}$ **4.** See Student Solutions Manual

8.6 Exercises **1.** 13 **3.** $2\sqrt{14}$ **5.** $7\sqrt{2}$ **7.** $3\sqrt{7}$ **9.** $2\sqrt{14}$ **11.** $\sqrt{57}$ **13.** 10.08 **15.** 9.4 ft **17.** 98.0 ft

19. Top cable = 145.40 ft; bottom cable = 140.77 ft **21.** $x = 7$ **23.** $x = 9$ **25.** $x = 7$ **27.** $x = \frac{81}{2}$ **29.** $x = 5$ only **31.** $y = 0, 3$

33. $x = 1$ only **35.** $y = 0, 1$ **37.** $y = \frac{1}{2}$ only **39.** $x = \frac{2}{3}$ only **41.** $x = 5$ **42.** $49x^2 - 28x + 4$ **43.** 20 **44.** Relation

Quick Quiz 8.6 **1.** $2\sqrt{7}$ **2.** $x = 14$ **3.** $x = 8$ **4.** See Student Solutions Manual

8.7 Exercises **1.** $y = 72$ **3.** $y = \frac{1029}{2}$ **5.** $y = 1296$ **7.** 225 calories **9.** 56 min **11.** $y = \frac{45}{4}$ **13.** $y = 1$ **15.** $y = \frac{40}{27}$

17. 3.45 min **19.** $444\frac{4}{9}$ lb **21. (a)** 15.0 km, 8.2 km, and 2.7 km **23. (a)** $k = 1$ **(b)** 248.25 Earth years **25.** 160

(b)

26. $S = 4.80h$

Quick Quiz 8.7 **1.** 6 in. **2.** 281.25 calories **3.** 40 cm^2 **4.** See Student Solutions Manual

Putting Your Skills to Work **1.** $\frac{8}{3}$ or $2\frac{2}{3}$ cups **2.** 56 oz **3.** 3.5 lb **4. (a)** $0.26/lb **(b)** 70% **5.** $2.20

6. Answers may vary **7.** Answers may vary

Chapter 8 Review Problems **1.** 6 **2.** 5 **3.** 13 **4.** 14 **5.** −9 **6.** 8 **7.** 7 **8.** −11 **9.** −12 **10.** 17 **11.** 0.2

12. 0.7 **13.** $\frac{1}{10}$ **14.** $\frac{8}{9}$ **15.** 10.247 **16.** 14.071 **17.** 8.775 **18.** 9.381 **19.** $3\sqrt{3}$ **20.** $2\sqrt{13}$ **21.** $2\sqrt{7}$ **22.** $5\sqrt{5}$

23. $2\sqrt{10}$ **24.** $4\sqrt{5}$ **25.** x^4 **26.** y^5 **27.** $x^2y^3\sqrt{x}$ **28.** $ab^2\sqrt{a}$ **29.** $4xy^2\sqrt{xy}$ **30.** $7x^2y^3\sqrt{2}$ **31.** $2x^2\sqrt{3x}$ **32.** $6x^4\sqrt{2x}$

33. $4x^4\sqrt{3}$ **34.** $5x^6\sqrt{5}$ **35.** $2ab^2c^2\sqrt{30ac}$ **36.** $11a^3b^2\sqrt{c}$ **37.** $2x^3y^4\sqrt{14xy}$ **38.** $3x^6y^3\sqrt{11xy}$ **39.** $3\sqrt{5}$ **40.** $9\sqrt{6}$ **41.** $7x\sqrt{3}$

42. $8a\sqrt{2} + 2a\sqrt{3}$ **43.** $-7\sqrt{5} + 2\sqrt{10}$ **44.** $-2\sqrt{10} + 6\sqrt{7}$ **45.** $6x^2$ **46.** $-10a\sqrt{b}$ **47.** $6ab^2\sqrt{b}$ **48.** $-8x^3$ **49.** −18

50. −12 **51.** $\sqrt{10} - \sqrt{6} - 4$ **52.** $\sqrt{30} - 10 + 5\sqrt{2}$ **53.** $20 + 3\sqrt{11}$ **54.** $27 + 8\sqrt{10}$ **55.** $8 - 4\sqrt{3} + 12\sqrt{6} - 18\sqrt{2}$

56. $15 - 3\sqrt{2} - 10\sqrt{3} + 2\sqrt{6}$ **57.** $66 + 36\sqrt{2}$ **58.** $74 - 40\sqrt{3}$ **59.** $\frac{\sqrt{3x}}{3x}$ **60.** $\frac{2y\sqrt{5}}{5}$ **61.** $\frac{x^2y\sqrt{2}}{4}$ **62.** $\frac{3a\sqrt{2b}}{2}$

63. $\frac{\sqrt{30}}{6}$ **64.** $\frac{\sqrt{66}}{11}$ **65.** $\frac{x\sqrt{5}}{5}$ **66.** $\frac{a^2\sqrt{6}}{6}$ **67.** $\sqrt{5} - \sqrt{2}$ **68.** $\frac{2(\sqrt{6} + \sqrt{3})}{3}$ **69.** $-7 + 3\sqrt{5}$ **70.** $\frac{3 - 2\sqrt{3}}{3}$

71. $c = \sqrt{89}$ **72.** $a = \sqrt{2}$ **73.** $b = \frac{\sqrt{51}}{2}$ **74.** $c = 3.124$ **75.** 30 m **76.** 186.7 mi **77.** 60 ft **78.** $x = 29$ **79.** $x = 13$

80. $x = -2$ **81.** $x = 6$ **82.** $x = 4$ **83.** $x = 7$ **84.** $x = 2$ **85.** $x = 7$ **86.** $y = 48$ **87.** $y = 1$ **88.** $y = \frac{1}{2}$

89. 5 lumens **90.** Approximately 134 ft **91.** It will be 8 times as much. **92.** $\frac{6}{11}$ **93.** 0.02 **94.** $7x^3y\sqrt{2y}$ **95.** $3\sqrt{3}$

96. $2\sqrt{15} + 3\sqrt{6} - 20 - 6\sqrt{10}$ **97.** $57 + 24\sqrt{3}$ **98.** $\frac{5\sqrt{3}}{6}$ **99.** $\frac{6 - \sqrt{6} + 6\sqrt{2} - 2\sqrt{3}}{10}$ **100.** $x = 42$ **101.** $x = -\frac{1}{2}, x = 2$

How Am I Doing? Chapter 8 Test **1.** 11 (obj. 8.1.1) **2.** $\frac{3}{10}$ (obj. 8.1.1) **3.** $4xy^3\sqrt{3y}$ (obj. 8.2.2) **4.** $10xz^2\sqrt{xy}$ (obj. 8.2.2)

5. $3\sqrt{3}$ (obj. 8.3.2) **6.** $8\sqrt{a} + 5\sqrt{2a}$ (obj. 8.3.2) **7.** $12ab$ (obj. 8.4.1) **8.** $5\sqrt{2} + 2\sqrt{15} - 15$ (obj. 8.4.2) **9.** $37 + 20\sqrt{3}$ (obj. 8.4.3)

10. $19 + \sqrt{10}$ (obj. 8.4.3) **11.** $\frac{\sqrt{5x}}{5}$ (obj. 8.5.2) **12.** $\frac{\sqrt{3}}{2}$ (obj. 8.5.2) **13.** $\frac{17 + \sqrt{3}}{22}$ (obj. 8.5.3) **14.** $a(\sqrt{5} - \sqrt{2})$ (obj. 8.5.3)

15. 12.49 (obj. 8.1.2) **16.** $x = \sqrt{133}$ (obj. 8.6.1) **17.** $x = 3\sqrt{3}$ (obj. 8.6.1) **18.** $x = \frac{35}{2}$ (obj. 8.6.2) **19.** $x = 9$ (obj. 8.6.2)

20. 16 in. (obj. 8.7.2) **21.** \$85.77 (obj. 8.7.1) **22.** 21.2 cm^2 (obj. 8.7.1)

Cumulative Test for Chapters 0–8 **1.** $\frac{2}{3}$ **2.** $7\frac{1}{12}$ **3.** $\frac{15}{16}$ **4.** 0.0007 **5.** 3 **6.** 31 **7.** $-3a^2bc^3 - 2abc^2$

8. $x^3 - x^2 - 16x + 16$ **9.** $16x^2 - 40x + 25$ **10.** $9x^2 - 121$ **11.** $\frac{4x + 4}{x + 2}$ **12.** $\frac{(x - 3)(x + 1)}{(x - 6)(x + 2)}$ **13.** $7x^2y^3\sqrt{2x}$ **14.** $7\sqrt{2}$

15. $9 - 6\sqrt{2}$ **16.** $5 + 2\sqrt{6}$ **17.** $10 + 22\sqrt{3}$ **18.** $-4 - 2\sqrt{5}$ **19.** -5 **20.** Not a real number **21.** $c = \sqrt{23}$

22. $b = 4\sqrt{5}$ **23.** $x = 1, 3$ **24.** $y = 6$ **25.** $y = 10$ **26.** 78.5 m^2

Chapter 9 9.1 Exercises

1. $a = 1$ $b = 8$ $c = 7$ **3.** $a = 8$ $b = -11$ $c = 0$ **5.** $a = 1$ $b = 3$ $c = -15$ **7.** $x = 0$ $x = \frac{1}{3}$ **9.** $x = 0$ $x = -2$ **11.** $x = 0$ $x = \frac{3}{2}$ **13.** $x = 7$ $x = -4$ **15.** $x = \frac{1}{2}$ $x = -6$ **17.** $x = \frac{1}{3}$ $x = 5$

19. $x = 7$ $x = -2$ **21.** $x = -\frac{3}{4}$ $x = -\frac{2}{3}$ **23.** $a = -8$ $a = -4$ **25.** $y = \frac{1}{3}$ $y = -\frac{3}{2}$ **27.** $x = \frac{6}{5}$ **29.** $x = 0$ $x = 3$ **31.** $x = 0$ $x = -\frac{14}{3}$ **33.** $y = -2$ $y = 5$

35. $x = 3$ $x = 7$ **37.** $x = 1$ **39.** $x = 4$ $x = 2$ **41.** $x = -7$ $x = 2$ **43.** $y = \frac{4}{3}$ $y = -1$ **45.** $x = \frac{5}{2}$ $x = -4$ **47.** $x = -\frac{10}{3}$ $x = 4$ **49.** You can always factor out x.

51. $a = \frac{14}{9}$ $c = -14$ **53.** $n = 1200$ $n = 2000$ **55.** Average = 1600 Revenue for 1600 = \$360, which is more than producing 1200 or 2000 labels **57.** Revenue for 1500 = \$357.50 Revenue for 1700 = \$357.50 It appears the maximum revenue is at 1600 labels.

59. $\frac{(2x - 1)(x + 2)}{2x}$ **60.** $\frac{2x^2 - 4x + 2}{3x^2 - 11x + 10}$

Quick Quiz 9.1 **1.** $x = \frac{5}{3}$ $x = -\frac{3}{2}$ **2.** $x = 8$ $x = 6$ **3.** $x = -3$ $x = \frac{2}{3}$ **4.** See Student Solutions Manual

9.2 Exercises **1.** $x = \pm 8$ **3.** $x = \pm 7\sqrt{2}$ **5.** $x = \pm 2\sqrt{7}$ **7.** $x = \pm 3$ **9.** $x = \pm 2\sqrt{5}$ **11.** $x = \pm 5\sqrt{5}$ **13.** $x = \pm 2\sqrt{3}$

15. $x = \pm\sqrt{5}$ **17.** $x = 3, x = 11$ **19.** $x = -4 \pm \sqrt{6}$ **21.** $x = \frac{-5 \pm \sqrt{2}}{2}$ **23.** $x = \frac{1 \pm \sqrt{7}}{3}$ **25.** $x = \frac{-1 \pm 3\sqrt{2}}{5}$

27. $x = \frac{5 \pm 3\sqrt{6}}{4}$ **29.** $x = 3 \pm 2\sqrt{5}$ **31.** $x = 2, x = -10$ **33.** $x = 6 \pm \sqrt{41}$ **35.** $x = 0, x = 7$ **37.** $x = 0, x = 5$

39. $x = \frac{9}{2}, x = -1$ **41.** $a = -4, b = -4$ **42.** $x = -1, x = 4$ **43.** 200 lb/in.^{2}

44. A total of 16 tires were defective. 6 had both kinds of defects. 7 had defects only in workmanship. 3 had defects only in materials.

Quick Quiz 9.2 **1.** $x = \pm\sqrt{6}$ **2.** $x = \frac{2 \pm 2\sqrt{10}}{3}$ **3.** $x = -4 \pm \sqrt{26}$ **4.** See Student Solutions Manual

9.3 Exercises **1.** $\sqrt{b^2 - 4ac} = \sqrt{100} = 10$ Therefore, there are two rational roots. **3.** No. Standard form is $4x^2 + 5x - 6 = 0$. Then $a = 4, b = 5$, and $c = -6$.

5. $x = 2, x = -5$ **7.** $x = \frac{3 \pm \sqrt{41}}{2}$ **9.** $x = \frac{1}{4}, x = -2$ **11.** $x = 4, x = -\frac{5}{2}$ **13.** $x = \frac{3 \pm \sqrt{33}}{12}$

15. $x = \frac{1 \pm \sqrt{19}}{6}$ **17.** $x = 5, x = 2$ **19.** No real solution **21.** $y = \frac{5 \pm \sqrt{13}}{4}$ **23.** $d = \frac{4}{3}, d = -3$ **25.** No real solution

27. $x = -6$ **29.** $x \approx 0.372, x \approx -5.372$ **31.** $x \approx 4.108, x \approx -0.608$ **33.** $x \approx -1.894, x \approx -0.106$ **35.** $x = \frac{3}{2}, x = \frac{2}{3}$

37. $x = 3, x = \frac{1}{3}$ **39.** $t = -1 \pm \sqrt{23}$ **41.** $y = \frac{1 \pm \sqrt{51}}{5}$ **43.** $x = \frac{\pm\sqrt{39}}{3}$ **45.** $x = 1 \pm 2\sqrt{2}$

47. 2 ft **49.** $x^2 + 5x + 2$ **50.** $y = -\frac{5}{13}x - \frac{14}{13}$

Quick Quiz 9.3 **1.** $x = \dfrac{-1 \pm \sqrt{2}}{3}$ **2.** $x = \dfrac{1 \pm 2\sqrt{5}}{2}$ **3.** $x = \dfrac{-1 \pm \sqrt{7}}{3}$ **4.** See Student Solutions Manual

How Am I Doing? Sections 9.1–9.3 **1.** $x = 16, x = -3$ (obj. 9.1.2) **2.** $x = 0, x = \dfrac{7}{5}$ (obj. 9.1.3) **3.** $x = \dfrac{2}{5}, x = 4$ (obj. 9.1.3)

4. $x = \dfrac{1}{3}, x = -\dfrac{1}{2}$ (obj. 9.1.3) **5.** $x = -8, x = 3$ (obj. 9.1.3) **6.** $x = 3, x = -\dfrac{9}{2}$ (obj. 9.1.3) **7.** $x = \pm 3\sqrt{2}$ (obj. 9.2.1)

8. $x = \pm 5$ (obj. 9.2.1) **9.** $x = -3 \pm 2\sqrt{2}$ (obj. 9.2.2) **10.** $x = \dfrac{-3 \pm \sqrt{65}}{4}$ (obj. 9.2.2) **11.** $x = \dfrac{-2 \pm \sqrt{14}}{2}$ (obj. 9.3.1)

12. $x = \dfrac{5}{4}, x = -1$ (obj. 9.3.1) **13.** $x = \dfrac{-4 \pm \sqrt{13}}{3}$ (obj. 9.3.1) **14.** No real solution (obj. 9.3.3)

9.4 Exercises **1.** It may have 2, 1, or none. **3.** If $b = 0$, then the parabola always has a vertex at $(0, c)$ which is on the y-axis.

5.
$y = x^2 + 2$

7.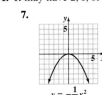
$y = -\dfrac{1}{3}x^2$

9.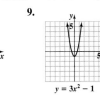
$y = 3x^2 - 1$

11.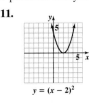
$y = (x - 2)^2$

13.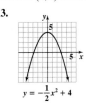
$y = -\dfrac{1}{2}x^2 + 4$

15.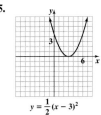
$y = \dfrac{1}{2}(x - 3)^2$

17.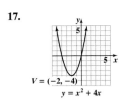
$V = (-2, -4)$
$y = x^2 + 4x$

19.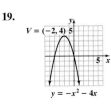
$V = (-2, 4)$
$y = -x^2 - 4x$

21.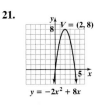
$V = (2, 8)$
$y = -2x^2 + 8x$

23.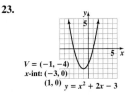
$V = (-1, -4)$
x-int: $(-3, 0)$
$(1, 0)$ $y = x^2 + 2x - 3$

25.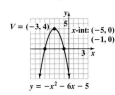
$V = (-3, 4)$ x-int: $(-5, 0)$
$(-1, 0)$
$y = -x^2 - 6x - 5$

27.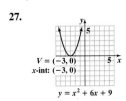
$V = (-3, 0)$
x-int: $(-3, 0)$
$y = x^2 + 6x + 9$

29. (a)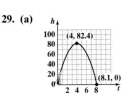
(4, 82.4)
(8.1, 0)
(b) 62.8 m
(c) 82.4 m
(d) About 8.1 sec

31. (a)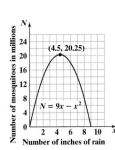
$N = 9x - x^2$
(4.5, 20.25)
Number of mosquitoes in millions
Number of inches of rain
(b) 4.5 in.
(c) 20.25 million

33. Vertex: $(-0.277, 1551.164)$
y-intercept: $(0, 1528)$
x-intercepts: $(1.993, 0)$ and $(-2.548, 0)$

34. $y = 1$ **35.** $y = 75$ **36.** 8.25 sec

Quick Quiz 9.4 **1.** $(3, -4)$ **2.** $(1, 0)$ and $(5, 0)$ **3.**

$y = x^2 - 6x + 5$

4. See Student Solutions Manual

9.5 Exercises **1.** Base = 11 cm **3.** 10 ft by 10 ft **5.** 21 people **7.** 80 members were expected. **9.** Width = 25 ft
altitude = 16 cm   100 members attended.   length = 50 ft

11. 600 mph **13.** 50 stores **15.** 15 stores **17.** 2002 and 2003 **19.** About 74 mi **21.**

2        $\dfrac{5}{2}$ 3

22. $\sqrt{15} - 2\sqrt{3}$ **23.** 0.75 ft

Quick Quiz 9.5 **1.** 12 yd, 5 yd **2.** 6 sec **3.** 61.25 m **4.** See Student Solutions Manual

Putting Your Skills to Work **1.** $16.40 **2.** $24.60 **3.** About $246 **4.** Answers may vary **5.** Answers may vary
6. $150 **7.** $2225 **8.** $178 **9.** $356 **10.** $62°$

Chapter 9 Review Problems
1. $a = 6, b = -5, c = 8$ **2.** $a = 14, b = 3, c = -16$ **3.** $a = 1, b = 3, c = -14$
4. $a = 11, b = 8, c = 0$ **5.** $a = -2, b = -6, c = 3$ **6.** $a = 4, b = 5, c = -14$ **7.** $x = -25, x = -1$ **8.** $x = -8$ **9.** $x = 5, x = -11$
10. $x = \frac{3}{2}, x = \frac{5}{2}$ **11.** $x = \frac{4}{3}$ **12.** $x = \frac{2}{5}, x = \frac{4}{3}$ **13.** $x = \frac{5}{3}, x = \frac{1}{2}$ **14.** $x = 1, x = -\frac{9}{4}$ **15.** $x = -\frac{3}{2}, x = \frac{4}{3}$ **16.** $x = \frac{1}{2}, x = 1$
17. $x = \frac{1}{5}, x = -\frac{1}{3}$ **18.** $x = \frac{1}{4}, x = -\frac{4}{3}$ **19.** $x = 0, x = -8$ **20.** $x = 0, x = 3$ **21.** $x = 0, x = -\frac{5}{3}$ **22.** $x = 0, x = 4$
23. $x = -\frac{2}{5}, x = 10$ **24.** $x = \frac{1}{3}, x = \frac{4}{3}$ **25.** $x = \pm 7$ **26.** $x = \pm 9$ **27.** $x = \pm\sqrt{22}$ **28.** $x = \pm\sqrt{39}$ **29.** $x = \pm 2\sqrt{2}$
30. $x = \pm 5\sqrt{2}$ **31.** $x = \pm 3\sqrt{2}$ **32.** $x = \pm 2\sqrt{6}$ **33.** $x = 4 \pm \sqrt{7}$ **34.** $x = 2 \pm \sqrt{3}$ **35.** $x = \frac{-2 \pm 2\sqrt{7}}{3}$ **36.** $x = \frac{1 \pm 4\sqrt{2}}{2}$
37. $x = -1, x = -7$ **38.** $x = -3, x = -11$ **39.** $x = 9, x = -5$ **40.** $x = 3 \pm \sqrt{2}$ **41.** $x = -1 \pm \sqrt{3}$ **42.** $x = \frac{-5 \pm \sqrt{31}}{2}$
43. $x = -2 \pm \sqrt{10}$ **44.** $x = -2 \pm 2\sqrt{3}$ **45.** $x = \frac{7 \pm \sqrt{17}}{4}$ **46.** $x = \frac{-5 \pm \sqrt{73}}{4}$ **47.** $x = \frac{4 \pm 2\sqrt{7}}{3}$ **48.** $x = \frac{1 \pm 3\sqrt{5}}{4}$
49. $x = \frac{1 \pm \sqrt{41}}{10}$ **50.** $x = \frac{-3 \pm \sqrt{21}}{6}$ **51.** $x = \frac{5}{2}, x = 2$ **52.** $x = \frac{3}{2}, x = -\frac{1}{2}$ **53.** $x = -\frac{1}{5}$ **54.** $x = 4, x = \frac{3}{2}$
55. $x = \frac{3 \pm \sqrt{3}}{3}$ **56.** $x = \frac{7 \pm \sqrt{209}}{10}$ **57.** $x = \frac{-2 \pm \sqrt{19}}{3}$ **58.** $x = \frac{-7 \pm \sqrt{29}}{10}$ **59.** $x = \frac{9 \pm \sqrt{93}}{2}$ **60.** $x = -\frac{1}{2}, x = 5$
61. $x = \frac{3}{4}, x = -2$ **62.** $x = \pm 3\sqrt{2}$ **63.** $y = -8$ only **64.** $y = -7$ only **65.** $x = \frac{-4 \pm \sqrt{31}}{3}$ **66.** $x = \frac{-3 \pm \sqrt{13}}{4}$
67. $y = -\frac{9}{8}, y = 1$ **68.** $y = -4, y = 2$ **69.** $y = -3, y = 5$ **70.** $x = 0, x = \frac{1}{6}$

71.
$y = 2x^2$

72.
$y = x^2 + 4$

73.
$y = x^2 - 3$

74.
$y = -\frac{1}{2}x^2$

75.
$y = x^2 - 3x - 4$

76.
$y = \frac{1}{2}x^2 - 2$

77.
$y = -2x^2 + 12x - 17$

78.
$y = -3x^2 - 2x + 4$

79. 7 ft by 4 ft or 8 ft by 3.5 ft   **80.** 16 cm, 12 cm

81. Width $= 18$ in.   **82.** $t = 30$ sec   **83.** 24 members   **84.** Going, 45 mph; returning, 60 mph   **85.** 60 pizza stores
length $= 24$ in.
86. 30 fewer pizza stores   **87.** 1990 and 2020   **88.** 2000 and 2010

How Am I Doing? Chapter 9 Test
1. $x = \frac{-7 \pm \sqrt{129}}{10}$ (obj. 9.3.1)   **2.** $x = -5, x = \frac{2}{3}$ (obj. 9.1.3)
3. No real solution (obj. 9.3.3)   **4.** $x = \frac{3}{4}, x = 5$ (obj. 9.2.2)   **5.** $x = -\frac{5}{4}, x = \frac{1}{3}$ (obj. 9.2.1)   **6.** $x = \frac{4}{3}$ (obj. 9.2.2)
7. $x = 0, x = 8$ (obj. 9.3.1)   **8.** $x = \pm 3$ (obj. 9.3.1)   **9.** $x = -\frac{1}{2}, x = 6$ (obj. 9.3.1)   **10.** $x = \frac{3}{2}, x = -\frac{1}{2}$ (obj. 9.3.1)

11.   (obj. 9.4.1)
$y = 3x^2 - 6x$

12.   (obj. 9.4.1)
$y = -x^2 + 8x - 12$

13. 9 m, 12 m (obj. 9.5.1)   **14.** 7 sec (obj. 9.5.1)

Practice Final Examination
1. $-2x + 21y + 18xy + 6y^2$ **2.** 14   **3.** $18x^5y^5$ **4.** $2x^2y - 8xy$ **5.** $x = \frac{5}{2}$ **6.** $x = \frac{8}{5}$
7. $x \geq 2.6$
8. $4x^2 + 12x + 9$ **9.** $2x^3 - 5x^2y + xy^2 + 2y^3$ **10.** $2(2x + 1)(x - 5)$
11. $3x(x - 5)(x + 2)$ **12.** $\frac{6x - 11}{(x + 2)(x - 3)}$ **13.** $\frac{11(x + 2)}{2x(x + 4)}$ **14.** $-\frac{12}{5}$ **15.** slope $= \frac{5}{2}$
$5x - 2y - 3 = 0$

16. $3x + 4y = 14$ or $y = -\dfrac{3}{4}x + \dfrac{7}{2}$ **17.** 38.13 in.^{2} **18.** $x = -5, y = 2$ **19.** $a = -2, b = -3$ **20.** $x\sqrt{5x}$ **21.** $6\sqrt{3} - 12 + 12\sqrt{2}$

22. $-\dfrac{\sqrt{15} + \sqrt{35} + \sqrt{21} + 7}{2}$ **23.** $x = \dfrac{2}{3}, x = -\dfrac{1}{4}$ **24.** $y = \dfrac{3 \pm \sqrt{7}}{2}$ **25.** $x = \pm 2$ **26.** $\sqrt{39}$ **27.** 5 **28.** Width 7 m, length 12 m

29. 200 reserved-seat tickets, 160 general admission tickets   **30.** Base 8 m, altitude 17 m

Appendix B

Exercises

1. $305.50   **3. (a)** $82\dfrac{1}{3}$ ft   **(b)** 90 feet; $17.90   **5.** jog $2\dfrac{2}{3}$ mi; walk $3\dfrac{1}{9}$ mi; rest $4\dfrac{4}{9}$ min; walk $1\dfrac{7}{9}$ mi

7. Betty; Melinda increases each activity by $\dfrac{2}{3}$ by day 3 but Betty increases each activity by $\dfrac{7}{9}$ by day 3.   **9.** $4\dfrac{1}{2}$ miles   **11.** $98,969

13. (a) $45,000   **(b)** $13,950   **15.** 18%   **17.** 69%

Appendix C

Addition Practice

1. 37   **2.** 75   **3.** 94   **4.** 99   **5.** 78   **6.** 29   **7.** 61   **8.** 81   **9.** 369   **10.** 277   **11.** 491   **12.** 600   **13.** 1421

14. 1052   **15.** 900   **16.** 401   **17.** 1711   **18.** 1319   **19.** 289   **20.** 399   **21.** 589   **22.** 887   **23.** 422   **24.** 792   **25.** 902

26. 930

Subtraction Practice

1. 21   **2.** 62   **3.** 22   **4.** 43   **5.** 68   **6.** 5   **7.** 29   **8.** 45   **9.** 531   **10.** 223   **11.** 726   **12.** 191   **13.** 378

14. 66   **15.** 509   **16.** 228   **17.** 256   **18.** 63   **19.** 183   **20.** 269   **21.** 873   **22.** 1892   **23.** 7479   **24.** 2868   **25.** 2066

26. 678

Multiplication Practice

1. 69   **2.** 26   **3.** 378   **4.** 603   **5.** 1554   **6.** 1643   **7.** 3680   **8.** 3640   **9.** 7790   **10.** 2691   **11.** 9116   **12.** 7502

13. 12,095   **14.** 30,168   **15.** 20,672   **16.** 24,180   **17.** 37,584   **18.** 53,692   **19.** 284,490   **20.** 506,566   **21.** 238,119

22. 375,820   **23.** 803,396   **24.** 785,354   **25.** 49,516,817   **26.** 62,526,175

Division Practice

1. 16   **2.** 56   **3.** 59 R2   **4.** 47 R5   **5.** 124   **6.** 296   **7.** 234   **8.** 567   **9.** 521 R6   **10.** 739 R2   **11.** 321 R11

12. 753 R6   **13.** 89   **14.** 91   **15.** 32   **16.** 41   **17.** 108 R29   **18.** 123 R25   **19.** 214   **20.** 342   **21.** 1134 R14

22. 1076 R65   **23.** 124   **24.** 153   **25.** 1125   **26.** 1532

Appendix D

Exercises

1. $y = -\dfrac{2}{3}x + 8$ **3.** $y = 5x + 33$ **5.** $y = -\dfrac{1}{5}x + \dfrac{6}{5}$ **7.** $y = \dfrac{5}{7}x + \dfrac{13}{7}$ **9.** $y = -\dfrac{2}{3}x - \dfrac{8}{3}$ **11.** $y = -3$

13. $5x - y = -10$ **15.** $x - 3y = 8$ **17.** $2x - 3y = 15$ **19.** $7x - y = -27$ **21.** Neither   **23.** Parallel   **25.** Perpendicular

Absolute value of a number (1.1)   The absolute value of a number x is the distance between 0 and the number x on the number line. It is written as $|x|$. $|x| = x$ if $x \geq 0$, but $|x| = -x$ if $x < 0$.

Altitude of a geometric figure (1.8)   The height of the geometric figure. In the three figures shown, the altitude is labeled a.

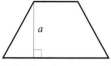

Altitude of a trapezoid   Altitude of a parallelogram

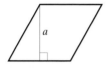

Altitude of a rhombus

Altitude of a triangle (1.8)   The height of any given triangle. In the three triangles shown, the altitude is labeled a.

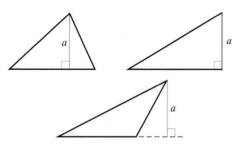

Associative property of addition (1.1)   If a, b, and c are real numbers, then

$$a + (b + c) = (a + b) + c.$$

This property states that if three numbers are added, it does not matter *which two numbers* are added first; the result will be the same.

Associative property of multiplication (1.3)   If a, b, and c are real numbers, then

$$a \times (b \times c) = (a \times b) \times c.$$

This property states that if three numbers are multiplied, it does not matter *which two numbers* are multiplied first; the result will be the same.

Base (1.4)   The number or variable that is raised to a power. In the expression 2^6, the number 2 is the base.

Base of a triangle (1.8)   The side of a triangle that is perpendicular to the altitude.

Binomial (5.3)   A polynomial of two terms. The expressions $a + 2b$, $6x^3 + 1$, and $5a^3b^2 + 6ab$ are all binomials.

Circumference of a circle (1.8)   The distance around a circle. The circumference of a circle is given by the formula $C = \pi d$ or $C = 2\pi r$, where d is the diameter of the circle and r is the radius of the circle.

Coefficient (5.1)   A coefficient is a factor or a group of factors in a product. In the term $4xy$ the coefficient of y is $4x$, but the coefficient of xy is 4. In the term $-5x^3y$ the coefficient of x^3y is -5.

Commutative property for addition (1.1)   If a and b are any real numbers, then $a + b = b + a$.

Commutative property for multiplication (1.3)   If a and b are any real numbers, then $ab = ba$.

Complex fraction (7.4)   A fraction that contains at least one fraction in the numerator or in the denominator or both. These three fractions are complex fractions:

$$\frac{7 + \dfrac{1}{x}}{x^2 + 2}, \qquad \frac{1 + \dfrac{1}{5}}{2 - \dfrac{1}{7}}, \qquad \text{and} \qquad \frac{\dfrac{1}{3}}{4}$$

Conjugates (8.5)   The expressions $a\sqrt{x} + b\sqrt{y}$ and $a\sqrt{x} - b\sqrt{y}$. The conjugate of $2\sqrt{3} + 5\sqrt{2}$ is $2\sqrt{3} - 5\sqrt{2}$. The conjugate of $4 - \sqrt{x}$ is $4 + \sqrt{x}$.

Constant (2.3)   Symbol or letter that is used to represent exactly one single quantity during a particular problem or discussion.

Coordinates of a point (3.1)   An ordered pair of numbers (x, y) that specifies the location of a point in a rectangular coordinate system.

Degree of a polynomial (5.3)   The degree of the highest-degree term of a polynomial. The degree of the polynomial $5x^3 + 2x^2 - 6x + 8$ is 3. The degree of the polynomial $5x^2y^2 + 3xy + 8$ is 4.

Degree of a term of a polynomial (5.3)   The sum of the exponents of the variables in the term. The degree of $3x^3$ is 3. The degree of $4x^5y^2$ is 7.

Denominator (0.1) and (7.1) The bottom number or algebraic expression in a fraction. The denominator of

$$\frac{3x - 2}{x + 4}$$

is $x + 4$. The denominator of $\frac{3}{7}$ is 7. The denominator of a fraction may not be zero.

Dependent equations (4.1)   Two equations are dependent if every value that satisfies one equation satisfies the other. A system of two dependent equations in two variables will not have a unique solution.

Difference-of-two-squares polynomial (6.5)   A polynomial of the form $a^2 - b^2$ that may be factored by using the formula

$$a^2 - b^2 = (a + b)(a - b).$$

Direct variation (8.7)   When a variable y varies directly with x, written $y = kx$, where k represents some real number that will stay the same over a range of x-values. This value k is called the *constant of variation*.

Distributive property (1.6)   For all real numbers a, b, and c, $a(b + c) = ab + ac$.

Dividend (0.4)   The number that is to be divided by another. In the problem $30 \div 5 = 6$, the three parts are as follows:

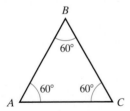

5 is the divisor
30 is the dividend
6 is the quotient

Divisor (0.4)   The number you divide into another.

Domain of a relation (3.6)   In any relation, the set of values that can be used for the independent variable is called its domain. This is the set of all the first coordinates of the ordered pairs that define the relation.

Equilateral triangle (1.8)   A triangle with three sides equal in length and three angles that each measure 60°. Triangle ABC is an equilateral triangle.

Even integers (1.3)   Integers that are exactly divisible by 2, such as $\ldots, -4, -2, 0, 2, 4, 6, \ldots$.

Exponent (1.4)   The number that indicates the power of a base. If the number is a positive integer, it indicates how many times the base is multiplied. In the expression 2^6, the exponent is 6.

Expression (5.3)   A mathematical expression is any quantity using numbers and variables. Therefore, $2x$, $7x + 3$, and $5x^2 + 6x$ are all mathematical expressions.

Extraneous solution (7.5) and (8.6)   An obtained solution to an equation that when substituted back into the original equation, does *not* yield an identity. $x = 2$ is an extraneous solution to the equation

$$\frac{x}{x - 2} - 4 = \frac{2}{x - 2}$$

An extraneous solution is also called an extraneous root.

Factor (0.1) and (6.1)   When two or more numbers, variables, or algebraic expressions are multiplied, each is called a factor. If we write $3 \cdot 5 \cdot 2$, the factors are 3, 5, and 2. If we write $2xy$, the factors are 2, x, and y. In the expression $(x - 6)(x + 2)$, the factors are $(x - 6)$ and $(x + 2)$.

Fractions

Algebraic fractions (7.1)   The indicated quotient of two algebraic expressions.

$$\frac{x^2 + 3x + 2}{x - 4} \quad \text{and} \quad \frac{y - 6}{y + 8}$$

are algebraic fractions. In these fractions the value of the denominator cannot be zero.

Numerical fractions (0.1)   A set of numbers used to describe parts of whole quantities. A numerical fraction can be represented by the quotient of two integers for which the denominator is not zero. The numbers

$$\frac{1}{5}, \ -\frac{2}{3}, \ \frac{8}{2}, \ -\frac{4}{31}, \ \frac{8}{1}, \ \text{and} \ -\frac{12}{1}$$

are all numerical fractions. The set of rational numbers can be represented by numerical fractions.

Function (3.6)   A relation in which no two different ordered pairs have the same first coordinate.

Graph of a function (3.6)   A graph in which a vertical line will never cross in more than one place. The following sketches represent the graphs of functions.

Hypotenuse of a right triangle (8.6)   The side opposite the right angle in any right triangle. The hypotenuse is always the longest side of a right triangle. In the following sketch the hypotenuse is side c.

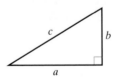

Improper fraction (0.1)   A numerical fraction whose numerator is larger than or equal to its denominator. $\frac{8}{3}, \frac{5}{2}$, and $\frac{7}{7}$ are improper fractions.

Inconsistent system of equations (4.1)   A system of equations that does not have a solution.

Independent equations (4.1)   Two equations that are not dependent are said to be independent.

Inequality (2.8) and (3.5)   A mathematical relationship between quantities that are not equal. $x \leq -3$, $w > 5$, and $x < 2y + 1$ are inequalities.

Integers (1.1) The set of numbers ..., $-5, -4, -3, -2,$ $-1, 0, 1, 2, 3, 4, 5, \ldots$.

Intercepts of an equation (3.2) The point or points where the graph of the equation crosses the x-axis or the y-axis or both. (*See* x-intercept, y-intercept.)

Inverse variation (8.7) When a variable y varies inversely with x, written $y = \dfrac{k}{x}$, where k is the constant of variation.

Irrational number (1.1) and (8.1) A real number that cannot be expressed in the form $\dfrac{a}{b}$, where a and b are integers and $b \neq 0$. $\sqrt{2}$, π, $5 + 3\sqrt{2}$, and $-4\sqrt{7}$ are irrational numbers.

Isosceles triangle (1.8) A triangle with two equal sides and two equal angles. Triangle ABC is an isosceles triangle. Angle BAC is equal to angle ACB. Side AB is equal in length to side BC.

Least common denominator of numerical fractions (0.2) The smallest whole number that is exactly divisible by all denominators of a group of fractions. The least common denominator (LCD) of $\dfrac{1}{6}, \dfrac{2}{3}$, and $\dfrac{3}{5}$ is 30. The least common denominator is also called the lowest common denominator.

Leg of a right triangle (8.6) One of the two shorter sides of a right triangle. In the following sketch, sides a and b are the legs of the right triangle.

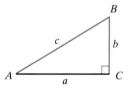

Like terms (1.7) Terms that have identical variables and exponents. In the expression $5x^3 + 2xy^2 + 6x^2 - 3xy^2$, the term $2xy^2$ and the term $-3xy^2$ are like terms.

Line of symmetry of a parabola (9.4) A line that can be drawn through a parabola such that, if the graph were folded on this line, the two halves of the curve would correspond exactly. The line of symmetry through the parabola formed by $y = ax^2 + bx + c$ is given by $x = \dfrac{-b}{2a}$. The line of symmetry of a parabola always passes through the vertex of the parabola.

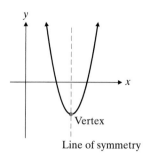

Linear equation in two variables (3.2) An equation of the form $Ax + By = C$, where A, B, and C are real numbers. The graph of a linear equation in two variables is a straight line.

Mixed number (0.1) A number that consists of an integer written next to a proper fraction. $2\dfrac{1}{3}$, $4\dfrac{6}{7}$, and $3\dfrac{3}{8}$ are all mixed numbers. Mixed numbers are sometimes called mixed fractions or mixed numerals.

Natural numbers (0.1) The set of numbers $1, 2, 3, 4, 5, \ldots$. This set is also called the set of counting numbers.

Numeral (0.1) The symbol used to describe a number.

Numerator (0.1) and (7.1) The top number or algebraic expression in a fraction. The numerator of
$$\frac{x + 3}{5x - 2}$$
is $x + 3$. The numerator of $\dfrac{12}{13}$ is 12.

Numerical coefficient (5.1) The number that is multiplied by a variable or a group of variables. The numerical coefficient in $5x^3y^2$ is 5. The numerical coefficient in $-6abc$ is -6. The numerical coefficient in x^2y is 1. A numerical coefficient of 1 is not usually written.

Odd integers (1.3) Integers that are not exactly divisible by 2, such as $\ldots, -3, -1, 1, 3, 5, 7, 9, \ldots$.

Opposite of a number (1.1) Two numbers that are the same distance from zero on the number line but lie on different sides of it are considered opposites. The opposite of -6 is 6. The opposite of $\dfrac{22}{7}$ is $-\dfrac{22}{7}$.

Ordered pair (3.1) A pair of numbers presented in a specified order. An ordered pair is often used to specify a location on a graph. Every point in a rectangular coordinate system can be represented by an ordered pair (x, y).

Origin (3.1) The point $(0, 0)$ in a rectangular coordinate system.

Parabola (9.4)   A curve created by graphing a quadratic equation. The curves shown are all parabolas. The graph of the equation $y = ax^2 + bx + c$, where $a \neq 0$, will always be a parabola.

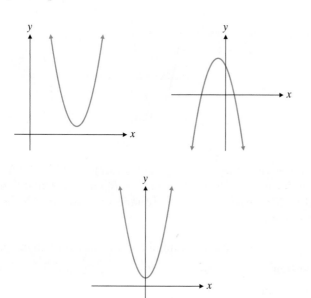

Parallel lines (3.3) and (4.1)   Two straight lines that never intersect. The graph of an inconsistent system of two linear equations in two variables will result in parallel lines.

Parallelogram (1.8)   A four-sided figure with opposite sides parallel. Figure $ABCD$ is a parallelogram.

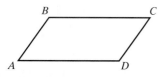

Percent (0.5)   Hundredths or "per one hundred"; indicated by the % symbol. Thirty-seven hundredths $\left(\dfrac{37}{100}\right) = 37\%$ (thirty-seven percent).

Perfect square number (6.5)   A number that is the square of an integer. The numbers 1, 4, 9, 16, 25, 36, 49, 64, 81, 100, 121, 144, . . . are perfect square numbers.

Perfect-square trinomial (6.5)   A polynomial of the form $a^2 + 2ab + b^2$ or $a^2 - 2ab + b^2$ that may be factored using one of the following formulas:

$$a^2 + 2ab + b^2 = (a + b)^2$$

or

$$a^2 - 2ab + b^2 = (a - b)^2.$$

Perimeter (1.8)   The distance around any plane figure. The perimeter of this triangle is 13. The perimeter of this rectangle is 20.

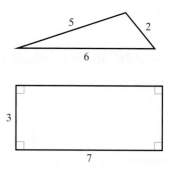

Pi (1.8)   An irrational number, denoted by the symbol π, that is approximately equal to 3.141592654. In most cases 3.14 can be used as a sufficiently accurate approximation for π.

Point-slope form of the equation of a straight line (Appendix D)   For a straight line passing through the point (x_1, y_1) and having slope m, $y - y_1 = m(x - x_1)$.

Polynomial (5.3)   An expression that only contains terms with nonnegative integer exponents. The expressions $5ab + 6$, $x^3 + 6x^2 + 3$, -12, and $x + 3y - 2$ are all polynomials. The expressions $x^{-2} + 2x^{-1}$, $2\sqrt{x} + 6$, and $\dfrac{5}{x} + 2x^2$ are not polynomials.

Prime number (0.1)   Any natural number greater than 1 whose only natural number factors are 1 and itself. The first eight prime numbers are 2, 3, 5, 7, 11, 13, 17, and 19.

Prime polynomial (6.6)   A prime polynomial is a polynomial that cannot be factored by the methods of elementary algebra. $x^2 + x + 1$ is a prime polynomial.

Principal square root (8.1)   For any given nonnegative number N, the principal square root of N (written $\sqrt{N}$) is the nonnegative number a if and only if $a^2 = N$. The principal square root of 25 $\left(\sqrt{25}\right)$ is 5.

Proper fraction (0.1)   A numerical fraction whose numerator is less than its denominator; $\dfrac{3}{7}$, $\dfrac{2}{5}$, and $\dfrac{8}{9}$ are proper fractions.

Proportion (7.6)   A proportion is an equation stating that two ratios are equal.

$$\frac{a}{b} = \frac{c}{d} \qquad \text{where } b, d \neq 0$$

is a proportion.

Pythagorean theorem (8.6) In any right triangle, if c is the length of the hypotenuse and a and b are the lengths of the two legs, then $c^2 = a^2 + b^2$.

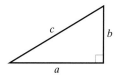

Quadratic equation in standard form (5.8) and (9.1) An equation of the form $ax^2 + bx + c = 0$, where a, b, and c are real numbers and $a \neq 0$. A quadratic equation is classified as a second-degree equation.

Quadratic formula (9.3) If $ax^2 + bx + c = 0$ and $a \neq 0$, then the roots to the equation are found by the formula

$$x = \frac{-b \pm \sqrt{b^2 - 4ac}}{2a}.$$

Quotient (0.4) The result of dividing one number or expression by another. In the problem $12 \div 4 = 3$, the quotient is 3.

Radical (8.1) An expression composed of a radical sign and a radicand. The expressions $\sqrt{5x}$, $\sqrt{\frac{3}{5}}$, $\sqrt{5x + b}$, and $\sqrt{10}$ are called radicals.

Radical equation (8.6) An equation that contains one or more radicals. $\sqrt{x + 4} = 12$, $\sqrt{x} = 3$, and $\sqrt{3x + 4} = x + 2$ are all radical equations.

Radical sign (8.1) The symbol $\sqrt{}$, which is used to indicate the root of a number.

Radicand (8.1) The expression beneath the radical sign. The radicand of $\sqrt{7x}$ is $7x$.

Range of a relation (3.6) In any relation, the set of values that represents the dependent variable is called its range. This is the set of all the second coordinates of the ordered pairs that define the relation.

Ratio (7.6) The ratio of one number a to another number b is the quotient $a \div b$ or $\frac{a}{b}$.

Rational numbers (1.1) and (8.1) A number that can be expressed in the form $\frac{a}{b}$, where a and b are integers and $b \neq 0$. $\frac{7}{3}$, $-\frac{2}{5}$, $\frac{7}{-8}$, $\frac{5}{1}$, 1.62, and 2.7156 are rational numbers.

Rationalizing the denominator (8.5) The process of transforming a fraction that contains one or more radicals in the denominator to an equivalent fraction that does not contain any radicals in the denominator. When we rationalize the denominator of $\frac{5}{\sqrt{3}}$, we obtain $\frac{5\sqrt{3}}{3}$. When we rationalize the denominator of $\frac{-2}{\sqrt{11} - \sqrt{7}}$, we obtain $-\frac{\sqrt{11} + \sqrt{7}}{2}$.

Real number (1.1) and (8.1) Any number that is rational or irrational. 2, 7, $\sqrt{5}$, $\frac{3}{8}$, π, $-\frac{7}{5}$, and $-3\sqrt{5}$ are all real numbers.

Rectangle (1.8) A four-sided figure with opposite sides parallel and each interior angle measuring $90°$. The opposite sides of a rectangle are equal.

Relation (3.6) A relation is any set of ordered pairs.

Rhombus (1.8) A parallelogram with four equal sides. Figure $ABCD$ is a rhombus.

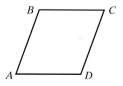

Right triangle (1.8) and (8.6) A triangle that contains one right angle (an angle that measures exactly 90 degrees). It is indicated by a small square at the corner of the angle.

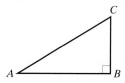

Root of an equation (2.1) and (6.7) A value of the variable that makes an equation into a true statement. The root of an equation is also called the solution of an equation.

Scientific notation (5.2) A positive number is written in scientific notation if it is in the form $a \times 10^n$, where $1 \leq a < 10$ and n is an integer.

Slope–intercept form (3.3) The equation of a line that has slope m and the y-intercept at $(0, b)$ is given by $y = mx + b$.

Slope of a line (3.3)   The ratio of change in y over the change in x for any two different points on a nonvertical line. The slope m is determined by

$$m = \frac{y_2 - y_1}{x_2 - x_1},$$

where $x_2 \neq x_1$ for any two points (x_1, y_1) and (x_2, y_2) on a nonvertical line.

Solution of an equation (2.1)   A number that, when substituted into a given equation, yields an identity. The solution of an equation is also called the root of an equation.

Solution of an inequality in two variables (3.5)   The set of all possible ordered pairs that, when substituted into the inequality, will yield a true statement.

Solution of a linear inequality (2.8)   The possible values that make a linear inequality true.

Solution to a system of two equations in two variables (4.2)   An ordered pair that can be substituted into each equation to obtain an identity in each case.

Square (1.8)   A rectangle with four equal sides.

Square root (8.1)   For any given nonnegative number N, the square root of N is the number a if $a^2 = N$. One square root of 16 is 4 since $(4)^2 = 16$. Another square root of 16 is -4 since $(-4)^2 = 16$. When we write $\sqrt{16}$, we want only the positive root, which is 4.

Standard form of a quadratic equation (6.7)   A quadratic equation that is in the form $ax^2 + bx + c = 0$.

System of equations (4.1)   A set of two or more equations that must be considered together. The solution is the value for each variable of the system that satisfies each equation.

$$x + 3y = -7 \qquad 4x + 3y = -1$$

is a system of two equations in two unknowns. The solution is $(2, -3)$, or the values $x = 2$, $y = -3$.

Term (1.7)   A number, a variable, or a product of numbers and variables. For example, in the expression $a^3 - 3a^2b + 4ab^2 + 6b^3 + 8$, there are five terms. They are a^3, $-3a^2b$, $4ab^2$, $6b^3$, and 8.

Trapezoid (1.8)   A four-sided figure with two sides parallel. The parallel sides are called the bases of the trapezoid. Figure $ABCD$ is a trapezoid.

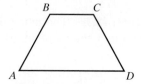

Trinomial (5.3)   A polynomial of three terms. The expressions $x^2 + 6x - 8$ and $a + 2b - 3c$ are trinomials.

Variable (1.4)   A letter that is used to represent a number or a set of numbers.

Variation (8.7)   An equation relating values of one variable to those of other variables. An equation of the form $y = kx$, where k is a constant, indicates *direct variation*.

An equation of the form $y = \dfrac{k}{x}$, where k is a constant, indicates *inverse variation*. In both cases, k is called the *constant of variation*.

Vertex of a parabola (9.4)   The lowest point on a parabola opening upward or the highest point on a parabola opening downward. The x-coordinate of the vertex of the parabola formed by the equation $y = ax^2 + bx + c$ is given by $\dfrac{-b}{2a}$.

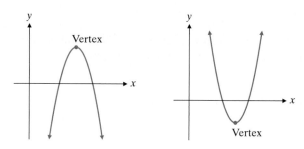

Vertical line test (3.6) If a vertical line can intersect the graph of a relation more than once, the relation is not a function.

Whole numbers (0.1)   The set of numbers 0, 1, 2, 3, 4, 5, ...

x-intercept (3.2)   The ordered pair $(a, 0)$ is the x-intercept of a line if the line crosses the x-axis at $(a, 0)$. The x-intercept of line l on the following graph is $(4, 0)$.

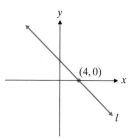

y-intercept (3.2)   The ordered pair $(0, b)$ is the y-intercept of a line if the line crosses the y-axis at $(0, b)$. The y-intercept of line p on the following graph is $(0, 3)$.

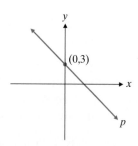

Subject Index

Applications Index

Photo Credits

CHAPTER 0   CO Comstock Complete   **p. 34** Getty Images - Stockbyte, Royalty Free   **p. 42** Adam Jones/Photo Researchers, Inc.
p. 51 Comstock Complete

CHAPTER 1   CO Comstock Complete   **p. 59** Peter Skinner/Photo Researchers, Inc.   **p. 92** AP Wide World Photos   **p. 107** John Coletti/Stock Boston   **p. 115** Jeanne White/Photo Researchers, Inc.

CHAPTER 2   CO Courtesy of www.istockphoto.com   **p. 135** Courtesy of www.istockphoto.com   **p. 135** Shutterstock   **p. 165** Courtesy of www.istockphoto.com   **p. 166** Townsend P. Dickinson/The Image Works   **p. 173** Courtesy of www.istockphoto.com   **p. 173** © Adrian Bradshaw/epa/CORBIS All Rights Reserved   **p. 177** NASA/Johnson Space Center   **p. 184** Photos.com.   **p. 188** Photos.com.   **p. 189** Javier Larrea/Pixtal/Superstock Royalty Free

CHAPTER 3   CO Courtesy of www.istockphoto.com

CHAPTER 4   CO Wong Tsu Shi/Shutterstock   **p. 304** Marcos Welsh/Art Life Images

CHAPTER 5   CO Courtesty of www.istockphoto.com   **p. 329** A. Riess (STSci)/NASA/John F. Kennedy Space Center   **p. 357** Courtesy of www.istockphoto.com

CHAPTER 6   CO Courtesy of www.istockphoto.com   **p. 370** Courtesy of www.istockphoto.com   **p. 395** Courtesy of www.istockphoto.com **p. 405** Philip H. Coblentz/World Travel Images, Inc. Royalty Free

CHAPTER 7   CO Comstock Complete   **p. 458** Comstock Complete

CHAPTER 8   CO Alexander Gordeyev/Shutterstock   **p. 470** David Madison/Getty Images Inc.—Stone Allstock   **p. 502** Comstock Complete **p. 511** Photos.com.

CHAPTER 9   CO Comstock Complete   **p. 533** Courtesy of www.istockphoto.com   **p. 557** Tetra Images/Superstock Royalty Free

PROPERTIES OF REAL NUMBERS

	Addition	**Multiplication**
Commutative Properties	$a + b = b + a$	$ab = ba$
Associative Properties	$(a + b) + c = a + (b + c)$	$(ab)c = a(bc)$
Identity Properties	$a + 0 = 0 + a = a$	$a \cdot 1 = 1 \cdot a = a$
Inverse Properties	$a + (-a) = -a + a = 0$	$a \cdot \dfrac{1}{a} = \dfrac{1}{a} \cdot a = 1 \ (a \neq 0)$
Distributive Property		$a(b + c) = ab + ac$

PROPERTIES OF EXPONENTS

If $x, y \neq 0$, then

$$x^a \cdot x^b = x^{a+b}$$

$$\frac{x^a}{x^b} = x^{a-b} \text{ if } a \geq b$$

$$\frac{x^a}{x^b} = \frac{1}{x^{b-a}} \text{ if } a < b$$

$$x^0 = 1$$

$$(x^a)^b = x^{ab}$$

$$(xy)^a = x^a y^a$$

$$\left(\frac{x}{y}\right)^a = \frac{x^a}{y^a}$$

ABSOLUTE VALUE

Definition

$$|x| = \begin{cases} x \text{ if } x \geq 0 \\ -x \text{ if } x < 0 \end{cases}$$

INEQUALITIES

If $a < b$, then for all values of c, $a + c < b + c$ and $a - c < b - c$.

If $a < b$ when c is a **positive number** $(c > 0)$, then $ac < bc$ and $\dfrac{a}{c} < \dfrac{b}{c}$.

If $a < b$ when c is a **negative number** $(c < 0)$, then $ac > bc$ and $\dfrac{a}{c} > \dfrac{b}{c}$.

FACTORING AND MULTIPLYING FORMULAS

Perfect-Square Trinomials	$a^2 + 2ab + b^2 = (a + b)^2$ $a^2 - 2ab + b^2 = (a - b)^2$
Difference of Two Squares	$a^2 - b^2 = (a + b)(a - b)$
Sum of Two Squares	$a^2 + b^2$ cannot be factored

PROPERTIES OF LINES AND SLOPES

The **slope** of any nonvertical line passing through (x_1, y_1) and (x_2, y_2) is $m = \dfrac{y_2 - y_1}{x_2 - x_1} (x_1 \neq x_2)$.

The **standard form** of the equation of a straight line is $Ax + By = C$.

The **slope–intercept form** of the equation of a straight line with slope m and y-intercept $(0, b)$ is $y = mx + b$.

A **horizontal line** has a slope of zero. The equation of a horizontal line can be written as $y = b$.

A **vertical line** has no slope (the slope is not defined for a vertical line).

The equation of a vertical line can be written as $x = a$.